(continued on back cover)

FINITE MATHEMATICS

Solving Problems in Business,
Economics, and the Social
and Behavioral Sciences

FINITE MATHEMATICS

Solving Problems in Business, Economics, and the Social and Behavioral Sciences

Bill Armstrong
Lakeland Community College

Don Davis
Lakeland Community College

Pearson Education, Inc.
Upper Saddle River, New Jersey 07458

Library of Congress Cataloging-in-Publication Data

Armstrong, Bill (William A.)
    Finite mathematics : solving problems in business, economics, and behavioral sciences
    / Bill Armstrong, Don Davis.
        p. cm.
    Includes index.
    ISBN 0-13-019958-3
        1. Mathematics. 2. Social sciences—Mathematics. I. Davis, Don (Donald Edgar)
    II. Title.

QA39.3 .A76 2003
510--dc21

                                                                    2002074991

Editor-in-Chief/Acquisitions Editor: Sally Yagan
Executive Project Manager: Ann Heath
Assistant Editor: Vince Janson
Vice President/Director of Production and Manufacturing: David W. Riccardi
Executive Managing Editor: Kathleen Schiaparelli
Senior Managing Editor: Linda Mihatov Behrens
Production Editor: Bob Walters
Manufacturing Buyer: Alan Fischer
Manufacturing Manager: Trudy Pisciotti
Marketing Manager: Patrice Lumumba Jones
Marketing Assistant: Rachel Beckman
Assistant Managing Editor, Math Media Production: John Matthews
Editorial Assistant/Supplements Editor: Joanne Wendelken
Art Director: Maureen Eide
Interior/Cover Designer: Joseph Sengotta
Creative Director: Carole Anson
Art Editor: Thomas Benfatti
Director of Creative Services: Paul Belfanti
Director, Image Resource Center: Melinda Reo
Manager, Rights and Permissions: Zina Arabia
Interior Image Specialist: Beth Boyd-Brenzel
Cover Image Specialist: Karen Sanatar
Image Permission Coordinator: Debbie Latronica
Cover Photo: David Lissy/EStock Photography LLC Leo De Wys
Art Studio: Laserwords

© 2003 Pearson Education, Inc.
Upper Saddle River, New Jersey 07458

Printed in the United States of America
10 9 8 7 6 5 4 3 2 1

ISBN: 0-13-019958-3

Pearson Education LTD., *London*
Pearson Education Australia PTY, Limited, *Sydney*
Pearson Education Singapore, Pte. Ltd
Pearson Education North Asia Ltd., *Hong Kong*
Pearson Education Canada, Ltd., *Toronto*
Pearson Educación de Mexico, S.A. de C.V.
Pearson Education – Japan, *Tokyo*
Pearson Education Malaysia, Pte. Ltd.

Our book is dedicated . . .

In memory of Mr. James Young.
His love of mathematics and the
teaching of mathematics continues
to inspire 20 years later.

and

To Dr. David Keck, whose
mathematical and teaching skills
were a true inspiration.

Contents

Preface

Audience

In preparing to write this text, we talked with many of our colleagues who teach a finite math course to find out if they experienced the same difficulties in teaching this course. What we learned is that while there is very little similarity in the topics covered or in how much time is spent on each area, there is remarkable uniformity in the needs of students who enroll in these diverse courses. Professors at Community Colleges, Universities, and Liberal Arts Colleges all told us that their students are generally unmotivated, unsure of their algebra skills, and uncomfortable with translating English into mathematics and knowing how to set up problems for solution. Armed with this knowledge, we set out to write a new Finite Math text that would address these fundamental needs.

As with other Finite Mathematics textbooks, our text may be used in either a one or two term course for students majoring in economics, business, social, or behavioral sciences. We have organized the topics for maximum flexibility so that the text may be adapted to any college or university's curriculum. However, that is where the similarity ends. We have crafted this book around six key principles designed to address student's needs:

- Present the mathematics in language that students can read and understand
- Teach good problem solving techniques and provide ample practice
- Use real data applications to keep it interesting
- Provide timely reinforcement of algebra and other essential skills
- Let Instructor's decide whether to incorporate technology
- Present Probability and Statistics with clarity and precision

Present the Mathematics in Language That Students Can Read and Understand

By writing this text in a conversational, easy-to-read style, we strive to evoke the one-on-one communication of a tutorial session. When students find that they can understand the clear presentation and follow the interesting, real world examples, we believe they will get into the habit of reading the text. Although we have written a text that is accessible, we have been careful not to sacrifice the proper depth of coverage and necessary rigor required of finite mathematics. We are confident both objectives have been met.

Teach Good Problem Solving Techniques and Provide Ample Practice

Problem Solving Sections

Before explaining how to solve an entire class of problems, we provide a special section that demonstrates how to apply the appropriate mathematical tools to analyze a given type of problem. For example, the first section of the book, *Problem Solving: Linear Equations*, introduces our general approach to problem solving and then addresses the challenge of analyzing a word problem and translating the words into mathematical expressions; after which we examine the more complex situation of writing a word problem as a system of linear equations. The other dedicated problem solving sections introduce strategies and techniques for solving each type of problem.

- Section 3.1, *Problem Solving: Linear Inequalities* reviews writing linear models and solving linear inequalities, both of which are vital skills for solving linear programming problems.
- Section 6.1, *Problem Solving: Probability and Statistics* explains some basic concepts in probability and statistics such as probability experiment, mutual exclusivity, and ordered samples.

Many of the exercises in these problem solving sections prepare students for subsequent sections because they introduce exercises that are solved later in the textbook. Reviewer feedback was enthusiastic on these sections which indicated to us that we were pursuing an approach that many felt was lacking in this area.

Problem Solving Method

To help students solve problems in finite mathematics, we developed a problem solving method that is used throughout the entire text. Many reviewers commented that this is the single most distinctive and user-friendly feature of our book that provides students a tool in their learning and understanding of the mathematics at hand. Frequently, applied mathematics instructors hear students comment that "I don't even know how to begin this problem." or "If the problem was just set up for me, I could solve it." Because skills such as setting up problems and writing the solution in its proper context can be a major challenge for finite mathematics students, we have integrated a **problem solving method** throughout the text. The steps of the problem solving method are used throughout the text. We use the phrases "Understand the Situation" and "Interpret the Solution" to identify two critical steps in problem solving.

- When solving an application example, we first need to **Understand the Situation.** In these clearly labeled paragraphs we state the quantity we are trying to determine and identify the method used to determine the solution. Frequently we reveal the thought processes necessary to attack the problem.
- After determining a numerical solution in a traditional solution step, we **Interpret the Solution.** In the interpretation step, we write a concluding sentence that conveys the meaning of the numerical answer in the context of the application. To reinforce these techniques, we often ask students to solve and interpret the solution to various applied problems in the exercise sets.

Exercises

A typical section exercise set contains numerous skill builder problems, a generous selection of applications from many different disciplines, and an abundant supply of real data modeling applications, which are denoted by the ⊕ symbol. We have included more applications based on real data in our textbook than in any other finite mathematics book on the market. Another lesson learned from our market research is that many finite mathematics textbooks fail to provide enough exercises for the student to grasp the course content. With more than 3200 exercises, we are confident our textbook has more than enough exercises to meet student's needs.

Use Real Data Applications to Keep It Interesting

Students in this course tend to be very pragmatic; they want to know why they must learn the mathematical content in this course. Including many **real data modeling** applications (denoted in text by ⊕) helps to answer their unstated question and provides motivation and interest. Many of the models, parameters, and scenarios in the examples and exercises are based on data gathered from the U.S. Statistical Abstract, the Census Bureau, and other reliable sources for which URLs are always given. For example, we include real data modeling applications such as determining planting schedules in California (*Source:* U.S. Department of Agriculture, www.usda.gov/nass) in the linear programming chapters. Other examples include:

- The out-of-pocket hospital care and physician services expenditures by patients and by Medicare used as an application of matrix operations (*Source:* U.S. Census Bureau, www.census.gov).
- The percentage of American adults who exercised for 30 minutes or more at least 5 times per week, classified by gender, as an application of conditional probability. (*Source:* National Center for Chronic Disease Prevention, www.cdc.gov/nccdphp).
- The typical amount of financial debt held by American families, classified by age group, as an application of frequency distributions (*Source:* U.S. Census Bureau, www.census.gov).

The quantity, quality, and variation of these types of applications are simply not found in other finite mathematics textbooks. We believe that real data modeling applications not only keep the course content relevant and fresh, but compel students to interpret the numerical solution in the context of the problem they have solved.

Provide Timely Reinforcement of Algebra and Other Essential Skills

One of the major challenges faced by students, and frustrations encountered by instructors, is weak preparation in algebra and other essential skills. Even students who have proficient algebra skills are often rusty and unsure of which algebraic tool to apply. Further, the content demands of a finite mathematics course does not allow for extensive time spent on review. In an effort to address this pervasive problem, we have developed the "From Your Toolbox" feature in the body of the text and also in the *MathPro for Finite Math* software.

From Your Toolbox

When appropriate, the "From Your Toolbox" feature directs students to read background material in the Algebra Review or other appendices. In addition, this feature is used to review previously introduced definitions, theorems, or properties as needed. By providing a brief review when it is needed, students stay on task and do not need to flip back to hunt through previous sections for key information.

MathPro5 for Finite Math.   Powered by Prentice Hall's celebrated *MathPro* engine, this unique new program is keyed section by section to our text. MathPro5 is online, customizable tutorial and assessment software that is fully integrated with *Finite Math* at the section level for anytime, anywhere learning. Students access multiple types of help while working algorithmically-generated tutorial exercises for review and concept mastery.

Let Instructor's Decide Whether to Incorporate Technology

As graduates and instructors of The Ohio State University, we began using graphing calculator technology in the classroom long before it was fashionable to do so. Based on our years of experience in this area, we have seen the strengths of using a graphing calculator and the drawbacks as well. Our philosophy is to let the instructor, rather than the textbook, determine how much or how little graphing calculators, spreadsheets, or other desktop applications are used in classroom. Since our goal was to write a text where graphing calculator technology was truly optional, we have developed the **Technology Option** in our text, which allows each instructor to decide whether or not to use technology in the curriculum. These shaded, optional parts are easy to find, or to skip, and typically follow examples. As a result, an instructor who desires to fully implement technology in the classroom can use this text, while at the same time, an instructor who desires to not use any technology in the classroom can also use this text. The content of a Technology Option box mirrors the traditional presentation but shows how the answer to a particular example may be found using a graphing calculator. Although keystroke commands are not given, we provide answers to general questions students may have. All screen shots are from the Texas Instruments TI-83 calculator. Keystrokes and commands for various models and brands of calculator are found at the online graphing calculator manual found at the companion website at www.prenhall.com/armstrong.

Exercises that assume the use of a graphing calculator are clearly marked with a ▦ symbol so they can be assigned or skipped as desired by the instructor. We recognize that the graphing calculator is simply a tool to be used in the understanding of mathematics. We have been very careful to introduce the technology only where it is appropriate and not to let its use overshadow the mathematics.

Present Probability and Statistics with Clarity and Precision

Our experiences in the classroom, and supported in the market research, indicated that probability is one of the most difficult topics for the student to comprehend.

Because probability is so challenging, we believe it is important to prepare students for the study of this topic. Students need to know the types of problems they will encounter and whether or not order is important. They also need to know when events are independent and when events are mutually exclusive. Addressing these basic concepts and introducing the language of probability in Section 6.1, *Problem Solving: Probability and Statistics*, provides students with a solid foundation. Additional strengths in the probability and statistics chapters include:

- Clear and well-paced development of permutations and combinations (Section 6.3). The *Flashback* to Section 6.1 helps students understand these often confusing topics.
- Introducing frequency distributions in connection with empirical probability in Section 6.4 helps students distinguish the difference between meanings of probability in various contexts.
- Chapter 7 incorporates more graphics and illustrations than competitors and focuses on interpretation. Students are encouraged to understand and interpret different representations of mathematical ideas by way of algebra, graphs, data, and verbal descriptions and to easily translate from one representation to another.
- Many Finite Math books incorrectly use a single notation, X or x, to represent both a random variable and the values that the random variable can assume. We believe that one need not sacrifice accuracy and precision to be accessible, therefore we use X to represent the random variable and x to represent the values that the random variable can assume. Students quickly become comfortable with this distinction.
- The coverage of random variables, expectation, and standard deviation are divided into two sections (8.1 and 8.2). This division allows for a slower paced development of these complex ideas and better prepares students for probability distributions.
- We have written these chapters in consultation with experts in this field to ensure that our goal of clarity and precision was being met. Saeed Ghahramani, Dean, School of Arts and Sciences, Western New England College, served as the expert reviewer for the Probability and Statistics chapters.

Content Features

In addition to exceptionally strong coverage of probability and statistics (Chapters 6, 7, and 8), the following sections have been praised by reviewers:

- Reviewers found the motivation in Section 2.3, Matrix Multiplication, to be particularly effective. Especially in the areas of interpreting the entries of a matrix product; determining when entries make sense, and when the entries are meaningless.
- Several reviewers commented that the explanations in Section 2.5, Leontief Input-Output Models, are exceptionally clear.
- The section on Applications of Linear Programming (Section 3.4) drew rave reviews from reviewers. Most liked the approach of the section in tying together all of Chapter 3 and summarizing the process. Some reviewers found the Flashback in this section to be a wonderful transition to applications of linear programming.
- Reviewers lauded the first two sections of Chapter 4 (Linear Programming and the Simplex Method) as being exceptionally clear and concise, yet thorough.

- Chapter 5 (Mathematics of Finance) was another favorite for our review panel. From the Chapter Opening page through the Chapter Project, these are topics that all reviewers said were necessary and appropriate for students.
- Reviewers who said they covered Markov Chains and Game Theory were extremely impressed with Chapters 9 and 10. All applauded the flow of material in these chapters and how students should find the material quite easy to follow.

Chapter Features

Chapter Openers. The first page of each chapter lists the sections included in that chapter, along with a photo and representative graphs or figures that fore-shadow the fundamental ideas presented in the chapter in the context of an application. "What We Know" reiterates what information has been learned in previous chapters and "Where Do We Go" explains what topics will be covered. The chapter opener helps create a roadmap to guide the student through the book and underscore the connections between topics.

Flashbacks. Selected examples used earlier in the textbook are revisited in the **Flashback** feature. The Flashback carries over an applied example from a previous section and then extends the content of the example by considering new questions. In this manner, new topics are introduced in a more natural way within a familiar context. Moreover, the Flashback often reviews the necessary skills and concepts from previous chapters. We believe that this pedagogical technique of using applications previously discussed allows students to concentrate on new topics using familiar applications.

Interactive Activities. Extensions to completed examples are included in the **interactive activities** that usually appear after examples. The interactive activities may ask students to solve a problem using a different method, to discover a pattern that can lead to a mathematical property, or to explore additional properties of recently introduced topics. Interactive Activities may be used in a number of ways in the classroom. Instructors may assign them as critical thinking exercises, they may be used as a springboard for classroom discussion, or they may provide a vehicle for collaborative activities. Interactive Activities that include the Web icon 🌐 have solutions provided on the companion website at www.prenhall.com/armstrong.

Checkpoints. At strategic points in each section, an example is followed by a **checkpoint**. Each checkpoint directs a student to complete a selected odd numbered problem in that section's exercise set. Guiding the student to a parallel exercise to the example problem helps to reinforce the topic at hand and ensures that the recently introduced skill or concept is practiced immediately. This pedagogical tool promotes interaction between the text and the student and helps students to develop good study habits. Students who use the checkpoints will quickly learn to take ownership of the course material.

Notes. Another popular feature is the **Notes** that appear after many definitions, theorems, and properties. Notes are used to clarify a mathematical idea verbally and to provide students with additional insight into the material. Many times these notes echo what a professor might state in the classroom to help the student understand the definition, theorem or property.

Section Projects.   At the end of each exercise set, a **Section Project** presents a series of questions that ask students to explore the idea that is presented. Some of the projects are based on real data and ask students to use the regression capabilities of their calculator to determine a model. Others give a step-by-step procedure that can be used to solve classic problems in finite mathematics. Instructors can use section projects as standard hand-in assignments, collaborative activities, or for demos in graphing calculator usage.

Section Summary.   At the end of each section, a short summary highlights the important concepts of the section. Section summaries are a convenient resource that students may use while completing exercises, or to offer a springboard for test review.

End of Chapter Material.   Each chapter concludes with three components; first is a summary called **Why We Learned It**. This narrative outlines the major topics of each chapter and describes how the topics are used in various careers. An extensive set of **Chapter Review Exercises** is designed to augment the exercises in each section of the chapter. Concluding the chapter is the **Chapter Project**. These extensive exercises use real data modeling to explore and extend the ideas discussed in each chapter. They may be assigned as individual homework or as a group project.

Supplements

Instructor's Solutions Manual (ISBN 0-13-065917-7).   Written by Denyse Kerr and accuracy checked by the authors and Priscilla Gathoni, this volume contains complete worked out solutions to all exercises as well as complete solutions to the section and chapter projects.

Student's Solutions Manual (ISBN 0-13-065918-5).   Written by Denyse Kerr, and accuracy checked by the authors and Priscilla Gathoni, this volume contains complete worked out solutions to all of the odd numbered section and review exercises in the textbook.

Companion Website www.prenhall.com/armstrong.   This site is designed to supplement the textbook by offering a variety of teaching and learning resources for each chapter. This includes a net search of topics relevant to the chapter, links to related chapter topics, solutions to selected interactive activities, quizzes, program downloads for various models of graphing calculators, and an online graphing calculator reference manual.

MathPro5 for Finite Math.   MathPro5 is online, customizable tutorial and assessment software that is fully integrated with *Finite Math* at the section. Loaded with many support tools, students work algorithmically-generated tutorial exercises for review and concept mastery and get step-by-step guidance as needed.

   The easy-to-use Course Management System enables instructors to track and assess student performance on tutorial work, quizzes, and tests. Instructors can create any number of algorithmic, multi-form quizzes, pre-tests, post-tests, chapter tests and cumulative reviews with the **integrated TestGen-EQ software**. A **robust**

reports wizard provides a grade book, individual student reports and class summaries. **A messaging system** enhances communication between instructors and students from wherever they are working in MathPro5. The **customizable course syllabus** allows instructors to remove and reorganize topics, and add website links and tests to maximize integration of the classroom and tutorial experience.

The combination of MathPro5's richly integrated tutorial and testing and robust course management tools provides an unparalleled tutorial experience for students, and new assessment and time-saving tools for instructors.

TestGen-EQ for Windows and Macintosh (ISBN 0-13-065910-X).   This algorithmic software-testing program allows instructors to create tests quickly and efficiently using the supplied questions or to personalize tests using the built-in editing features.

Test Item File (ISBN 0-13-065919-3).   Hard copy of the algorithmic computerized testing materials provides a quick reference for the testing software.

Acknowledgments

We owe a debt of gratitude to many individuals who helped us shape and refine *Finite Mathematics: Problem Solving in Business, Economics and the Social and Behavioral Sciences.* The reviewers included:

Brenda Ammons, Pellissippi State Tech and CC; Paul Boisvert, Oakton Community College; Dale Buske—St. Cloud State; Elias Deeba—University of Houston—Downtown; Jennifer Fowler, UT Knoxville; Lou Hoelzlel—Bucks County Community College; Frank Hummer—Iowa State University; Stephanie Kurtz, Louisiana State University; Mike McCraith, Joliet Junior College; David Miller—William Patterson University; and Bill Siebler.

We owe special thanks to Saeed Ghahramani, Dean, School of Arts and Sciences Western New England College who served as our expert reviewer for the Probability and Statistics chapters. Ed Bolger, Miami University of Ohio, was our expert reviewer for the Markov Process and Game Theory chapters. Sarah Street served as an additional reviewer for Chapter 8.

A special thank you to Priscilla Gathoni, our accuracy checker, and to Mike McCraith, who provided the Chapter Projects and solutions to Interactive Activities.

We are indebted to Denyse Kerr, who is an extremely valuable team member. Her professional, high-quality work on creating both the Instructor's and Student's Solutions Manuals is valued for its precision and careful proofreading.

We owe special thanks to Tony Palermino. Besides serving as our Developmental Editor, Tony was also our mentor, coach, advisor, advocate, and valued team member. Tony's top quality professional work was critical in preparing the manuscript. Tony deserves much credit for helping create this text.

Many people involved in the management and production of the text deserve recognition. Bob Walters has provided his high quality, professional production abilities. Maureen Eide worked through several iterations of the design. Vince Jansen contributed his expertise on MathPro5 for Finite Math and the TestGen EQ. We appreciate the quality services provided by Laserwords in preparing the graphs, tables, and art in the text and of Laurel Technical Services in writing the review and test bank exercises. Patrice Jones continues to provide effective marketing techniques. Thanks to Joanne Wendelken for routing our numerous telephone calls. A very special thanks to the Prentice Hall sales staff for their efforts and enthusiasm.

Special thanks to Sally Yagan for her commitment to this project. Sally continues to impress us in her level of dedication and her can-do attitude is refreshing and contagious.

Ann Heath deserves special recognition for the many hats she wears. Without Ann on our team, this project would not be at the high level of quality it currently is. Ann is a master of crisis management as well as providing an ear at any time to hear our frustrations.

Finally, we owe personal thanks to our families. Our wives, Lisa and Melissa, were supportive and patient during the entire process. To our children Austin and Dylan; Randy, Rusty, and Ronnie who understood why sometimes Daddy could not come out and play in the backyard.

Bill Armstrong

Don Davis

A Guide to Using This Text

Opening Application

These engaging real-world applications appear at the beginning of every chapter. They include compelling photos and graphs that illustrate how the math introduced and covered in the chapter is used outside the classroom.

Sets and the Fundamentals of Probability

Odds to Probability

$$P(A) = \frac{a}{a+b}$$

Probability of Tiger Woods Winning the 2001 British Open

$$P(A) = \frac{a}{a+b} = \frac{3}{3+2} = \frac{3}{5}$$

We frequently hear about the odds of something occurring. But how do odds relate to probability? In 2001, golf pro Tiger Woods was "hot" and the odds of his winning the British Open were 3 to 2. Using the formula for converting odds to probability, we can translate odds into the probability of a given event occurring. In this case, the probability of Tiger Woods winning the Open was 3/5. Although the probability of a win was high, Tiger actually suffered a stunning defeat. Probability is a valuable tool for predicting likely outcomes, but it does not insure that a given event will occur.

What We Know

So far in this text, we have seen many applications involving matrices and linear programming, as well as the mathematics of finance. We have seen how an understanding of these topics can be important for individuals in many diverse areas.

Page 307

Section 1.1   Problem Solving: Linear Equations

Most of us use or hear the word *problem* in our daily conversation. We may say or hear statements such as

- "I'm having a problem loading this software."
- "The problem is that management doesn't communicate with us."
- "I had a problem getting tickets for the concert."

Many problems that we encounter can be solved easily, but others require a more thoughtful approach. In finite mathematics we are often confronted with the challenge of solving problems that require a thoughtful approach. As a result, we need to review a general problem-solving strategy and the accompanying skills that will allow us to solve a variety of problems. The skills reviewed in this section will be used throughout the text.

English to Mathematics

From prior mathematics classes we have seen how a **variable** can be used to represent an unknown quantity. For example, suppose that we hear the following report on the local news:

"CleanAir produces 1475 of two types of vacuums, Mod I and Mod II, each week."

Page 2

Problem Solving

Teaching problem-solving skills is central to the approach of the text. The pedagogy and organization encourage students to apply sound problem-solving principles and gives them practice applying them.

Example 4 **Using the Addition Rule with Tabular Data**

The source of payment for patients discharged from U.S. hospitals during 1996 for two age groups is shown in Table 6.5.3. The number of discharges is given in thousands. For the labeled events A, B, C, and D, determine the following probabilities and interpret.

(a) $P(B \cap D)$ (b) $P(A \cup C)$

Table 6.5.3

Age Group	Private (A)	Government (B)	Total
Less than 44 years old (C)	6,090	6,442	12,532
45 years or older (D)	4,636	13,337	17,973
Total	10,726	19,779	30,505

SOURCE: U.S. National Center for Health Statistics, www.cdc.gov/nchs

Solution

(a) Understand the Situation: First notice that event B is the event that the source of payment for a discharged patient in 1996 was the government. Event D is the event that a discharged patient was 45 years old or older. So when we compute $P(B \cap D)$, we are computing the probability that a randomly selected patient discharged from a hospital in 1996 was 45 years old or older and the source of payment was the government.

By inspecting Table 6.5.3, we see that if we follow down the event B column and across the event D row they intersect at the cell with the value 13,337 thousand. Since the total number of patients discharged in 1996 was 30,505 thousand, we have

$$P(B \cap D) = \frac{\text{value where the event } B \text{ column and event } D \text{ row intersect}}{\text{total patients discharged}}$$

$$= \frac{13,337}{30,505} \approx 0.44$$

Interpret the Solution: This means that if we select a discharged patient at random the probability that he or she had government health insurance **and** was 45 year or older is about 0.44 or about 44%.

(b) Understand the Situation: Since event A is the event that the source of payment was private and event C is the event that the discharged patient was less than 44 years old, we are being asked to determine the probability that a randomly selected patient has private health insurance, is 44 years old or younger, or both. Table 6.5.3 indicates that

$$P(A) = \frac{\text{total in event } A \text{ column}}{\text{total patients discharged}} = \frac{10,726}{30,505},$$

$$P(C) = \frac{\text{total in event } C \text{ row}}{\text{total patients discharged}} = \frac{12,532}{30,505},$$

and where event A column and event C row intersect gives us

$$P(A \cap C) = \frac{6090}{30,505}.$$

Examples Reinforce Good Problem-Solving Skills

The solutions to many examples are broken into three steps.

- The first step is to **Understand the Situation.** Using straightforward language, the problem situation is analyzed. The quantity to be determined is identified, the method used to find the solution is discussed, and the thought process necessary to attack the problem is presented.

- Then the numerical solution is found in a traditional solution step.

- Finally, students see how to **Interpret the Solution** in the concluding sentence that conveys the meaning of the numerical answer in the context of the application.

Many of the exercises and examples in the text use **Real Data** to make them more interesting and relevant to students.

Technology Option

The text's technology coverage is thorough yet optional. Presented in shaded boxes, technology coverage is easy to find, or to skip, and typically follows examples. The contents of these boxes mirrors the traditional presentation but shows how the answer to a particular example may be found using a TI-83 graphing calculator. For keystroke level instruction on a variety of different calculators, students are directed to the online graphing calculator manuals on the companion website at **www.prenhall.com/armstrong**

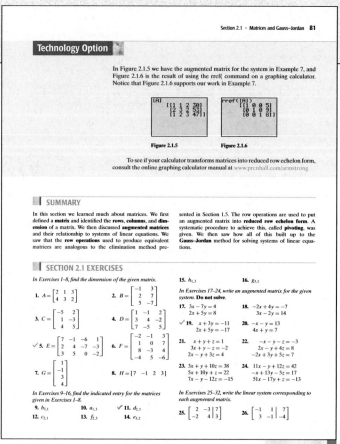

Section 2.1 · Matrices and Gauss–Jordan **81**

Technology Option

In Figure 2.1.5 we have the augmented matrix for the system in Example 7, and Figure 2.1.6 is the result of using the rref(command on a graphing calculator. Notice that Figure 2.1.6 supports our work in Example 7.

Figure 2.1.5   Figure 2.1.6

To see if your calculator transforms matrices into reduced row echelon form, consult the online graphing calculator manual at www.prenhall.com/armstrong

▌ SUMMARY

In this section we learned much about matrices. We first defined a **matrix** and identified the **rows, columns,** and **dimension** of a matrix. We then discussed **augmented matrices** and their relationship to systems of linear equations. We saw that the **row operations** used to produce equivalent matrices are analogous to the elimination method pre-sented in Section 1.5. The row operations are used to put an augmented matrix into **reduced row echelon form**. A systematic procedure to achieve this, called **pivoting**, was given. We then saw how all of this built up to the **Gauss–Jordan** method for solving systems of linear equations.

▌ SECTION 2.1 EXERCISES

In Exercises 1–8, find the dimension of the given matrix.

1. $A = \begin{bmatrix} 2 & 1 & 3 \\ 4 & 3 & 2 \end{bmatrix}$ 2. $B = \begin{bmatrix} -1 & 3 \\ 2 & 7 \\ 5 & -7 \end{bmatrix}$

3. $C = \begin{bmatrix} -5 & 2 \\ 1 & -3 \\ 4 & 5 \end{bmatrix}$ 4. $D = \begin{bmatrix} 1 & -1 & 2 \\ 3 & 4 & -2 \\ 7 & -5 & 5 \end{bmatrix}$

5. $E = \begin{bmatrix} 7 & -1 & -6 & 1 \\ 2 & 4 & -7 & -3 \\ 3 & 5 & 0 & -2 \end{bmatrix}$ 6. $F = \begin{bmatrix} -2 & -1 & 3 \\ 1 & 0 & 7 \\ 8 & -3 & 4 \\ -4 & 5 & -6 \end{bmatrix}$

7. $G = \begin{bmatrix} 1 \\ -1 \\ 3 \\ 4 \end{bmatrix}$ 8. $H = [7 \ -1 \ 2 \ 3]$

In Exercises 9–16, find the indicated entry for the matrices given in Exercises 1–8.

9. $b_{3,1}$ 10. $a_{1,3}$ 11. $d_{2,3}$
12. $c_{2,1}$ 13. $f_{2,3}$ 14. $e_{3,2}$
15. $h_{1,3}$ 16. $g_{3,1}$

In Exercises 17–24, write an augmented matrix for the given system. **Do not solve.**

17. $3x - 7y = 4$
$2x + 5y = 8$

18. $-2x + 4y = -7$
$3x - 2y = 14$

19. $x + 3y = -11$
$2x + 5y = -17$

20. $-x - y = 13$
$4x + y = 7$

21. $x + y + z = 1$
$3x + y - z = -2$
$2x - y + 3z = 4$

22. $-x - y - z = -3$
$2x - y + 4z = 8$
$-2x + 3y + 5z = 7$

23. $3x + y + 10z = 38$
$5x + 10y + z = 22$
$7x - y - 12z = -15$

24. $11x - y + 12z = 42$
$-x + 13y - 5z = 17$
$51x - 17y + z = -13$

In Exercises 25–32, write the linear system corresponding to each augmented matrix.

25. $\begin{bmatrix} 2 & -3 & | & 7 \\ -2 & 4 & | & 3 \end{bmatrix}$ 26. $\begin{bmatrix} -1 & 1 & | & 7 \\ 3 & -1 & | & -4 \end{bmatrix}$

Simple Interest

 **From Your Toolbox**

Interest rates should be converted to decimals before being used in any formulas. For more review on working with percents, consult the review in Appendix A.1.

When financial institutions advertise **interest rates** on certificates of deposit (CDs) or on loans, it usually means **rate per year**. In general, rates are expressed as percents. Since the word percent means **per hundred**, we can quickly convert from percents to decimals, and vice versa. For our purposes in this chapter, remember the information in the Toolbox to the left.

Simple interest is earned on the amount borrowed and not on any past interest. The **time** that the money is earning interest is given in years. **Simple interest** is simply the product of the principal, rate, and time.

▌ **Simple Interest**

To compute **simple interest**, use the formula $I = Prt$, where

P = principal (amount invested or borrowed)
r = interest rate per year (as a decimal)
t = time in years
I = interest (earned or paid)

Example 1   Calculating Simple Interest

Lisa borrows \$10,000 from her parents for a down payment on a house. She agrees to pay back the loan in 5 years at the rate of 6.5% simple interest. How much interest will Lisa pay?

Solution

Understand the Situation: For this situation, we have that the principal P is \$10,000, the interest rate r is $6.5\% = 0.065$, and the time t is 5 years.

From Your Toolbox

The "From Your Toolbox" makes study time more efficient. The feature pulls the appropriate review material into a given discussion or example. This timely review helps students stay on task, reducing the need to flip back through the text, hunting through previous sections for key information.

■ **Formula for Permutation**

The number of ways in which r objects can be selected in a specific order from among n objects is given by the **permutation** $_nP_r$ (read "n select r"), where

$$_nP_r = \frac{n!}{(n-r)!}$$

▶ **Note:** Permutations have two important characteristics:

1. Order is important
2. There is no repetition.

Let's take a Flashback to an example first presented in Section 6.1 to practice our newest counting technique.

 Flashback

Computer Club Revisited

There are 24 members in the New Strawsburg Junior High School Computer club. How many ways can the club pick a president, a vice-president, a secretary, and a treasurer?

Flashback Solution

In Section 6.1 we saw that order is important and there is no repetition. The names that we associate with the president, vice-president, secretary, and treasurer make a difference. So the total number of ways that the 4 club officers can be chosen is a 24 select 4 permutation, which is calculated by

$$_{24}P_4 = \frac{24!}{(24-4)!} = \frac{24!}{20!} = 24 \cdot 23 \cdot 22 \cdot 21 = 255{,}024$$

Interpret the Solution: So the Computer club officers can be selected in 255,024 different ways.

Notes

synthesize the ideas presented into everyday language—similar to how an instructor would explain in class.

Flashbacks

allow new topics to be introduced in a more natural way within a familiar context. The Flashback carries over an applied example from a previous section and then extends the content of the example by considering new questions.

Interactive Activities

accompany many examples and provide an opportunity for students to explore the example in greater depth. When the icon appears, a worked solution to the Interactive Activity may be found on the companion website. Interactive Activities are ideal for sparking classroom discussion or assigning as an independent or group project.

Example 2 **Applying the Multiplication Rule for Independent Events**

In a July 2000 Harris poll, it was found that 47% of the American public had independently read or heard something about the missile defense system that the Pentagon was developing. If two Americans are selected at random, determine the probability that both had heard about the missile defense system. (*Source:* www.harrisinteractive.com.)

Solution

Understand the Situation: Let A represent the event that the first randomly selected American had heard about the missile defense system and B represent the event that the second randomly selected American had heard about the missile defense system. The events A and B are independent since whether the first person has heard about the system does not affect whether the second person has heard about the system.

Since events A and B are independent, we can use the Multiplication Rule for Independent Events. This gives us

$$P(A \cap B) = P(A) \cdot P(B) = (0.47)(0.47) \approx 0.22$$

Interpretation: This means that if two Americans are selected at random there is about a 22% probability that both of them have read or heard something about the missile defense system. ■

Conditional Probability

We have seen that events are independent if the occurrence of one event does not affect the occurrence of the other. Our next step is to determine the probability $P(A \cap B)$ if the events are dependent. If the events are dependent, we must compute a new type of probability called **conditional probability** denoted by $P(B \mid A)$. We read $P(B \mid A)$ as "the probability of B given A." This means that we compute the probability of B given that A has occurred.

Exercise Sets

The section exercise set contains numerous skill builder problems, applications from many different disciplines, and an abundant supply of real data modeling applications, denoted by this symbol ⊕. The text contains over 3200 exercises.

- Exercises marked with a ✓ are called out in the body of the text as "Checkpoint" problems that relate closely to a worked in-text example.

- Building on the "Connections" approach, we use these symbols ⇐⇒ to indicate a problem that uses a scenario or concept that has been or will be revisited.

- The symbol ▦ denotes problems that are best solved using a graphing calculator.

In Exercises 29–34, use a calculator to determine the following:

29. $_{50}P_{20}$ **30.** $_{30}P_{15}$

31. $_{40}P_{15}$ **32.** $_{45}P_{28}$

33. $_{90}P_{30}$ **34.** $_{50}P_{41}$

35. The board of directors at the SafeSide Company consists of 12 members. In how many different ways can a chief executive officer, a vice-president and a chief financial officer be selected?

36. The Chess Club at Dyeston High School has 15 members. In how many different ways can a president, vice-president, and treasurer be selected?

37. The Miss Farrago County Fair queen contest has 19 entrants. In how many different ways can a winner, first runner-up, and second runner-up be selected?

38. Traditionally, 33 cars start the Indianapolis 500. In how many different ways can the drivers place first, second, and third?

39. Food critic Hero Maramoto has eaten at 25 restaurants this year. To introduce the top 4 restaurants in his newspaper column, how many choices does he have?

40. In a customer preference survey, a person is asked to try 16 different brands of applesauce and rank the first 5 in order of preference. How many different rankings are possible?

41. Cabin cruisers can communicate with each other by lofting 5 from among 12 flags up a mast. If the message is determined by the order in which the flags are raised, how many different messages are possible?

42. A researcher has found that 13 drugs are effective in treating a certain ailment. She conjectures that the sequencing of the administration of the drugs can affect their effectiveness. In how many different ways can 6 of the drugs be administered?

43. The starting lineup for the Jasper Comets softball team consists of 9 members. In how many ways can the manager arrange the batting order?

44. The 1921 class of Stringsbrug High School consists of 7 surviving members. In how many ways can they be arranged in 7 chairs, in a row, for a class photo?

In Exercises 45–56, evaluate the expression without using a calculator.

45. $_5C_3$ **46.** $_6C_2$

47. $_{10}C_6$ **48.** $_9C_5$

49. $_{12}C_5$ **50.** $_{11}C_4$

51. $_7C_1$ **52.** $_9C_1$

53. $_{13}C_0$ **54.** $_{17}C_0$

55. $_{12}C_{12}$ **56.** $_{14}C_{14}$

In Exercises 57–62, use a calculator to determine the following:

57. $_{50}C_{20}$ **58.** $_{30}C_{15}$

59. $_{40}C_{15}$ **60.** $_{45}C_{28}$

61. $_{90}C_{30}$ **62.** $_{50}C_{41}$

63. In Crawfords County, 6 judges are running for 3 county judge positions. In how many different ways can the 3 county judges be elected?

64. The central office of the Higgens Corporation has to choose 3 of its 8 regional offices for an audit. In how many different ways can the regional offices be chosen?

65. A group of 20 professors in an Archeology department wants to choose a research team of 5 to study a new archeological site. In how many different ways can the research team be formed?

66. For a literature class, a student has to read 4 books from a list of 21. In how many different ways can the 4 books be chosen?

67. There are 29 companies bidding for the construction of a new building. In how many different ways can 5 of the companies be chosen for hire?

68. How many different 5-card hands can be dealt from a standard deck of 52 cards?

69. In the Maine State lottery called *Megabucks*, 6 balls numbered 1 to 42 are drawn from a hopper. In how many different ways can the 6 numbers be chosen? (*Source:* www.mainelottery.com.)

70. In the Washington State lottery called Lotto, 6 balls numbered 1 to 49 are drawn from a hopper. In how many different ways can the 6 numbers be chosen? (*Source:* www.wa.gov/lot/home.htm.)

71. Jelly Belly manufactures 40 different flavors of jelly beans. If a person goes to a Jelly Belly store, they can get a sampler cup of 8 jelly beans in which the 8 jelly beans are all different flavors. How many different sampler cups are possible? (*Source:* www.jellybelly.com.)

72. Sweet Dreams has 34 different flavors of ice cream. One of its ice cream cones is a triple dip of 3 different flavors for which the order of the 3 different flavors does not matter. How many different kinds of triple-dip ice cream cone can be made?

73. A city council has 5 Democrats and 7 Republicans. In how many different ways can a subcommittee be formed with 2 Democrats and 4 Republicans?

74. The Euratal Company has 4 women and 8 men on its board of directors. In how many different ways can a gender-equality subcommittee of 4 be formed so that the numbers of men and women are equal?

■ SECTION PROJECT

We began this section by stating that Wassily Leontief organized the 1958 U.S. economy into an 81×81 matrix. He also aggregated these 81 sectors into six groups. This section project deals with four of these six groups. The technological matrix for the four groups is

$$T = \begin{matrix} & \text{FM} & \text{BM} & \text{BN} & \text{E} \\ \text{FM} \\ \text{BM} \\ \text{BN} \\ \text{E} \end{matrix} \begin{bmatrix} 0.295 & 0.018 & 0.002 & 0.004 \\ 0.173 & 0.46 & 0.007 & 0.011 \\ 0.037 & 0.021 & 0.403 & 0.011 \\ 0.001 & 0.039 & 0.025 & 0.358 \end{bmatrix}$$

FM represents final metals, such as construction machinery and household goods.
BM represents basic metal, such as mining and machine shop products.
BN represents basic nonmetal, such as glass, wood, and textiles.

E represents energy, such as coal, petroleum, electricity, and gas.

Matrix D gives the estimated external demands for the 1958 U.S. economy, in millions of dollars.

$$D = \begin{bmatrix} 75{,}548 \\ 14{,}444 \\ 33{,}501 \\ 23{,}527 \end{bmatrix} \begin{matrix} \text{FM} \\ \text{BM} \\ \text{BN} \\ \text{E} \end{matrix}$$

(a) Use a graphing calculator to determine the total output for each sector.
(b) If an energy crisis increased the cost of energy by 20% in each sector, what would the total output for each sector need to be?

■ Why We Learned It

In this chapter we learned that matrices are common in business, life science, and other occupations. Knowing the characteristics of matrices, as well as operations used with matrices, is vital for applying them to the many diversified applications. For example, we saw that a business may use matrices to determine the maximum production capacity of a product, and we saw that a farmer can use matrix multiplication to determine the amount of nutrients that have to be added to crops for a growing season. Another example of matrix operations is that matrices can be used to keep track of inventories and the sales of certain items and the revenue generated from these sales at different stores. Matrix multiplication quickly led us to learn about inverse matrices. Inverse matrices were used to solve systems of equations. For example, we saw how a commercial baker could use matrix inverses to determine how much of each ingredient is needed to fill an order. Finally, Leontief input–output models are an application of matrices and inverse matrices, frequently used by economists, to analyze the relationship between production and demands for various sectors in an economy.

■ CHAPTER REVIEW EXERCISES

In Exercises 1 and 2, determine the dimension of the given matrix.

1. $A = \begin{bmatrix} 0 & 5 \\ -1 & 7 \\ 2 & 8 \end{bmatrix}$

2. $B = \begin{bmatrix} -3 & 4 & 6 & -1 \end{bmatrix}$

In Exercises 3 and 4, determine the indicated entry for the matrices given in Exercises 1 and 2.

3. $a_{2,1}$
4. $b_{1,3}$

5. Write an augmented matrix for the given system. **Do not solve.**

$$\begin{aligned} -2x + 3y - z &= 5 \\ x + y + z &= 13 \\ 4x - 5y + 2z &= -1 \end{aligned}$$

6. Write the linear system corresponding to the augmented matrix.

$$\begin{bmatrix} -2 & 5 & | & -7 \\ 1 & 4 & | & 3 \end{bmatrix}$$

Section Projects

conclude each exercise set and provide an opportunity for students to explore the main chapter ideas in greater depth. Instructors can use section projects as standard hand-in assignments, collaborative activities, or as demos for graphing calculator usage.

Why We Learned It

parallels the chapter-opening "road map" and explains how the chapter topic is used in the real world.

Chapter Review Exercises

conclude each chapter and offer the same range of skill builder and applied problems as do the section projects. These exercises also serve to tie together ideas from several sections.

Chapter Project

A Chapter Project rounds out each chapter. These extensive exercises use real data modeling to explore and extend the ideas discussed in each chapter. They may be assigned as individual homework or as a group project.

This project about the computer game, *The Sims,* is a fun way for students to use math in a practical setting.

■ CHAPTER 4 PROJECT

SOURCE: www.thesims.com and *The Sims: Prima's Official Strategy Guide* by Rick Barba

Maxis is a computer game company known for its Sim games. One of the more recent and very popular computer games is called "The Sims." In the game, the player controls people (called Sims) and helps them to get a job and establish relationships and shows them the way to the bathroom. One of the more strategic points of the game is to make the Sims happy. To do this, the player must purchase accessories for the home, such as a couch or a bubble blower. Most objects will affect the Sims' motive(s): those being Hunger, Comfort, Hygiene, Bladder, Energy, Fun, Social, and Room. Some objects do a better job at satisfying needs than others. For instance, would you rather watch T.V. on a 13-inch black and white screen or on a 53-inch HDTV? Unfortunately, the objects that do a better job cost a lot more. The point of this project is to minimize the amount of money spent while satisfying the given needs.

We will consider six objects and only three motives: Fun, Room, and Comfort. Table 1 gives information about these six objects and the needs that they satisfy.

Suppose that a Sim needed at least 16 Fun points, 6 Room points, and 15 Comfort points.

4. Referring to the objective function in 1 and the results of 3, why does it make sense that these items were purchased?

5. Referring to 3, for some items more than one is needed. Suppose that we don't want to purchase more than two items. Repeat 1 and 2 and construct new constraints for those items so that no more than two are purchased. Do not solve the corresponding linear programming problem.

Table 1   Items and Corresponding Price, Fun, Room, and Comfort Values

Item	Price § (in Simoleans)	Fun	Room	Comfort
Super Schlooper Bubble Blower	§710	3	0	2
Liebefunkenmann Piano	§5399	4	6	0
Spazmatronic Plasmatronic Go-Go Cage	§1749	7	3	0

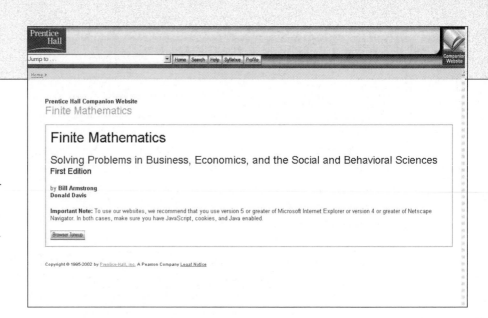

Companion Website

complements the text by offering a variety of teaching and learning resources including: a chapter quiz, solutions to many of the interactive activities, the online technology manuals referenced in the Technology Options, links to other useful resources, and a wealth of additional support material.

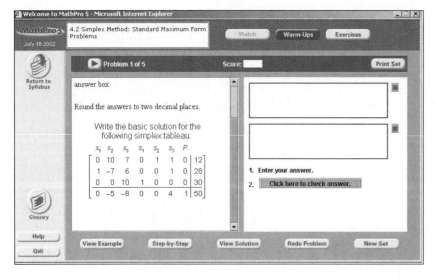

MathPro5 for Finite Math

MathPro5 is a online, customizable tutorial and assessment software that is fully integrated with *Finite Mathematics* at the section level for anytime, anywhere learning. This powerful and easy-to-use program allows students access to multiple types of help while working algorithmically-generated tutorial exercises for review and concept mastery.

Available Spring 2003

About the Authors

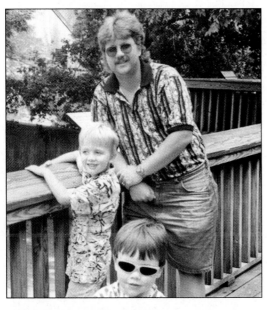

Bill Armstrong

Bill is a native of Ohio and became a hard-core Buckeyes fan after earning both his Bachelors and Masters degrees in Mathematics at The Ohio State University.

Bill taught at numerous colleges including Ohio State and Phoenix College, before taking his current position at Lakeland Community College near Cleveland, Ohio. Bill enjoys working with students at all levels and teaches courses ranging from Algebra to Differential Equations. He employs various teaching strategies to interest and motivate students including using humor to lighten the subject matter and inviting comments from students. His enjoyment of teaching and constant interaction with students has earned him a reputation as an innovative, enthusiastic, and effective teacher.

When Bill is not teaching, tutoring during office hours, or writing, he enjoys playing pool and golf, coaching his sons' little league baseball teams, and hanging out at home with his wife... not necessarily in that order!

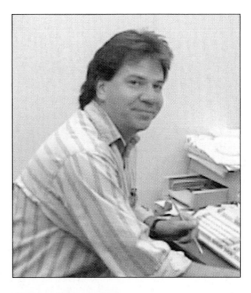

Don Davis

Also a native of Ohio, Don earned a Bachelor of Science degree in Education, specializing in Political Science, Economics, and Mathematics from Bowling Green State University in Bowling Green, Ohio. After teaching at the high school level, Don received his Master of Science degree in Mathematics from Ohio University. He taught at Ohio State University – Newark before joining the Lakeland Community College faculty. With his background in Economics, Don is always searching for ways to apply mathematics to the "real world" and to connect mathematical concepts to the various courses his students are taking. By using technology, along with interesting, practical applications, Don brings his many mathematics classes to life for students. Like Bill, one typically finds Don in his office helping students understand the concepts of his courses. His dedication to his students makes him a sought-after teacher.

In his free time, Don is often found playing with his three children or scurrying after one of the many pets currently in residence at his home including a rat, two parakeets, a turtle, a lizard, a dog, and two cats. Occasionally, he enjoys a quiet night at home.

FINITE MATHEMATICS

Solving Problems in Business,
Economics, and the Social
and Behavioral Sciences

Linear Models and Systems of Linear Equations

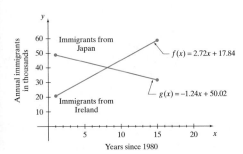

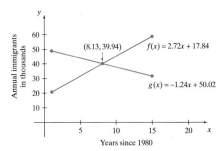

The Statue of Liberty has been a beacon of hope for immigrants to the United States for many years. Although the flow of immigrants slowed by the end of the 20th century, America continued to open her doors to men and women from across the globe. By using linear models and graphing techniques we can compare patterns of immigration and predict future trends, as seen in these graphs of Japanese and Irish immigrants. In the graph to the right, the intersection point shows that the number of immigrants from Japan and Ireland were equal in 1988.

What We Know

We begin our study of finite mathematics having learned the basic skills in algebra. These skills include solving, factoring, and graphing equations.

Where Do We Go

In this chapter we will review the concept of problem solving, explore properties of linear functions, and study the average rate of change. Then we will see how to solve two linear equations simultaneously using graphic and algebraic techniques.

Section 1.1   Problem Solving: Linear Equations

Most of us use or hear the word *problem* in our daily conversation. We may say or hear statements such as

- "I'm having a problem loading this software."
- "The problem is that management doesn't communicate with us."
- "I had a problem getting tickets for the concert."

Many problems that we encounter can be solved easily, but others require a more thoughtful approach. In finite mathematics we are often confronted with the challenge of solving problems that require a thoughtful approach. As a result, we need to review a general problem-solving strategy and the accompanying skills that will allow us to solve a variety of problems. The skills reviewed in this section will be used throughout the text.

English to Mathematics

From prior mathematics classes we have seen how a **variable** can be used to represent an unknown quantity. For example, suppose that we hear the following report on the local news:

"CleanAir produces 1475 of two types of vacuums, Mod I and Mod II, each week."

From this statement we do not know how many of each vacuum is made each week, but from the report we do know that

The number of Mod I produced plus the number of Mod II produced is 1475.

Now, if we let x represent the number of Mod I produced and let y represent the number of Mod II produced, it is more convenient to write the prior English sentence as the mathematical sentence

$$x + y = 1475$$

Converting English to mathematics is an important skill. An essential part of this skill is called **defining the variable**. This is when we state what the variable represents. Defining the variable occurs only after we carefully read the problem and understand what we are asked to do. Example 1 eases us into this process.

Example 1   Translating English to Mathematics

A certain number tripled plus seven is nineteen. Translate this English sentence to a mathematical sentence.

Solution

After reading the first sentence, we realize that we need to define a variable. So we let x represent the certain number. By "a certain number tripled," we mean the product of 3 and a certain number. Thus $3x$ represents this number tripled.

So we have

A certain number tripled plus seven is nineteen.	English sentence
$3x + 7 = 19$	Mathematical sentence

■

✓ **Checkpoint 1**

Now work Exercise 5.

The mathematical sentence we came up with in Example 1,

$$3x + 7 = 19$$

is more commonly called an **equation**. Once we have an equation, we can solve it and answer the original problem. Solving the equation in Example 1 gives us

$3x + 7 = 19$	Given equation
$3x + 7 - 7 = 19 - 7$	Subtract 7 from both sides
$3x = 12$	Simplify
$x = 4$	Divide both sides by 3

So the "certain number" in Example 1 was $x = 4$. This can be checked by going back to the original English sentence.

A certain number tripled plus seven is nineteen.

We found that the certain number is $x = 4$. Four tripled is 12; $12 + 7$ is 19.

Example 1 and the discussion following Example 1 illustrate the general strategy for analyzing and solving word problems.

■ **Strategy for Solving Word Problems**

1. Carefully **read** the problem and **understand the situation**. As you analyze a problem, you may need to

   • **Define the variables** to represent all the unknowns.
   • **Construct** a mathematical model to represent the problem or write an equation (or equations) to represent the problem.

2. **Solve** the equation. Make sure to check the solution in the original equation.

3. **Interpret the solution**.

Example 2   Solving a Word Problem

The Dirt Rocket Company produces spokes for off road motorcycles. Each spoke costs $3 to produce. The company also spends $1250 for advertisements in a certain motorcycle enthusiast magazine.

(a) Determine the total cost for advertising and producing 400 spokes.

(b) Write an equation for the costs incurred for advertisement and the production of any number of spokes.

Solution

(a) **Understand the Situation:** If The Dirt Rocket Company produces no spokes, they still have to pay the $1250 for advertising. Such a cost is called a **fixed cost**. Since each spoke costs $3 to produce, the cost to produce 400 spokes is given by

$$\text{(cost per spoke)} \cdot \text{(number of spokes)}$$
$$3(400) = \$1200$$

Letting C represent the total cost of producing 400 spokes and paying for the advertisements gives us

$$C = \text{(cost per spoke)} \cdot \text{(number of spokes)} + \text{(advertisement cost)}$$
$$= 3 \cdot 400 + 1250$$
$$= 1200 + 1250$$
$$= \$2450$$

Interpret the Solution: The total cost, C, of producing 400 spokes and the advertisements is $2450.

(b) **Understand the Situation:** Let C represent the total cost of advertising and producing any number of spokes. Since we do not know how many spokes are produced, we let x represent the number of spokes produced.

Using the variables C and x, we can write an equation to model the problem. This gives us

$$C = \text{(cost per spoke)} \cdot \text{(number of spokes)} + \text{(advertisement cost)}$$
$$= 3x + 1250 \qquad \blacksquare$$

✓**Checkpoint 2**

Now work Exercise 13.

Systems of Linear Equations

In the first part of this text we often need to represent a situation with a **system of linear equations**. (A system of linear equations is basically two or more linear equations with at least one variable.) We will review the techniques for solving systems of linear equations in Sections 1.4 and 1.5. For now, we focus on representing a word problem with a system of linear equations. Example 3 illustrates how to set up linear equations that form a system. You will solve this system in Exercise 67 of Section 2.1.

Example 3   Writing Multiple Linear Equations

Max George is a craftsman who makes end tables and night stands. To make one end table takes 5 hours of his labor, and to make one night stand takes 4 hours of his labor. The materials for one end table cost him $200 and the materials for one night stand cost him $80. Suppose that for 1 week Max works 40 hours and spends $1200 on materials.

(a) Write an equation that represents Max's costs.
(b) Write an equation that represents Max's time (labor).

Solution

(a) **Understand the Situation:** We know that materials for each end table cost Max $200 and materials for each night stand cost Max $80. We also know that Max spent $1200 on materials. What we don't know is how many end tables and how many night stands he made. So we let x represent the number of end tables made in 1 week and let y represent the number of night stands made in 1 week.

This means that he spent

$$\underset{200}{(\text{cost per end table})} \cdot \underset{x}{(\text{number of end tables})} = 200x$$

dollars on end tables. We also know that he spent

$$\underset{80}{(\text{cost per night stand})} \cdot \underset{y}{(\text{number of night stands})} = 80y$$

dollars on night stands. Since he spent a total of $1200 on materials, an equation that represents Max's costs is

$$\left(\begin{array}{c}\text{dollars spent}\\\text{on end tables}\end{array}\right) + \left(\begin{array}{c}\text{dollars spent}\\\text{on night stands}\end{array}\right) = \left(\begin{array}{c}\text{total dollars}\\\text{spent on materials}\end{array}\right)$$

$$200x + 80y = 1200$$

(b) **Understand the Situation:** We know that to make each end table takes Max 5 hours and to make each night stand takes Max 4 hours. We also know that Max spent a total of 40 hours working on these items in 1 week. What we don't know is how many end tables and how many night stands he made. As in part (a), we let x represent the number of end tables made in 1 week and let y represent the number of night stands made in 1 week.

Then we know that he spent

$$\underset{5x}{(\text{hours per end table})} \cdot (\text{number of end tables})$$

hours on end tables. We also know that he spent

$$\underset{4y}{(\text{hours per night stand})} \cdot (\text{number of night stands})$$

hours on night stands. Since he spent a total of 40 hours working on these items, an equation that represents Max's time is

$$\left(\begin{array}{c}\text{hours spent}\\\text{on end tables}\end{array}\right) + \left(\begin{array}{c}\text{hours spent}\\\text{on night stands}\end{array}\right) = \left(\begin{array}{c}\text{total hours}\\\text{spent on both}\end{array}\right)$$

$$5x + 4y = 40 \qquad \blacksquare$$

✓ **Checkpoint 3**   Now work Exercise 15.

In Example 3 notice that in both parts (a) and (b) we let x represent the number of end tables made in 1 week and let y represent the number of night stands made in 1 week. Since each equation had the same variables defined to represent the same unknown quantities, we can write the equations as the **linear system**

$$200x + 80y = 1200$$
$$5x + 4y = 40$$

Solving this linear system requires finding a value for x and a value for y that makes both equations true. As stated earlier, you will learn how to solve systems later in Chapter 1.

Example 4   Writing Multiple Linear Equations

Gonzalez Office Supplies manufactures three types of office chairs: Secretarial, Executive, and Presidential. Each chair requires time on three different machines. These times and the total available time for each machine are given in Table 1.1.1.

Table 1.1.1

	Secretarial	Executive	Presidential	Total Weekly Hours
Hours on Machine A	1	1	2	30
Hours on Machine B	2	3	2	53
Hours on Machine C	1	2	3	47

Assuming that all available time for each machine must be used each week, do the following:

(a) Write an equation that represents the total number of hours that all three chairs spend on Machine A.

(b) Write an equation that represents the total number of hours that all three chairs spend on Machine B.

(c) Write an equation that represents the total number of hours that all three chairs spend on Machine C.

Solution

(a) **Understand the Situation:** Let x represent the number of Secretarial chairs made in 1 week, y represent the number of Executive chairs made in 1 week, and z represent the number of Presidential chairs made in 1 week. Since x represents the number of Secretarial chairs made in 1 week, and it takes 1 hour of machine time on Machine A, the weekly hours on Machine A for x Secretarial chairs is given by

$$\left(\begin{array}{c}\text{hours per}\\\text{Secretarial chair}\end{array}\right) \cdot \left(\begin{array}{c}\text{number of}\\\text{Secretarial chairs}\end{array}\right) = \left(\begin{array}{c}\text{hours on Machine A}\\\text{spent on Secretarial chairs}\end{array}\right)$$

$$1 \cdot x = x$$

Similarly, the weekly hours on Machine A for y Executive chairs is given by

$$\left(\begin{array}{c}\text{hours per}\\\text{Executive chair}\end{array}\right) \cdot \left(\begin{array}{c}\text{number of}\\\text{Executive chairs}\end{array}\right) = \left(\begin{array}{c}\text{hours on Machine A}\\\text{spent on Executive chairs}\end{array}\right)$$

$$1 \cdot y = y$$

We know that it takes 2 hours on Machine A for each Presidential chair. So for z Presidential chairs, we have

$$\left(\begin{array}{c}\text{hours per}\\\text{Presidential chair}\end{array}\right) \cdot \left(\begin{array}{c}\text{number of}\\\text{Presidential chairs}\end{array}\right) = \left(\begin{array}{c}\text{hours on Machine A}\\\text{spent on Presidential chairs}\end{array}\right)$$

$$2 \cdot z = 2z$$

So if we have a total of 30 hours available on Machine A each week that must be used, we get the equation

$$\left(\begin{array}{c}\text{hours spent on}\\\text{Secretarial chairs}\end{array}\right)+\left(\begin{array}{c}\text{hours spent on}\\\text{Executive chairs}\end{array}\right)+\left(\begin{array}{c}\text{hours spent on}\\\text{Presidential chairs}\end{array}\right)=\left(\begin{array}{c}\text{total weekly hours}\\\text{on Machine A}\end{array}\right)$$

$$x+y+2z=30$$

(b) For Machine B, we know that each Secretarial chair takes 2 hours, each Executive chair takes 3 hours, and each Presidential chair takes 2 hours. So for a total of 53 hours available on Machine B that must be used, we get the equation

$$\left(\begin{array}{c}\text{hours spent on}\\\text{Secretarial chairs}\end{array}\right)+\left(\begin{array}{c}\text{hours spent on}\\\text{Executive chairs}\end{array}\right)+\left(\begin{array}{c}\text{hours spent on}\\\text{Presidential chairs}\end{array}\right)=\left(\begin{array}{c}\text{total weekly hours}\\\text{on Machine B}\end{array}\right)$$

$$2x+3y+2z=53$$

(c) For Machine C, we know that each Secretarial chair takes 1 hour, each Executive chair takes 2 hours and each Presidential chair takes 3 hours. So for a total of 47 hours available on Machine C that must be used, we get the equation

$$\left(\begin{array}{c}\text{hours spent on}\\\text{Secretarial chairs}\end{array}\right)+\left(\begin{array}{c}\text{hours spent on}\\\text{Executive chairs}\end{array}\right)+\left(\begin{array}{c}\text{hours spent on}\\\text{Presidential chairs}\end{array}\right)=\left(\begin{array}{c}\text{total weekly hours}\\\text{on Machine C}\end{array}\right)$$

$$x+2y+3z=47 \qquad ■$$

Since x, y, and z represent the number of Secretarial, Executive, and Presidential chairs, respectively, for each of the three linear equations, we can express the time used on all three machines with the system of linear equations

$$x+y+2z=30$$
$$2x+3y+2z=53$$
$$x+2y+3z=47$$

We will solve and interpret this system in a Flashback in Section 1.5.

SUMMARY

In this section, we learned how to translate English sentences into mathematical sentences. Regardless of the application, we use the **Strategy for Solving Word Problems**:

1. Carefully **read** the problem and **understand the situation**. As you analyze a problem, you may need to

   • **Define the variables** to represent all the unknowns.

   • **Construct** a mathematical model to represent the problem or write an equation (or equations) to represent the problem.

2. **Solve** the equation. Make sure to check the solution in the original equation.

3. **Interpret the solution**.

SECTION 1.1 EXERCISES

In our exercise sets, we often revisit applications. When a number appears under the ⇒ symbol, it means that the application will be seen again or was previously seen in that section.

In Exercises 1–10, translate the English sentence into a mathematical sentence.

1. A certain number increased by ten is 14.

2. A certain number increased by four is six.

3. A certain number decreased by seven is 28.

4. A certain number decreased by one is five.

✓ **5.** The product of five and a certain number is 45.

6. The product of two and a certain number is 112.

7. The product of six and a certain number minus one is 29.

8. The product of four and a certain number minus three is five.

9. The product of three and a certain number plus 11 is 44.

10. The product of eight and a certain number plus nine is 57.

⇒
1.4 **11.** Adel starts a business of manufacturing picture frames for small 3 by 5 inch photographs. He determines the fixed cost to be $252 and the cost to produce each frame is $1.50.

(a) Determine the total cost for producing 10 picture frames.

(b) Determine an equation for the total cost of producing any number of picture frames.

⇒
1.4 **12.** Sally makes dream catchers as a hobby and wants to sell them at craft shows. She knows that the fixed costs are $90 and the cost to produce each dream catcher is $2.50.

(a) Determine the total cost for producing 18 dream catchers.

(b) Determine an equation for the total cost of producing any number of dream catchers.

⇒
1.4 ✓ **13.** The Lullaby Company knows that the fixed costs to manufacture their portable crib are $250 and the cost to produce each crib is $100.

(a) Determine the total cost for producing 110 portable cribs.

(b) Determine an equation for the total cost of producing any number of portable cribs.

⇒
1.4 **14.** Clark and Florence make bumper stickers. They know that their fixed costs are $200 and the cost of each bumper sticker is $3.

(a) Determine the total cost for producing 200 bumper stickers.

(b) Determine an equation for the total cost of producing any number of bumper stickers.

⇒
2.1 ✓ **15.** Martha George is a craftswoman who makes coffee tables and night stands. To make 1 coffee table takes 6 hours of her labor, and to make 1 night stand takes 4 hours of her labor. The materials for 1 coffee table cost her $220, and the materials for 1 night stand cost her $80. Suppose that for 1 week Martha works 40 hours and spends $1400 on materials. Let x represent the number of coffee tables made in 1 week and let y represent the number of night stands made in 1 week.

(a) Write an equation that represents Martha's costs.

(b) Write an equation that represents Martha's time (labor).

⇒
2.1 **16.** The WeDo Wood Canoes Company makes two types of canoes: a two-person model and a four-person model. Each two-person model requires 1 hour in the cutting department and 1.5 hours in the assembly department. Each four-person model requires 1 hour in the cutting department and 2.75 hours in the assembly department. The cutting department has a maximum of 640 hours available each week, while the assembly department has a maximum of 1080 hours available each week. Let x represent the number of two-person canoes made each week and let y represent the number of four-person canoes made each week.

(a) Write an equation that represents the number of hours that both models of canoe spend in the cutting department each week, assuming that all available hours are used.

(b) Write an equation that represents the number of hours that both models of canoe spend in the assembly department each week, assuming that all available hours are used.

⇒
2.1 **17.** (*continuation of Exercise 16*) Suppose that a breakthrough in the assembly process decreases the time required in the assembly department for each two-person model to $\frac{3}{4}$ hour and for each four-person model to 2 hours. The cutting department still has a maximum of 640 hours available each week, while the assembly department has a maximum of 1080 hours available each week. Let x represent the number of two-person canoes made each week and let y represent the number of four-person canoes made each week.

(a) Write an equation that represents the number of hours that both models of canoe spend in the cutting department each week, assuming that all available hours are used.

(b) Write an equation that represents the number of hours that both models of canoe spend in the assembly department each week, assuming that all available hours are used.

18. The SnackPower company has two ingredients to combine to make their new ZapSnack energy bar. Ingredient I contains 3 grams of carbohydrate and 2 grams of protein per ounce. Ingredient II contains 5 grams of carbohydrate and 3 grams of protein per ounce. They want to combine the two ingredients so that the energy bar has 73 grams of carbohydrate and 46 grams of protein per ounce. Let x represent the number of ounces of ingredient I and let y represent the number of ounces of ingredient II.

(a) Write an equation that represents the number of ounces of carbohydrates in the two ingredients.

(b) Write an equation that represents the number of ounces of protein in the two ingredients.

⇒
2.4 **19.** The Drama Department in a local high school is performing a play for three consecutive nights. The auditorium at the high school has 1600 seats, and tickets are broken into two categories: $3 student tickets and $5 adult tickets. Table 1.1.2 gives the data for ticket revenue and total tickets sold for each play night.

Table 1.1.2

	Thursday	Friday	Saturday
Tickets sold	1600	1600	1600
Revenue	$5760	$5980	$5500

(a) Define the variables and write an equation representing the number of tickets sold on Thursday night, and write an equation representing the revenue generated on Thursday night.

(b) Define the variables and write an equation representing the number of tickets sold on Friday night, and write an equation representing the revenue generated on Friday night.

(c) Define the variables and write an equation representing the number of tickets sold on Saturday night, and write an equation representing the revenue generated on Saturday night.

$\overset{\Rightarrow}{2.4}$ **20.** (*continuation of Exercise 19*) Because each night was a sellout, the next semester the Drama Department moved the play to a local community college. The auditorium at the community college has 2800 seats. The ticket prices remain the same. Table 1.1.3 gives the data for ticket revenue and total tickets sold for each play night.

Table 1.1.3

	Thursday	Friday	Saturday
Tickets sold	2800	2800	2800
Revenue	$11,400	$10,180	$10,760

(a) Define the variables and write an equation representing the number of tickets sold on Thursday night, and write an equation representing the revenue generated on Thursday night.

(b) Define the variables and write an equation representing the number of tickets sold on Friday night, and write an equation representing the revenue generated on Friday night.

(c) Define the variables and write an equation representing the number of tickets sold on Saturday night, and write an equation representing the revenue generated on Saturday night.

21. The Second City Bank has two types of checking accounts. The City account charges $5 per month plus $0.15 for each check that is written. The Urban account charges $4 per month plus $0.20 for each check that is written.

(a) Define the variables and write an equation that represents the monthly cost of the City checking account.

(b) Define the variables and write an equation that represents the monthly cost of the Urban checking account.

22. Membership at the Iron Wrist Tennis Club costs $1000 per year plus $3 per hour of court usage. The Line-Out! Tennis Club has an $800 yearly fee with $5 per hour court fees.

(a) Define the variables and write an equation that represents the total yearly cost of playing at the Iron Wrist Tennis Club.

(b) Define the variables and write an equation that represents the total yearly cost of playing at the Line-Out! Tennis Club.

23. The Frank & Earnst Law Firm charges a $1500 flat fee plus 25% of the settlement for litigating a civil case. The Lopez & Mackenzie Law Firm charges a $2000 flat fee plus 20% of the settlement for litigating a civil case.

(a) Define the variables and write an equation that represents the cost that the Frank & Earnst Law Firm charges for litigating a civil case.

(b) Define the variables and write an equation that represents the cost that the Lopez & Mackenzie Law Firm charges for litigating a civil case.

24. The Yorkville School District has two property tax proposals. Under proposal D, a homeowner pays $500 in addition to 15% of the assessed value of the house. Under proposal E, a homeowner pays $2000 in addition to 4% of the assessed value of the house.

(a) Define the variables and write an equation for the property tax under proposal D.

(b) Define the variables and write an equation for the property tax under proposal E.

$\overset{\Rightarrow}{2.1}$ **25.** The Stitz Tent Company makes three types of tents: two- , four- , and six-person models. Each tent requires the services of three departments as shown in Table 1.1.4. The fabric cutting, assembly, and packaging departments have available a maximum of 450, 390, and 190 hours each week, respectively.

Table 1.1.4

	Two-person (hr)	Four-person (hr)	Six-person (hr)
Fabric cutting	0.5	0.8	1.1
Assembly	0.6	0.7	0.9
Packaging	0.2	0.3	0.6

Let x represent the number of two-person tents made in 1 week, y represent the number of four-person tents made in 1 week, and z represent the number of six-person tents made in 1 week.

(a) Write an equation that represents the number of hours that the three models spend in the fabric cutting department per week, assuming that all available hours are used.

(b) Write an equation that represents the number of hours that the three models spend in the assembly department per week, assuming that all available hours are used.

(c) Write an equation that represents the number of hours that the three models spend in the packaging department per week, assuming that all available hours are used.

$\overset{\Rightarrow}{2.1}$ **26.** Rework Exercise 25(a), (b), and (c), except here assume that the fabric cutting, assembly, and packaging departments have available a maximum of 420, 390, and 185 hours each week, respectively.

$\overset{\Rightarrow}{2.1}$ **27.** Last year the Hannus Corporation took out three loans totaling $50,000 to fund the purchase of new equipment. Some of the money was financed at 6%, $1000 more than $\frac{1}{2}$ the amount of the 6% loan was financed at 8%, and the rest was financed at 7%. The total annual interest on the loans was $3410. Let x represent the amount borrowed at 6%, y represent the amount borrowed at 7%, and z represent the amount borrowed at 8%.

(a) Write an equation that represents the first sentence.

(b) Write an equation that represents "$1000 more than $\frac{1}{2}$ the amount of the 6% loan was financed at 8%."

(c) Write an equation that represents the third sentence.

28. Rework Exercise 27(a), (b), and (c), except here assume that the three loans totaled $60,000 and the total annual interest on the loans was $4085.

⇒ 2.1 🌐 **29.** The data in Table 1.1.5 give the amounts of nitrogen, phosphate, and potash (all in lb/acre) needed to grow asparagus, cauliflower, and celery for an entire growing season in California.

Table 1.1.5

	Asparagus	Cauliflower	Celery
Nitrogen	240	215	336
Phosphate	148	98	171
Potash	42	69	147

SOURCE: U.S. Department of Agriculture, www.usda.gov/nass

Juan has 280 acres, 65,790 pounds of nitrogen, 31,860 pounds of phosphate and 21,630 pounds of potash. Let x represent the number of acres planted in asparagus, y represent the number of acres planted in cauliflower, and z represent the number of acres planted in celery. Assume that Juan wants to use all his resources completely.

(a) Write an equation that represents the total amount of nitrogen needed for the three crops.

(b) Write an equation that represents the total amount of phosphate needed for the three crops.

(c) Write an equation that represents the total amount of potash needed for the three crops.

⇒ 2.1 🌐 **30.** The data in Table 1.1.6 give the amounts of nitrogen, phosphate, and potash (all in lb/acre) needed to grow head lettuce, cantaloupe, and bell peppers for an entire growing season in California.

Table 1.1.6

	Head lettuce	Cantaloupe	Bell peppers
Nitrogen	173	175	212
Phosphate	122	161	108
Potash	76	39	72

SOURCE: U.S. Department of Agriculture, www.usda.gov/nass

Zach has 200 acres, 38,430 pounds of nitrogen, 27,845 pounds of phosphate, and 12,985 pounds of potash. Let x represent the number of acres planted in head lettuce,

y represent the number of acres planted in cantaloupe, and z represent the number of acres planted in bell peppers. Assume that Zach wants to use all his resources completely.

(a) Write an equation that represents the total amount of nitrogen needed for the three crops.

(b) Write an equation that represent the total amount of phosphate needed for the three crops.

(c) Write an equation that represent the total amount of potash needed for the three crops.

⇒ 2.1 **31.** Chris O'Neill, a dietitian in a local hospital, is to arrange a special diet for a patient that is made up of three basic foods. The diet is to include 340 units of calcium, 205 units of vitamin A, and 225 units of vitamin C. The number of units per ounce for each basic food is given in Table 1.1.7.

Table 1.1.7

	Basic!	Vaca	Secra
Calcium	20	10	30
Vitamin A	10	5	20
Vitamin C	20	15	10

Assume that the dietitian wants to exactly meet the diet requirements.

(a) Define the variables and write an equation that represents the total amount of calcium from the three foods.

(b) Define the variables and write an equation that represents the total amount of vitamin A from the three foods.

(c) Define the variables and write an equation that represents the total amount of vitamin C from the three foods.

⇒ 2.1 **32.** Rework Exercise 31 if the patient's diet is changed so that it includes 480 units of calcium, 285 units of vitamin A, and 325 units of vitamin C.

⇒ 2.4 **33.** Davram manufactures bicycle frames at three different plants. Due to different technologies at each plant, the production rate (in frames/hr) differs for each plant, as shown in Table 1.1.8.

Table 1.1.8

	Plant I	Plant II	Plant III
Racing frame	18	20	16
Hybrid frame	12	9	11
Mountain frame	22	18	17

Shannon, the production manager at Davram, has received an order to produce 5900 racing frames, 3580 hybrid frames, and 6350 mountain frames. Shannon needs to fill this order exactly.

(a) Define the variables and write an equation that represents the total number of racing frames from the three plants.

(b) Define the variables and write an equation that represents the total number of hybrid frames from the three plants.

(c) Define the variables and write an equation that represents the total number of mountain frames from the three plants.

$\overset{\Rightarrow}{2.4}$ **34.** (*continuation of Exercise 33*) Shannon has received another order requesting 6040 racing frames, 3780 hybrid frames, and 6695 mountain frames. With everything else from Exercise 33 remaining the same, rework parts (a), (b), and (c).

$\overset{\Rightarrow}{2.1}$ ▌ ## SECTION PROJECT

During rush hour, traffic congestion may be encountered at certain intersections. For a network of four one-way streets in a certain city, the rush hour flow is as shown in Figure 1.1.1. To help to speed the flow of traffic, traffic engineers need to coordinate the signals at these intersections. To do this, data

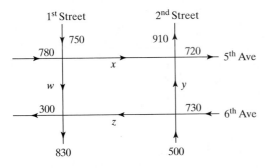

Figure 1.1.1

are collected giving the hourly flow of traffic during rush hour at certain locations. Figure 1.1.1 shows the data that were collected. Notice that the figure indicates that a total of 1530 cars per hour enter the intersection of 1st Street and 5th Avenue. Of this 1530 cars, x of these cars leave the intersection on 5th Avenue and w of these cars leave the intersection on 1st Street. Since the total number of cars leaving this (or any) intersection equals the number entering, we have

$$x + w = 1530$$

This equation gives the flow at the intersection of 1st Street and 5th Avenue.

(a) Determine the equations for the traffic flow at the other three intersections. Remember that the total entering the intersection equals the total leaving the intersection, and vice versa.

(b) Write a system of four equations representing the traffic flow for this network of four one-way streets.

Section 1.2   Introduction to Linear Functions

Many times in life, what people desire is some measure of predictability. There is a certain solace in knowing that if today is Tuesday, then it is garbage day. This same kind of idea is also sought in mathematics. We want to know that if we are given a value x we can determine a value for y, without ambiguity. This, in part, is what the **function concept** is about. In this section we will learn how functions are defined, what makes up functions, and the usefulness of functions. We will also explore properties of a special type of function called the **linear function**.

The Function Concept

Suppose that Manuel takes a typing class and, during the course of the semester, takes a series of tests to determine how fast he is typing. His results are shown in Table 1.2.1.

Table 1.2.1

Week of class	0	3	5	7	9	12
Typing speed in words per minute	10	20	28	45	50	54

To make a picture of these data, we can use the **Cartesian plane**. This rectangular coordinate system is made up of two perpendicular number lines called **axes**. Usually, the horizontal number line is called the **x-axis**, and the vertical number line is called the **y-axis**. See Figure 1.2.1.

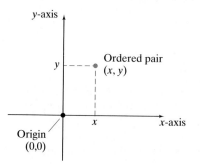

Figure 1.2.1

Since we want to investigate the relationship between the two sets of data, we can plot the data in the table by making **ordered pairs** consisting of x- and y-**coordinates** in the form (x, y). For Manuel, if we let x represent the week of typing class and let y represent typing speed in words per minute, we can construct ordered pairs of the form (x, y) as shown in Table 1.2.2.

Table 1.2.2

Week of class, x	0	3	5	7	9	12
Typing speed in words per minute, y	10	20	28	45	50	54
Ordered pair (x, y)	(0, 10)	(3, 20)	(5, 28)	(7, 45)	(9, 50)	(12, 54)

Now we can plot each ordered pair or **point** by moving to the right x units and up y units. The graph of the typing speed data is given in Figure 1.2.2. Note the labels on the axes.

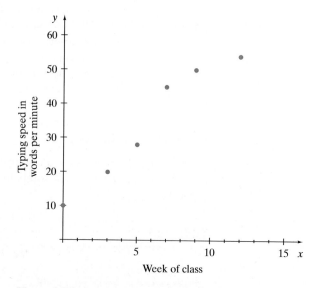

Figure 1.2.2

The choice of letting x represent the number of weeks in class and y the typing speed was not done arbitrarily. We commonly call x the **independent** variable and y the **dependent** variable. Since the typing speed presumably *depends* on the length of time that Manuel is in the typing class, we assigned y to the typing speed.

Notice that if we know what week of class it is we can determine the typing speed. This is much of what the function concept is all about! Now let's apply a mathematical definition to the function concept.

Function

A **function** is a rule, or a set of ordered pairs, that assigns a unique y-value to each x-value. The set of all possible x-values is called the **domain**, and the set of all possible y-values is called the **range**.

Example 1   **Plotting Ordered Pairs and Identifying the Domain and Range**

Plot the data in Table 1.2.3 on the Cartesian plane and write the sets for the domain and range.

Table 1.2.3

x	−2	−1	0	1	2
y	3	−2	4	5	0

Solution

Notice here that some of the x-values are negative, so we must include negative numbers on the axes on our Cartesian plane. The data are plotted in Figure 1.2.3.

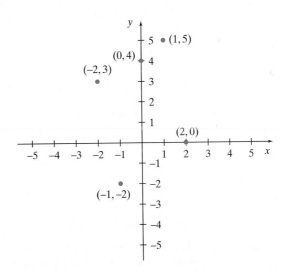

Figure 1.2.3

Since the domain is the set of all possible x-values, the domain is the set $\{-2, -1, 0, 1, 2\}$. Since the range is the set of all possible y-values, the range is the set $\{-2, 0, 3, 4, 5\}$. Note that we write the elements of these sets in ascending order, that is, from smallest to largest.   ■

✓ **Checkpoint 1**   Now work Exercise 11.

We often desire to write a convenient algebraic expression to represent a function. We denote a function by writing $f(x)$, read "f of x," where f is the name of the function and x is the independent variable. This notation allows us to write functions such as

$$f(x) = 3x - 4 \qquad g(x) = \frac{9x + 6}{4} \qquad h(t) = t^2 - 2t$$

Notice that we can designate functions with names like f, g, and h. The letters x and t are used to denote independent variables. **This means that $f(x)$, $g(x)$, and $h(t)$ are all different names for the dependent variable y.** To get ordered pairs from these functions, we must substitute (or "plug in") a value for the independent variable and compute the corresponding value of the dependent variable. This process of substituting values is called **evaluating** a function. When we write $f(4)$, for example, this means to replace every occurrence of the independent variable with 4 and then simplify. The process of evaluating functions is shown in Example 2.

Example 2   **Evaluating a Function**

For the function $f(x) = 2x - 3$, evaluate $f(-1)$, $f(0)$, $f(2)$, and $f(4)$ and then make a table of values including ordered pairs.

Solution

Evaluating this function, we get

$$f(-1) = 2(-1) - 3 = -2 - 3 = -5$$
$$f(0) = 2(0) - 3 = 0 - 3 = -3$$
$$f(2) = 2(2) - 3 = 4 - 3 = 1$$
$$f(4) = 2(4) - 3 = 8 - 3 = 5$$

The values and ordered pairs are shown in Table 1.2.4.

Table 1.2.4

x	-1	0	2	4
y	-5	-3	1	5
Ordered pairs (x, y)	$(-1, -5)$	$(0, -3)$	$(2, 1)$	$(4, 5)$

Linear Functions

Many of the topics that we study in finite mathematics focus on the study of **linear functions**. When graphed, a linear function has the shape of a line. But in what ways are linear functions used? Let's say that Ken has a pickup truck with a 12-gallon gas tank. From experience, he knows that the truck consumes $\frac{1}{20}$ of a gallon of gas for each mile driven. (This is another way to say that the truck gets 20 miles per gallon.) This means, for example, that if he drives the truck 60 miles he uses $60\left(\frac{1}{20}\right) = 3$ gallons of gas. We can say that the amount of gas in the tank has **decreased** by 3 gallons. Now let's suppose that Ken starts a road trip with a full tank of gas and wants to know how many gallons of gas remain in the tank as he takes his trip. To solve this problem, let's begin by **defining the variables**.

> **Define the Variables:** Let x represent the number of miles traveled and $y = f(x)$ represent the number of gallons of gas remaining in the tank.

To see how a function can be written, we make a table to represent the number of gallons left in the tank as various mileages are driven. See Table 1.2.5.

Table 1.2.5

Miles driven, x	Calculation	Gallons remaining, $y = f(x)$
0	$12 - \frac{1}{20}(0)$	12
60	$12 - \frac{1}{20}(60)$	9
120	$12 - \frac{1}{20}(120)$	6
180	$12 - \frac{1}{20}(180)$	3
240	$12 - \frac{1}{20}(240)$	0

The general form for the function is

$$\left(\begin{array}{c}\text{gallons remaining}\\ \text{in the tank}\end{array}\right) = \left(\begin{array}{c}\text{gas tank}\\ \text{capacity}\end{array}\right) - \left(\begin{array}{c}\text{rate of gas}\\ \text{consumption}\end{array}\right)\left(\begin{array}{c}\text{miles}\\ \text{driven}\end{array}\right)$$

In terms of our defined variables, we have

$$f(x) = 12 - \frac{1}{20}x, \qquad 0 \le x \le 240$$

The graph of this function is shown in Figure 1.2.4. Notice that the graph only applies for values of x between 0 and 240, inclusive. This is why we wrote $0 \le x \le 240$ to the right of the function. We call this the **reasonable domain** of the function.

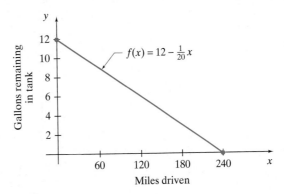

Figure 1.2.4

We know that Ken's truck uses gas at a **rate** of $\frac{1}{20}$ gallon per mile. This tells us how quickly the fuel in the tank decreases. We call this the **slope**, or **average rate of change**, of the linear function. Now let's define the linear function.

■ **Linear Function**

A **linear function** has the form

$$y = f(x) = mx + b$$

where m is the slope or the average rate of change of the line and b is a constant.

We could have written the linear function for the gas remaining in the tank in the form $f(x) = -\dfrac{1}{20}x + 12$. In this form we observe that the slope is $m = -\dfrac{1}{20}$, which tells us that the amount of gas in the tank is **decreasing** at a rate of $\dfrac{1}{20}$ gallon per mile, on average. Also, we observe that $b = 12$. This is simply the capacity of the fuel tank or, equivalently, the number of gallons remaining when 0 miles have been driven.

Often we are given ordered pairs and are not explicitly given the slope. To compute the slope, we use the following formula.

■ **Slope**

The **slope** of a line, denoted by m, is a measurement of the steepness of the line. Given two ordered pairs (x_1, y_1) and (x_2, y_2), the slope is determined by computing

$$m = \frac{y_2 - y_1}{x_2 - x_1}$$

The slope of a line gives the **average rate of change** of y with respect to x.

▶ **Note:** The units of measure associated with slope are $\dfrac{\text{dependent units}}{\text{independent units}}$.

A graphical description of the slope is given in Figure 1.2.5. In addition to an average rate of change, the following are ways to interpret the slope of a line.

$$\text{slope} = m = \frac{\text{difference in } y\text{-values}}{\text{difference in } x\text{-values}} = \frac{\text{change in } y}{\text{change in } x}$$

$$= \frac{\text{rise}}{\text{run}} = \frac{\Delta y}{\Delta x} = \text{steepness of the line}$$

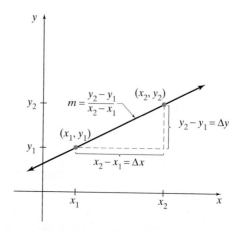

Figure 1.2.5

Now let's see how we can compute the slope if we are given some ordered pairs.

Example 3   Computing the Slope between Two Points

For each of the following, compute the slope m of the line passing through the points that represent the data. Also plot the points and sketch the line through the points.

(a) Table 1.2.6

x	y
-2	-3
3	1

(b) $(0, 4)$ and $(5, 1)$

Solution

(a) For the values in Table 1.2.6, we let $(x_1, y_1) = (-2, -3)$ and $(x_2, y_2) = (3, 1)$. Then the slope is

$$m = \frac{y_2 - y_1}{x_2 - x_1} = \frac{1 - (-3)}{3 - (-2)} = \frac{1 + 3}{3 + 2} = \frac{4}{5}$$

See Figure 1.2.6 for the graph of the line through the two points.

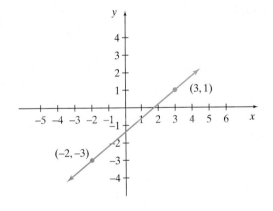

Figure 1.2.6

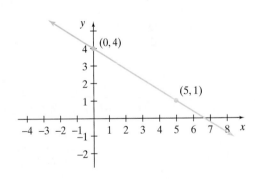

Figure 1.2.7

(b) If we let $(x_1, y_1) = (0, 4)$ and $(x_2, y_2) = (5, 1)$ we get

$$m = \frac{1 - 4}{5 - 0} = \frac{-3}{5}$$

The line through these points is shown in Figure 1.2.7.

Interactive Activity

Recompute the slope for part (b) of Example 3, letting $(x_1, y_1) = (5, 1)$ and $(x_2, y_2) = (0, 4)$. This illustrates that **the slope of a line is unique**.

Before we leave the slope concept, let's discuss a few of its properties. Note in part (a) of Example 3 that the slope is a positive number. We can see in Figure 1.2.6 that the line is rising or **increasing** as we follow the graph from the left to the right. In part (b) of Example 3 the slope has a negative value. We can see in Figure 1.2.7 that the line is falling or **decreasing** as we follow it from left to right. It appears that there is an association between the value of the slope and the type of line that we have. This is summarized in Figures 1.2.8 through 1.2.11.

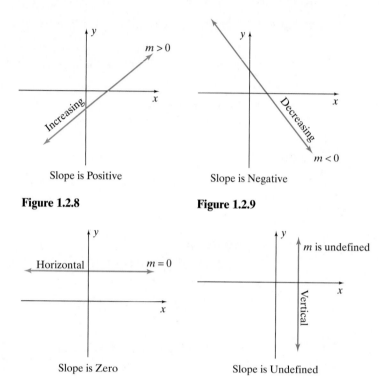

Slope is Positive

Figure 1.2.8

Slope is Negative

Figure 1.2.9

Slope is Zero

Figure 1.2.10

Slope is Undefined

Figure 1.2.11

Now let's see how we can interpret slope in an application.

Example 4   Applying Slope

The per capita bottled water consumption, measured in gallons per day, in the United States from 1990 to 1998 is shown in Table 1.2.7. Compute the slope of the water consumption with respect to time and interpret the result.

Solution

Understand the Situation: We need to use the formula $m = \dfrac{y_2 - y_1}{x_2 - x_1}$ to compute the slope.

If we let $(x_1, y_1) = (1990, 8.0)$ and $(x_2, y_2) = (1998, 13.1)$, we get

$$m = \frac{\text{change in consumption}}{\text{change in years}} = \frac{13.1 - 8.0}{1998 - 1990} = \frac{5.1}{8} \approx 0.64 \, \frac{\text{gallon}}{\text{year}}$$

Interpret the Solution: This means that from 1990 to 1998 the per capita consumption of bottled water in the United States was increasing at an average rate of 0.64 gallon per day each year. ∎

Table 1.2.7

Year	Per Capita Water Consumption
1990	8.0
1998	13.1

SOURCE: U.S. Census Bureau, www.census.gov

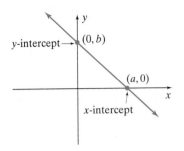

Figure 1.2.12

Intercepts

Two important points on the graph of a linear function are its **intercepts**. These are the points where the graph crosses the x (independent variable) and y (dependent variable) axes, as shown in Figure 1.2.12. The figure also suggests how to find the intercepts.

- To determine the y-intercept, we can replace x with zero in the equation of the line and solve the resulting equation for y.
- To determine the x-intercept, we replace y with zero in the equation of the line and solve the resulting equation for x.

Example 5 **Finding the Intercepts of a Linear Function**

Determine the x- and y-intercepts of the graph of the linear function $y = \frac{2}{3}x + \frac{11}{3}$.

Solution

To find the y-intercept, we replace x with zero and solve for y. This gives

$$y = \frac{2}{3}(0) + \frac{11}{3} = 0 + \frac{11}{3} = \frac{11}{3}$$

So the y-intercept is $\left(0, \frac{11}{3}\right)$.

To get the x-intercept, we replace y with zero and solve for x to get

$$
\begin{aligned}
0 &= \frac{2}{3}x + \frac{11}{3} \\
0 &= 2x + 11 \qquad &\text{Multiply both sides by 3.} \\
-11 &= 2x \qquad &\text{Subtract 11 from both sides.} \\
\frac{-11}{2} &= x \qquad &\text{Divide both sides by 2.}
\end{aligned}
$$

> ### From Your Toolbox
>
> For more on how to solve linear equations, consult Appendix B.3.

Thus the x-intercept is $\left(\frac{-11}{2}, 0\right)$. These points are shown on the graph in Figure 1.2.13.

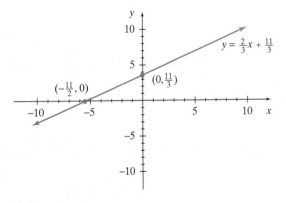

Figure 1.2.13

✓ **Checkpoint 2**

Now work Exercise 49.

Linear Cost Functions

In business, managers are usually very interested in the cost of producing items. This is often expressed as a **linear cost function**. A cost function consists of two parts. The first part is called the **fixed cost**. A fixed cost is what the business pays no matter how many items are produced. Utilities, rent, and other monthly costs are fixed costs. The second part changes as the number of items produced changes. Economists call this part the **variable costs**. We can put fixed costs together with the variable costs and get the cost function.

■ **Linear Cost Function**

The cost $C(x)$ of producing x units of a product is given by the **linear cost function**

$$C(x) = \left(\begin{matrix} \text{cost per} \\ \text{unit} \end{matrix}\right) \cdot x + \left(\begin{matrix} \text{fixed} \\ \text{costs} \end{matrix}\right)$$

▶ **Note:** Some cost functions are not linear because the cost per unit is a variable itself. In the applications that we will study in this text, all cost functions will be linear.

Example 6 **Determining a Cost Function**

A printing job has fixed costs of $5 and per unit costs of $0.10 per copy.

(a) Determine the cost function for the printing job.

(b) Use the result of part (a) to evaluate and interpret $C(100)$.

Solution

(a) **Understand the Situation:** Let x represent the number of copies made and $C(x)$ represent the cost of the printing job.

Then, using the model given in the definition of linear cost function, we have

$$C(x) = \left(\begin{matrix} \text{cost per} \\ \text{unit} \end{matrix}\right) \cdot x + \left(\begin{matrix} \text{fixed} \\ \text{costs} \end{matrix}\right)$$
$$= 0.10x + 5$$

(b) To evaluate $C(100)$, we simply substitute 100 for each occurrence of x. This gives

$$C(x) = 0.10x + 5$$
$$C(100) = 0.10(100) + 5 = 15$$

Interpret the Solution: The total cost of printing 100 copies is $15. ■

✓ **Checkpoint 3**

Now work Exercise 61.

SUMMARY

In this section we examined the definition of function and explored the properties of lines, including the slope of a line.

- A **function** is a rule, or a set of ordered pairs, that assigns a unique y-value to each x-value. The set of all possible x-values is called the **domain**, and the set of all possible y-values is called the **range**.
- A **linear function** has the form $y = f(x) = mx + b$.
- Given two ordered pairs (x_1, y_1) and (x_2, y_2), the **slope** of the line through the two points is determined by computing $m = \dfrac{y_2 - y_1}{x_2 - x_1}$.
- We also found that the value of the slope indicates the behavior of the line.

If $m > 0$, then the line is **increasing**.
If $m < 0$, then the line is **decreasing**.
If $m = 0$, then the line is **horizontal**.
If m is undefined, then the line is **vertical**.

- The **intercepts** are the points where a graph crosses the x- (independent variable) and y- (dependent variable) axes.
- The cost $C(x)$ of producing x units of a product is given by the **linear cost function**

$$C(x) = \left(\begin{array}{c}\text{cost per} \\ \text{unit}\end{array}\right) \cdot x + \left(\begin{array}{c}\text{fixed} \\ \text{costs}\end{array}\right)$$

SECTION 1.2 EXERCISES

In Exercises 1–8, use the table to construct ordered pairs (x, y). Plot the ordered pairs on the Cartesian plane.

1.

x	y
0	1
3	5
6	7

2.

x	y
1	3
4	7
7	10

3.

x	y
−2	3
−1	5
4	−4

4.

x	y
−2	0
−1	−3
−2	−5

5.

x	−4	−3.5	−2	2
y	−5	−6	−8	−11

6.

x	0	−1	−2	−3
y	0	1	5	6

7.

x	0	1	2	3
y	1.1	1.5	1.8	2

8.

x	1	2	3	4
y	1.5	2	2.5	3.5

For Exercises 9–16, write the sets for the domain and range for the values in the table:

9. In Exercise 1.   **10.** In Exercise 2.

✓ **11.** In Exercise 3.   **12.** In Exercise 4.

13. In Exercise 5.   **14.** In Exercise 6.

15. In Exercise 7.   **16.** In Exercise 8.

For the functions in Exercises 17–26, complete the following:

(a) Identify the independent and dependent variables.

(b) Complete the table. Note that in Exercises 25 and 26 the dependent variable value is given.

17. $f(x) = x - 5$

x	f(x)
2	
−3	

18. $f(x) = 3 - x$

x	f(x)
1	
−4	

19. $g(x) = 3x + \dfrac{1}{2}$

x	g(x)
−1	
0	
5	

20. $g(x) = 2x + \dfrac{1}{4}$

x	g(x)
−2	
1	
4	

21. $y = 0.3x + 6$

x	y
0	
5	
10	

22. $y = 0.2x - 1.1$

x	y
1	
6	
11	

23. $f(t) = \dfrac{1}{2}(7 - 0.3t)$

t	f(t)
0	
1	
3	
6	

24. $f(t) = \dfrac{1}{3}(10 - 0.4t)$

t	f(t)
0	
4	
8	
12	

25. $f(x) = \dfrac{2x+3}{5}$

x	$f(x)$
1	
4	
	3

26. $f(x) = \dfrac{3x-1}{4}$

x	$f(x)$
3	
4	
	5

In Exercises 27–30, determine which of the tables represents function.

27.

Domain	Range
0	1
2	1
3	2
4	5

28.

Domain	Range
−5	−2
−3	1
−2	5
0	10

29.

Domain	Range
−3	1
−2	0
−2	3
−1	4
0	5
1	6

30.

Domain	Range
2	1
−2	3
0	0
−1	2
−3	7
2	4

In Exercises 31–34, determine if the graph represents a function.

31.

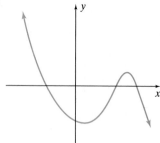

32.

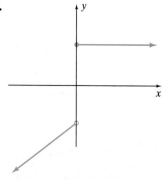

33.

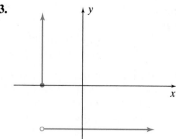

34.

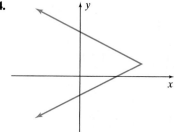

In Exercises 35–38, a table of values is given.

(a) Complete the table.

(b) Write a linear function for the data in the table.

35.

x	expression	y
0		
2	$3(2)+1$	
		4

36.

x	expression	y
	$2(0)+4$	4
		8
3		

37.

x	expression	$f(x)$
0		6
	$6-3(1)$	
		0
3		

38.

x	expression	$f(x)$
	$10-\frac{1}{2}(0)$	10
2		
		8
	$10-\frac{1}{2}(6)$	

In Exercises 39–48, compute the slope of the line passing through the two points. Then classify the line as increasing, decreasing, horizontal, or vertical.

39.

x	y
−6	1
4	4

40.

x	y
−4	−2
−1	7

41.

x	$f(x)$
−1	−4
−3	2

42.

x	$f(x)$
−2	9
1	−3

43. $(3, 8)$ and $(−3, 8)$

44. $(−4, 7)$ and $(3, 7)$

45. $(2, −3)$ and $(2, 4)$

46. $(5, −2)$ and $(5, 3)$

47.

x	0.5	2.5
y	4.1	1.1

48.

x	1.6	4.8
y	0.3	1.5

In Exercises 49–58, determine the x- and y-intercepts of the graph of the linear function.

✓ **49.** $y = 2x - 10$

50. $y = 3x - 9$

51. $2x - 3y = 6$

52. $x - 4y = 16$

53. $y = \dfrac{1}{2}x - 4$

54. $y = \dfrac{1}{3}x - 12$

55. $g(x) = \dfrac{10x - 5}{2}$

56. $g(x) = \dfrac{15x - 3}{5}$

57. $f(x) = 0.2x + 10.3$

58. $f(x) = 1.4x + 6.6$

Applications

In Exercises 59–62, write the linear cost function for the given information in the form $C(x) = mx + b$.

59. The cost per unit is $70 and the fixed costs are $10,000.

60. The cost per unit is $147.75 and the fixed costs are $2700.

✓ **61.** The fixed costs are $14,500 and the cost per unit is $134.

62. The fixed costs are $3580 and the cost per unit is $80.

63. The Dirt Rocket Company produces spokes for offroad motorcycles and determines that the daily cost in dollars of producing x quick-replace spokes is $C(x) = 1250 + 3x$.

(a) What are the fixed and per unit costs for producing the spokes?

(b) Evaluate and interpret $C(400)$.

(c) How many spokes can be produced for $1520? That is, what is x when $C(x) = 1520$?

64. The Fontana Vitamin Company determines that the cost of producing x bottles of a new supplement is given by the linear cost function $C(x) = 1.25x + 550$.

(a) What are the fixed and per unit costs for producing the supplement?

(b) Evaluate and interpret $C(70)$.

(c) How many bottles of the supplement can be produced for $693.75? That is, what is x when $C(x) = 693.75$?

65. The Get-It-Now florist determines that the cost, in dollars, of arranging and delivering x get-well bouquets is given by the cost function $C(x) = 10 + 32x$.

(a) What are the fixed and per units costs for producing the bouquets?

(b) Evaluate and interpret $C(15)$.

(c) How many bouquets can be produced and delivered for $8010?

66. The Never Die Light Bulb Company determines that the cost, in dollars, of producing x units of their new brand of bulb is $C(x) = 2.5x + 1050$.

(a) What are the fixed and per units costs for producing the light bulbs?

(b) Evaluate and interpret $C(260)$.

(c) How many bulbs can be produced for $1480?

🌐 **67.** The data in Table 1.2.8 represent the public expenditures at private universities and colleges in the United States from 1989 to 1999.

Table 1.2.8	
Year	**Public Expenditures**
1989	$68.4 billion
1999	$92.6 billion

SOURCE: U.S. Census Bureau, www.census.gov

Compute the average rate of change in public expenditures at private universities and colleges from 1989 to 1999 and interpret. Be sure to include proper units.

🌐 **68.** The data in Table 1.2.9 represent the number of visits to national parks from 1990 to 1998.

Table 1.2.9	
Year	**Number of Visits**
1990	58.7 million
1998	64.5 million

SOURCE: U.S. Census Bureau, www.census.gov

Compute the average rate of change in the number of visits to national parks from 1990 to 1998 and interpret. Be sure to include proper units.

▌ SECTION PROJECT

Determine if each of the following can be represented by a function. Explain your reasoning.

(a) A person's astrological sign is a function of his or her birth date.

(b) A person's income is a function of how long the person works each day.

(c) The voter turnout in an election is a function of the weather.

(d) The distance a person is from a pizzeria is a function of the deliveries made.

(e) The number of U.S. congress members from a state is a function of its population.

Section 1.3   Linear Models

In Section 1.2 we defined the function and explored the properties of the average rate of change for linear functions. In this section we will see the different ways in which we can write linear functions and examine the concept of **linear models**.

Equations of Lines

When we defined the linear function, we wrote it in the form $f(x) = mx + b$. But there are other ways to mathematically express a line. One way that we will see frequently is the **standard form** of a line.

> ■ **Standard Form**
>
> In the Cartesian plane, the **standard form** of a line is given by
> $$ax + by = c$$
> where a, b, and c are real numbers, $a \geq 0$, and a and b are not both zero.

▶ **Note:** If a, b, and c are rational numbers, they are often converted to integers by multiplying both sides of the equation by the least common multiple of the denominators.

Example 1   **Writing a Linear Equation in Standard Form**

For the linear equation $y = \dfrac{2}{3}x + \dfrac{11}{3}$, write the equation in the standard form $ax + by = c$, where a, b, and c are integers.

Solution

First, we collect the terms with x and y on the left side and leave the constant on the right.

$$-\frac{2}{3}x + y = \frac{11}{3}$$

Since we need $a \geq 0$, we multiply both sides of the equation by -1 to get

$$\frac{2}{3}x - y = -\frac{11}{3}$$

Finally, we can multiply both sides by 3 so that no fractions are present.

$$2x - 3y = -11$$ ■

In addition to standard form, there are other ways to write an equation of a line. Let's say that we know the slope of a line and the coordinates of one point on the line. For example, suppose that a line has a slope of $m = -2$ and passes through the point $(-3, 5)$. Now call some other point on the line (x, y). See

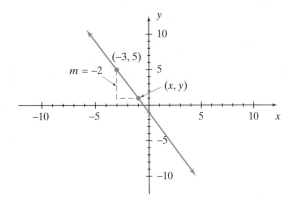

Figure 1.3.1 The slope of the line through $(-3, 5)$ and (x, y) is $m = -2$.

Figure 1.3.1. Using the definition of slope, $m = \dfrac{y_2 - y_1}{x_2 - x_1}$, we can write

$$-2 = \frac{y - 5}{x - (-3)} \quad \text{or} \quad -2 = \frac{y - 5}{x + 3}$$

where $(x_1, y_1) = (-3, 5)$ and $(x_2, y_2) = (x, y)$. Multiplying both sides of the equation by $x + 3$ gives us

$$y - 5 = -2(x + 3)$$

This is an equation of the line with slope $m = -2$ passing through the point $(-3, 5)$. Since this equation uses a known slope and a known point, we call it the **point–slope form**.

> **Point–Slope Form**
>
> An equation of the line with known slope m and known point (x_1, y_1) can be written in the **point–slope form** given by
>
> $$y - y_1 = m(x - x_1)$$

We can use this form to write an equation of any line if we are given either one point on the line and its slope or if we are given two points on the line. This is illustrated in Example 2.

Example 2 **Writing an Equation in Point–Slope Form**

For the points $(2, 4)$ and $(3, 7)$, complete the following:

(a) Determine the slope m of the line passing through the two points.

(b) Use the point $(2, 4)$ to write an equation for the line passing through the points.

(c) Rewrite the equation in standard form.

Solution

(a) If we let $(x_1, y_1) = (2, 4)$ and let $(x_2, y_2) = (3, 7)$ we get

$$m = \frac{y_2 - y_1}{x_2 - x_1} = \frac{7 - 4}{3 - 2} = \frac{3}{1} = 3$$

(b) Using $m = 3$ from part (a) and knowing that $(x_1, y_1) = (2, 4)$, we get

$$y - 4 = 3(x - 2)$$

(c) Here we need to rewrite the equation in the standard form $ax + by = c$, where $a \geq 0$ (that is, a is nonnegative).

$y - 4 = 3(x - 2)$	Point–slope form
$y - 4 = 3x - 6$	Distribute the 3.
$-3x + y - 4 = -6$	Subtract $3x$ from both sides.
$-3x + y = -2$	Add 4 to both sides.
$3x - y = 2$	Multiply both sides by -1. ∎

Let's say that we want to write the equation of a line in point–slope form with slope m and y-intercept $(0, b)$. Using $(x_1, y_1) = (0, b)$, we get

$$y - b = m(x - 0)$$

Solving this equation for y gives us

$$y = m(x - 0) + b \quad \text{or} \quad y = mx + b$$

This is another way to write the equation of a line and is called the **slope–intercept form**.

■ Slope Intercept Form

An equation of a line with slope m and y-intercept $(0, b)$ can be written in the **slope–intercept form** given by

$$y = mx + b$$

▶ **Note:** The slope–intercept form is how we defined the linear function in Section 1.2. The slope–intercept form is useful when we are given an average rate of change (that is, a slope) along with some y-value when $x = 0$. This is sometimes called an **initial value**. Now let's practice writing a linear equation in slope–intercept form.

Interactive Activity*

Redo Example 2, but let $(x_1, y_1) = (3, 7)$. Do you get the same result as was found in part (c)? This is why $ax + by = c$ is called the *standard* form.

From Your Toolbox

In part (c) of Example 2, we used the distributive property. For more practice on using this property, consult the Appendix B.1.

Example 3 Writing a Linear Equation in Slope–Intercept Form

A line passes through the points $(-2, 6)$ and $(0, 1)$.

(a) Write an equation of the line in slope–intercept form.

(b) Rewrite the equation in standard form.

---

*Interactive Activities that have the 💻 icon have worked out solutions at the companion website at www.prenhall.com/armstrong.

Solution

(a) We begin by computing the slope of the line using the two given points.

$$m = \frac{y_2 - y_1}{x_2 - x_1} = \frac{1 - 6}{0 - (-2)} = \frac{-5}{2} = -\frac{5}{2}$$

Knowing that the y-intercept is $(0, b) = (0, 1)$, we have $b = 1$. So we get the slope–intercept form

$$y = mx + b$$

$$y = -\frac{5}{2}x + 1$$

(b) Here we need to write the equation found in part (a) in the form $ax + by = c$, where $a \geq 0$.

$$y = -\frac{5}{2}x + 1 \qquad \text{Given equation}$$

$$\frac{5}{2}x + y = 1 \qquad \text{Add } \frac{5}{2}x \text{ to both sides.}$$

$$5x + 2y = 2 \qquad \text{Multiply both sides by 2.} \qquad \blacksquare$$

✔ **Checkpoint 1**

Now work Exercise 3.

Linear Models

Often, in our study of finite mathematics, we wish to take real data and use them to write a function that shows the general trends of these data. We call this process **modeling**. Many times, this model starts with a given value and then either steadily increases or decreases. We call these types of models **linear models**.

> ▨ **Linear Model**
>
> A **linear model** is a type of linear function that is derived from a real-world situation or from real-world data.

Let's start by examining a linear model in an example. In this section we will assume that all the applications discussed can be modeled by a linear function.

Example 4   Evaluating and Interpreting a Linear Model

The average annual pay of Americans from 1990 to 2000 can be represented by a linear model that contains the points shown in Table 1.3.1. In the table, x represents the number of years since 1990 and y represents the average annual pay in thousands of dollars.

(a) Compute the average rate of change and interpret.

(b) Write an equation of the line containing the two points in point–slope form.

(c) Write a linear model in slope–intercept form

$$f(x) = mx + b, \qquad 0 \leq x \leq 10$$

for the average annual pay of Americans from 1990 to 2000.

(d) Evaluate $f(5)$ and interpret.

Table 1.3.1

x	y
1	24.28
3	26.40

SOURCE: U.S. Census Bureau, www.census.gov

Solution

(a) Understand the Situation: Recall that the average rate of change is simply the slope, m, of the line through two points.

Using the ordered pairs $(x_1, y_1) = (1, 24.28)$ and $(x_2, y_2) = (3, 26.40)$, we get

$$m = \frac{y_2 - y_1}{x_2 - x_1} = \frac{26.40 - 24.28}{3 - 1} = \frac{2.12}{2} = 1.06$$

Interpret the Solution: This means that from 1990 to 2000 the average annual pay for Americans increased at an average rate of about 1.06 thousand dollars per year.

(b) Using the point $(x_1, y_1) = (1, 24.28)$, we get

$$y - y_1 = m(x - x_1)$$
$$y - 24.28 = 1.06(x - 1)$$

A graph of this equation is given in Figure 1.3.2.

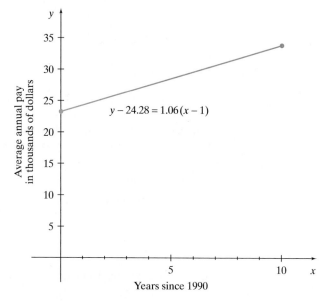

Figure 1.3.2

(c) To get the result of part (b) in the form $f(x) = mx + b$, we solve the equation for y and then replace y with $f(x)$.

$y - 24.28 = 1.06(x - 1)$	Given equation
$y = 1.06(x - 1) + 24.28$	Add 24.28 to both sides.
$y = 1.06x - 1.06 + 24.28$	Distribute 1.06.
$y = 1.06x + 23.22$	Simplify.
$f(x) = 1.06x + 23.22$	Replace y with $f(x)$.

(d) Understand the Situation: Since x represents the number of years since 1990, we know that $x = 5$ corresponds to the year $1990 + 5 = 1995$.

Now we need to evaluate the linear model when $x = 5$ and interpret the result.

$$f(5) = 1.06(5) + 23.22 = 5.3 + 23.22 = 28.52$$

Interactive Activity

Write the result of part (b) in Example 4 in standard form. (*Hint:* Multiply both sides of the equation by 100 to eliminate the decimals.)

Interpret the Solution: This means that in 1995 the average annual pay of Americans was about 28.52 thousand dollars (that is, \$28,520). ■

✓ **Checkpoint 2**   Now work Exercise 51.

Another application of linear modeling involves the **linear depreciation model**. When the value of a product decreases as time goes on, we say it depreciates. The amount of value that the product loses each year is called the **depreciation rate**. Many companies consider linear depreciation models in their fiscal plans and for planning upgrades in equipment. Example 5 illustrates how to construct a linear depreciation model.

Example 5   **Writing a Linear Depreciation Model**

The Clark Clipboard Company recently purchased a metal press for \$15,000. The press is estimated to depreciate at a rate of $2500\dfrac{\text{dollars}}{\text{year}}$.

  (a) Write a linear model in the form $f(x) = mx + b$ for the value of the metal press.

  (b) Determine the x-intercept and interpret.

Solution

  (a) **Understand the Situation:** Since the value of the press appears to depend on how old it is, we let x represent the age of the press in years and $y = f(x)$ represent the value of the press in dollars.

   We know that when the press is new, meaning that the age is $x = 0$, its value is \$15,000. So $f(0) = 15,000$, which means that the point $(0, 15,000)$ is on the graph. Notice that this point is the y-intercept. (Recall that this is also known as the initial value.) Since the rate at which the value of the press decreases is $2500\dfrac{\text{dollars}}{\text{year}}$, the value of the press has an average rate of change of $-2500\dfrac{\text{dollars}}{\text{year}}$. So knowing that $m = -2500$ and the initial value is $b = 15,000$, we get the model

$$f(x) = mx + b$$
$$f(x) = -2500x + 15,000$$

  (b) To determine the x-intercept, we let $f(x) = 0$ and solve for x.

$0 = -2500x + 15,000$	Set $f(x) = 0$.
$2500x = 15,000$	Add 2500x to each side.
$x = \dfrac{15,000}{2500}$	Divide by 2500.
$= 6$	Simplify.

Interpret the Solution: This means that after 6 years the metal press will have no value. ■

Interactive Activity

Determine the reasonable domain for the model in Example 5a.

✓ **Checkpoint 3**   Now work Exercise 37.

Technology Option

Calculator-Friendly Form

Many times, we want to use our calculator to graph a line with a known point and known slope. We can do this by solving the point–slope form $y - y_1 = m(x - x_1)$ for y. This gives us the calculator-friendly (or C-F) form.

> ■ **Calculator-Friendly or C-F Form**
>
> A linear function with a known slope m and known point (x_1, y_1) can be written in the **calculator-friendly form** given by
>
> $$y = m(x - x_1) + y_1$$

This form is particularly useful in applications, as is illustrated in Example 6.

Example 6 **Using the Calculator-Friendly Form**

In 1994 there were 25.27 million children enrolled in the National School Lunch Program. In 1999, this number had risen to 26.82 million. Assume that the number of children enrolled in the lunch program grew linearly from 1990 to 1999 (*Source:* U.S. Census Bureau, www.census.gov).

(a) Let x represent the number of years since 1990 and y represent the number of children enrolled in the National School Lunch Program in millions. Determine the average rate of change in the number of enrollees from 1990 to 1999.

(b) Write a linear model in the **calculator-friendly form** given by

$$y = m(x - x_1) + y_1, \qquad 0 \le x \le 9$$

to represent the number of children enrolled in the National School Lunch Program in millions from 1990 to 1999.

(c) During what year did the number of enrollees exceed 25 million?

Solution

(a) **Understand the Situation:** If we let x represent the number of years since 1990, we see that 1994 corresponds to $x = 4$ and 1999 corresponds to $x = 9$. So we have to determine the average rate of change, or slope, between the points $(x_1, y_1) = (4, 25.27)$ and $(x_2, y_2) = (9, 26.82)$.
   This gives us

$$m = \frac{26.82 - 25.27}{9 - 4} = \frac{1.55}{5} = 0.31$$

Interpret the Solution: This means that from 1990 to 1999 the number of children enrolled in the National School Lunch Program increased at an average rate of 0.31 million enrollees per year.

(b) Knowing that $m = 0.31$ and $(x_1, y_1) = (4, 25.27)$, we get

$$y = 0.31(x - 4) + 25.27$$

Figure 1.3.3 is a graph of the model.

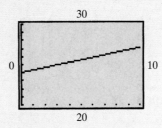

Figure 1.3.3

(c) Here we need the x-values so that $y = 25$. This gives us the equation

$$y = 0.31(x - 4) + 25.27$$
$$25 = 0.31(x - 4) + 25.27$$
$$-0.27 = 0.31(x - 4)$$
$$\frac{-0.27}{0.31} = x - 4$$
$$\frac{-0.27}{0.31} + 4 = x$$
$$3.13 \approx x$$

So we get $x \approx 3.13$. Since x represents the number of years since 1990, we see that during 1993 the number of children enrolled in the National School Lunch Program exceeded 25 million. Figure 1.3.4 shows how to check our algebraic solution by using the TRACE command. For more information on using the TRACE command, consult the online calculator manual at www.prenhall.com/armstrong.

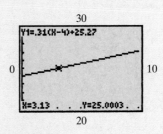

Figure 1.3.4

SUMMARY

In this section we learned four ways to write linear equations.

- **Standard form:** $ax + by = c$
- **Point–slope form:** $y - y_1 = m(x - x_1)$
- **Slope–intercept form:** $y = mx + b$
- **Calculator-friendly form:** $y = m(x - x_1) + y_1$

SECTION 1.3 EXERCISES

In Exercises 1–14, write an equation of the line in standard form (if possible).

1. A line with a slope of $m = \dfrac{2}{3}$ that passes through the point $(2, -5)$.

2. A line with a slope of $m = -\dfrac{1}{2}$ that passes through the point $(2, -2)$.

✓ **3.** The line that passes through the points $(2, 6)$ and $(5, -3)$.

4. The line that passes through the points $(3, 2)$ and $(-1, -2)$.

5. The line is given by the values in Table 1.3.2.

Table 1.3.2

x	y
3	5
5	13

6. The line is given by the values in Table 1.3.3.

Table 1.3.3

x	y
2	4
0	4

7. The line is given by the values in Table 1.3.4.

Table 1.3.4

x	1	5
$f(x)$	5.1	10.6

8. The line is given by the values in Table 1.3.5.

Table 1.3.5

x	0	3
$f(x)$	8.8	4.3

9. The line has a y-intercept at -3 and passes through the origin.

10. The line has an x-intercept at 2 and a y-intercept at 4.

11. The line is vertical and passes through the point $(-2, 9)$.

12. The line is horizontal and passes through the point $(-3, 1)$.

13. The line has an average rate of change of $-\dfrac{2}{3}$ and an x-intercept at 3.

14. The line has an average rate of change of 0.3 and an x-intercept at -4.

In Exercises 15–30, write a function for the line in $f(x) = mx + b$ form.

15. A line with a slope of $m = 3$ that passes through the point $(8, 4)$.

16. A line with a slope of $m = \dfrac{5}{6}$ that passes through the point $(-1, 6)$.

17. A linear function has a slope of $m = -\dfrac{5}{8}$ and $f(1) = 3$.

18. A linear function has a slope of $m = 4$ and $f(-2) = 4$.

19. A linear function has a slope of $m = 5$ and a y-intercept at $(0, -13)$.

20. A linear function has a slope of $m = \dfrac{3}{5}$ and a y-intercept at $(0, -10)$.

21. A linear function with $f(1) = 3$ and $f(-10) = 1$.

22. A linear function with $f(7) = 3$ and $f(-3) = 6$.

23. A linear function $f(x)$ has function values of $f(5) = 1$ and $f(-7) = -1.19$.

24. A linear function $g(x)$ has function values of $g(0) = 6$ and $g(4) = 1.3$.

25. A linear function contains the points in Table 1.3.6.

Table 1.3.6

x	$f(x)$
-1	2
3	1

26. A linear function contains the points in Table 1.3.7.

Table 1.3.7

x	$f(x)$
0	2
-3	4

27. A line has an x-intercept at $(-12, 0)$ and y-intercept at $\left(0, \dfrac{3}{2}\right)$.

28. A line has an x-intercept at $(-5, 0)$ and a y-intercept at $(0, 9)$.

29. A linear function contains the points in Table 1.3.8.

Table 1.3.8

x	1	3
$f(x)$	25.3	10.8

30. A linear function contains the points in Table 1.3.9.

Table 1.3.9		
x	1	6
$f(x)$	13.7	16.9

In Exercises 31–34, write an equation in slope–intercept form for the given graph.

31.

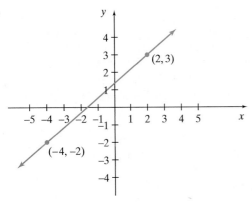

32.

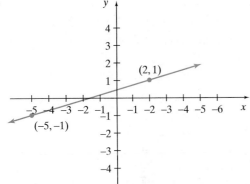

33.

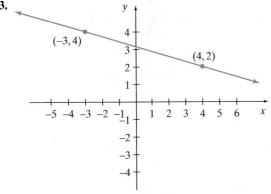

34.

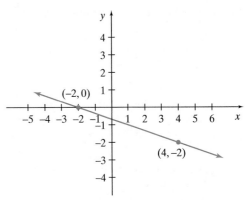

Applications

In Exercises 35–38, answer by using a linear depreciation model.

35. A new photocopier initially costs $25,000 and depreciates at a rate of $1250 per year.

(a) Define the variables and determine the depreciation function.

(b) In how many years will the photocopier be worth $10,000?

36. A new multimedia computer costs $4000 and depreciates at a rate of $150 \dfrac{\text{dollars}}{\text{year}}$.

(a) Define the variables and determine the depreciation function.

(b) When will the computer depreciate to the point that it has no worth?

✓ **37.** The Westbrook Company invests $90,000 into a new machine press that depreciates at a rate of $6000 per year.

(a) Define the variables and determine the depreciation function.

(b) How long will it take for the press to be worth half of its original cost?

38. The XQ-383 sports cars costs $40,000 and depreciates $1500 each year.

(a) Define the variables and determine the depreciation function.

(b) How much will the car be worth in 5 years?

🌐 **39.** The annual total expenditures on pollution abatement in the United States can be modeled by

$$f(x) = 2x + 48.17, \qquad 0 \le x \le 20$$

where x represents the number of years since 1972 and $f(x)$ represents the expenditures in billions of constant 1987 dollars (*Source:* U.S. Bureau of Economic Analysis, www.bea.doc.gov).

(a) Is the amount of money spent on pollution abatement generally increasing or decreasing? At what rate?

(b) Evaluate and interpret $f(12)$.

(c) In what year did the annual expenditures approach 90 billion dollars?

40. The percentage of waste generated in the United States that is wood can be modeled by

$$f(x) = 0.23x + 5.95, \qquad 0 \le x \le 9$$

where x represents the number of years since 1990 and $f(x)$ represents the percentage of waste generated that is wood (*Source:* U.S. Bureau of Economic Analysis, www.bea.doc.gov).

(a) Is the percentage of waste generated that is wood generally increasing or decreasing? At what rate?

(b) Evaluate and interpret $f(6)$.

(c) In what year did the percentage approach 7.4%?

41. The population of the United States can be modeled by

$$f(x) = 1.92x + 67.15, \qquad 0 \le x \le 100$$

where x represents the number of years since 1900 and $f(x)$ represents the U.S. population in millions (*Source:* U.S. Census Bureau, www.census.gov).

(a) Evaluate and interpret $f(50)$.

(b) Solve $f(x) = 150$ and interpret.

42. The percentage of pleasure trips taken by U.S. households with incomes above \$40,000 can be modeled by

$$f(x) = 2.66x + 23.62, \qquad 1 \le x \le 14$$

where x represents the number of years since 1984 and $f(x)$ represents the percentage of pleasure trips taken by U.S. households with incomes above \$40,000 (*Source:* U.S. Census Bureau, www.census.gov).

(a) Evaluate and interpret $f(10)$.

(b) Solve $f(x) = 40$ and interpret.

43. The amount spent annually in college bookstores in the United States can be modeled by

$$f(x) = 0.19x + 1.67, \qquad 0 \le x \le 15$$

where x represents the number of years since 1982 and $f(x)$ represents the amount spent in billions of dollars (*Source:* U.S. National Center for Education Statistics, www.nces.ed.gov).

(a) On average, by how much does the amount of spending increase in college bookstores each year?

(b) According to the model, how much was spent in college bookstores in 1985?

(c) According to the model, how much was spent in college bookstores in 1990?

(d) How much more was spent in college bookstores in 1990 compared to 1985?

44. The average annual salary of counselors in the U.S. public school system can be modeled by

$$f(x) = 1557.29x + 19{,}647.80, \qquad 0 \le x \le 20$$

where x represents the number of years since 1980 and $f(x)$ represents the average salary in dollars (*Source:* U.S. National Center for Education Statistics, www.nces.ed.gov).

(a) On average, by how much does a counselor's average salary increase each year?

(b) According to the model, what was a counselor's average salary in 1985?

(c) According to the model, what was a counselor's average salary in 1995?

(d) How much more was a counselor's average salary in 1995 compared to 1985?

45. In 1990 there were 4.24 million children in the United States with physical disabilities and the number was increasing at an average rate of 0.12 million per year (*Source:* U.S. Census Bureau, www.census.gov).

(a) Let x represent the number of years since 1990. Write a linear model of the form

$$g(x) = y = mx + b, \qquad 0 \le x \le 10$$

to represent the number of children in the United States with physical disabilities from 1990 to 2000.

(b) Evaluate and interpret $g(4)$.

46. In 1980 there were 28.3 million Americans enrolled in the Medicare system. The number of enrollees was increasing at an average rate of 0.61 enrollees per year (*Source:* U.S. Census Bureau, www.census.gov).

(a) Let x represent the number of years since 1980. Write a linear model of the form

$$g(x) = y = mx + b, \qquad 0 \le x \le 20$$

to represent the number of Medicare enrollees from 1980 to 2000.

(b) Evaluate and interpret $g(15)$.

47. From 1990 to 1999 the amount spent to purchase a car exported to the United States increased at an average rate of \$1776.71 per year. In 1995, the average amount spent to purchase a car exported to the United States was \$25,483.95 (*Source:* U.S. Census Bureau, www.census.gov).

(a) Let x represent the number of years since 1990. Write a linear model in the calculator-friendly form

$$y = m(x - x_1) + y_1, \qquad 0 \le x \le 9$$

to represent the average amount spent to purchase a car exported to the United States from 1990 to 1999.

(b) Use the model to determine in what year the average price exceeded \$30,000.

48. From 1990 to 2000 the percentage of married women in the U.S. labor force was increasing at an average rate of 0.23% each year. In 1996, 48.1% of the labor force was comprised of married women (*Source:* U.S. Census Bureau, www.census.gov).

(a) Let x represent the number of years since 1990. Write a linear model in the calculator-friendly form

$$y = m(x - x_1) + y_1, \qquad 0 \le x \le 10$$

to represent the percentage of married women in the U.S. labor force from 1990 to 2000.

(b) Use the model to determine in what year the percentage of married women in the U.S. labor force exceeded 48%.

49. In 1989 the median price of a single-family house in the southern United States was $96.4 thousand. In 1999 the median price had risen to $140 thousand (*Source:* U.S. Census Bureau, www.census.gov).

(a) Let x represent the number of years since 1989 and y represent the median price of a single-family house in the southern United States in thousands. Determine the average rate of change in the median price from 1989 to 1999.

(b) Write a linear model in the slope–intercept form

$$y = mx + b, \qquad 0 \le x \le 10$$

to represent the median price of a single-family house in the southern United States from 1989 to 1999.

(c) According to the model, during what year was the median price of a single-family house in the southern United States $118.2 thousand?

50. The annual amount of Social Security payments from 1990 to 2000 can be represented by a linear model that contains the points shown in Table 1.3.10, where x represents the number of years since 1990 and $f(x)$ represents the annual amount of Social Security payments measured in billions of dollars (*Source:* U.S. Census Bureau, www.census.gov).

Table 1.3.10

x	$f(x)$
0	250.63
1	266.18

(a) Compute the slope m and interpret.

(b) Use the information in the table to write a linear model in the slope–intercept form

$$f(x) = mx + b, \qquad 0 \le x \le 10$$

for the annual amount of Social Security payments from 1990 to 2000.

(c) Evaluate and interpret $f(5)$.

51. The size of the adult civilian labor force in the United States from 1990 to 2000 can be represented by a linear model that contains the points shown in Table 1.3.11, where x represents the number of years since 1990 and $f(x)$ represents the civilian labor force in millions (*Source:* U.S. Census Bureau, www.census.gov).

Table 1.3.11

x	$f(x)$
0	125.50
1	127.02

(a) Compute the slope m and interpret.

(b) Use the information in the table to write a linear model in slope–intercept form

$$f(x) = mx + b, \qquad 0 \le x \le 10$$

for the size of the adult civilian labor force in the United States from 1990 to 2000.

(c) Evaluate and interpret $f(4)$.

52. The average annual pay of Americans from 1990 to 2000 can be represented by a linear model that contains the points shown in Table 1.3.12, where x represents the number of years since 1990 and $f(x)$ represents the average annual pay in thousands of dollars (*Source:* U.S. Census Bureau, www.census.gov).

Table 1.3.12

x	$f(x)$
1	24.28
3	26.40

(a) Compute the slope m and interpret.

(b) Use the information in the table to write a linear model in slope–intercept form

$$f(x) = mx + b, \qquad 0 \le x \le 10$$

for the average annual pay of Americans from 1990 to 2000.

(c) Determine the y-intercept and interpret.

53. The number of U.S. motor vehicle registrations from 1990 to 2000 can be represented by a linear model that contains the points shown in Table 1.3.13, where x represents the number of years since 1990 and $f(x)$ represents the number of motor vehicle registrations in millions (*Source:* U.S. Census Bureau, www.census.gov).

Table 1.3.13

x	$f(x)$
1	191.28
4	200.1

(a) Compute the slope m and interpret.

(b) Use the information in the table to write a linear model in slope–intercept form

$$f(x) = mx + b, \qquad 0 \le x \le 10$$

for the number of U.S. motor vehicle registrations from 1990 to 2000.

(c) Determine the y-intercept and interpret.

54. The number of students enrolled in elementary (K–8) school annually in the United States from 1990 to 2000 can be represented by a linear model that contains the points shown in Table 1.3.14, where x represents the number of years since 1990 and y represents the number of students enrolled in elementary (K–8) school in millions (*Source:* U.S. Census Bureau, www.census.gov).

Table 1.3.14

x	y
2	35.09
6	37.01

(a) Use the information in the table to write a linear model in standard form

$$ax + by = c, \qquad 0 \le x \le 10$$

for the number of students enrolled in elementary (K–8) school annually in the United States from 1990 to 2000. (*Hint:* Multiply both sides of the equation by 100 so that a, b, and c are integers.)

(b) Determine x when $y = 37.97$ and interpret.

55. The number of students enrolled in higher education institutions annually in the United States from 1990 to 2000 can be represented by a linear model that contains the points shown in Table 1.3.15, where x represents the number of years since 1990 and y represents the number of students enrolled in higher education institutions in millions (*Source:* U.S. Census Bureau, www.census.gov).

Table 1.3.15

x	y
2	13.97
6	14.41

(a) Use the information in the table to write a linear model in standard form

$$ax + by = c, \qquad 0 \le x \le 10$$

for the number of students enrolled in higher education institutions from 1990 to 2000. (*Hint:* Multiply both sides of the equation by 100 so that a, b, and c are integers.)

(b) Determine x when $y = 14.3$ and interpret.

SECTION PROJECT

Consider the data in Table 1.3.16, where the values in the right column show the number of Americans covered by private or government health insurance, measured in millions of people.

Table 1.3.16

Year	Number Insured
1987	241
1989	246
1991	251
1993	260
1995	264
1997	269

SOURCE: U.S. Healthcare Financing Administration, www.hcfa.gov

Let x represent the number of years since 1985 and let $f(x)$ represent the number of Americans covered by private or government health insurance, in millions. Then the data can be represented by the linear model

$$f(x) = 2.9x + 234.87, \qquad 2 \le x \le 12$$

(a) Evaluate and interpret $f(6)$.

(b) Determine the difference between the value in the table in 1991 and the answer in part (a) by computing

$$(\text{value in table}) - [\text{answer to part (a)}]$$

This difference is called the **residual**.

(c) Now graph the scatterplot of the data in the table along with the given linear model. Is the data point for $x = 6$ (1991) above or below the graph of the linear model? How does this relate to the result of part (b)? Explain.

Section 1.4   Solving Systems of Linear Equations Graphically

There are instances when we need more than just one equation with one variable to find a solution. For example, let's say that an amusement park charges $6 for children under 6 and $22 for anyone 7 years or older. On a certain day, the cashier reports that the park took in $140,000 in gate receipts and that 8000 people visited the park. If we want to know how many child and how many adult tickets were sold, the best way to find the solution is by solving a **system of**

equations. In this section we will learn how to solve systems of equations using graphing techniques and how these systems are used in applications.

Fundamentals of Systems of Linear Equations

A **system of linear equations** is a collection of two or more linear equations with each having at least one variable. The following are examples of systems of equations common in this text.

$$2x + y = 6 \qquad 2x + y - z = 2 \qquad y = 3x - 5$$
$$x - y = 6 \qquad x + 3y + 2z = 1 \qquad y = 1 - 4x$$
$$x + y + z = 2$$

Note that the exponents on all the variables in these systems are 1. This is why they are called **linear systems** of equations. Linear systems are our focus in the remainder of this chapter and in Chapter 2.

A **solution** to a system of linear equations is the values of the variables that make each equation true. Solutions to systems of equations with two or three variables are written as ordered pairs (x, y) or ordered triples (x, y, z). Let's see how we can verify that an ordered pair is a solution to a system of equations.

Example 1   **Checking a Solution to a System of Equations**

Determine if the ordered pair $(4, -2)$ is a solution to the system

$$2x + y = 6$$
$$x - y = 6$$

Solution

Here we need to show that if we substitute $x = 4$ and $y = -2$ into each of the equations both of the equations are true. Substituting into the first equation gives

$$2x + y = 6$$
$$2(4) + (-2) = 6$$
$$8 + (-2) = 6$$
$$6 = 6$$

which is true. In the second equation, we have

$$x - y = 6$$
$$4 - (-2) = 6$$
$$6 = 6$$

which is also true. Thus $(4, -2)$ is a solution to the system.   ∎

✓ Checkpoint 1

Now work Exercise 7.

To see why we want to write solutions as ordered pairs, let's graph each equation in Example 1. We see in the first equation $2x + y = 6$ that the x-intercept is at $(3, 0)$ and the y-intercept is at $(0, 6)$. In the second equation, $x - y = 6$, the x- and y-intercepts are at $(6, 0)$ and $(0, -6)$, respectively. The graphs are shown in

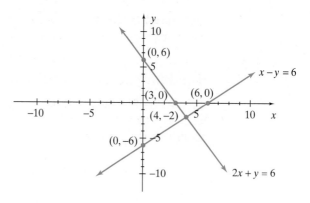

Figure 1.4.1

Figure 1.4.1. Notice that the solution $(4, -2)$ is the point of intersection of the two graphs. This illustrates a fundamental fact behind solving systems of equations.

> **The solution to a linear system of two equations in two variables corresponds to the point(s) of intersection of their graphs.**

Let's use this fact to solve a system of equations graphically.

Example 2   Solving a System of Equations by Graphing

For the system

$$x + 3y = 18$$
$$2x - 6y = 12$$

determine the intercepts of each equation, graph the equations, and write the coordinates of the point of intersection.

Interactive Activity

Numerically verify that $(12, 2)$ is an exact solution in Example 2 by using the technique outlined in Example 1.

Solution

For the first equation, $x + 3y = 18$, the x-intercept is $(18, 0)$ and the y-intercept is $(0, 6)$. For the second equation, $2x - 6y = 12$, the x and y intercepts are at $(6, 0)$ and $(0, -2)$, respectively. The graphs of these equations are shown in Figure 1.4.2. We can see by inspecting the graph that the point of intersection appears to be $(12, 2)$. So the ordered pair $(12, 2)$ is an approximate solution to the system.

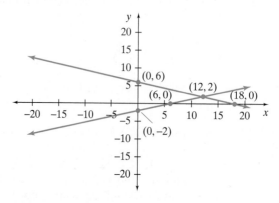

Figure 1.4.2

Solutions to Systems

We were able to find a solution to the system in Example 2 because the two equations had exactly **one point of intersection**. To see what else can happen, consider the system

$$x - y = 2$$
$$3x - 3y = -9$$

The graph of this system is shown in Figure 1.4.3. It appears from the graph that the lines are **parallel**, meaning that they never intersect. If the lines are parallel and never intersect, then the system has **no solution**. But how do we know for certain that the lines are parallel? First, we write each of the equations in slope–intercept form (that is, solve each equation for y) and get

$$x - 2 = y$$
$$\frac{-9 - 3x}{-3} = y$$

or in slope–intercept form

$$y = x - 2$$
$$y = x + 3$$

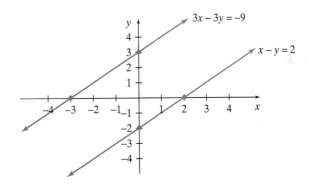

Figure 1.4.3

Notice that each of these equations has a slope of $m = 1$. It stands to reason that if the steepness of the lines are the same then they will not intersect, making the lines parallel. This illustrates that *parallel lines have equal slopes*. Linear systems that do not have an ordered pair that satisfy both equations are called **inconsistent**.

■ **Inconsistent Linear Systems**

Linear systems that have no solution are called **inconsistent systems**.

Let's now take a look at another type of system. The system

$$y = -2x + 1$$
$$6x + 3y = 3$$

is graphed in Figure 1.4.4. Notice that the two equations really represent the same line! We say that the two lines **coincide**. Thus any ordered pair that satisfies the first equation also satisfies the second equation. This means that the system has **infinitely many solutions**. If we write the first equation in standard form, we get

$$2x + y = 1$$
$$6x + 3y = 3$$

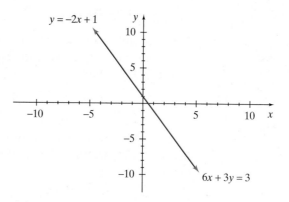

Figure 1.4.4

Notice that we can obtain the second equation by multiplying the first equation by 3. In mathematics, we say that the equations are **linearly dependent**.

■ **Dependent Linear Systems**

Linear systems whose graphs coincide are called **dependent systems**. Ordered pairs that satisfy either equation are solutions to the system.

Table 1.4.1 summarizes the types of solutions possible for systems of linear equations.

Table 1.4.1

Position of Graphs	Number of Solutions	Characteristics of Equations
Lines intersect at one point	One (see Figure 1.4.5(a))	Slopes are not equal
Lines are parallel	None (see Figure 1.4.5(b))	Slopes are equal and y-intercepts are not equal
Lines coincide	Infinite (see Figure 1.4.5(c))	Slopes are equal and y-intercepts are equal

From Table 1.4.1 we see that we can determine the system type if it is written in slope–intercept form. In slope–intercept form we can compare the values of the slopes and the y-intercepts.

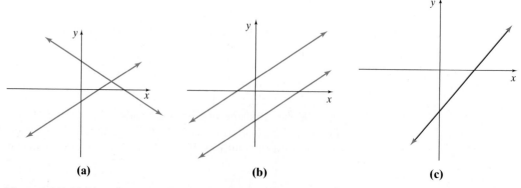

(a) (b) (c)

Figure 1.4.5 **(a)** Lines intersect at one point. The system has one solution. **(b)** Lines are parallel. The system has no solution. **(c)** Lines coincide. The system has infinitely many solutions.

Example 3 **Determining the Number of Solutions of a Linear System**

Determine the number of solutions for each system and describe the position of the lines without graphing the system.

(a) $2y - x = 4$
$2x + 8 = 4y$

(b) $2x + 2y = 4$
$2x - y = 3$

(c) $2x + 3y = 6$
$4x + 6y = 24$

Solution

Understand the Situation: To determine the system type without graphing, we will write each equation in slope–intercept form and compare the values of m and b.

(a) Solving each equation for y, we get

$$y = \frac{x+4}{2} \qquad\qquad y = \frac{1}{2}x + 2$$
$$\text{or}$$
$$\frac{2x+8}{4} = y \qquad\qquad y = \frac{1}{2}x + 2$$

Since each equation has a slope of $m = \frac{1}{2}$ and y-intercept at $b = 2$, we see that the graphs of the lines coincide, and this system has an infinite number of solutions.

(b) In the slope–intercept form, this system is

$$y = \frac{4 - 2x}{2} \qquad\qquad y = -x + 2$$
$$\text{or}$$
$$2x - 3 = y \qquad\qquad y = 2x - 3$$

Here the slopes are not equal. This means that the graphs of the lines intersect at a point and the system has one solution.

(c) Isolating y gives us the slope–intercept form.

$$y = \frac{6 - 2x}{3} \qquad\qquad y = -\frac{2}{3}x + 2$$
$$\text{or}$$
$$y = \frac{24 - 4x}{6} \qquad\qquad y = -\frac{2}{3}x + 4$$

In this case the slopes are the same at $m = -\frac{2}{3}$, yet the y-intercepts are different. This means that the graphs of the lines are parallel and this linear system has no solution. ∎

✓ **Checkpoint 2**

Now work Exercise 41.

Applications of Linear Systems

We can see that graphing systems of linear equations gives us a visualization of the solution. However, it has drawbacks. For example, we saw that in part (b) of Example 3 the system is dependent and has one solution. However, this solution turns out to be $\left(\frac{5}{3}, \frac{1}{3}\right)$. It is very difficult to determine the exact solution just by graphing. For now, we will approximate the coordinates of the point of intersection. Let's take a look at a few of these solutions in the context of applications.

One application of systems of equations common to economics is the **Law of Supply and Demand**.

> ■ **Law of Supply and Demand**
>
> As the price of a product increases,
>
> • the **demand** for this product decreases and
> • the **supply** of this product increases.

The demand for the product decreases when the price increases because some consumers are not willing to pay higher prices. The supply of a product increases because the manufacturers of that product are willing to supply more units of the product, allowing more to be sold. A simplified version of this relationship between the supply, given by the function s and the demand, denoted by the function d, is shown in Figure 1.4.6. Notice that x represents the number of units of the product sold.

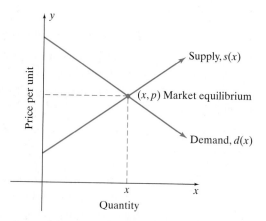

Figure 1.4.6

The point of intersection of the graphs is called the **market equilibrium**. This represents the point at which the quantity produced is equal to the quantity demanded. Let's see how this law works in an application.

Example 4 Applying the Law of Supply and Demand

The Dress-4-Success Company determines that the demand for their new office-appropriate skirt is given by

$$d(x) = 90 - 0.5x$$

while the related supply function is given by

$$s(x) = 0.45x$$

where x is the daily quantity and d and s are measured in dollars for each skirt.

(a) Determine the x- and y-intercepts of the graph of the demand function d.

(b) Determine the supply s when $x = 0$ and $x = 160$ and then graph d and s on the same Cartesian plane.

(c) Approximate the point of intersection rounded to the nearest 5 skirts and 5 dollars and interpret.

Solution

(a) The y-intercept of d can be found by letting $x = 0$ to get

$$d(0) = 90 - 0.5(0) = 90$$

So the y-intercept is at $(0, 90)$. The x-intercept is found by setting $d(x) = 0$ and solving for x.

$$0 = 90 - 0.5x$$
$$0.5x = 90$$
$$x = \frac{90}{0.5} = 180$$

So the x-intercept is at the point $(180, 0)$.

(b) When $x = 0$, the supply is $s(0) = 0.45(0) = 0$, and when $x = 160$, the supply is $s(160) = 0.45(160) = 72$. The graphs of the demand and supply functions are shown in Figure 1.4.7.

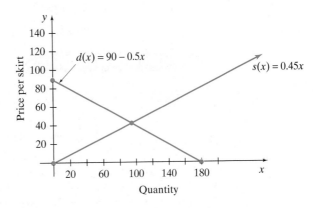

Figure 1.4.7

(c) From inspecting the graph, we can see that the market equilibrium is at about the point $(95, 40)$.

Interpret the Solution: This means that the daily supply will meet the demand when 95 skirts are supplied at a price of $40 per skirt. ∎

Technology Option

By using the INTERSECT command on the calculator, we can get a more accurate approximation of the point of intersection. By graphing the demand and supply functions in Example 4 and using the INTERSECT command, we get an intersection point of about $(94.74, 42.63)$. See Figure 1.4.8. This means that at a daily supply of about 95 skirts the price that consumers are willing

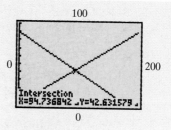

Figure 1.4.8

to pay is about $42.63. For more on how to use the INTERSECT command on your calculator, consult the online graphing calculator manual at www.prenhall.com/armstrong.

Another type of application of systems in economics is **break-even analysis**. This compares the cost of producing a product to the revenue that the product generates. To see how this process works, let's start by reviewing the **cost function** in the Toolbox below.

 From Your Toolbox

The cost of producing x units of a product is given by

$$\left(\begin{array}{c}\text{cost of}\\\text{producing } x \text{ units}\end{array}\right) = \left(\begin{array}{c}\text{cost per}\\\text{unit}\end{array}\right) \cdot x + \left(\begin{array}{c}\text{fixed}\\\text{costs}\end{array}\right)$$

For x units produced at cost per unit m and fixed costs b, the **cost function**, denoted by C, is $C(x) = mx + b$. Recall that fixed costs are costs incurred no matter how many units are produced.

We define the **revenue function** as follows:

■ **Revenue Function**

The revenue generated by producing and selling x units of a product at a certain price is

$$\left(\begin{array}{c}\text{revenue from producing}\\\text{and selling } x \text{ units}\end{array}\right) = \left(\begin{array}{c}\text{quantity}\\\text{sold}\end{array}\right) \cdot \left(\begin{array}{c}\text{unit}\\\text{price}\end{array}\right)$$

For x units sold at a price p, the **revenue function** R is

$$R(x) = x \cdot p$$

When the revenue is equal to the costs (that is, when money coming in is equal to money going out), we have reached the break-even point. Mathematically, this is the point where $R(x) = C(x)$. See Figure 1.4.9. Notice in the figure that when $R(x) > C(x)$ a profit occurs. When $R(x) < C(x)$, a loss occurs. Now let's try an application.

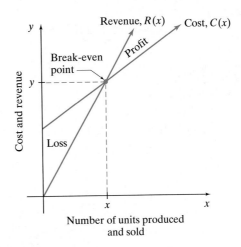

Figure 1.4.9

Example 5   Determining the Break-even Point

The Outdoors Always Company determines that the fixed costs for their new pop-up mini canopy are $2500 and the cost to produce each canopy is $25. If they plan to sell the mini canopies for $50 each, what is the break-even point?

Solution

Understand the Situation: Since the fixed costs are $2500 and the per unit cost is $25, we get the cost function $C(x) = 25x + 2500$. The selling price for the mini canopy is $50, so the revenue is given by the function $R(x) = 50x$.

  This means that we must solve the system

$$C(x) = y = 25x + 2500$$
$$R(x) = y = 50x$$

  The graphs of these functions are shown in Figure 1.4.10. From inspecting the figure, we believe that the point of intersection is at about $(100, 5000)$.

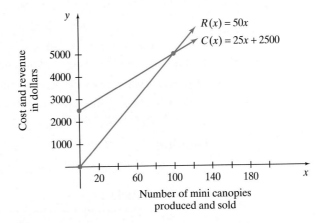

Figure 1.4.10

Interpret the Solution: This means that the company will have to produce and sell 100 mini canopies for $5000 in revenue in order to break even.   ∎

Checkpoint 3   Now work Exercise 51.

Technology Option

If we graph the cost and revenue functions given in Example 5 and use the INTERSECT command, we get the result shown in Figure 1.4.11.

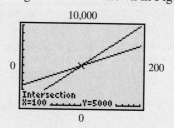

Figure 1.4.11

We see that the point of intersection is at (100, 5000). This confirms our approximation for the break-even point found in Example 5.

We can also use graphical methods when examining **systems of linear models**. These are pairs of models much like the ones we studied in Section 1.3. One of these systems is given in Example 6.

Example 6   Solving a System of Linear Models

From 1990 to 1997, the percent of low-birth-weight babies (less than 5 pounds, 8 ounces) born in Indiana can be modeled by

$$f(x) = -0.038x + 7.621, \qquad 0 \leq x \leq 7$$

where x represents the number of years since 1990 and $f(x)$ represents the percentage of low-birth-weight babies born. During the same time period, the percent of low-birth-weight babies born in Missouri can be modeled by

$$g(x) = 0.088x + 7.113, \qquad 0 \leq x \leq 7$$

where x represents the number of years since 1990 and $g(x)$ represents the percentage of low-birth-weight babies born.

(a) Evaluate each model at $x = 0$ and $x = 7$ and graph the system of equations.

(b) Approximate the point of intersection rounded to the nearest half-unit and interpret each coordinate.

Solution

(a) Table 1.4.2 shows the values of each model evaluated at $x = 0$ and $x = 7$.

Table 1.4.2

x	$f(x)$	$g(x)$
0	$-0.038(0) + 7.621 = 7.621$	$0.088(0) + 7.113 = 7.113$
7	$-0.038(7) + 7.621 = 7.355$	$0.088(7) + 7.113 = 7.729$

Using the coordinates in Table 1.4.2, we get the graph of the system shown in Figure 1.4.12.

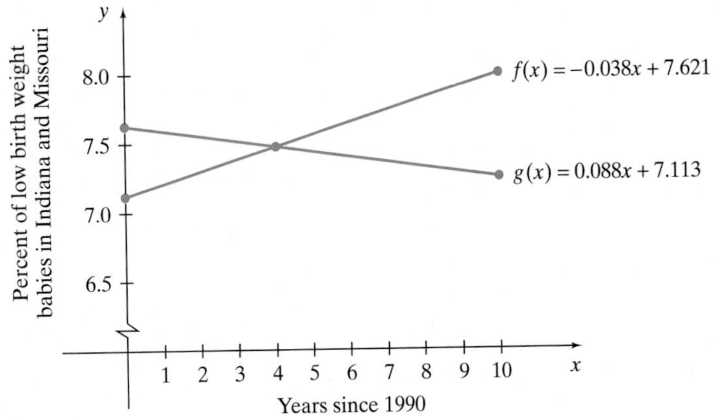

Figure 1.4.12

(b) By inspecting the graph, we approximate the point of intersection to be about $(4, 7.5)$.

Interpret the Solution: This means that in 1994 both Indiana and Missouri had low-birth-weight percentages of about 7.5%. ∎

SUMMARY

In this section, we were introduced to systems of linear equations. We learned that one way we could determine the solution to a system of linear equations is by **graphing** each equation in the system. The **point of intersection** of the graphs gives us the solution. We also found that we could have three types of solution for linear systems of equations. Linear systems can produce

- a point of intersection and **one solution**,
- parallel lines and **no solution**,
- the same line and **infinitely many solutions**.

We also learned that some of the applications of linear systems of equations included

- Law of Supply and Demand
- Break-even analysis
- Systems of linear models

SECTION 1.4 EXERCISES

In Exercises 1–10, determine if the given ordered pair is a solution to the system.

1. $(1, -1)$
$x + 4y = -3$
$5x - y = 6$

2. $(2, 2)$
$2x - y = 2$
$x + 3y = 8$

3. $(2, -1)$
$2x + y = 5$
$x + y = 3$

4. $(3, 2)$
$x - y = 5$
$2x + y = 10$

5. $(-1, 1)$
$2x + y = 1$
$x - 2y = 3$

6. $(-2, 2)$
$3x + 2y = -2$
$2x - y = -6$

✓ **7.** $(0, 2)$
$-2x + y = 0$
$7x - 2y = 3$

8. $(3, -3)$
$x - 4y = 15$
$3x - 2y = 15$

9. $(3, 13)$
$x + y = 16$
$x - y = 10$

10. $(-1, 0)$
$x - y = 1$
$3x + y = -1$

In Exercises 11–20, solve the system by determining the x- and y-intercepts of each equation. Then graph the system on the same Cartesian plane. If the system has no solution or infinite solutions, then say so and state the reason why.

11. $x + y = 2$
$x - y = 4$

12. $x + y = 6$
$x - y = 2$

13. $y = 4 - x$
$y = x + 4$

14. $y = 1 - x$
$y = x + 3$

15. $-x + y = -3$
$3x - y = 3$

16. $4x + 3y = 12$
$2x - 3y = 6$

17. $6x + 2y = 12$
$-3x - y = -3$

18. $x + y = 6$
$-x - y = -5$

19. $-4x - 2y = -8$
$2x + y = 4$

20. $-x + y = 3$
$3x - 3y = -9$

In Exercises 21–24, solve the system by inspecting the graph and approximating the point of intersection.

21. $2x + y = 3$
$-x + y = -3$

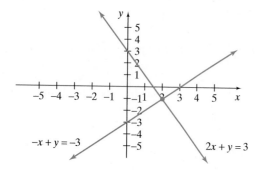

22. $4x - y = -3$
$x - y = 0$

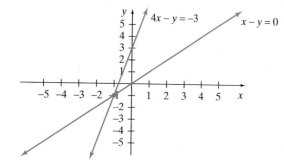

23. $4x + y = 4$
$3x - y = 3$

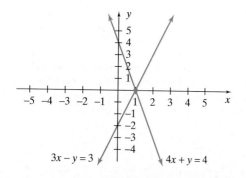

24. $2x + y = 4$
$5x - y = 10$

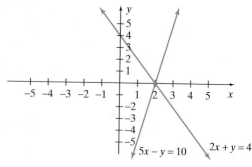

In Exercises 25–32, graph the equations on the same Cartesian plane and then use the graphs to estimate the solutions.

25. $x + y = 4$
$y = -2$

26. $x + y = 10$
$y = 5$

27. $x - y = 8$
$x = 4$

28. $x - y = 6$
$x = 3$

29. $x = 3$
$y = -2$

30. $y = -1$
$x = 4$

31. $y = 3 - 2x$
$y = 1 - x$

32. $y = 3x - 4$
$y = 1 - 2x$

In Exercises 33–38, solve the system by graphing each equation on your calculator and using the INTERSECT command. If the system has no solution or infinite solutions, then say so.

33. $3y = 2x$
$y - x = 1$

34. $2x + y = 3$
$2y = 3x + 6$

35. $3x + 2y = 8$
$3x = 10 - 2y$

36. $3x + 4y = 5$
$10x - 6y = 3$

37. $2x + y = 3$
$2y - 3x = 6$

38. $2x + 3y = 6$
$4x + 2y = 4$

In Exercises 39–44, determine the value of k that satisfies the given condition for each system.

39. k so that $\begin{aligned} y &= kx + 8 \\ y &= 3x - 1 \end{aligned}$ has no solution.

40. k so that $\begin{aligned} 2x - y &= -2 \\ kx - y &= 1 \end{aligned}$ has no solution.

✓ **41.** k so that $\begin{aligned} 4x - 2y &= 7 \\ 12x - ky &= 21 \end{aligned}$ has infinitely many solutions.

42. k so that $\begin{aligned} y &= 3x - 4 \\ 4y &= kx - 16 \end{aligned}$ has infinitely many solutions.

43. k so that $\begin{aligned} y &= -6x - 4 \\ y &= -3x - k \end{aligned}$ has no solution.

44. k so that $\begin{aligned} y &= 8x + 5 \\ y &= 4x + k \end{aligned}$ has no solution.

Applications

45. The Say-It-Now Company develops communications software and determines that the demand for their teleconference mini suite is given by $d(x) = 80 - \dfrac{3}{5}x$ with a supply function given by $s(x) = \frac{2}{5}x$, where x is the daily quantity and $d(x)$ and $s(x)$ are measured in dollars per minisuite.

(a) Determine the x- and y-intercepts for the graph of the demand function d.

(b) Determine the supply s when $x = 0$ and $x = 130$. Graph d and s on the same Cartesian plane.

(c) Approximate the point of intersection and interpret.

46. The SunMaker Company has determined that the demand for their patio sets is given by $d(x) = 630 - 0.75x$, while the supply is given by $s(x) = 0.75x$, where x is the quantity and $d(x)$ and $s(x)$ are in dollars per patio set.

(a) Determine the x- and y-intercepts for the graph of the demand function d.

(b) Determine the supply s when $x = 0$ and $x = 830$. Graph d and s on the same Cartesian plane.

(c) Approximate the point of intersection and interpret.

47. The Omniview Company makes hand-held color televisions and determines that the demand for their entry-level TK-2 model is $d(x) = 150 - 0.2x$, while the supply is given by $s(x) = 0.4x$, where x is the quantity and $d(x)$ and $s(x)$ are in dollars per TK-2 television.

(a) Determine the x- and y-intercepts for the graph of the demand function d.

(b) Determine the supply s when $x = 0$ and $x = 400$. Graph d and s on the same Cartesian plane.

(c) Approximate the point of intersection and interpret.

48. The Big Sound Company has determined that the demand for their Faster Blaster boom box is $d(x) = 450 - 0.3x$, while the supply is given by $s(x) = 0.15x$, where x is the quantity and $d(x)$ and $s(x)$ are measured in dollars per boom box.

(a) Determine the x- and y-intercepts for the graph of the demand function d.

(b) Determine the supply s when $x = 0$ and $x = 2000$. Graph d and s on the same Cartesian plane.

(c) Approximate the point of intersection and interpret.

49. The Ryovic Company determines that the demand for their single-person pup tent is given by $d(x) = 75 - \frac{1}{4}x$, while the supply is given by $s(x) = \frac{1}{2}x$, where x is the quantity and $d(x)$ and $s(x)$ are measured in dollars per tent.

(a) Write a system of equations to determine the equilibrium point and graph the system in the viewing window $[0, 200]$ by $[0, 100]$.

(b) Determine the point of intersection via the INTERSECT command and interpret each coordinate.

50. The TechnoPic Company determines that the demand for their new picture-in-a-picture TV set is given by $d(x) = 1800 - \frac{3}{5}x$, while the supply is given by $s(x) = \frac{3}{10}x$, where x is the quantity and $d(x)$ and $s(x)$ are measured in dollars per TV set.

(a) Write a system of equations to determine the equilibrium point and graph the system in the viewing window $[0, 3000]$ by $[0, 1200]$.

(b) Determine the point of intersection via the INTERSECT command and interpret each coordinate.

51. Adel starts a business of manufacturing picture frames for small 3 by 5 inch photographs. He determines that the fixed cost is $252 and the unit cost is $1.50. He sells each frame for $5.50.

(a) Determine a system of linear equations for the cost and the revenue that Adel makes from the picture frames.

(b) Use Figure 1.4.13 to approximate the break-even point and interpret each coordinate. Round the cost and revenue to the nearest $10 and the number of frames sold to the nearest 10.

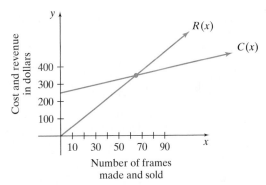

Figure 1.4.13

52. Sally makes dream catchers as a hobby and wants to sell them at craft shows. She knows that the fixed costs are $90 and the cost to produce each dream catcher is $2.50. She plans to set a price of $3 for each dream catcher.

(a) Determine a system of equations for the cost and the revenue that Sally makes from the dream catchers.

(b) Use Figure 1.4.14 to approximate the break-even point and interpret each coordinate.

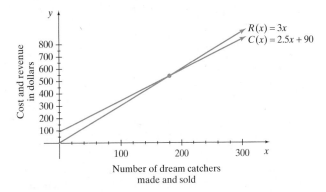

Figure 1.4.14

53. The Lullaby Company knows that the fixed costs to manufacture their portable crib are $250 and the cost to produce each crib is $100. The sale price for each crib is $125. Define the variables and write a system of equations for the cost and the revenue that the Lullaby Company makes from the portable cribs.

54. For the system determined in Exercise 53, graph C and R in the same viewing window. Use the INTERSECT command to determine the break-even point and interpret each coordinate.

1.1 **55.** Clark and Florence make bumper stickers. They know that their fixed costs are $200 and the per unit cost is $3. They sell the bumper stickers for $5 each. Define the variables and write a system of equations for the cost and the revenue that Clark and Florence make from the bumper stickers.

56. For the system determined in Exercise 55, graph C and R in the same viewing window. Use the INTERSECT command to determine the break-even point and interpret each coordinate.

57. The Stay-Connected Company determines that the fixed costs for their Family Radio Service radios are $20,000 and it costs $30 to manufacture each unit. They sell each radio for $50. Define the variables and write a system of equations for the cost and the revenue that the Stay-Connected Company makes from the Family Radio Service radios.

58. For the system determined in Exercise 57, graph C and R in the same viewing window. Use the INTERSECT command to determine the break-even point and interpret each coordinate.

59. The Squeaky Clean Company has fixed costs of $100,000 and unit costs of $150 for their pressure washers. Each pressure washer sells for $250. Define the variables and write a system of equations for the cost and the revenue that the Squeaky Clean Company makes from the pressure washers.

60. For the system determined in Exercise 59, graph C and R in the same viewing window. Use the INTERSECT command to determine the break-even point and interpret each coordinate.

61. From 1990 to 1998, the number of U.S. children under 5 years of age who were of Hispanic origin can be modeled by

$$f(x) = 0.12x + 2.51, \qquad 0 \le x \le 8$$

where x represents the number of years since 1990 and $f(x)$ represents the number of children under 5 years of age who were of Hispanic origin, in millions. During the same time period, the number of U.S. children under 5 years of age who were African American can be modeled by

$$g(x) = -0.02x + 2.84, \qquad 0 \le x \le 8$$

where x represents the number of years since 1990 and $g(x)$ represents the number of children under 5 years of age who were African American, in millions (*Source:* U.S. Census Bureau, www.census.gov).

(a) Evaluate each model at $x = 0$ and $x = 8$ and graph the system of equations.

(b) Approximate the point of intersection rounded to the nearest half-unit and interpret each coordinate.

62. For the models given in Exercise 61, graph f and g in the viewing window [0, 8] by [2, 4]. Determine the point of intersection via the INTERSECT command and interpret each coordinate.

63. From 1990 to 1998, the number of people 25 to 29 years of age who were of Hispanic origin in the United States can be modeled by

$$f(x) = 0.024x + 2.335, \qquad 0 \le x \le 8$$

where x represents the number of years since 1990 and $f(x)$ represents the number of people 25 to 29 years of age who were of Hispanic origin, in millions. During the same time period, the number of people 25 to 29 years of age who were African American in the United States can be modeled by

$$g(x) = -0.023x + 2.629, \qquad 0 \le x \le 8$$

where x represents the number of years since 1990 and $g(x)$ represents the number of people 25 to 29 years of age who were African American, in millions (*Source:* U.S. Census Bureau, www.census.gov).

(a) Evaluate each model at $x = 0$ and $x = 8$ and graph the system of equations.

(b) Approximate the point of intersection rounded to the nearest half-unit and interpret each coordinate.

64. For the models given in Exercise 63, graph f and g in the viewing window [0, 8] by [2, 3]. Determine the point of intersection via the INTERSECT command and interpret each coordinate.

SECTION PROJECT

The populations (measured in thousands) of the metropolitan areas of the cities of Austin, Texas, and Dayton, Ohio, are shown in Table 1.4.3.

Table 1.4.3

City	1990 Population	1998 Population
Austin	846	1106
Dayton	951	949

SOURCE: U.S. Census Bureau, www.census.gov

(a) Write linear models in slope–intercept form for the populations of Austin and Dayton.

(b) State the average rate of change for the populations of each city.

(c) Graph each of the models by hand, and estimate the point of intersection and interpret each coordinate.

(d) Graph each model with your calculator, determine the point of intersection, and interpret each coordinate.

Section 1.5   Solving Systems of Linear Equations Algebraically

In the previous section we saw how we could solve systems of linear equations using the **graphical method**. This method provides a good visualization of the solution, but we have seen that the method has some drawbacks. If the solution contains a fraction, for example, the graphing method makes it difficult to pinpoint the solution. In this section we will learn two algebraic methods for solving systems of linear equations. We will also examine more applications of systems of linear equations.

The Substitution Method

The challenge in solving systems of linear equations is that there are two equations with two unknowns (that is, variables). Much of what we studied in algebra focused on how to solve one equation with one variable. With the **substitution method**, we can alter the system in such a way that there is only one linear equation written in terms of one variable. To see how this works, let's Flashback to an example from Section 1.4.

Flashback

Revisiting Systems of Equations

In Example 3, part (b) of the previous section, we were given the system

$$2x + 2y = 4$$
$$2x - y = 3$$

Graph the system and verify that the solution is $\left(\dfrac{5}{3}, \dfrac{1}{3}\right)$.

Flashback Solution

The graph of the system is shown in Figure 1.5.1. If the solution to the system is $\left(\dfrac{5}{3}, \dfrac{1}{3}\right)$, then the lines intersect at $\left(\dfrac{5}{3}, \dfrac{1}{3}\right)$, as shown in Figure 1.5.1.

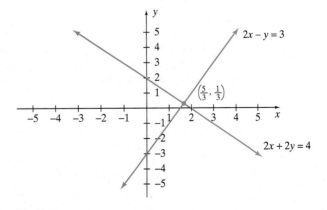

Figure 1.5.1

To verify the solution, we replace x with $\frac{5}{3}$ and y with $\frac{1}{3}$ in each equation and show that the equations are true. In the first equation we get

$$2x + 2y = 4$$
$$2\left(\frac{5}{3}\right) + 2\left(\frac{1}{3}\right) = 4$$
$$\frac{10}{3} + \frac{2}{3} = 4$$
$$\frac{12}{3} = 4$$

which is true. The second equation yields

$$2x - y = 3$$
$$2\left(\frac{5}{3}\right) - \left(\frac{1}{3}\right) = 3$$
$$\frac{10}{3} - \frac{1}{3} = 3$$
$$\frac{9}{3} = 3$$

which is true. So the solution is verified.

In the Flashback we verified that the solution to the system is $\left(\frac{5}{3}, \frac{1}{3}\right)$, although it is difficult to see this from looking at the graph. Now let's try an algebraic approach. For the system in the Flashback

$$2x + 2y = 4$$
$$2x - y = 3$$

let's begin by solving the second equation for y. This gives us

$$2x + 2y = 4$$
$$2x - 3 = y$$

Since the solution to a system of linear equations is made up of the values that make each equation true, we can replace or **substitute** the expression $2x - 3$ for every occurrence of y in the first equation. This gives us

$$2x + 2(2x - 3) = 4 \qquad \text{First equation, with } 2x - 3 \text{ replacing } y$$

Now we have one equation written in terms of one variable, x. This allows us to solve for x to get

$$2x + 4x - 6 = 4 \qquad \text{Distribute the 2.}$$
$$6x = 10 \qquad \text{Combine like terms.}$$
$$x = \frac{10}{6} = \frac{5}{3} \qquad \text{Divide both sides by 6.}$$

To find y, we can replace $\frac{5}{3}$ for x in either equation of the system. If we let $x = \frac{5}{3}$ in the system's first equation, we get

$$2\left(\frac{5}{3}\right) + 2y = 4 \qquad \text{Substitute } x \text{ with } \frac{5}{3}.$$

$$\frac{10}{3} + 2y = 4 \qquad \text{Simplify.}$$

$$2y = 4 - \frac{10}{3} = \frac{2}{3} \qquad \text{Subtract } \frac{10}{3} \text{ from both sides.}$$

$$y = \frac{1}{3} \qquad \text{Divide both sides by 2.}$$

So the solution to the system is $(x, y) = \left(\frac{5}{3}, \frac{1}{3}\right)$.

We have just illustrated an algebraic technique known as the **substitution method**. We now list the steps used in the substitution method.

■ **Substitution Method**

Given a system of linear equations and two variables, we can solve the system using the substitution method by the following steps.

1. Choose one of the equations and solve for one of the variables, x or y. A good choice is a variable with a coefficient of 1 or −1.
2. Substitute the resulting expression from step 1 into the other equation for the appropriate variable.
3. Solve the resulting equation.
4. Find the value of the remaining variable by substituting the value of the variable found in step 3 into either equation containing both variables.

Example 1 **Solving a System of Linear Equations by the Substitution Method**

Solve the system using the substitution method.

$$2x + 3y = 1$$
$$x - y = 3$$

Solution

For this example, let's follow the four steps in the substitution process.

Step 1: Since the coefficient of x in the second equation is 1, we will solve the second equation for x to get

$$x - y = 3$$
$$x = y + 3 \qquad \text{Add } y \text{ to both sides.}$$

Step 2: Substituting the expression $y + 3$ for x in the first equation gives us

$$2x + 3y = 1$$
$$2(y + 3) + 3y = 1$$

Step 3: Now we want to solve the equation found in step 2 for y to get

$$2y + 6 + 3y = 1 \qquad \text{Distribute the 2.}$$
$$5y + 6 = 1 \qquad \text{Combine like terms.}$$
$$5y = -5 \qquad \text{Subtract 6 from both sides.}$$
$$y = -1 \qquad \text{Divide both sides by 5.}$$

Step 4: We can now substitute $y = -1$ into either equation to determine the x-value.

Using the first equation in the system gives us

$$2x + 3(-1) = 1$$
$$2x - 3 = 1$$
$$2x = 4$$
$$x = 2$$

So the solution to the system is $(x, y) = (2, -1)$. ∎

✓ Checkpoint 1

Now work Exercise 5.

Many of our systems of linear models are written in terms of the dependent variable, whether it be f, g, y, and so on. We can use the substitution method to solve these systems of linear models by simply setting the expressions containing the independent variables equal to each other. Some call this method of using substitution the **comparison method**.

Example 2   **Solving a System of Linear Models**

The number of immigrants from Ireland to the United States from 1981 to 1995 can be modeled by

$$y = f(x) = 2.72x + 17.84, \qquad 1 \le x \le 15$$

where x represents the number of years since 1980 and $f(x)$ represents the annual number of immigrants from Ireland to the United States, in thousands. During the same time period, the number of immigrants from Japan to the United States can be modeled by

$$y = g(x) = -1.24x + 50.02, \qquad 1 \le x \le 15$$

where x represents the number of years since 1980 and $g(x)$ represents the annual number of immigrants from Japan to the United States, in thousands. Solve the corresponding system of equations and interpret the meaning of each coordinate (*Source:* U.S. Immigration and Naturalization Service, www.ins.usdoj.gov).

Solution

Understand the Situation: Here we have to solve the system of equations

$$y = f(x) = 2.72x + 17.84$$
$$y = g(x) = -1.24x + 50.02$$

Since each equation is solved for y, we can set the expressions involving x equal to each other to get

$$2.72x + 17.84 = -1.24x + 50.02$$
$$3.96x = 32.18$$
$$x = \frac{32.18}{3.96}$$
$$x \approx 8.13$$

Substituting this result into the second model gives us

$$y = -1.24(8.13) + 50.02 \approx 39.94$$

So the solution to the system is about $(8.13, 39.94)$.

Interpret the Solution: This means that during the year 1988 the number of immigrants from Ireland and from Japan to the United States was approximately equal at about 39.94 thousand, or about 39,940.   ∎

Interactive Activity

Using your calculator, enter $y_1 = 2.72x + 17.84$ and $y_2 = -1.24x + 50.02$. Graph y_1 and y_2 in the viewing window [0,15] by [10,50] and use the INTERSECT command to verify the result of Example 2.

The Elimination Method

The **elimination method** is a powerful method used to solve systems of equations and works well when none of the coefficients of the variables is 1 or -1. The method will also lead us to the method for solving systems of equations using matrices in Chapter 2. The fundamental idea behind the elimination method is to continue replacing the original systems of equations by a system of **equivalent equations** until the solution is reached.

Consider the system

$$3x - 5y = 2$$
$$2x + 5y = 13$$

Notice that the coefficients of the variable y in the two equations are opposite, so if we simply add the like terms of the two equations, we get

$$
\begin{array}{r}
3x - 5y =\ \ 2 \\
\underline{2x + 5y = 13} \\
5x = 15
\end{array}
$$

Solving for x gives us

$$5x = 15$$
$$x = 3$$

If we substitute $x = 3$ into the first equation, we get

$$3(3) - 5y = 2$$
$$9 - 5y = 2$$
$$-5y = -7$$
$$y = \frac{7}{5}$$

So the solution is $(x, y) = \left(3, \dfrac{7}{5}\right)$.

From Your Toolbox

A skill from algebra that is necessary for solving systems by the elimination method is combining monomial terms. For more practice on this skill, see Appendix B.

We have just outlined the elimination method. But what if the coefficients of a variable are not opposites? Then we use one of the following three operations for obtaining equivalent equations.

■ **Equivalent Equation Operations**

We can transform a system of linear equations into an equivalent system of linear equations using these three operations.

1. Interchange two equations (this is sometimes called *swapping* equations).
2. Multiply an equation by a nonzero constant.
3. Add a multiple of one equation to another equation.

Usually, the system's equations are written in the standard form $ax + by = c$ before performing any equivalent equation operations. We will see that the third of the operations is used often. Now let's use these operations to solve another system.

Example 3   **Solving a System of Equations by the Elimination Method**

Solve the system using the elimination method.

$$3x + 2y = 11$$
$$-x + y = 3$$

Solution

We can make the coefficients of x opposites if we multiply the second equation by 3. This gives us the equivalent system

$$3x + 2y = 11$$
$$-3x + 3y = 9$$

Now we can add the two equations to get

$$
\begin{array}{r}
3x + 2y = 11 \\
-3x + 3y = 9 \\
\hline
5y = 20 \\
y = 4
\end{array}
$$

If we replace the y-value in our given system's first equation, we have

$$3x + 2(4) = 11$$
$$3x + 8 = 11$$
$$3x = 3$$
$$x = 1$$

So the solution is (1, 4). This is illustrated in Figure 1.5.2.

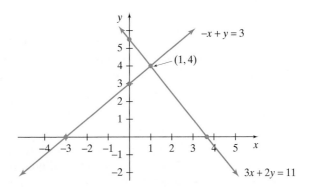

Figure 1.5.2

In Section 1.4, we saw that a system of equations can have either one solution, no solution, or infinitely many solutions. If a system has an infinite number of solutions, we can express the solution by using what is called a **parameter**. Let's see how we can express a solution to a system with an infinite number of solutions by using a parameter.

Example 4 **Using a Parameter in Writing a Linear System's Solution**

Solve the system by elimination.

$$x - y = 3$$
$$-2x + 2y = -6$$

Solution

We begin by multiplying the first equation by 2 and adding the equations to get

$$2x - 2y = 6$$
$$\underline{-2x + 2y = -6}$$
$$0 = 0$$

Notice that all the variables have disappeared and we are left with the equation $0 = 0$. This is sometimes called an **identity** and indicates that the system has an infinite number of solutions. We will express the solutions in terms of y, where y can represent any real number. In this case, y is the parameter. (We could have also chosen x to be our parameter.) When we solve the first equation for x, we get

$$x - y = 3$$
$$x = y + 3$$

So we can say that all ordered pairs of the form

$$(y + 3, y)$$

are solutions to the original system.

Example 4 illustrated that when using an algebraic technique to solve a system and all variables disappear, the system has an infinite number of solutions if we are left with an equation that is an identity. To see what happens when we have no solution, consider the following system:

$$x - y = 4$$
$$-2x + 2y = -6$$

We begin by multiplying the first equation by 2 and then add the equations to get

$$
\begin{aligned}
2x - 2y &= 8 \\
-2x + 2y &= -6 \\
\hline
0 &= 2
\end{aligned}
$$

Notice that all the variables have disappeared and we are left with the equation $0 = 2$. This is sometimes called a **contradiction** and indicates that the system has no solution.

Solving a System with Three Linear Equations and Three Variables

One reason that elimination is so powerful is that we can use it to solve systems of three linear equations and three variables. This is sometimes called a **3 by 3 system**. Similarly, a system of two equations and two unknowns is called a **2 by 2 system**. The general strategy behind solving these systems is to reduce the system to a system of two equations and two variables. Then we can use the techniques we have learned so far to determine the variable's values.

■ **Solving a System of Three Equations and Three Variables**
1. Choose any two equations and eliminate a variable.
2. Choose a different pair of equations and eliminate the same variable.
3. Use the results of steps 1 and 2 to write a system of two equations and two variables.
4. Solve the system from step 3.
5. Replace the value found in step 4 into the system of two equations and two variables to determine the second variable's value.
6. Replace the known values into any equation in the original system to determine the third variable's value.

Notice that, since the system has three variables, we will get an **ordered triple** for a solution. Let's look at an example of how to solve one of these systems.

Example 5   **Solving a System of Three Linear Equations and Three Variables**

Solve the 3 by 3 system using the elimination method.

$$x + 2y + 2z = 9$$
$$2x + y + 4z = 12$$
$$3x - 3y - 2z = 1$$

Solution

In this example, let's follow the six steps for solving 3 by 3 systems.

Step 1: Notice that if we add the first and third equations, we can eliminate z.

$$x + 2y + 2z = 9$$
$$\underline{3x - 3y - 2z = 1}$$
$$4x - y = 10$$

Step 2: Now we need to eliminate z from another pair of equations in our given system. If we multiply the third equation by 2 and add it to the second equation, we get

$$2x + y + 4z = 12 \quad \text{Multiply bottom} \qquad 2x + y + 4z = 12$$
$$3x - 3y - 2z = 1 \quad \text{equation by 2 and add.} \rightarrow \underline{6x - 6y - 4z = 2}$$
$$8x - 5y = 14$$

Step 3: This gives us the 2 by 2 system

$$4x - y = 10$$
$$8x - 5y = 14$$

Step 4: We can now eliminate y by multiplying the top equation by -5 and adding the equations. This yields

$$4x - y = 10 \quad \text{Multiply top equation} \qquad -20x + 5y = -50$$
$$8x - 5y = 14 \quad \text{by } -5 \text{ and add.} \rightarrow \underline{8x - 5y = 14}$$
$$-12x = -36$$
$$x = 3$$

Step 5: We can now determine the y-value by replacing x by 3 in either equation of the 2 by 2 system. Substituting x with 3 in the second equation gives us

$$8(3) - 5y = 14$$
$$24 - 5y = 14$$
$$-5y = -10$$
$$y = 2$$

Step 6: We can now determine the z-value by substituting $x = 3$ and $y = 2$ into any equation in our original 3 by 3 system. If we use the first equation, we find

$$x + 2y + 2z = 9$$
$$3 + 2(2) + 2z = 9$$
$$3 + 4 + 2z = 9$$
$$7 + 2z = 9$$
$$2z = 2$$
$$z = 1$$

So the solution to the system is the ordered triple $(x, y, z) = (3, 2, 1)$. ∎

✔ **Checkpoint 2**   Now work Exercise 41.

We learned in Section 1.4 that the graph of a system of two linear equations and two variables was generally two lines in the Cartesian plane. But what is the graph of three linear equations with three variables? The graph of an equation of the form $ax + by + cz = d$ is a surface in three dimensions called a **plane**. Just like 2 by 2 systems, the 3 by 3 system can have either one solution, no solution, or infinitely many solutions. This depends on how the three planes given by the three equations intersect. Figure 1.5.3 shows this relationship.

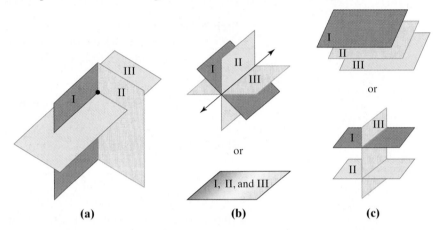

(a)   **(b)**   **(c)**

Figure 1.5.3 (a) Three planes intersect at a point. System has one solution.
(b) Three planes intersect in a line or are the same plane. System has infinitely many solutions. **(c)** Three parallel planes or two parallel planes intersected by a third plane. No point or points in common. System has no solution.

Example 6   Solving a 3 by 3 System of Linear Equations

Solve the system by elimination.

$$2x + 3y - 2z = 4$$
$$3x - 3y + 2z = 16$$
$$6x - 12y + 8z = 5$$

Solution

If we multiply the first equation by 4 and add it to the third equation, we get

$$
\begin{array}{ll}
2x + 3y - 2z = 4 & \text{Multiply top equation} \\
6x - 12y + 8z = 5 & \text{by 4 and add.}
\end{array}
\rightarrow
\begin{array}{r}
8x + 12y - 8z = 16 \\
\underline{6x - 12y + 8z = 5} \\
14x = 21 \\
x = \dfrac{3}{2}
\end{array}
$$

Now if we add the first and second equations of our original system, we get

$$
\begin{array}{r}
2x + 3y - 2z = 4 \\
\underline{3x - 3y + 2z = 16} \\
5x = 20 \\
x = 4
\end{array}
$$

So we see that two different values are obtained for x, which is a contradiction. This indicates that the system has no solution.   ∎

In Example 4 we saw that the solutions to a 2 by 2 system of linear equations with an infinite number of solutions can be expressed by a parameter. Parameters can also be written to express infinite solutions to 3 by 3 systems of linear equations.

Application

Systems of three equations and three variables can also be used in applications, as is shown in the following Flashback.

Flashback

System of Three Equations and Three Variables Revisited

In Section 1.1 we saw that Gonzalez Office Supplies manufactures three types of office chairs: Secretarial, Executive, and Presidential. Each chair requires time on three different machines. These times and the total available time for each machine are given in Table 1.5.1. If all available time for each machine must be used each week, determine how many of each type of chair can be made each week.

Table 1.5.1

Type:	Secretarial	Executive	Presidential	Total Weekly Hours
Hours on Machine A	1	1	2	30
Hours on Machine B	2	3	2	53
Hours on Machine C	1	2	3	47

Flashback Solution

Understand the Situation: Here we let x represent the number of Secretarial chairs made each week, y represent the number of Executive chairs made each week, and z represent the number of Presidential chairs made each week. Then we have to solve the system

$$x + y + 2z = 30$$
$$2x + 3y + 2z = 53$$
$$x + 2y + 3z = 47$$

We can eliminate x by multiplying the first equation by -1 and adding it to the third equation.

$$
\begin{array}{l}
x + y + 2z = 30 \\
x + 2y + 3z = 47
\end{array}
\quad
\begin{array}{c}
\text{Multiply top equation} \\
\text{by } -1 \text{ and add.}
\end{array}
\;\rightarrow\;
\begin{array}{r}
-x - y - 2z = -30 \\
\underline{x + 2y + 3z = 47} \\
y + z = 17
\end{array}
$$

We can eliminate x from the second and third equations by multiplying the third equation by -2 and adding it to the second.

$$
\begin{array}{l}
2x + 3y + 2z = 53 \\
x + 2y + 3z = 47
\end{array}
\quad
\begin{array}{c}
\text{Multiply bottom equation} \\
\text{by } -2 \text{ and add.}
\end{array}
\;\rightarrow\;
\begin{array}{r}
2x + 3y + 2z = 53 \\
\underline{-2x - 4y - 6z = -94} \\
-y - 4z = -41
\end{array}
$$

This gives us the 2 by 2 system

$$y + z = 17$$
$$-y - 4z = -41$$

We can now eliminate y just by adding the two equations. This gives

$$
\begin{aligned}
y + z &= 17 \\
-y - 4z &= -41 \\
\hline
-3z &= -24 \\
z &= 8
\end{aligned}
$$

If we replace z with 8 in the first equation of our 2 by 2 system, we find

$$y + 8 = 17$$
$$y = 9$$

If we replace z with 8 and y with 9 in the first equation of our given 3 by 3 system, we have

$$x + y + 2z = 30$$
$$x + 9 + 2(8) = 30$$
$$x + 9 + 16 = 30$$
$$x = 5$$

So we get the ordered triple $(5, 9, 8)$.

Interpret the Solution: This means that if we use all the available time for each machine each week we can make 5 Secretarial chairs, 9 Executive chairs, and 8 Presidential chairs.

SUMMARY

In this section we learned two new ways to solve systems of equations. The first was the substitution method and the second was the elimination method. We can now examine the advantages and disadvantages of each method.

Method	Advantages	Disadvantages
Graphical	• We can visualize the solution	• Difficult to determine the solution if the coordinates are not integers
Substitution	• Gives exact solutions • Can be used for systems of linear models	• Solutions cannot be visualized
Elimination	• Can be used to solve 3 by 3 systems • Gives exact solutions	• Difficult to use with systems of linear models with decimal coefficients. • Solutions cannot be visualized

SECTION 1.5 EXERCISES

In Exercises 1–20, solve the system by the substitution method. If the system has no solution, then say so. If the system has an infinite number of solutions, use a parameter to represent the infinite number of solutions as shown in Example 4.

1. $y = x$
$x + y = 4$

2. $x = y$
$x - y = 6$

3. $x + y = 4$
$3x - y = 0$

4. $x + y = 6$
$2x - y = 0$

✓ **5.** $2x - y = 9$
$x + 3y = 8$

6. $2x - y = -7$
$2x - 3y = -13$

7. $2x + 3y = 11$
$x = 2 + 2y$

8. $2x + y = 20$
$x = -2y - 12$

9. $4x - 3y = 2$
$y = -2x - 4$

10. $5x + 6y = 14$
$y = 4x - 17$

11. $x + 2y = 13$
$x = -2y + 9$

12. $2x - y = -1$
$y = 2x + 3$

13. $y = 2x + 1$
$y = x + 4$

14. $y = 4x - 7$
$y = 2x + 1$

15. $-3x + y = -5$
$21x - 7y = 35$

16. $-3x + y = -1$
$12x - 4y = 4$

17. $x - 2y = -2$
$2x - y = 3$

18. $-2x + y = 3$
$x + y = 4$

19. $2x - 3y = 1.3$
$-x + y = -0.5$

20. $1.5x + y = 4.05$
$2x - 3y = -3.7$

In Exercises 21–40, solve the system by the elimination method. If the system has no solution, then say so. If the system has an infinite number of solutions, use a parameter to represent the infinite number of solutions as shown in Example 4.

21. $x - y = -2$
$x + y = 6$

22. $x - y = 3$
$x + y = 1$

23. $x - y = 3$
$x + y = 7$

24. $x - y = 7$
$x + y = 1$

25. $2x + y = -10$
$2x - y = -6$

26. $x + 2y = -9$
$x - 2y = -1$

27. $2x + y = 4$
$3x + 2y = 3$

28. $x - y = -3$
$2x - 5y = -21$

29. $7x - y = 4$
$3x - 2y = -3$

30. $2x + y = 1$
$-3x + y = -4$

31. $5x - 10y = -2$
$-5x + 10y = -2$

32. $3x - y = 1$
$-3x + y = -2$

33. $2x + 2y = 3$
$-5x + 4y = 15$

34. $x + 4y = 2$
$x - 8y = -7$

35. $3x - 2y = 6$
$x + y = 3$

36. $4x - y = 3$
$x - 3y = -4$

37. $4x - 7y = -13$
$-7x + 3y = -5$

38. $9x + 7y = 21$
$x + 12y = 36$

39. $3x - 4y - 1 = 0$
$4x + 3y - 1 = 0$

40. $4x - 2y - 5 = 0$
$2x + 3y - 4 = 0$

In Exercises 41–50, solve the system of three equations and three unknowns by the elimination method. If the system has no solution, then say so. If the system has an infinite number of solutions, use a parameter to represent the infinite number of solutions as shown in Example 4.

✓ **41.** $-x + 3y + z = 10$
$3x + 2y - 2z = 3$
$2x - y - 4z = -7$

42. $x + y - 3z = 8$
$3x + 4y - 2z = 20$
$2x - 3y + z = -6$

43. $x + y + z = 1$
$x + 3y + 7z = 13$
$x + 2y + 3z = 4$

44. $x + y + z = 2$
$2x + y + 2z = 3$
$3x - y + z = 4$

45. $x + 2y + 3z = 9$
$-3x + 5y - 4z = -7$
$3x - y + 2z = -1$

46. $2x + 2y + 3z = 10$
$3x + y - z = 0$
$x + y + 2z = 6$

47. $2x + y - z = 1$
$x + 2y + 2z = 2$
$4x + 5y + 3z = 3$

48. $2x + 3y + 4z = 6$
$2x - 5y - 4z = -4$
$4x + 6y + 8z = 12$

49. $x - 2y = 14$
$y + 2z = 11$
$2x + z = 16$

50. $y + z = -3$
$x - 2z = -5$
$3x + 2y = -5$

Applications

51. Victor purchased 2 coats and 1 pair of slacks for $120 at the Men's Suit Palace. The next week, he returned and purchased 1 coat and 2 pairs of slacks also for $120.

(a) Define the variables and write a system of equations for the price of the coat and slacks.

(b) Solve the system in part (a) and interpret.

52. An office assistant purchased 5 cups of coffee and 6 bagels for $24 from the DaMatta Deli. The next day the assistant purchased 3 cups of coffee and 5 bagels for $18.

(a) Define the variables and write a system of equations for the price of the coffee and bagels.

(b) Solve the system in part (a) and interpret.

53. From 1980 to 1995 the number of households in the District of Columbia can be modeled by

$$f(x) = -1.14x + 255.31, \qquad 0 \le x \le 15$$

where x represents the number of years since 1980 and $f(x)$ represents the number of households in thousands. During the same time period, the number of households in South Dakota can be modeled by

$$g(x) = 1.75x + 242.58, \qquad 0 \le x \le 15$$

where x represents the number of years since 1980 and $g(x)$ represents the number of households in thousands. Determine the point of intersection and interpret each coordinate (*Source:* U.S. Census Bureau, www.census.gov).

54. From 1980 to 1997 the number of Americans aged 25 to 29 who owned a home can be modeled by

$$f(x) = -0.06x + 9.81, \qquad 0 \le x \le 17$$

where x represents the number of years since 1980 and $f(x)$ represents the number of Americans, in millions, aged 25 to 29 who owned a home. During the same time period, the number of Americans aged 30 to 34 who owned a home can be modeled by

$$g(x) = 0.09x + 9.60, \qquad 0 \le x \le 17$$

where x represents the number of years since 1980 and $g(x)$ represents the number of Americans, in millions, aged 30 to 34 who owned a home. Determine the point of intersection and interpret each coordinate (*Source:* U.S. Census Bureau, www.census.gov).

55. From 1990 to 2000 the population of the District of Columbia can be modeled by

$$f(x) = -10.30x + 604.64, \qquad 0 \le x \le 10$$

where x represents the number of years since 1990 and $f(x)$ represents the District of Columbia population in thousands. During the same time period, the population of Alaska can be modeled by

$$g(x) = 33.58x + 394.78, \qquad 0 \le x \le 10$$

where x represents the number of years since 1990 and $g(x)$ represents the Alaska population in thousands. Determine the point of intersection and interpret each coordinate (*Source:* U.S. Census Bureau, www.census.gov).

56. From 1990 to 2000 the population of Utah can be modeled by

$$f(x) = 46.49x + 1732.49, \qquad 0 \le x \le 10$$

where x represents the number of years since 1990 and $f(x)$ represents the Utah population in thousands. During the same time period, the population of West Virginia can be modeled by

$$g(x) = 1.76x + 1802.56, \qquad 0 \le x \le 10$$

where x represents the number of years since 1990 and $g(x)$ represents the West Virginia population in thousands. Determine the point of intersection and interpret each coordinate (*Source:* U.S. Census Bureau, www.census.gov).

57. The Clip-it Company manufacturers three types of paper clips: standard, large, and jumbo. The daily cost of manufacturing a box of each type of paper clip, along with the sales price for each box, is shown in Table 1.5.2.

Table 1.5.2

	Standard	Large	Jumbo
Cost per box	$4	$6	$7
Sales price per box	$6	$9	$12

The company knows that the daily cost of manufacturing 100 total boxes is $520 and the daily revenue from the sales is $810. Let x, y, and z represent the daily number of standard, large, and jumbo boxes manufactured, respectively.

(a) Write an equation for the total number of boxes manufactured daily.

(b) Write equations for the daily total cost and for the sales revenue.

(c) Solve the system and interpret the result.

58. In a blind study the number of grams of nutrients per serving of three foods is as shown in Table1.5.3.

Table 1.5.3

	I	II	III	Total
Fat	1	2	2	11
Carbohydrate	1	1	1	6
Protein	2	1	2	10

Let x, y, and z represent the number of servings of foods I, II, and III, respectively. The study is intended to see how many servings of each food must be used to provide 11 grams of fat, 6 grams of carbohydrate, and 10 grams of protein.

(a) Write a system of equations to determine the number of servings of each food that is needed to provide the necessary number of grams of fat, carbohydrate, and protein.

(b) Solve the system and interpret the result.

59. In a blind study the number of grams of nutrients per serving of three foods is as shown in Table 1.5.4.

Table 1.5.4

	A	B	C	Total
Fat	2	3	1	14
Carbohydrate	1	2	1	9
Protein	2	1	2	9

Let x, y, and z represent the number of servings of foods A, B, and C, respectively. The study is intended to see how many servings of each food must be used to provide 14 grams of fat, 9 grams of carbohydrate, and 9 grams of protein.

(a) Write a system of equations to determine the number of servings of each food that is needed to provide the necessary number of grams of fat, carbohydrate, and protein.

(b) Solve the system and interpret the result.

60. The Crash-em-Up body repair shop fixes three types of cars: compact, midsize, and luxury. If a damaged car fender arrives, the times required, in hours, to fill, sand, and paint the fender for the three types of cars is as shown in Table 1.5.5.

Table 1.5.5

	Compact	Midsize	Luxury	Time Available
Filling	2	2	1	14
Sanding	1	2	2	15
Painting	3	2	2	21

Let x, y, and z, represent the number of compact, midsize, and luxury car fenders repaired, respectively. The total time available to complete each job is in the right column.

(a) Write a system of equations to determine the number of fenders of each type that can be repaired using all available labor hours.

(b) Solve the system and interpret the result.

SECTION PROJECT

Systems of equations can also be used to perform curve fitting, meaning that if we are given some coordinates of a function we can determine the function itself. Let's say that a line of the form $y = ax + b$ contains the coordinates $(-1, 4)$ and $(2, 6)$. We can write the following system of equations to determine the constants a and b.

$$\begin{aligned} 4 &= a(-1) + b \\ 6 &= a(2) + b \end{aligned} \quad \text{or in simplified form} \quad \begin{aligned} 4 &= -a + b \\ 6 &= 2a + b \end{aligned}$$

(a) Solve the given system to determine the constants a and b and write the equation in slope–intercept form. Graph and verify that the two given points are on the line.

(b) Determine a and b so that the line $y = ax + b$ contains the points $(1, 1)$ and $(-2, 4)$.

We can also use a system to perform curve fitting on parabolas resulting from quadratic equations. Let's say that

we have a quadratic equation that is symmetric about the vertical line $x = 2$. Then the quadratic equation has the form $y = a(x - 2)^2 + k$. Suppose that we know that the points $(1, -1)$ and $(4, 8)$ are on the parabola. Then, to determine the values of a and k, we can solve the system

$$\begin{aligned} -1 &= a(1 - 2)^2 + k \\ 8 &= a(4 - 2)^2 + k \end{aligned} \quad \text{or in simplified form} \quad \begin{aligned} -1 &= a + k \\ 8 &= 4a + k \end{aligned}$$

(c) Solve the given system to determine the constants a and k and write the quadratic equation in the form $y = a(x - 2)^2 + k$. Graph and verify that the two given points are on the parabola.

(d) Determine a and k so that the parabola $y = a(x - 2)^2 + k$ contains the points $(1, -3)$ and $(4, 0)$. Note that this parabola is also symmetric about the line $x = 2$.

■ Why We Learned It

In Chapter 1 we reviewed the properties of linear functions and linear systems of equations. The study of linear functions is very important to the study of finite mathematics. We saw that linear models could be used to express applications in the business, social, and life sciences. For example, a researcher could examine trends in immigration or birth weight over time. The amount that these values increase or decrease over time is given by the average rate of change. We saw that systems of equations are often used in business, particularly in applications such as the Law of Supply and Demand and for determining break-even points of cost and revenue functions. Finally, we saw that we could use graphical or algebraic techniques to solve systems of linear equations.

CHAPTER REVIEW EXERCISES

In Exercises 1–5, translate the English sentence into a mathematical sentence.

1. A certain number increased by 7 is 12.

2. A certain number decreased by 4 is five.

3. The product of three and a certain number is 36.

4. The product of two and a certain number minus three is seven.

5. The product of nine and a certain number plus five is 41.

6. Molly makes wool sweaters and sells them at craft shows. She knows that her fixed costs are $65 and the cost to produce each sweater is $14.

(a) Determine the total cost for producing 15 sweaters.

(b) Determine an equation for the total cost of producing any number of sweaters.

7. The Drink Power company has two ingredients, zap and wham, to combine to make their new power drink. Zap contains 7 grams of carbohydrate and 4 grams of protein per ounce. Wham contains 5 grams of carbohydrate and 4 grams of protein per ounce. They want to combine the two ingredients so that the power drink has 70 grams of carbohydrates and 48 grams of protein. Let x represent the number of ounces of zap and y represent the number of ounces of wham.

(a) Write an equation that represents the number of grams of carbohydrate in the power drink.

(b) Write an equation that represents the number of grams of protein in the power drink.

8. A local ski shop sells discounted lift tickets for a nearby ski resort. Tickets cost $30 per day for children and $42 per day for adults. Table R.1 gives the data for the number of tickets sold and revenue generated over one weekend.

Table R.1

	Friday	Saturday	Sunday
Tickets sold	20	30	15
Revenue	$696	$720	$570

(a) Define the variables and write an equation representing the number of tickets sold on Friday, and write an equation representing the revenue generated on Friday.

(b) Define the variables and write an equation representing the number of tickets sold on Saturday, and write an equation representing the revenue generated on Saturday.

(c) Define the variables and write an equation representing the number of tickets sold on Sunday, and write an equation representing the revenue generated on Sunday.

9. Harold invested a total of $20,000 in three different investments. Some of the money was invested in a certificate of deposit earning 6%, $8000 more than $\frac{1}{4}$ the amount invested in the certificate of deposit earning 6% was invested in a mutual fund earning 7%, and the rest was placed in a savings account earning 5%. The total annual return on Harold's investment was $1280. Let x represent the amount invested in the certificate of deposit earning 6%, y represent the amount invested in the mutual fund earning 7%, and z represent the amount invested in the savings account earning 5%.

(a) Write an equation that represents the first sentence.

(b) Write an equation that represents "$8000 more than $\frac{1}{4}$ the amount invested in the certificate of deposit earning 6% was invested in a mutual fund earning 7%."

(c) Write an equation that represents the third sentence.

10. The Tasty Snack company makes three brands of snack mixes made up of three different kinds of dried fruit. The proportions of the ingredients of the three brands are shown in Table R.2.

Table R.2

	Cranberry Lovers (%)	Banana Lovers (%)	Apricot Lovers (%)
Cranberries	70	20	10
Bananas	10	70	20
Apricots	20	10	70

Tasty Snack wants to create a new snack mix that contains 6.6 ounces of cranberries, 3.4 ounces of bananas, and 6 ounces of apricots.

(a) Define the variables and write an equation that represents the total amount of each snack mix that can

be combined to produce 6.6 ounces of cranberries in the new mix.

(b) Define the variables and write an equation that represents the total amount of each snack mix that can be combined to produce 3.4 ounces of bananas in the new mix.

(c) Define the variables and write an equation that represents the total amount of each snack mix that can be combined to produce 6 ounces of apricots in the new mix.

For Exercises 11 and 12, use the given table to construct ordered pairs and plot them on the Cartesian plane.

11.

x	y
−2	4
−1	0
2	3

12.

x	y
−1	−2
1	3
−3	4

For Exercises 13 and 14, write the sets for the domain and range.

13. For the values in the table in Exercise 11.

14. For the values in the table in Exercise 12.

For the functions in Exercises 15 and 16, complete the following.

(a) Identify the independent and dependent variables.

(b) Complete the table.

15. $g(x) = 2x + \dfrac{3}{4}$

x	$g(x)$
−2	
4	
10	

16. $f(x) = \dfrac{2x - 1}{5}$

x	$f(x)$
5	
0	
	7

In Exercises 17 and 18, determine which of the tables represent functions.

17.

Domain	Range
2	5
−1	6
3	4
2	0

18.

Domain	Range
−3	4
3	2
0	1
5	1

In Exercises 19 and 20, determine if the graph represents a function.

19.

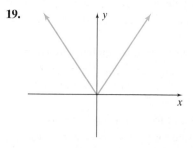

20.

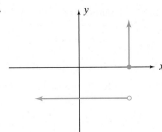

In Exercises 21 and 22, a table of values is given.

(a) Complete the table.

(b) Write a linear function for the data in the table.

21.

x	Expression	y
0		
4	2(4) − 1	
		0
2		

22.

x	Expression	f(x)
	3 + 4(−2)	−5
2		
		15
	3 + 4(0)	

In Exercises 23 and 24, compute the slope of the line passing through the two ordered pairs. Then classify the line as increasing, decreasing, horizontal, or vertical.

23.

x	y
−3	4
5	6

24. $(-3, 4)$ and $(-3, 7)$

In Exercises 25 and 26, determine the x- and y-intercepts of the graph of the linear function.

25. $y = 2x - 14$

26. $g(x) = \dfrac{5x + 4}{2}$

In Exercises 27 and 28, write the linear cost function for the given information in the form $C(x) = mx + b$.

27. The cost per unit is $17.33 and the fixed costs are $210.

28. The fixed costs are $35,200 and the cost per unit is $410.

29. The Spencer Baseball Company determines that the cost of producing x baseballs is given by the linear cost function $C(x) = 1.50x + 700$.

(a) What are the fixed and per unit costs for producing the baseballs?

(b) Evaluate and interpret $C(50)$.

(c) How many baseballs can be produced for $812.50? That is, what is x when $C(x) = 812.50$?

30. The data in Table R.3 show the enrollment of a small college in 1995 and 2000.

Table R.3

Year	Enrollment
1995	3210
2000	5600

Compute the average rate of change in enrollment at the small college from 1995 to 2000 and interpret. Be sure to include proper units.

In Exercises 31–34, write an equation of the line in standard form (if possible).

31. The line has slope $m = \dfrac{3}{4}$ and passes through the point $(-1, 4)$.

32. The line is given by the values in Table R.4.

Table R.4

x	0	4
f(x)	−3	−19

33. The line is vertical and passes through the point $(-5, 12)$.

34. The line has an average rate of change of $\dfrac{1}{5}$ and an x-intercept at 3.

In Exercises 35–38, write a function for the line in slope–intercept form.

35. The line has a slope of $m = 2$ and passes through the point $(1, -4)$.

36. The linear function $f(x)$ has a slope of $m = -\dfrac{3}{7}$ and $f(7) = 1$.

37. The linear function $g(x)$ has function values of $g(0) = -5$ and $g(2) = -11$.

38. The line has an x-intercept at $(-3, 0)$ and a y-intercept at $(0, 4)$.

In Exercises 39 and 40, write an equation in slope–intercept form for the given graph.

39.

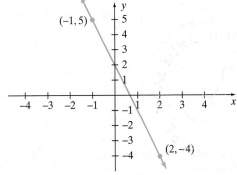

40.

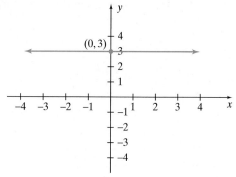

In Exercises 41 and 42, answer using a depreciation model.

41. A new $15,525 car depreciates $800 each year.

(a) Define the variables and determine the depreciation function.

(b) How much will the car be worth in 4 years?

42. A new computer costs $3200 and depreciates at a rate of $400 per year.

(a) Define the variables and determine the depreciation function.

(b) When will the computer depreciate to the point that it has no worth?

43. The total pounds of garbage produced per year in a growing housing community since 1995 can be modeled by

$$f(x) = 50x + 5200, \qquad 0 \le x \le 5$$

where x represents the number of years since 1995 and $f(x)$ represents the total amount of garbage in pounds.

(a) Is the total pounds of garbage produced per year increasing or decreasing and by how much?

(b) How many pounds of garbage were produced in 1995?

(c) Evaluate and interpret $f(5)$.

44. Suppose that the population of the housing community, from Exercise 43, stabilizes in the year 2000 and the community begins to recycle. Since the recycling concept takes some time to catch on, over the next 5 years the community will reduce the amount of garbage that they produce by 375 pounds per year.

(a) Assuming that the community produced 5450 pounds of garbage in the year 2000, write a linear model of the form

$$g(x) = y = mx + b, \qquad 0 \le x \le 5$$

to represent the predicted amount of garbage that the community will produce per year x years after 2000.

(b) Evaluate and interpret $g(5)$.

45. From 1990 to 2000 the average price of a resident fishing license was increasing at an average rate of $2.07 per year. In 1992 the average cost for a license was $20.15.

(a) Let x represent the number of years since 1990. Write a linear model in the calculator-friendly form

$$y = m(x - x_1) + y_1, \qquad 0 \le x \le 10$$

to represent the average cost of a license from 1990 to 2000.

(b) At the beginning of what year did the average cost per license exceed $24.50?

In Exercises 46 and 47, determine if the given ordered pair is a solution to the system.

46. $(2, -3)$
$$2x - 3y = 13$$
$$x + 5y = -2$$

47. $(1, 4)$
$$3x - 5y = -17$$
$$x + 4y = 17$$

In Exercises 48 and 49, solve the system by determining the x- and y-intercepts of the equation and then graphing the equations to determine their point of intersection. If the system has no solution or infinite solutions, then say so and state a reason why.

48. $y = 3x - 2$
$y = 5x - 11$

49. $3x + 4y = 1$
$9x + 12y = 3$

In Exercises 50 and 51, graph the equations on the same Cartesian plane and then use the graphs to estimate the solutions.

50. $x - y = 3$
$x = 4$

51. $y = -3$
$x = 2$

In Exercises 52 and 53, solve the system by graphing each equation on the calculator and using the INTERSECT command. If the system has no solution or infinite solutions, then say so.

52. $4x - 3y = 2$
$2y = 4x$

53. $x + 2y = -3$
$3x + 6y = 4$

In Exercises 54 and 55, determine the value of k that satisfies the given condition for each system.

54. k so that $\begin{array}{l} y = kx + 5 \\ y = 2x - 1 \end{array}$ has no solution.

55. k so that $\begin{array}{l} 3x - 4y = 6 \\ 9x - ky = 18 \end{array}$ has infinitely many solutions.

56. The Quinn Outdoor Company determines that the demand for their new dog backpacks is given by $d(x) = 80 - 0.75x$, with a supply function given by $s(x) = 0.34x$, where x is the daily quantity and $d(x)$ and $s(x)$ are measured in dollars per backpack.

(a) Determine the x- and y-intercepts for the graph of the demand function d.

(b) Determine the supply s when $x = 0$ and $x = 50$ and then graph d and s on the same Cartesian plane.

(c) Approximate the point of intersection and interpret.

57. The All Gear Company determines that the demand for their new adventure running shoe is given by $d(x) = 100 - \frac{1}{5}x$, while the supply is given by $s(x) = \frac{3}{5}x$, where x is the quantity and $d(x)$ and $s(x)$ are measured in dollars per shoe.

(a) Write a system of equations to determine the equilibrium point and graph the system in the viewing window [0, 500] by [0, 100].

(b) Determine the point of intersection via the INTERSECT command and interpret each coordinate.

58. The All Gear Company determines that the demand function for their new kayak passes through the points (0, 800) and (7, 797), and the supply function passes through the points (0, 0) and (14, 12).

(a) Use the given information to find demand and supply functions of the form $d(x) = mx + b$ and $s(x) = mx + b$, where x is the quantity and $d(x)$ and $s(x)$ are measured in dollars per kayak.

(b) Write a system of equations to determine the equilibrium point and graph the system in the viewing window [0, 2000] by [0, 1000].

(c) Determine the point of intersection via the INTERSECT command and interpret each coordinate. Round each coordinate to the nearest hundredth.

59. Molly makes wool sweaters and sells them at craft shows. She knows that her fixed costs are $65 and the cost to produce each sweater is $14. She sells each sweater for $30.

(a) Define the variables and determine a system of linear equations for the cost and the revenue that Molly makes from the sweaters.

(b) Graph the system determined in part (a) in the same viewing window, use the INTERSECT command to determine the break-even point, and interpret each coordinate. Round each coordinate to the nearest whole number.

60. From July 1984 to July 1999 the number of U.S. children under 5 years of age can be modeled by

$$f(x) = -151.6x + 19{,}700, \qquad 0 \le x \le 5$$

where x represents the number of years since July 1994 and $f(x)$ represents the number of U.S. children under 5 in thousands. During the same time period, the number of U.S. children who were 10 to 14 years old can be modeled by

$$g(x) = 166.4x + 18{,}716, \qquad 0 \le x \le 5$$

where x represents the number of years since July 1994 and $g(x)$ represents the number of U.S. children 10 to 14 years old in thousands (*Source:* U.S. Census Bureau, www.census.gov).

(a) Evaluate each model at $x = 0$ and $x = 5$.

(b) Graph f and g in the viewing window [0, 5] by [18,000, 20,000] and determine the point of intersection via the INTERSECT command. Round each coordinate to the nearest thousandth.

(c) Interpret each of the coordinates from your result in part (b).

In Exercises 61–64, solve the system by the substitution method. If the system has no solution, then say so. If the

system has an infinite number of solutions, use a parameter to represent the infinite number of solutions.

61. $3x + y = -5$
$x + 4y = 2$

62. $x + 3y = 10$
$x = -3y + 4$

63. $-2x + 4y = -12$
$3x - 2y = 6$

64. $y = 4x + 1$
$y = x + 13$

In Exercises 65–68, solve the system by the elimination method. If the system has no solution, then say so. If the system has an infinite number of solutions, use a parameter to represent the infinite number of solutions.

65. $x + y = 8$
$x - y = 12$

66. $x + 5y = -9$
$4x + y = -17$

67. $x - 3y = 7$
$-4x + 12y = -28$

68. $2x - 5y + 14 = 0$
$4x + 3y - 24 = 0$

In Exercises 69 and 70, solve the system of three equations and three unknowns by the elimination method. If the system has no solution, then say so. If the system has an infinite number of solutions, use a parameter to represent the infinite number of solutions.

69. $-x + 2y + z = -1$
$3x + 2y - 2z = 9$
$x + 4y + 3z = -1$

70. $x + 3y = 1$
$y = 3z - 8$
$2x + z = -1$

71. Alicia purchased 3 T-shirts and 2 pairs of jeans for $90 at Casual Wear. The next week she purchased 1 T-shirt and 4 pairs of jeans for $130.

(a) Define the variables and write a system of equations for the price of the T-shirts and jeans.

(b) Solve the system in part (a) and interpret.

72. An office assistant purchased 5 boxes of pens and 2 boxes of staples for $9. The next day the assistant purchased 2 boxes of pens and 8 boxes of staples for $9.

(a) Define the variables and write a system of equations for the boxes of pens and boxes of staples.

(b) Solve the system in part (a) and interpret.

73. From July 1990 to July 1999 the population of Colorado can be modeled by

$$f(x) = 0.084x + 3.30, \qquad 0 \le x \le 9$$

where x represents the number of years since July 1990 and $f(x)$ represents the population of Colorado in millions. During the same time period, the population of Kentucky can be modeled by

$$g(x) = 0.030x + 3.69, \qquad 0 \le x \le 9$$

where x represents the number of years since July 1990 and $g(x)$ represents the population of Kentucky in millions. Determine the point of intersection and interpret each coordinate (*Source:* U.S. Census Bureau, www.census.gov).

74. Papa Harold's sandwich shop makes three sizes of sandwiches: large, extra-large, and Papa size. The cost of making each sandwich and the sales price for each sandwich are shown in Table R.5.

Table R.5

| | Type | | |
	Large	Extra-large	Papa size
Cost per sandwich	$1	$2	$3
Sales price per sandwich	$4	$5	$7

The shop knows that the daily cost of making 200 sandwiches is $360 and the daily revenue from the sales is $990. Let x, y, and z represent the daily number of large, extra-large, and Papa size sandwiches made, respectively.

(a) Write an equation for the total number of sandwiches made daily.

(b) Write equations for the daily cost and for the sales revenue.

(c) Solve the system and interpret the results.

75. In a blind study the number of grams of nutrients per serving of three foods is as shown in Table R.6.

Table R.6

| | Food ID | | | |
	A	B	C	Total
Fat	4	2	1	14
Carbohydrate	3	1	2	15
Protein	2	2	3	18

Let x, y, and z represent the number of servings of foods A, B, and C, respectively. The study is intended to see how many servings of each food must be used to provide grams of fat, grams of carbohydrates, and grams of protein.

(a) Write a system of equations to determine the number of servings of each food that is needed to provide the necessary number of grams of fat, carbohydrates, and protein.

(b) Solve the system and interpret the results.

CHAPTER 1 PROJECT

SOURCE: Centers for Disease Control and Prevention, www.cdc.gov

As you have seen in this chapter, there are many examples of two quantities having a linear relationship. However, some relationships can only be *approximated* by a linear equation. To compute this correlation, we can use regression.

Table 1 offers data collected by the Centers for Disease Control and Prevention concerning deaths by race and sex from 1985 to 1997 in the United States. The selected races are Native Americans[1] and Asian or Pacific Islanders[2] and the sex is female for both groups.

1. Let x represent the number of years since 1984. Let y_1 be the number of deaths of American Indian females. Enter the data into your calculator and find a linear regression model of the form $y_1 = ax + b$ to model the data. Round coefficients to two decimal places. Interpret the slope of the line.

2. Again, let x be as defined in question 1 and let y_2 be the number of deaths of Asian or Pacific Islander females. Enter the data into your calculator and find a linear regression model of the form $y_2 = ax + b$ to model the data. Round coefficients to two decimal places. Interpret the slope of the line.

3. Plot the data along with the regression lines found in questions 1 and 2 in the same viewing window. According to the regression lines, during what year did the number of deaths for American Indian females equal the number of deaths for Asian or Pacific Islander females?

4. Confirm your answer from question 3 by using the substitution method.

5. According to the models, predict for each race the number of deaths in the years 2000, 2005, and 2010.

6. For the year 1986, compute the residual. That is, compute the difference between the value in the table and the value given by the linear regression line from questions 1 and 2.

Table 1 Crude Rates on an Annual Basis per 100,000 Population

Year	American Indian Females	Asian or Pacific Islander Females
1985	2,973	6,646
1986	2,936	6,719
1987	3,170	7,193
1988	3,300	7,808
1989	3,548	8,354
1990	3,439	8,916
1991	3,673	9,446
1992	3,772	10,092
1993	4,145	10,854
1994	4,140	11,695
1995	4,423	12,364
1996	4,564	12,958
1997	4,591	13,696

---
[1]Includes Aleuts and Eskimos (Inuits).
[2]Includes Chinese, Filipino, Hawaiian, Japanese, and other Asian or Pacific Islanders.

2

Systems of Linear Equations and Matrices

Although the concept of a matrix may seem complex, the fact is that matrices are part of our everyday experience. Matrices are used in such diverse areas as agriculture and urban design. Farmers use systems of equations and matrices to keep track of crops and to determine planting schedules. Traffic engineers help control gridlock in busy cities by using matrices and systems of equations to sequence traffic signals.

What We Know

In Chapter 1 we reviewed linear functions and their graphs. Solving systems of linear equations was discussed using both graphical and analytical methods. The analytical methods included the substitution method and the elimination method.

Where Do We Go

In this chapter we will learn how to use *matrices* to solve systems of linear equations. We begin with *augmented matrices* and the *Gauss–Jordan* method and end with *inverse matrices* and matrix equations. Along the way, we will learn operations with matrices and encounter many applications.

Section 2.1   Matrices and Gauss–Jordan

In this section we will learn what a **matrix**, an **augmented matrix**, **reduced row echelon form**, and the **Gauss–Jordan method** are all about. The Gauss–Jordan method is simply a generalization of the elimination method studied in Section 1.5. Its main advantage is that it works well for systems of any size. Before we learn the power of the Gauss–Jordan method, we must first analyze the matrix.

Basic Matrix Vocabulary

A **matrix** (plural is **matrices**) is a rectangular arrangement of numbers enclosed by brackets. Table 2.1.1 gives the predicted high temperature for selected cities in Texas for four days in June.

Table 2.1.1

	Saturday	Sunday	Monday	Tuesday
Amarillo	86	85	88	91
Austin	87	88	87	91
Brownsville	88	89	89	91
Dallas	84	88	88	90
Houston	84	88	86	89

SOURCE: *The Plain Dealer*

These data can be written as the matrix T as follows:

$$T = \begin{bmatrix} 86 & 85 & 88 & 91 \\ 87 & 88 & 87 & 91 \\ 88 & 89 & 89 & 91 \\ 84 & 88 & 88 & 90 \\ 84 & 88 & 86 & 89 \end{bmatrix}$$

The horizontal listings of any matrix are called **rows**, and the vertical listings are called **columns**. In any matrix each number is called an **element** or an **entry**.

The number of rows and the number of columns give the **dimension** of a matrix. If a matrix has m rows and n columns, we say that its dimension is $m \times n$ (read "m by n"). Hence, since matrix T has 5 rows and 4 columns, we say that it has dimension 5×4. We can also say that matrix T is a 5×4 matrix.

It is a convention to use uppercase letters to name a matrix and lowercase letters with a double subscript to label entries within a matrix. For example, $t_{4,3}$ refers to the entry in matrix T that is in the 4th row and 3rd column. Notice that $t_{4,3}$ refers to the number 88.

Example 1   Determining the Dimension and Entries of a Matrix

For the given matrix A, determine the dimension of A and determine $a_{2,3}$.

$$A = \begin{bmatrix} -3 & 2 & 0 \\ 4 & -1 & 3 \end{bmatrix}$$

Solution

Since A has 2 rows and 3 columns, it is a 2×3 matrix. Also, $a_{2,3}$ refers to the entry in the 2nd row and the 3rd column. This means that $a_{2,3} = 3$. ∎

Checkpoint 1

Now work Exercises 5 and 11.

Technology Option

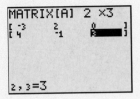

Figure 2.1.1

Matrices can be entered on a graphing calculator. Figure 2.1.1 shows a screen shot of what matrix A from Example 1 looks like on a graphing calculator. Note that the entry $a_{2,3}$ is highlighted and the lower-left corner of the screen shows the value of this entry.

To learn how to enter matrices on your calculator, consult the online graphing calculator manual at www.prenhall.com/armstrong

Augmented Matrices and Solving Systems of Equations

At the beginning of this section we stated that the **Gauss–Jordan** method is a generalization of the elimination method studied in Section 1.5. To aid our transition to this new method for solving systems of linear equations, let's take a Flashback to a problem encountered in Sections 1.1 and 1.5.

 Flashback

Gonzalez Office Supplies Revisited

In Chapter 1 we learned that Gonzalez Office Supplies manufactures three types of office chairs: Secretarial, Executive, and Presidential. Each chair requires time on three different machines. These times and the total available time for each machine are given in Table 2.1.2. If all available time for each machine must be used each week, set up a system of equations that can be solved to determine how many of each type can be made each week.

Table 2.1.2

	Secretarial	Executive	Presidential	Total hours each week
Hours on Machine A	1	1	2	30
Hours on Machine B	2	3	2	53
Hours on Machine C	1	2	3	47

Flashback Solution

We let x represent the number of Secretarial chairs made each week, y represent the number of Executive chairs made each week, and z represent the number of Presidential chairs made each week. The total number of hours each week for all three chairs on Machine A is given by $x + y + 2z$ and must equal 30. This gives the equation

$$x + y + 2z = 30$$

In a similar fashion, the total number of hours for Machine B is given by

$$2x + 3y + 2z = 53$$

and the total number of hours for Machine C is given by

$$x + 2y + 3z = 47$$

These three equations form the system of equations

$$\begin{aligned} x + y + 2z &= 30 \\ 2x + 3y + 2z &= 53 \\ x + 2y + 3z &= 47 \end{aligned}$$

In Section 1.5 we solved the system in the Flashback solution using the elimination method. Our task at this time is to develop a new method. For any linear system the matrix having as its entries the coefficients of the variables is called the **coefficient matrix** of the system. For example, the system of equations in the Flashback solution has the coefficient matrix

$$\begin{bmatrix} 1 & 1 & 2 \\ 2 & 3 & 2 \\ 1 & 2 & 3 \end{bmatrix}$$

If we add another column to the coefficient matrix whose entries are the constants from the right side of each equation in the system, we form what is called an **augmented matrix** of the system. The augmented matrix from the Flashback solution is

$$\left[\begin{array}{ccc|c} 1 & 1 & 2 & 30 \\ 2 & 3 & 2 & 53 \\ 1 & 2 & 3 & 47 \end{array}\right]$$

We use a vertical line to separate the constants in the last column from the coefficients of the variables.

Example 2   **Writing an Augmented Matrix**

Write an augmented matrix for the system

$$\begin{aligned} 2x - 3y &= 7 \\ 5x + y &= 8 \end{aligned}$$

Solution

The augmented matrix is

$$\left[\begin{array}{cc|c} 2 & -3 & 7 \\ 5 & 1 & 8 \end{array}\right]$$

✓ **Checkpoint 2**   Now work Exercise 19.

Since an augmented matrix is just a compact form of a system of equations, we can manipulate the rows of an augmented matrix in the same way as we did the equations in the system in Section 1.5. In fact, the manipulations of systems of equations given in Section 1.5 are now given as **row operations** for an augmented matrix.

■ Row Operations

For an augmented matrix, the following row operations produce an augmented matrix of an equivalent system:

1. Two rows may be interchanged (denoted $R_i \leftrightarrow R_j$).
2. A row is multiplied by a nonzero constant (denoted $cR_i \rightarrow R_i$).
3. Add a constant multiple of one row to another (denoted $cR_i + R_j \rightarrow R_j$).

▷ **Note:** We use the arrow $\rightarrow$ to mean *replaces*. For example, $3R_1 + R_2 \rightarrow R_2$ means "3 times row 1 plus row 2 replaces row 2."

Example 3   Solving a System Using Augmented Matrices

Solve the system

$$8x + 9y = 69$$
$$2x + 3y = 21$$

using augmented matrices and the row operations.

Solution

We begin by writing the augmented matrix corresponding to the system

$$\begin{bmatrix} 8 & 9 & | & 69 \\ 2 & 3 & | & 21 \end{bmatrix}$$

Our goal is to use the row operations to transform our augmented matrix into an augmented matrix of the form

$$\begin{bmatrix} 1 & 0 & | & c \\ 0 & 1 & | & d \end{bmatrix}$$

where c and d are real numbers. This form gives the solution $x = c$ and $y = d$ because **a row in an augmented matrix corresponds to an equation in a linear system**. In other words,

The first row of $\begin{bmatrix} \mathbf{1} & \mathbf{0} & | & c \\ 0 & 1 & | & d \end{bmatrix}$ corresponds to $x + 0y = c$ or $x = c$.

The second row of $\begin{bmatrix} 1 & 0 & | & c \\ \mathbf{0} & \mathbf{1} & | & d \end{bmatrix}$ corresponds to $0x + y = d$ or $y = d$.

We proceed by getting a 1 in the upper-left corner. We multiply R_1 by $\frac{1}{8}$ (row operation 2) to get

$$\begin{bmatrix} 8 & 9 & | & 69 \\ 2 & 3 & | & 21 \end{bmatrix} \frac{1}{8}R_1 \rightarrow R_1 \begin{bmatrix} \frac{1}{8}(8) & \frac{1}{8}(9) & | & \frac{1}{8}(69) \\ 2 & 3 & | & 21 \end{bmatrix} \rightarrow \begin{bmatrix} 1 & \frac{9}{8} & | & \frac{69}{8} \\ 2 & 3 & | & 21 \end{bmatrix}$$

To get a 0 in the lower-left corner, we multiply $-2R_1$ and add to R_2 (row operation 3). This gives

$$\begin{bmatrix} 1 & \frac{9}{8} & \Big| & \frac{69}{8} \\ 2 & 3 & \Big| & 21 \end{bmatrix} \begin{array}{c} -2R_1 + R_2 \to R_2 \end{array} \begin{bmatrix} 1 & \frac{9}{8} & \Big| & \frac{69}{8} \\ -2(1)+2 & -2\left(\frac{9}{8}\right)+3 & \Big| & -2\left(\frac{69}{8}\right)+21 \end{bmatrix}$$

$$\to \begin{bmatrix} 1 & \frac{9}{8} & \Big| & \frac{69}{8} \\ 0 & \frac{3}{4} & \Big| & \frac{15}{4} \end{bmatrix}$$

Notice how the last row operation left row 1 unaffected. We now get a 1 in the 2nd row, 2nd column by multiplying R_2 by $\frac{4}{3}$. This gives

$$\begin{bmatrix} 1 & \frac{9}{8} & \Big| & \frac{69}{8} \\ 0 & \frac{3}{4} & \Big| & \frac{15}{4} \end{bmatrix} \begin{array}{c} \frac{4}{3}R_2 \to R_2 \end{array} \begin{bmatrix} 1 & \frac{9}{8} & \Big| & \frac{69}{8} \\ \frac{4}{3}(0) & \frac{4}{3}\left(\frac{3}{4}\right) & \Big| & \frac{4}{3}\left(\frac{15}{4}\right) \end{bmatrix} \to \begin{bmatrix} 1 & \frac{9}{8} & \Big| & \frac{69}{8} \\ 0 & 1 & \Big| & 5 \end{bmatrix}$$

Finally, to get a 0 in the 1st row, 2nd column, we multiply $-\frac{9}{8}R_2$ and add to R_1. This yields

$$\begin{bmatrix} 1 & \frac{9}{8} & \Big| & \frac{69}{8} \\ 0 & 1 & \Big| & 5 \end{bmatrix} \begin{array}{c} -\frac{9}{8}R_2 + R_1 \to R_1 \end{array} \begin{bmatrix} -\frac{9}{8}(0)+1 & -\frac{9}{8}(1)+\frac{9}{8} & \Big| & -\frac{9}{8}(5)+\frac{69}{8} \\ 0 & 1 & \Big| & 5 \end{bmatrix}$$

$$\to \begin{bmatrix} 1 & 0 & \Big| & 3 \\ 0 & 1 & \Big| & 5 \end{bmatrix}$$

This last augmented matrix corresponds to the linear system

$$x = 3$$
$$y = 5$$

So the solution to our original system is $x = 3$ and $y = 5$. ■

Interactive Activity

Verify that $x = 3$ and $y = 5$ is the solution to the original system given in Example 3.

www

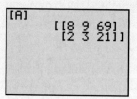
Checkpoint 3

Now work Exercise 43.

Technology Option

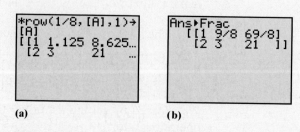
Figure 2.1.2

The row operations in Example 3 can be performed on most graphing calculators.

Figure 2.1.2 shows the augmented matrix for the original system in Example 3.

Figure 2.1.3(a) shows how to get a 1 in the upper-left corner by using the *row(command, and Figure 2.1.3(b) shows the resulting matrix with fractions instead of decimals.

(a)   (b)

Figure 2.1.3

Figure 2.1.4(a) shows how to get a 0 in the lower-left corner by using the *row+(command, and Figure 2.1.4(b) shows the resulting matrix with fractions instead of decimals.

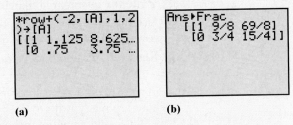

(a)   **(b)**

Figure 2.1.4

To learn how row operations are done on your calculator, consult the online graphing calculator manual at www.prenhall.com/armstrong

Reduced Row Echelon Form and Gauss–Jordan

We found the final matrix in Example 3 to be $\begin{bmatrix} 1 & 0 & | & 3 \\ 0 & 1 & | & 5 \end{bmatrix}$. This is an example of a matrix that is in **reduced row echelon form**. Let's see how to determine when a matrix is in this form.

Reduced Row Echelon Form

For an $m \times n$ matrix to be in **reduced row echelon form**, it must satisfy the following:

1. All rows containing only zeros are at the bottom of the matrix.
2. The first (leftmost) nonzero entry in a nonzero row is a 1. (This is called the **leading 1** of its row).
3. The leading 1 of each nonzero row lies left of the leading 1 of any lower row.
4. For any column that has a leading 1, the other entries in the column are zeros.

From the preceding definition we can see that matrices such as $\begin{bmatrix} 1 & 0 & | & 5 \\ 0 & 1 & | & 6 \end{bmatrix}$

and $\begin{bmatrix} 1 & 0 & 0 & | & 6 \\ 0 & 1 & 0 & | & 9 \\ 0 & 0 & 0 & | & 0 \end{bmatrix}$ are in reduced row echelon form. On the other hand,

matrices such as $\begin{bmatrix} 0 & 1 & | & -2 \\ 1 & 0 & | & 6 \end{bmatrix}$ and $\begin{bmatrix} 1 & 0 & 0 & | & 7 \\ 0 & 0 & 0 & | & 0 \\ 0 & 0 & 1 & | & 8 \end{bmatrix}$ are not in reduced row echelon form. These matrices do not meet requirements 3 and 1, respectively. The process we went through in Example 3 to solve the system of linear equations was really our first example of using the **Gauss–Jordan method**. Let's examine the procedure used to transform a matrix into reduced row echelon form using the Gauss–Jordan method.

■ **Gauss–Jordan Method**

1. **Write the system** so that all variables (in the same order) are on the left side of the equations.
2. **Write the augmented matrix** that corresponds to the system.
3. **Use the row operations** to transform the augmented matrix into reduced row echelon form. (This is called **pivoting**.)
4. **Read the solution** once in reduced row echelon form.

There are many ways to do step 3 in the Gauss–Jordan method. In Example 3 we illustrated a routine procedure that always works. It is called **pivoting** and basically it effectively produces the requisite 1s and 0s in reduced row echelon form.

■ **Pivoting Procedure**

1. Change the appropriate entry of the 1st column to a 1 by multiplying the row containing that entry by a nonzero constant (row operation 2).
2. Once the 1 is secured, change all entries in the column to a 0 by adding a multiple of the row containing the 1 to the row that we wish to have a 0 (row operation 3).
3. Repeat the above steps with the next column.

Example 4   Using the Gauss–Jordan Method

Solve the following system by using the Gauss–Jordan method.

$$x - y = 2$$
$$5x + 2y = 17$$

Solution

Understand the Situation: We begin by writing the augmented matrix for the system and use the row operations to transform the augmented matrix into reduced row echelon form.

$$\begin{bmatrix} 1 & -1 & | & 2 \\ 5 & 2 & | & 17 \end{bmatrix} -5R_1 + R_2 \rightarrow R_2 \begin{bmatrix} 1 & -1 & | & 2 \\ 0 & 7 & | & 7 \end{bmatrix}$$

$$\frac{1}{7}R_2 \rightarrow R_2 \begin{bmatrix} 1 & -1 & | & 2 \\ 0 & 1 & | & 1 \end{bmatrix}$$

$$R_2 + R_1 \rightarrow R_1 \begin{bmatrix} 1 & 0 & | & 3 \\ 0 & 1 & | & 1 \end{bmatrix}$$

So the solution to the system is $x = 3$ and $y = 1$.   ■

In Section 1.5 we solved some systems of linear equations that had no solutions and some that had an infinite number of solutions. In Examples 5 and 6 we analyze these two scenarios using the Gauss-Jordan method.

Example 5   **Using the Gauss–Jordan Method (No Solution)**

Use Gauss–Jordan to solve the linear system

$$x - y = 3$$
$$2x - 2y = 7$$

Solution

The augmented matrix for the system is

$$\begin{bmatrix} 1 & -1 & | & 3 \\ 2 & -2 & | & 7 \end{bmatrix}$$

We try to put this matrix into reduced row echelon form as follows:

$$\begin{bmatrix} 1 & -1 & | & 3 \\ 2 & -2 & | & 7 \end{bmatrix} \quad -2R_1 + R_2 \rightarrow R_2 \quad \begin{bmatrix} 1 & -1 & | & 3 \\ 0 & 0 & | & 1 \end{bmatrix}$$

Observe that this transformed augmented matrix corresponds to the system

$$x - y = 3$$
$$0x + 0y = 1$$

Since the second equation states $0 = 1$, which is not true, we know that this system is inconsistent and has no solution. Notice how the last row $[0 \; 0 \,|\, 1]$ of the matrix is our clue that the system is **inconsistent**.   ■

Example 6   **Using the Gauss–Jordan Method (Infinite Number of Solutions)**

Use the Gauss–Jordan method to solve the linear system

$$x + 2y + z = 3$$
$$2x - 3y + z = 2$$
$$-2x - 4y - 2z = -6$$

Solution

We first write the augmented matrix for the system. Since the first entry in row 1 is a 1, we begin with row operations in order to get 0s for all the other entries in the first column.

$$\begin{bmatrix} 1 & 2 & 1 & | & 3 \\ 2 & -3 & 1 & | & 2 \\ -2 & -4 & -2 & | & -6 \end{bmatrix} \quad \begin{array}{l} -2R_1 + R_2 \rightarrow R_2 \\ 2R_1 + R_3 \rightarrow R_3 \end{array} \quad \begin{bmatrix} 1 & 2 & 1 & | & 3 \\ 0 & -7 & -1 & | & -4 \\ 0 & 0 & 0 & | & 0 \end{bmatrix}$$

$$-\tfrac{1}{7}R_2 \rightarrow R_2 \quad \begin{bmatrix} 1 & 2 & 1 & | & 3 \\ 0 & 1 & \tfrac{1}{7} & | & \tfrac{4}{7} \\ 0 & 0 & 0 & | & 0 \end{bmatrix}$$

$$-2R_2 + R_1 \rightarrow R_1 \quad \begin{bmatrix} 1 & 0 & \tfrac{5}{7} & | & \tfrac{13}{7} \\ 0 & 1 & \tfrac{1}{7} & | & \tfrac{4}{7} \\ 0 & 0 & 0 & | & 0 \end{bmatrix}$$

Although the first two columns are in the correct form, we cannot get a zero in row 1, column 3 without changing the structure of the 1st two columns. So we write the system that corresponds to this last matrix:

$$x + \tfrac{5}{7}z = \tfrac{13}{7}$$
$$y + \tfrac{1}{7}z = \tfrac{4}{7}$$
$$0 = 0$$

Interactive Activity

Give three solutions to the system in Example 6 corresponding to $z = 2$, $z = -3$, and $z = 5$.

As in Section 1.5, since the first two equations contain a z, we choose z to be the parameter. Notice that the system has an infinite number of solutions for the infinite number of values of z that could be selected. The last row of the matrix, $[0 \ \ 0 \ \ 0 \,|\, 0]$, is our clue to this. Solving the first two equations for x and y, respectively, yields

$$x = \frac{13 - 5z}{7} \quad \text{and} \quad y = \frac{4 - z}{7}$$

We can write the solution as $\left(\frac{13-5z}{7}, \frac{4-z}{7}, z\right)$.

Application

We conclude this section by using the Gauss–Jordan method to solve the application in the Flashback from earlier in the section.

Example 7 **Applying the Gauss–Jordan Method**

Solve the system

$$x + y + 2z = 30$$
$$2x + 3y + 2z = 53$$
$$x + 2y + 3z = 47$$

that was set up in the Flashback solution earlier in this section.

Solution

Again, we write the augmented matrix for the system and use row operations.

$$\begin{bmatrix} 1 & 1 & 2 & 30 \\ 2 & 3 & 2 & 53 \\ 1 & 2 & 3 & 47 \end{bmatrix} \begin{array}{c} -2R_1 + R_2 \to R_2 \\ -1R_1 + R_3 \to R_3 \end{array} \begin{bmatrix} 1 & 1 & 2 & 30 \\ 0 & 1 & -2 & -7 \\ 0 & 1 & 1 & 17 \end{bmatrix}$$

$$\begin{array}{c} -1R_2 + R_1 \to R_1 \\ -1R_2 + R_3 \to R_3 \end{array} \begin{bmatrix} 1 & 0 & 4 & 37 \\ 0 & 1 & -2 & -7 \\ 0 & 0 & 3 & 24 \end{bmatrix}$$

$$\tfrac{1}{3}R_3 \to R_3 \begin{bmatrix} 1 & 0 & 4 & 37 \\ 0 & 1 & -2 & -7 \\ 0 & 0 & 1 & 8 \end{bmatrix}$$

$$\begin{array}{c} 2R_3 + R_2 \to R_2 \\ -4R_3 + R_1 \to R_1 \end{array} \begin{bmatrix} 1 & 0 & 0 & 5 \\ 0 & 1 & 0 & 9 \\ 0 & 0 & 1 & 8 \end{bmatrix}$$

So the solution is $x = 5$, $y = 9$, and $z = 8$.

Interpret the Solution: This means that Gonzalez Office Supplies should manufacture 5 Secretarial, 9 Executive, and 8 Presidential chairs each week.

Technology Option

In Figure 2.1.5 we have the augmented matrix for the system in Example 7, and Figure 2.1.6 is the result of using the rref(command on a graphing calculator. Notice that Figure 2.1.6 supports our work in Example 7.

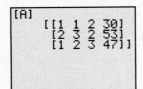

Figure 2.1.5

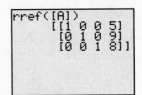

Figure 2.1.6

To see if your calculator transforms matrices into reduced row echelon form, consult the online graphing calculator manual at www.prenhall.com/armstrong

SUMMARY

In this section we learned much about matrices. We first defined a **matrix** and identified the **rows**, **columns**, and **dimension** of a matrix. We then discussed **augmented matrices** and their relationship to systems of linear equations. We saw that the **row operations** used to produce equivalent matrices are analogous to the elimination method presented in Section 1.5. The row operations are used to put an augmented matrix into **reduced row echelon form**. A systematic procedure to achieve this, called **pivoting**, was given. We then saw how all of this built up to the **Gauss–Jordan** method for solving systems of linear equations.

SECTION 2.1 EXERCISES

In Exercises 1–8, find the dimension of the given matrix.

1. $A = \begin{bmatrix} 2 & 1 & 3 \\ 4 & 3 & 2 \end{bmatrix}$

2. $B = \begin{bmatrix} -1 & 3 \\ 2 & 7 \\ 5 & -7 \end{bmatrix}$

3. $C = \begin{bmatrix} -5 & 2 \\ 1 & -3 \\ 4 & 5 \end{bmatrix}$

4. $D = \begin{bmatrix} 1 & -1 & 2 \\ 3 & 4 & -2 \\ 7 & -5 & 5 \end{bmatrix}$

✓ **5.** $E = \begin{bmatrix} 7 & -1 & -6 & 1 \\ 2 & 4 & -7 & -3 \\ 3 & 5 & 0 & -2 \end{bmatrix}$

6. $F = \begin{bmatrix} -2 & -1 & 3 \\ 1 & 0 & 7 \\ 8 & -3 & 4 \\ -4 & 5 & -6 \end{bmatrix}$

7. $G = \begin{bmatrix} 1 \\ -1 \\ 3 \\ 4 \end{bmatrix}$

8. $H = \begin{bmatrix} 7 & -1 & 2 & 3 \end{bmatrix}$

In Exercises 9–16, find the indicated entry for the matrices given in Exercises 1–8.

9. $b_{3,1}$

10. $a_{1,3}$

✓ **11.** $d_{2,3}$

12. $c_{2,1}$

13. $f_{2,3}$

14. $e_{3,2}$

15. $h_{1,3}$

16. $g_{3,1}$

In Exercises 17–24, write an augmented matrix for the given system. **Do not solve**.

17. $\begin{aligned} 3x - 7y &= 4 \\ 2x + 5y &= 8 \end{aligned}$

18. $\begin{aligned} -2x + 4y &= -7 \\ 3x - 2y &= 14 \end{aligned}$

✓ **19.** $\begin{aligned} x + 3y &= -11 \\ 2x + 5y &= -17 \end{aligned}$

20. $\begin{aligned} -x - y &= 13 \\ 4x + y &= 7 \end{aligned}$

21. $\begin{aligned} x + y + z &= 1 \\ 3x + y - z &= -2 \\ 2x - y + 3z &= 4 \end{aligned}$

22. $\begin{aligned} -x - y - z &= -3 \\ 2x - y + 4z &= 8 \\ -2x + 3y + 5z &= 7 \end{aligned}$

23. $\begin{aligned} 3x + y + 10z &= 38 \\ 5x + 10y + z &= 22 \\ 7x - y - 12z &= -15 \end{aligned}$

24. $\begin{aligned} 11x - y + 12z &= 42 \\ -x + 13y - 5z &= 17 \\ 51x - 17y + z &= -13 \end{aligned}$

In Exercises 25–32, write the linear system corresponding to each augmented matrix.

25. $\left[\begin{array}{cc|c} 2 & -3 & 7 \\ -2 & 4 & 3 \end{array} \right]$

26. $\left[\begin{array}{cc|c} -1 & 1 & 7 \\ 3 & -1 & -4 \end{array} \right]$

27. $\begin{bmatrix} 1 & 0 & | & 8 \\ 0 & 1 & | & -3 \end{bmatrix}$ **28.** $\begin{bmatrix} 1 & 0 & | & -2 \\ 0 & 1 & | & 4 \end{bmatrix}$

29. $\begin{bmatrix} 3 & -1 & 4 & | & 7 \\ 2 & 8 & -3 & | & 2 \\ -1 & 4 & 5 & | & -3 \end{bmatrix}$ **30.** $\begin{bmatrix} 4 & -4 & -3 & | & -2 \\ -5 & 18 & 4 & | & 22 \\ 7 & 31 & -2 & | & 37 \end{bmatrix}$

31. $\begin{bmatrix} 1 & 0 & 0 & | & 8 \\ 0 & 1 & 0 & | & -2 \\ 0 & 0 & 1 & | & 7 \end{bmatrix}$ **32.** $\begin{bmatrix} 1 & 0 & 0 & | & -3 \\ 0 & 1 & 0 & | & 15 \\ 0 & 0 & 1 & | & 9 \end{bmatrix}$

In Exercises 33–40, perform the indicated row operations to change each augmented matrix.

33. $-1R_1 \rightarrow R_1$

$\begin{bmatrix} -1 & 1 & | & 7 \\ 3 & -1 & | & -4 \end{bmatrix}$

34. $\frac{1}{2}R_1 \rightarrow R_1$

$\begin{bmatrix} 2 & -3 & | & 7 \\ -2 & 4 & | & 3 \end{bmatrix}$

35. $2R_1 + R_2 \rightarrow R_2$

$\begin{bmatrix} 1 & 3 & | & -2 \\ -2 & 4 & | & 3 \end{bmatrix}$

36. $-5R_1 + R_2 \rightarrow R_2$

$\begin{bmatrix} 1 & -8 & | & 7 \\ 5 & 2 & | & -3 \end{bmatrix}$

37. $\frac{1}{4}R_1 \rightarrow R_1$

$\begin{bmatrix} 4 & -4 & -3 & | & -2 \\ -5 & 18 & 4 & | & 22 \\ 7 & 31 & -2 & | & 37 \end{bmatrix}$

38. $\frac{1}{3}R_1 \rightarrow R_1$

$\begin{bmatrix} 3 & 8 & -2 & | & 5 \\ -7 & -1 & 1 & | & 8 \\ 2 & 5 & 8 & | & 13 \end{bmatrix}$

39. $-2R_1 + R_3 \rightarrow R_3$

$\begin{bmatrix} 1 & -1 & -1 & | & -2 \\ 0 & 3 & 2 & | & 5 \\ 2 & -8 & 1 & | & 5 \end{bmatrix}$

40. $3R_1 + R_3 \rightarrow R_3$

$\begin{bmatrix} 1 & 4 & 5 & | & 3 \\ 0 & -1 & 4 & | & -7 \\ -3 & 1 & 4 & | & 8 \end{bmatrix}$

In Exercises 41–58, use the Gauss–Jordan method to solve the given system of linear equations.

41.
$x - y = 2$
$-2x + 3y = 7$

42.
$x + y = 4$
$-3x - 2y = -5$

✓ **43.**
$3x - 5y = 8$
$-2x + 3y = -7$

44.
$-4x + 5y = 9$
$3x - 4y = -2$

45. $-5x + 3y = -11$
$2x - y = 5$

46. $-3x + 5y = -13$
$2x - 2y = 5$

47. $-7x - 2y = 9$
$4x + 5y = -2$

48.
$7x - 8y = -9$
$-2x - 5y = 6$

49. $x + 3y - 6z = 7$
$2x - y + 2z = 0$
$x + y + 2z = -1$

50. $2x + 3y + 5z = 21$
$x - y - 5z = -2$
$2x + y - z = 11$

51. $x + 3y - z = -3$
$2x + y - 2z = 4$
$3x + 4y - z = 7$

52. $x + 2y + 2z = 11$
$x - y - z = -4$
$2x + 5y + 9z = 39$

53.
$x - 3y + z = 5$
$-2x + 7y - 6z = -9$
$x + 2y - 3z = 6$

54. $3x - 6y + 9z = 0$
$4x - 6y + 8z = -4$
$-2x - y + z = 7$

55. $x + 2y - 4z = 5$
$3x - y + z = 3$
$x + 2y + z = -3$

56. $-2x + 3y + 4z = 10$
$3x - 5y + 2z = 7$
$4x + 2y - 3z = -2$

57.
$x + 3y + 2z - w = 2$
$-x - y - 4z + 2w = 1$
$2x + y - z + 3w = -4$
$x - 2y - z - 3w = 4$

58.
$x - 3y + z + 2w = 3$
$-2x + y + 3z + w = 0$
$-x + 2y - 3z + 2w = -3$
$x + 2y - z + 3w = 4$

In Exercises 59–66, use the Gauss–Jordan method to solve the given system of equations using your calculator's matrix capabilities (either step-by-step row operations or rref(command). Round answers to four decimal places.

59.
$2.3x - 5.7y = 12.3$
$-13.1x - 18.2y = 7.1$

60. $-3.1x + 2.9y = -5.8$
$6.1x + 3.5y = -15.2$

61. $1.53x + 3.57y = 2.68$
$7.28x + 9.85y = 13.41$

62. $-3.75x - 4.23y = 7.82$
$7.28x + 8.65y = 4.86$

63. $1.1x - 3.4y + 7.2z = 5.9$
$-8.9x - 4.7y - 3.1z = -7.1$
$0.8x + 0.3y + 1.2z = 0.9$

64. $-2.3x + 3.5y - 2.1z = 5.7$
$3.6x - 0.5y - 0.3z = -5.3$
$-8.2x - 1.1y + 3.7z = -2.2$

65. $-2.31x + 4.21y - 7.28z + 3.42w = 8.17$
$7.91x + 8.99y + 4.01z - 4.21w = -2.71$
$-13.11x - 5.07y - 0.39z + 1.22w = 0.77$
$8.71x + 3.72y - 2.73z - 4.81w = 3.14$

66. $0.72x - 1.13y + 4.27z - 3.12w = 5.77$
$-1.11x + 0.93y - 2.21z + 5.19w = -2.29$
$-0.04x + 1.98y + 7.61z + 4.23w = 5.13$
$1.92x - 5.03y + 3.41z - 2.71w = 8.28$

Applications

67. Max George is a craftsman who makes end tables and night stands. To make one end table takes 5 hours of his labor, and to make one night stand takes 4 hours of his labor. The materials for one end table cost him $200, and the materials for one night stand cost him $80. Suppose that for 1 week Max works 40 hours and spends $1200 on materials.

(a) Let x represent the number of end tables made in 1 week and y represent the number of night stands made in 1 week. Write a system of two linear equations in which one equation relates costs and the other relates time.

(b) Solve the system in part (a) to determine how many end tables and how many night stands Max can make in 1 week, assuming that he works 40 hours and spends $1200 on materials.

$\Leftarrow$ **1.1 68.** Martha George is a craftswoman who makes coffee tables and night stands. To make one coffee table takes 6 hours of her labor, and to make one night stand takes 4 hours of her labor. The materials for one coffee table cost her $220 and the materials for one night stand cost her $80. Suppose that for 1 week Martha works 40 hours and spends $1400 on materials.

(a) Let x represent the number of coffee tables made in 1 week and y represent the number of night stands made in 1 week. Write a system of two linear equations in which one equation relates costs and the other relates time.

(b) Solve the system in part (a) to determine how many coffee tables and how many night stands Martha can make in 1 week, assuming that she works 40 hours and spends $1400 on materials.

$\Leftarrow$ **1.1 69.** The WeDo Wood Canoes Company makes two types of canoes: a two-person model and a four-person model. Each two-person model requires 1 hour in the cutting department and 1.5 hours in the assembly department. Each four-person model requires 1 hour in the cutting department and 2.75 hours in the assembly department. The cutting department has a maximum of 640 hours available each week, while the assembly department has a maximum of 1080 hours available each week. How many of each type of canoe should be produced each week for the company to operate at full capacity, that is, use all available hours in each department?

$\Leftarrow$ **1.1 70.** (*continuation of Exercise 69*) Suppose that a breakthrough in the assembly process decreases the time required in the assembly department for each two-person model to $\frac{3}{4}$ hour and for each four-person model to 2 hours. Now how many of each type of canoe should be produced each week for the company to operate at full capacity, that is, use all available hours in each department?

71. Each order from the TidyTrader wholesaler contains 10 boxes of lighters and 36 boxes of coffee stirrers, while each order from the Trade-it-Now wholesaler contains 30 boxes of lighters and 12 boxes of coffee stirrers. How many orders should the Gas-N-Go convenient store get from each wholesaler to get exactly 120 boxes of lighters and 240 boxes of coffee stirrers?

72. The Java-Go-Go coffee shop carries two blends of coffee. The first blend contains 20% Mexican and 80% Colombian coffee. The second blend contains 60% Mexican and 40% Colombian coffee. How many ounces of each blend should be combined to get a new blend that has 100 ounces of Mexican and 100 ounces of Colombian coffee?

$\Leftarrow$ **1.1 73.** The Stitz Tent Company makes three types of tents: two-person, four-person and, six-person models. Each tent requires the services of three departments, as shown in Table 2.1.3. The fabric cutting, assembly, and packaging departments have available a maximum of 450, 390, and 190 hours each week, respectively.

Table 2.1.3

	Two-person	Four-person	Six-person
Fabric cutting (hr)	0.5	0.8	1.1
Assembly (hr)	0.6	0.7	0.9
Packaging (hr)	0.2	0.3	0.6

(a) Let x represent the number of two-person tents made in 1 week, y represent the number of four-person tents made in 1 week, and z represent the number of six-person tents made in 1 week. Write a system of three linear equations using this information and the given department maximum weekly times.

(b) Solve the system in part (a) to determine how many two-, four-, and six-person tents can be made in 1 week, assuming that the company operates at full capacity.

$\Leftarrow$ **1.1 74.** Rework Exercise 73(a) and (b), except here assume that the fabric cutting, assembly, and packaging departments have available a maximum of 420, 390, and 185 hours each week, respectively.

$\Leftarrow$ **1.1 75.** Last year the Hannus Corporation took out three loans totaling $50,000 to fund the purchase of new equipment. Some of the money was at 6%, $1000 more than $\frac{1}{2}$ the amount of the 6% loan was at 8%, and the rest was at 7%. The total annual interest on the loans was $3410.

(a) Let x represent the amount borrowed at 6%, y represent the amount borrowed at 7%, and z represent the amount borrowed at 8%. Write a system of three linear equations that can be solved using the Gauss–Jordan method to determine the amount borrowed at each rate.

(b) Solve the system in part (a) to determine the amount borrowed at each rate.

76. Debbie Ashton, a certified financial planner, is to invest $80,000 for a client in three different areas: certificates of deposit earning 6%, a mutual fund earning 12%, and speculative junk bonds earning 17%. The client is a conservative investor and stipulates that four times as much be invested in CDs as in junk bonds. The client also wishes that $7700 be produced each year on these investments. Assuming these rates and adhering to her client's wishes, how much should Debbie place in each area to produce exactly $7700 each year on the investments?

77. Tim Kling, chairman of the Chemistry Department at a local community college, wants to make 2 quarts of an 18% acid solution. He has available a solution that is 21% acid and a solution that is 14% acid. How many quarts of each type should he use to produce the 18% solution?

78. Rework Exercise 77 assuming that Tim made an error and really desired 2 quarts of a 16% acid solution.

← 1.1 **79.** Chris O'Neill, a dietitian in a local hospital, is to arrange a special diet for a patient that is made up of three basic foods, Basic!, Vaca, and Secra. The diet is to include 340 units of calcium, 205 units of vitamin A, and 225 units of vitamin C. The number of units per ounce for each basic food is given in Table 2.1.4. How many ounces of each food must be used to exactly meet the diet requirements?

Table 2.1.4

	Basic!	Vaca	Secra
Calcium	20	10	30
Vitamin A	10	5	20
Vitamin C	20	15	10

← 1.1 **80.** Rework Exercise 79 if the patient's diet is changed so that it includes 480 units of calcium, 285 units of vitamin A, and 325 units of vitamin C.

← 1.1 🌐 📊 **81.** The data in Table 2.1.5 give the amounts of nitrogen, phosphate, and potash (all in lb/acre) needed to grow asparagus, cauliflower, and celery for an entire growing season in California.

Table 2.1.5

	Asparagus	Cauliflower	Celery
Nitrogen	240	215	336
Phosphate	148	98	171
Potash	42	69	147

SOURCE: U.S. Department of Agriculture, www.usda.gov/nass

(a) If Juan has 280 acres, 20,000 pounds of nitrogen, 31,860 pounds of phosphate, and 21,630 pounds of potash, is it possible for him to use all resources completely. If so, how many acres should he allot for each crop?

(b) Suppose that everything is the same as in part (a), except now Juan has 65,790 pounds of nitrogen. Is it possible for him to use all resources completely? If so, how many acres should he allot for each crop?

← 1.1 🌐 📊 **82.** The data in Table 2.1.6 give the amounts of nitrogen, phosphate, and potash (all in lb/acre) needed to grow head lettuce, cantaloupe, and bell peppers for an entire growing season in California.

Table 2.1.6

	Head Lettuce	Cantaloupe	Bell Peppers
Nitrogen	173	175	212
Phosphate	122	161	108
Potash	76	39	72

SOURCE: U.S. Department of Agriculture, www.usda.gov/nass

(a) If Zach has 200 acres, 38,430 pounds of nitrogen, 27,845 pounds of phosphate, and 6125 pounds of potash, is it possible for him to use all resources completely. If so, how many acres should he allot for each crop?

(b) Suppose that everything is the same as in part (a), except now Zach has 12,985 pounds of potash. Is it possible for him to use all resources completely? If so, how many acres should he allot for each crop?

📊 **83.** Tammy K's has a 1-month intensive training program for receptionist/secretaries for three types of office positions: medical field, small business, and executive secretarial. Each individual receives training in three departments at Tammy K's: word processing, telephone and LAN (local area network), and ethics. The data in Table 2.1.7 show the number of units of training that each type of receptionist/secretary receives.

Table 2.1.7

	Medical Field	Small Business	Executive Secretary
Word processing	0.8	1.1	1.7
Telephone and LAN	0.6	1.3	1.5
Ethics	1.2	0.8	1.3

Each month at Tammy K's, word processing can give 32.1 units of word processing, telephone and LAN can give 29.7 units, and ethics can give 33 units. To completely utilize their training capacity, how many of each type of receptionist/secretary should Tammy K's admit to their training program each month?

📊 **84.** Due to a strong demand for their services, Tammy K's from Exercise 83 has added more trainers. This results in Tammy K's being able to deliver each month 41.9 units of word processing, 38.9 units of telephone and LAN training, and 41 units of ethics training. To completely utilize their training capacity, how many of each type of receptionist/secretary should Tammy K's admit to their training program each month?

←
1.1

SECTION PROJECT

During rush hour, traffic congestion may be encountered at certain intersections. For a network of four one-way streets in a certain city, the rush hour flow is as shown in Figure 2.1.7.

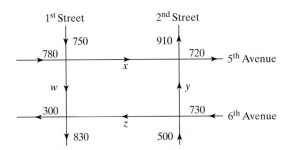

Figure 2.1.7

To help to speed the flow of traffic, traffic engineers need to coordinate the signals at these intersections. To do this, data are collected giving the hourly flow of traffic during rush hour at certain locations. Figure 2.1.7 shows the data that were collected. Notice that the figure indicates that a total of 1530 cars per hour enter the intersection of 1st Street and

5th Avenue. Of these 1530 cars, x cars leave the intersection on 5th Avenue and w cars leave the intersection on 1st Street. Since the total number of cars leaving this (or any) intersection equals the number entering, we have

$$x + w = 1530$$

This equation gives the flow at the intersection of 1st Street and 5th Avenue.

(a) Determine the equations for the traffic flow at the other three intersections. Remember that the total entering the intersection equals the total leaving the intersection, and vice versa.

(b) Write a system of four equations representing the traffic flow for this network of four one-way streets.

(c) Write the augmented matrix for the system in part (b). Note that some entries must be zero.

(d) Use the Gauss–Jordan method to solve. Here we use w as the parameter. If necessary, see Example 6.

(e) Using the result of part (d), what is the smallest and largest value for the parameter w.

(f) Using the results of parts (d) and (e), determine the smallest and largest values for the variables x, y, and z.

Section 2.2   Matrices and Operations

In Section 2.1 we introduced matrices as a means of helping us solve systems of linear equations using the Gauss–Jordan method. In the next three sections we take a more in-depth look at matrices and operations with matrices. We will also see how matrices can be used to solve problems in the business, life, and social sciences.

Matrix Arithmetic

One beauty of matrices is its ability to represent data. For example, Staminly's has three furniture stores, East, West and Metro, in a large metropolitan area. Tables 2.2.1, 2.2.2, and 2.2.3 give the sales of their Country Style Contemporary sofa for each store during July.

Table 2.2.1 Staminly's East

	60″	72″	84″
Two cushion	8	13	5
Three cushion	9	16	7

Table 2.2.2 Staminly's Metro

	60″	72″	84″
Two cushion	13	7	3
Three cushion	6	8	4

Table 2.2.3 Staminly's West

	60″	72″	84″
Two cushion	4	11	13
Three cushion	6	9	8

With the understanding that rows refer to the sofa type (two or three cushion) and columns refer to the lengths (60″, 72″, or 84″), we can use matrices to represent the information in Tables 2.2.1 through 2.2.3 as follows:

$$E = \begin{bmatrix} 8 & 13 & 5 \\ 9 & 16 & 7 \end{bmatrix} \qquad M = \begin{bmatrix} 13 & 7 & 3 \\ 6 & 8 & 4 \end{bmatrix} \qquad W = \begin{bmatrix} 4 & 11 & 13 \\ 6 & 9 & 8 \end{bmatrix}$$

Matrix for Table 2.2.1 Matrix for Table 2.2.2 Matrix for Table 2.2.3

Now suppose that we wanted to know the total number of two-cushion 60″ sofas sold at Staminly's East and Metro. Since 8 were sold at East and 13 were sold at Metro, we conclude that a total of $8 + 13 = 21$ two-cushion 60″ sofas were sold at Staminly's East and Metro. Notice that $e_{1,1} + m_{1,1} = 21$. If we continue to sum corresponding entries for E and M, we get a new matrix, call it S, that is given by

$$S = E + M = \begin{bmatrix} 8 & 13 & 5 \\ 9 & 16 & 7 \end{bmatrix} + \begin{bmatrix} 13 & 7 & 3 \\ 6 & 8 & 4 \end{bmatrix}$$

$$= \begin{bmatrix} 8+13 & 13+7 & 5+3 \\ 9+6 & 16+8 & 7+4 \end{bmatrix} = \begin{bmatrix} 21 & 20 & 8 \\ 15 & 24 & 11 \end{bmatrix}$$

Matrix S gives the total sales of Staminly's Country Style Contemporary sofa for both styles and all three lengths for their East and Metro stores during July. The manner in which E and M were added illustrates the following definition for the addition of matrices.

Interactive Activity

For the matrices M and W in Example 1, determine $W + M$. Compare the result with the result of Example 1. This illustrates that matrix addition is **commutative**; that is, $M + W = W + M$.

Addition of Matrices

The **sum** of two (or more) $m \times n$ matrices A and B is the $m \times n$ matrix $A + B$, where each entry in $A + B$ is the sum of the corresponding entries of A and B.

▶ **Note:** We can only add matrices that have the same dimension.

Example 1 Adding Matrices

Use matrix addition to find the total sales of Staminly's Country Style Contemporary sofa for their Metro and West stores during July.

Solution

The total sales for Metro and West are

$$M + W = \begin{bmatrix} 13 & 7 & 3 \\ 6 & 8 & 4 \end{bmatrix} + \begin{bmatrix} 4 & 11 & 13 \\ 6 & 9 & 8 \end{bmatrix}$$

$$= \begin{bmatrix} 13+4 & 7+11 & 3+13 \\ 6+6 & 8+9 & 4+8 \end{bmatrix} = \begin{bmatrix} 17 & 18 & 16 \\ 12 & 17 & 12 \end{bmatrix}$$

✓ **Checkpoint 1** Now work Exercise 1.

Example 2 Adding Matrices and Interpreting an Entry

For Staminly's, define matrix $T = E + M + W$. Find matrix T and interpret $t_{2,3}$.

Interactive Activity

For matrices E, M, and W from Example 2, do the following: Find $(E+M)+W$ and compare to $E+(M+W)$. This will illustrate that matrix addition is **associative**; that is, $(E+M)+W = E+(M+W)$.

Solution

Matrix T is given by

$$T = E + M + W = \begin{bmatrix} 8 & 13 & 5 \\ 9 & 16 & 7 \end{bmatrix} + \begin{bmatrix} 13 & 7 & 3 \\ 6 & 8 & 4 \end{bmatrix} + \begin{bmatrix} 4 & 11 & 13 \\ 6 & 9 & 8 \end{bmatrix}$$

$$= \begin{bmatrix} 25 & 31 & 21 \\ 21 & 33 & 19 \end{bmatrix}$$

By looking in the 2nd row, 3rd column we have $t_{2,3} = 19$.

Interpret the Solution: This entry means that the total number of three-cushion 84″ Country Style Contemporary sofas sold in July at all three Staminly stores was 19. ■

Technology Option

Figure 2.2.1

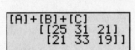

Matrix operations can be performed on graphing calculators as well. Figure 2.2.1 shows the result of $E + M + W$ in Example 2. Matrices E, M, and W were entered into the graphing calculator as A, B, and C, respectively.

The way that matrices are added and subtracted can vary from calculator to calculator. To learn how matrix operations are performed on your calculator, consult the online graphing calculator manual at www.prenhall.com/armstrong

Scalar Multiplication

The product of a real number k and a matrix A, denoted by kA, is a matrix in which each entry is the product of each entry in A by k. When a matrix is multiplied by a real number, we call this **scalar multiplication**.

Scalar Multiplication

For a matrix $A = \begin{bmatrix} a & b \\ c & d \end{bmatrix}$ and a real number k, the scalar product kA is given by the **scalar multiplication**

$$kA = k\begin{bmatrix} a & b \\ c & d \end{bmatrix} = \begin{bmatrix} ka & kb \\ kc & kd \end{bmatrix}$$

Note: This definition can be extended to a matrix of any dimension.

Example 3 Performing a Scalar Multiplication

Given matrix $A = \begin{bmatrix} -1 & 2 & 2 \\ 4 & 3 & -7 \end{bmatrix}$, find $3A$.

Solution

We multiply each entry in A by 3 to get

$$3A = 3\begin{bmatrix} -1 & 2 & 2 \\ 4 & 3 & -7 \end{bmatrix} = \begin{bmatrix} 3(-1) & 3(2) & 3(2) \\ 3(4) & 3(3) & 3(-7) \end{bmatrix} = \begin{bmatrix} -3 & 6 & 6 \\ 12 & 9 & -21 \end{bmatrix}$$ ■

Technology Option

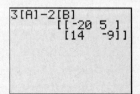

Figure 2.2.2

Figure 2.2.2 shows the scalar multiplication in Example 3 on a graphing calculator.

Matrix subtraction is defined similarly to real number subtraction. This means that if A and B are two matrices with the same dimension then $A - B = A + (-B)$.

Example 4   **Subtracting Scalar Multiplied Matrices**

Given $A = \begin{bmatrix} -2 & 3 \\ 4 & -1 \end{bmatrix}$ and $B = \begin{bmatrix} 7 & 2 \\ -1 & 3 \end{bmatrix}$, find $3A - 2B$.

Solution

We have

$$
\begin{aligned}
3A - 2B &= 3\begin{bmatrix} -2 & 3 \\ 4 & -1 \end{bmatrix} - 2\begin{bmatrix} 7 & 2 \\ -1 & 3 \end{bmatrix} \\
&= 3\begin{bmatrix} -2 & 3 \\ 4 & -1 \end{bmatrix} + (-2)\begin{bmatrix} 7 & 2 \\ -1 & 3 \end{bmatrix} \\
&= \begin{bmatrix} -6 & 9 \\ 12 & -3 \end{bmatrix} + \begin{bmatrix} -14 & -4 \\ 2 & -6 \end{bmatrix} \\
&= \begin{bmatrix} -20 & 5 \\ 14 & -9 \end{bmatrix}
\end{aligned}
$$

✓**Checkpoint 2**   Now work Exercise 13.

Technology Option

Figure 2.2.3

Figure 2.2.3 shows the result of $3A - 2B$ from Example 4 on a graphing calculator.

In Example 4 matrix A and matrix B are examples of **square matrices**. In general, an $n \times n$ matrix, that is, a matrix having the same number of rows as columns, is called a **square matrix**. A matrix with one row is called a **row matrix**, and a matrix with one column is called a **column matrix**.

Equality of Matrices

We have to satisfy two conditions in order for two matrices A and B to be equal.

> ■ **Equality of Matrices**
>
> Two matrices are **equal** if they have the same dimension and if corresponding entries are equal.

Example 5   **Determining Matrix Equality**

Determine the values for x, y, z, and w that make the following matrices equal.

(a) $\begin{bmatrix} -2 & x & 3 \\ y+2 & 4 & 7 \end{bmatrix} = \begin{bmatrix} z & 1 & 3 \\ -5 & 4 & w+1 \end{bmatrix}$

(b) $\begin{bmatrix} 3 & x & 1 \\ y & -7 & 4 \end{bmatrix} = \begin{bmatrix} 5 & w \\ z & -1 \\ 3 & 4 \end{bmatrix}$

Solution

(a) Both matrices are 2×3, so they are equal if corresponding entries are equal. Hence

$$x = 1, \qquad y + 2 = -5, \qquad z = -2, \qquad w + 1 = 7$$

This means that $x = 1$, $y = -7$, $z = -2$, and $w = 6$ makes the matrices equal.

(b) The matrix on the left is a 2×3, and the one on the right is a 3×2; hence they cannot be equal no matter the values for x, y, z, and w.   ■

✓ **Checkpoint 3**

Now work Exercise 35.

Zero Matrix and Transpose of a Matrix

In the real number system, the number zero is called the **additive identity**, because any real number plus zero produces the same real number. There is a matrix that behaves the same way, the **zero matrix**. The zero matrix is simply a matrix in which all entries are zero. Given a matrix A, the zero matrix O has the property that

$$A + O = O + A = A$$

as long as A and matrix O have the same dimension. For example, if matrix A is any 2×3 matrix, then the zero matrix for any 2×3 matrix is

$$O = \begin{bmatrix} 0 & 0 & 0 \\ 0 & 0 & 0 \end{bmatrix}$$

In Section 4.3, we will use the **transpose** of a matrix to solve some applications. For a given matrix A, if we interchange the rows and columns we produce what is called the **transpose of A** and denote it with A^T.

> ■ **Transpose of a Matrix**
>
> If A is a $m \times n$ matrix with entries $a_{i,j}$, then the **transpose** of A, denoted A^T, is an $n \times m$ matrix with entries $a_{j,i}$.

Example 6   **Finding a Transpose**

For matrix A, $A = \begin{bmatrix} 3 & -1 & 2 \\ 4 & 7 & -2 \end{bmatrix}$, determine A^T and the dimension of A^T.

Solution

Interchanging the rows and columns gives us

$$A^T = \begin{bmatrix} 3 & 4 \\ -1 & 7 \\ 2 & -2 \end{bmatrix}$$

The dimension of matrix A is 2×3, and the dimension of A^T is 3×2.   ■

✓ **Checkpoint 4**   Now work Exercise 23.

Technology Option

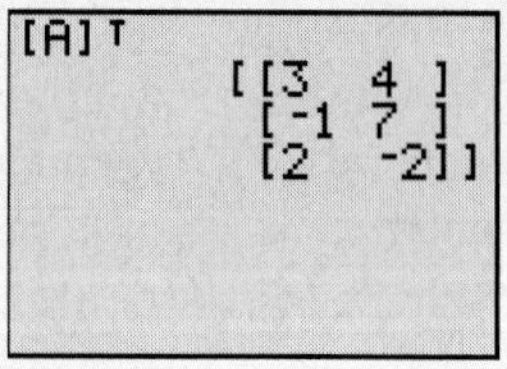

Figure 2.2.4 shows the result of Example 6 on a graphing calculator.

Figure 2.2.4

Our final example returns us to the Staminly furniture stores from the beginning of the section.

Example 7   **Predicting Monthly Sales**

Joyce, the CEO of Staminly's, wants to see an across the board increase in sales for the month of August. She wants the East and West stores to increase monthly sales by 15% and the Metro store to increase monthly sales by 10%. Determine a matrix giving the total sales that Joyce expects for all three stores in August. Round all entries to the nearest whole number.

Solution

Recall that the July sales for the East, West, and Metro stores, given by matrices E, W, and M, respectively, are

$$E = \begin{bmatrix} 8 & 13 & 5 \\ 9 & 16 & 7 \end{bmatrix} \qquad W = \begin{bmatrix} 4 & 11 & 13 \\ 6 & 9 & 8 \end{bmatrix} \qquad M = \begin{bmatrix} 13 & 7 & 3 \\ 6 & 8 & 4 \end{bmatrix}$$

The matrix for expected August sales at the East store is computed to be

$$E + 0.15E = \begin{bmatrix} 8 & 13 & 5 \\ 9 & 16 & 7 \end{bmatrix} + 0.15 \begin{bmatrix} 8 & 13 & 5 \\ 9 & 16 & 7 \end{bmatrix}$$

$$= \begin{bmatrix} 8 & 13 & 5 \\ 9 & 16 & 7 \end{bmatrix} + \begin{bmatrix} 1.2 & 1.95 & 0.75 \\ 1.35 & 2.4 & 1.05 \end{bmatrix} = \begin{bmatrix} 9.2 & 14.95 & 5.75 \\ 10.35 & 18.4 & 8.05 \end{bmatrix}$$

Similarly, the expected August sales at the West store are

$$W + 0.15W = \begin{bmatrix} 4 & 11 & 13 \\ 6 & 9 & 8 \end{bmatrix} + 0.15 \begin{bmatrix} 4 & 11 & 13 \\ 6 & 9 & 8 \end{bmatrix}$$

$$= \begin{bmatrix} 4 & 11 & 13 \\ 6 & 9 & 8 \end{bmatrix} + \begin{bmatrix} 0.6 & 1.65 & 1.95 \\ 0.9 & 1.35 & 1.2 \end{bmatrix} = \begin{bmatrix} 4.6 & 12.65 & 14.95 \\ 6.9 & 10.35 & 9.2 \end{bmatrix}$$

Finally, the expected August sales at the Metro store are

$$M + 0.10M = \begin{bmatrix} 13 & 7 & 3 \\ 6 & 8 & 4 \end{bmatrix} + 0.10 \begin{bmatrix} 13 & 7 & 3 \\ 6 & 8 & 4 \end{bmatrix}$$

$$= \begin{bmatrix} 13 & 7 & 3 \\ 6 & 8 & 4 \end{bmatrix} + \begin{bmatrix} 1.3 & 0.7 & 0.3 \\ 0.6 & 0.8 & 0.4 \end{bmatrix} = \begin{bmatrix} 14.3 & 7.7 & 3.3 \\ 6.6 & 8.8 & 4.4 \end{bmatrix}$$

So the matrix giving the total sales that Joyce expects for all three stores in August is the sum of

$$(E + 0.15E) + (W + 0.15W) + (M + 0.10M)$$

$$= \begin{bmatrix} 9.2 & 14.95 & 5.75 \\ 10.35 & 18.4 & 8.05 \end{bmatrix} + \begin{bmatrix} 4.6 & 12.65 & 14.95 \\ 6.9 & 10.35 & 9.2 \end{bmatrix} + \begin{bmatrix} 14.3 & 7.7 & 3.3 \\ 6.6 & 8.8 & 4.4 \end{bmatrix}$$

$$= \begin{bmatrix} 28.1 & 35.3 & 24 \\ 23.85 & 37.55 & 21.65 \end{bmatrix} \approx \begin{bmatrix} 28 & 35 & 24 \\ 24 & 38 & 22 \end{bmatrix}$$ ∎

Interactive Activity

For the matrices given in Example 7, determine $1.15E + 1.15W + 1.1M$ and compare to the final result.

SUMMARY

In this section we saw how to add, subtract, and multiply matrices by a scalar. We also learned how to determine when two matrices are equal. The zero matrix was presented along with how to determine the transpose of a matrix. We saw how matrices can be used to represent data and analyze problems in an applied setting.

SECTION 2.2 EXERCISES

In Exercises 1–16, refer to the following matrices to perform the indicated operations.

$$A = \begin{bmatrix} -2 & 0 & 3 \\ 1 & -3 & 4 \end{bmatrix} \qquad B = \begin{bmatrix} 1 & -3 \\ 4 & -2 \end{bmatrix}$$

$$C = \begin{bmatrix} -4 & 1 \\ 3 & 5 \end{bmatrix} \qquad D = \begin{bmatrix} 4 & -1 & -3 \\ 0 & 7 & 1 \end{bmatrix}$$

$$E = \begin{bmatrix} 1 & -1 & 4 & -3 \end{bmatrix} \qquad F = \begin{bmatrix} -2 \\ 1 \\ 0 \\ 3 \end{bmatrix}$$

$$G = \begin{bmatrix} -1 & 4 \\ 3 & -7 \\ -5 & -2 \end{bmatrix} \qquad H = \begin{bmatrix} 4 & -2 \\ 3 & -1 \\ -2 & 5 \end{bmatrix}$$

✓ **1.** $A + D$ **2.** $G + H$

3. $A + B$ **4.** $D + G$

5. $B - C$ **6.** $C - B$

7. $-2A$ **8.** $-3B$

9. $4E$ **10.** $2F$

11. $3A + 2D$ **12.** $-4A + 3D$

✓ **13.** $-3G - 4H$ **14.** $-2G - 7H$

15. $4H + 2G$ **16.** $5H - 3G$

To answer Exercises 17–30, refer to the matrices A through H given for Exercises 1–16.

17. Which matrices are square matrices?

18. Identify the column matrix.

19. Identify the row matrix.

20. Explain why $A + B$ cannot be done.

21. Explain why $D + G$ cannot be done.

22. Determine A^T, the transpose of A, and give the dimension of A^T.

✓ **23.** Determine D^T, the transpose of D, and give the dimension of D^T.

24. Determine B^T, the transpose of B, and give the dimension of B^T.

25. Determine C^T, the transpose of C, and give the dimension of C^T.

26. Determine $(A + D)^T$, the transpose of $A + D$.

27. Write the zero matrix for A.

28. Write the zero matrix for B.

29. Write the zero matrix for G.

30. Write the zero matrix for E.

In Exercises 31–38, find the values of the variables in each equation.

31. $\begin{bmatrix} 3 \\ x \\ y \end{bmatrix} = \begin{bmatrix} z \\ -1 \\ 4 \end{bmatrix}$

32. $\begin{bmatrix} -4 \\ x \\ 3 \end{bmatrix} = \begin{bmatrix} y \\ 2 \\ z \end{bmatrix}$

33. $\begin{bmatrix} -2 & x \\ y & 1 \end{bmatrix} = \begin{bmatrix} -2 & 5 \\ 3 & z \end{bmatrix}$

34. $\begin{bmatrix} x & 1 \\ y & 4 \end{bmatrix} = \begin{bmatrix} -9 & 1 \\ 3 & z \end{bmatrix}$

✓ **35.** $\begin{bmatrix} x+3 & y-2 \\ 4 & 3 \end{bmatrix} = \begin{bmatrix} 8 & 11 \\ 4 & z \end{bmatrix}$

36. $\begin{bmatrix} 3 & 8 \\ x & 1 \end{bmatrix} = \begin{bmatrix} y-4 & z+3 \\ 2 & 1 \end{bmatrix}$

37. $\begin{bmatrix} -3+x & 4y & 3z \\ 5w & 3 & 4 \end{bmatrix} + \begin{bmatrix} -6 & 3y & 2 \\ 7 & 2 & 9 \end{bmatrix}$
$= \begin{bmatrix} 18 & 28 & 8 \\ 22 & 5 & 26p \end{bmatrix}$

38. $\begin{bmatrix} x+1 & 3y-1 & 4z \\ -7w & -3 & 5 \end{bmatrix} + \begin{bmatrix} 2x & 7y & 5z-1 \\ -3w & 8 & 13 \end{bmatrix}$
$= \begin{bmatrix} 16 & 91 & 8 \\ 20 & -11 & 54p \end{bmatrix}$

In Exercises 39–48, use a graphing calculator and its matrix capabilities, and refer to the given matrices to perform the indicated operations.

$A = \begin{bmatrix} 1.32 & 4.67 & 1.19 \\ 3.14 & 2.78 & 5.61 \end{bmatrix}$ $B = \begin{bmatrix} 8.11 & 9.23 & 1.41 \\ 4.15 & 3.14 & 4.59 \end{bmatrix}$

39. $A + B$

40. $A - B$

41. $2A + 3B$

42. $4A - 2B$

43. $1.3A$

44. $2.4B$

45. $2.37A - 0.17B$

46. $7.81B + 2.19A$

47. Verify that $1.7(A + B) = 1.7A + 1.7B$.

48. Verify that $(6.3 \cdot 1.2)A = 6.3(1.2A)$.

Applications

49. Kirzan Auto has the following data for sales of the popular Blink sedan during June, July, and August. The data are from Kirzan's Northside dealership. (Manual and automatic refer to transmission type.)

June Sales

	Manual	Automatic
Two door	13	7
Four door	5	12

July Sales

	Manual	Automatic
Two door	12	9
Four door	7	14

August Sales

	Manual	Automatic
Two door	14	11
Four door	3	19

(a) If the rows refer to the sedan type (two door/four door) and the columns refer to the transmission type (manual/automatic), write three matrices to represent the data. Let J, Y, and A be the matrices for June, July, and August, respectively.

(b) Determine a matrix giving the total sales for Kirzan's Northside dealership during June, July, and August.

50. Kirzan Auto has the following data for sales of the popular Blink sedan during June, July, and August. The data are from Kirzan's Southside dealership. (Manual and automatic refer to transmission type.)

June Sales

	Manual	Automatic
Two door	11	8
Four door	6	9

July Sales

	Manual	Automatic
Two door	15	6
Four door	7	11

August Sales	Manual	Automatic
Two door	16	13
Four door	2	13

(a) If the rows refer to the sedan type (two door/four door) and the columns refer to the transmission type (manual/automatic), write three matrices to represent the data. Let J, Y, and A be the matrices for June, July, and August, respectively.

(b) Determine a matrix giving the total sales for Kirzan's Southside dealership during June, July, and August.

51. For J, Y, and A in Exercise 49, compute $\frac{1}{3}(J + Y + A)$ and interpret.

52. For J, Y, and A in Exercise 50, compute $\frac{1}{3}(J + Y + A)$ and interpret.

53. The Basich Company has two plants, one on the East Coast and one on the West Coast, that manufacture 2D and 3D graphing calculators. The production costs, in dollars, for each calculator are summarized in the following matrices:

$$E = \begin{bmatrix} 25 & 35 \\ 10 & 25 \end{bmatrix} \begin{matrix} \text{Materials} \\ \text{Labor} \end{matrix} \quad W = \begin{bmatrix} 28 & 39 \\ 15 & 27 \end{bmatrix} \begin{matrix} \text{Materials} \\ \text{Labor} \end{matrix}$$

East Coast   West Coast

(a) Find and interpret $\frac{1}{2}(E + W)$.

(b) If both labor and materials at the East Coast plant increase by 10% and both labor and materials at the West Coast plant are decreased by 5%, determine new matrices to represent the new production costs.

54. Shawna Wells has two types of IRA retirement accounts, certificates of deposit (CD) and Grade A municipal bonds (GMB), held in two banks, First Century (FC) and Money National (MN). The dollar amounts in each at the beginning of 2002 are given in matrix A.

$$A = \begin{bmatrix} 15{,}000 & 25{,}000 \\ 40{,}000 & 35{,}000 \end{bmatrix} \begin{matrix} \text{FC} \\ \text{MN} \end{matrix}$$

(a) Assuming a 7% growth rate on each account, determine a matrix I giving the earnings on the various accounts during 2002.

(b) Determine a matrix that represents the value of each account at the end of 2002.

55. Nick Dominic has two types of IRA retirement accounts, treasury notes (TN) and municipal bonds (MB), held in two banks, Valley Bank (VB) and Hilltop Bank (HB). The dollar amounts in each at the beginning of 2002 are given in matrix A.

$$A = \begin{bmatrix} 20{,}000 & 25{,}000 \\ 15{,}000 & 35{,}000 \end{bmatrix} \begin{matrix} \text{VB} \\ \text{HB} \end{matrix}$$

(a) Assuming an 8% growth rate on each account, determine a matrix I giving the earnings on the various accounts during 2002.

(b) Determine a matrix that represents the value of each account at the end of 2002.

56. The hospital care and physician services expenditures, in billions of dollars, by patients (out of pocket) and Medicare are given in Table 2.2.4.

Table 2.2.4

	Hospital Care		Physician Services	
	Out of pocket	Medicare	Out of pocket	Medicare
1980	5.3	26.4	14.7	8.0
1990	11.1	68.7	32.2	29.2
1995	11.5	108.9	30.0	39.9
1996	11.8	116.3	31.1	42.8
1997	12.4	123.7	34.1	46.4

SOURCE: U.S. Census Bureau, www.census.gov

The data for hospital care are summarized in matrix H.

$$H = \begin{bmatrix} 5.3 & 26.4 \\ 11.1 & 68.7 \\ 11.5 & 108.9 \\ 11.8 & 116.3 \\ 12.4 & 123.7 \end{bmatrix} \begin{matrix} 1980 \\ 1990 \\ 1995 \\ 1996 \\ 1997 \end{matrix}$$

with columns Out of pocket, Medicare.

(a) Write a matrix M to summarize the data for physician services.

(b) Find $H + M$ and interpret the entries.

57. The number of people, in millions, covered by private or government health insurance is given in Table 2.2.5.

Table 2.2.5

Year	Private	Government	
		Medicare	Medicaid
1990	182.1	32.3	24.3
1992	181.5	33.2	29.4
1993	182.4	33.1	31.7
1994	184.3	33.9	31.6
1995	185.9	34.7	31.9
1996	187.4	35.2	31.5
1997	188.5	35.6	29.0

SOURCE: U.S. Census Bureau, www.census.gov

The data for people with private insurance is summarized in matrix P.

$$P = \begin{bmatrix} 182.1 \\ 181.5 \\ 182.4 \\ 184.3 \\ 185.9 \\ 187.4 \\ 188.5 \end{bmatrix} \begin{matrix} 1990 \\ 1992 \\ 1993 \\ 1994 \\ 1995 \\ 1996 \\ 1997 \end{matrix}$$

(a) Write a matrix MCR to summarize the data for people with Medicare.

(b) Write a matrix MCD to summarize the data for people with Medicaid.

(c) Let $TG = MCR + MCD$. Determine TG and interpret the entries.

🌐 **58.** (*continuation of Exercise 57*)

(a) Let $TC = TG + P$. Matrix TC gives the total covered by private or government insurance.

(b) Matrix F gives the U.S. population in millions (*Source:* U.S. Census Bureau, www.census.gov).

$$\text{Population}$$

$$F = \begin{bmatrix} 248.9 \\ 256.8 \\ 259.8 \\ 262.1 \\ 264.3 \\ 266.8 \\ 269.1 \end{bmatrix} \begin{matrix} 1990 \\ 1992 \\ 1993 \\ 1994 \\ 1995 \\ 1996 \\ 1997 \end{matrix}$$

Compute $F - TC$ and interpret the entries.

🌐 **59.** Matrix M gives the percentage of U.S. male smokers, and matrix F gives the percentage of U.S. female smokers (*Source:* U.S. Census Bureau, www.census.gov).

$$\begin{matrix} & 18\text{--}24 & 25\text{--}34 & 35\text{--}44 & 45\text{--}64 & 65\text{ and over} \\ M = & \begin{bmatrix} 28.0 & 38.2 & 37.6 & 33.4 & 19.6 \\ 26.6 & 31.6 & 34.5 & 29.3 & 14.6 \\ 29.8 & 31.4 & 33.2 & 28.3 & 13.2 \end{bmatrix} & \begin{matrix} 1985 \\ 1990 \\ 1994 \end{matrix} \end{matrix}$$

$$\begin{matrix} & 18\text{--}24 & 25\text{--}34 & 35\text{--}44 & 45\text{--}64 & 65\text{ and over} \\ F = & \begin{bmatrix} 30.4 & 32.0 & 31.5 & 29.9 & 13.5 \\ 22.5 & 28.2 & 24.8 & 24.8 & 11.5 \\ 25.2 & 28.8 & 26.8 & 22.8 & 11.1 \end{bmatrix} & \begin{matrix} 1985 \\ 1990 \\ 1994 \end{matrix} \end{matrix}$$

(a) Interpret the meaning of entry $m_{3,2}$ and entry $f_{3,2}$.

(b) Using M and F, write a matrix in which each entry shows how much more (or less) smoking is done by males than females.

🌐 **60.** The data in Table 2.2.6 give the educational attainment in the United States of people of Hispanic origin 25 years old and over (entries are measured in percentages). HS means high school and CG means college.

Table 2.2.6

	Puerto Rican		Cuban	
	4 yr HS	**4 yr CG or more**	**4 yr HS**	**4 yr CG or more**
1990	55.5	9.7	63.5	20.2
1995	61.3	10.7	64.7	19.4
1996	60.4	11.0	63.8	18.8
1997	61.1	10.7	65.2	19.7
1998	63.8	11.9	67.8	22.2

SOURCE: U.S. Census Bureau, www.census.gov

(a) Write a matrix P for the educational attainment of Puerto Ricans and a matrix C for the educational attainment of Cubans.

(b) Use the matrices from part (a) to write a matrix in which each entry shows how much more (or less) education Cubans have attained than Puerto Ricans.

🌐 **61.** The data in Table 2.2.7 give the percentage of U.S. students in grades 9–12 that participated on a sports team in 1997.

Table 2.2.7

	Male		Female	
Run by:	**School**	**Other organization**	**School**	**Other organization**
Grade 9	57.2	51.3	48.5	36.8
Grade 10	58.0	47.3	45.0	34.7
Grade 11	54.0	41.6	40.7	26.4
Grade 12	53.4	42.6	35.7	21.9

SOURCE: U.S. Census Bureau, www.census.gov

(a) Write a 4×2 matrix M for the participation of males on a sports team.

(b) Write a 4×2 matrix F for the participation of females on a sports team.

(c) Use the matrices from parts (a) and (b) to write a matrix in which each entry shows how much more (or less) participation on sports teams males have than females.

SECTION PROJECT

In the text two Interactive Activities illustrated that matrix addition is commutative and associative. In this project you will discover and verify another property of matrices. Given matrices A and B, do the following:

$$A = \begin{bmatrix} -2 & 3 \\ 4 & 7 \end{bmatrix} \qquad B = \begin{bmatrix} 8 & -1 \\ 2 & 3 \end{bmatrix}$$

(a) $4A$

(b) $4B$

(c) $4A + 4B$

(d) $A + B$

(e) $4(A + B)$

(f) Based on the results to parts (c) and (e), do you think scalar multiplication is distributive?

(g) To verify that scalar multiplication is distributive in the general case, consider

$$A = \begin{bmatrix} a & b \\ c & d \end{bmatrix} \qquad B = \begin{bmatrix} e & f \\ g & h \end{bmatrix}$$

and real number k. Also, a, b, c, d, e, f, g, and h are real numbers. To show that

$$k(A + B) = kA + kB$$

determine a matrix for $k(A + B)$ and a matrix for $kA + kB$. Then use the equality of matrices definition to verify that the matrices are equal.

Section 2.3 Matrix Multiplication

In Section 2.2 we saw how to do scalar multiplication, that is, multiplying a real number k times a matrix. In this section we learn how to multiply two matrices. **Matrix multiplication** has several applications, as we will see in this section. Matrix multiplication will also reappear later in this chapter, as well as in later chapters.

Matrix Multiplication

Multiplying two matrices is more involved than scalar multiplication. To understand the meaning behind matrix multiplication, it may be helpful to take a Flashback to the Staminly's furniture store from Section 2.2.

 Flashback

Staminly's Furniture Revisited

In Section 2.2 we were introduced to Staminly's and its three furniture stores. In Example 2 from Section 2.2, we found a matrix T that gave the total number of two-cushion and three-cushion Country Style Contemporary sofas (of lengths 60″, 72″, and 84″) sold in July for all three stores. This matrix is

$$T = \begin{matrix} & \begin{matrix} 60'' & 72'' & 84'' \end{matrix} & \\ & \begin{bmatrix} 25 & 31 & 21 \\ 21 & 33 & 19 \end{bmatrix} & \begin{matrix} \text{Two cushion} \\ \text{Three cushion} \end{matrix} \end{matrix}$$

From matrix T, write a matrix A for the total number of two-cushion models of each length that were sold and a matrix B for the total number of three-cushion models of each length that were sold.

Flashback Solution

The total number of two-cushion models sold is given by the matrix

$$A = \begin{matrix} 60'' \quad 72'' \quad 84'' \\ \begin{bmatrix} 25 & 31 & 21 \end{bmatrix} \end{matrix} \qquad \text{(a } 1 \times 3 \text{ row matrix)}$$

and the total number of three-cushion models sold is given by the matrix

$$B = \begin{matrix} 60'' \quad 72'' \quad 84'' \\ \begin{bmatrix} 21 & 33 & 19 \end{bmatrix} \end{matrix} \qquad \text{(a } 1 \times 3 \text{ row matrix)}$$

Now suppose that the selling price of the two-cushion models of lengths 60″, 72″, and 84″ is $600, $800, and $1100, respectively. We can represent this information in the following column matrix, P:

$$P = \begin{bmatrix} 600 \\ 800 \\ 1100 \end{bmatrix} \begin{matrix} 60'' \\ 72'' \\ 84'' \end{matrix} \quad \text{(a } 3 \times 1 \text{ column matrix)}$$

The first entry in the row matrix A in the Flashback gives the number of 60″ two-cushion sofas sold, and the first entry in the column matrix P gives the selling price for each 60″ two-cushion sofa. The product of these entries, $25 \cdot 600$, gives the revenue generated from the sale of the 60″ two-cushion sofas in July. Similarly, the second entry in matrix A multiplied by the second entry in matrix P gives the revenue realized from the sale of 72″ two-cushion sofas. Also, the third entry in matrix A multiplied by the third entry in P yields the revenue made from the sale of 84″ two-cushion sofas. So the revenue generated for each type of two-cushion sofa is

$$\begin{aligned} 60'': & \quad 25 \cdot 600 = \$15{,}000 \\ 72'': & \quad 31 \cdot 800 = \$24{,}800 \\ 84'': & \quad 21 \cdot 1100 = \$23{,}100 \end{aligned}$$

The total revenue for July for all three types of two-cushion sofas is the sum of these products, which gives us

$$(25 \cdot 600) + (31 \cdot 800) + (21 \cdot 1100) = \$62{,}900$$

Notice that the revenue, $62,900, was simply found by taking the entries in matrix A and multiplying them by the entries in matrix P and then adding the products. We have actually computed the following matrix multiplication:

$$A \cdot P = \begin{bmatrix} 25 & 31 & 21 \end{bmatrix} \cdot \begin{bmatrix} 600 \\ 800 \\ 1100 \end{bmatrix}$$

$$= (25 \cdot 600) + (31 \cdot 800) + (21 \cdot 1100) = 62{,}900$$

This suggests that if we have a **row matrix** A with dimension $1 \times n$,

$$A = \begin{bmatrix} a_1 & a_2 & a_3 & a_4 & \cdots & a_n \end{bmatrix}$$

and a **column matrix** B with dimension $n \times 1$,

$$B = \begin{bmatrix} b_1 \\ b_2 \\ b_3 \\ b_4 \\ \vdots \\ b_n \end{bmatrix}$$

then the **matrix product** of A and B, AB, is given by

$$AB = \begin{bmatrix} a_1 & a_2 & a_3 & a_4 & \cdots & a_n \end{bmatrix} \begin{bmatrix} b_1 \\ b_2 \\ b_3 \\ b_4 \\ \vdots \\ b_n \end{bmatrix}$$

$$= a_1 b_1 + a_2 b_2 + a_3 b_3 + a_4 b_4 + \cdots + a_n b_n$$

Example 1 Multiplying a Row Matrix and a Column Matrix

The total number of three-cushion Country Style Contemporary sofas (lengths $60''$, $72''$, and $84''$) sold in July by Staminly's is given by matrix B, and the selling price for each of these sofas is given by matrix C.

$$B = \begin{matrix} 60'' & 72'' & 84'' \\ [21 & 33 & 19] \end{matrix} \qquad C = \begin{bmatrix} 650 \\ 850 \\ 1200 \end{bmatrix} \begin{matrix} 60'' \\ 72'' \\ 84'' \end{matrix}$$

Determine the total revenue for July for all three types of three-cushion sofas.

Solution

The total revenue is given by

$$BC = [21 \quad 33 \quad 19] \begin{bmatrix} 650 \\ 850 \\ 1200 \end{bmatrix} = 21(650) + 33(850) + 19(1200) = 64{,}500$$

So the total revenue for July for all three types of three-cushion sofas is $64,500. ∎

If we look closer at the matrix product BC in Example 1, we notice that the number of columns of matrix B is equal to the number of rows of matrix C. Also, the matrix product BC has dimension 1×1 (we can think of a real number as being a 1×1 matrix). See Figure 2.3.1. In general, if matrix A has dimension $m \times n$ and matrix B has dimension $n \times p$, then the matrix product of A and B, AB, is defined and has dimension $m \times p$. See Figure 2.3.2.

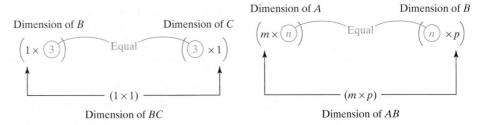

Figure 2.3.1	Figure 2.3.2

Example 2 Determining the Dimension of a Matrix Product

Given matrices A, B, and C, determine if the following matrix products are defined. If so, give the dimension of the matrix product.

$$A = \begin{bmatrix} 2 & 3 & -1 & 7 \\ 5 & 6 & 2 & 1 \end{bmatrix} \qquad B = \begin{bmatrix} 1 & 3 \\ -1 & 4 \\ 2 & -2 \end{bmatrix} \qquad C = \begin{bmatrix} 2 & 1 \\ -1 & 3 \\ 4 & -2 \\ 7 & -3 \end{bmatrix}$$

(a) AB **(b)** BA **(c)** AC

✓ **Checkpoint 1**

Solution

(a) A is a 2×4 matrix and B is a 3×2 matrix. Since the number of columns of A *does not equal* the number of rows of B, the matrix product AB does not exist.

(b) B is a 3×2 matrix and A is a 2×4 matrix. Since the number of columns of B *does equal* the number of rows of A, the matrix product BA is defined and would yield a 3×4 matrix.

(c) A is a 2×4 matrix and C is a 4×2 matrix. As in part (b), we see that the matrix product AC is defined and would yield a 2×2 matrix. ∎

Now work Exercise 5.

Example 2(b) indicates that BA is defined and yields a 3×4 matrix. If we call the matrix product D, we have

$$BA = D = \begin{bmatrix} d_{1,1} & d_{1,2} & d_{1,3} & d_{1,4} \\ d_{2,1} & d_{2,2} & d_{2,3} & d_{2,4} \\ d_{3,1} & d_{3,2} & d_{3,3} & d_{3,4} \end{bmatrix}$$

$$\begin{bmatrix} 1 & 3 \\ -1 & 4 \\ 2 & -2 \end{bmatrix} \begin{bmatrix} 2 & 3 & -1 & 7 \\ 5 & 6 & 2 & 1 \end{bmatrix} = D = \begin{bmatrix} d_{1,1} & d_{1,2} & d_{1,3} & d_{1,4} \\ d_{2,1} & d_{2,2} & d_{2,3} & d_{2,4} \\ d_{3,1} & d_{3,2} & d_{3,3} & d_{3,4} \end{bmatrix}$$

To compute the entries of D, we do the following:

$$d_{1,1} = [\,b_{1,1} \quad b_{1,2}\,] \begin{bmatrix} a_{1,1} \\ a_{2,1} \end{bmatrix} = [1 \quad 3] \begin{bmatrix} 2 \\ 5 \end{bmatrix} = 1(2) + 3(5) = 17$$

$$d_{1,2} = [\,b_{1,1} \quad b_{1,2}\,] \begin{bmatrix} a_{1,2} \\ a_{2,2} \end{bmatrix} = [1 \quad 3] \begin{bmatrix} 3 \\ 6 \end{bmatrix} = 1(3) + 3(6) = 21$$

The other entries in D are computed in a similar fashion. Matrix products are summarized in the following definition.

■ Matrix Product

If A is an $m \times n$ matrix and B is an $n \times p$ matrix, then the **matrix product**, AB, is an $m \times p$ matrix. To find the entry of the ith row and jth column of AB, multiply each entry in the ith row of A by the corresponding entry in the jth column of B and add the products.

▶ **Note:** Observe that determining the entries in a matrix product is analogous to taking a row matrix times a column matrix, as we did in Example 1.

Example 3 Computing a Matrix Product

Determine AB given

$$A = \begin{bmatrix} 2 & -1 & -3 \\ 4 & 1 & 6 \end{bmatrix} \qquad B = \begin{bmatrix} 3 & 1 \\ -1 & 4 \\ 5 & 6 \end{bmatrix}$$

Solution

The first thing that we should always check is whether the matrix product is defined. Since A is a 2×3 and B is a 3×2, the matrix product is defined and yields a 2×2 matrix. If we call the matrix product C, we have

$$AB = C$$

$$\begin{bmatrix} 2 & -1 & -3 \\ 4 & 1 & 6 \end{bmatrix} \begin{bmatrix} 3 & 1 \\ -1 & 4 \\ 5 & 6 \end{bmatrix} = C = \begin{bmatrix} c_{1,1} & c_{1,2} \\ c_{2,1} & c_{2,2} \end{bmatrix}$$

$c_{1,1}$ is found by taking the 1st row of A and multiplying by the 1st column of B.

$$c_{1,1} = \begin{bmatrix} 2 & -1 & -3 \end{bmatrix} \begin{bmatrix} 3 \\ -1 \\ 5 \end{bmatrix} = 2(3) + -1(-1) + -3(5) = -8$$

$c_{1,2}$ is found by taking the 1st row of A and multiplying by the 2nd column of B.

$$c_{1,2} = \begin{bmatrix} 2 & -1 & -3 \end{bmatrix} \begin{bmatrix} 1 \\ 4 \\ 6 \end{bmatrix} = 2(1) + -1(4) + -3(6) = -20$$

$c_{2,1}$ is found by taking the 2nd row of A and multiplying by the 1st column of B.

$$c_{2,1} = \begin{bmatrix} 4 & 1 & 6 \end{bmatrix} \begin{bmatrix} 3 \\ -1 \\ 5 \end{bmatrix} = 4(3) + 1(-1) + 6(5) = 41$$

$c_{2,2}$ is found by taking the 2nd row of A and multiplying by the 2nd column of B.

$$c_{2,2} = \begin{bmatrix} 4 & 1 & 6 \end{bmatrix} \begin{bmatrix} 1 \\ 4 \\ 6 \end{bmatrix} = 4(1) + 1(4) + 6(6) = 44$$

So we have

$$AB = \begin{bmatrix} -8 & -20 \\ 41 & 44 \end{bmatrix}$$ ∎

Technology Option

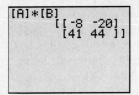

Figure 2.3.3 shows the matrix product from Example 3 on a graphing calculator. Many calculators perform matrix multiplication. To learn how your calculator multiplies matrices, consult the online graphing calculator manual at www.prenhall.com/armstrong

Figure 2.3.3

✓ **Checkpoint 2**   Now work Exercise 15.

Example 4   **Computing a Matrix Product**

Determine the matrix products AB and BA for the matrices

$$A = \begin{bmatrix} 2 & 3 & 1 \\ -1 & 2 & 3 \\ 3 & 4 & 1 \end{bmatrix} \qquad B = \begin{bmatrix} 4 & -1 & 3 \\ 2 & 1 & 4 \\ 1 & 3 & 2 \end{bmatrix}$$

Solution

We have

$$AB = \begin{bmatrix} 2 & 3 & 1 \\ -1 & 2 & 3 \\ 3 & 4 & 1 \end{bmatrix} \begin{bmatrix} 4 & -1 & 3 \\ 2 & 1 & 4 \\ 1 & 3 & 2 \end{bmatrix}$$

$$= \begin{bmatrix} 2(4)+3(2)+1(1) & 2(-1)+3(1)+1(3) & 2(3)+3(4)+1(2) \\ -1(4)+2(2)+3(1) & -1(-1)+2(1)+3(3) & -1(3)+2(4)+3(2) \\ 3(4)+4(2)+1(1) & 3(-1)+4(1)+1(3) & 3(3)+4(4)+1(2) \end{bmatrix}$$

$$= \begin{bmatrix} 15 & 4 & 20 \\ 3 & 12 & 11 \\ 21 & 4 & 27 \end{bmatrix}$$

For BA we have

$$BA = \begin{bmatrix} 4 & -1 & 3 \\ 2 & 1 & 4 \\ 1 & 3 & 2 \end{bmatrix} \begin{bmatrix} 2 & 3 & 1 \\ -1 & 2 & 3 \\ 3 & 4 & 1 \end{bmatrix}$$

We save the details of this multiplication for you in the Interactive Activity to the left. The matrix product is

$$BA = \begin{bmatrix} 18 & 22 & 4 \\ 15 & 24 & 9 \\ 5 & 17 & 12 \end{bmatrix}$$

Interactive Activity

By hand, perform the details of computing BA in Example 4. Check AB and BA using a graphing calculator.

Example 5   **Applying Matrix Multiplication**

Leslie, the director of KiddieTime a local day-care center, needs to get the monthly billing under control. Children at the center are there either 2, 4, or 8 hours. Some attend daily, for different hours each day, and some are there only certain days of the week. To further complicate matters, months do not have the same number of Mondays, Tuesdays, and so on, and this is further complicated for months with holidays. Leslie's task of creating a monthly billing system must account for the individual schedule of each child, the number of Mondays, Tuesdays, and so on, in a given month, and the cost per hour for day care. (The latter will probably rise over time.) Leslie believes that matrices can help her, and she selected three children and their schedules for the month of November 2000 to see. Matrix C is the children's daily schedule in which the entries represent the number of hours spent in day care for a particular day. Matrix M gives the number of Mondays, Tuesdays, and so on, that the day care was open in November

2000. Currently, the cost is $4 per hour. How can Leslie use this information to create a billing system?

$$C = \begin{array}{c} \\ \end{array} \begin{array}{ccccc} \text{M} & \text{T} & \text{W} & \text{Th} & \text{F} \end{array}$$

$$C = \begin{bmatrix} 4 & 0 & 4 & 0 & 2 \\ 8 & 4 & 8 & 4 & 2 \\ 0 & 8 & 2 & 8 & 0 \end{bmatrix} \begin{array}{l} \text{Don} \\ \text{Ron} \\ \text{Sarah} \end{array} \qquad M = \begin{bmatrix} 4 \\ 4 \\ 5 \\ 4 \\ 3 \end{bmatrix} \begin{array}{l} \text{M} \\ \text{T} \\ \text{W} \\ \text{Th} \\ \text{F} \end{array}$$

Solution

Understand the Situation: Leslie needs to use the given information and determine the total number of hours that each child was in day care. Then she can multiply by $4 to find the bill for each child.

The matrix product CM will give the total hours in November 2000 for each child.

$$CM = \begin{bmatrix} 4 & 0 & 4 & 0 & 2 \\ 8 & 4 & 8 & 4 & 2 \\ 0 & 8 & 2 & 8 & 0 \end{bmatrix} \begin{bmatrix} 4 \\ 4 \\ 5 \\ 4 \\ 3 \end{bmatrix}$$

$$= \begin{bmatrix} 4(4) + 0(4) + 4(5) + 0(4) + 2(3) \\ 8(4) + 4(4) + 8(5) + 4(4) + 2(3) \\ 0(4) + 8(4) + 2(5) + 8(4) + 0(3) \end{bmatrix} = \begin{bmatrix} 42 \\ 110 \\ 74 \end{bmatrix} \begin{array}{l} \text{Don} \\ \text{Ron} \\ \text{Sarah} \end{array}$$

To determine the November 2000 bill for each child, we simply multiply the matrix product CM by $4. This gives

$$4CM = 4 \begin{bmatrix} 42 \\ 110 \\ 74 \end{bmatrix} = \begin{bmatrix} 168 \\ 440 \\ 296 \end{bmatrix} \begin{array}{l} \text{Don} \\ \text{Ron} \\ \text{Sarah} \end{array}$$

Interpret the Solution: For November 2000, Don owes $168, Ron owes $440, and Sarah owes $296. It appears that Leslie can enter into matrix C all the children in the day-care center and their daily hours of attendance. In matrix M she can put the number of Mondays, Tuesdays, and so on, that the center is open. The hourly rate multiplied by CM will produce the monthly bill for each child. ∎

In our next example we compute a matrix product in which some entries are meaningful while other entries are meaningless. It is important to always remember the definition of matrix multiplication so that we know when a matrix product has meaning in an application.

Example 6 Applying Matrix Multiplication

Recall that for Staminly's three furniture stores, the total sales in July for its two- and three-cushion Country Style Contemporary sofa (lengths of 60″, 72″, and 84″) are represented in matrix T.

$$T = \begin{array}{c} \begin{array}{ccc} 60'' & 72'' & 84'' \end{array} \\ \begin{bmatrix} 25 & 31 & 21 \\ 21 & 33 & 19 \end{bmatrix} \begin{array}{l} \text{Two cushion} \\ \text{Three cushion} \end{array} \end{array}$$

The selling price for each is given in matrix S.

$$S = \begin{bmatrix} 600 & 650 \\ 800 & 850 \\ 1100 & 1200 \end{bmatrix} \begin{matrix} 60'' \\ 72'' \\ 84'' \end{matrix}$$

(Two cushion, Three cushion)

Compute TS and interpret the entries.

Solution

The matrix product TS is

$$TS = \begin{bmatrix} 25 & 31 & 21 \\ 21 & 33 & 19 \end{bmatrix} \begin{bmatrix} 600 & 650 \\ 800 & 850 \\ 1100 & 1200 \end{bmatrix}$$

$$= \begin{bmatrix} 62,900 & 67,800 \\ 59,900 & 64,500 \end{bmatrix}$$

Interpret the Solution: $ts_{1,1}$ gives the total revenue for the three lengths of the two-cushion model to be $62,900. Now $ts_{1,2}$ is the product of the number of two-cushion models sold times the prices for the three-cushion models; hence this entry has no meaning. For the same reason, $ts_{2,1}$ is also meaningless. Finally, $ts_{2,2}$ gives the total revenue for the three lengths of the three-cushion model to be $64,500. ∎

Technology Option

```
[A]*[B]
[[62900 67800]
 [59900 64500]]
```

In Example 6, if we use a graphing calculator to do the matrix multiplication, with matrix T called A and matrix S called B, we get the result in Figure 2.3.4.

Figure 2.3.4

Whether matrix multiplication is done by hand or on a graphing calculator, it is imperative to know the definition of matrix multiplication so that we can interpret the entries meaningfully!

Representing Systems with a Matrix Equation

Our final topic in this section is to lay the groundwork for Section 2.4. In Example 7 we show how to use matrices to rewrite a system of linear equations in a more compact form called a **matrix equation**.

Example 7   Writing a Matrix Equation for a System of Equations

Write the following system of linear equations as a matrix equation.

$$2x + y + 3z = 4$$
$$x - 2y + z = -3$$
$$4x - y - z = -1$$

Solution

We start by writing

$$A = \begin{bmatrix} 2 & 1 & 3 \\ 1 & -2 & 1 \\ 4 & -1 & -1 \end{bmatrix} \qquad X = \begin{bmatrix} x \\ y \\ z \end{bmatrix} \qquad B = \begin{bmatrix} 4 \\ -3 \\ -1 \end{bmatrix}$$

Observe that A is a 3×3 matrix of coefficients, X is a 3×1 column matrix of the variables, and B is a 3×1 column matrix of the constants. The matrix equation for this system has the form

$$AX = B$$

$$\begin{bmatrix} 2 & 1 & 3 \\ 1 & -2 & 1 \\ 4 & -1 & -1 \end{bmatrix} \begin{bmatrix} x \\ y \\ z \end{bmatrix} = \begin{bmatrix} 4 \\ -3 \\ -1 \end{bmatrix}$$

To verify that this is true, we perform the matrix product AX and get

$$AX = \begin{bmatrix} 2 & 1 & 3 \\ 1 & -2 & 1 \\ 4 & -1 & -1 \end{bmatrix} \begin{bmatrix} x \\ y \\ z \end{bmatrix} = \begin{bmatrix} 2x + y + 3z \\ x - 2y + z \\ 4x - y - z \end{bmatrix}$$

Equating this product with matrix B gives

$$\begin{bmatrix} 2x + y + 3z \\ x - 2y + z \\ 4x - y - z \end{bmatrix} = \begin{bmatrix} 4 \\ -3 \\ -1 \end{bmatrix}$$

which via matrix equality is obviously the same as the original system. ■

✓ **Checkpoint 3**   Now work Exercise 39.

SUMMARY

The section started with multiplying a row matrix by a column matrix, and that lead us to defining **matrix multiplication** in general. It is important to know when matrix multiplication is defined and the dimension of the resulting product. For example, if A is $m \times n$ and B is $n \times p$, then the product AB is defined and has dimension $m \times p$, whereas the product BA is not defined. We saw several applications of matrix multiplication, and the section concluded with a preview of the next section: representing systems of linear equations with a matrix equation.

SECTION 2.3 EXERCISES

In Exercises 1–8, the dimension of matrices A and B are given.

(a) Find the dimension of the matrix product AB, if it exists.

(b) Find the dimension of the matrix product BA, if it exists.

1. A is a 3×2 and B is a 2×3

2. A is a 2×4 and B is a 4×2

3. A is a 2×4 and B is a 4×2

4. A is a 5×2 and B is a 2×3

✓ **5.** A is a 3×4 and B is a 5×3

6. A is a 3×2 and B is a 4×3

7. A is a 5×2 and B is a 3×3

8. A is a 6×2 and B is a 4×2

In Exercises 9–24, find the matrix product, if it exists.

9. $\begin{bmatrix} 3 & 1 \\ 4 & 2 \end{bmatrix} \begin{bmatrix} -2 \\ 3 \end{bmatrix}$

10. $\begin{bmatrix} -3 & 2 \\ 4 & 1 \end{bmatrix} \begin{bmatrix} 2 \\ 3 \end{bmatrix}$

11. $\begin{bmatrix} 1 & 3 & -2 \\ -1 & 2 & 4 \end{bmatrix} \begin{bmatrix} -2 \\ 2 \\ 1 \end{bmatrix}$

12. $\begin{bmatrix} 6 & -1 & 2 \\ 3 & 4 & 1 \end{bmatrix} \begin{bmatrix} 0 \\ 4 \\ 5 \end{bmatrix}$

13. $\begin{bmatrix} 2 & -2 \\ 1 & -3 \end{bmatrix} \begin{bmatrix} 4 & -3 \\ -1 & 2 \end{bmatrix}$

14. $\begin{bmatrix} -1 & 2 \\ 3 & -4 \end{bmatrix} \begin{bmatrix} -2 & 3 \\ -1 & 5 \end{bmatrix}$

✓ **15.** $\begin{bmatrix} 3 & 2 & 0 \\ -1 & -2 & 4 \end{bmatrix} \begin{bmatrix} 6 & 1 \\ 5 & -1 \\ 2 & 4 \end{bmatrix}$

16. $\begin{bmatrix} 1 & 6 \\ -1 & -5 \\ 2 & -4 \end{bmatrix} \begin{bmatrix} -3 & 1 & -1 \\ 1 & 4 & -2 \end{bmatrix}$

17. $\begin{bmatrix} 2 & -3 \\ 4 & 1 \\ 0 & -4 \\ -1 & 5 \end{bmatrix} \begin{bmatrix} 2 \\ 1 \\ 4 \\ 3 \end{bmatrix}$

18. $\begin{bmatrix} -1 & 3 & 2 & -4 \end{bmatrix} \begin{bmatrix} -1 & 1 \\ 2 & 3 \end{bmatrix}$

19. $\begin{bmatrix} 5 & -6 \\ -2 & 1 \end{bmatrix} \begin{bmatrix} 1 & 0 \\ 0 & 1 \end{bmatrix}$

20. $\begin{bmatrix} 1 & 0 \\ 0 & 1 \end{bmatrix} \begin{bmatrix} 5 & -6 \\ -2 & 1 \end{bmatrix}$

21. $\begin{bmatrix} 6 & -1 & 4 \\ 0 & 3 & 2 \\ 1 & 4 & 5 \end{bmatrix} \begin{bmatrix} 1 & 0 & 0 \\ 0 & 1 & 0 \\ 0 & 0 & 1 \end{bmatrix}$

22. $\begin{bmatrix} 1 & 0 & 0 \\ 0 & 1 & 0 \\ 0 & 0 & 1 \end{bmatrix} \begin{bmatrix} 6 & -1 & 4 \\ 0 & 3 & 2 \\ 1 & 4 & 5 \end{bmatrix}$

23. $\begin{bmatrix} 10 & 20 & 15 & 5 \\ 6 & 8 & 6 & 4 \\ 5 & 5 & 15 & 30 \\ 20 & 18 & 22 & 16 \end{bmatrix} \begin{bmatrix} 5 \\ 8 \\ 11 \\ 13 \end{bmatrix}$

24. $\begin{bmatrix} 6 & 9 & 18 \\ 15 & 12 & 10 \\ 8 & 6 & 9 \end{bmatrix} \begin{bmatrix} 4 \\ 6 \\ 7 \end{bmatrix}$

In Exercises 25–32, use a graphing calculator to find the matrix products, if possible.

25. $\begin{bmatrix} 1.3 & 1.4 \\ 2.8 & 3.2 \end{bmatrix} \begin{bmatrix} 5.4 & 6.1 \\ 8.9 & 3.5 \end{bmatrix}$

26. $\begin{bmatrix} 0.7 & 1.3 \\ 0.4 & 9.1 \end{bmatrix} \begin{bmatrix} 8.4 & 7.3 \\ 6.1 & 5.9 \end{bmatrix}$

27. $\begin{bmatrix} \frac{4}{11} & -\frac{3}{22} \\ -\frac{3}{11} & \frac{5}{22} \end{bmatrix} \begin{bmatrix} 5 & 3 \\ 6 & 8 \end{bmatrix}$

28. $\begin{bmatrix} -1.5 & 2 \\ -0.5 & 1 \end{bmatrix} \begin{bmatrix} -2 & 4 \\ -1 & 3 \end{bmatrix}$

29. $\begin{bmatrix} 21 & 33 & 58 \\ 17 & 25 & 13 \\ 42 & 20 & 37 \end{bmatrix} \begin{bmatrix} 700 \\ 800 \\ 900 \end{bmatrix}$

30. $\begin{bmatrix} 19 & 17 & 24 \\ 64 & 33 & 36 \\ 56 & 90 & 18 \end{bmatrix} \begin{bmatrix} 20 \\ 50 \\ 82 \end{bmatrix}$

31. $\begin{bmatrix} 2 & 3 & 2 \\ -1 & 5 & 6 \\ 2 & -3 & 4 \end{bmatrix} \begin{bmatrix} \frac{19}{55} & -\frac{9}{55} & \frac{4}{55} \\ \frac{8}{55} & \frac{2}{55} & -\frac{7}{55} \\ -\frac{7}{110} & \frac{6}{55} & \frac{13}{110} \end{bmatrix}$

32. $\begin{bmatrix} 1 & 5 & -2 \\ 2 & 3 & 4 \\ -1 & -3 & 1 \end{bmatrix} \begin{bmatrix} -\frac{5}{3} & -\frac{1}{9} & -\frac{26}{9} \\ \frac{2}{3} & \frac{1}{9} & \frac{8}{9} \\ \frac{1}{3} & \frac{2}{9} & \frac{7}{9} \end{bmatrix}$

33. Consider $A = \begin{bmatrix} 0.65 & 0.35 \\ 0.71 & 0.29 \end{bmatrix}$

(a) Use a graphing calculator to compute A^2, A^5, A^{10}, and A^{20}.

(b) Describe any pattern observed in part (a)

34. Consider $B = \begin{bmatrix} 0.37 & 0.63 \\ 0.25 & 0.75 \end{bmatrix}$

(a) Use a graphing calculator to compute B^2, B^5, B^{10}, and B^{20}.

(b) Describe any pattern observed in part (a)

In Exercises 35–42 write the system of linear equations as a matrix equation. See Example 7.

35. $\begin{aligned} 3x - 2y &= 14 \\ 2x - 5y &= 8 \end{aligned}$

36. $\begin{aligned} x + 3y &= -11 \\ 2x - 7y &= 4 \end{aligned}$

37. $\begin{aligned} x + y &= 4 \\ -3x - 2y &= -5 \end{aligned}$

38. $\begin{aligned} 7x - 8y &= -9 \\ -2x - 5y &= 6 \end{aligned}$

✓ **39.** $\begin{aligned} x + 2y + 2z &= 11 \\ x - y - z &= -4 \\ 2x + 5y + 9z &= 39 \end{aligned}$

40. $\begin{aligned} x - 3y + z &= 5 \\ -2x + 7y - 6z &= -9 \\ x + 2y - 3z &= 6 \end{aligned}$

41. $\begin{aligned} -2x + 3y + 4z &= 10 \\ 3x - 5y + 2z &= 7 \\ 4x + 2y - 3z &= -2 \end{aligned}$

42. $\begin{aligned} x + 2y - 4z &= 5 \\ 3x - y + z &= 3 \\ x + 2y + z &= -3 \end{aligned}$

Applications

43. Hilltop Motors has two dealerships, East and West. The inventory for its three most popular selling sports utility vehicles (simply called A, B, and C) at each dealership is given in matrix M.

$$M = \begin{matrix} & \begin{matrix} \text{A} & \text{B} & \text{C} \end{matrix} & \\ \begin{bmatrix} 15 & 22 & 19 \\ 17 & 18 & 23 \end{bmatrix} & \begin{matrix} \text{East} \\ \text{West} \end{matrix} \end{matrix}$$

The wholesale (W) and retail (R) values of each model of vehicle are given in matrix P.

$$P = \begin{bmatrix} 15{,}000 & 22{,}000 \\ 17{,}500 & 26{,}100 \\ 18{,}700 & 28{,}000 \end{bmatrix} \begin{matrix} A \\ B \\ C \end{matrix}$$

with column headers W, R.

(a) Find MP.

(b) Interpret the entries in MP.

44. Holobinko's Auto has three dealerships, North, South, and Metro. The inventory for its four most popular selling vehicles (we will call them simply A, B, C, and D) at each dealership is given in matrix M.

$$M = \begin{bmatrix} 10 & 12 & 8 & 15 \\ 13 & 7 & 11 & 12 \\ 17 & 11 & 5 & 13 \end{bmatrix} \begin{matrix} \text{North} \\ \text{South} \\ \text{Metro} \end{matrix}$$

with column headers A, B, C, D.

The wholesale (W) and retail (R) values of each model of car are given in matrix P.

$$P = \begin{bmatrix} 8000 & 11{,}000 \\ 8900 & 12{,}200 \\ 10{,}100 & 15{,}600 \\ 9500 & 13{,}000 \end{bmatrix} \begin{matrix} A \\ B \\ C \\ D \end{matrix}$$

with column headers W, R.

(a) Find MP.

(b) Interpret the entries in MP.

45. The Stumpf Company has two manufacturing plants, one on the East Coast and one on the West Coast, that produce three types of tents. The labor hours and wage requirements for the manufacture of these three types of tents are given in matrices L and W, respectively.

$$L = \begin{bmatrix} 0.5 & 0.6 & 0.2 \\ 0.8 & 0.7 & 0.3 \\ 1.1 & 0.9 & 0.6 \end{bmatrix} \begin{matrix} \text{Two person} \\ \text{Four person} \\ \text{Six person} \end{matrix}$$

with column headers Fabric cutting, Assembly, Packaging.

and

$$W = \begin{bmatrix} 13 & 14 \\ 15 & 15 \\ 10 & 11 \end{bmatrix} \begin{matrix} \text{Fabric cutting} \\ \text{Assembly} \\ \text{Packaging} \end{matrix}$$

with column headers East Coast, West Coast.

The entries in L are in hours and the entries in W are in dollars.

(a) Find LW.

(b) Determine the labor costs for a two-person tent manufactured on the West Coast.

(c) Determine the labor costs for a six-person tent manufactured on the East Coast.

(d) Interpret the entries in LW.

46. (*continuation of Exercise 45*) Dave, the CEO of the Stumpf Company, decides to give all employees a 10% raise.

(a) Use this information to construct a new wage matrix W_2.

(b) Find LW_2.

(c) Determine the labor costs for a two-person tent manufactured on the West Coast.

(d) Determine the labor costs for a six-person tent manufactured on the East Coast.

47. Aida, a dietitian in a local hospital, plans a special diet for a patient that is made up of three basic foods, Guhde, Vita, and Zaca. The number of units of calcium, vitamin A, and vitamin C, per ounce, is given in matrix F.

$$F = \begin{bmatrix} 20 & 10 & 30 \\ 10 & 5 & 20 \\ 20 & 15 & 10 \end{bmatrix} \begin{matrix} \text{Calcium} \\ \text{Vitamin A} \\ \text{Vitamin C} \end{matrix}$$

with column headers Guhde, Vita, Zaca.

At lunch the patient had 8 ounces of Guhde, 5 ounces of Vita, and 9 ounces of Zaca. For dinner the patient had 4 ounces of Guhde, 5 ounces of Vita, and 7 ounces of Zaca.

(a) Write a 3×1 matrix A that represents the amount of each food consumed (in ounces) at lunch.

(b) Write a 3×1 matrix B that represents the amount of each food consumed (in ounces) at dinner.

(c) Find FA and interpret the entries.

(d) Find FB and interpret the entries.

48. Manuel, a dietitian in a local hospital, plans a special diet for a patient that is made up of three basic foods, Veni, Gala, and Mogi. The number of units of calcium, vitamin A, vitamin C, and fiber, per ounce, is given in matrix F.

$$F = \begin{bmatrix} 10 & 15 & 30 \\ 5 & 20 & 10 \\ 10 & 30 & 15 \\ 10 & 5 & 5 \end{bmatrix} \begin{matrix} \text{Calcium} \\ \text{Vitamin A} \\ \text{Vitamin C} \\ \text{Fiber} \end{matrix}$$

with column headers Veni, Gala, Mogi.

At lunch the patient had 8 ounces of Veni, 9 ounces of Gala, and 5 ounces of Mogi. For dinner the patient had 3 ounces of Veni, 5 ounces of Gala, and 7 ounces of Mogi.

(a) Write a 3×1 matrix A that represents the amount of each food consumed (in ounces) at lunch.

(b) Write a 3×1 matrix B that represents the amount of each food consumed (in ounces) at dinner.

(c) Find FA and interpret the entries.

(d) Find FB and interpret the entries.

49. The amounts of nitrogen, phosphate, and potash needed to grow artichokes, cabbage, and carrots for an entire growing season in California are given in matrix F, in lb/acre (*Source:* U.S. Department of Agriculture, www.usda.gov/nass).

$$F = \begin{bmatrix} & \text{Arti-} & & \\ & \text{chokes} & \text{Cabbage} & \text{Carrots} \\ 197 & 203 & 249 \\ 123 & 86 & 228 \\ 123 & 78 & 67 \end{bmatrix} \begin{matrix} \text{Nitrogen} \\ \text{Phosphates} \\ \text{Potash} \end{matrix}$$

Matrix P represents the number of acres of each crop that Francis wishes to plant.

$$P = \begin{bmatrix} 30 \\ 40 \\ 130 \end{bmatrix} \begin{matrix} \text{Artichokes} \\ \text{Cabbage} \\ \text{Carrots} \end{matrix}$$

(a) Find FP.

(b) How many pounds each of nitrogen, phosphate, and potash will Francis need for the entire growing season?

50. (*continuation of Exercise 49*) Matrix T gives the number of acres of each crop planted in 1990 in California (*Source:* U.S. Department of Agriculture, www.usda.gov/nass).

$$T = \begin{bmatrix} 10{,}000 \\ 11{,}200 \\ 56{,}100 \end{bmatrix} \begin{matrix} \text{Artichokes} \\ \text{Cabbage} \\ \text{Carrots} \end{matrix}$$

(a) Find FT.

(b) In 1990, how many pounds each of nitrogen, phosphate, and potash were used in California for an entire growing season to grow these three crops?

51. The amounts of three pollutants, measured in thousands of tons, released into the atmosphere during 1985, 1990, and 1995 by the United States are given in matrix A (*Source:* U.S. Census Bureau, www.census.gov).

$$A = \begin{bmatrix} & 1985 & 1990 & 1995 \\ 45{,}584 & 29{,}844 & 26{,}760 \\ 23{,}230 & 23{,}678 & 19{,}189 \\ 23{,}488 & 23{,}436 & 23{,}768 \end{bmatrix} \begin{matrix} \text{Particulates} \\ \text{Sulfur dioxide} \\ \text{Nitrogen oxides} \end{matrix}$$

Matrix B gives the percent, as a decimal, of these pollutants due to electric utilities.

$$\begin{matrix} \text{Parti-} & \text{Sulfur} & \text{Nitrogen} \\ \text{culates} & \text{dioxide} & \text{oxides} \end{matrix}$$
$$B = [\,0.01 \quad 0.64 \quad 0.26\,]$$

Here for example, 0.01 means 1% of particulate air pollution is caused by electric companies. Compute BA and interpret the entries.

52. The amounts of waste generated, in million of tons, for various items in the United States during 1990, 1992, 1994, and 1996 is given in matrix A (*Source:* U.S. Census Bureau, www.census.gov).

$$A = \begin{bmatrix} 1990 & 1992 & 1994 & 1996 \\ 72.7 & 74.3 & 80.8 & 79.7 \\ 2.8 & 2.9 & 3.0 & 3.0 \\ 13.1 & 13.1 & 13.4 & 12.3 \\ 17.1 & 18.4 & 19.3 & 19.8 \end{bmatrix} \begin{matrix} \text{Paper and paper board} \\ \text{Aluminum} \\ \text{Glass} \\ \text{Plastics} \end{matrix}$$

Matrix B gives the percent, as a decimal, of these items that were recovered or recycled over these 4 years.

$$\begin{matrix} P_P & A & G & Pl \end{matrix}$$
$$B = [\,0.35 \quad 0.37 \quad 0.23 \quad 0.04\,]$$

Here 0.35 means 35% of paper and paperboard was recovered. Compute BA and interpret the entries.

53. Denyse and Jamie's stock holdings for the companies CGC, BA, and DD are given in matrix A. The entries in A represent the number of shares owned.

$$A = \begin{bmatrix} \text{CGC} & \text{BA} & \text{DD} \\ 250 & 300 & 150 \\ 300 & 350 & 200 \end{bmatrix} \begin{matrix} \text{Denyse} \\ \text{Jamie} \end{matrix}$$

At the close of trading on December 31, 2000, the prices in dollars per share of the stocks are as given in matrix P.

$$P = \begin{bmatrix} 27 \\ 81 \\ 67 \end{bmatrix} \begin{matrix} \text{CGC} \\ \text{BA} \\ \text{DD} \end{matrix}$$

(a) Find AP.

(b) Use the entries of AP to determine the value of Denyse's stock holdings and the value of Jamie's stock holdings on December 31, 2000.

54. (*continuation of Exercise 53*) Suppose that on December 31, 2001, Denyse and Jamie noticed that the prices of their stocks had increased by 15%.

(a) Determine a new matrix P representing the prices in dollars per share of their stocks.

(b) Assuming A from Exercise 53 and P from part (a), determine AP.

(c) Use the result from part (b) to determine the value of Denyse's stock holdings and the value of Jamie's stock holdings on December 31, 2001.

55. Peggy, a mathematics professor, has given three tests to a class of five students. Peggy records their scores in matrix S.

$$S = \begin{bmatrix} & \text{Test} & \\ 1 & 2 & 3 \\ 88 & 78 & 82 \\ 89 & 81 & 91 \\ 92 & 93 & 89 \\ 96 & 66 & 98 \\ 72 & 92 & 81 \end{bmatrix} \begin{matrix} \text{Josh} \\ \text{Juan} \\ \text{Maria} \\ \text{Diane} \\ \text{Zach} \end{matrix}$$

Let $C = \begin{bmatrix} 1 \\ 1 \\ 1 \end{bmatrix}$ and $R = [\,1 \quad 1 \quad 1 \quad 1 \quad 1\,]$.

(a) Compute $\frac{1}{3}(SC)$ and interpret the entries.

(b) Compute $\frac{1}{5}(RS)$ and interpret the entries.

56. Susan, a biology professor, has given three tests to a class of six students. Susan records their scores in matrix S.

$$S = \begin{bmatrix} 82 & 78 & 91 \\ 80 & 76 & 82 \\ 94 & 92 & 89 \\ 82 & 92 & 98 \\ 76 & 86 & 96 \\ 90 & 76 & 72 \end{bmatrix} \begin{matrix} \text{Paul} \\ \text{Sue} \\ \text{Brent} \\ \text{Megan} \\ \text{Teri} \\ \text{Steve} \end{matrix}$$

with column headings "Test 1 2 3".

Let $C = \begin{bmatrix} 1 \\ 1 \\ 1 \end{bmatrix}$ and $R = [1 \quad 1 \quad 1 \quad 1 \quad 1 \quad 1]$.

(a) Compute $\frac{1}{3}(SC)$ and interpret the entries.

(b) Compute $\frac{1}{6}(RS)$ and interpret the entries.

57. Bill is a beagle breeder. He uses two feeds, Octav and Zoso, to make three different mixtures of the perfect beagle food. The three mixtures are shown in matrix F (the entries in F are in ounces).

$$F = \begin{bmatrix} 30 & 15 & 10 \\ 10 & 15 & 20 \end{bmatrix} \begin{matrix} \text{Octav} \\ \text{Zoso} \end{matrix}$$

with column headings "Mix I Mix II Mix III".

The nutritional value of each of the two feeds is given in matrix N (the entries are in grams/ounce).

$$N = \begin{bmatrix} 6 & 4 \\ 15 & 12 \\ 20 & 15 \end{bmatrix} \begin{matrix} \text{Protein} \\ \text{Carbohydrates} \\ \text{Fat} \end{matrix}$$

with column headings "Octav Zoso".

(a) Find NF.

(b) Use part (a) to determine the amount of protein in mix II.

(c) Use part (a) to determine the amount of carbohydrates in mix I.

(d) Interpret the entries in NF.

58. (*continuation of Exercise 57*) Bill, the beagle breeder, introduces a new feed, Zipz, and creates a fourth mixture. The four mixtures are shown in matrix F (the entries are in ounces).

$$F = \begin{bmatrix} 15 & 10 & 10 & 20 \\ 10 & 10 & 10 & 5 \\ 5 & 10 & 10 & 5 \end{bmatrix} \begin{matrix} \text{Octav} \\ \text{Zoso} \\ \text{Zipz} \end{matrix}$$

with column headings "Mix I Mix II Mix III Mix IV".

The nutritional values of each of the three feeds are given in matrix N (the entries are in grams/ounce).

$$N = \begin{bmatrix} 6 & 4 & 8 \\ 15 & 12 & 15 \\ 20 & 15 & 12 \end{bmatrix} \begin{matrix} \text{Protein} \\ \text{Carbohydrates} \\ \text{Fat} \end{matrix}$$

with column headings "Octav Zoso Zipz".

(a) Find NF.

(b) Use part (a) to determine the amount of carbohydrates in mix III.

(c) Use part (a) to determine the amount of fat in mix II.

(d) Interpret the entries in NF.

59. Matrix A gives the percentage, as a decimal, of eligible voters in the United States in 1994 classified according to party affiliation and age group. (*Note:* Rows may not sum to 100% due to not including parties such as Libertarian, Reform, Green, and the like.) (*Source:* U.S. Census Bureau, www.census.gov.)

$$A = \begin{bmatrix} 0.51 & 0.10 & 0.37 \\ 0.44 & 0.12 & 0.43 \\ 0.47 & 0.07 & 0.44 \\ 0.53 & 0.08 & 0.385 \end{bmatrix} \begin{matrix} 18\text{--}24 \\ 25\text{--}44 \\ 45\text{--}64 \\ \text{Over } 65 \end{matrix}$$

with column headings "Democratic Independent Republican".

The population of eligible voters, in millions, in 1994 in the United States according to age group is given in matrix B.

$$B = [25.2 \quad 83 \quad 50.9 \quad 31.1]$$

with column headings "18–24 25–44 45–64 Over 65".

If all eligible voters vote along party lines, find a matrix giving the total number of eligible voters in the United States in 1994 who will vote Democratic, Independent, and Republican.

60. Matrix A gives the percentage, as a decimal, of the total vote for five states for the Democratic, Republican, and Reform party candidates in the 1996 presidential election (*Source:* U.S. Census Bureau, www.census.gov).

$$A = \begin{bmatrix} 0.511 & 0.382 & 0.07 \\ 0.48 & 0.423 & 0.091 \\ 0.474 & 0.41 & 0.107 \\ 0.44 & 0.498 & 0.056 \\ 0.498 & 0.373 & 0.089 \end{bmatrix} \begin{matrix} \text{California} \\ \text{Florida} \\ \text{Ohio} \\ \text{South Carolina} \\ \text{Washington} \end{matrix}$$

with column headings "Democratic Republican Reform".

Matrix B gives the total number of votes, in thousands, cast in each state.

$$B = [10{,}019 \quad 5304 \quad 4534 \quad 1152 \quad 2254]$$

with column headings "California Florida Ohio South Carolina Washington".

Compute BA and interpret each entry.

SECTION PROJECT

For this section project, let

$$A = \begin{bmatrix} 2 & -1 \\ 3 & 4 \end{bmatrix} \qquad B = \begin{bmatrix} -1 & 3 \\ 1 & 2 \end{bmatrix} \qquad C = \begin{bmatrix} 4 & -2 \\ -1 & 3 \end{bmatrix}$$

(a) Let $D = AB$. Find D.

(b) Find DC.

(c) Let $E = BC$. Find E.

(d) Find AE.

(e) Compare the results of parts (b) and (d). They should be equal. This illustrates the **associative law** of matrix multiplication, $(AB)C = A(BC)$. Of course, the associative law is true only if the matrix products are defined. Using the letters $m, n, p,$ and q and given that matrix A (not the A given above, but any matrix A) has dimensions $m \times n$, determine

the dimensions of B and C (not the B and C given above) so that the associative law is true.

For parts (f) through (i), use the matrices given above.

(f) Let $S = B + C$. Find S.

(g) Find AS.

(h) Let $D = AB$ and $P = AC$. Find $D + P$.

(i) Compare the results of parts (g) and (h). They should be equal, illustrating the **distributive law** of matrix multiplication, $A(B + C) = AB + AC$. Of course, the distributive law is true only if the matrix sums and products are defined. For any matrix A having dimensions $m \times n$, what must be true about the dimensions of any matrices B and C?

Section 2.4 Matrix Inverses and Solving Systems of Linear Equations

From Your Toolbox

If $b \neq 0$, then $b^{-1} = \frac{1}{b}$ is the multiplicative inverse (reciprocal) of b, and $b \cdot b^{-1} = b \cdot \frac{1}{b} = 1$.

Sections 2.2 and 2.3 have laid the groundwork for matrix arithmetic. In this section, we learn how to determine the **inverse of a matrix** and how to use the inverse matrix to solve a **matrix equation** that represents a system of linear equations. We will see that a **matrix inverse** is analogous to the reciprocal of a real number. See the Toolbox to the left.

Since $b \cdot 1 = b$ for any real number b, the number 1 is called the **multiplicative identity**. Before studying the inverse of a matrix, we need to discuss another important matrix called the **identity matrix**.

Identity Matrix

From Your Toolbox

A matrix is a **square matrix** if the number of rows equals the number of columns. That is, its dimension is $n \times n$.

An **identity matrix** is a matrix that behaves like the multiplicative identity 1. We know that, for any real number b, $b \cdot 1 = 1 \cdot b = b$. If I is an identity matrix and A is another matrix, then the matrix products (assuming that they exist) AI and IA must equal A. For this reason, an identity matrix exists only for square matrices.[1] (See the Toolbox to the left for a review of square matrices.) In general, the identity matrix of dimension $n \times n$ has 1s along the diagonal from upper left to lower right (called the **main diagonal**) and 0s elsewhere.

■ **Identity Matrix**

The **identity matrix** with dimension $n \times n$ is given by

$$I = \begin{bmatrix} 1 & 0 & \cdots & 0 \\ 0 & 1 & \cdots & 0 \\ \vdots & \vdots & \ddots & \vdots \\ 0 & 0 & \cdots & 1 \end{bmatrix}$$

---

[1]Although some nonsquare matrices may have an identity matrix, only square matrices satisfy $AI = IA = A$.

Example 1   **Verifying an Identity Matrix**

Given the matrices

$$A = \begin{bmatrix} 2 & 1 & 3 \\ -1 & 3 & -2 \\ 4 & 1 & 7 \end{bmatrix} \qquad I = \begin{bmatrix} 1 & 0 & 0 \\ 0 & 1 & 0 \\ 0 & 0 & 1 \end{bmatrix}$$

show that $AI = A$ and $IA = A$. (We can express these equalities as $AI = IA = A$.)

Solution

$$AI = \begin{bmatrix} 2 & 1 & 3 \\ -1 & 3 & -2 \\ 4 & 1 & 7 \end{bmatrix} \begin{bmatrix} 1 & 0 & 0 \\ 0 & 1 & 0 \\ 0 & 0 & 1 \end{bmatrix} = \begin{bmatrix} 2 & 1 & 3 \\ -1 & 3 & -2 \\ 4 & 1 & 7 \end{bmatrix} = A$$

$$IA = \begin{bmatrix} 1 & 0 & 0 \\ 0 & 1 & 0 \\ 0 & 0 & 1 \end{bmatrix} \begin{bmatrix} 2 & 1 & 3 \\ -1 & 3 & -2 \\ 4 & 1 & 7 \end{bmatrix} = \begin{bmatrix} 2 & 1 & 3 \\ -1 & 3 & -2 \\ 4 & 1 & 7 \end{bmatrix} = A$$

So $AI = IA = A$. This means that I is an identity matrix.   ∎

The identity matrix in Example 1 is the identity matrix for 3×3 matrices. The identity matrix for 2×2 matrices and 4×4 matrices are, respectively

$$I = \begin{bmatrix} 1 & 0 \\ 0 & 1 \end{bmatrix} \quad \text{and} \quad I = \begin{bmatrix} 1 & 0 & 0 & 0 \\ 0 & 1 & 0 & 0 \\ 0 & 0 & 1 & 0 \\ 0 & 0 & 0 & 1 \end{bmatrix}$$

Inverse of a Square Matrix

From Your Toolbox

To see more review on negative exponents, consult Section A.3 in Appendix A.

As stated in the Toolbox on page 108, the **multiplicative inverse** (reciprocal) of a real number b ($b \neq 0$) is the unique real number b^{-1} such that

$$b \cdot b^{-1} = b^{-1} \cdot b = 1$$

For example, 3^{-1} is the multiplicative inverse of 3 since $3^{-1} \cdot 3 = \frac{1}{3} \cdot 3 = 1$. We use this same idea to define the **inverse of a matrix**.

■ **Inverse of a Matrix**

Let A be a square matrix of dimension $n \times n$. The square matrix A^{-1} of dimension $n \times n$, where
$$A^{-1} \cdot A = A \cdot A^{-1} = I$$
is called the **inverse of A**. Note that I is the identity matrix of dimension $n \times n$.

▶ **Note:** A^{-1} does *not* mean $\frac{1}{A}$. When working with matrices, A^{-1} is simply notation for the multiplicative inverse, or simply inverse, of matrix A.

Example 2   **Verifying an Inverse Matrix**

Show that matrix A has as its inverse matrix A^{-1}.

$$A = \begin{bmatrix} 2 & 3 \\ -1 & 4 \end{bmatrix} \qquad A^{-1} = \begin{bmatrix} \frac{4}{11} & -\frac{3}{11} \\ \frac{1}{11} & \frac{2}{11} \end{bmatrix}$$

Solution

We need to verify that $A \cdot A^{-1} = A^{-1} \cdot A = I$. We have

$$A \cdot A^{-1} = \begin{bmatrix} 2 & 3 \\ -1 & 4 \end{bmatrix} \begin{bmatrix} \frac{4}{11} & -\frac{3}{11} \\ \frac{1}{11} & \frac{2}{11} \end{bmatrix}$$

$$= \begin{bmatrix} \frac{8}{11} + \frac{3}{11} & -\frac{6}{11} + \frac{6}{11} \\ -\frac{4}{11} + \frac{4}{11} & \frac{3}{11} + \frac{8}{11} \end{bmatrix} = \begin{bmatrix} 1 & 0 \\ 0 & 1 \end{bmatrix} = I$$

$$A^{-1} \cdot A = \begin{bmatrix} \frac{4}{11} & -\frac{3}{11} \\ \frac{1}{11} & \frac{2}{11} \end{bmatrix} \begin{bmatrix} 2 & 3 \\ -1 & 4 \end{bmatrix}$$

$$= \begin{bmatrix} \frac{8}{11} + \frac{3}{11} & \frac{12}{11} + \left(-\frac{12}{11}\right) \\ \frac{2}{11} + \left(-\frac{2}{11}\right) & \frac{3}{11} + \frac{8}{11} \end{bmatrix} = \begin{bmatrix} 1 & 0 \\ 0 & 1 \end{bmatrix} = I$$

Since $AA^{-1} = A^{-1}A = I$, we see that A^{-1} is the inverse of A. ∎

✓ Checkpoint 1

Now work Exercise 1.

Before we outline a process to determine inverse matrices, we must first address whether or not every square matrix has an inverse matrix. We know that every nonzero real number has an inverse (reciprocal); however, unlike real numbers, this is not true for matrices. A matrix that does not have an inverse matrix is said to be **singular**. For example, matrix A does not have an inverse matrix; hence it is singular.

$$A = \begin{bmatrix} 0 & 0 \\ 2 & 0 \end{bmatrix}$$

If A had an inverse, it would be of the form

$$A^{-1} = \begin{bmatrix} a & b \\ c & d \end{bmatrix}$$

where a, b, c, and d are real numbers. By the definition of an inverse, we must have $AA^{-1} = I$, that is,

$$\begin{bmatrix} 0 & 0 \\ 2 & 0 \end{bmatrix} \begin{bmatrix} a & b \\ c & d \end{bmatrix} = \begin{bmatrix} 1 & 0 \\ 0 & 1 \end{bmatrix}$$

Multiplying on the left side of the equation gives

$$\begin{bmatrix} 0a + 0c & 0b + 0d \\ 2a + 0c & 2b + 0d \end{bmatrix} = \begin{bmatrix} 1 & 0 \\ 0 & 1 \end{bmatrix}$$

$$\begin{bmatrix} 0 & 0 \\ 2a & 2b \end{bmatrix} = \begin{bmatrix} 1 & 0 \\ 0 & 1 \end{bmatrix}$$

Recalling that in order for two matrices to be equal they must have the same dimension and corresponding entries must be equal, we see that if these two matrices are equal, then, by examining their entries in the 1st row, 1st column, we must have $0 = 1$, which is not possible. This impossibility tells us that A does not have an inverse; that is, A is singular.

Finding the Inverse of a Square Matrix

To discover a step-by-step procedure to find the inverse of a matrix, let's find the inverse of matrix B

$$B = \begin{bmatrix} 1 & -3 \\ -1 & 4 \end{bmatrix}$$

Let's assume that B^{-1} exists and has the form

$$B^{-1} = \begin{bmatrix} a & b \\ c & d \end{bmatrix}$$

where a, b, c, and d are real numbers. By definition we know that $BB^{-1} = I$, which gives us the equation

$$\begin{bmatrix} 1 & -3 \\ -1 & 4 \end{bmatrix} \begin{bmatrix} a & b \\ c & d \end{bmatrix} = \begin{bmatrix} 1 & 0 \\ 0 & 1 \end{bmatrix}$$

Performing the multiplication on the left side of the equation yields

$$\begin{bmatrix} a - 3c & b - 3d \\ -a + 4c & -b + 4d \end{bmatrix} = \begin{bmatrix} 1 & 0 \\ 0 & 1 \end{bmatrix}$$

From equality of matrices, this equation is the same as the following two systems of linear equations.

$$a - 3c = 1, \qquad b - 3d = 0$$
$$-a + 4c = 0, \qquad -b + 4d = 1$$

From Section 2.1 we can write each system as an augmented matrix.

$$\begin{bmatrix} 1 & -3 & | & 1 \\ -1 & 4 & | & 0 \end{bmatrix} \quad \text{and} \quad \begin{bmatrix} 1 & -3 & | & 0 \\ -1 & 4 & | & 1 \end{bmatrix}$$

Notice that the matrices of coefficients (that is, each matrix left of the vertical line) for both augmented matrices are identical. This means that we would use the same row operations on each augmented matrix to transform it into reduced row echelon form. So we can save some time and effort by combining the two augmented matrices into one augmented matrix.

$$\begin{bmatrix} 1 & -3 & | & 1 & 0 \\ -1 & 4 & | & 0 & 1 \end{bmatrix}$$

We now use the Gauss–Jordan method to get

$$\begin{bmatrix} 1 & -3 & | & 1 & 0 \\ -1 & 4 & | & 0 & 1 \end{bmatrix} R_1 + R_2 \rightarrow R_2 \begin{bmatrix} 1 & -3 & | & 1 & 0 \\ 0 & 1 & | & 1 & 1 \end{bmatrix}$$

$$3R_2 + R_1 \rightarrow R_1 \begin{bmatrix} 1 & 0 & | & 4 & 3 \\ 0 & 1 & | & 1 & 1 \end{bmatrix}$$

The numbers in the 1st column to the right of the vertical bar give the values of a and c, and the 2nd column gives the values of b and d. Hence $a = 4$, $b = 3$, $c = 1$, and $d = 1$, which means that B^{-1} is

$$B^{-1} = \begin{bmatrix} 4 & 3 \\ 1 & 1 \end{bmatrix}$$

We use the prior discussion as motivation for the following algorithm, which is used to compute the inverse of a square matrix.

Interactive Activity

For B and B^{-1} in the above discussion, multiply BB^{-1} and $B^{-1}B$ to show that both products equal I. This verifies that B^{-1} is the inverse of B.

■ **Finding an Inverse Matrix**

Given an $n \times n$ matrix A, we find A^{-1} (when it exists) by doing the following:

1. Make the augmented matrix $[A \mid I]$, where I is the $n \times n$ identity matrix.
2. Use row operations to reduce $[A \mid I]$ to the form $[I \mid B]$, if possible.
3. Matrix B is the inverse of A; that is, $B = A^{-1}$.

Example 3   **Finding an Inverse Matrix**

Given $A = \begin{bmatrix} 1 & 1 & 2 \\ 2 & 1 & 3 \\ 1 & -1 & 2 \end{bmatrix}$, find A^{-1}.

Solution

We write the augmented matrix $[A \mid I]$.

$$\left[\begin{array}{ccc|ccc} 1 & 1 & 2 & 1 & 0 & 0 \\ 2 & 1 & 3 & 0 & 1 & 0 \\ 1 & -1 & 2 & 0 & 0 & 1 \end{array}\right]$$

We now use the Gauss–Jordan method to transform it to the form $[I \mid B]$.

$$\left[\begin{array}{ccc|ccc} 1 & 1 & 2 & 1 & 0 & 0 \\ 2 & 1 & 3 & 0 & 1 & 0 \\ 1 & -1 & 2 & 0 & 0 & 1 \end{array}\right] \begin{array}{c} -2R_1 + R_2 \rightarrow R_2 \\ -1R_1 + R_3 \rightarrow R_3 \end{array} \left[\begin{array}{ccc|ccc} 1 & 1 & 2 & 1 & 0 & 0 \\ 0 & -1 & -1 & -2 & 1 & 0 \\ 0 & -2 & 0 & -1 & 0 & 1 \end{array}\right]$$

$$-2R_2 + R_3 \rightarrow R_3 \left[\begin{array}{ccc|ccc} 1 & 1 & 2 & 1 & 0 & 0 \\ 0 & -1 & -1 & -2 & 1 & 0 \\ 0 & 0 & 2 & 3 & -2 & 1 \end{array}\right]$$

$$\tfrac{1}{2}R_3 \rightarrow R_3 \left[\begin{array}{ccc|ccc} 1 & 1 & 2 & 1 & 0 & 0 \\ 0 & -1 & -1 & -2 & 1 & 0 \\ 0 & 0 & 1 & \frac{3}{2} & -1 & \frac{1}{2} \end{array}\right]$$

$$R_2 + R_1 \rightarrow R_1 \left[\begin{array}{ccc|ccc} 1 & 0 & 1 & -1 & 1 & 0 \\ 0 & -1 & -1 & -2 & 1 & 0 \\ 0 & 0 & 1 & \frac{3}{2} & -1 & \frac{1}{2} \end{array}\right]$$

$$\begin{array}{c} -1R_3 + R_1 \rightarrow R_1 \\ R_3 + R_2 \rightarrow R_2 \end{array} \left[\begin{array}{ccc|ccc} 1 & 0 & 0 & -\frac{5}{2} & 2 & -\frac{1}{2} \\ 0 & -1 & 0 & -\frac{1}{2} & 0 & \frac{1}{2} \\ 0 & 0 & 1 & \frac{3}{2} & -1 & \frac{1}{2} \end{array}\right]$$

$$-1R_2 \rightarrow R_2 \left[\begin{array}{ccc|ccc} 1 & 0 & 0 & -\frac{5}{2} & 2 & -\frac{1}{2} \\ 0 & 1 & 0 & \frac{1}{2} & 0 & -\frac{1}{2} \\ 0 & 0 & 1 & \frac{3}{2} & -1 & \frac{1}{2} \end{array}\right]$$

So the inverse of A is

$$A^{-1} = \begin{bmatrix} -\frac{5}{2} & 2 & -\frac{1}{2} \\ \frac{1}{2} & 0 & -\frac{1}{2} \\ \frac{3}{2} & -1 & \frac{1}{2} \end{bmatrix}$$

Interactive Activity

For A in Example 3, compute AA^{-1} and $A^{-1}A$.

We leave the verification of this result to you in the Interactive Activity to the left.

■

Checkpoint 2

Now work Exercise 13.

Technology Option

Inverses of matrices can be found using a graphing calculator. In Figure 2.4.1(a), we have entered matrix A from Example 3 into the calculator and used the calculator to find A^{-1}. Figure 2.4.1(b) shows the entries as fractions.

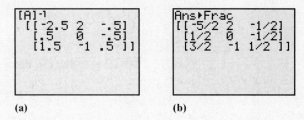

(a)   (b)

Figure 2.4.1

Example 4   Finding an Inverse Matrix

Given $A = \begin{bmatrix} 1 & 2 & 2 \\ 2 & 1 & 2 \\ 3 & 3 & 4 \end{bmatrix}$, find A^{-1}.

Solution

We form the augmented matrix

$$\begin{bmatrix} 1 & 2 & 2 & | & 1 & 0 & 0 \\ 2 & 1 & 2 & | & 0 & 1 & 0 \\ 3 & 3 & 4 & | & 0 & 0 & 1 \end{bmatrix}$$

and use the Gauss–Jordan method as in Example 3.

$$\begin{bmatrix} 1 & 2 & 2 & | & 1 & 0 & 0 \\ 2 & 1 & 2 & | & 0 & 1 & 0 \\ 3 & 3 & 4 & | & 0 & 0 & 1 \end{bmatrix} \begin{matrix} -2R_1 + R_2 \to R_2 \\ -3R_1 + R_3 \to R_3 \end{matrix} \begin{bmatrix} 1 & 2 & 2 & | & 1 & 0 & 0 \\ 0 & -3 & -2 & | & -2 & 1 & 0 \\ 0 & -3 & -2 & | & -3 & 0 & 1 \end{bmatrix}$$

$$-1R_2 + R_3 \to R_3 \begin{bmatrix} 1 & 2 & 2 & | & 1 & 0 & 0 \\ 0 & -3 & -2 & | & -2 & 1 & 0 \\ 0 & 0 & 0 & | & -1 & -1 & 1 \end{bmatrix}$$

Since the last row has all zeros to the left of the vertical bar, there is no way we can continue and get the reduced form $[I \mid B]$. This means that matrix A is singular and has no inverse. ∎

Whenever we encounter a matrix like the final augmented matrix in Example 4 that has all zeros in a row to the left of the vertical bar, this tells us that the matrix has no inverse.

Formula for the Inverse of 2 × 2 Matrices

Before we look at how inverse matrices can be used to solve matrix equations, let's examine an alternative way to find the inverse of a 2 × 2 matrix. The derivation of this formula is outlined in Exercise 27.

> ### ■ Formula for the Inverse of a 2 × 2 Matrix
>
> Let $A = \begin{bmatrix} a & b \\ c & d \end{bmatrix}$. Suppose that $\det(A) = ad - bc$ and $\det(A) \neq 0$. Then A^{-1} is given by
>
> $$A^{-1} = \frac{1}{\det(A)} \begin{bmatrix} d & -b \\ -c & a \end{bmatrix}$$

▶ **Note:** $\det(A)$ is called the **determinant of A**.

Example 5   **Using the Formula to Find A^{-1} for a 2 × 2 Matrix**

Given $A = \begin{bmatrix} 3 & -1 \\ 4 & -2 \end{bmatrix}$, find A^{-1} by using the formula.

Solution

First we compute $\det(A)$. Since $a = 3$, $b = -1$, $c = 4$, and $d = -2$, we have

$$\det(A) = 3(-2) - (-1)4 = -6 - (-4) = -2$$

Next we write the matrix

$$\begin{bmatrix} d & -b \\ -c & a \end{bmatrix} = \begin{bmatrix} -2 & 1 \\ -4 & 3 \end{bmatrix}$$

Finally, we multiply this matrix by $\frac{1}{\det(A)} = \frac{1}{-2}$ to get A^{-1}.

$$A^{-1} = \frac{1}{-2} \begin{bmatrix} -2 & 1 \\ -4 & 3 \end{bmatrix} = \begin{bmatrix} 1 & -\frac{1}{2} \\ 2 & -\frac{3}{2} \end{bmatrix}$$

■

Interactive Activity

Use a graphing calculator to find A^{-1} for matrix A in Example 5.

✔ **Checkpoint 3**

Now work Exercise 19.

Using Inverses to Solve Systems of Linear Equations

When we have a system of linear equations in which the number of equations equals the number of variables, we can write a corresponding matrix equation and use the inverse of the coefficient matrix to solve the system. For example, consider the system

$$3x - y = 6$$
$$4x - 2y = 7$$

If we let A be the coefficient matrix, X be the variable matrix, and B be the constants matrix, we have

$$A = \begin{bmatrix} 3 & -1 \\ 4 & -2 \end{bmatrix} \qquad X = \begin{bmatrix} x \\ y \end{bmatrix} \qquad B = \begin{bmatrix} 6 \\ 7 \end{bmatrix}$$

Notice that

$$AX = B$$

$$\begin{bmatrix} 3 & -1 \\ 4 & -2 \end{bmatrix} \begin{bmatrix} x \\ y \end{bmatrix} = \begin{bmatrix} 6 \\ 7 \end{bmatrix}$$

In Example 5 we found that $A^{-1} = \begin{bmatrix} 1 & -\frac{1}{2} \\ 2 & -\frac{3}{2} \end{bmatrix}$. We can use A^{-1} to solve as follows.

(We will show both the numerical and general solution simultaneously.)

$$\begin{bmatrix} 3 & -1 \\ 4 & -2 \end{bmatrix} \begin{bmatrix} x \\ y \end{bmatrix} = \begin{bmatrix} 6 \\ 7 \end{bmatrix} \qquad AX = B$$

$$\begin{bmatrix} 1 & -\frac{1}{2} \\ 2 & -\frac{3}{2} \end{bmatrix} \begin{bmatrix} 3 & -1 \\ 4 & -2 \end{bmatrix} \begin{bmatrix} x \\ y \end{bmatrix} = \begin{bmatrix} 1 & -\frac{1}{2} \\ 2 & -\frac{3}{2} \end{bmatrix} \begin{bmatrix} 6 \\ 7 \end{bmatrix} \qquad A^{-1}AX = A^{-1}B$$

$$\begin{bmatrix} 1 & 0 \\ 0 & 1 \end{bmatrix} \begin{bmatrix} x \\ y \end{bmatrix} = \begin{bmatrix} \frac{5}{2} \\ \frac{3}{2} \end{bmatrix} \qquad IX = A^{-1}B$$

$$\begin{bmatrix} x \\ y \end{bmatrix} = \begin{bmatrix} \frac{5}{2} \\ \frac{3}{2} \end{bmatrix} \qquad X = A^{-1}B$$

So the solution to the system is $x = \frac{5}{2}$ and $y = \frac{3}{2}$. The generalization on the right side is summarized as follows:

■ **Using Matrix Inverses to Solve Systems of Equations**

Given a system of linear equations $AX = B$, where A is the coefficient matrix, X is the variable matrix, and B is the constants matrix, we solve the system by first finding A^{-1}. Then $X = A^{-1}B$ gives the solution.

Example 6 Using a Matrix Inverse to Solve a System

Write the given system in the form $AX = B$ and use A^{-1} to solve the system.

$$x + y + 2z = -3$$
$$2x + y + 3z = 4$$
$$x - y + 2z = 5$$

Solution
We have

$$A = \begin{bmatrix} 1 & 1 & 2 \\ 2 & 1 & 3 \\ 1 & -1 & 2 \end{bmatrix} \qquad X = \begin{bmatrix} x \\ y \\ z \end{bmatrix} \qquad B = \begin{bmatrix} -3 \\ 4 \\ 5 \end{bmatrix}$$

So

$$AX = B$$

$$\begin{bmatrix} 1 & 1 & 2 \\ 2 & 1 & 3 \\ 1 & -1 & 2 \end{bmatrix} \begin{bmatrix} x \\ y \\ z \end{bmatrix} = \begin{bmatrix} -3 \\ 4 \\ 5 \end{bmatrix}$$

In Example 3, we determined that $A^{-1} = \begin{bmatrix} -\frac{5}{2} & 2 & -\frac{1}{2} \\ \frac{1}{2} & 0 & -\frac{1}{2} \\ \frac{3}{2} & -1 & \frac{1}{2} \end{bmatrix}$,

so the solution is given by

$$X = A^{-1}B = \begin{bmatrix} -\frac{5}{2} & 2 & -\frac{1}{2} \\ \frac{1}{2} & 0 & -\frac{1}{2} \\ \frac{3}{2} & -1 & \frac{1}{2} \end{bmatrix} \begin{bmatrix} -3 \\ 4 \\ 5 \end{bmatrix} = \begin{bmatrix} 13 \\ -4 \\ -6 \end{bmatrix}$$

So $x = 13$, $y = -4$, and $z = -6$ is the solution to the system. ∎

✓ **Checkpoint 4**

Now work Exercise 41.

Applications

For our first application, let's revisit the Flashback from Section 2.1.

Flashback

Gonzalez Office Supplies Revisited

In Section 2.1 we revisited Gonzalez Office Supplies and saw that they manufacture three types of office chairs: Secretarial, Executive, and Presidential. Each chair requires time on three different machines. These times and the total available time for each machine are given in Table 2.4.1.

Table 2.4.1

	Secretarial	Executive	Presidential	Total hours each week
Hours on Machine A	1	1	2	30
Hours on Machine B	2	3	2	53
Hours on Machine C	1	2	3	47

We assume that all available time for each machine must be used each week. If x represents the number of Secretarial chairs made each week, y represents the number of Executive chairs made each week, and z represents the number of Presidential chairs made each week, then the system of equations that can be solved to determine how many of each type of chair can be made each week is

$$x + y + 2z = 30$$
$$2x + 3y + 2z = 53$$
$$x + 2y + 3z = 47$$

Solve this system of linear equations by using the inverse of the coefficient matrix.

Flashback Solution

For this system we have

$$A = \begin{bmatrix} 1 & 1 & 2 \\ 2 & 3 & 2 \\ 1 & 2 & 3 \end{bmatrix} \qquad X = \begin{bmatrix} x \\ y \\ z \end{bmatrix} \qquad B = \begin{bmatrix} 30 \\ 53 \\ 47 \end{bmatrix}$$

To find A^{-1}, we start with

$$[A \mid I] = \begin{bmatrix} 1 & 1 & 2 & | & 1 & 0 & 0 \\ 2 & 3 & 2 & | & 0 & 1 & 0 \\ 1 & 2 & 3 & | & 0 & 0 & 1 \end{bmatrix}$$

and transform it to get $[I \mid A^{-1}]$. Omitting the details, the result is

$$A^{-1} = \begin{bmatrix} \frac{5}{3} & \frac{1}{3} & -\frac{4}{3} \\ -\frac{4}{3} & \frac{1}{3} & \frac{2}{3} \\ \frac{1}{3} & -\frac{1}{3} & \frac{1}{3} \end{bmatrix}$$

The solution is given by

$$X = A^{-1}B = \begin{bmatrix} \frac{5}{3} & \frac{1}{3} & -\frac{4}{3} \\ -\frac{4}{3} & \frac{1}{3} & \frac{2}{3} \\ \frac{1}{3} & -\frac{1}{3} & \frac{1}{3} \end{bmatrix} \begin{bmatrix} 30 \\ 53 \\ 47 \end{bmatrix} = \begin{bmatrix} 5 \\ 9 \\ 8 \end{bmatrix}$$

So the solution to the system is $x = 5$, $y = 9$, and $z = 8$.

Interpret the Solution: Five Secretarial, nine Executive, and eight Presidential chairs should be made each week to use all available time on each machine.

Our final example utilizes the formula for determining an inverse for a 2×2 matrix.

Example 7   Determining a Planting Schedule

The data in Table 2.4.2 give the amounts of nitrogen and phosphate needed to grow cauliflower and head lettuce for an entire growing season in California (all entries are in lb/acre). If Fernando has 150 acres and uses 28,890 pounds of nitrogen and 16,620 pounds of phosphate, how many acres did he allot for each crop to use all the resources completely?

Table 2.4.2

	Cauliflower	Head Lettuce
Nitrogen	215	173
Phosphate	98	122

SOURCE: U.S. Department of Agriculture, www.usda.gov/nass

Solution

Understand the Situation: We need to determine the number of acres of cauliflower and the number of acres of head lettuce that Fernando planted. To do this, we let x represent the number of acres of cauliflower and y represent the number of acres of head lettuce. The number of acres of cauliflower times the number of pounds per acre of nitrogen plus the number of acres of head lettuce times the number of pounds per acre of nitrogen must equal 28,890. This gives

$$215x + 173y = 28{,}890$$

Similarly, for phosphates we get

$$98x + 122y = 16{,}620$$

So the system is

$$215x + 173y = 28{,}890$$
$$98x + 122y = 16{,}620$$

Letting $A = \begin{bmatrix} 215 & 173 \\ 98 & 122 \end{bmatrix}$, $X = \begin{bmatrix} x \\ y \end{bmatrix}$, and $B = \begin{bmatrix} 28{,}890 \\ 16{,}620 \end{bmatrix}$, we have the system in the form $AX = B$. Since $X = A^{-1}B$, we need to find A^{-1}. Here A is a 2×2 matrix, so we find A^{-1} by using the formula

$$A^{-1} = \tfrac{1}{\det(A)} \begin{bmatrix} d & -b \\ -c & a \end{bmatrix}$$

We have $\det(A) = 215(122) - 173(98) = 9276$, so

$$A^{-1} = \tfrac{1}{9276} \begin{bmatrix} 122 & -173 \\ -98 & 215 \end{bmatrix} = \begin{bmatrix} \frac{122}{9276} & -\frac{173}{9276} \\ -\frac{98}{9276} & \frac{215}{9276} \end{bmatrix}$$

Therefore,

$$X = A^{-1}B = \begin{bmatrix} \frac{122}{9276} & -\frac{173}{9276} \\ -\frac{98}{9276} & \frac{215}{9276} \end{bmatrix} \begin{bmatrix} 28{,}890 \\ 16{,}620 \end{bmatrix} = \begin{bmatrix} 70 \\ 80 \end{bmatrix}$$

So the solution to the system is $x = 70$ and $y = 80$.

Interpret the Solution: This means that Fernando planted 70 acres of cauliflower and 80 acres of head lettuce to use all his resources completely.   ■

▍ SUMMARY

We began this section with a brief discussion on an **identity matrix**, I, and quickly proceeded to determining the **inverse of a square matrix**. We saw how the Gauss–Jordan method is key in finding an inverse matrix by hand. We also learned a formula for the inverse of a 2×2 matrix that uses the **determinant**. The section concluded with a discussion on using inverse matrices to solve systems of equations, provided the number of equations equals the number of variables, and using this new method of solving systems in an applied setting.

- **Finding an Inverse Matrix**   Given an $n \times n$ matrix A, we find A^{-1} (when it exists) by:

  1. Writing the augmented matrix $[A \mid I]$, where I is the $n \times n$ identity matrix.

2. Using row operations to reduce $[A\,|\,I]$ to the form $[I\,|\,B]$, if possible.

3. Matrix B is the inverse of A; that is, $B = A^{-1}$.

- **Formula for the Inverse of a 2×2 Matrix** Let $A = \begin{bmatrix} a & b \\ c & d \end{bmatrix}$. Suppose that $\det(A) = ad - bc$ and $\det(A) \neq 0$. Then A^{-1} is given by

$$A^{-1} = \tfrac{1}{\det(A)} \begin{bmatrix} d & -b \\ -c & a \end{bmatrix}.$$

- **Using Matrix Inverses to Solve Systems of Equations** Given a system of linear equations $AX = B$, where A is the coefficient matrix, X is the variable matrix, and B is the constants matrix, we solve the system by first finding A^{-1}. Then $X = A^{-1}B$ gives the solution.

SECTION 2.4 EXERCISES

In Exercises 1–8, determine if the given matrices are inverses of each other. (Hint: See if their product is the identity matrix I.)

✓ **1.** $\begin{bmatrix} -2 & -3 \\ 1 & 1 \end{bmatrix}$ and $\begin{bmatrix} 1 & 3 \\ -1 & -2 \end{bmatrix}$

2. $\begin{bmatrix} 2 & -1 \\ 3 & 1 \end{bmatrix}$ and $\begin{bmatrix} \frac{1}{5} & \frac{1}{5} \\ -\frac{3}{5} & \frac{2}{5} \end{bmatrix}$

3. $\begin{bmatrix} 1 & 1 \\ 2 & 3 \end{bmatrix}$ and $\begin{bmatrix} -3 & 1 \\ 2 & -1 \end{bmatrix}$

4. $\begin{bmatrix} -3 & 1 \\ 5 & 1 \end{bmatrix}$ and $\begin{bmatrix} \frac{1}{8} & -\frac{1}{8} \\ -\frac{5}{8} & -\frac{3}{8} \end{bmatrix}$

5. $\begin{bmatrix} 1 & 3 & 2 \\ 4 & -2 & -1 \\ 0 & 4 & 3 \end{bmatrix}$ and $\begin{bmatrix} \frac{1}{3} & \frac{1}{6} & -\frac{1}{6} \\ 2 & -\frac{1}{2} & -\frac{3}{2} \\ -\frac{8}{3} & \frac{2}{3} & \frac{7}{3} \end{bmatrix}$

6. $\begin{bmatrix} -1 & 2 & 3 \\ 0 & 2 & -1 \\ 2 & 4 & 5 \end{bmatrix}$ and $\begin{bmatrix} -\frac{7}{15} & -\frac{1}{15} & \frac{4}{15} \\ \frac{1}{15} & \frac{11}{30} & \frac{1}{30} \\ \frac{2}{15} & -\frac{4}{15} & \frac{1}{15} \end{bmatrix}$

7. $\begin{bmatrix} -1 & 0 & 2 \\ 0 & -1 & 4 \\ 3 & 0 & 5 \end{bmatrix}$ and $\begin{bmatrix} \frac{5}{11} & 0 & \frac{2}{11} \\ \frac{12}{11} & -1 & \frac{4}{11} \\ \frac{3}{11} & 0 & -\frac{1}{11} \end{bmatrix}$

8. $\begin{bmatrix} 1 & 2 & 7 \\ 0 & -1 & 1 \\ 2 & 1 & 2 \end{bmatrix}$ and $\begin{bmatrix} -\frac{1}{5} & \frac{1}{5} & -\frac{3}{5} \\ \frac{2}{15} & -\frac{4}{5} & -\frac{1}{15} \\ -\frac{2}{15} & \frac{1}{5} & -\frac{1}{15} \end{bmatrix}$

In Exercises 9–18, find the inverse for each matrix. See Example 3.

9. $\begin{bmatrix} 1 & 3 \\ 4 & 2 \end{bmatrix}$ **10.** $\begin{bmatrix} -1 & -2 \\ 3 & 2 \end{bmatrix}$

11. $\begin{bmatrix} -2 & 7 \\ -4 & -1 \end{bmatrix}$ **12.** $\begin{bmatrix} 3 & 4 \\ -2 & -3 \end{bmatrix}$

✓ **13.** $\begin{bmatrix} 1 & -1 & 2 \\ 0 & 2 & 3 \\ 3 & -1 & 4 \end{bmatrix}$ **14.** $\begin{bmatrix} 1 & 2 & 7 \\ 0 & -1 & 1 \\ 2 & 1 & 2 \end{bmatrix}$

15. $\begin{bmatrix} 2 & 1 & 0 \\ -1 & 1 & 2 \\ 4 & 2 & 3 \end{bmatrix}$ **16.** $\begin{bmatrix} 3 & 2 & 0 \\ -2 & -1 & -2 \\ 1 & 0 & 1 \end{bmatrix}$

17. $\begin{bmatrix} 1 & 2 & -2 & 3 \\ 0 & 2 & -3 & 1 \\ 2 & 0 & 1 & 2 \\ -3 & 1 & 2 & 5 \end{bmatrix}$ **18.** $\begin{bmatrix} 1 & -2 & 3 & 4 \\ 0 & 1 & -1 & 2 \\ -2 & 1 & 3 & -2 \\ 3 & 2 & 1 & 5 \end{bmatrix}$

In Exercises 19–26, use the formula for the inverse of a 2×2 matrix to find the inverse for the given matrix, if it exists.

✓ **19.** $\begin{bmatrix} 2 & -3 \\ 4 & 8 \end{bmatrix}$ **20.** $\begin{bmatrix} -3 & 4 \\ 6 & 3 \end{bmatrix}$

21. $\begin{bmatrix} 2 & -4 \\ -2 & 4 \end{bmatrix}$ **22.** $\begin{bmatrix} 3 & 7 \\ -6 & -14 \end{bmatrix}$

23. $\begin{bmatrix} 7 & 3 \\ -1 & -2 \end{bmatrix}$ **24.** $\begin{bmatrix} -1 & -2 \\ 3 & 2 \end{bmatrix}$

25. $\begin{bmatrix} 2 & -1 \\ -4 & 7 \end{bmatrix}$ **26.** $\begin{bmatrix} -3 & -2 \\ 1 & 4 \end{bmatrix}$

27. Given that $A = \begin{bmatrix} a & b \\ c & d \end{bmatrix}$ is a 2×2 matrix, where $a, b, c,$ and d are real numbers, $a \neq 0$, derive the formula, $A^{-1} = \tfrac{1}{\det(A)} \begin{bmatrix} d & -b \\ -c & a \end{bmatrix}$, $\det(A) \neq 0$, by doing the following:

(a) Write the matrix $[A\,|\,I]$ and use row operations until you have $[I\,|\,A^{-1}]$.

(b) From A^{-1}, factor out $\frac{1}{ad-bc}$. This should leave you with $\frac{1}{ad-bc} \begin{bmatrix} d & -b \\ -c & a \end{bmatrix}$.

28. Given that $A = \begin{bmatrix} 3 & 2 \\ -1 & 1 \end{bmatrix}$ and $B = \begin{bmatrix} 1 & -2 \\ 3 & 4 \end{bmatrix}$:

(a) Find A^{-1} and B^{-1}.

(b) Show that $(A^{-1})^{-1} = A$ and $(B^{-1})^{-1} = B$.

(c) Find $(AB)^{-1}$.

(d) Find $A^{-1}B^{-1}$ and $B^{-1}A^{-1}$.

(e) Which is true? $(AB)^{-1} = A^{-1}B^{-1}$ or $(AB)^{-1} = B^{-1}A^{-1}$.

In Exercises 29–36, use a graphing calculator to find the inverse for each matrix, if it exists. Where possible, express entries as fractions.

29. $\begin{bmatrix} 1.1 & 2.1 & 1.1 \\ 3.1 & 2.1 & 2.1 \\ 4.1 & 3.1 & 4.1 \end{bmatrix}$

30. $\begin{bmatrix} 2.2 & 3.2 & 1.2 \\ 4.2 & 1.2 & 5.2 \\ 1.2 & 0.2 & 3.2 \end{bmatrix}$

31. $\begin{bmatrix} 1 & 3 & -1 & 2 \\ 4 & 7 & -4 & -1 \\ 0 & 3 & 8 & 1 \\ 2 & 1 & 4 & 3 \end{bmatrix}$

32. $\begin{bmatrix} 1 & -1 & -2 & 3 \\ 7 & 1 & 8 & 2 \\ 3 & -2 & -1 & -4 \\ 0 & 3 & 5 & 7 \end{bmatrix}$

33. $\begin{bmatrix} 7 & -1 & 3 & 4 \\ 5 & 3 & -2 & 5 \\ 0 & 2 & -3 & -1 \\ 1 & -1 & 4 & 3 \end{bmatrix}$

34. $\begin{bmatrix} 2 & 1 & 7 & 8 \\ -3 & -1 & -4 & -1 \\ 5 & 9 & -1 & 3 \\ 0 & 1 & 2 & 1 \end{bmatrix}$

35. $\begin{bmatrix} 1 & 5 & 4 & 3 & 2 \\ 0 & 4 & -1 & 4 & 5 \\ -1 & -2 & 6 & -1 & 0 \\ 7 & -3 & 3 & 2 & 1 \\ 8 & 1 & 4 & 7 & 9 \end{bmatrix}$

36. $\begin{bmatrix} 1 & 0 & -1 & 7 & 8 \\ 5 & 4 & -2 & -3 & 1 \\ 4 & -1 & 6 & 3 & 4 \\ 3 & 4 & -1 & 2 & 7 \\ 2 & 5 & 0 & 1 & 9 \end{bmatrix}$

In Exercises 37–46, solve each system of linear equations by using the inverse of the coefficient matrix. See Example 6. (All inverses were found in Exercises 9–18.)

37. $x + 3y = 7$
$4x + 2y = 9$
(See Exercise 9)

38. $-x - 2y = 8$
$3x + 2y = -2$
(See Exercise 10)

39. $-2x + 7y = 3$
$-4x - y = -5$
(See Exercise 11)

40. $3x + 4y = -1$
$-2x - 3y = 5$
(See Exercise 12)

✓ 41. $x - y + 2z = 3$
$2y + 3z = -1$
$3x - y + 4z = 2$
(See Exercise 13)

42. $x + 2y + 7z = -2$
$-y + z = -1$
$2x + y + 2z = -3$
(See Exercise 14)

43. $2x + y = 3$
$-x + y + 2z = -2$
$4x + 2y + 3z = 4$
(See Exercise 15)

44. $3x + 2y = -1$
$-2x - y - 2z = 2$
$x + z = 3$
(See Exercise 16)

45. $x + 2y - 2z + 3w = 3$
$2y - 3z + w = -1$
$2x + z + 2w = 1$
$-3x + y + 2z + 5w = 4$
(See Exercise 17)

46. $x - 2y + 3z + 4w = -2$
$y - z + 2w = 3$
$-2x + y + 3z - 2w = 1$
$3x + 2y + z + 5w = -3$
(See Exercise 18)

In Exercises 47–52, solve each system of linear equations by using the inverse of the coefficient matrix and a graphing calculator. (All inverse were found in Exercises 29–36.)

47. $1.1x + 2.1y + 1.1z = 13.1$
$3.1x + 2.1y + 2.1z = 5.1$
$4.1x + 3.1y + 4.1z = 8.1$
(See Exercise 29)

48. $2.2x + 3.2y + 1.2z = 4.2$
$4.2x + 1.2y + 5.2z = 1.2$
$1.2x + 0.2y + 3.2z = 6.2$
(See Exercise 30)

49. $7x - y + 3z + 4w = 8$
$5x + 3y - 2z + 5w = -1$
$2y - 3z - w = 2$
$x - y + 4z + 3w = 3$
(See Exercise 33)

50. $2x + y + 7z + 8w = 2$
$-3x - y - 4z - w = -4$
$5x + 9y - z + 3w = 5$
$y + 2z + w = 3$
(See Exercise 34)

51. $x + 5y + 4z + 3v + 2w = 1$
$4y - z + 4v + 5w = -2$
$-x - 2y + 6z - v = 2$
$7x - 3y + 3z + 2v + w = -3$
$8x + y + 4z + 7v + 9w = 4$
(See Exercise 35)

52.
$$x - z + 7v + 8w = 4$$
$$5x + 4y - 2z - 3v + w = -2$$
$$4x - y + 6z + 3v + 4w = 3$$
$$3x + 4y - z + 2v + 7w = 1$$
$$2x + 5y + v + 9w = 2$$
(See Exercise 36)

Applications

53. Mary Lou Smith makes three types of bread, lemon poppy seed glaze (Type I), pumpkin glaze (Type II), and banana walnut glaze (Type III). Each type of bread requires the same basic ingredients as shown in matrix A.

$$A = \begin{matrix} & \text{I} & \text{II} & \text{III} & \\ & \begin{bmatrix} 1 & 2 & 2 \\ 3 & 2 & 3 \\ 2 & 2 & 1 \end{bmatrix} & \begin{matrix} \text{Flour (in cups)} \\ \text{Sugar (in cups)} \\ \text{Eggs} \end{matrix} \end{matrix}$$

Mary Lou has regular customers, so she makes these breads daily. To fill her daily orders for these three breads, the amount of basic ingredients used daily is given in matrix B.

$$B = \begin{bmatrix} 24 \\ 45 \\ 27 \end{bmatrix} \begin{matrix} \text{Flour (in cups)} \\ \text{Sugar (in cups)} \\ \text{Eggs} \end{matrix}$$

(a) Let the number of daily orders for the three bread types (Types I, II, and III) be represented by x, y, and z, respectively. Write a 3×1 matrix X with entries x, y, and z that represents the daily orders. Write a matrix equation in the form $AX = B$.

(b) Use A^{-1} to solve the equation in part (a).

54. During the holiday season, Mary Lou Smith from Exercise 53 makes more bread on a daily basis than usual. While matrix A remains the same, matrix B changes to

$$B = \begin{bmatrix} 30 \\ 56 \\ 34 \end{bmatrix} \begin{matrix} \text{Flour (in cups)} \\ \text{Sugar (in cups)} \\ \text{Eggs} \end{matrix}$$

(a) Let the number of daily orders for the three bread types (Types I, II, and III) be represented by x, y, and z, respectively. Write a 3×1 matrix X with entries x, y, and z that represents the daily orders. Write a matrix equation in the form $AX = B$.

(b) Use A^{-1} to solve the equation in part (a).

$\overset{\Leftarrow}{1.1}$ **55.** The Drama Department in a local high school is performing a play for three consecutive nights. The auditorium at the high school has 1600 seats, and tickets are broken into two categories: $3 student tickets and $5 adult

tickets. Table 2.4.3 gives the data for ticket revenue and total tickets sold for each play night.

Table 2.4.3

	Thursday	**Friday**	**Saturday**
Tickets sold	1600	1600	1600
Revenue	$5760	$5980	$5500

Let x represent the number of $3 student tickets sold and let y represent the number of $5 adult tickets sold.

(a) Write a system of equations for each night, Thursday, Friday, and Saturday.

(b) For each system in part (a), let matrix A be the coefficient matrix, matrix X the variable matrix, and matrix B the constants matrix. Write a matrix equation in the form $AX = B$ for each system in part (a).

(c) Use A^{-1} to solve each system in part (b) and interpret. (Notice that A^{-1} is the same for all three equations.)

$\overset{\Leftarrow}{1.1}$ **56.** (*continuation of Exercise 55*) Because each night was a sellout, the next semester the Drama Department moved the play to a local community college. The auditorium at the community college has 2800 seats. The ticket prices remain the same. Table 2.4.4 gives the data for ticket revenue and total tickets sold for each play night.

Table 2.4.4

	Thursday	**Friday**	**Saturday**
Tickets sold	2,800	2,800	2,800
Revenue	$11,400	$10,180	$10,760

(a) Write a system of equations for each night, Thursday, Friday, and Saturday.

(b) For each system in part (a), let matrix A be the coefficient matrix, matrix X the variable matrix, and matrix B the constants matrix. Write a matrix equation in the form $AX = B$ for each system in part (a).

(c) Use A^{-1} to solve each system in part (b) and interpret. (Notice that A^{-1} is the same for all three equations.)

$\overset{\Leftarrow}{1.1}$ **57.** Davram manufactures bicycle frames at three different plants. Due to different technologies at each plant, the production rate (in frames/hour) differs for each plant, as shown in Table 2.4.5.

Table 2.4.5

	Plant I	**Plant II**	**Plant III**
Racing frame	18	20	16
Hybrid frame	12	9	11
Mountain frame	22	18	17

Shannon, the production manager at Davram, has received an order to produce 5900 racing frames, 3580 hybrid frames, and 6350 mountain frames. To fill this order exactly, Shannon lets x, y, and z represent the number of hours that plants I, II, and III, respectively, should be scheduled to operate.

(a) Write a system of equations to model this situation.

(b) Write a matrix equation in the form $AX = B$ for the system in part (a).

(c) Use A^{-1} to solve the equation in part (b) and interpret.

$\overleftarrow{1.1}$ **58.** (*continuation of Exercise 57*) Shannon has received another order requesting 6040 racing frames, 3780 hybrid frames, and 6695 mountain frames. With everything else from Exercise 57 remaining the same, rework parts (a), (b), and (c). Notice that A^{-1} is the same as it was in Exercise 57.

59. The Basich Company makes three types of tents: two-person, four-person, and six-person models. Each tent requires the services of three departments, as shown in Table 2.4.6. (The entries in the table are in hours.) The fabric cutting, assembly, and packaging departments have available a maximum of 450, 390, and 190 hours each week, respectively.

Table 2.4.6

Department:	Two-person	Four-person	Six-person
Fabric cutting	0.5	0.8	1.1
Assembly	0.6	0.7	0.9
Packaging	0.2	0.3	0.6

Let x, y, and z represent the number of two-, four-, and six-person tents made in 1 week, respectively. Assuming that the company operates at full capacity, write a matrix equation in the form $AX = B$ that models this situation. Use

$$A^{-1} = \begin{bmatrix} -6 & 6 & 2 \\ \frac{36}{5} & -\frac{16}{5} & -\frac{42}{5} \\ -\frac{8}{5} & -\frac{2}{5} & \frac{26}{5} \end{bmatrix}$$ to solve the matrix equation and

interpret.

60. Rework Exercise 59, except here assume that the fabric cutting, assembly, and packaging departments have available a maximum of 420, 390, and 185 hours each week, respectively.

61. Last year the Yagan Corporation took out three loans totaling $120,000 to fund the purchase of new computer equipment. Some of the money was at 6%, $5000 more than $\frac{1}{3}$ the amount of the 6% loan was at 8.5%, and the rest was at 7.5%. The total annual interest on the loans was $8350. Let x represent the amount borrowed at 6%, y the amount borrowed at 7.5%, and z the amount borrowed at 8.5%.

(a) Write a matrix equation in the form $AX = B$ that models this scenario.

(b) Use A^{-1} to solve the matrix equation and interpret.

62. A dog breeder has available three commercial dog food mixes, Grava, Zita, and Blando. These mixes containing the percentages, as a decimal, of protein, carbohydrates, and fat are summarized in Table 2.4.7.

Table 2.4.7

	Grava	Zita	Blando
Protein	0.25	0.10	0.20
Carbohydrates	0.10	0.15	0.05
Fat	0.20	0.15	0.15

How many ounces of each mix should be used to prepare a diet for a cocker spaniel that requires 10.25 ounces of protein, 4.75 ounces of carbohydrates, and 8.75 ounces of fat?

63. Mona is a dietitian for a national weight loss program. Using only three basic foods, Lacto, Fracto, and Hima, she needs to arrange a restrictive diet for a new patient. The diet is to include 280 units of calcium, 175 units of vitamin B_{12}, and 225 units of vitamin B_6. The number of units per ounce for each basic food is given in Table 2.4.8. How many ounces of each food must be used to exactly meet the diet requirements?

Table 2.4.8

	Lacto	Fracto	Hima
Calcium	20	10	20
Vitamin B_{12}	10	5	15
Vitamin B_6	15	20	10

64. The data in Table 2.4.9 give the amounts of nitrogen, phosphate, and potash (all in lb/acre) needed to grow three crops for an entire growing season in California.

Table 2.4.9

	Asparagus	Cauliflower	Celery
Nitrogen	240	215	336
Phosphate	148	98	171
Potash	42	69	147

SOURCE: U.S. Department of Agriculture, www.usda.gov/nass

If George has 280 acres and uses 65,790 pounds of nitrogen, 31,860 pounds of phosphate, and 21,630 pounds of potash, how many acres did he allot for each crop in order to use all his resources completely?

65. The data in Table 2.4.10 give the amounts of nitrogen, phosphate, and potash (all in lb/acre) needed to grow three crops for an entire growing season in California.

Table 2.4.10

	Head Lettuce	Cantaloupe	Bell Peppers
Nitrogen	173	175	212
Phosphate	122	161	108
Potash	76	39	72

SOURCE: U.S. Department of Agriculture, www.usda.gov/nass

If Zach has 300 acres and uses 55,220 pounds of nitrogen, 39,380 pounds of phosphate, and 18,780 pounds of potash, how many acres did he allot for each crop in order to use all his resources completely?

SECTION PROJECT

This section project illustrates how matrices and inverse matrices can be used in **cryptography**, that is, encoding and decoding messages. To start, we assign the numbers 1 through 26 to the letters of the alphabet (A = 1, B = 2, C = 3, ..., Y = 25, Z = 26) and assign the number 27 to a blank so that we can put a space between words. For example, the message GET LOST corresponds to

$$7 \quad 5 \quad 20 \quad 27 \quad 12 \quad 15 \quad 19 \quad 20$$

Now any square matrix whose entries are positive integers (and whose inverse exists) can be used to encode the message. Such a matrix is called an **encoding matrix**. For example, we can use encoding matrix E:

$$E = \begin{bmatrix} 1 & 2 \\ 3 & 4 \end{bmatrix}$$

to encode the message. To do this, we put our sequence of numbers into matrix A.

$$A = \begin{bmatrix} 7 & 20 & 12 & 19 \\ 5 & 27 & 15 & 20 \end{bmatrix}$$

Notice that we entered the sequence going down each column. Matrix product EA encodes the message.

$$EA = \begin{bmatrix} 1 & 2 \\ 3 & 4 \end{bmatrix} \begin{bmatrix} 7 & 20 & 12 & 19 \\ 5 & 27 & 15 & 20 \end{bmatrix} = \begin{bmatrix} 17 & 74 & 42 & 59 \\ 41 & 168 & 96 & 137 \end{bmatrix}$$

Hence the original message has been coded as

$$17 \quad 41 \quad 74 \quad 168 \quad 42 \quad 96 \quad 59 \quad 137$$

To decode the message, the receiver of the message needs E^{-1}, the inverse matrix of the encoding matrix. E^{-1} is called the **decoding matrix**. Here E^{-1} is

$$E^{-1} = \begin{bmatrix} -2 & 1 \\ \frac{3}{2} & -\frac{1}{2} \end{bmatrix}$$

To decode, put the coded message into matrix form and multiply on the left by E^{-1}.

(a) You have received the following sequence of a coded message

$$31 \quad 70 \quad 21 \quad 62 \quad 36 \quad 81 \quad 46 \quad 119 \quad 37 \quad 90$$

It was encoded with matrix $E = \begin{bmatrix} 1 & 1 \\ 2 & 3 \end{bmatrix}$. Decode the message.

(b) Use the encoding matrix in part (a) to encode the message MEET ME AT NOON.

(c) Determine an encoding matrix E and encode any mes-[...] E to a friend and see if he/she can decode it.

Section 2.5   Leontief Input–Output Models

In this concluding section of Chapter 2, we introduce one of the most important applications of matrices and their inverses and systems of equations to economics. It is called the **Leontief input–output** model, named after Wassily Leontief. Wassily Leontief was awarded the Nobel prize in economics in 1973 because of this matrix theory applied to economics. What Leontief did was organize the 1958 U.S. economy into an 81×81 matrix. The 81 sectors of the U.S. economy, such as steel, agriculture, manufacturing, and transportation, each represented resources that relied on input from the output of other resources. For example, production of agriculture requires an input of manufacturing, transportation, and even some agriculture. Our discussion will not be nearly as in depth as Leontief's; however, the ideas that we present can be generalized to an economy of any size.

Input–Output Models

Input–output analysis deals with the relationship between the inputs and outputs of different sectors of an economy. To help us understand the concepts at hand, let's consider a simple model of an economy in which only three goods are produced: petroleum, transportation, and chemicals. These three industries are called **sectors** in a three-sector economy. Any industry that does not produce petroleum, transportation, or chemicals is said to be outside the economy. However, industries (or governments) outside the economy may have demands on the outputs of these three sectors. These types of demands are called **external demands**. Now the situation becomes a little more complicated because the three sectors tend to use their own outputs. For example, petroleum may demand as an input the output from transportation. These types of demands among the three sectors are referred to as **internal demands**.

Internal Demands

To get a handle on internal demands, and the most important matrix that it leads to, consider the following for our three-sector economy:

- One unit of petroleum (P) requires 0.1 unit of itself, 0.2 unit of transportation, and 0.4 unit of chemicals.
- One unit of transportation (T) requires 0.6 unit of petroleum, 0.1 unit of itself, and 0.25 unit of chemicals.
- One unit of chemicals (C) requires 0.2 unit of petroleum, 0.3 unit of transportation, and 0.2 unit of chemicals.

Units are usually measured in dollars. Hence we can view the above as saying that the production of \$1 worth of petroleum requires \$0.10 of petroleum, \$0.20 of transportation, and \$0.40 of chemicals. The information above can be nicely summarized in what is the central idea of input–output analysis: the input–output or **technological matrix**, matrix T.

$$T = \text{Input (amount used in production)} \quad \begin{array}{c} \text{Petr} \\ \text{Tran} \\ \text{Chem} \end{array} \begin{bmatrix} 0.1 & 0.6 & 0.2 \\ 0.2 & 0.1 & 0.3 \\ 0.4 & 0.25 & 0.2 \end{bmatrix}$$

Output (amount produced): Petr, Tran, Chem

Notice that in matrix T the 1st column read top to bottom tells us the requirements to produce 1 unit of petroleum.

Example 1   Determining an Internal Demand

If the three-sector economy produces \$900 million of petroleum, \$850 million of transportation, and \$800 million of chemicals, determine how much of this production is consumed internally. This internal consumption is also known as the **internal demand**.

Solution

We begin by organizing the information in a 3×1 matrix P.

$$P = \begin{bmatrix} 900 \\ 850 \\ 800 \end{bmatrix} \begin{array}{l} \text{Petr} \\ \text{Tran} \\ \text{Chem} \end{array}$$

Matrix P is called the **production matrix** or the **total output matrix**. The internal demand is simply the matrix product TP. This gives

$$TP = \begin{bmatrix} 0.1 & 0.6 & 0.2 \\ 0.2 & 0.1 & 0.3 \\ 0.4 & 0.25 & 0.2 \end{bmatrix} \begin{bmatrix} 900 \\ 850 \\ 800 \end{bmatrix} = \begin{bmatrix} 760 \\ 505 \\ 732.5 \end{bmatrix} \begin{matrix} \text{Petr} \\ \text{Tran} \\ \text{Chem} \end{matrix}$$

Interpret the Solution: The entries of TP tell us that the internal demand for petroleum, transportation, and chemicals is $760 million, $505 million, and $732.5 million, respectively. ∎

✓ Checkpoint 1

Now work Exercise 5.

External Demands and Leontief

In Example 1 we saw that the production (or total output) matrix was $P = \begin{bmatrix} 900 \\ 850 \\ 800 \end{bmatrix}$ and the internal demand was $\begin{bmatrix} 760 \\ 505 \\ 732.5 \end{bmatrix}$. To determine the **external demand** matrix D, we simply subtract internal demand from matrix P. That is,

$$\text{external demand} = D = \begin{bmatrix} 900 \\ 850 \\ 800 \end{bmatrix} - \begin{bmatrix} 760 \\ 505 \\ 732.5 \end{bmatrix} = \begin{bmatrix} 140 \\ 345 \\ 67.5 \end{bmatrix}$$

Observe that this leads to the following:

$$\text{production} = \text{internal demand} + \text{external demand}$$

$$\begin{bmatrix} 900 \\ 850 \\ 800 \end{bmatrix} = \begin{bmatrix} 760 \\ 505 \\ 732.5 \end{bmatrix} + \begin{bmatrix} 140 \\ 345 \\ 67.5 \end{bmatrix}$$

This is a fundamental observation of input–output analysis.

Production = Internal demand + External demand

or

Total output = Internal demand + External demand

Using the result of Example 1 and the discussion following Example 1, we can use matrices and write the prior boxed equation as

$$\text{Total output} = \text{Internal demand} + \text{External demand}$$
$$P = TP + D \tag{I}$$

This simple model will be our guide as we develop a technique to attack the major goal of input–output analysis. We now present this major goal.

■ **Basic Input–Output Problem**

Given the technological matrix for an economy and the external demands, determine the total output matrix (production matrix) of each sector that will meet both the internal and external demands.

Our basic input–output problem suggests that usually we are not given matrix P in equation (I). This is what we need to determine! Since these are usually our unknowns, let's work through our three-sector economy and see how matrices (and matrix inverses) will aid us. First let's call the output of the three sectors

$x_1 = $ dollar value of output of petroleum

$x_2 = $ dollar value of output of transportation

$x_3 = $ dollar value of output of chemicals

This means that matrix P now has the form

$$P = \begin{bmatrix} x_1 \\ x_2 \\ x_3 \end{bmatrix}$$

Since this is a variable matrix, let's rename it X. This means equation (I) becomes

$$X = TX + D \qquad \text{(II)}$$

Example 2 Solving for Total Output

Using the technological matrix T for our three-sector economy and external demand matrix D, use equation (II) and solve for X.

Solution

Substituting the given information into equation (II) and solving for X gives

$$\begin{bmatrix} x_1 \\ x_2 \\ x_3 \end{bmatrix} = \begin{bmatrix} 0.1 & 0.6 & 0.2 \\ 0.2 & 0.1 & 0.3 \\ 0.4 & 0.25 & 0.2 \end{bmatrix} \begin{bmatrix} x_1 \\ x_2 \\ x_3 \end{bmatrix} + \begin{bmatrix} 140 \\ 345 \\ 67.5 \end{bmatrix} \qquad X = TX + D$$

$$\begin{bmatrix} x_1 \\ x_2 \\ x_3 \end{bmatrix} - \begin{bmatrix} 0.1 & 0.6 & 0.2 \\ 0.2 & 0.1 & 0.3 \\ 0.4 & 0.25 & 0.2 \end{bmatrix} \begin{bmatrix} x_1 \\ x_2 \\ x_3 \end{bmatrix} = \begin{bmatrix} 140 \\ 345 \\ 67.5 \end{bmatrix} \qquad X - TX = D$$

$$\left(\begin{bmatrix} 1 & 0 & 0 \\ 0 & 1 & 0 \\ 0 & 0 & 1 \end{bmatrix} - \begin{bmatrix} 0.1 & 0.6 & 0.2 \\ 0.2 & 0.1 & 0.3 \\ 0.4 & 0.25 & 0.2 \end{bmatrix} \right) \begin{bmatrix} x_1 \\ x_2 \\ x_3 \end{bmatrix} = \begin{bmatrix} 140 \\ 345 \\ 67.5 \end{bmatrix} \qquad (I - T)X = D$$

$$\begin{bmatrix} 0.9 & -0.6 & -0.2 \\ -0.2 & 0.9 & -0.3 \\ -0.4 & -0.25 & 0.8 \end{bmatrix} \begin{bmatrix} x_1 \\ x_2 \\ x_3 \end{bmatrix} = \begin{bmatrix} 140 \\ 345 \\ 67.5 \end{bmatrix} \qquad \text{Simplification}$$

We now multiply both sides of the equation, on the left, by the inverse of the coefficient matrix.

$$\begin{bmatrix} \frac{1290}{661} & \frac{1060}{661} & \frac{720}{661} \\ \frac{560}{661} & \frac{1280}{661} & \frac{620}{661} \\ \frac{820}{661} & \frac{930}{661} & \frac{1380}{661} \end{bmatrix} \begin{bmatrix} 0.9 & -0.6 & -0.2 \\ -0.2 & 0.9 & -0.3 \\ -0.4 & -0.25 & 0.8 \end{bmatrix} \begin{bmatrix} x_1 \\ x_2 \\ x_3 \end{bmatrix} = \begin{bmatrix} \frac{1290}{661} & \frac{1060}{661} & \frac{720}{661} \\ \frac{560}{661} & \frac{1280}{661} & \frac{620}{661} \\ \frac{820}{661} & \frac{930}{661} & \frac{1380}{661} \end{bmatrix} \begin{bmatrix} 140 \\ 345 \\ 67.5 \end{bmatrix}$$

$$\begin{bmatrix} 1 & 0 & 0 \\ 0 & 1 & 0 \\ 0 & 0 & 1 \end{bmatrix} \begin{bmatrix} x_1 \\ x_2 \\ x_3 \end{bmatrix} = \begin{bmatrix} 900 \\ 850 \\ 800 \end{bmatrix}$$

$$\begin{bmatrix} x_1 \\ x_2 \\ x_3 \end{bmatrix} = \begin{bmatrix} 900 \\ 850 \\ 800 \end{bmatrix}$$

Interpret the Solution: An output of $900 million in petroleum, $850 million in transportation, and $800 million in chemicals will satisfy internal and external demands. ∎

The result of Example 2 is no surprise since the results were given to us in Example 1. There are two reasons we went through the calculations in Example 2. The first is to illustrate that we *can* find the outputs x_1, x_2, and x_3, and the second reason is to show numerically how to solve for matrix X. We now generalize our work in Example 2 by algebraically solving $X = TX + D$ for X.

$$X = TX + D \qquad \text{Given equation}$$
$$X - TX = D \qquad \text{Subtract } TX \text{ from both sides.}$$
$$IX - TX = D \qquad I \text{ is the identity matrix.}$$
$$(I - T)X = D \qquad \text{Factor out } X.$$
$$(I - T)^{-1}(I - T)X = (I - T)^{-1}D \qquad \text{Provided } I - T \text{ has an inverse}$$
$$IX = (I - T)^{-1}D \qquad (I - T)^{-1}(I - T) = I$$
$$X = (I - T)^{-1}D \qquad \text{Simplification}$$

We now summarize these results.

> ### Leontief Input–Output Model
>
> In a Leontief input–output model, the matrix equation giving the total output of each sector that meets the internal and external demands is
> $$X = TX + D$$
> where X is the **total output matrix**, T is the **technological matrix**, and D is the **external demand matrix**. The solution of this equation is
> $$X = (I - T)^{-1}D \qquad \text{(provided } I - T \text{ has an inverse)}$$

Example 3   Determining Total Output

Bucksville has three main industries: a coal-mining operation, an electric power-generating plant, and a local railroad. To mine $1 of coal, the mining operation must purchase $0.35 of electricity and $0.15 of transportation. To produce $1 of electricity, the generating plant requires $0.60 of coal, $0.10 of its own electricity, and $0.05 of transportation. To provide $1 of transportation, the railroad needs $0.50 of coal and $0.15 of electricity. In a certain week, the coal-mining operation receives an order for $50,000 of coal outside Bucksville and the electric generating plant receives an order for $30,000 of electricity outside Bucksville. For this week, there is no external demand for the local railroad. How much must each industry produce in this week to exactly meet the internal and external demands?

Solution

Understand the Situation: We begin by determining the technological matrix T and the external demand matrix D.

$$T = \text{Input} \begin{array}{c} \text{Coal} \\ \text{Elec.} \\ \text{RR} \end{array} \overset{\begin{array}{ccc} \text{Coal} & \text{Elec.} & \text{RR} \end{array}}{\begin{bmatrix} 0 & 0.60 & 0.50 \\ 0.35 & 0.10 & 0.15 \\ 0.15 & 0.05 & 0 \end{bmatrix}} \qquad D = \begin{bmatrix} 50,000 \\ 30,000 \\ 0 \end{bmatrix} \begin{array}{c} \text{Coal} \\ \text{Elec.} \\ \text{RR} \end{array}$$

with Output above.

We need to find

$$X = \begin{bmatrix} x_1 \\ x_2 \\ x_3 \end{bmatrix}$$

where x_1, x_2, and x_3 represent the value of coal products, electricity, and the railroad, respectively.

From the Leontief input–output model we know that

$$X = (I - T)^{-1} D$$

To determine $(I - T)^{-1}$, we first compute $I - T$.

$$I - T = \begin{bmatrix} 1 & 0 & 0 \\ 0 & 1 & 0 \\ 0 & 0 & 1 \end{bmatrix} - \begin{bmatrix} 0 & 0.60 & 0.50 \\ 0.35 & 0.10 & 0.15 \\ 0.15 & 0.05 & 0 \end{bmatrix} = \begin{bmatrix} 1 & -0.6 & -0.5 \\ -0.35 & 0.9 & -0.15 \\ -0.15 & -0.05 & 1 \end{bmatrix}$$

Using the methods in Section 2.4, we compute $(I - T)^{-1}$ to be

$$(I - T)^{-1} = \begin{bmatrix} \frac{3570}{2371} & \frac{2500}{2371} & \frac{2160}{2371} \\ \frac{1490}{2371} & \frac{3700}{2371} & \frac{1300}{2371} \\ \frac{610}{2371} & \frac{560}{2371} & \frac{2760}{2371} \end{bmatrix}$$

Finally, we compute X to be

$$X = (I - T)^{-1} D = \begin{bmatrix} \frac{3570}{2371} & \frac{2500}{2371} & \frac{2160}{2371} \\ \frac{1490}{2371} & \frac{3700}{2371} & \frac{1300}{2371} \\ \frac{610}{2371} & \frac{560}{2371} & \frac{2760}{2371} \end{bmatrix} \begin{bmatrix} 50{,}000 \\ 30{,}000 \\ 0 \end{bmatrix}$$

$$\approx \begin{bmatrix} 106{,}916.91 \\ 78{,}237.03 \\ 19{,}949.39 \end{bmatrix} \quad \text{(entries rounded to nearest hundredth)}$$

Interactive Activity

Determine the internal demand of the three industries in Example 3.

Interpret the Solution: To fulfill internal and external demand, $106{,}916.91 worth of coal, $78{,}237.03 worth of electricity, and $19{,}949.39 worth of railroad transportation should be produced. ∎

✓ **Checkpoint 2**   Now work Exercise 25.

SUMMARY

This section saw the presentation of one of the most important applications of matrices and their inverses to economics: the Leontief input–output model. Our presentation revolved around a simple three-sector economy; however, the concepts in this section easily extend to any n-sector economy. This includes the 81 sectors that Wassily Leontief analyzed in the 1958 U.S. economy. (You would want to use a computer and a spreadsheet for 81 sectors.) The most important part of Leontief input–output is the **technological matrix**. Other important components include **internal demands** and **external demands**.

• **Basic input–output problem**   Given the technological matrix for an economy and the external demands,

determine the total output matrix (production matrix) of each sector that will meet both the internal and external demands.

• **Leontief input–output model**   In a Leontief input–output model, the matrix equation giving the total output of each sector that meets the internal and external demands is

$$X = TX + D$$

where X is the total output matrix, T is the technological matrix, and D is the external demand matrix. The solution of this equation is

$$X = (I - T)^{-1} D \quad \text{(provided } I - T \text{ has an inverse)}$$

SECTION 2.5 EXERCISES

Exercises 1–4 use the following technological matrix T. Here agr, manu, trd, and srv represent agriculture, manufacturing, trade, and service, respectively.

$$T = \begin{array}{c} \text{agr} \\ \text{manu} \\ \text{trd} \\ \text{srv} \end{array} \begin{array}{cccc} \overset{\text{agr}}{} & \overset{\text{manu}}{} & \overset{\text{trd}}{} & \overset{\text{srv}}{} \\ \begin{bmatrix} 0.27 & 0.39 & 0 & 0.02 \\ 0.05 & 0.15 & 0 & 0.01 \\ 0.06 & 0.07 & 0.36 & 0.15 \\ 0.12 & 0.13 & 0.2 & 0.17 \end{bmatrix} \end{array}$$

Determine the number of units of agriculture, manufacturing, trade, and service required to produce:

1. One unit of agriculture

2. One unit of manufacturing

3. One unit of trade

4. One unit of service

In Exercises 5–8, a technological matrix T and a total output matrix X for each sector is given. Determine the internal demand. Assume that all entries represent dollars. See Example 1.

✓ **5.** $T = \begin{array}{c} \text{Util} \\ \text{Trans} \end{array} \begin{array}{cc} \overset{\text{Util}}{} & \overset{\text{Trans}}{} \\ \begin{bmatrix} 0.25 & 0.40 \\ 0.30 & 0.10 \end{bmatrix} \end{array}$ $X = \begin{bmatrix} 25{,}000 \\ 30{,}000 \end{bmatrix} \begin{array}{c} \text{Util} \\ \text{Trans} \end{array}$

6. $T = \begin{array}{c} \text{Util} \\ \text{Trans} \end{array} \begin{array}{cc} \overset{\text{Util}}{} & \overset{\text{Trans}}{} \\ \begin{bmatrix} 0.20 & 0.45 \\ 0.30 & 0.05 \end{bmatrix} \end{array}$ $X = \begin{bmatrix} 30{,}000 \\ 40{,}000 \end{bmatrix} \begin{array}{c} \text{Util} \\ \text{Trans} \end{array}$

7. $T = \begin{array}{c} \text{Agr} \\ \text{Cons} \\ \text{Util} \end{array} \begin{array}{ccc} \overset{\text{Agr}}{} & \overset{\text{Cons}}{} & \overset{\text{Util}}{} \\ \begin{bmatrix} 0.05 & 0.03 & 0.04 \\ 0.01 & 0.23 & 0.09 \\ 0.15 & 0.17 & 0.17 \end{bmatrix} \end{array}$

$X = \begin{bmatrix} 30{,}000 \\ 50{,}000 \\ 45{,}000 \end{bmatrix} \begin{array}{c} \text{Agr} \\ \text{Cons} \\ \text{Util} \end{array}$

8. $T = \begin{array}{c} \text{Agr} \\ \text{Cons} \\ \text{Util} \end{array} \begin{array}{ccc} \overset{\text{Agr}}{} & \overset{\text{Cons}}{} & \overset{\text{Util}}{} \\ \begin{bmatrix} 0.04 & 0.02 & 0.01 \\ 0.02 & 0.29 & 0.16 \\ 0.23 & 0.21 & 0.19 \end{bmatrix} \end{array}$

$X = \begin{bmatrix} 25{,}000 \\ 60{,}000 \\ 50{,}000 \end{bmatrix} \begin{array}{c} \text{Agr} \\ \text{Cons} \\ \text{Util} \end{array}$

Exercises 9–14 utilize the following Leontief input–output model: An economy is based on two sectors, construction (C) and services (S). The technological matrix T and the external demand matrices D_1, D_2, and D_3, in millions of dollars, are

$$T = \begin{array}{c} \text{C} \\ \text{S} \end{array} \begin{array}{cc} \overset{\text{C}}{} & \overset{\text{S}}{} \\ \begin{bmatrix} 0.2 & 0.1 \\ 0.4 & 0.6 \end{bmatrix} \end{array} \qquad D_1 = \begin{bmatrix} 5 \\ 3 \end{bmatrix} \begin{array}{c} \text{C} \\ \text{S} \end{array}$$

$$D_2 = \begin{bmatrix} 6 \\ 5 \end{bmatrix} \begin{array}{c} \text{C} \\ \text{S} \end{array} \qquad D_3 = \begin{bmatrix} 8 \\ 6 \end{bmatrix} \begin{array}{c} \text{C} \\ \text{S} \end{array}$$

9. Determine the number of units (dollars) of C and S required to produce $1 worth of C.

10. Determine the number of units (dollars) of C and S required to produce $1 worth of S.

11. Determine $I - T$ and $(I - T)^{-1}$.

12. Determine the total output for each sector that is required to satisfy D_1. (In other words, find the total output matrix X.)

13. Determine the total output for each sector that is required to satisfy D_2. (In other words, find the total output matrix X.)

14. Determine the total output for each sector that is required to satisfy D_3. (In other words, find the total output matrix X.)

Exercises 15–20 utilize the following Leontief input–output model: An economy is based on three sectors, agriculture (A), manufacturing (M), and transportation (T). The technological matrix T and the external demand matrices D_1 and D_2, in millions of dollars, are

$$T = \begin{array}{c} \text{A} \\ \text{M} \\ \text{T} \end{array} \begin{array}{ccc} \overset{\text{A}}{} & \overset{\text{M}}{} & \overset{\text{T}}{} \\ \begin{bmatrix} 0.2 & 0.2 & 0.1 \\ 0.4 & 0.5 & 0.2 \\ 0.1 & 0 & 0.2 \end{bmatrix} \end{array}$$

$$D_1 = \begin{bmatrix} 9 \\ 12 \\ 16 \end{bmatrix} \begin{array}{c} \text{A} \\ \text{M} \\ \text{T} \end{array} \qquad D_2 = \begin{bmatrix} 8 \\ 10 \\ 15 \end{bmatrix} \begin{array}{c} \text{A} \\ \text{M} \\ \text{T} \end{array}$$

15. Determine the number of units (dollars) of A, M, and T required to produce $1 worth of M.

16. Determine the number of units (dollars) of A, M, and T required to produce $1 worth of T.

17. Find $(I - T)$.

18. Given $(I - T)^{-1} = \begin{bmatrix} \frac{400}{247} & \frac{160}{247} & \frac{90}{247} \\ \frac{340}{247} & \frac{630}{247} & \frac{200}{247} \\ \frac{50}{247} & \frac{20}{247} & \frac{320}{247} \end{bmatrix}$, determine the

total output for each sector required to satisfy D_1.

19. Repeat Exercise 18 for D_2.

20. Explain why the entries in the columns of any technological matrix should not sum to greater than 1.

Applications

21. An economy has two main industries: agriculture and energy. Production of $1 worth of agriculture requires $0.10 from the agriculture sector and $0.35 from the energy sector. Production of $1 worth of energy requires $0.05 from the agriculture sector and $0.25 from the energy sector. Find the total outputs for each sector that are

needed to meet the external demand of $5 billion for agriculture and $12 billion for energy.

22. An economy has two main industries: agriculture and chemicals. Production of $1 worth of agriculture requires $0.10 from the agriculture sector and $0.45 from the chemicals sector. Production of $1 worth of chemicals requires $0.05 from the agriculture sector and $0.35 from the chemicals sector. Find the total outputs for each sector that are needed to meet the external demand of $5 billion for agriculture and $11 billion for chemicals.

23. A much simplified version of the U.S. economy in 1996, which deals with only two sectors, has the following technological matrix:

$$\begin{array}{cc} & \text{Agri} \quad \text{Manu} \\ T = \begin{array}{c} \text{Agri} \\ \text{Manu} \end{array} & \begin{bmatrix} 0.24 & 0.04 \\ 0.17 & 0.36 \end{bmatrix} \end{array}$$

The external demand (in millions of dollars) as given by exports is

$$D = \begin{bmatrix} 27{,}066 \\ 465{,}357 \end{bmatrix} \begin{array}{c} \text{Agri} \\ \text{Manu} \end{array}$$

Find the total outputs of each sector that are needed to meet the external demands. (*Source:* Bureau of Economic Analysis, www.bea.gov.)

24. A much simplified version of the U.S. economy in 1996, which deals with only two sectors, has the following technological matrix:

$$\begin{array}{cc} & \text{Minerals} \quad \text{Const} \\ T = \begin{array}{c} \text{Minerals} \\ \text{Const} \end{array} & \begin{bmatrix} 0.18 & 0.007 \\ 0.02 & 0.0008 \end{bmatrix} \end{array}$$

The external demand (in millions of dollars) as given by exports is

$$D = \begin{bmatrix} 8123 \\ 97 \end{bmatrix} \begin{array}{c} \text{Minerals} \\ \text{Const} \end{array}$$

Find the total outputs of each sector that are needed to meet the external demands. (*Source:* Bureau of Economic Analysis, www.bea.gov.)

25. An economy has three main industries: agriculture (A), manufacturing (M), and trade (T). Production of $1 worth of agriculture requires $0.27 from the agriculture sector, $0.05 from the manufacturing sector, and $0.06 from the trade sector. Production of $1 worth of manufacturing requires $0.39 from the agriculture sector, $0.15 from the manufacturing sector, and $0.07 from the trade sector. Production of $1 worth of trade requires nothing from agriculture and manufacturing and $0.36 from the trade sector.

(a) Write the technological matrix T for this economy.

(b) The external demand is $8 billion for agriculture, $23 billion for manufacturing, and $10 billion for trade. Write the external demand matrix D.

(c) Given that $(I - T)^{-1} = \begin{bmatrix} 1.41 & 0.65 & 0 \\ 0.08 & 1.21 & 0 \\ 0.14 & 0.19 & 1.56 \end{bmatrix}$, (entries

rounded to the nearest hundredth), determine the total outputs that are needed to meet the external demand.

26. An economy has three main industries: manufacturing (M), trade (T), and service (S). Production of $1 worth of manufacturing requires $0.15 from the manufacturing sector, $0.07 from the trade sector, and $0.13 from the service sector. Production of $1 worth of trade requires nothing from manufacturing, $0.36 from the trade sector, and $0.20 from the service sector. Production of $1 worth of service requires $0.01 from the manufacturing sector, $0.15 from the trade sector, and $0.17 from the service sector.

(a) Write the technological matrix T for this economy.

(b) The external demand is $12 billion for manufacturing, $20 billion for trade, and $8 billion for service. Write the external matrix D.

(c) Given that $(I - T)^{-1} = \begin{bmatrix} 1.18 & 0.05 & 0.02 \\ 0.18 & 1.66 & 0.30 \\ 0.23 & 0.40 & 1.28 \end{bmatrix}$, (entries

rounded to the nearest hundredth), determine the total outputs that are needed to meet the external demand.

27. A much simplified version of the U.S. economy in 1996, which deals with only three sectors, has the following technological matrix:

$$\begin{array}{cccc} & \text{Agri} & \text{Manu} & \text{Trade} \\ T = \begin{array}{c} \text{Agri} \\ \text{Manu} \\ \text{Trade} \end{array} & \begin{bmatrix} 0.24 & 0.04 & 0.0009 \\ 0.17 & 0.36 & 0.05 \\ 0.05 & 0.06 & 0.02 \end{bmatrix} \end{array}$$

The external demand (in millions of dollars) as given by exports is

$$D = \begin{bmatrix} 27{,}066 \\ 465{,}357 \\ 66{,}786 \end{bmatrix} \begin{array}{c} \text{Agri} \\ \text{Manu} \\ \text{Trade} \end{array}$$

Find the total output of each sector that is needed to meet the external demands. (*Source:* Bureau of Economic Analysis, www.bea.gov.)

28. A much simplified version of the U.S. economy in 1996, which deals with only three sectors, has the following technological matrix:

$$\begin{array}{cccc} & \text{Minerals} & \text{Const} & \text{Manu} \\ T = \begin{array}{c} \text{Minerals} \\ \text{Const} \\ \text{Manu} \end{array} & \begin{bmatrix} 0.18 & 0.007 & 0.03 \\ 0.02 & 0.0008 & 0.007 \\ 0.08 & 0.30 & 0.36 \end{bmatrix} \end{array}$$

The external demand (in millions of dollars) as given by exports is

$$D = \begin{bmatrix} 8123 \\ 97 \\ 465{,}357 \end{bmatrix} \begin{array}{c} \text{Minerals} \\ \text{Const} \\ \text{Manu} \end{array}$$

Find the total output of each sector that is needed to meet the external demands. (*Source:* Bureau of Economic Analysis, www.bea.gov.)

SECTION PROJECT

We began this section by stating that Wassily Leontief organized the 1958 U.S. economy into an 81×81 matrix. He also aggregated these 81 sectors into six groups. This section project deals with four of these six groups. The technological matrix for the four groups is

$$T = \begin{array}{c} \\ \text{FM} \\ \text{BM} \\ \text{BN} \\ \text{E} \end{array} \overset{\begin{array}{cccc} \text{FM} & \text{BM} & \text{BN} & \text{E} \end{array}}{\begin{bmatrix} 0.295 & 0.018 & 0.002 & 0.004 \\ 0.173 & 0.46 & 0.007 & 0.011 \\ 0.037 & 0.021 & 0.403 & 0.011 \\ 0.001 & 0.039 & 0.025 & 0.358 \end{bmatrix}}$$

FM represents final metals, such as construction machinery and household goods.

BM represents basic metal, such as mining and machine shop products.

BN represents basic nonmetal, such as glass, wood, and textiles.

E represents energy, such as coal, petroleum, electricity, and gas.

Matrix D gives the estimated external demands for the 1958 U.S. economy, in millions of dollars.

$$D = \begin{bmatrix} 75,548 \\ 14,444 \\ 33,501 \\ 23,527 \end{bmatrix} \begin{array}{l} \text{FM} \\ \text{BM} \\ \text{BN} \\ \text{E} \end{array}$$

(a) Use a graphing calculator to determine the total output for each sector.

(b) If an energy crisis increased the cost of energy by 20% in each sector, what would the total output for each sector need to be?

■ Why We Learned It

In this chapter we learned that matrices are common in business, life science, and other occupations. Knowing the characteristics of matrices, as well as operations used with matrices, is vital for applying them to the many diversified applications. For example, we saw that a business may use matrices to determine the maximum production capacity of a product, and we saw that a farmer can use matrix multiplication to determine the amount of nutrients that have to be added to crops for a growing season. Another example of matrix operations is that matrices can be used to keep track of inventories and the sales of certain items and the revenue generated from these sales at different stores. Matrix multiplication quickly led us to learn about inverse matrices. Inverse matrices were used to solve systems of equations. For example, we saw how a commercial baker could use matrix inverses to determine how much of each ingredient is needed to fill an order. Finally, Leontief input–output models are an application of matrices and inverse matrices, frequently used by economists, to analyze the relationship between production and demands for various sectors in an economy.

CHAPTER REVIEW EXERCISES

In Exercises 1 and 2, determine the dimension of the given matrix.

1. $A = \begin{bmatrix} 0 & 5 \\ -1 & 7 \\ 2 & 8 \end{bmatrix}$

2. $B = \begin{bmatrix} -3 & 4 & 6 & -1 \end{bmatrix}$

In Exercises 3 and 4, determine the indicated entry for the matrices given in Exercises 1 and 2.

3. $a_{2,1}$

4. $b_{1,3}$

5. Write an augmented matrix for the given system. **Do not solve.**

$$\begin{aligned} -2x + 3y - z &= 5 \\ x + y + z &= 13 \\ 4x - 5y + 2z &= -1 \end{aligned}$$

6. Write the linear system corresponding to the augmented matrix.

$$\begin{bmatrix} -2 & 5 & | & -7 \\ 1 & 4 & | & 3 \end{bmatrix}$$

In Exercises 7–10, use the Gauss–Jordan method to solve the given system of linear equations.

7. $-3x + 4y = 11$
$x + 5y = 9$

8. $-2x + 5y - z = 6$
$x - 10y + 4z = 0$
$3x + y - 2z = -11$

9. $x + 3y - z = 1$
$2x + 3y + 4z = 5$
$-3x - 9y + 3z = -3$

10. $x - 2y + z - 3w = -10$
$3x - y + 2z + w = -1$
$-x + 2y - z - 3w = -8$
$2x - 3y + 3z - w = -4$

In Exercises 11 and 12, use the Gauss–Jordan method to solve the given system of equations using your calculator's matrix capabilities (either step-by-step row operators or rref(command). Round answers to four decimal places.

11. $-3.7x + 5.1y = 12.1$
$0.5x - 14.2y = 3.8$

12. $-2.17x + 5.13y - 0.42z + 4.12w = 5.67$
$5.89x - 3.24y + 8.53z + 1.02w = -3.29$
$-0.05x + 1.07y - 0.43z + 8.59w = 0.78$
$1.14x + 2.29y + 1.92z - 7.38w = 1.11$

13. Grace Richards makes 4-weight and 5-weight fly-fishing rods. To make a 4-weight rod takes 6 hours of labor and to make a 5-weight rod takes 8 hours of labor. The materials for one 4-weight rod cost her $60 and the materials for one 5-weight rod cost her $80. Suppose that for 1 week Grace works 40 hours and spends $400 on materials.

(a) Let x represent the number of 4-weight rods made in 1 week and y represent the number of 5-weight rods made in 1 week. Write a system of two linear equations in which one equation relates costs and the other relates time.

(b) Solve the system in part (a) to determine how many 4- and 5-weight rods Grace can make in 1 week, assuming that she works 40 hours and spends $400 on materials.

14. The Smoothie-Go-Go drink stand carries two blends of smoothies. The first blend contains 70% fruit and 30% yogurt. The second blend contains 85% fruit and 15% yogurt. How many ounces of each blend should be combined to get a 16-ounce drink that is 75% fruit and 25% yogurt?

15. The Quinn Outdoor Company makes three types of day backpacks: small, medium, and large. Each size requires the resources of three departments, as shown in Table R.1. The fabric cutting, assembly, and packing departments have available a maximum of 310, 405, and 150 hours each week, respectively.

Table R.1

	Small (hr)	Medium (hr)	Large (hr)
Fabric cutting	0.3	0.4	0.6
Assembly	0.4	0.5	0.8
Packaging	0.1	0.2	0.3

(a) Let x represent the number of small day backpacks made in 1 week, y represent the number of medium day backpacks made in 1 week, and z represent the number of large day backpacks made in 1 week. Write a system of three linear equations using this information and the given department maximum weekly times.

(b) Solve the system in part (a) to determine how many small, medium, and large day backpacks can be made in 1 week, assuming that the company operates at full capacity.

16. Pat wishes to invest $15,000 in three different areas: a savings account earning 3%, a certificate of deposit earning 5%, and a mutual fund earning 8%. She decides that twice as much should be invested in the mutual fund as in the savings account. Assuming these rates, how much should she place in each area to produce exactly $3000 each year on the investments?

17. A dietitian arranged a special diet for a patient that is made up of three basic foods, Alive, Gday, and Cardio. The diet is to include 270 units of iron, 340 units of vitamin C, and 126 units of vitamin B. The number of units per ounce for each basic food is given in Table R.2. How many ounces of each food must be used to exactly meet the diet requirements?

Table R.2

	Alive	Gday	Cardio
Iron	15	10	8
Vitamin C	25	5	10
Vitamin B	5	7	4

Refer to the following matrices to perform the indicated operations in Exercises 18–23.

$$A = \begin{bmatrix} -2 & 0 \\ 1 & 7 \end{bmatrix} \quad B = \begin{bmatrix} 4 & 5 \\ -1 & 3 \end{bmatrix}$$

$$C = \begin{bmatrix} 2 & 1 & 5 \\ -1 & 0 & 4 \\ 3 & 1 & 2 \end{bmatrix} \quad D = \begin{bmatrix} -1 & -3 & 5 \\ 0 & 8 & 4 \\ -2 & 1 & 3 \end{bmatrix}$$

18. $A + B$ **19.** $C - D$

20. $2A$ **21.** $-3D$

22. $4A + 2B$ **23.** $-2C - 3D$

Refer to matrices A through D given for Exercises 18–23 to answer Exercises 24 and 25.

24. Explain why $A + D$ cannot be done.

25. Determine $(C + D)^T$, the transpose of $C + D$.

In Exercises 26 and 27, find the values of the variables in each equation.

26. $\begin{bmatrix} 3 & 4 \\ y & x \end{bmatrix} = \begin{bmatrix} z+3 & 4 \\ 7 & -3 \end{bmatrix}$

27. $\begin{bmatrix} -x+1 & 3y & z+2 \\ -2w & 5 & 2 \end{bmatrix} + \begin{bmatrix} 3x & 4y & 1 \\ w-1 & 3 & 4 \end{bmatrix} = \begin{bmatrix} 11 & 21 & -4 \\ -8 & 8 & 24p \end{bmatrix}$

In Exercises 28 and 29, use a graphing calculator and its matrix capabilities and refer to the given matrices to perform the indicated operations.

$$A = \begin{bmatrix} 5.9 & 3.8 \\ 1.2 & -4.1 \\ -0.4 & 2.1 \end{bmatrix} \qquad B = \begin{bmatrix} -2.6 & -1.1 \\ 7.3 & 2.7 \\ -0.5 & 1.6 \end{bmatrix}$$

28. $3.8A + 0.6B$ **29.** $-2.2A - 4.7B$

30. Coulombe's Snowboards has the following data for sales of a popular board during December, January, and February (150″ and 160″ refer to the size of the board and Step-in and Traditional refer to two different types of bindings).

December Sales

	Step-in	Traditional
150″	5	8
160″	4	6

January Sales

	Step-in	Traditional
150″	10	12
160″	7	9

February Sales

	Step-in	Traditional
150″	9	7
160″	7	12

(a) If the rows refer to the length of the board (150″ or 160″) and the columns refer to the binding type (Step-in or Traditional), write three matrices to represent the data. Let D, J, and F be the matrices for December, January, and February, respectively.

(b) Determine a matrix giving the total sales for Coulombe's Snowboards during December, January, and February.

31. Quin Jackson has two types of IRA retirement accounts, certificates of deposit (CD) and treasury notes (TN), held in two banks, First Century (FC) and Hilltop

Bank (HB). The dollar amounts in each at the beginning of 2001 are given in matrix A.

$$A = \begin{matrix} & \text{CD} \quad\;\; \text{TD} \\ \begin{bmatrix} 25{,}000 & 20{,}000 \\ 15{,}000 & 30{,}000 \end{bmatrix} & \begin{matrix} \text{FC} \\ \text{HB} \end{matrix} \end{matrix}$$

(a) Assuming a 9% growth rate on each account, determine a matrix I giving the earnings on the various accounts during 2001.

(b) Determine a matrix that represents the value of each account at the end of 2001.

(c) Assuming that the CDs and TNs at both banks have the same growth rate in 2002, what would those growth rates have to be for Quin Jackson to earn \$3379 in interest from both accounts at FC and \$3597 in interest from both accounts at HB at the end of 2002?

32. For D, J, and F in Exercise 30, compute $\frac{1}{3}(D + J + F)$ and interpret.

33. For matrices A and I in Exercise 31, find $a_{2,2}$ and $i_{1,2}$ and interpret.

34. The data in Table R.3 give the percentage of U.S. male smokers and the percentage of U.S. female smokers.

Table R.3

	18–24	25–34	35–44	45–64	65 and over
			Male		
1985	28.0	38.2	37.6	33.4	19.6
1990	26.6	31.6	34.5	29.3	14.6
1994	29.8	31.4	33.2	28.3	13.2
			Female		
1985	30.4	32.0	31.5	29.9	13.5
1990	22.5	28.2	24.8	24.8	11.5
1994	25.2	28.8	26.8	22.8	11.1

SOURCE: U.S. Census Bureau, www.census.gov

(a) Write a matrix M and a matrix F showing the percentage of U.S. male smokers and U.S. female smokers, respectively.

(b) Find and interpret $m_{2,3}$ and $f_{1,4}$.

(c) Using M and F, write a matrix S in which each entry shows how much more (or less) smoking is done by females than males.

(d) During which year and in what age group did a higher percentage of females than males smoke?

In Exercises 35 and 36, the dimension of matrices A and B are given.

(a) Find the dimension of the matrix product AB, if it exists.

(b) Find the dimension of the matrix product BA, if it exists.

35. A is a 5×2 and B is a 2×5.

36. A is a 3×4 and B is a 2×3.

In Exercises 37–41 find the matrix product.

37. $\begin{bmatrix} -3 & 1 \\ 7 & 4 \end{bmatrix}\begin{bmatrix} -2 \\ 3 \end{bmatrix}$

38. $\begin{bmatrix} -1 & 4 \\ 5 & 3 \end{bmatrix}\begin{bmatrix} 1 & 0 \\ 0 & 1 \end{bmatrix}$

39. $\begin{bmatrix} -2 & 1 \\ 4 & 4 \\ 3 & -2 \end{bmatrix}\begin{bmatrix} 7 & 6 & -2 \\ 1 & 4 & 3 \end{bmatrix}$

40. $[-2 \ \ 0 \ \ 5]\begin{bmatrix} 7 & -8 \\ 2 & -6 \end{bmatrix}$

41. $\begin{bmatrix} 1 & -2 & 4 & 3 \\ 7 & 8 & -5 & 2 \\ 16 & -1 & -12 & 4 \\ 8 & 9 & -13 & 7 \end{bmatrix}\begin{bmatrix} 2 \\ -3 \\ 5 \\ 7 \end{bmatrix}$

42. Use a graphing calculator to find the matrix product.

$$\begin{bmatrix} 27 & 35 & 80 \\ 42 & 51 & 17 \\ 14 & 37 & 63 \end{bmatrix}\begin{bmatrix} 200 \\ 500 \\ 700 \end{bmatrix}$$

43. Consider $A = \begin{bmatrix} 0.45 & 0.55 \\ 0.34 & 0.66 \end{bmatrix}$.

(a) Use a graphing calculator to compute A^2, A^5, A^{10}, and A^{20}.

(b) Describe any pattern observed in part (a).

In Exercises 44 and 45, write the system of linear equations as a matrix equation.

44. $4x - 5y = 8$
$-2x + y = -3$

45. $x + 3y - 2z = 4$
$-2x + 4y - 5z = -1$
$-x - y + z = 7$

46. The total sales for January for three models of snowboards (lengths of $160''$ and $170''$), Racing, All-mountain, and Backcountry, at Coulombe's Snowboards is represented in matrix T.

$$T = \begin{bmatrix} 5 & 4 \\ 20 & 10 \\ 7 & 10 \end{bmatrix}\begin{matrix} \text{Racing} \\ \text{All-Racing} \\ \text{Backcountry} \end{matrix}$$

with columns $160''$ $160''$.

The selling price in dollars for each is given in the matrix S.

$$S = \begin{bmatrix} 600 & 500 & 550 \\ 700 & 600 & 650 \end{bmatrix}\begin{matrix} 160'' \\ 170'' \end{matrix}$$

with columns Racing, All-mountain, Backcountry.

(a) Find TS.

(b) Interpret the entries in TS.

47. Kris, a dietitian in a local hospital, plans a special diet for a patient that is made up of three basic foods, Zor, Daan, and Cretin. The number of units of iron, vitamin C, and vitamin B, per ounce, is given in matrix F.

$$F = \begin{bmatrix} 15 & 10 & 8 \\ 25 & 5 & 10 \\ 5 & 7 & 4 \end{bmatrix}\begin{matrix} \text{Iron} \\ \text{Vitamin C} \\ \text{Vitamin B} \end{matrix}$$

with columns Zor, Daan, Cretin.

At breakfast the patient had 5 ounces of Zor, 8 ounces of Daan, and 12 ounces of Cretin. For lunch the patient had 6 ounces of Zor, 10 ounces of Daan, and 16 ounces of Cretin.

(a) Write a 3×1 matrix A that represents the amount of each food consumed (in ounces) at breakfast.

(b) Write a 3×1 matrix B that represents the amount of each food consumed (in ounces) at lunch.

(c) Find FA and interpret the entries.

(d) Find FB and interpret the entries.

48. The data in matrix F give the amounts of nitrogen, phosphate, and potash (in lb/acre) needed to grow celery, cabbage, and carrots for an entire growing season in California. (*Source:* U.S. Department of Agriculture, www.usda.gov/nass.)

$$F = \begin{bmatrix} 336 & 203 & 249 \\ 171 & 86 & 228 \\ 147 & 78 & 67 \end{bmatrix}\begin{matrix} \text{Nitrogen} \\ \text{Phosphate} \\ \text{Potash} \end{matrix}$$

with columns Celery, Cabbage, Carrots.

Matrix P represents the number of acres of each crop that Rebecca wishes to plant.

$$P = \begin{bmatrix} 20 \\ 40 \\ 15 \end{bmatrix}\begin{matrix} \text{Celery} \\ \text{Cabbage} \\ \text{Carrots} \end{matrix}$$

(a) Find FP.

(b) How many pounds of nitrogen, phosphate, and potash will Rebecca need for the entire growing season?

49. Matrix B gives the percentage of people who live in three different buildings of an apartment complex, A, B, and C, who recycle and don't recycle.

$$R = \begin{bmatrix} 0.80 & 0.40 & 0.50 \\ 0.20 & 0.60 & 0.50 \end{bmatrix}\begin{matrix} \text{Recycle} \\ \text{Don't recycle} \end{matrix}$$

with columns A, B, C.

(a) If 500 people live in building A, 400 in building B, and 600 in building C, write a 3×1 matrix P that represents the amount of people in each building.

(b) Find RP.

(c) How many people total, among the three buildings, recycle?

(d) How many people total, among the three buildings, don't recycle?

50. Lindsay, a mathematics teacher, has given three tests to two different geometry sections, Sections 1 and 2. Each section consist of five students. Lindsay records their scores in matrices S and T.

$$S = \begin{array}{c} \begin{array}{ccc} & \text{Test} & \\ 1 & 2 & 3 \end{array} \\ \begin{bmatrix} 89 & 95 & 97 \\ 92 & 99 & 100 \\ 84 & 76 & 79 \\ 85 & 87 & 88 \\ 74 & 78 & 81 \end{bmatrix} \begin{array}{l} \text{Paul} \\ \text{Beth} \\ \text{Dave} \\ \text{Kris} \\ \text{Joe} \end{array} \end{array}$$

Section 1

$$T = \begin{array}{c} \begin{array}{ccc} & \text{Test} & \\ 1 & 2 & 3 \end{array} \\ \begin{bmatrix} 90 & 92 & 87 \\ 71 & 93 & 74 \\ 81 & 87 & 78 \\ 96 & 99 & 100 \\ 72 & 83 & 85 \end{bmatrix} \begin{array}{l} \text{Teri} \\ \text{Cindy} \\ \text{Mika} \\ \text{Lera} \\ \text{Rob} \end{array} \end{array}$$

Section 2

Let $C = \begin{bmatrix} 1 \\ 1 \\ 1 \end{bmatrix}$ and $R = \begin{bmatrix} 1 & 1 & 1 & 1 & 1 \end{bmatrix}$. In parts (a) through (e), round all entries to the nearest whole number.

(a) Compute $\frac{1}{3}(SC)$ and interpret the entries.

(b) Compute $\frac{1}{3}(TC)$ and interpret the entries.

(c) Compute $\frac{1}{5}(RS)$ and interpret the entries.

(d) Compute $\frac{1}{5}(RT)$ and interpret the entries.

(e) Compute $\frac{1}{10}(RS + RT)$ and interpret the entries.

In Exercises 51 and 52, determine if the given matrices are inverses of each other.

51. $\begin{bmatrix} -1 & 2 \\ 7 & 6 \end{bmatrix}$ and $\begin{bmatrix} 3 & 5 \\ 2 & -1 \end{bmatrix}$

52. $\begin{bmatrix} -1 & 2 & 4 \\ 2 & 6 & 0 \\ 0 & 3 & 3 \end{bmatrix}$ and $\begin{bmatrix} -3 & -1 & 4 \\ 1 & \frac{1}{2} & -\frac{4}{3} \\ -1 & -\frac{1}{2} & \frac{5}{3} \end{bmatrix}$

In Exercises 53 and 54, find the inverse of each matrix.

53. $\begin{bmatrix} -1 & 0 & 3 \\ 2 & 1 & 2 \\ 0 & -2 & 4 \end{bmatrix}$

54. $\begin{bmatrix} 1 & 0 & 3 & -1 \\ 0 & -2 & 1 & 4 \\ 0 & 0 & 1 & -3 \\ -2 & -3 & 1 & 4 \end{bmatrix}$

In Exercises 55 and 56, use the formula for the inverse of a 2×2 matrix to find the inverse for the given matrix, if it exists.

55. $\begin{bmatrix} 3 & 4 \\ -2 & 6 \end{bmatrix}$

56. $\begin{bmatrix} 2 & -3 \\ -2 & 3 \end{bmatrix}$

In Exercises 57 and 58, use a graphing calculator to find the inverse for each matrix, if it exists. Where possible, express entries as fractions.

57. $\begin{bmatrix} 3.3 & 2.3 & 1.3 \\ 0.3 & 3.3 & 1.3 \\ 4.3 & 6.3 & 5.3 \end{bmatrix}$

58. $\begin{bmatrix} 1 & 0 & 2 & -2 & 3 \\ 1 & 5 & -2 & 6 & -1 \\ 0 & 0 & 5 & -3 & -2 \\ 8 & 4 & -4 & -2 & 5 \\ 1 & 0 & 1 & -1 & 2 \end{bmatrix}$

In Exercises 59 and 60, solve each system of linear equations by using the inverse of the coefficient matrix.

59. $\begin{aligned} 3x + 4y &= 13 \\ -2x + 6y &= 26 \end{aligned}$
(See Exercise 55)

60. $\begin{aligned} -x + 3z &= -5 \\ 2x + y + 2z &= 10 \\ -2y + 4z &= 20 \end{aligned}$
(See Exercise 53)

In Exercises 61 and 62, solve each system of linear equations by using the inverse of the coefficient matrix and a graphing calculator.

61. $\begin{aligned} 3.3x + 2.3y + 1.3z &= 4.3 \\ 0.3x + 3.3y + 1.3z &= 5.3 \\ 4.3x + 6.3y + 5.3z &= 8.3 \end{aligned}$
(See Exercise 57)

62. $\begin{aligned} x + 2z - 2v + 3w &= 1 \\ x + 5y - 2z + 6v - w &= -2 \\ 5z - 3v - 2w &= 7 \\ 8x + 4y - 4z - 2v + 5w &= 4 \\ x + z - v + 2w &= 3 \end{aligned}$
(See Exercise 58)

63. A small pizza shop makes three different sizes of pizza: small, medium, and large. Each pizza requires a certain amount (in cups) of flour, cheese, and sauce, as shown in matrix A.

$$A = \begin{array}{c} \begin{array}{ccc} \text{Small} & \text{Medium} & \text{Large} \end{array} \\ \begin{bmatrix} 1 & 3 & 3 \\ 2 & 3 & 4 \\ 1 & 2 & 3 \end{bmatrix} \begin{array}{l} \text{Flour} \\ \text{Cheese} \\ \text{Sauce} \end{array} \end{array}$$

To fill the daily orders for the three sizes of pizza, the amounts of ingredients used daily are given in matrix B.

$$B = \begin{bmatrix} 135 \\ 180 \\ 125 \end{bmatrix} \begin{array}{l} \text{Flour} \\ \text{Cheese} \\ \text{Sauce} \end{array}$$

(a) Let the number of daily orders for small, medium, and large pizzas be represented by x, y, and z, respectively. Write a 3×1 matrix X with entries x, y, and z that represents the daily orders. Write a matrix equation $AX = B$.

(b) Use A^{-1} to solve the equation in part (a).

(c) How many large pizzas does the shop sell daily?

(d) Using X from part (b), verify that $AX = B$.

64. Freeman's Mountain, a local ski resort, sells seasons passes for children and adults at $350 and $500, respectively. Table R.4 gives the revenue generated for the 1998–1999, 1999–2000, and 2000–2001 seasons and the number of season passes sold.

Table R.4

	1998–1999	1999–2000	2000–2001
Tickets sold	1200	1300	1550
Revenue	$525,000	$560,000	$670,000

Let x represent the number of $350 children season passes sold, and let y represent the number of $500 adult season passes sold.

(a) Write a system of equations for each season: 1998–1999, 1999–2000, and 2000–2001.

(b) For each system in part (a), let matrix A be the coefficient matrix, matrix X the variable matrix, and matrix B the constant matrix. Write a matrix equation in the form $AX = B$ for each system in part (a).

(c) Use A^{-1} to solve each system in part (b) and interpret. (Notice that A^{-1} is the same for all three equations.)

65. Snowsports manufactures snowshoes at three different plants. Due to different technologies at each plant, the production rate (in snowshoes/hour) differs for each plant, as shown in Table R.5.

Table R.5

	Plant I	Plant II	Plant III
Running snowshoes	10	12	8
Hiking snowshoe	15	7	12
Backpacking snowshoes	6	12	15

Kerri, the production manager at Snowsports, has received an order to produce 2800 running snowshoes, 2880 hiking snowshoes, and 2970 backpacking snowshoes. To fill this order exactly, Kerri lets x, y, and z represent the number of hours that plants I, II, and III, respectively, should be scheduled to operate.

(a) Write a system of equations to model this situation.

(b) Write a matrix equation in the form $AX = B$ for the system in part (a).

(c) Use A^{-1} to solve the equation in part (b) and interpret.

66. (*continuation of Exercise 65*) Kerri has received another order requesting 2980 running snowshoes, 3370 hiking snowshoes, and 3390 backpacking snowshoes. With everything else from Exercise 65 remaining the same, do the following.

(a) Write a system of equations to model this situation.

(b) Set up an augmented matrix that represents the system in part (a).

(c) Use your graphing calculator to find the row reduced echelon form of the matrix in part (b) and interpret.

(d) Write a matrix equation in the form $AX = B$ for the system in part (a).

(e) Use A^{-1} to solve the equation in part (d) and compare the results to those in part (c).

67. Matrix A gives the percentage, as a decimal, of the total vote for three states, California, Florida, and Ohio, for the Democratic, Republican, and Reform party candidates in the 1996 presidential election (*Source:* U.S. Census Bureau, www.census.gov).

$$A = \begin{bmatrix} 0.511 & 0.48 & 0.474 \\ 0.382 & 0.423 & 0.41 \\ 0.07 & 0.091 & 0.107 \end{bmatrix} \begin{matrix} \text{Democratic} \\ \text{Republican} \\ \text{Reform} \end{matrix}$$

with columns CA, FL, OH.

Matrix B represents the total number of Democratic, Republican, and Reform party votes, in thousands, for the three states combined.

$$B = \begin{bmatrix} 9815 \\ 7930 \\ 1669 \end{bmatrix} \begin{matrix} \text{Democratic} \\ \text{Republican} \\ \text{Reform} \end{matrix}$$

(a) Let the total number of votes in California, Florida, and Ohio be represented by x, y, and z, respectively. Write a matrix equation $AX = B$.

(b) Use your graphing calculator to solve the equation in part (a) and interpret. Round entries to the nearest whole number.

Exercises 68–70 use the following technology matrix T. Here A, E, and C represent agriculture, energy, and chemicals, respectively.

$$T = \begin{matrix} A \\ E \\ C \end{matrix} \begin{bmatrix} 0.10 & 0.18 & 0.05 \\ 0.25 & 0.15 & 0.20 \\ 0.34 & 0 & 0.20 \end{bmatrix}$$

with columns A, E, C.

68. Determine the number of units of agriculture, energy, and chemicals required to produce 1 unit of agriculture.

69. Determine the number of units of agriculture, energy, and chemicals required to produce 1 unit of energy.

70. Determine the number of units of agriculture, energy, and chemicals required to produce 1 unit of chemicals.

In Exercises 71 and 72 you are given a technology matrix T and a total output matrix X for each sector. Determine the internal demand. Assume that all entries represent dollars.

71. $T = \begin{matrix} \text{Util} \\ \text{Trans} \end{matrix} \begin{bmatrix} 0.15 & 0.20 \\ 0.35 & 0.40 \end{bmatrix}$ (cols Util Trans) $\quad X = \begin{bmatrix} 30,000 \\ 45,000 \end{bmatrix} \begin{matrix} \text{Util} \\ \text{Trans} \end{matrix}$

72. $T = \begin{matrix} \text{Agr} \\ \text{Cons} \\ \text{Util} \end{matrix} \begin{bmatrix} 0.10 & 0.01 & 0.08 \\ 0.04 & 0.30 & 0.09 \\ 0.23 & 0.15 & 0.24 \end{bmatrix}$ (cols Agr Cons Util) $\quad X = \begin{bmatrix} 15,000 \\ 25,000 \\ 20,000 \end{bmatrix} \begin{matrix} \text{Agr} \\ \text{Cons} \\ \text{Util} \end{matrix}$

Exercises 73–77 utilize the following Leontief input–output model: An economy is based on three sectors, construction (C), manufacturing (M), and transportation (T). The technology matrix T and the external demand matrices D_1 and D_2, in millions of dollars, are

$$T = \begin{array}{c} \\ C \\ M \\ T \end{array} \begin{array}{ccc} C & M & T \\ \begin{bmatrix} 0.1 & 0.3 & 0.1 \\ 0.2 & 0.1 & 0.3 \\ 0.4 & 0.1 & 0.5 \end{bmatrix} \end{array} \qquad D_1 = \begin{bmatrix} 6 \\ 8 \\ 13 \end{bmatrix} \begin{array}{c} C \\ M \\ T \end{array}$$

$$D_2 = \begin{bmatrix} 7 \\ 4 \\ 10 \end{bmatrix} \begin{array}{c} A \\ M \\ T \end{array}$$

73. Determine the number of units (dollars) of C, M, and T required to produce $1 worth of C.

74. Determine the number of units (dollars) of C, M, and T required to produce $1 worth of M.

75. Find $(I - T)$.

76. Given $(I - T)^{-1} = \begin{bmatrix} \frac{210}{137} & \frac{80}{137} & \frac{90}{137} \\ \frac{110}{137} & \frac{205}{137} & \frac{145}{137} \\ \frac{190}{137} & \frac{105}{137} & \frac{375}{137} \end{bmatrix}$, determine

the total output for each sector required to satisfy D_1.

77. Repeat Exercise 76 for D_2.

78. An economy has two main industries: agriculture and manufacturing. Production of $1 worth of agriculture requires $0.15 from the agriculture sector and $0.45 from the manufacturing sector. Production of $1 worth of manufacturing requires $0.20 from the agriculture sector and $0.25 from the manufacturing sector. Find the total output for each sector needed to meet the external demand of $4 billion for agriculture and $10 billion for manufacturing.

79. A simplified version of an economy that deals with two sectors, agriculture (A) and energy (E), has the following technology matrix:

$$T = \begin{array}{c} A \\ E \end{array} \begin{array}{cc} A & E \\ \begin{bmatrix} 0.10 & 0.14 \\ 0.34 & 0.21 \end{bmatrix} \end{array}$$

The external demand (in millions of dollars) is given by

$$D = \begin{bmatrix} 8 \\ 15 \end{bmatrix} \begin{array}{c} A \\ E \end{array}$$

Find the total output of each sector needed to meet the external demands.

80. (continuation of Exercise 79) Repeat Exercise 79 if the external demand for agriculture is $10 million and $15 million for energy.

81. An economy has three main industries: agriculture (A), chemicals (C), and energy (E). Production of $1 worth of agriculture requires $0.15 from the agriculture sector, $0.04 from the chemical sector, and $0.05 from the energy sector. Production of $1 worth of chemicals requires $0.02 from the agriculture sector, $0.12 from the chemical sector, and $0.25 from the energy sector. Production of $1 worth of energy requires $0.04 from the agriculture sector, $0.18 from the chemical sector, and $0.10 from the energy sector.

(a) Write the technological matrix T for this economy.

(b) The external demand is $12 billion for agriculture, $8 billion for chemicals, and $15 billion for energy. Write the external demand matrix D.

(c) Given $(I - T)^{-1} = \begin{bmatrix} 1.18 & 0.04 & 0.06 \\ 0.07 & 1.21 & 0.24 \\ 0.09 & 0.34 & 1.18 \end{bmatrix}$ (entries to the

nearest hundredth), determine the total output needed to meet the external demand.

82. An economy that deals with three sectors has the following technological matrix:

$$T = \begin{array}{c} \\ Agri \\ Manu \\ Trade \end{array} \begin{array}{ccc} Agri & Manu & Trade \\ \begin{bmatrix} 0.18 & 0.24 & 0.03 \\ 0.12 & 0.16 & 0.14 \\ 0.05 & 0.20 & 0.22 \end{bmatrix} \end{array}$$

The total output of each sector, in millions of dollars, needed to meet the external demand is given by

$$X = \begin{bmatrix} 16.2 \\ 4.2 \\ 8.7 \end{bmatrix}$$

(a) Find $I - T$.

(b) Determine the external demand matrix D. Round entries to the nearest tenth.

(c) Verify the given external demand matrix X by using your calculator to compute $(I - T)^{-1} D$, where D is the matrix found in part (b).

CHAPTER 2 PROJECT

SOURCE: Palm, Inc., www.palm.com

Tech Mart, Electrifun, Gadgets-4-U, and Plug-n-Play are selling the same type of Palm Pilots: the m100, m105, m500, and m505. The retail prices of these models are $129.00, $199.00, $399.00, and $449.00, respectively.

Suppose that Tech Mart sells 2 m100s, 3 m105s, 7 m500s, and 6 m505s; Electrifun sells 4 m100s, 4 m105s, 2 m500s, and 5 m505s; Gadgets-4-U sells 3 m100s, 5 m105s, 8 m500s, and 9 m505s; and Plug-n-Play sells 3 m100s, 3 m105s, 5 m500s, and 5 m505s. All sales are for 1 month's time.

1. Create a 4 × 4 matrix A for the amount sold by each store in which the rows represent the name of the store and the columns represent the model name.

2. Create a 4 × 1 matrix B for the price of each model.

3. Find AB and interpret the results. Which store generated the most revenue? By referring to the number of Palms sold and the price per model, can you explain why this store generated the most revenue?

Suppose that a manufacturer makes the Palm m100, m500, and m505. The m100 has a volume of 10.47 cubic inches, weighs 4.4 ounces, and contains 2 MB of memory. The m500 has a volume of 5.58 cubic inches, weighs 4 ounces, and contains 8 MB of memory. The m505 has a volume of 6.98 cubic inches, weighs 4.9 ounces, and contains 8 MB of memory.

Consider a store, McMikes, that will sell a specified number of each type of Palm Pilot in one bundled package.

Suppose that the bundled package holds 72.58 cubic inches and can carry 45.3 ounces, and that there's a tag on the box that says "a total of 68 MB inside."

4. Create a 3 × 3 matrix C where the rows represent the size, weight and amount of memory and the columns represent Palm m100, m500, and m505, respectively.

5. Create a 3 × 1 matrix D for the limits of the package.

6. Find $C^{-1}D$ and interpret the results.

7. Using the result of question 6 and the price of each Palm Pilot as listed above, how much revenue can McMikes hope to generate from selling all the Palms in this shipping box?

8. Repeat questions 1 through 7 if a different shipping box was used that could hold 70.49 cubic inches and carry 34.9 ounces, and the limit of memory is 34 MB.

3

Linear Programming: The Graphical Method

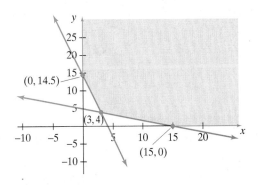

Minimize
$$C = 2.09x + 1.59y$$

Subject to :
$$28x + 8y \geq 116$$
$$2x + 6y \geq 30$$
$$x \geq 0, \ y \geq 0$$

Hospital dietitians are often charged with minimizing food costs while meeting certain nutrient requirements. One technique for meeting both goals is to develop a mathematical model involving constraints and then to use linear programming methods to solve the model. Individuals can employ the same process even when eating at a fast food restaurant such as McDonald's. The figure shows the graph of the linear inequalities that form the problem constraints. One of the corner points of the feasible region minimizes the objective function.

What We Know

In Chapter 1 we learned how to solve systems of linear equations graphically and algebraically. In Chapter 2 we analyzed how matrices could be used to solve systems of linear equations.

Where Do We Go

In this chapter we will discuss how to graph linear inequalities and systems of linear inequalities and how this relates to a powerful tool called *linear programming*. Linear programming is a relatively new field of mathematics, and we will use it to solve a variety of applications.

Problem Solving: Linear Inequalities

In Section 1.1 we examined techniques for solving linear equations and writing systems of equations. Now we turn our attention to the problem-solving techniques associated with **linear inequalities**. This is a skill necessary for solving linear programming problems.

Expressing Inequalities

We begin our study of **inequalities** with a review of the inequality symbols used most frequently in this text.

■ Inequality Symbols

Symbol	*Meaning*
$<$	Less than
$>$	Greater than
$\leq$	Less than or equal to
$\geq$	Greater than or equal to

▶ **Note:** The symbols $<$ and $>$ are sometimes called **strict inequalities**.

Any value that makes an inequality a true statement is a **solution** to the inequality. Example 1 illustrates how to determine if a value is a solution to an inequality.

Example 1   Checking Solutions to Inequalities

For the following inequalities, complete Tables 3.1.1 and 3.1.2 by stating whether the inequality is *true* or *false*.

(a) $x > -7$ **(b)** $x \leq 12$

Table 3.1.1

x	$x > -7$	Statement
-10		
-7		
0		
3		

Table 3.1.2

x	$x \leq 12$	Statement
0		
9		
12		
15		

Solution

(a) The inequality statements are listed in Table 3.1.3.

(b) The inequality statements are shown in Table 3.1.4.

Notice that $12 \leq 12$ is a true statement since $12 = 12$.

Table 3.1.3

x	$x > -7$	Statement
-10	$-10 > -7$	False
-7	$-7 > -7$	False
0	$0 > -7$	True
3	$5 > -7$	True

Table 3.1.4

x	$x \leq 12$	Statement
0	$0 \leq 12$	True
9	$9 \leq 12$	True
12	$12 \leq 12$	True
15	$15 \leq 12$	False

Since the solution to an inequality is the set of all values that make the inequality true, one of the most efficient ways to represent solutions to inequalities is by graphing them on a number line. To aid us in graphing, we utilize the following:

- $a < b$ means that a is to the left of b on a number line.
- $a > b$ means that a is to the right of b on a number line.

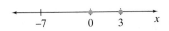

Figure 3.1.1

Figure 3.1.2

In Example 1 (a) we verified that $x = 0$ and $x = 3$ are two solutions to the inequality $x > -7$. These solutions are shown in Figure 3.1.1. These are not the only solutions to $x > -7$. Any value for x *to the right of* -7 on the number line is a solution to $x > -7$. So we graph all solutions to $x > -7$ as shown in Figure 3.1.2. We use parentheses, (or,) to show when an endpoint is not a solution. We use brackets, [or,] to show when an endpoint is a solution. Inequalities and their graphs are summarized below.

■ **Inequalities and Graphs**

For a real number a:

Inequality	Graph
$x < a$	
$x > a$	

Inequality	Graph
$x \leq a$	
$x \geq a$	

Solving Linear Inequalities

Before we discuss how to solve a linear inequality, let's see how we can translate an English sentence into a mathematical sentence using an inequality symbol.

Example 2   **Translating an English Sentence into an Inequality**

The product of five and a certain number plus two is greater than 37. Define the variable and translate the English sentence into a mathematical sentence.

Solution

Understand the Situation: Let x represent the certain number. The phrase "the product of five and a certain number" means to multiply 5 by x. This gives the term $5x$. So we get

The product of five and a certain number plus two	English sentence
is greater than 37.	
$5x + 2 > 37$	Mathematical sentence ∎

✓ **Checkpoint 1**

Now work Exercise 9.

The mathematical sentence we came up with in Example 2 is a **linear inequality**. In general, a linear inequality is defined as follows:

■ **Linear Inequality**

An inequality of the form

$$ax + b \leq c$$

is called a **linear inequality**, where a, b, and c are constants.

▶ **Note:** When working with inequalities, we tend generally to use the symbol $\leq$ in definitions and theorems. However, remember that the symbols $\geq$, $<$, or $>$ could be used in place of $\leq$.

Solving linear inequalities is much like solving linear equations. The following inequality properties are used in solving linear inequalities.

■ **Inequality Properties**

When given an inequality, an equivalent inequality is produced and the inequality symbol **remains the same** if

1. Each side of the original inequality has the same real number added to or subtracted from it.
2. Each side of the original inequality is multiplied or divided by the same positive real number.

An equivalent inequality is produced and the inequality symbol **reverses direction** if

3. Each side of the original inequality is multiplied or divided by the same negative real number.

▶ **Note:** 1. As with solving linear equations, multiplying by zero and dividing by zero are not allowed.
2. These properties illustrate that we solve linear inequalities in the same manner as we solve linear equations. The exception is when multiplying or dividing the inequality by a negative coefficient. In this case, reverse the inequality symbol.

Example 3 Solving Linear Inequalities

Solve and graph the solutions to the following.

(a) $2x - 3 \leq 7$ **(b)** $3 - 7x < 17$

Solution

(a) To solve this inequality, we isolate the variable x. This gives us

$$2x - 3 \le 7 \qquad \text{Given inequality}$$
$$2x \le 10 \qquad \text{Add 3 to both sides.}$$
$$x \le 5 \qquad \text{Divide both sides by 2.}$$

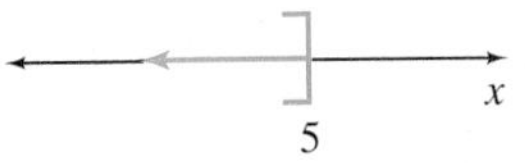

So any value of x that is less than or equal to 5 is a solution to the inequality. The solution is graphed, using a bracket on the endpoint 5, in Figure 3.1.3.

(b) Proceeding as in part (a) yields

$$3 - 7x < 17 \qquad \text{Given inequality}$$
$$-7x < 14 \qquad \text{Subtract 3 from both sides.}$$
$$x > 2 \qquad \text{Divide by } -7 \text{ and reverse inequality symbol.}$$

Notice in the final step that we had to reverse the inequality sign from $<$ to $>$ because we divided by a negative number. The solution is shown graphically in Figure 3.1.4 using a parenthesis on the endpoint 2. ∎

Figure 3.1.3

Figure 3.1.4

✔ **Checkpoint 2**   Now work Exercise 23.

Writing Linear Inequalities with More Than One Variable

Many applications in this text require us to work with *linear inequalities with more than one variable*. This means that many times we will need to write inequalities to represent a situation. Typically, linear inequalities have many solutions. We will study solutions of linear inequalities, as well as systems of linear inequalities, later in Chapter 3 and in Chapter 4. For now our goal is to represent situations with linear inequalities. Example 4 illustrates how to set up linear inequalities. Before we proceed with Example 4, let's recall the Strategy for Solving Word Problems first presented in Section 1.1.

> ■ **Strategy for Solving Word Problems**
>
> 1. Carefully **read** the problem and **understand the situation**. As you analyze a problem, you may need to:
>   - **Define the variables** to represent all of the unknowns.
>   - **Construct** a mathematical model or write one or more inequalities to represent the problem.
> 2. **Solve** the inequality. Be sure to check the solution in the original inequality.
> 3. **Interpret the solution**.

▷ **Note:** The only difference we see in the above strategy from the one presented in Section 1.1 is that we are now working with inequalities.

Example 4   **Writing Inequalities**

Table 3.1.5 gives nutritional information in grams for selected items from McDonald's menu.

Table 3.1.5

Item	Protein (g)	Fat (g)
Small French fries	3	10
Garden salad	7	6

SOURCE: McDonald's, www.mcdonalds.com

Julia wishes to eat a lunch consisting of small French fries and garden salad. She wants the lunch to contain at least 12 grams of protein. Let x represent the number of orders of small French fries and let y represent the number of orders of garden salad. Write inequalities that give appropriate restrictions on x and y.

Solution

Here we have already had our variables defined for us. Since x represents the number of orders of small French fries, and Julia cannot eat a negative number of small French fries, we have the restriction of x being greater than or equal to zero. This English sentence can be expressed as an inequality by writing

$$x \geq 0$$

Likewise, since y represents the number of orders of garden salad, we have

$$y \geq 0$$

Since each order of small French fries has 3 grams of protein, if Julia has x orders of small French fries, she would consume

$$3x$$

grams of protein. Since each order of garden salad has 7 grams of protein, if Julia has y orders of garden salad, she would consume

$$7y$$

grams of protein. So the number of grams of protein in a meal consisting of small French fries and garden salad is given by

$$3x + 7y$$

But Julia wants **at least** 12 grams of protein in her meal. Her meal could have 12, 14, or even 20 grams of protein, but not 11, 9, or 7 grams of protein. She wants at least 12 grams! The equation $3x + 7y = 12$ represents a meal with *exactly* 12 grams of protein. However, this equation excludes a meal of 14 or 20 grams. The inequality $3x + 7y > 12$ represents a meal having *more than* 12 grams of protein. However, it excludes a meal of exactly 12 grams. Combining the equation and the inequality gives us

$$3x + 7y \geq 12$$

which satisfies the condition that her meal have 12 or more grams of protein. ■

✓ **Checkpoint 3**

Now work Exercise 29.

Example 5   Writing Inequalities with More than One Variable

Using the information from Table 3.1.5 in Example 4, let's say that Julia wishes to eat a lunch consisting of orders of small French fries and orders of garden salad that contains at most 25 grams of fat. Let x represent the number of small French fries and let y represent the number of garden salads. Write inequalities that give appropriate restrictions on x and y.

Solution

In this case we still have the restrictions $x \geq 0$ and $y \geq 0$, since Julia cannot eat a negative number of French fries or garden salads. Since each order of small French fries has 10 grams of fat, if Julia has x orders of small French fries, she would consume

$$10x$$

grams of fat. Since each order of garden salad has 6 grams of fat, if Julia has y orders of garden salad, she would consume

$$6y$$

grams of fat. So the number of grams of fat in a meal consisting of small French fries and garden salad is given by

$$10x + 6y$$

Since Julia wants **at most** 25 grams of fat, this means that she could have any number of fat grams that are *less than or equal to* 25. This gives us the inequality

$$10x + 6y \leq 25$$ ■

We can see in Examples 4 and 5 that we get a total of four inequalities. In all these inequalities, x represents the number of orders of small French fries and y represents the number of orders of garden salad. If we want to express the number of orders of small French fries and garden salads that Julia can order so that she consumes both at least 12 grams of protein and at most 25 grams of fat, we can write the **system of inequalities**

$$x \geq 0$$
$$y \geq 0$$
$$3x + 7y \geq 12$$
$$10x + 6y \leq 25$$

In Section 3.2 we will study how to graph systems of linear inequalities.

Examples 4 and 5 illustrated how inequalities can be used to place restrictions on the variables so that we only consider reasonable values for the variables. This is true of many applications that we will encounter in the next two chapters, particularly in linear programming. In Example 6 we continue to practice how to use inequalities to place restrictions on the variables.

Example 6   Using Inequalities to Express Reasonable Values

Dub, Inc., is a small manufacturing firm that produces two microwave switches, switch A and switch B. Each switch passes through two departments: assembly and testing. Switch A requires 4 hours of assembly and 1 hour of testing, and switch B requires 3 hours of assembly and 2 hours of testing. The maximum number of hours available each week for assembly and testing is 240 hours and 140 hours, respectively.

(a) Define the variables and write inequalities to express reasonable values for the variables.

(b) Write an inequality to express the time restriction of the assembly department.

(c) Write an inequality to express the time restriction of the testing department.

Solution

(a) **Understand the Situation:** Since we do not know how many units of switch A and switch B are made each week, we let x represent the number of units of switch A made each week and let y represent the number of units of switch B made each week.

We know that there cannot be a negative number of switches made. This means that x and y must be either zero or greater than zero. This gives us the two inequalities

$$x \geq 0 \quad \text{and} \quad y \geq 0$$

(b) **Understand the Situation:** To see what kind of time restriction is made, we carefully examine the assembly department. It takes 4 hours to assemble each switch A and 3 hours to assemble each switch B. So the total amount of time that it takes to assemble x switch A's and y switch B's is given by the expression

$$\begin{pmatrix} \text{hours to assemble} \\ \text{each switch A} \end{pmatrix} \begin{pmatrix} \text{number of switch} \\ \text{A's assembled} \end{pmatrix} + \begin{pmatrix} \text{hours to assemble} \\ \text{each switch B} \end{pmatrix} \begin{pmatrix} \text{number of switch} \\ \text{B's assembled} \end{pmatrix}$$

$$4x + 3y$$

We know that a maximum of 240 hours is available each week for assembly. Thus the total number of hours cannot be more than 240. This is another way to say that the number of hours available is less than or equal to 240. So the restriction on the assembly department is given by

$$4x + 3y \leq 240$$

(c) **Understand the Situation:** For the testing department, we know that it takes 1 hour to test each switch A and 2 hours to test each switch B. Using the same kind of reasoning that we used for the assembly department in part (b), the total amount of time that it takes to test x switch A's and y switch B's is given by

$$x + 2y$$

We know that the maximum number of hours available each week for testing is 140 hours. So the restriction on the testing department is given by

$$x + 2y \leq 140$$ ■

The restrictions in Example 6 (a), $x \geq 0$ and $y \geq 0$, are commonly called the **nonnegative constraints**. We will study constraints in more detail as we learn the techniques of linear programming in Chapters 3 and 4. We conclude this section with an example that involves four variables.

Example 7   Using Inequalities to Express Restrictions

Skinner Bikes sends bicycles from its two plants, plant A and plant B, to its bicycle retailers I and II located in Columbus, Ohio. Plant A has a total of 210 bicycles to send and plant B has 60. Retailer I needs at least 170 bicycles and retailer II needs at least 100.

(a) Define variables and express the nonnegative constraints.

(b) Write inequalities to express restrictions on the availability and transportation costs for the bicycles.

Solution

(a) Understand the Situation: Since we do not know how many bicycles are sent from each plant to each retailer, we let x represent the number of bicycles sent from plant A to retailer I, y represent the number of bicycles sent from plant A to retailer II, z represent the number of bicycles sent from plant B to retailer I, and w represent the number of bicycles sent from plant B to retailer II.

Since all these variables represent quantities of bicycles, all must be greater than or equal to zero. This gives the nonnegative constraints

$$x \geq 0, \qquad y \geq 0, \qquad z \geq 0, \qquad w \geq 0$$

(b) Since plant A has 210 bicycles to ship and we know that the total number of bicycles sent from plant A to retailer I is x and that the total of the number of bicycles sent from plant A to retailer II is y, we get the inequality

$$\begin{pmatrix} \text{bicycles sent from} \\ \text{plant A to retailer I} \end{pmatrix} + \begin{pmatrix} \text{bicycles sent from} \\ \text{plant A to retailer II} \end{pmatrix} \begin{pmatrix} \text{cannot} \\ \text{exceed} \end{pmatrix} \begin{pmatrix} \text{total shipped from} \\ \text{plant A} \end{pmatrix}$$

$$x + y \leq 240$$

Similarly, since plant B has a total of 60 bicycles to ship, we get the inequality

$$\begin{pmatrix} \text{bicycles sent from} \\ \text{plant B to retailer I} \end{pmatrix} + \begin{pmatrix} \text{bicycles sent from} \\ \text{plant B to retailer II} \end{pmatrix} \begin{pmatrix} \text{cannot} \\ \text{exceed} \end{pmatrix} \begin{pmatrix} \text{total shipped from} \\ \text{plant B} \end{pmatrix}$$

$$z + w \leq 60$$

Since retailer I needs at least 170 bicycles, this means that no less than 170 bicycles must be delivered. So we get

$$\begin{pmatrix} \text{bicycles sent from} \\ \text{plant A to retailer I} \end{pmatrix} + \begin{pmatrix} \text{bicycles sent from} \\ \text{plant B to retailer I} \end{pmatrix} \begin{pmatrix} \text{must be at} \\ \text{least} \end{pmatrix} \begin{pmatrix} \text{number of} \\ \text{bicycles needed} \end{pmatrix}$$

$$x + z \geq 170$$

Likewise, we know that retailer II needs at least 100 bicycles, giving us the inequality

$$\begin{pmatrix} \text{bicycles sent from} \\ \text{plant A to retailer II} \end{pmatrix} + \begin{pmatrix} \text{bicycles sent from} \\ \text{plant B to retailer II} \end{pmatrix} \begin{pmatrix} \text{must be at} \\ \text{least} \end{pmatrix} \begin{pmatrix} \text{number of} \\ \text{bicycles needed} \end{pmatrix}$$

$$y + w \geq 100$$

Notice that if we collect the solutions from Example 7 we get the **system of inequalities**

$$x + y \leq 210$$
$$z + w \leq 60$$
$$x + z \geq 170$$
$$y + w \geq 100$$

$$x \geq 0, \quad y \geq 0, \quad z \geq 0, \quad w \geq 0$$

We will study how to solve systems of inequalities like this one in Chapter 4.

SUMMARY

• **Inequality Symbols**

Symbol	Meaning
$<$	Less than
$>$	Greater than
$\leq$	Less than or equal to
$\geq$	Greater than or equal to

• A **linear inequality** has the form $ax + b \leq c$, where $a, b,$ and c are constants.

• When **solving linear inequalities**, we use the same process as for solving linear equations. The exception is when multiplying or dividing the inequality by a negative coefficient. In this case, reverse the inequality symbol.

• When translating English sentences into mathematical sentences using inequalities, we use the Strategy for Solving Word Problems.

SECTION 3.1 EXERCISES

For the inequalities in Exercises 1–6, complete the tables by stating whether the inequality is true *or* false.

1. $x > 6$

x	$x > 6$	Statement
-12		
2		
6		
10		

2. $x > 0$

x	$x > 0$	Statement
-1		
0		
1		
6		

3. $x \leq -4$

x	$x \leq -4$	Statement
-8		
-6		
-4		
0		

4. $x \leq 5$

x	$x \leq 5$	Statement
-5		
0		
5		
15		

5. $x \geq -1$

x	$x \geq -1$	Statement
-2		
-1		
0		
1		

6. $x \geq -10$

x	$x \geq -10$	Statement
-15		
-12		
-10		
0		

In Exercises 7–16, translate the English sentence into a mathematical sentence.

7. A certain number increased by five is greater than four.

8. A certain number increased by seven is greater than six.

✓ **9.** A certain number decreased by eight is less than −18.

10. A certain number decreased by two is less than negative six.

11. The product of nine and a certain number is greater than or equal to 54.

12. The product of three and a certain number is greater than or equal to one.

13. The product of three and a certain number minus 15 is less than or equal to −20.

14. The product of four and a certain number minus four is less than or equal to −10.

15. The product of eight and a certain number plus 1 is not equal to 14.

16. The product of six and a certain number plus one is not equal to 97.

In Exercises 17–28, solve and graph the solutions.

17. $x - 3 > 4$ **18.** $x - 1 > 5$

19. $x + 4 < 9$ **20.** $x + 6 < 4$

21. $2x + 2 \le 18$ **22.** $3x + 5 \le 20$

✓ **23.** $2x - 3 \ge 7$ **24.** $3x + 2 \ge 14$

25. $-2x - 3 > 7$ **26.** $-4x + 3 > 15$

27. $14 - 3x \le 5$ **28.** $3 - 7x \le 17$

Applications

$\Rightarrow$ 3.2 ✓ **29.** McSavaney's trains administrative assistants for two types of office positions: medical field and executive secretarial. Each individual receives training in two departments at McSavaney's: word processing along with telephone and LAN (local area networks). An individual in the medical field program receives 2 units of word processing and 2.5 units of telephone and LAN training. An individual in the executive secretarial program receives 3 units of word processing and 2.5 units of telephone and LAN training. Each month at McSavaney's, corporate training can give no more than 32 units of word processing training and no more than 30 units of telephone and LAN training. Let x represent the number admitted each month to the medical field program, and let y represent the number admitted each month to the executive secretarial program. Write linear inequalities that give appropriate restrictions on x and y.

$\Rightarrow$ 3.2 **30.** (*continuation of Exercise 29*) Feedback from employers has indicated that McSavaney's needs to provide

another area of training, ethics. McSavaney's adds an ethics training department. It has been determined that an individual in the medical field program receive 3.5 units of ethics training and an individual in the executive secretarial program receive 2.5 units of ethics training. Each month at McSavaney's, corporate training can give no more than 35 units of ethics training. Assuming that everything else from Exercise 29 remains the same, write linear inequalities that give appropriate restrictions on x and y.

$\Rightarrow$ 3.2 **31.** Sandkuhl, Inc., is a small manufacturing firm that produces two microwave switches, switch A and switch B. Each switch passes through two departments: assembly and testing. Switch A requires 4 hours of assembly and 1 hour of testing; switch B requires 3 hours of assembly and 2 hours of testing. The maximum number of hours available each week for assembly and testing is 240 hours and 140 hours, respectively. Let x represent the number of units of switch A manufactured each week, and let y represent the number of units of switch B manufactured each week. Write linear inequalities that give appropriate restrictions on x and y.

$\Rightarrow$ 3.2 **32.** (*continuation of Exercise 31*) Due to a new contract commitment, Sandkuhl, Inc., must manufacture at least 10 units of switch A each week. Assuming that everything else from Exercise 31 remains the same, write linear inequalities that give appropriate restrictions on x and y.

$\Rightarrow$ 3.2 **33.** Klosterman's owns 200 acres of tillable farmland that is used to grow corn and wheat. On average, each acre of corn and wheat yields 110 and 35 bushels, respectively. At least 11,000 bushels of corn and 2205 bushels of wheat are required due to prior contracts. Let x represent the number of acres of corn planted and let y represent the number of acres of wheat planted. Write linear inequalities that give appropriate restrictions on x and y.

$\Rightarrow$ 3.2 **34.** Mix for Chix, Inc., a local feed store, manufactures chicken feed by mixing two ingredients. Each ingredient contains two key nutrients: protein and fat. The number of units of each nutrient in 1 pound of the two basic ingredients is shown in Table 3.1.6.

Table 3.1.6

Ingredient	Protein	Fat
1	25	11
2	40	8

A local chicken farmer desires at most 500 pounds of a mix in which each pound of feed has a maximum of 35 units of protein and a maximum of 10 units of fat. Let x represent the number of pounds of ingredient 1 used and let y represent the number of pounds of ingredient 2 used. Write linear inequalities that give appropriate restrictions on x and y.

$\Rightarrow$ 3.2 **35.** The Stitz Tent Company makes two types of tents: two-person and six-person models. Each tent requires the services of two departments, in hours, as shown in Table 3.1.7.

Table 3.1.7

Department	Two-person	Six-person
Fabric cutting	6	8
Assembly and packaging	2	4

Fabric cutting has a maximum of 580 hours available each week and assembly and packaging has a maximum of 260 hours available.

(a) Define the variables and write linear inequalities to express reasonable values for the variables.

(b) Write linear inequalities to express the time restrictions of the fabric cutting and of the assembly and packaging departments.

⇒ 4.1 **36.** Due to new regulations approved by the Environmental Protection Agency (EPA), the Henderson Chemical Company was required to implement a new process to reduce the pollution created when making a particular chemical. The older process releases 4 grams of sulfur and 3 grams of particulates for each gallon of chemical produced. The newer process releases 2 grams of sulfur and 1 gram of particulates for each gallon of chemical produced. The new EPA regulations do not allow for more than 10,000 grams of sulfur and 7000 grams of particulates to be released daily.

(a) Define the variables and write linear inequalities to express reasonable values for the variables.

(b) Write linear inequalities to express the EPA restrictions on sulfur and particulates.

⇒ 4.1 **37.** Randy and Rusty have decided to enter the field of personal digital assistants (PDAs). Initially, they plan to design two types of PDAs, the WOW 1000 and the WOW 1500. Because of interest in PDAs, they can sell all that they can produce. However, they need to determine a production rate so as to satisfy various limits with a small production crew. These include a maximum of 150 hours per week for assembly and 100 hours per week for testing. The WOW 1000 requires 3 hours of assembly and 3 hours of testing, and the WOW 1500 requires 6 hours of assembly and 3.5 hours of testing.

(a) Define the variables and write linear inequalities to express reasonable values for the variables.

(b) Write linear inequalities to express the time restrictions of the assembly and testing departments.

⇒ 4.1 🌐 **38.** Table 3.1.8 gives data for growing artichokes and asparagus for an entire growing season in California (entries are in lb/acre).

Table 3.1.8

	Artichokes	Asparagus
Nitrogen	197	240
Potash	123	42

SOURCE: U.S. Department of Agriculture, www.usda.gov/nass

Franco has available 150 acres suitable for cultivating crops. He has purchased a contract for a total of 31,880 pounds of nitrogen and 9120 pounds of potash.

(a) Define the variables and write linear inequalities to express reasonable values for the variables.

(b) Write linear inequalities to express the restrictions of land, nitrogen, and potash.

⇒ 4.1 🌐 **39.** Table 3.1.9 gives data for growing cauliflower and head lettuce for an entire growing season in California (entries are in lb/acre).

Table 3.1.9

	Cauliflower	Head Lettuce
Nitrogen	215	173
Phosphate	98	122

SOURCE: U.S. Department of Agriculture, www.usda.gov/nass

Sonja has 210 acres available for cultivating crops. She has purchased a contract for a total of 38,380 pounds of nitrogen and 22,240 pounds of phosphate.

(a) Define the variables and write linear inequalities to express reasonable values for the variables.

(b) Write linear inequalities to express the restrictions of land, nitrogen, and phosphate.

⇒ 4.1 **40.** The Smolko Tent Company makes three types of tents: two-person, four-person and six-person models. Each tent requires the services, in hours, of three departments, as shown in Table 3.1.10. The fabric cutting, assembly, and packaging departments have available a maximum of 420, 310, and 180 hours each week, respectively.

Table 3.1.10

	Two-person	Four-person	Six-person
Fabric cutting	0.5	0.8	1.1
Assembly	0.6	0.7	0.9
Packaging	0.2	0.3	0.6

(a) Define the variables and write linear inequalities to express reasonable values for the variables.

(b) Write linear inequalities to express the time restrictions of the fabric cutting, assembly, and packing departments.

⇒ 4.1 **41.** Kathy's Secretarial Services has a 1-month intensive training program for receptionist/secretaries for three types of office positions: medical field, small business, and executive secretarial. Each individual receives training in all three departments at Kathy's: word processing, telephone and LAN (local area network), and ethics. Table 3.1.11 shows the number of units of training that each type of receptionist/secretary receives.

Table 3.1.11

	Medical Field	Small Business	Executive Secretary
Word processing	0.8	1.1	1.7
Telephone and LAN	0.6	1.3	1.5
Ethics	1.2	0.8	1.3

Table 3.1.12

	Asparagus	Cauliflower	Celery
Nitrogen	240	215	336
Phosphate	148	98	171
Potash	42	69	147

SOURCE: U.S. Department of Agriculture,
www.usda.gov/nass

Each month, word processing, telephone and LAN, and ethics have available a maximum of 32.1 units, 29.7 units, and 33 units, respectively.

(a) Define the variables and write linear inequalities to express reasonable values for the variables.

(b) Write linear inequalities to express the time restrictions of the word processing, telephone and LAN, and ethics departments.

⇒ 4.1 🌐 **42.** The data in Table 3.1.12 give the amount of nitrogen, phosphate, and potash (all in lb/acre) needed to

grow asparagus, cauliflower, and celery for an entire growing season in California.

Eldrick has 280 acres and can get no more than 65,970 pounds of nitrogen, 31,860 pounds of phosphate, and 21,630 pounds of potash.

(a) Define the variables and write linear inequalities to express reasonable values for the variables.

(b) Write linear inequalities to express the restrictions on nitrogen, phosphate, and potash.

⇒ 3.2 ▮ **SECTION PROJECT**

The Sierra Refining Company produces a grade of unleaded gasoline, grade A, which it supplies to its chain of service stations. The grade A unleaded gasoline is blended from Sierra's inventory of gasoline components and must meet the specifications given in the table.

	Minimum Octane Rating	Maximum Demand (barrels/week)	Minimum Deliveries (barrels/week)
Grade A	87	80,000	60,000

The characteristics of the components in inventory are given in the next table.

Gasoline Component	Octane Rating	Inventory (barrels)
1	86	70,000
2	96	60,000

Let x represent the number of barrels of component 1 blended into grade A gasoline per week, and let y represent the number of barrels of component 2 blended into grade A gasoline per week.

(a) Verify that the total weekly amount of grade A gasoline is given by $x + y$.

(b) Determine two linear inequality restrictions for $x + y$.

Section 3.2 Graphing Systems of Linear Inequalities

Many realistic applications involve **inequalities** rather than equations. For example, a dietitian may need to plan a diet for a patient that has *at least* 30 grams of calcium. Let's say that one food provides 15 grams per serving while another provides 7 grams per serving. If the dietitian has two (or more) foods to select from and two (or more) dietary requirements, he can use a **system of linear inequalities** to find the right amount of each food. One of the most convenient ways to represent a solution to a system of linear inequalities with two variables is with a graph. Before we can do this, however, we must first learn how to graph a single linear inequality.

Graphing Linear Inequalities

In Chapter 1 we stated that the **standard form** of a line is given by

$$ax + by = c$$

where a, b, and c are constants, $a \geq 0$, and a and b are not both zero. We can use this standard form to define a **linear inequality**.

> ### Linear Inequality
>
> A **linear inequality** in two variables is given by
>
> $$ax + by \leq c$$
>
> where a, b, and c are constants, with a and b not both zero.

Example 1 Graphing a Linear Inequality

Graph the linear inequality $2x + y \leq 6$.

Solution

Recall that the symbol $\leq$ is read "less than **or** equal to." The coordinates of any point (x, y) in the plane must satisfy $2x + y < 6$ **or** $2x + y = 6$ to be a solution to the inequality, $2x + y \leq 6$. Using the techniques from Chapter 1, we begin by graphing $2x + y = 6$. To graph $2x + y = 6$, we determine its intercepts. (Consult the Toolbox to the left.) The x-intercept is at $(3, 0)$ and the y-intercept is at $(0, 6)$. We use these points to graph $2x + y = 6$, as shown in Figure 3.2.1.

From Your Toolbox

To determine the x-intercept, we let $y = 0$, and to determine the y-intercept, we let $x = 0$.

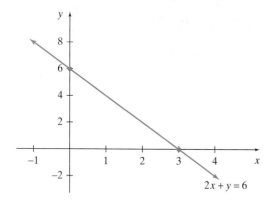

Figure 3.2.1

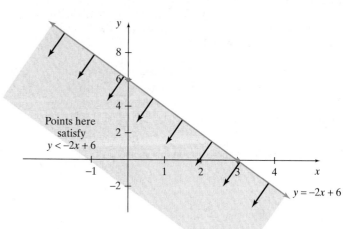

Figure 3.2.2

Before continuing, we must emphasize that each point on the line is a solution to $2x + y = 6$. Now, to determine the points that satisfy (or are solutions to) the inequality $2x + y < 6$, we can solve the inequality for y to get

$$2x + y < 6$$
$$y < -2x + 6$$

Figure 3.2.2 shows that points below the line $2x + y = 6$ (or in slope–intercept form, $y = -2x + 6$) satisfy the inequality $y < -2x + 6$. Since points on the line satisfy $y = -2x + 6$, visually it appears that when below the line the values of y

decrease the farther we move away from the line. What this means is that our inequality $2x + y \leq 6$ has as its solution **all points on or below the line** $2x + y = 6$. We indicate the points below the line by shading, as shown in Figure 3.2.3. The line and shaded region in Figure 3.2.3 give us the graph of $2x + y \leq 6$.

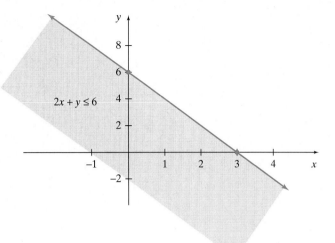

Figure 3.2.3 Graph of $2x + y \leq 6$ ■

The line $2x + y = 6$ in Figure 3.2.3 is called the **boundary**. Quite simply, a boundary is a line that separates the points in the solution from those **not** in the solution. Also, the line $2x + y = 6$ has split the Cartesian plane into two planes called the **lower half-plane** and the **upper half-plane**. Figure 3.2.4 summarizes these important features that we find in graphing linear inequalities.

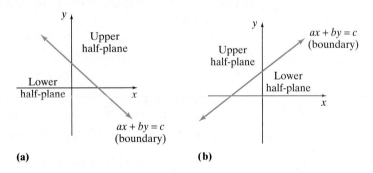

Figure 3.2.4

Example 2   Graphing a Linear Inequality

Graph the linear inequality $2x + y > 6$.

Solution

We have the same boundary as in Example 1, $2x + y = 6$. However, since our inequality symbol ($>$) indicates that points on the line itself do **not** satisfy the inequality, we use a dashed line. As in Example 1, to determine the points that satisfy $2x + y > 6$, we solve for y and get $y > -2x + 6$. Since points on the line satisfy $y = -2x + 6$, from Figure 3.2.2 it appears that, when above the line, the values of y increase the farther we move away from the line. This means that we must shade the upper half-plane. Figure 3.2.5 shows the graphical solution to $2x + y > 6$.

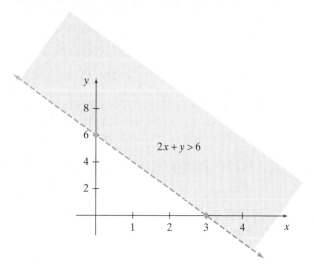

Figure 3.2.5

There is an alternative method for finding the correct region to shade when graphing a linear inequality. After graphing the boundary, select a **test point** that is **not** on the boundary. For example, in Examples 1 and 2 such a point is the origin $(0, 0)$. We can then substitute the coordinates of the test point into the given inequality. If, after substituting, the resulting inequality is true, shade the region containing the test point. If the resulting inequality is false, shade the region not containing the test point. For example, using the test point $(0, 0)$ in Example 1 would give us

$$2x + y \leq 6 \qquad \text{Given inequality}$$
$$2(0) + 0 \leq 6 \qquad \text{Substitute } (0, 0) \text{ for } (x, y).$$
$$0 \leq 6 \qquad \text{Inequality is true.}$$

Since the result is a true statement, we would shade the region containing the test point $(0, 0)$. This discussion, along with Examples 1 and 2, supplies the following procedure for graphing linear inequalities.

■ **Graphing a Linear Inequality**

1. Draw the graph of the boundary line. If the inequality involves $\leq$ or $\geq$, make the line solid. If the inequality involves $<$ or $>$, make the line dashed.
2. Pick a test point (x, y) that is in one of the half-planes. Substitute the values of x and y into the given inequality and simplify. (*Hint:* Whenever possible, pick the origin.)
3. If the test point satisfies the original inequality (gives a true statement), shade the region containing the test point. If the test point does not satisfy the original inequality (gives a false statement), shade the region that does not contain the test point.

▶ **Note:** When choosing a test point (x, y), be sure it is not on the boundary line.

Example 3   **Graphing a Linear Inequality**

Graph the following linear inequalities.

(a) $3x + 2y > 6$ **(b)** $x \geq -2$

Solution

(a) We begin by graphing the boundary line $3x + 2y = 6$. The x-intercept is $(2, 0)$ and the y-intercept is $(0, 3)$. Since the inequality symbol is $>$, the boundary line is dashed, as shown in Figure 3.2.6. Our test point is $(0, 0)$. Substituting into the original inequality gives

$$3x + 2y > 6 \qquad \text{Given inequality}$$
$$3(0) + 2(0) > 6 \qquad \text{Substitute } (0, 0) \text{ for } (x, y).$$
$$0 > 6 \qquad \text{Inequality is false.}$$

Since $0 > 6$ is false, we shade the half-plane that does *not* contain the test point, as shown in Figure 3.2.6.

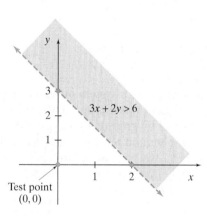

Figure 3.2.6

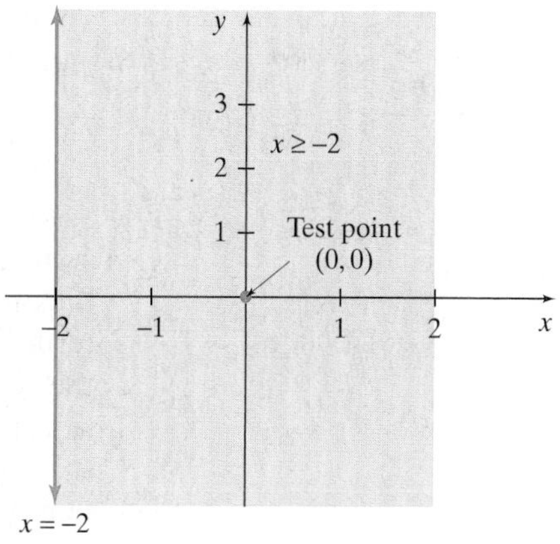

Figure 3.2.7

(b) The graph of $x = -2$ is a vertical line. Since the inequality symbol is $\geq$, we make the boundary line solid, as shown in Figure 3.2.7. Our test point is $(0, 0)$, and substituting into the given inequality yields

$$x \geq -2 \qquad \text{Given inequality}$$
$$0 \geq -2 \qquad \text{Inequality is true when substituting } 0 \text{ for } x.$$

Since it is true, we shade the half-plane containing the test point, as shown in Figure 3.2.7.   ■

✓ **Checkpoint 1**

Now work Exercise 15.

Graphing Systems of Linear Inequalities

In Chapter 1 we saw that a solution to a system of linear equations with two variables was a pair of values for the variables that made all the equations in the

Technology Option

Graphing calculators can shade regions showing solutions to linear inequalities. Figure 3.2.8 gives two different shadings options for Example 3a. Figure 3.2.8a uses the DRAW SHADE command, and Figure 3.2.8b uses a feature on the $Y =$ editor screen.

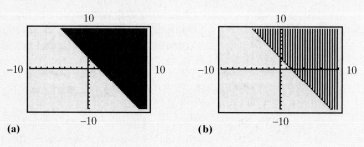

(a)   (b)

Figure 3.2.8

To learn how to shade inequalities on your calculator, consult the online graphing calculator manual at www.prenhall.com/armstrong

system true. A solution to a **system of linear inequalities** is the set of all points in the plane that satisfies (makes true) all the inequalities in the system. We can determine these points by graphing all inequalities in the system and determining the region that is common to all the graphs of the inequalities. To avoid confusion (and a mess!), we will use arrows on each boundary line to indicate the shading for a single linear inequality and will shade only the solution to the system. Example 4 illustrates this process.

Example 4   **Graphing a Solution to a System of Linear Inequalities**

Graph the system

$$2x + y \geq 6$$
$$x - y \leq 0$$

Solution

We begin by graphing each boundary line, $2x + y = 6$ and $x - y = 0$, as shown in Figure 3.2.9. Using the test point $(0, 0)$ for $2x + y \geq 6$ gives us $2(0) + 0 \geq 6$ or $0 \geq 6$, which is *false*. Hence we shade the region not containing $(0, 0)$. This is indicated by the arrows on the boundary line. Since $(0, 0)$ is on the boundary line $x - y = 0$ and not in a half-plane, we cannot use $(0, 0)$ as a test point. Choosing $(0, 1)$, which is not on the boundary line, as our test point for $x - y \leq 0$ gives us $0 - 1 \leq 0$ or $-1 \leq 0$, which is *true*. Hence we shade the region containing $(0, 1)$. This is indicated by the arrows on the boundary line. The region common to both inequalities is shaded in Figure 3.2.9.

Interactive Activity

Pick three points in the solution set (shaded region) in Example 4 and verify that each point satisfies both inequalities.

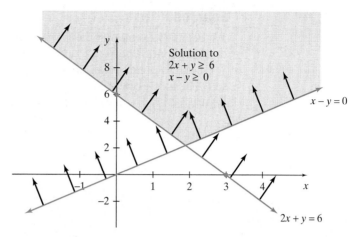

Figure 3.2.9

Technology Option

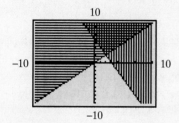

Figure 3.2.10

The graph of the system in Example 4 is shown in Figure 3.2.10. Notice that the solution to the system is given by the region where the two shaded regions overlap.

✓ **Checkpoint 2**

Now work Exercise 27.

The shaded region in Figure 3.2.9 from Example 4, which gives the solutions to the system of linear inequalities, is also called the **feasible region**. It has this name since it gives all points that satisfy all inequalities in the system. We will use this phrase frequently in the remainder of Chapter 3.

■ **Feasible Region**

The set of all points that satisfies a system of linear inequalities is called the **feasible region**.

Example 5   Graphing a Feasible Region

Graph the feasible region for the system

$$x + y \leq 6$$
$$x - y \leq 2$$
$$x \geq 0, \; y \geq 0$$

Solution

We begin by examining the last two inequalities. Observe that $x \geq 0$ corresponds to all points to the right of the y-axis plus all points on the y-axis itself. Also, $y \geq 0$ corresponds to all points above the x-axis plus all points on the x-axis itself. The half-plane solutions for the first two inequalities are shown by arrows in Figure 3.2.11. The graph of the feasible region is shaded in Figure 3.2.11.`

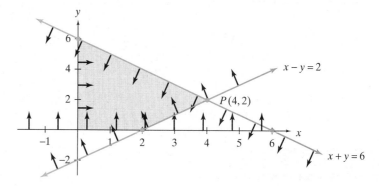

Figure 3.2.11

The feasible region in Example 5, the shaded region, is said to be **bounded**. This simply means that it is enclosed or it could be enclosed by some circle. If we draw a circle centered at the origin with a radius of, say, 15, then the shaded region would be entirely inside the circle. The feasible region in Example 4 is said to be **unbounded**. This means that it cannot be enclosed by a circle.

Applications of Systems of Linear Inequalities

In this section we focus on writing systems of linear inequalities to represent a problem situation and graphing and interpreting the corresponding feasible region. In the remaining two sections of Chapter 3 we learn how to apply this necessary skill to solving applications.

Example 6 Writing a System of Linear Inequalities

The WeDo Wood Company makes two types of canoes: a two-person model and a four-person model. Each two-person model requires 1 hour in the cutting department and 1.5 hours in the assembly department. Each four-person model requires 1 hour in the cutting department and 2.75 hours in the assembly department. The cutting department has a maximum of 640 hours available each week, and the assembly department has a maximum of 1080 hours available each week. Define the variables and write a system of linear inequalities that expresses these restrictions. Graph the feasible region.

Solution

Understand the Situation: We do not know how many two-person canoes or four-person canoes are made each week. So we let x represent the number of two-person canoes made each week and let y represent the number of four-person canoes made each week.

Since we cannot make a negative number of canoes, we immediately have the linear inequalities

$$x \geq 0 \quad \text{and} \quad y \geq 0$$

Now x units of two-person canoes requires $1 \cdot x = x$ hours in the cutting department, and y units of four-person canoes requires $1 \cdot y = y$ hours in the cutting department. Since the cutting department has available a maximum of 640 hours, we have the linear inequality

$$x + y \leq 640 \qquad \text{Cutting department restrictions}$$

Similarly, x units of two-person canoes requires $1.5x$ hours in the assembly department, and y units of four-person canoes requires $2.75y$ hours in the assembly department. Since the assembly department has available a maximum of 1080 hours, this gives the inequality

$$1.5x + 2.75y \leq 1080 \qquad \text{Assembly department restrictions}$$

So the system of linear inequalities that expresses the restrictions is

$$x \geq 0$$
$$y \geq 0$$
$$x + y \leq 640$$
$$1.5x + 2.75y \leq 1080$$

A graph of the feasible region is given in Figure 3.2.12. Observe that every point in the feasible region satisfies all inequalities in the system. Thus every point in the feasible region is a feasible solution to the problem.

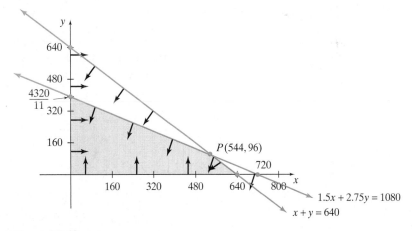

Figure 3.2.12

✓ **Checkpoint 3**   Now work Exercise 51.

The linear inequalities in Example 6 are also known as the problem **constraints**. Each point in the feasible region in Figure 3.2.12 is a feasible solution given the problem constraints. We will see more on constraints in the last two sections of Chapter 3.

SUMMARY

In this section we learned what a linear inequality is and what its graph looks like. The latter gave rise to a **boundary line** as well as **half-planes**. A three-step procedure was given to graph linear inequalities. We then discussed graphing systems of linear inequalities. We saw that the common region of the graph of a system gives the solution to the system. We called this region the **feasible region**. We concluded the section with an application on writing a system of linear inequalities, given some restrictions.

- A **linear inequality** in two variables is given by $ax + by \leq c$, where a, b, and c are constants, with a and b not both zero.
- **Graphing a Linear Inequality**

1. Draw the graph of the boundary line. If the inequality involves $\leq$ or $\geq$, make the line solid. If the inequality involves $<$ or $>$, make the line dashed.
2. Pick a test point (x, y) that is in one of the half-planes. Substitute the values of x and y into the given inequality and simplify. (*Hint:* Whenever possible, pick the origin.)
3. If the test point satisfies the original inequality (gives a true statement), shade the region containing the test point. If the test point does not satisfy the original inequality (gives a false statement), shade the region that does not contain the test point.

SECTION 3.2 EXERCISES

In Exercises 1–4 match the graph with the correct inequality.

1. $x + y < 4$ **2.** $x + y > 4$ **3.** $x + y \geq 4$ **4.** $x + y \leq 4$

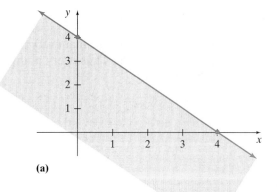

(a)

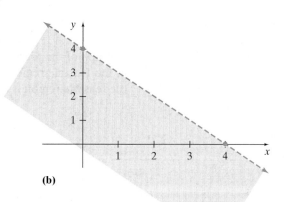

(b)

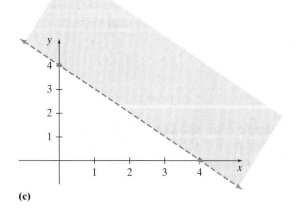

(c)

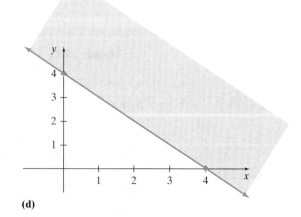

(d)

In Exercises 5–24, graph each linear inequality.

5. $y \leq x + 2$

6. $y \leq x - 3$

7. $y \geq -2x + 1$

8. $y \geq -3x + 2$

9. $y < x - 2$

10. $y < x + 5$

11. $y > 3x - 2$

12. $y > -2x + 5$

13. $3x - y \leq 7$

14. $4x - y \leq 6$

✔ **15.** $2x + 4y \geq 8$

16. $3x + 6y \geq 12$

17. $x \geq -1$

18. $y \geq -3$

19. $-2x - y \leq 12$

20. $-3x - y \leq 15$

21. $2x + 3y \leq 0$

22. $4x + 5y \leq 0$

23. $3x - 2y \geq 0$

24. $5x - 4y \geq 0$

In Exercises 25–28, use Figure 3.2.13 to match the correct feasible region with the corresponding system of linear inequalities.

25. $x - y \leq 0$
$x + 2y \geq 6$

26. $x - y \geq 0$
$x + 2y \geq 6$

✔ **27.** $x - y \leq 0$
$x + 2y \leq 6$

28. $x - y \geq 0$
$x + 2y \leq 6$

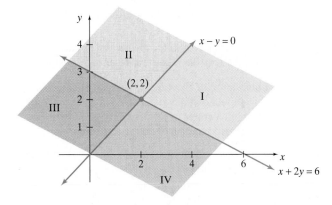

Figure 3.2.13

In Exercises 29–42, graph the feasible region for each system of linear inequalities. Indicate if the region is bounded or unbounded.

29. $x + y \leq 2$
$x - y \geq 3$

30. $x - y \leq 4$
$x + y \geq -2$

31. $2x + 3y \leq 6$
$-x + 2y < 4$

32. $3x + 2y > 6$
$x + 3y \geq 9$

33. $2x - y \geq 4$
$4x - 2y \leq -2$

34. $3x - y \leq 9$
$2x + y \geq 8$

35. $x + y \leq 7$
$2x - y \leq 4$
$x \geq 0, \ y \geq 0$

36. $x + y \leq 7$
$2x - y \geq 4$
$x \geq 0, \ y \geq 0$

37. $-2x + y \leq 3$
$5x + 2y \leq 10$
$x \geq 0, \ y \geq 0$

38. $-2x + y \leq 3$
$5x + 2y \geq 10$
$x \geq 0, \ y \geq 0$

39. $3x + y \leq 7$
$4x + y \geq 6$
$x \geq 0, \ y \geq 0$

40. $3x + 6y \leq 12$
$5x + 4y \geq 8$
$x \geq 0, \ y \geq 0$

41. $5x - y \leq 10$
$2x + y \geq 3$
$x + 2y \leq 4$
$x \geq 0, \ y \geq 0$

42. $6x + 7y \leq 42$
$4x - 9y \leq 18$
$6x - 7y \leq 14$
$x \geq 0, \ y \geq 0$

In Exercises 43–48, use a graphing calculator to graph the following.

43. $y \leq 6 - x$

44. $y \geq 3 + x$

45. $2x + 3y \geq 7$

46. $4x - y \leq 8$

47. $3x + 2y \geq 6$
$x - 3y \geq 9$

48. $2x - y \geq 4$
$4x - 2y \leq -2$

Applications

49. McSavaney's trains administrative assistants for two types of office positions: medical field and executive secretarial. Each individual receives training in two departments at McSavaney's: word processing along with telephone and LAN (local area networks). An individual in the medical field program receives 2 units of word processing and 2.5 units of telephone and LAN training. An individual in the executive secretarial program receives 3 units of word processing and 2.5 units of telephone and LAN training. Each month at McSavaney's, corporate training can give no more than 32 units of word processing training and no more than 30 units of telephone and LAN training. Let x represent the number admitted each month to the medical field program, and let y represent the number admitted each month to the executive secretarial program.

(a) Write a system of linear inequalities that gives appropriate restrictions on x and y.

(b) Graph the feasible region for the system in part (a). Interpret what the feasible region represents.

50. (*continuation of Exercise 49*) Feedback from employers has indicated that McSavaney's needs to provide another area of training, ethics. McSavaney's adds an ethics training department. It has been determined that an individual in the medical field program receives 3.5 units of ethics training and an individual in the executive secretarial program receives 2.5 units of ethics training. Each month at McSavaney's, corporate training can give no more than 35 units of ethics training. Assuming that everything else from Exercise 49 remains the same, complete the following:

(a) Write a system of linear inequalities that gives appropriate restrictions on x and y.

(b) Graph the feasible region for the system in part (a). Interpret what the feasible region represents.

⇐ 3.1 ✓ **51.** Sandkuhl, Inc., is a small manufacturing firm that produces two microwave switches, switch A and switch B. Each switch passes through two departments: assembly and testing. Switch A requires 4 hours of assembly and 1 hour of testing, and switch B requires 3 hours of assembly and 2 hours of testing. The maximum number of hours available each week for assembly and testing is 240 hours and 140 hours, respectively.

(a) Define variables and write a system of linear inequalities that gives appropriate restrictions on x and y.

(b) Graph the feasible region for the system in part (a). Interpret what the feasible region represents.

⇐ 3.1 **52.** (*continuation of Exercise 51*) Due to a new contract commitment, Sandkuhl, Inc., must manufacture at least 10 units of switch A each week. Assuming that everything else from Exercise 51 remains the same, complete the following:

(a) Write a system of linear inequalities that gives appropriate restrictions on x and y.

(b) Graph the feasible region for the system in part (a). Interpret what the feasible region represents.

⇐ 3.1 **53.** Klosterman's owns 200 acres of tillable farmland that is used to grow corn and wheat. On average, each acre of corn and wheat yields 110 and 35 bushels, respectively. At least 11,000 bushels of corn and 2205 bushels of wheat are required due to prior contracts.

(a) Define the variables and write a system of linear inequalities that gives appropriate restrictions on x and y.

(b) Graph the feasible region for the system in part (a). Interpret what the feasible region represents.

⇐ 3.1 **54.** Mix for Chix, Inc., a local feed store, manufactures chicken feed by mixing two ingredients. Each ingredient contains two key nutrients: protein and fat. The number of units of each nutrient in 1 pound of the two basic ingredients is shown in Table 3.2.1.

Table 3.2.1

Ingredient	Protein	Fat
1	25	11
2	40	8

A local chicken farmer desires at most 500 pounds of a mix in which each pound of the feed has a maximum of 35 units of protein and a maximum of 10 units of fat.

(a) Define the variables and write a system of linear inequalities that gives appropriate restrictions on x and y.

(b) Graph the feasible region for the system in part (a). Interpret what the feasible region represents.

🌐 **55.** Table 3.2.2 gives nutritional information, in grams, for selected items from McDonald's menu.

Table 3.2.2

Item	Protein	Fat
Small French fries	3	10
Garden salad	7	6

SOURCE: McDonald's, www.mcdonalds.com

Julia wishes to eat a lunch consisting of small French fries and garden salad. She wants the lunch to contain at least 12 grams of protein.

(a) Define the variables and write a system of linear inequalities that gives appropriate restrictions on x and y.

(b) Graph the feasible region for the system in part (a). Interpret what the feasible region represents.

🌐 **56.** (*continuation of Exercise 55*) Now suppose that Julia wishes to eat a lunch consisting of small French fries and garden salad that contains no more than 25 grams of fat. Use the data in Table 3.2.2 to answer the following:

(a) Define the variables and write a system of linear inequalities that gives appropriate restrictions on x and y.

(b) Graph the feasible region for the system in part (a). Interpret what the feasible region represents.

🌐 **57.** (*continuation of Exercises 55 and 56*) Julia decides that her lunch should contain at least 12 grams of protein, but no more than 25 grams of fat. Use Table 3.2.2 and the results of Exercises 55 and 56 to do the following:

(a) Write a system of linear inequalities that gives appropriate restrictions on x and y.

(b) Graph the feasible region for the system in part (a). Interpret what the feasible region represents.

⇐ 3.1 **58.** The Stitz Tent Company makes two types of tents: two-person and six-person models. Each tent requires the services, in hours, of two departments as shown in Table 3.2.3.

Table 3.2.3

Department	Two-person	Six-person
Fabric cutting	6	8
Assembly and packaging	2	4

Fabric cutting has a maximum of 580 hours available each week, and assembly and packaging has a maximum of 260 hours available.

(a) Define the variables and write a system of linear inequalities that gives appropriate restrictions on x and y.

(b) Graph the feasible region for the system in part (a). Interpret what the feasible region represents.

59. (*continuation of Exercise 58*) Suppose that the Stitz Tent Company makes a profit of $40 on each two-person model and a profit of $60 on each six-person model.

(a) Verify that the point (20, 20) is in the feasible region found in Exercise 58.

(b) If the company makes and sells 20 two-person models and 20 six-person models, the weekly profit is $2000. The

equation $40x + 60y = 2000$ gives *all* production schedules that result in a $2000 weekly profit. Graph this line in the feasible region from Exercise 58 and verify that the point (20, 20) is on the line.

(c) Determine an equation that gives all production schedules that result in a $2400 weekly profit.

(d) Verify that the graph of the equation found in part (c) is parallel to the graph of the equation given in part (b).

⇐
3.1 ■ **SECTION PROJECT**

The Sierra Refining Company produces a grade of unleaded gasoline, grade A, which it supplies to its chain of service stations. The grade A unleaded gasoline is blended from Sierra's inventory of gasoline components and must meet the specifications in Table 3.2.4.

Table 3.2.4

	Minimum Octane Rating	Maximum Demand (barrels/week)	Minimum Deliveries (barrels/week)
Grade A	87	80,000	60,000

The characteristics of the components in inventory are given in Table 3.2.5

Table 3.2.5

Gasoline Component	Octane Rating	Inventory (barrels)
1	86	70,000
2	96	60,000

Let x represent the number of barrels of component 1 blended into grade A gasoline per week, and let y represent the number of barrels of component 2 blended into grade A gasoline per week.

(a) Verify that the total weekly amount of grade A gasoline is given by $x + y$.

(b) Use Table 3.2.4 to determine two linear inequality restrictions for $x + y$.

(c) The minimum octane rating is obtained by computing the weighted average of the octane in the mixture. This means to divide the total octane in the mixture by the number of barrels in the mixture. Mathematically, this gives

$$\frac{86x + 96y}{x + y} \geq 87 \qquad \text{Restriction for minimum octane rating for grade A}$$

Multiply both sides by $(x + y)$ and simplify so that 0 is on one side of the inequality. (How do we know that $x + y$ is not a negative number?)

(d) Determine any other restrictions on x and y.

(e) Write the system of linear inequalities that gives all restrictions on x and y. (*Hint:* You should have seven total inequalities.)

(f) Graph the feasible region for the system in part (e).

Section 3.3 Solving Linear Programming Problems Graphically

In Section 3.2 we laid the groundwork for our next task, solving the linear programming problem. In the areas of management science, quantitative analysis, and operations research, linear programming is the most used and best known mathematical process. Linear programming allows us to maximize the profit of a product or maximize the protein in a diet, given certain restrictions. It also allows us to minimize the costs in the manufacture of a product or minimize the calories in a diet, given certain restrictions. For example, a company that makes two types of canoes may have restrictions as given by the number of labor hours available in the assembly and cutting departments. If the company knows the profit made on each type of canoe, it would want to determine how to utilize the available labor hours so that the profit is maximized. In Section 3.4 we will address applications such as these. In this section we will focus exclusively on how to solve linear programming problems graphically so that this skill can be used effectively in the next section.

Linear Programming

Many applications in the real world ask us to determine an **optimal value** (maximum or minimum) of a function that is subject to certain **constraints**. For example, the agribusiness in Chang Qing County in China is made up of crop farming, livestock husbandry, forestry, fisheries, and food processing. County officials wanted to increase the net profit while having no adverse effects on the environment. (*Source:* A Case Write Up: Agriculture in China, www.informs.org). In a **linear programming** problem the function that we are trying to find an optimal value for is called an **objective function**. In Chang Qing County, the objective function was the profit function for the agribusinesses, and the constraints were the restrictions on environmental impact. The constraints that the objective function is subject to are usually given by linear inequalities. When only two variables are involved, we can use the techniques of Section 3.2. Before solving applications, let's examine the method behind linear programming.

Example 1   **Maximizing an Objective Function**

Maximize the objective function $P = 2x + y$ subject to the constraints

$$x + y \leq 6$$
$$x - y \leq 2$$
$$x \geq 0, \ y \geq 0$$

Solution

The first thing we do is **graph the feasible region** as given by the system of linear inequalities using the techniques discussed in Section 3.2. See Figure 3.3.1. The points labeled $A, B, C,$ and D occur where boundary lines intersect. Such points are called **corner points**. Soon we will see that corner points are crucial in maximizing (or minimizing) an objective function. Hence we need to **determine the coordinates** of the corner points. $A, B,$ and C are fairly easy to determine since

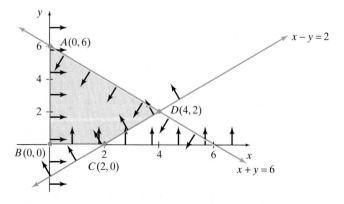

Figure 3.3.1

they are simply intercepts. We determine the coordinates of point A to be $(4, 2)$ by solving the system

$$x - y = 2$$
$$x + y = 6$$

Recall that this system can be solved by using the substitution method (Chapter 1) or the elimination method (Chapter 1), graphically on a calculator, and by using the INTERSECT command (Chapter 1 and 2), Gauss–Jordan (Chapter 2), or matrix inverses (Chapter 2).

Now every point in the feasible region satisfies all the constraints given by the system of linear inequalities. However, we want to **find the point that gives the largest value for the objective function**. To see how we can do this, we will add lines to Figure 3.3.1 that represent our objective function $P = 2x + y$ for various values of P. We arbitrarily pick $0, 4, 8$, and 12 for P. This produces the equations

$$0 = 2x + y, \qquad 4 = 2x + y, \qquad 8 = 2x + y, \qquad 12 = 2x + y$$

We now add the graphs of these four lines to Figure 3.3.1 as shown in Figure 3.3.2. The four lines are parallel since they all have a slope of -2. Observe that P cannot have a value of 12 since the graph for $P = 12$ is outside the feasible region. Every point on the line $P = 8$ that is within the feasible region gives a value for x and y such that $2x + y = 8$. It appears that a line parallel to these four lines, somewhere between the lines representing the objective function when $P = 8$ and $P = 12$, will yield the maximum value for P. The values of x and y that maximize P **and** still satisfy all constraints will be coordinates (x, y) of a point on this parallel line **if** the line just touches the feasible region. We observe that this happens at the **corner point** $(4, 2)$. The value of P at this point is

$$P = 2x + y = 2(4) + 2 = 10$$

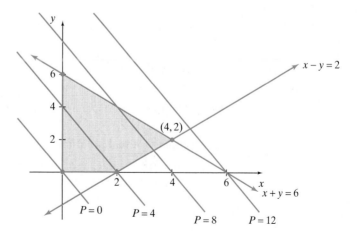

Interactive Activity

Conjecture what point minimizes the objective function in Example 1. What is the minimum value?

Figure 3.3.2

Before doing another example, we want to stress that of all the points in the feasible region in Example 1, the corner point $(4, 2)$ gives the largest possible value of P. Also, recall from Section 3.2 that the feasible region in Example 1 is said to be **bounded**.

Example 2   Minimizing an Objective Function

Minimize the objective function $C = 2x + 2y$ subject to the constraints

$$x + y \geq 8$$
$$5x + 7y \geq 50$$
$$x \geq 0, \ y \geq 0$$

Solution

Figure 3.3.3 shows the feasible region along with lines when C in the objective function is replaced with 0, 8, 16, and 24. The smallest value of z for which a line touches the feasible region is $C = 16$. We have two corner points, $(0, 8)$ and $(3, 5)$, on this line. This means that both corner points, $(0, 8)$ and $(3, 5)$, as well as **all points on the boundary line between them**, yield the same minimum value for C. So there are an infinite number of values for x and y that produce the minimum value of 16 for the objective function.

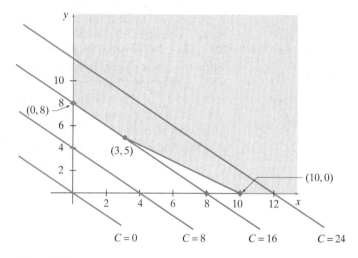

Figure 3.3.3

Interactive Activity

Explain why the lines for $C = 0$ and $C = 8$ in Example 2 are not candidates for minimizing C. Also, do you believe that there is a maximum value of the objective function in Example 2? Explain your reasoning.

Example 2 is an example of a **multiple optimal solution**. In general,

▪ Multiple Optimal Solution

If two corner points are both optimal solutions to a linear programming problem, then any point that is on the line segment joining them is also an optimal solution. We call the collection of these points the **multiple optimal solution**.

Recall from Section 3.2 that the feasible region in Example 2 is said to be **unbounded**. The feasible region in Example 1 was bounded and, as Example 1 showed, along with the Interactive Activity following Example 1, linear programming problems with bounded regions always have optimal solutions (a maximum and a minimum). This is not true for the unbounded region in Example 2. It had a minimum, but there is no value for x and y that will maximize the objective function.

We now summarize the prior discussion as follows:

◼ Existence of Optimal Solution

- If the feasible region for a linear programming problem is bounded, then both a maximum value and a minimum value of the objective function exist.
- If the feasible region for a linear programming problem is unbounded, then carefully examine the feasible region and objective function to see if the maximum or minimum value exists.

Example 2 illustrates bullet item 2. Examples 1 and 2 illustrate that the optimal value of an objective function occurs at a corner point. These examples suggests the following theorem.

◼ Corner Point Theorem

If the optimal value (either maximum or minimum) of the objective function exists, then it will occur at one (or more) of the corner points of the feasible region.

The Corner Point Theorem greatly simplifies the task of finding an optimal value. We now supply a general strategy for locating optimal values.

◼ Solving Linear Programming Problems with Two Variables Graphically

Step 1: Graph the feasible region. Take note if the region is bounded or unbounded.

Step 2: Determine the coordinates of all corner points.

Step 3: Evaluate each corner point in the objective function.

Step 4: Determine the optimal solution from the results of step 3.

Example 3   Solving a Linear Programming Problem

$$\text{Maximize:} \qquad z = x - 3y$$
$$\text{Subject to:} \qquad 2x - 3y \leq 6$$
$$-x + 4y \leq 4$$
$$x \geq 0, \ y \geq 0$$

Solution

We will utilize the step-by-step procedure given prior to Example 3.

Step 1: The graph of the feasible region is shown in Figure 3.3.4. Notice that the feasible region is bounded, so the existence of a maximum value for the objective function is guaranteed.

Step 2: Corner point A has coordinates $(3, 0)$, corner point B has coordinates $(0, 0)$, and corner point C has coordinates $(0, 1)$. The coordinates for corner point D are found by solving the system

$$2x - 3y = 6$$
$$-x + 4y = 4$$

We leave it to you to decide which method to use to solve the system and verify that the coordinates for corner point D are $\left(\frac{36}{5}, \frac{14}{5}\right)$.

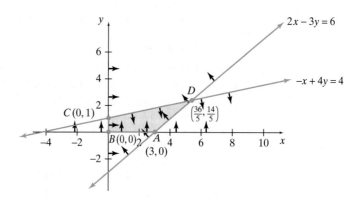

Figure 3.3.4

Step 3: Here we evaluate each corner point in the objective function. We use Table 3.3.1 to organize our work.

Table 3.3.1

Corner Point	Value of $z = x - 3y$	
$(3, 0)$	$3 - 3(0) = 3$	**Maximum**
$(0, 0)$	$0 - 3(0) = 0$	
$(0, 1)$	$0 - 3(1) = -3$	
$\left(\frac{36}{5}, \frac{14}{5}\right)$	$\frac{36}{5} - 3\left(\frac{14}{5}\right) = -\frac{6}{5}$	

Step 4: The maximum value of $z = x - 3y$ is 3 at the corner point $(3, 0)$.

■

✓ **Checkpoint 1**

Now work Exercise 9.

In the next example we determine both a maximum and a minimum.

Example 4 **Solving a Linear Programming Problem**

Maximize and minimize: $z = 2x + y$

Subject to: $x + y \geq 2$

$6x + 4y \leq 36$

$4x + 2y \leq 20$

$x \geq 0, y \geq 0$

Solution

As in Example 3, we utilize the step-by-step procedure.

Step 1: Figure 3.3.5 is a graph of the feasible region.

Step 2: The corner points are shown in Figure 3.3.5. Notice that four corner points are intercepts, while the fifth corner point is the solution to the system

$$6x + 4y = 36$$
$$4x + 2y = 20$$

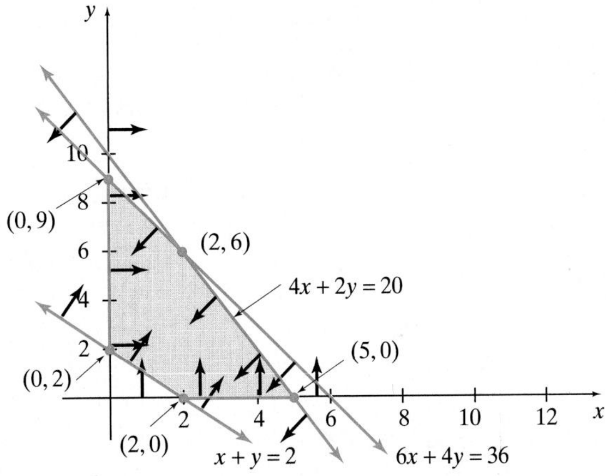

Figure 3.3.5

Step 3: We evaluate each corner point in the objective function. As in Example 3, we use a table to organize our work. See Table 3.3.2.

Table 3.3.2

Corner Point	Value of $z = 2x + y$	
$(5, 0)$	$2(5) + 0 = 10$	**Maximum**
$(2, 0)$	$2(2) + 0 = 4$	
$(0, 2)$	$2(0) + 2 = 2$	**Minimum**
$(0, 9)$	$2(0) + 9 = 9$	
$(2, 6)$	$2(2) + 6 = 10$	**Maximum**

Step 4: The minimum value of $z = 2x + y$ is 2 at the corner point $(0, 2)$. The maximum value of $z = 2x + y$ is 10 at the corner points $(5, 0)$ and $(2, 6)$. As stated earlier, this is a multiple optimal solution, and any point on the line segment connecting $(5, 0)$ and $(2, 6)$ also produces a maximum value of $z = 2x + y$. ■

✔ **Checkpoint 2**

Now work Exercise 15.

Our final example supplies a general strategy when dealing with an unbounded feasible region.

Example 5 **Solving a Linear Programming Problem**

Maximize and minimize: $z = x + y$

Subject to: $9x - 3y \leq 6$
$4x + 2y \geq 8$
$x \geq 0, y \geq 0$

Solution

Using our four-step process yields:

Step 1: Figure 3.3.6 is a graph of the feasible region. Notice that the feasible region is unbounded.

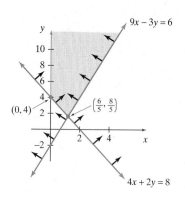

Figure 3.3.6

Step 2: The corner points are shown in Figure 3.3.6.

Step 3: We evaluate each corner point in the objective function and record the results in Table 3.3.3.

Table 3.3.3

Corner Point	Value of $z = x + y$	
$(0, 4)$	$0 + 4 = 4$	
$\left(\frac{6}{5}, \frac{8}{5}\right)$	$\frac{6}{5} + \frac{8}{5} = \frac{14}{5} = 2.8$	**Minimum**

Step 4: The minimum value of $z = x + y$ is 2.8 at the corner point $\left(\frac{6}{5}, \frac{8}{5}\right)$. For a maximum value, we need to take care since the feasible region is unbounded. To help us, we will add two lines to Figure 3.3.6, representing our objective function $z = x + y$ for $z = 2.8$ and $z = 4$, the values from Table 3.3.3. See Figure 3.3.7, which suggests that as z increases the lines move farther away from the origin. Since the feasible region is unbounded, the region continues indefinitely; there is no line that produces a maximum value of z.

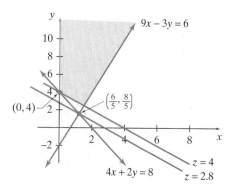

Figure 3.3.7

Example 5 illustrates how we need to take care when dealing with an unbounded feasible region. It also illustrates the following fact when confronted with an unbounded feasible region.

■ **Optimal Values of an Unbounded Feasible Region**

If the feasible region is unbounded and all the coefficients of the objective function are positive, then a minimum value of the objective function exists, but a maximum value does not.

▶ **Note:** This fact is true only when the unbounded feasible region is in the first quadrant. In this text, all applications considered will be restricted to the first quadrant.

SUMMARY

In this section we introduced the methods of solving a **linear programming problem** graphically. We determined the **optimal value** of an **objective function**, given certain **constraints**. We saw how **corner points** were the key to the linear programming problem.

- **Corner Point Theorem** If the optimal value (either maximum or minimum) of the objective function exists, then it will occur at one (or more) of the corner points of the feasible region.

- **Solving Linear Programming Problems with Two Variables Graphically**

  Step 1: Graph the feasible region. Take note if it is bounded or unbounded.

  Step 2: Determine the coordinates of all corner points.

  Step 3: Evaluate each corner point in the objective function.

  Step 4: Determine the optimal solution from the results of step 3.

SECTION 3.3 EXERCISES

In Exercises 1–8, the graph of a feasible region is given. Use these regions to determine the maximum and minimum values of the given objective function.

1. $z = 2x + 3y$

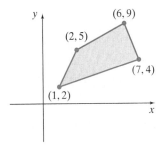

2. $z = 3x + 2y$

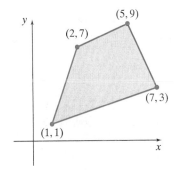

3. $z = 2x + 5y$

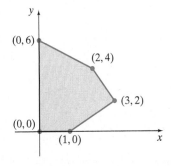

4. $z = 4x + y$

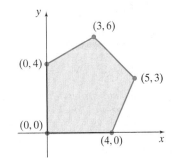

5. $z = x + 4y$

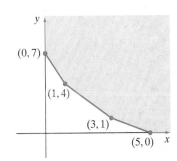

6. $z = 2x + 4y$

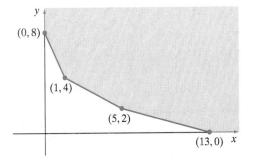

7. $z = 0.4x + 0.3y$

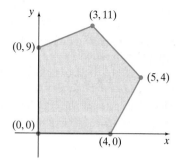

8. $z = 0.2x + 0.7y$

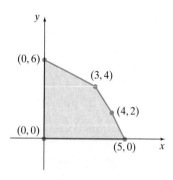

In Exercises 9–22, use the graphical and corner point method to solve the given linear programming problem.

 **9.** Maximize: $z = 4x + 3y$
Subject to: $3x - y \leq 5$
$x + 2y \leq 4$
$x \geq 0, \ y \geq 0$

10. Maximize: $z = 3x + y$
Subject to: $4x + 5y \leq 9$
$2x - y \geq 1$
$x \geq 0, \ y \geq 0$

11. Minimize: $z = x + 2y$
Subject to: $x + y \leq 7$
$2x + 3y \leq 16$
$x \geq 0, \ y \geq 0$

12. Minimize: $z = 2x + 5y$
Subject to: $x + y \leq 7$
$2x + 3y \geq 16$
$x \geq 0, \ y \geq 0$

13. Maximize: $z = 2x + 2y$
Subject to: $x + y \leq 7$
$x - 2y \leq 16$
$x \geq 0, \ y \geq 0$

14. Maximize: $z = x + 3y$
Subject to: $x + y \leq 3$
$x - 2y \geq 0$
$x \geq 0, \ y \geq 0$

15. Minimize and maximize: $z = 3x + 2y$
Subject to: $x + y \leq 5$
$3x - 2y \leq 0$
$2x + y \leq 6$
$x \geq 0, \ y \geq 0$

16. Minimize and maximize: $z = 2x + 3y$
Subject to: $x + 2y \leq 2$
$3x - y \leq 3$
$x \geq 0, \ y \geq 0$

17. Minimize and maximize: $z = x + 4y$
Subject to: $2x + y \leq 32$
$x + y \leq 18$
$x + 3y \leq 36$
$x \geq 0, \ y \geq 0$

18. Minimize and maximize: $z = 5x + y$
Subject to: $-x + y \geq 0$
$x + 3y \leq 9$
$4x + 3y \leq 12$
$x \geq 0, \ y \geq 0$

19. Minimize and maximize: $z = 2x + 5y$
Subject to: $x + y \geq 2$
$2x + 3y \leq 12$
$3x + y \leq 12$
$x \geq 0, \ y \geq 0$

20. Minimize and maximize: $z = 5x + 2y$
Subject to: $x + y \geq 2$
$x + y \leq 8$
$2x + y \leq 10$
$x \geq 0, \ y \geq 0$

21. Minimize and maximize: $z = 3x + 5y$
Subject to: $-x + 2y \leq 10$
$2x + y \leq 15$
$5x + y \geq 8$
$-x + y \geq -1$
$x \geq 0, \ y \geq 0$

22. Minimize and maximize: $z = 5x + y$
Subject to: $-x + 2y \leq 10$
$2x + y \leq 15$
$5x + y \geq 8$
$-x + y \geq -5$
$x \geq 0, \ y \geq 0$

SECTION PROJECT

Minimize and maximize: $z = \frac{1}{2}x + \frac{1}{3}y$

Subject to:
$$2x + 3y \le 18$$
$$3x + 2y \le 18$$
$$8x + 4y \le 32$$
$$9x + 2y \ge 16$$
$$0 \le x \le 5$$
$$0 \le y \le 6$$

Section 3.4   Applications of Linear Programming

This section combines the skills learned in the first three sections of this chapter. This includes writing systems of linear inequalities, representing problem constraints as inequalities, graphing the feasible region, and maximizing (or minimizing) an objective function for given constraints. In other words, in this section we will solve applications using linear programming. We'll start with a Flashback.

Flashback

WeDo Wood Canoes Company Revisited

In Section 3.2 we saw that the WeDo Wood Canoes Company makes two types of canoes: a two-person model and a four-person model. Each two-person model requires 1 hour in the cutting department and 1.5 hours in the assembly department. Each four-person model requires 1 hour in the cutting department and 2.75 hours in the assembly department. The cutting department has a maximum of 640 hours available each week, and the assembly department has a maximum of 1080 hours available each week. We saw in Section 3.2 that if we let x represent the number of two-person canoes made each week and let y represent the number of four-person canoes made each week then the following system of linear inequalities expresses the constraints.

$$x + y \le 640$$
$$1.5x + 2.75y \le 1080$$
$$x \ge 0, \; y \ge 0$$

A graph of this system, showing the feasible region and corner points, is given in Figure 3.4.1.

A profit of $40 on each two-person canoe and $60 on each four-person canoe is realized by WeDo Wood Canoes. Using the definition of the variables given, write an equation to represent the weekly profit and determine values for x and y that maximize weekly profit.

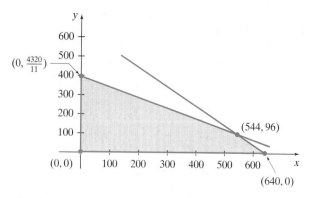

Figure 3.4.1

Flashback Solution

Understand the Situation: We have already defined x to represent the number of two-person canoes produced and sold and to let y represent the number of four-person canoes produced and sold. Since each two-person canoe yields a profit of \$40, WeDo Wood Canoes realizes a profit of $40x$ dollars on the sale of x two-person canoes. Similarly, a profit of $60y$ dollars is realized on the sale of y four-person canoes. So the weekly profit is represented by

$$P = 40x + 60y$$

Determining the values for x and y that maximize weekly profit sounds similar to what we did in Section 3.3. Actually, it is identical to what we did in Section 3.3! We are really solving the following linear programming problem:

Maximize: $P = 40x + 60y$
Subject to: $x + y \leq 640$
$1.5x + 2.75y \leq 1080$
$x \geq 0, y \geq 0$

Figure 3.4.1 shows the feasible region along with the corner points, so all we need to do is evaluate the objective function at each corner point, as shown in Table 3.4.1.

Table 3.4.1

Corner Point	Value of $P = 40x + 60y$	
$(0, 0)$	$40(0) + 60(0) = 0$	
$(0, \frac{4320}{11})$	$40(0) + 60\left(\frac{4320}{11}\right) \approx 23{,}563.64$	
$(544, 96)$	$40(544) + 60(96) = 27{,}520$	**Maximum**
$(640, 0)$	$40(640) + 60(0) = 25{,}600$	

Interpret the Solution: Table 3.4.1 indicates that producing and selling 544 two-person canoes and 96 four-person canoes each week maximizes the weekly profit at \$27,520.

In the Flashback, the variables x and y are known as **decision variables**. Our goal was to determine values of the decision variables that maximized the **objective function**, $P = 40x + 60y$, given the **problem constraints**. Our problem constraints, as given by the system of linear inequalities, were

Cutting department constraint:	$x + y \leq 640$
Assembly department constraint:	$1.5x + 2.75y \leq 1080$
Nonnegative constraints:	$x \geq 0, y \geq 0$

Since our decision variables, x and y, represented the number of items produced, the feasible region in Figure 3.4.1 actually gave us the **feasible region for production schedules**. In other words, we could choose any production schedule (x, y) in the feasible region and use the objective function $P = 40x + 60y$ to compute the profit at production level (x, y). In the Flashback we used the techniques from Section 3.3 to find the production level that maximized profit.

Before we consider another example, let's combine our solving strategy from Section 3.3 with some new information to get a strategy for solving applied linear programming problems.

■ **Solving an Applied Linear Programming Problem with Two Decision Variables**

Step 1: Construct a mathematical model for the problem. This includes:

- Define the decision variables.
- Write a linear objective function.
- Write the problem constraints.

Step 2: Graph the feasible region. Note if it is bounded or unbounded.

Step 3: Determine the coordinates of all corner points.

Step 4: Evaluate each corner point in the objective function.

Step 5: Determine the optimal solution from the result of step 4.

Step 6: Interpret the optimal solution in terms of the original problem.

▶ **Note:** Steps 2 through 5 are taken from the steps for solving linear programming problems in Section 3.3.

Example 1   **Solving an Applied Linear Programming Problem**

Dub, Inc., is a small manufacturing firm that produces two microwave switches, switch A and switch B. Each switch passes through two departments: assembly and testing. Switch A requires 4 hours of assembly and 1 hour of testing, and switch B requires 3 hours of assembly and 2 hours of testing. The maximum number of hours available each week for assembly and testing are 240 and 140 hours, respectively. The per unit profit for switch A is $20, whereas the per unit profit for switch B is $30. How many units of each switch should be made and sold each week to maximize profit?

Solution

We use the six-step strategy outlined prior to Example 1.

Step 1: **Understand the Situation:** Since we do not know how many of each switch is made, we let x represent the number of units of switch A made each week and let y represent the number of units of switch B made each week. Since the profit must be maximized, we determine an objective function for profit. The objective function is

$$P = 20x + 30y$$

The problem constraints are

Assembly department constraint:	$4x + 3y \leq 240$
Testing department constraint:	$x + 2y \leq 140$
Nonnegative assembly and testing constraints:	$x \geq 0, \ y \geq 0$

So the mathematical model is

Maximize: $P = 20x + 30y$

Subject to: $4x + 3y \leq 240$

$x + 2y \leq 140$

$x \geq 0, \ y \geq 0$

Step 2: The feasible region is shown in Figure 3.4.2. Since the region is bounded, a maximum value is guaranteed to exist.

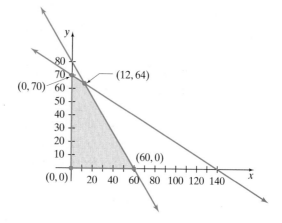

Figure 3.4.2

Step 3: The corner points are shown in Figure 3.4.2.

Step 4: Table 3.4.2 shows the result of each corner point evaluated in the objective function.

Table 3.4.2

Corner Point	Value of $P = 20x + 30y$	
$(0, 0)$	$20(0) + 30(0) = 0$	
$(0, 70)$	$20(0) + 30(70) = 2100$	
$(12, 64)$	$20(12) + 30(64) = 2160$	**Maximum**
$(60, 0)$	$20(60) + 30(0) = 1200$	

Step 5: The optimal (maximum) value shown in Table 3.4.2 is 2160 and occurs at the corner point $(12, 64)$.

Step 6: **Interpret the Solution:** To maximize weekly profit, Dub, Inc., should produce and sell 12 units of switch A and 64 units of switch B. This generates a maximum weekly profit of $2160.   ∎

✓ **Checkpoint 1**

Now work Exercise 1.

In Example 2 we turn our attention to minimizing costs.

Example 2   Solving an Applied Linear Programming Problem

A beagle breeder can buy a special food, DogsLife, at $0.30 per pound and another special food, King Kibble, at $0.40 per pound. Each pound of DogsLife has 210 units of protein and 240 units of fat. Each pound of King Kibble has 150 units of protein and 380 units of fat. If the minimum daily requirements for the beagles (collectively) are 2115 units of protein and 3460 units of fat, how many pounds of each food mix should be used daily to minimize daily food costs while still meeting (or even exceeding) the minimum daily requirements?

Solution

Step 1: **Understand the Situation:** Since we do not know how many pounds of each special food are used, we let x represent the number of pounds of DogsLife used daily and let y represent the number of pounds of King Kibble used daily. Since the daily food costs must be minimized, we determine an objective function for daily food costs. The objective function is

$$C = 0.30x + 0.40y$$

The problem constraints are

Protein constraint:	$210x + 150y \geq 2115$
Fat constraint:	$240x + 380y \geq 3460$
Nonnegative food constraints:	$x \geq 0, y \geq 0$

So the mathematical model is

Minimize: $C = 0.30x + 0.40y$

Subject to: $210x + 150y \geq 2115$
$240x + 380y \geq 3460$
$x \geq 0, \ y \geq 0$

Step 2: The feasible region is shown in Figure 3.4.3. The region is unbounded. However, since all coefficients in the objective function are positive, a minimum optimal value exists. (Consult the Toolbox on p. 178.)

Step 3: The corner points are shown in Figure 3.4.3.

Step 4: The results of the corner points being evaluated in the objective function are shown in Table 3.4.3.

Step 5: The optimal (minimum) value shown in Table 3.4.3 is 3.95 and occurs at corner point (6.5, 5).

From Your Toolbox

If the feasible region is unbounded and all the coefficients of the objective function are positive, then a minimum value of the objective function exists, but a maximum value does not.

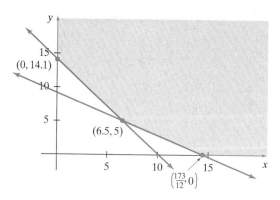

Figure 3.4.3

Table 3.4.3

Corner Point	Value of $C = 0.30x + 0.40y$	
$(0, 14.1)$	$0.30(0) + 0.40(14.1) = 5.64$	
$(6.5, 5)$	$0.30(6.5) + 0.40(5) = 3.95$	**Minimum**
$\left(\frac{173}{12}, 0\right)$	$0.30\left(\frac{173}{12}\right) + 0.40(0) = 4.325 \approx 4.33$	

Step 6: Interpret the Solution: To minimize daily food costs, while still meeting the minimum daily nutritional requirements, the beagle breeder should use 6.5 pounds of DogsLife and 5 pounds of King Kibble. The daily minimum cost is $3.95.   ■

✔ **Checkpoint 2**

Now work Exercise 19.

Example 3   **Solving an Applied Linear Programming Problem**

Table 3.4.4 gives nutritional information, in grams, and current prices for selected items from Wendy's menu.

Table 3.4.4

Item	Protein	Fiber	Price
Breaded Chicken Sandwich	28	2	$2.89
Biggie® Fries	7	6	$1.19

SOURCE: Wendy's, www.wendys.com

Joshua desires to eat nothing but Breaded Chicken Sandwiches and Biggie® Fries. He wants at least 112 grams of protein and at least 30 grams of fiber daily. If Joshua eats nothing but Breaded Chicken Sandwiches and Biggie® Fries and wishes to meet his minimum nutrient requirements, how much of each food should he eat to minimize daily food costs?

Solution

Step 1: **Understand the Situation:** Let x represent the number of orders of Breaded Chicken Sandwiches and let y represent the number of orders of Biggie® Fries. The objective function is

$$C = 2.89x + 1.19y$$

The problem constraints are

Protein constraint:	$28x + 7y \geq 112$
Fiber constraint:	$2x + 6y \geq 30$
Nonnegative constraints:	$x \geq 0, y \geq 0$

So the mathematical model is

Minimize:	$C = 2.89x + 1.19y$
Subject to:	$28x + 7y \geq 112$
	$2x + 6y \geq 30$
	$x \geq 0, \ y \geq 0$

Step 2: Figure 3.4.4 shows the feasible region. The region is unbounded but, as in Example 2, since all coefficients of the objective function are positive, a minimum optimal value exists.

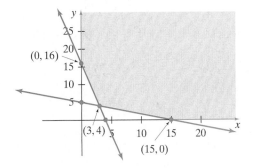

Figure 3.4.4

Step 3: The corner points are shown in Figure 3.4.4.

Step 4: Table 3.4.5 shows the results of each corner point evaluated in the objective function.

Table 3.4.5

Corner Point	Value of $C = 2.89x + 1.19y$	
$(0, 16)$	$2.89(0) + 1.19(16) = 19.04$	
$(3, 4)$	$2.89(3) + 1.19(4) = 13.43$	**Minimum**
$(15, 0)$	$2.89(15) + 1.19(0) = 43.35$	

Step 5: The optimal (minimum) value shown in Table 3.4.5 is 13.43 and occurs at the corner point $(3, 4)$.

Step 6: Interpret the Solution: To minimize daily food costs, while still meeting the daily nutrient requirements, Joshua should eat 3 Breaded Chicken Sandwiches and 4 orders of Biggie® Fries. His daily minimum cost is $13.43.   ∎

✓ **Checkpoint 3**

Now work Exercise 21.

Before we consider our final example, we hope that the first three examples have illustrated that the skills learned in Sections 3.1 and 3.2 (writing a system of linear inequalities and graphing the feasible region), along with the skills learned in Section 3.3 (solving linear programming problems using corner points), are simply being combined in this section.

Example 4   Solving an Applied Linear Programming Problem

The Hoofer Mining Company owns two different mines that produce ore, which is crushed and graded into three classes: high, medium, and low grade. The company has a contract to provide a smelting plant with at least 13 tons of high-grade, 9 tons of medium-grade, and 24 tons of low-grade ore per week. The two mines have different operating characteristics as given in Table 3.4.6.

Table 3.4.6

Mine	Cost per Day (in 1000s)	Production (tons per day)		
		High	Medium	Low
I	12	3	3	4
II	10	2	1	6

Each mine operates at most 5 days per week. How many days per week should each mine operate to meet (or exceed) the smelting plant contract and minimize the total cost?

Solution

Step 1: Understand the Situation: Let x represent the number of days per week that mine I is operated, and let y represent the number of days per week that mine II is operated. The objective function is

$$C = 12x + 10y \qquad \text{In thousands}$$

The problem constraints are

High-grade ore constraint:	$3x + 2y \geq 13$
Medium-grade ore constraint:	$3x + y \geq 9$
Low-grade ore constraint:	$4x + 6y \geq 24$
Operating days constraint:	$x \leq 5, y \leq 5$
Nonnegative constraints:	$x \geq 0, y \geq 0$

So the mathematical model is

$$\text{Minimize:} \quad C = 12x + 10y$$
$$\text{Subject to:} \quad 3x + 2y \geq 13$$
$$3x + y \geq 9$$
$$4x + 6y \geq 24$$
$$0 \leq x \leq 5, 0 \leq y \leq 5$$

Step 2: Figure 3.4.5 shows the feasible region.

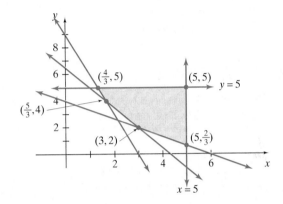

Figure 3.4.5

Step 3: The corner points are shown in Figure 3.4.5.

Step 4: Table 3.4.7 shows the results of each corner point evaluated in the objective function.

Table 3.4.7

Corner Point	Value of $C = 12x + 10y$	
$(5, \frac{2}{3})$	$12(5) + 10\left(\frac{2}{3}\right) = 66\frac{2}{3}$	
$(3, 2)$	$12(3) + 10(2) = 56$	**Minimum**
$(\frac{5}{3}, 4)$	$12\left(\frac{5}{3}\right) + 10(4) = 60$	
$(\frac{4}{3}, 5)$	$12\left(\frac{4}{3}\right) + 10(5) = 66$	
$(5, 5)$	$12(5) + 10(5) = 110$	

Interactive Activity

Determine how many total tons of high-, medium-, and low-grade ore are produced for a week by the Hoofer Mining Company at the optimal value found in Example 4.

Step 5: The optimal (minimum) value in Table 3.4.7 is 56 and occurs at the corner point $(3, 2)$.

Step 6: **Interpret the Solution:** To minimize the total cost, while still meeting the smelting plant contract, Hoofer Mining Company should operate mine I 3 days per week and mine II 2 days per week. The minimum weekly cost is $56,000. ∎

SUMMARY

In this section we learned how to solve applications of linear programming problems. Basically, this combined the skills acquired in the first three sections of the chapter. We also used a six-step process to solve these applications.

- **Solving an Applied Linear Programming Problem with Two Decision Variables**

  Step 1: Construct a mathematical model for the problem. This includes:

  - Define the decision variables.
  - Write a linear objective function.
  - Write the problem constraints.

Step 2: Graph the feasible region. Note if it is bounded or unbounded.

Step 3: Determine the coordinates of all corner points.

Step 4: Evaluate each corner point in the objective function.

Step 5: Determine the optimal solution from the result of step 4.

Step 6: Interpret the optimal solution in terms of the original problem.

SECTION 3.4 EXERCISES

✓ **1.** The Yuen Tent Company makes two types of tents: two-person and six-person models. Each tent requires the services of two departments, as shown in Table 3.4.8.

Table 3.4.8

Department	Labor (hours per unit)	
	Two-person	**Six-person**
Fabric cutting	6	8
Assembly and packaging	2	4

Fabric cutting has a maximum of 580 hours available each week and assembly and packaging has a maximum of 260 hours available each week. Each two-person tent yields $40 in profit and each six-person tent yields $60 in profit.

(a) Define the variables and write the objective function and the problem constraints.

(b) How many of each type of tent should be made and sold to maximize weekly profit?

2. (*continuation of Exercise 1*) Irene, the head of research and development at the Yuen Tent Company, has discovered a way to reduce the amount of time needed for the six-person tent in the fabric cutting department to 7 hours. Assume that everything else from Exercise 1 remains the same.

(a) Define the variables and write the objective function and the problem constraints.

(b) How many of each type of tent should be made and sold to maximize weekly profit?

3. Denyse and Mike, two recent graduates, have decided to enter the field of personal digital assistants. Initially, they plan to make two types of personal digital assistants, Zap 1000 and Zap 1200. Because of the interest in personal digital assistants, they can sell all that they can produce. However, they need to determine a production rate so as to satisfy various limits with a small production crew.

These limits include a maximum of 150 hours per week for assembly and 100 hours per week for testing. The Zap 1000 requires 3 hours of assembly and 3 hours of testing, and the Zap 1200 requires 6 hours of assembly and 3.5 hours of testing. Profit for the Zap 1000 is estimated at $300 per unit; profit for the Zap 1200 is estimated at $450 per unit.

(a) Define the variables and write the objective function and the problem constraints.

(b) How many of each type of microcomputer should be made and sold to maximize weekly profit?

4. (*continuation of Exercise 3*) Due to a favorable response from the market, Denyse and Mike decide to hire more employees and increase production. By doing so, they now have a maximum of 210 hours per week for assembly and 160 hours per week for testing. Everything else from Exercise 3 remains the same.

(a) Define the variables and write the objective function and the problem constraints.

(b) How many of each type of microcomputer should be made and sold to maximize weekly profit?

5. HotRubber, Inc., produces two types of tires for racing cars: slicks and rain. Each slick tire requires 4 minutes to make and costs $50; each rain tire requires 12 minutes to make and costs $60. The company has contracts that mandate that they must make a total of at least 2000 tires each day, of which at least 800 must be rain tires. The labor union contract with HotRubber dictates that they use at least 300 hours each day making these two types of tires.

(a) How many of each type of tire should be made to meet the conditions and minimize costs?

(b) What are the minimum costs?

6. (*continuation of Exercise 5*) HotRubber, Inc., has a machine that is crucial to the production process break down. As a result it now takes 5 minutes to make a slick tire and 15 minutes to make a rain tire. The costs per tire are still the same as in Exercise 5. The company still has to make a total of at least 2000 tires each day, but now at least 600 must be rain tires. Given that the labor union contract is the same as in Exercise 5, answer the following:

(a) How many of each type of tire should be made to meet the conditions and minimize costs?

(b) What are the minimum costs?

For Exercises 7 and 8, use the fact that

return on an investment
= (amount invested) · (percentage rate)

7. Kevan has received a $50,000 bonus from his employer and wishes to invest it in a growth-oriented mutual fund and an income mutual fund. The growth mutual fund has a historical average yearly return of 12%, and the income fund has a historical average yearly return of 9%. Each fund has a minimum investment of $5000. Kevan desires that at least twice as much be invested in the income fund as in the growth fund. Assuming the historical average yearly returns, how much should Kevan invest in each fund to maximize his return?

8. Brendan has received a $30,000 bonus from his employer and wishes to invest it in an international mutual fund and a growth and income mutual fund. The international fund has a historical average yearly return of 8% and the growth and income fund has a historical average yearly return of 11%. The international fund has a minimum investment of $2500 and the growth and income fund has a minimum investment of $1000. Brendan wishes that at least three times as much be invested in the growth and income fund as in the international fund. Assuming the historical average yearly returns, how much should Brendan invest in each fund to maximize his return?

9. Hoofer Oil has two refineries, one in Findlay, Ohio, and one in Ft. Wayne, Indiana. Each refinery can refine the amounts of fuel, in barrels, in a 24-hour period as shown in Table 3.4.9.

Table 3.4.9

Maximum Production of:	Findlay	Ft. Wayne
Low-octane gasoline	800	1000
High-octane gasoline	500	600
Diesel	1000	600

Daily operating costs for the Findlay refinery are $28,000, and daily operating costs for the Ft. Wayne refinery are $33,000. Zachary, the production manager of Hoofer Oil, receives an order for 60,000 barrels of low-octane gasoline, 120,000 barrels of high-octane gasoline, and 150,000 barrels of diesel.

(a) How many days should Hoofer Oil operate each refinery to fulfill the order while minimizing cost?

(b) What is the minimum cost?

10. Shannon Oil has two refineries, one in Lima, Ohio, and one in Detroit, Michigan. Each refinery can refine the amounts of fuel, in barrels, in a 24-hour period as shown in Table 3.4.10.

Table 3.4.10

Maximum Production of:	Lima	Detroit
Low-octane gasoline	900	1100
High-octane gasoline	600	700
Diesel	1100	700

Daily operating costs for the Lima refinery are $35,000, and daily operating costs for the Detroit refinery are $30,000. Joshua, the production manager of Shannon Oil, receives an order for 50,000 barrels of low-octane gasoline, 90,000 barrels of high-octane gasoline, and 130,000 barrels of diesel.

(a) How many days should Shannon Oil operate each refinery to fulfill the order while minimizing cost?

(b) What is the minimum cost?

11. Due to new regulations approved by the Environmental Protection Agency (EPA), the Roskos Chemical Company was required to implement a new process to reduce the pollution created when making a particular chemical. The older process releases 4 grams of sulfur and 3 grams of particulates for each gallon of chemical produced. The newer process releases 2 grams of sulfur and 1 gram of particulates for each gallon of chemical produced. The company realizes a profit of $0.40 for each gallon produced under the old process and $0.16 for each gallon produced under the new process. The new EPA regulations do not allow for more than 13,000 grams of sulfur and 7000 grams of particulates to be released daily.

(a) How many gallons of this chemical should Roskos Chemical Company produce using the old process and how many gallons should be produced using the new process to maximize profits and still satisfy the new EPA regulations?

(b) What is the maximum daily profit?

12. (*continuation of Exercise 11*) Phil, the plant manager for the Roskos Chemical Company, received word that the EPA is going to change the new regulations so that no more than 10,000 grams of sulfur can be released daily. Assuming that all other data remain the same as in Exercise 11, answer the following:

(a) How many gallons of this chemical should Roskos Chemical Company produce using the old process and how many gallons should be produced using the new process to maximize profits and still satisfy the new EPA regulations?

(b) What is the maximum daily profit?

13. Michael Ende, a famous German author, wrote *The Neverending Story*. Amazon.com currently provides two versions of the book: hardback and softback. The hardback has a volume of 69.90 in.^{3}, and the softback has a volume of

25.87 in.^{3} Suppose that a shipping box that holds 5184 in.^{3} will be used to ship the books to stores and that Amazon.com wants to ship at most 39 hardback versions. Currently, Amazon.com sells the hardback version for $21.99 and the softback version for $6.99 (*Source:* www.amazon.com).

(a) How many books of each should be shipped to maximize revenue? (Round to the nearest integer so that the shipping box restrictions are not violated.)

(b) What is the maximum revenue?

14. (*continuation of Exercise 13*) Suppose that Amazon.com desires to ship at most 56 hardback versions. Assuming that all other data from Exercise 13 remain the same, answer the following:

(a) How many books of each should be shipped to maximize revenue? (Round to the nearest integer so that the shipping box restrictions are not violated.)

(b) What is the maximum revenue?

15. Compaq makes many types of laptops. Currently, the 1200 Series sells for $1099, while the 1700 Series sells for $1499. The 1200 Series weighs 12.3 pounds and the 1700 Series weighs 11.0 pounds (both in their respective packaging). Suppose that the company was planning to ship the laptops in a box, but the shipper requires the package to weigh no more than 89.3 pounds. Also, suppose that Compaq wants at least one of each laptop in the package (*Source:* www.compaq.com).

(a) How many of each laptop should be packaged to maximize the potential revenue in a package?

(b) What is the maximum potential revenue?

16. (*continuation of Exercise 15*) Suppose that Compaq has a new shipper that requires the package to weigh no more than 78.3 pounds. With this new shipper, Compaq desires to ship at least one 1200 Series and two 1700 Series in a package. Assuming that all the other data in Exercise 15 remain the same, answer the following:

(a) How many of each laptop should be packaged to maximize the potential revenue in a package?

(b) What is the maximum potential revenue?

17. Dominic plans to make pancakes and crepes for breakfast at his restaurant. Pancakes require 1 egg, $\frac{3}{4}$ cup milk, 1 cup flour, and $3\frac{1}{2}$ teaspoons baking powder. Crepes require 2 eggs, 2 cups milk, $1\frac{1}{2}$ cups flour, and $\frac{1}{2}$ teaspoon baking powder. Pancakes sell for $3.00 and crepes sell for $5.50. Dominic only has 3 dozen eggs, 4 gallons of milk (64 cups), 40 cups of flour, and 35 teaspoons of baking powder.

(a) How many pancakes and how many crepes can Dominic make to maximize his revenue?

(b) What is the maximum revenue?

18. (*continuation of Exercise 17*) If Dominic makes the amount of pancakes and crepes determined in Exercise 17(a), how much of each ingredient is left over?

19. Anita, a dietitian at a local hospital, is to prepare a diet to contain at least 400 units of vitamins C and E, 500 units of minerals (such as zinc and iron), and 1400 calories. She has two foods, HiGlo and Sheen, available. An ounce of HiGlo contains 2 units of vitamin C and E, 1 unit of minerals, and 4 calories and costs $0.05. An ounce of Sheen contains 1 unit of vitamins C and E, 2 units of minerals, and 4 calories and costs $0.03.

(a) How many ounces of each food should be used to minimize daily food cost?

(b) What is the minimum daily food cost?

20. Lisa, a dietitian at a local hospital, is to prepare two foods to meet certain nutritional requirements. Each pound of JumpStart contains 200 units of vitamin C, 80 units of vitamin B_{12}, and 20 units of vitamin E and costs $0.40. Each pound of MainTain contains 20 units of vitamin C, 160 units of vitamin B_{12}, and 10 units of vitamin E and costs $0.30. To meet the patients' nutritional needs, the combination of the two foods must have at least 520 units of vitamin C, 640 units of vitamin B_{12}, and 100 units of vitamin E.

(a) How many pounds of each food should be used to minimize daily food cost?

(b) What is the minimum daily food cost?

21. Table 3.4.11 gives nutritional information, in grams, and current prices for selected items from Burger King's menu.

Table 3.4.11

Item	Protein	Fiber	Price
Cheeseburger	22	2	$1.19
Small onion rings	3	2	$0.99

SOURCE: Burger King, www.burgerking.com

Mike desires to eat nothing but cheeseburgers and orders of small onion rings. He wants at least 96 grams of protein and at least 26 grams of fiber daily. Also, he wishes to eat at least 1 cheeseburger and at least 1 order of small onion rings. If Mike eats nothing but cheeseburgers and small onion rings and meet his minimum nutrient requirements, answer the following:

(a) How much of each food should Mike eat to minimize his daily food cost?

(b) What is his minimum daily food cost?

22. (*continuation of Exercise 21*). Mike from Exercise 21 found out that Burger King is having a promotion where cheeseburgers sell for $0.49 on Wednesdays. Assuming all other data in Exercise 21 remains the same, answer the following:

(a) How much of each food should Mike eat to minimize his daily food cost?

(b) What is his minimum daily food cost?

23. Table 3.4.12 gives nutritional information, in grams, for selected items from McDonald's menu.

Table 3.4.12

Item	Protein	Fiber	Total Fat
Filet-o-fish™	15	1	26
Large French fries	8	6	26

SOURCE: McDonald's, www.mcdonalds.com

Austin wants to eat nothing but Filet-o-fish and large French fries. He wants at least 122 grams of protein and at least 30 grams of fiber daily. If Austin eats nothing but Filet-o-fish and large French fries and meets his minimum nutrient requirements, answer the following:

(a) How much of each food should Austin eat to minimize his daily total fat intake?

(b) What is his minimum daily total fat intake?

24. (*continuation of Exercise 23*) A Filet-o-fish contains 470 calories and a large French fries has 540 calories.

(a) If Austin is to consume no more than 2500 calories in a day, would he meet this restriction with the optimal solution found in Exercise 23(a)? If yes, by how many calories is he below 2500? If no, by how many calories would he exceed this restriction?

(b) Is it possible for Austin to meet the calorie restriction and still satisfy his nutrient requirements in Exercise 23? Explain. (*Hint:* Write an inequality for his calorie constraint and graph with the constraints in Exercise 23. Is there a feasible region?)

25. Table 3.4.13 gives data (in lb/acre) for growing artichokes and asparagus for an entire growing season in California.

Table 3.4.13

	Artichokes	Asparagus
Nitrogen	197	240
Potash	123	42

SOURCE: U.S. Department of Agriculture, www.usda.gov/nass

Frances has 150 acres available suitable for cultivating crops. She has purchased a contract for a total of 31,880 pounds of nitrogen and 9120 pounds of potash. She expects to make a profit of $140/acre on artichokes and $160/acre on asparagus.

(a) How many acres of each should she plant to maximize her profit?

(b) What is the maximum profit?

26. Table 3.4.14 gives data (in lb/acre) for growing cauliflower and head lettuce for an entire growing season in California.

Table 3.4.14

	Cauliflower	Head lettuce
Nitrogen	215	173
Phosphate	98	122

SOURCE: U.S. Department of Agriculture, www.usda.gov/nass

Sonja has 210 acres available for cultivating crops. She has purchased a contract for a total of 38,380 pounds of nitrogen and 22,240 pounds of phosphate. She expects to make a profit of $90/acre on cauliflower and $100/acre on head lettuce.

(a) How many acres of each should she plant to maximize her profit?

(b) What is the maximum profit?

SECTION PROJECT

Table 3.4.15 gives nutritional information and current prices for selected items from McDonald's menu.

Table 3.4.15

Item	Calories	Cholesterol (mg)	Price
Garden salad	100	75	$2.29
Small French fries	210	0	$1.29

SOURCE: McDonald's, www.mcdonalds.com

Frank is on a diet of 2500 (or fewer) calories per day, and it is recommended that his daily total cholesterol intake be less than 300 mg. Frank wants to eat nothing but fries and garden salad and not exceed the calorie and cholesterol requirement.

(a) Form the mathematical model for this situation in which the objective function is cost.

(b) Graph the feasible region. Label the corner points.

(c) Evaluate the objective function at each corner point.

(d) From part (c), determine the optimal (minimum) value. Interpret including the number of garden salads, small French fries, calories, and milligrams of cholesterol that Frank would consume.

(e) From part (c), determine the optimal (maximum) value. Interpret including the number of garden salads, small French fries, calories, and milligrams of cholesterol that Frank would consume.

(f) Based on parts (d) and (e), what would you tell Frank? In other words, how many small French fries and how many garden salads should he expect to eat each day and what is the daily cost?

■ Why We Learned It

In this chapter we learned how to graph systems of linear inequalities and a graphical method for solving linear programming problems. We saw that many managers in business can use linear programming to maximize profit, to minimize costs, or to determine optimum manufacturing times. For example, suppose that a company has three departments that its products must pass through. Due to certain restrictions, each department has a maximum number of hours available each week. Assuming that the company knows the profit it receives on each product that it produces and sells, it would want to determine how many units of each product to make in a week (given the time constraints of each department) in order to maximize profit. We also learned that dietitians can use linear programming to maximize the healthy nutritional elements of a diet, such as vitamins. Dietitians can also use linear programming to minimize elements of a diet, such as fat intake. We also saw that farmers can use linear programming to maximize profit, given restrictions of land and the availability of various fertilizers.

▌CHAPTER REVIEW EXERCISES

For the inequalities in Exercises 1 and 2, complete the tables by stating whether the inequality is true *or* false.

1. $x > 5$

x	$x > 5$	Statement
-14		
5		
7		
10		

2. $x \leq -3$

x	$x \leq -3$	Statement
-7		
-5		
-3		
0		

In Exercises 3–6, translate the English sentence into a mathematical sentence.

3. A certain number increased by eight is twelve.

4. A certain number decreased by three is greater than negative one.

5. The product of five and a certain number minus seven is less than or equal to nine.

6. The product of three and a certain number plus two is not equal to ten.

In Exercises 7–10, solve and graph the solution to the following.

7. $x - 5 > 3$

8. $x - 2 < 6$

9. $4x - 3 \geq 13$

10. $5 - 2x \leq 3$

11. Music Notes offers introductory music workshops for drums and guitar. Each student attends an individual music theory session and a music lesson. Drum students attend a 2-hour music theory session and a 3-hour music lesson. Guitar students attend a 3-hour music theory session and a 4-hour music lesson. Each month instructors can offer no more than 37 hours of music theory sessions and 51 hours of music lessons. Let x represent the number of students admitted each month for the drum workshop, and let y represent the number of students admitted each month for the guitar workshop. Write linear inequalities that give appropriate restrictions on x and y.

12. (*continuation of Exercise 11*) Feedback from students has indicated that Music Notes needs to provide another workshop: music recording. Guitar and drum students will both receive 1 hour of instruction in music recording. Each month instructors can give no more than 14 hours of instruction in music recording. Assuming that everything else from Exercise 11 remains the same, write linear inequalities that give appropriate restrictions on x and y.

13. Sopris Mountaineering makes two types of backpacks: a 3500-cubic-inch day pack and a 5500-cubic-inch multiday pack. Each backpack requires the services, in hours, of two departments as shown in Table R.1.

Table R.1

Department	Day Pack	Multiday Pack
Fabric cutting	4	8
Assembly and packaging	2	6

Fabric cutting has a maximum of 340 hours available each week and assembly and packaging has a maximum of 230 hours available.

(a) Define the variables and write linear inequalities to express reasonable values for the variables.

(b) Write linear inequalities to express the time restrictions of the fabric cutting and assembly and packaging departments.

⊕ 14. The data in Table R.2 give the amount of nitrogen, phosphate, and potash (all in lb/acre) needed to grow head lettuce, cantaloupe, and bell peppers for an entire growing season.

Table R.2

	Head Lettuce	Cantaloupe	Bell Peppers
Nitrogen	173	175	212
Phosphate	122	161	108
Potash	76	39	72

SOURCE: U.S. Department of Agriculture, www.usda.gov/nass

Mika has 250 acres and can get no more than 46,140 pounds of nitrogen, 32,640 pounds of phosphate, and 15,760 pounds of potash.

(a) Define the variables and write linear inequalities to express reasonable values for the variables.

(b) Write linear inequalities to express the restrictions on nitrogen, phosphate, and potash.

15. Blue's Music offers 1-month introductory music workshops for drums, guitar, and piano. Each student attends an individual music theory session, music lesson, and recording session. Each student attends each of the three workshops. Table R.3 shows the number of hours of instruction that each student receives.

Table R.3

	Drum Student	Guitar Student	Piano Student
Music theory	2	3	5
Music lesson	4	5	2
Recording session	1	3	1

Each month, instructors can give a maximum of 57 hours, 54 hours, and 24 hours of instruction in music theory, music lessons, and recording, respectively.

(a) Define the variables and write linear inequalities to express reasonable values for the variables.

(b) Write linear inequalities to express the time restrictions of the music theory, music lessons, and recording workshops.

In Exercises 16 and 17, match the graph with the correct inequality.

16. $x + y \leq 5$ **17.** $x + y \geq 5$

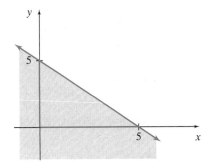

(a)

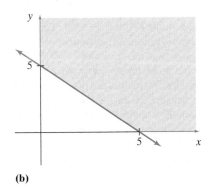

(b)

In Exercises 18–21, graph each linear inequality.

18. $y \leq 2x + 6$ **19.** $y > -4x + 2$

20. $-2x - y \geq 5$ **21.** $x > -2$

In Exercises 22 and 23, use Figure R.1 to match the correct feasible region with the corresponding system of linear inequalities.

22. $x - 2y \leq 0$ **23.** $x - 2y \geq 0$
   $2x + y \geq 5$ $2x + y \geq 5$

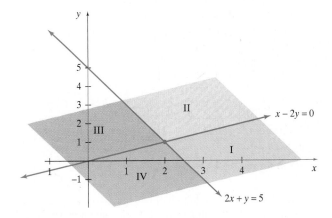

Figure R.1

In Exercises 24–27, graph the feasible region for each system of linear inequalities. Indicate if the region is bounded or unbounded.

24. $x + y \leq 4$

$x - y \geq 2$

25. $3x + 2y \geq 5$

$x + 3y \leq 4$

26. $x + y \leq 5$

$3x - y \geq 2$

$x \geq 0, y \geq 0$

27. $2x - y \leq 6$

$x + 2y \geq 4$

$3x + y \geq 1$

$x \geq 0, y \geq 0$

In Exercises 28 and 29, use a graphing calculator to graph the following:

28. $y \leq 7 - 2x$

29. $2x - 3y \geq 4$

$4x + y \geq 3$

30. Music Notes offers introductory music workshops for drums and guitar. Each student attends an individual music theory session and a music lesson. Drum students attend a 2-hour music theory session and a 3-hour music lesson. Guitar students attend a 3-hour music theory session and a 4-hour music lesson. Each month, instructors can offer no more than 37 hours of music theory sessions and 51 hours of music lessons. Let x represent the number of students admitted each month for the drum workshop, and let y represent the number of students admitted each month for the guitar workshop.

(a) Write a system of linear inequalities that gives appropriate restrictions on x and y.

(b) Graph the feasible region for the system in part (a). Interpret what the feasible region represents.

31. (*continuation of Exercise 30*) Feedback from students has indicated that Music Notes needs to provide another workshop: music recording. Guitar and drum students will both receive 1 hour of instruction in music recording. Each month, instructors can give no more than 14 hours of instruction in music recording. Assuming that everything else from Exercise 30 remains the same, complete the following:

(a) Write a system of linear inequalities that gives appropriate restrictions on x and y.

(b) Graph the feasible region for the system in part (a). Interpret what the feasible region represents.

32. Sopris Mountaineering makes two types of backpacks: a 3500-cubic-inch day pack and a 5500-cubic-inch multiday pack. Each backpack requires the services, in hours, of two departments as shown in Table R.4.

Table R.4

Department	Day Pack	Multiday Pack
Fabric cutting	4	8
Assembly and packaging	2	6

Fabric cutting has a maximum of 340 hours available each week, and assembly and packaging has a maximum of 230 hours available.

(a) Define the variables and write a system of linear inequalities that give appropriate restrictions on x and y.

(b) Graph the feasible region for the system in part (a). Interpret what the feasible region represents.

33. Table R.5 gives nutritional information, in grams, for selected items from Subway's menu.

Table R.5

Item	Protein	Fat
6″ Turkey	16	3.5
6″ Veggie Delight	7	2.5

SOURCE: Subway, www.subway.com

Lera wishes to eat a lunch consisting of turkey subs and veggie subs. She wants the lunch to contain at least 25 grams of protein.

(a) Define the variables and write a system of linear inequalities that gives appropriate restrictions on x and y.

(b) Graph the feasible region for the system in part (a). Interpret what the feasible region represents.

34. (*continuation of Exercise 33*) Now suppose that Lera wishes to eat a lunch consisting of turkey subs and veggie subs that contains no more than 8 grams of fat. Use the data in Table R.5 to answer the following:

(a) Define the variables and write a system of linear inequalities that gives appropriate restrictions on x and y.

(b) Graph the feasible region for the system in part (a). Interpret what the feasible region represents.

35. (*continuation of Exercises 33 and 34*) Lera decides that her lunch should contain at least 25 grams of protein, but no more than 8 grams of fat. Use Table R.5 and the results of Exercises 33 and 34 to do the following:

(a) Write a system of linear inequalities that gives appropriate restrictions on x and y.

(b) Graph the feasible region for the system in part (a). Interpret what the feasible region represents.

In Exercises 36–39, the graph of a feasible region is given. Use the regions to determine the maximum and minimum values of the given objective function.

36. $z = 3x + 4y$

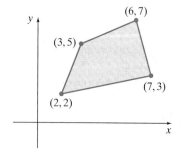

37. $z = 4x + 3y$

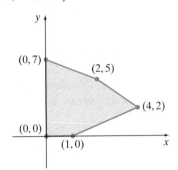

38. $z = x + 2y$

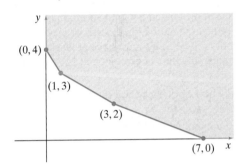

39. $z = 2x + 3y$

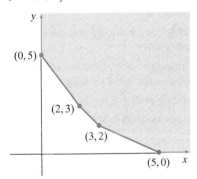

In Exercises 40–45, use the graphical and corner point method to solve the given linear programming problem.

40. Maximize: $z = 2x + y$
   Subject to: $2x - 3y \le 1$
   $x + y \le 3$
   $x \ge 0, y \ge 0$

41. Maximize: $z = 3x + y$
   Subject to: $6x - 2y \ge 4$
   $2x + 2y \le 4$
   $x \ge 0, y \ge 0$

42. Minimize: $z = x + 4y$
   Subject to: $x - y \le 5$
   $x + 2y \ge 8$
   $x \ge 0, y \ge 0$

43. Minimize: $z = 2x + 5y$
   Subject to: $2x + y \ge 6$
   $3x - y \le 2$
   $x \ge 0, y \ge 0$

44. Minimize and maximize: $z = x + 6y$
   Subject to: $x + y \ge 4$
   $2x - 3y \le 8$
   $2x - y \ge 2$
   $x \ge 0, y \ge 0$

45. Minimize and maximize: $z = 4x + 5y$
   Subject to: $-x + 4y \le 6$
   $x + y \le 13$
   $2x + y \ge 3$
   $-x + y \ge -2$
   $x \ge 0, y \ge 0$

46. Sopris Mountaineering makes two types of sleeping bags: +15 sleeping bag and −15 sleeping bag (+15 and −15 indicate the temperature ratings in degrees Fahrenheit). Each bag requires the services, in hours, of two departments as shown in Table R.6.

Table R.6		
Department	**+15 Bag**	**−15 Bag**
Fabric cutting	4	6
Assembly and packaging	2	4

Fabric cutting has a maximum of 320 hours available each week, and assembly and packaging has a maximum of 200 hours available each week. Each +15 bag yields $30 in profit and each −15 bag yields $50 in profit.

(a) Define the variables and write the objective function and problem constraints.

(b) How many of each type of bag should be made and sold to maximize weekly profit?

47. Music Notes offers introductory music workshops for drums and guitar. Each student attends an individual music theory session and a music lesson. Drum students attend a 2-hour music theory session and a 3-hour music lesson. Guitar students attend a 3-hour music theory session and a 4-hour music lesson. Each month, instructors can offer no more than 37 hours of music theory sessions and 51 hours of music lessons. Music notes makes a profit of $25 for each drum student and $30 for each guitar student.

(a) How many of each type of student should be admitted each month to maximize profit?

(b) What is the maximum profit?

48. (*continuation of Exercise 47*) Since Music Notes would really like to offer guitar lessons and students have been demanding a music recording workshop, Music Notes has decided to offer 1 hour of instruction in music recording to both drum and guitar students. Instructors can give no more than 14 hours of instruction in music recording. Assuming that everything else from Exercise 47 remains the same, answer the following questions:

(a) How many of each type of student should be admitted each month to maximize profit?

(b) What is the maximum profit?

(c) Compare your results from part (b) of Exercises 47 and 48. What is your recommendation for Music Notes?

49. Henderson Oil has two refineries, one in Aztec, New Mexico, and one in Buena Vista, Colorado. Each refinery can refine the amounts of fuel, in barrels, in a 24-hour period as shown in Table R.7.

Table R.7

Maximum Production of:	Aztec	Buena Vista
Low-octane gasoline	700	900
High-octane gasoline	400	500
Diesel	900	500

Daily operating costs for the Aztec refinery are $32,000, and daily operating costs for the Buena Vista refinery are $30,000. Lissa, the production manager at Henderson Oil, receives an order for 90,000 barrels of low-octane gasoline, 145,000 barrels of high-octane gasoline, and 150,000 barrels of diesel.

(a) How many days should Henderson Oil operate each refinery to fulfill the order while minimizing cost?

(b) What is the minimum cost?

50. Carolyn, a dietitian for a sports team, is to prepare two foods to meet certain nutritional requirements. Each ounce of PowerSource contains 40 units of carbohydrates, 8 units of protein, and 4 units of total fat and costs $0.12. Each ounce of EnergyMax contains 50 units of carbohydrates, 10 units of protein, and 4 units of total fat and costs $0.14. To meet the nutritional needs of players, the combination of the two foods must have at least 600 units of carbohydrates, 450 units of protein, and 220 units of total fat.

(a) How many ounces of each food should be used to minimize daily food cost?

(b) What is the minimum daily food cost?

CHAPTER 3 PROJECT

SOURCE: EAS, Inc., www.eas.com

EAS produces supplements for anyone who enjoys athletic activities. Two of their products are the SimplyProtein Nutrition Bar and the Myoplex Deluxe Shake. The nutritional information for each of these products is provided in Table 1.

Table 1   Information Is for One Serving

	SimplyProtein Nutrition Bar	Myoplex Deluxe Shake
Protein	33 g	42 g
Calories	300	300
Sugar	13 g	2 g
Sodium	45 mg	450 mg

Suppose that Dave is getting ready to run the Shamrock Shuffle 8k race in Chicago. He is planning on using the SimplyProtein Nutrition Bar and the Myoplex Deluxe Shake in his training. Dave wants to maximize his protein intake while keeping his calories under 1500, sugar under 32 g, and sodium under 1845 mg.

1. Declare the variables and write the objective function and the problem constraints.

2. Rewrite each nonnegative constraint in calculator-friendly form and graph in the same viewing window.

3. Determine the corner points of the feasible region.

4. Use the corner point method to solve the linear programming problem.

5. Using the optimal solution from question 4, determine how many calories, how much sugar, and how much sodium are being consumed.

6. Determine the difference between the amount of calories from question 5 and the limit on calories. Do the same for sugar and sodium. This result is called *slack*.

7. Referring to question 6, explain why some of the differences were approximately 0. (Note the graph and the answer you got for question 4.)

Now suppose that Dave's requirements change as follows: no more than 1800 calories, 56 g of sugar, and 1485 mg of sodium.

8. Repeat questions 1 through 7 for the new constraints.

4

Linear Programming: The Simplex Method

Maximize
$P = 40x_1 + 80x_2 + 70x_3$
Subject to:
$x_1 + x_2 + x_3 \leq 30$
$2x_1 + 3x_2 + 2x_3 \leq 53$
$x_1 + 2x_2 + 3x_3 \leq 47$
$x_1 \geq 0, x_2 \geq 0, x_3 \geq 0$

$$
\begin{array}{ccccccc|c}
x_1 & x_2 & x_3 & s_1 & s_2 & s_3 & P & \\
\hline
\frac{2}{5} & 0 & 0 & 1 & -\frac{1}{5} & -\frac{1}{5} & 0 & 10 \\
\frac{4}{5} & 1 & 0 & 0 & \frac{3}{5} & -\frac{2}{5} & 0 & 13 \\
-\frac{1}{5} & 0 & 1 & 0 & -\frac{2}{5} & \frac{3}{5} & 0 & 7 \\
\hline
10 & 0 & 0 & 0 & 20 & 10 & 1 & 1530
\end{array}
$$

Modern factories employ specialized machines to manufacture goods. Consider the case of a chair manufacturer who produces several different types of chairs. To maximize profit, the manufacturer must decide how many of each chair to produce each week. Several variables come into play for each chair including the amount of time spent on each machine, the total available time each machine can be run, and the profit per chair. When two or more decision variables must be considered the simplex method is a useful tool. The final tableau of the simplex method gives the solution to the linear programming problem.

What We Know

In Chapter 3 we solved linear programming problems graphically by sketching the feasible region, locating corner points, and then evaluating each corner point in the objective function.

Where Do We Go

In this chapter, we learn a method called the simplex method that allows us to solve linear programming problems without graphing the feasible region. This new method is also used when we have more than two decision variables.

Section 4.1   Introduction to the Simplex Method

In practice, applications of linear programming have more than two decision variables. Some applications may have dozens (or even hundreds) of variables. In 1947, George Dantzig developed an algebraic method to solve problems with many variables called the **simplex method**. In this section we introduce the vocabulary and skills necessary to convert a problem into a form so that the simplex method can be used. We also learn the beginning steps of the method. In Section 4.2 we learn how to complete the method and solve applications. The last two sections of this chapter address special situations involving the simplex method.

Standard Maximum Form

Since the simplex method is used for problems with many variables, it is now more convenient for us to use subscripts to name variables. This means that we will use x_1 (read "x sub one"), x_2, x_3, x_4, and so on, to name the unknowns. Let's quickly practice this new convention through a Flashback.

Flashback

A Maximization Problem Revisited

In Section 3.3, Example 3, we graphically solved the linear programming problem

$$\begin{aligned} \text{Maximize:} \quad & z = x - 3y \\ \text{Subject to:} \quad & 2x - 3y \le 6 \\ & -x + 4y \le 4 \\ & x \ge 0,\, y \ge 0 \end{aligned}$$

Rewrite this problem using x_1 and x_2 instead of x and y.

Flashback Solution

Using the new notation, this linear programming problem is written as

$$\begin{aligned} \text{Maximize:} \quad & z = x_1 - 3x_2 \\ \text{Subject to:} \quad & 2x_1 - 3x_2 \le 6 \\ & -x_1 + 4x_2 \le 4 \\ & x_1 \ge 0,\, x_2 \ge 0 \end{aligned}$$

Look carefully at the Flashback solution. In this section and the next, we will use the simplex method for problems written in this form. Problems written in the form shown in the Flashback solution are said to be in **standard maximum form**.

■ **Standard Maximum Form**

A linear programming problem is said to be in **standard maximum form** if the following are true:

1. The objective function is linear and is to be **maximized**.
2. All variables are nonnegative.
3. All constraints, except the nonnegative constraints, are written as a linear expression less than or equal to a nonnegative constant. Mathematically, this means

$$a_1 x_1 + a_2 x_2 + a_3 x_3 + \cdots + a_n x_n \leq b, \qquad \text{with } b \geq 0$$

▶ **Note:** In Sections 4.3 and 4.4 we address problems that do not meet these conditions.

Slack Variables and the Initial Simplex Tableau

The first step in using the simplex method to solve a standard maximum form problem, and the key in allowing us to view a linear programming problem algebraically is to turn each constraint (inequality) into an equation. If we can write equations, we can then use matrices to solve the problem. To turn an inequality into an equation requires us to introduce a nonnegative variable called a **slack variable**.

For example, to turn $2x_1 + x_2 \leq 12$ into an equation, we add the slack variable s_1 to the left side of the inequality to get

$$2x_1 + x_2 + s_1 = 12, \qquad \text{where } s_1 \geq 0$$

As x_1 and x_2 change, the variable s_1 "picks up the slack" to guarantee that an equality exists. If $2x_1 + x_2$ equals 9, then s_1 is 3. If $2x_1 + x_2$ equals 12, then s_1 is 0. In Example 1 we illustrate how to use slack variables to write constraints as equations.

Example 1 **Adapting a Linear Programming Problem**

Use slack variables to restate the linear programming problem from the Flashback into equational form.

$$\begin{aligned} \text{Maximize:} \qquad & z = x_1 - 3x_2 \\ \text{Subject to:} \qquad & 2x_1 - 3x_2 \leq 6 \\ & -x_1 + 4x_2 \leq 4 \\ & x_1 \geq 0, \ x_2 \geq 0 \end{aligned}$$

Solution

We begin by observing that this problem is in standard maximum form. We can rewrite the two constraints as equations by using the slack variables s_1 and s_2. Using these slack variables, we have the equational form as

$$\begin{aligned} \text{Maximize:} \qquad & z = x_1 - 3x_2 \\ \text{Subject to:} \qquad & 2x_1 - 3x_2 + s_1 \quad\;\; = 6 \\ & -x_1 + 4x_2 \quad\;\; + s_2 = 4 \end{aligned}$$

where $x_1 \geq 0, \ x_2 \geq 0, \ s_1 \geq 0, \text{ and } s_2 \geq 0$ ■

Before continuing, let's make some observations about slack variables.

■ Slack Variables

- Slack variables are not added to the nonnegative constraints $x_1 \geq 0$, $x_2 \geq 0$.
- A different slack variable is needed for each constraint (inequality).
- A slack variable "takes up the slack" to ensure that an equality exists.
- For any point in the feasible region of a linear programming problem, the value of each slack variable is either positive or zero.

To gain some insight into the last bullet item, consult Figure 4.1.1, which shows the feasible region of the linear programming problem considered in the Flashback and again in Example 1.

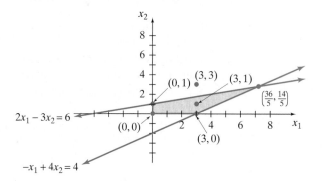

Figure 4.1.1

First observe that when $s_1 = 0$ and $s_2 = 0$, we get the graphs of the lines $2x_1 - 3x_2 = 6$ and $-x_1 + 4x_2 = 4$, respectively. We now pick a point in the feasible region, $(3, 1)$, and a point **not** in the feasible region, $(3, 3)$. These points are of the form (x_1, x_2). Let's use these points to determine s_1 and s_2 as shown in Table 4.1.1.

Table 4.1.1

Point	First Equation	Slack Value, s_1	Second Equation	Slack Value, s_2
	$2x_1 - 3x_2 + s_1 = 6$		$-x_1 + 4x_2 + s_2 = 4$	
$(3, 1)$	$2(3) - 3(1) + s_1 = 6$	$s_1 = 3$	$-3 + 4(1) + s_2 = 4$	$s_2 = 3$
$(3, 3)$	$2(3) - 3(3) + s_1 = 6$	$s_1 = 9$	$-3 + 4(3) + s_2 = 4$	$s_2 = -5$

As Table 4.1.1 indicates, for point $(3, 1)$ in the feasible region, the value of each slack variable is positive. For the point $(3, 3)$ which is **not** in the feasible region, there is at least one slack variable (s_2) that is **negative**. This illustrates that the value of each slack variable is either positive or zero for any point in the feasible region.

Back in Example 1, the addition of slack variables converted the constraints into equational form. Hence the slack variables actually converted the linear programming problem into a system of linear equations. From Chapters 1 and 2, we know that for a system of linear equations we should have all variables on one side of the equal sign and the constants on the other side. In Example 1 the only equation that does not satisfy this condition is the objective function, $z = x_1 - 3x_2$. However, we can write this objective function as

$$-x_1 + 3x_2 + z = 0$$

So the standard maximum form from the Flashback

Maximize: $z = x_1 - 3x_2$

Subject to: $2x_1 - 3x_2 \leq 6$

$-x_1 + 4x_2 \leq 4$

$x_1 \geq 0, x_2 \geq 0$

has the **equational form**

Maximize: $-x_1 + 3x_2 + z = 0$

Subject to: $2x_1 - 3x_2 + s_1 \quad\quad = 6$

$-x_1 + 4x_2 \quad\quad + s_2 = 4$

We can use an **augmented matrix** to write the equational form as follows:

$$\begin{array}{c} \overbrace{}^{\text{Decision variables}} \quad \overbrace{}^{\text{Slack variables}} \\ \begin{array}{ccccc} x_1 & x_2 & s_1 & s_2 & z \end{array} \\ \left[\begin{array}{ccccc|c} 2 & -3 & 1 & 0 & 0 & 6 \\ -1 & 4 & 0 & 1 & 0 & 4 \\ \hline -1 & 3 & 0 & 0 & 1 & 0 \end{array} \right] \\ \underbrace{}_{\text{Indicators}} \end{array}$$

The numbers in the bottom row are from the objective function. These numbers, except for the rightmost 1 and 0, are called **indicators**. This matrix is called the **initial simplex tableau**. Since forming the initial simplex tableau is the first step in the simplex method, let's do an example practicing this necessary skill.

Example 2   Forming an Initial Simplex Tableau

Form the initial simplex tableau for the following standard maximum form linear programming problem.

Maximize: $z = 2x_1 - x_2 + x_3$

Subject to: $2x_1 + x_2 + 3x_3 \leq 5$

$-x_1 + 3x_2 \quad\quad \leq 7$

$x_1 \geq 0, x_2 \geq 0, x_3 \geq 0$

Solution

Understand the Situation: Since we have two constraints (ignoring the non-negative constraints), we need to introduce two slack variables, s_1 and s_2, one for each constraint.

Rewriting the constraints in equational form gives

$$2x_1 + x_2 + 3x_3 + s_1 \quad\quad = 5$$

$$-x_1 + 3x_2 \quad\quad\quad + s_2 = 7$$

Next we write the objective function as

$$-2x_1 + x_2 - x_3 + z = 0$$

We now have the initial simplex tableau as

$$
\begin{array}{c}
\begin{matrix} x_1 & x_2 & x_3 & s_1 & s_2 & z \end{matrix} \\
\left[
\begin{array}{cccccc|c}
2 & 1 & 3 & 1 & 0 & 0 & 5 \\
-1 & 3 & 0 & 0 & 1 & 0 & 7 \\
\hline
-2 & 1 & -1 & 0 & 0 & 1 & 0
\end{array}
\right]
\end{array}
$$

$$\underbrace{}_{\text{Indicators}}$$

∎

✔ **Checkpoint 1**

Now work Exercise 15.

Basic Variables, Nonbasic Variables, and Basic Feasible Solutions

In Example 2 we have seen the system of linear inequalities rewritten as the system of linear equations

$$
\begin{aligned}
2x_1 + x_2 + 3x_3 + s_1 \phantom{{}+s_2} &= 5 \\
-x_1 + 3x_2 \phantom{{}+ 3x_3 + s_1} + s_2 &= 7
\end{aligned}
$$

This system of two equations in five variables has an infinite number of solutions. For example, if we solve each equation for s_1 and s_2, respectively, we get

$$
\begin{aligned}
s_1 &= 5 - 2x_1 - x_2 - 3x_3 \\
s_2 &= 7 + x_1 - 3x_2
\end{aligned}
$$

Notice that each choice of values for x_1, x_2, and x_3 yields values for s_1 and s_2 that give a solution to the system. However, a solution to the system is said to be **feasible** only if all variables are nonnegative.

When a linear system is written in the form

$$
\begin{aligned}
s_1 &= 5 - 2x_1 - x_2 - 3x_3 \\
s_2 &= 7 + x_1 - 3x_2
\end{aligned}
$$

s_1 and s_2 are called **basic variables** and x_1, x_2, and x_3 are called **nonbasic variables**. In general, we consider the nonbasic variables to be free and set them equal to zero. Doing this for the system under discussion gives us

$$
\begin{aligned}
s_1 &= 5 - 2(0) - (0) - 3(0) = 5 \\
s_2 &= 7 + (0) - 3(0) = 7
\end{aligned}
$$

The solution obtained when setting all nonbasic variables equal to zero is called a **basic solution**. In fact, the basic solution for this system is $x_1 = 0$, $x_2 = 0$, $x_3 = 0$, $s_1 = 5$, and $s_2 = 7$. Since all values are nonnegative, we can call it a **basic feasible solution**.

Since the identification of nonbasic variables and basic variables will be important, we offer the following procedure on how to determine basic and nonbasic variables from a tableau.

When a variable heads a column containing exactly one 1 and all other entries are 0, that variable is a **basic variable**. A variable that does not head such a column is a **nonbasic variable** and may be set equal to zero.

Example 3   **Identifying Basic and Nonbasic Variables and Reading a Solution**

Given the following tableau:

(a) Determine the basic and nonbasic variables.

(b) Determine the solution given by the tableau.

$$
\begin{array}{ccccccc}
x_1 & x_2 & x_3 & s_1 & s_2 & z & \\
\end{array}
$$
$$
\left[
\begin{array}{cccccc|c}
3 & 1 & -2 & 3 & 0 & 0 & 13 \\
1 & 0 & -1 & 1 & 1 & 0 & 22 \\
\hline
4 & 0 & 7 & 0 & 0 & 1 & 87
\end{array}
\right]
$$

Solution

(a) Since x_2, s_2, and z head columns containing exactly one 1 and all other entries are 0, we have the variables x_2, s_2, and z as our basic variables. This means that x_1, x_3, and s_1 are the nonbasic variables.

(b) Since x_1, x_3, and s_1 are the nonbasic variables, they may be set equal to 0. Recall that the first row of the tableau represents the equation

$$3x_1 + x_2 - 2x_3 + 3s_1 = 13$$

Since x_1, x_3, and s_1 are the nonbasic variables (and may be set equal to zero), this equation becomes

$$3(0) + x_2 - 2(0) + 3(0) = 13$$
$$x_2 = 13$$

Similarly, the second row of the tableau gives us $1 \cdot s_2 = 22$ or simply $s_2 = 22$. The bottom row gives us $1 \cdot z = 87$ or simply $z = 87$. So the solution represented by the tableau is

$$x_1 = 0, \quad x_2 = 13, \quad x_3 = 0, \quad s_1 = 0, \quad s_2 = 22, \quad z = 87$$

Observe how we can determine the solution directly from the tableau by reading down the column that is a basic variable until we hit the 1 and then turning right and reading the number in the last column.   ∎

✓ **Checkpoint 2**

Now work Exercise 33.

Pivot

Rarely is the solution read from the initial simplex tableau the optimal solution. Almost always, we need to locate other solutions. To determine these other solutions, we use the row operations for matrices (see the Toolbox below) and **pivot**. Example 4 reviews how to pivot. Keep in mind that pivoting produces another matrix (tableau) that gives another solution to the system of equations. Also, pivoting is the key to using the simplex method, as we will see in Section 4.2.

From Your Toolbox

1. Two rows may be interchanged (denoted $R_i \leftrightarrow R_j$).
2. A row is multiplied by a nonzero constant (denoted $cR_i \leftrightarrow R_i$).
3. Add a constant multiple of one row to another (denoted $cR_i + R_j \rightarrow R_j$).

▶ **Note:** It will not be necessary to use row operation 1 in this chapter.

Example 4 Pivoting

Pivot about the indicated 2.

$$
\begin{array}{ccccc}
x_1 & x_2 & s_1 & s_2 & z \\
\end{array}
$$

$$
\left[\begin{array}{ccccc|c}
② & -3 & 1 & 0 & 0 & 6 \\
\hline
-1 & 4 & 0 & 1 & 0 & 4 \\
\hline
-1 & 3 & 0 & 0 & 1 & 0
\end{array}\right]
$$

Solution

We get a 1 where the 2 is by using row operation 2.

$$
\begin{array}{ccccc}
x_1 & x_2 & s_1 & s_2 & z \\
\end{array}
$$

$$
\left[\begin{array}{ccccc|c}
2 & -3 & 1 & 0 & 0 & 6 \\
\hline
-1 & 4 & 0 & 1 & 0 & 4 \\
\hline
-1 & 3 & 0 & 0 & 1 & 0
\end{array}\right]
\quad \tfrac{1}{2}R_1 \to R_1 \quad
\left[\begin{array}{ccccc|c}
1 & -\tfrac{3}{2} & \tfrac{1}{2} & 0 & 0 & 3 \\
\hline
-1 & 4 & 0 & 1 & 0 & 4 \\
\hline
-1 & 3 & 0 & 0 & 1 & 0
\end{array}\right]
$$

We now get zeros in the first column with the pivot by using row operation 3.

Interactive Activity

Is the solution in Example 4 feasible? Explain.

$$
\begin{array}{ccccc}
x_1 & x_2 & s_1 & s_2 & z \\
\end{array}
$$

$$
\left[\begin{array}{ccccc|c}
1 & -\tfrac{3}{2} & \tfrac{1}{2} & 0 & 0 & 3 \\
\hline
-1 & 4 & 0 & 1 & 0 & 4 \\
\hline
-1 & 3 & 0 & 0 & 1 & 0
\end{array}\right]
\begin{array}{l}
R_1 + R_2 \to R_2 \\
R_1 + R_3 \to R_3
\end{array}
\left[\begin{array}{ccccc|c}
1 & -\tfrac{3}{2} & \tfrac{1}{2} & 0 & 0 & 3 \\
\hline
0 & \tfrac{5}{2} & \tfrac{1}{2} & 1 & 0 & 7 \\
\hline
0 & \tfrac{3}{2} & \tfrac{1}{2} & 0 & 1 & 3
\end{array}\right]
$$

From this last matrix we can read the solution as $x_1 = 3$, $x_2 = 0$, $s_1 = 0$, $s_2 = 7$, and $z = 3$. ∎

✓ **Checkpoint 3**

Now work Exercise 37.

The matrix given in Example 4,

$$
\begin{array}{ccccc}
x_1 & x_2 & s_1 & s_2 & z \\
\end{array}
$$

$$
\left[\begin{array}{ccccc|c}
2 & -3 & 1 & 0 & 0 & 6 \\
\hline
-1 & 4 & 0 & 1 & 0 & 4 \\
\hline
-1 & 3 & 0 & 0 & 1 & 0
\end{array}\right]
$$

was the initial simplex tableau for the equational form of the linear programming problem presented in the Flashback. We notice from this tableau that x_1 and x_2 are the nonbasic variables (and may be set equal to zero), and s_1, s_2, and z are the basic variables. Observe that this means that the tableau corresponds to the solution $x_1 = 0$, $x_2 = 0$, $s_1 = 6$, $s_2 = 4$, and $z = 0$. In particular, the solution corresponds to the corner point $(0, 0)$. See Figure 4.1.2.

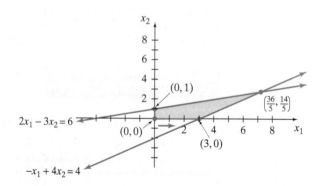

Figure 4.1.2

In Example 4 we actually performed a step in the simplex method (discussed in detail in Section 4.2). Observe that the solution read from the final tableau corresponds to the corner point $(3, 0)$ because $x_1 = 3$ and $x_2 = 0$. This illustrates that the simplex method seeks out an optimal solution by moving from corner point to corner point in the feasible region in a strategic fashion. In Section 4.2 we learn the strategy used to find the optimal solution.

Technology Option

In Chapter 2 we saw that row operations can be performed on a graphing calculator. Figure 4.1.3 shows the initial tableau from Example 4. Notice that the right side of the matrix is not visible, however, it can be seen by pressing the right arrow button. Figure 4.1.4 shows $\frac{1}{2}R_1 \rightarrow R_1$. Figures 4.1.5 and 4.1.6 show $R_1 + R_2 \rightarrow R_2$ and $R_1 + R_3 \rightarrow R_3$, respectively.

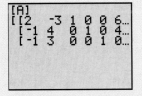

Figure 4.1.3

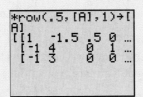

Figure 4.1.4

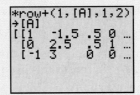

Figure 4.1.5

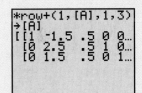

Figure 4.1.6

For more information on how to do matrix operations on your calculator, consult the online graphing calculator manual at www.prenhall.com/armstrong.

Applications

The main thrust of the applications in this section will be to use slack variables to adapt standard maximum form linear programming problems into equational form and then write the initial simplex tableau.

Example 5   Writing an Initial Simplex Tableau

Dub, Inc., is a small manufacturing firm that produces two microwave switches, switch A and switch B. Each switch passes through two departments: assembly and testing. Switch A requires 4 hours of assembly and 1 hour of testing and switch B requires 3 hours of assembly and 2 hours of testing. The maximum numbers of hours available each week for assembly and testing are 240 hours and 140 hours, respectively. The per unit profit for switch A is $20, whereas the per unit profit of switch B is $30. To determine how many units of each switch should be made and sold each week to maximize profit, do the following:

(a) Define the variables and write the mathematical model for this problem.

(b) Use slack variables to adapt the mathematical model to equational form.

(c) Write the initial simplex tableau.

Solution

(a) **Understand the Situation:** Since we do not know how many of each switch are made each week, we let x_1 represent the number of units of switch A made each week and x_2 represent the number of units of switch B made each week.

The objective function is

$$P = 20x_1 + 30x_2$$

where P is the profit in dollars. The problem constraints are

Assembly department constraint: $\quad 4x_1 + 3x_2 \leq 240$

Testing department constraint: $\quad x_1 + 2x_2 \leq 140$

Nonnegative constraints: $\quad x_1 \geq 0, x_2 \geq 0$

So the mathematical model is

$$\begin{aligned}
\text{Maximize:} \quad & P = 20x_1 + 30x_2 \\
\text{Subject to:} \quad & 4x_1 + 3x_2 \leq 240 \\
& x_1 + 2x_2 \leq 140 \\
& x_1 \geq 0, x_2 \geq 0
\end{aligned}$$

(b) Since we have two constraints, we need two slack variables, s_1 and s_2. Our constraints in equational form are

$$\begin{aligned}
4x_1 + 3x_2 + s_1 \quad\quad &= 240 \\
x_1 + 2x_2 \quad\quad + s_2 &= 140
\end{aligned}$$

Our objective function is

$$-20x_1 - 30x_2 + P = 0$$

(c) The initial simplex tableau is

$$\begin{array}{ccccc}
x_1 & x_2 & s_1 & s_2 & P \\
\end{array}$$
$$\left[\begin{array}{ccccc|c}
4 & 3 & 1 & 0 & 0 & 240 \\
1 & 2 & 0 & 1 & 0 & 140 \\
\hline
-20 & -30 & 0 & 0 & 1 & 0
\end{array}\right]$$

In the next section, we will learn a method that will transform this tableau into one that provides the optimal solution to the original problem. ∎

✓ **Checkpoint 4**   Now work Exercise 51.

SUMMARY

In this section we presented many necessary preliminaries pertaining to the simplex method. These included the **standard maximum form** of a linear programming problem, **slack variables**, how to adapt a standard maximum form linear programming problem into **equational form**, the **initial simplex tableau**, **basic variables**, **nonbasic variables**, **basic** **feasible solution**, and a review of matrix row operations and the **pivot**. The ability to read a simplex tableau was discussed and is very important. It gives a solution (optimal or otherwise) and indicates which variables are basic and which nonbasic.

SECTION 4.1 EXERCISES

In Exercises 1–8, convert the system of linear inequalities into a system of equations by using slack variables.

1. $x_1 + x_2 \leq 8$
$2x_1 - x_2 \leq 5$

2. $3x_1 + 2x_2 \leq 12$
$5x_1 - 3x_2 \leq 35$

3. $2x_1 + x_2 - x_3 \leq 5$
$-3x_1 + 2x_2 + x_3 \leq 7$
$4x_1 - x_2 + 2x_3 \leq 9$

4. $-x_1 + 2x_2 + x_3 \leq 1$
$7x_1 + 4x_2 - 2x_3 \leq 8$
$2x_1 - 3x_2 - 4x_3 \leq 7$

5. $2x_1 - x_2 \leq 4$
$-3x_1 + 2x_2 \leq 7$
$x_1 - 4x_2 \leq 8$

6. $-3x_1 + 2x_2 \leq 16$
$4x_1 - 7x_2 \leq 14$
$x_1 - x_2 \leq 7$

7. $2x_1 + 3x_2 - x_3 \leq 8$
$-4x_1 + x_2 + x_3 \leq 5$

8. $-2x_1 + x_2 - 3x_3 \leq 12$
$3x_1 - 2x_2 - 4x_3 \leq 31$

In Exercises 9–26, for the given standard maximum form linear programming problem, introduce slack variables and set up the initial simplex tableau.

9. Maximize: $z = 2x_1 + x_2$
Subject to: $x_1 + x_2 \leq 6$
$x_1 - x_2 \leq 2$
$x_1 \geq 0, x_2 \geq 0$

10. Maximize: $z = 3x_1 + x_2$
Subject to: $4x_1 + 5x_2 \leq 9$
$2x_1 - x_2 \leq 1$
$x_1 \geq 0, x_2 \geq 0$

11. Maximize: $z = x_1 + 2x_2$
Subject to: $x_1 + x_2 \leq 7$
$2x_1 + 3x_2 \leq 16$
$x_1 \geq 0, x_2 \geq 0$

12. Maximize: $z = 2x_1 + 3x_2$
Subject to: $3x_1 + 2x_2 \leq 8$
$x_1 - x_2 \leq 7$
$x_1 \geq 0, x_2 \geq 0$

13. Maximize: $z = 2x_1 + 2x_2$
Subject to: $x_1 + x_2 \leq 3$
$x_1 - 2x_2 \leq 0$
$x_1 \geq 0, x_2 \geq 0$

14. Maximize: $z = 5x_1 + 7x_2$
Subject to: $3x_1 - x_2 \leq 8$
$2x_1 + 4x_2 \leq 5$
$x_1 \geq 0, x_2 \geq 0$

✓ **15.** Maximize: $z = 2x_1 + 2x_2$
Subject to: $3x_1 - 2x_2 \leq 0$
$x_1 + x_2 \leq 5$
$2x_1 + x_2 \leq 6$
$x_1 \geq 0, x_2 \geq 0$

16. Maximize: $z = x_1 + 4x_2$
Subject to: $2x_1 + x_2 \leq 32$
$x_1 + x_2 \leq 18$
$x_1 + 3x_2 \leq 36$
$x_1 \geq 0, x_2 \geq 0$

17. Maximize: $z = 5x_1 + x_2$
Subject to: $-x_1 + 2x_2 \leq 10$
$2x_1 + x_2 \leq 15$
$5x_1 + x_2 \leq 8$
$-x_1 + x_2 \leq 1$
$x_1 \geq 0, x_2 \geq 0$

18. Maximize: $z = 3x_1 + 5x_2$
Subject to: $-x_1 + 2x_2 \leq 10$
$4x_1 + 2x_2 \leq 15$
$x_1 + 3x_2 \leq 8$
$6x_1 - x_2 \leq 4$
$x_1 \geq 0, x_2 \geq 0$

19. Maximize: $z = 2x_1 + x_2 + 3x_3$
Subject to: $x_1 - x_2 + 2x_3 \leq 4$
$2x_1 + x_2 - 3x_3 \leq 7$
$x_1 \geq 0, x_2 \geq 0, x_3 \geq 0$

20. Maximize: $z = 3x_1 + 2x_2 + 4x_3$
Subject to: $2x_1 + x_2 + x_3 \leq 10$
$5x_1 - x_2 + 2x_3 \leq 8$
$x_1 \geq 0, x_2 \geq 0, x_3 \geq 0$

21. Maximize: $z = 3x_1 + x_2 + 5x_3$
Subject to: $2x_1 + x_2 + 4x_3 \leq 8$
$4x_1 - x_2 + 3x_3 \leq 15$
$12x_1 + 2x_2 + 7x_3 \leq 12$
$x_1 \geq 0, x_2 \geq 0, x_3 \geq 0$

22. Maximize: $z = 2x_1 + 5x_2 + x_3$
Subject to: $5x_1 + 2x_2 + x_3 \leq 20$
$x_1 + 3x_2 + 5x_3 \leq 13$
$4x_1 + x_2 + 3x_3 \leq 15$
$x_1 \geq 0, x_2 \geq 0, x_3 \geq 0$

23. Maximize: $z = 4x_1 + 2x_2 + 3x_3$

Subject to: $2x_1 \qquad + 3x_3 \le 8$

$3x_1 + 2x_2 \qquad \le 5$

$x_1 \ge 0, x_2 \ge 0, x_3 \ge 0$

24. Maximize: $z = 9x_1 + x_2 + x_3$

Subject to: $4x_1 \qquad + x_3 \le 7$

$2x_1 + 3x_2 \qquad \le 8$

$x_1 \ge 0, x_2 \ge 0, x_3 \ge 0$

25. Maximize: $z = 5x_1 + x_2 + 3x_3$

Subject to: $15x_1 + 3x_2 \qquad \le 9$

$6x_2 + x_3 \le 7$

$x_1 \ge 0, x_2 \ge 0, x_3 \ge 0$

26. Maximize: $z = 4x_1 + 3x_2 + x_3$

Subject to: $6x_1 + x_2 \qquad \le 9$

$3x_2 + 4x_3 \le 24$

$x_1 \ge 0, x_2 \ge 0, x_3 \ge 0$

In Exercises 27–36, for each given matrix:

(a) Identify the basic and nonbasic variables.

(b) Determine the solution given by the matrix.

27.
$$\begin{array}{ccccc} x_1 & x_2 & s_1 & s_2 & z \\ \end{array}$$
$$\left[\begin{array}{ccccc|c} 1 & 1 & 0 & 3 & 0 & 5 \\ 0 & 2 & 1 & 1 & 0 & 2 \\ \hline 0 & -5 & 0 & 4 & 1 & 10 \end{array}\right]$$

28.
$$\begin{array}{ccccc} x_1 & x_2 & s_1 & s_2 & z \\ \end{array}$$
$$\left[\begin{array}{ccccc|c} 0 & 2 & 1 & 6 & 0 & 5 \\ 1 & 5 & 0 & 11 & 0 & 11 \\ \hline 0 & -3 & 0 & 4 & 1 & 15 \end{array}\right]$$

29.
$$\begin{array}{ccccc} x_1 & x_2 & s_1 & s_2 & z \\ \end{array}$$
$$\left[\begin{array}{ccccc|c} 2 & -3 & 1 & 0 & 0 & 3 \\ 4 & 6 & 0 & 1 & 0 & 7 \\ \hline -1 & 5 & 0 & 0 & 1 & 9 \end{array}\right]$$

30.
$$\begin{array}{ccccc} x_1 & x_2 & s_1 & s_2 & z \\ \end{array}$$
$$\left[\begin{array}{ccccc|c} 1 & 3 & 0 & 6 & 0 & 4 \\ 0 & -1 & 1 & 2 & 0 & 7 \\ \hline 0 & -2 & 0 & 7 & 1 & 13 \end{array}\right]$$

31.
$$\begin{array}{cccccc} x_1 & x_2 & s_1 & s_2 & s_3 & z \\ \end{array}$$
$$\left[\begin{array}{cccccc|c} 2 & 5 & 1 & 0 & 0 & 0 & 3 \\ 3 & 6 & 0 & 1 & 0 & 0 & 8 \\ 4 & 2 & 0 & 0 & 1 & 0 & 15 \\ \hline -1 & -2 & 0 & 0 & 0 & 1 & 0 \end{array}\right]$$

32.
$$\begin{array}{cccccc} x_1 & x_2 & s_1 & s_2 & s_3 & z \\ \end{array}$$
$$\left[\begin{array}{cccccc|c} -1 & 2 & 0 & 1 & 0 & 0 & 3 \\ 3 & 8 & 1 & 0 & 0 & 0 & 6 \\ 7 & 5 & 0 & 0 & 1 & 0 & 8 \\ \hline -2 & -7 & 0 & 0 & 0 & 1 & 0 \end{array}\right]$$

✓ **33.**
$$\begin{array}{cccccc} x_1 & x_2 & x_3 & s_1 & s_2 & z \\ \end{array}$$
$$\left[\begin{array}{cccccc|c} 1 & 0 & -3 & 5 & -1 & 0 & 7 \\ 0 & 1 & 2 & 8 & 2 & 0 & 12 \\ \hline 0 & 0 & 4 & 3 & 9 & 1 & 27 \end{array}\right]$$

34.
$$\begin{array}{cccccc} x_1 & x_2 & x_3 & s_1 & s_2 & z \\ \end{array}$$
$$\left[\begin{array}{cccccc|c} 0 & 1 & -2 & 1 & 8 & 0 & 9 \\ 1 & 0 & 3 & 8 & 2 & 0 & 17 \\ \hline 0 & 0 & 1 & 9 & 3 & 1 & 38 \end{array}\right]$$

35.
$$\begin{array}{cccccccc} x_1 & x_2 & x_3 & x_4 & s_1 & s_2 & s_3 & z \\ \end{array}$$
$$\left[\begin{array}{cccccccc|c} 2 & 1 & 0 & 2 & 2 & 0 & 5 & 0 & 15 \\ 4 & 0 & 1 & 10 & 1 & 0 & 1 & 0 & 20 \\ 2 & 0 & 0 & 0 & 2 & 1 & 0 & 0 & 10 \\ \hline -4 & 0 & 0 & -2 & 3 & 0 & 3 & 1 & 100 \end{array}\right]$$

36.
$$\begin{array}{cccccccc} x_1 & x_2 & x_3 & x_4 & s_1 & s_2 & s_3 & z \\ \end{array}$$
$$\left[\begin{array}{cccccccc|c} 1 & 5 & 0 & 0 & 2 & 5 & 3 & 0 & 9 \\ 0 & 3 & 0 & 1 & -1 & 4 & 1 & 0 & 13 \\ 0 & 2 & 1 & 0 & 3 & 7 & 9 & 0 & 27 \\ \hline 0 & -2 & 0 & 0 & 1 & 0 & 4 & 1 & 80 \end{array}\right]$$

In Exercises 37–46:

(a) Pivot about the circled entry.

(b) Read the solution after pivoting.

✓ **37.**
$$\begin{array}{ccccc} x_1 & x_2 & s_1 & s_2 & z \\ \end{array}$$
$$\left[\begin{array}{ccccc|c} 1 & 1 & 0 & 3 & 0 & 5 \\ 0 & ② & 1 & 1 & 0 & 2 \\ \hline 0 & -5 & 0 & 4 & 1 & 10 \end{array}\right]$$

38.
$$\begin{array}{ccccc} x_1 & x_2 & s_1 & s_2 & z \\ \end{array}$$
$$\left[\begin{array}{ccccc|c} 0 & 2 & 1 & 6 & 0 & 5 \\ 1 & ⑤ & 0 & 11 & 0 & 11 \\ \hline 0 & -3 & 0 & 4 & 1 & 15 \end{array}\right]$$

39.
$$\begin{array}{ccccc} x_1 & x_2 & s_1 & s_2 & z \\ \end{array}$$
$$\left[\begin{array}{ccccc|c} ② & -3 & 1 & 0 & 0 & 3 \\ 4 & 6 & 0 & 1 & 0 & 7 \\ \hline -1 & 5 & 0 & 0 & 1 & 9 \end{array}\right]$$

40.
$$\begin{array}{ccccc} x_1 & x_2 & s_1 & s_2 & z \\ \end{array}$$
$$\left[\begin{array}{ccccc|c} 1 & ③ & 0 & 6 & 0 & 4 \\ 0 & -1 & 1 & 2 & 0 & 7 \\ \hline 0 & -2 & 0 & 7 & 1 & 13 \end{array}\right]$$

41.
$$\begin{array}{cccccc} x_1 & x_2 & s_1 & s_2 & s_3 & z \\ \end{array}$$
$$\left[\begin{array}{cccccc|c} -1 & 2 & 0 & 1 & 0 & 0 & 3 \\ 3 & ⑧ & 1 & 0 & 0 & 0 & 6 \\ 7 & 5 & 0 & 0 & 1 & 0 & 8 \\ \hline -2 & -7 & 0 & 0 & 0 & 1 & 0 \end{array}\right]$$

42.
$$\begin{array}{cccccc} x_1 & x_2 & s_1 & s_2 & s_3 & z \\ \end{array}$$
$$\left[\begin{array}{cccccc|c} 2 & ⑤ & 1 & 0 & 0 & 0 & 3 \\ 3 & 6 & 0 & 1 & 0 & 0 & 8 \\ 4 & 2 & 0 & 0 & 1 & 0 & 15 \\ \hline -1 & -2 & 0 & 0 & 0 & 1 & 0 \end{array}\right]$$

43.
$$\begin{array}{ccccccc} x_1 & x_2 & x_3 & s_1 & s_2 & z \\ \end{array}$$

$$\left[\begin{array}{cccccc|c} 10 & 4 & 2 & 1 & 0 & 0 & 90 \\ 6 & 3 & ④ & 0 & 1 & 0 & 40 \\ \hline -2 & -3 & -8 & 0 & 0 & 1 & 0 \end{array} \right]$$

44.
$$\begin{array}{ccccccc} x_1 & x_2 & x_3 & s_1 & s_2 & z \\ \end{array}$$

$$\left[\begin{array}{cccccc|c} 3 & 6 & ① & 1 & 0 & 0 & 13 \\ 5 & 2 & 2 & 0 & 1 & 0 & 15 \\ \hline -8 & -4 & -10 & 0 & 0 & 1 & 0 \end{array} \right]$$

45.
$$\begin{array}{cccccccc} x_1 & x_2 & x_3 & x_4 & s_1 & s_2 & s_3 & z \\ \end{array}$$

$$\left[\begin{array}{cccccccc|c} 1 & 5 & 0 & 0 & 2 & 5 & 3 & 0 & 9 \\ 0 & 3 & 0 & 1 & -1 & 4 & 1 & 0 & 13 \\ 0 & ② & 1 & 0 & 3 & 7 & 9 & 0 & 27 \\ \hline 0 & -2 & 0 & 0 & 1 & 0 & 4 & 1 & 80 \end{array} \right]$$

46.
$$\begin{array}{cccccccc} x_1 & x_2 & x_3 & x_4 & s_1 & s_2 & s_3 & z \\ \end{array}$$

$$\left[\begin{array}{cccccccc|c} 6 & 0 & 1 & 2 & 2 & 0 & 5 & 0 & 5 \\ ② & 1 & 0 & 10 & 1 & 0 & 2 & 0 & 10 \\ 3 & 0 & 0 & 0 & 2 & 1 & 0 & 0 & 8 \\ \hline -10 & 0 & 0 & -2 & 3 & 0 & 3 & 1 & 20 \end{array} \right]$$

In Exercises 47–50, use a graphing calculator to:

(a) Pivot about the circled entry.

(b) Read the solution after pivoting.

47. Matrix in Exercise 38.

48. Matrix in Exercise 42.

49. Matrix in Exercise 44.

50. Matrix in Exercise 46.

Applications

In Exercises 51–60, perform the setup so that the solution can be determined by the simplex method as illustrated in Example 5. In the Section 4.2 exercises, you will determine optimal solutions for these exercises.

51. Due to new regulations approved by the Environmental Protection Agency (EPA), the Henderson Chemical Company was required to install a new process to reduce the pollution created when making a particular chemical. The older process releases 4 grams of sulfur and 3 grams of particulates for each gallon of chemical produced. The newer process releases 2 grams of sulfur and 1 gram of particulates for each gallon of chemical produced. The company realizes a profit of $0.40 for each gallon produced under the old process and $0.16 for each gallon produced under the new process. The new EPA regulations do not allow for more than 10,000 grams of sulfur and 7000 grams of particulates to be released daily. To determine how many gallons of this chemical should be produced using the old process to maximize profit, do the following:

(a) Define the variables and write the mathematical model for this problem.

(b) Use slack variables to adapt the mathematical model to equational form.

(c) Write the initial simplex tableau.

52. (*continuation of Exercise 51*) Barb, the plant manager for Henderson Chemical Company, has learned that the EPA is going to relax the new regulations so that no more than 13,000 grams of sulfur can be released daily. Assuming that all other data remain the same as in Exercise 51, to determine how many gallons of this chemical should be produced using the old process to maximize profit, do the following:

(a) Define the variables and write the mathematical model for this problem.

(b) Use slack variables to adapt the mathematical model to equational form.

(c) Write the initial simplex tableau.

53. Randy and Rusty have decided to enter the field of personal digital assistants (PDAs). Initially, they plan to design two types of PDAs, the WOW 1000 and the WOW 1500. Because of interest in PDAs, they can sell all that they can produce. However, they need to determine a production rate to satisfy various limits with a small production crew. These include a maximum of 150 hours per week for assembly and 100 hours per week for testing. The WOW 1000 requires 3 hours of assembly and 3 hours of testing, and the WOW 1500 requires 6 hours of assembly and 3.5 hours of testing. Profit for the WOW 1000 is estimated at $100 per unit, and profit for the WOW 1500 is estimated at $150 per unit. To determine how many of each type of PDA should be made and sold to maximize weekly profit, do the following:

(a) Define the variables and write the mathematical model for this problem.

(b) Use slack variables to adapt the mathematical model to equational form.

(c) Write the initial simplex tableau.

54. (*continuation of Exercise 53*) Randy and Rusty hire some new employees and can, as a result, increase production. By doing so, they now have a maximum of 210 hours per week for assembly and 160 hours per week for testing. Assuming that all other data remain the same as in Exercise 53, to determine how many of each type of PDA should be made and sold to maximize weekly profit, do the following:

(a) Define the variables and write the mathematical model for this problem.

(b) Use slack variables to adapt the mathematical model to equational form.

(c) Write the initial simplex tableau.

55. Table 4.1.2 gives data for growing artichokes and asparagus for an entire growing season in California (entries are in lb/acre).

Table 4.1.2

	Artichokes	Asparagus
Nitrogen	197	240
Potash	123	42

SOURCE: U.S. Department of Agriculture, www.usda.gov/nass

Franco has 150 acres available suitable for cultivating crops. He has purchased a contract for a total of 31,880 pounds of nitrogen and 9120 pounds of potash. He expects to make a profit of $140 per acre on artichokes and $160 per acre on asparagus. To determine how many acres of each he should plant to maximize his profit, do the following:

(a) Define the variables and write the mathematical model for this problem.

(b) Use slack variables to adapt the mathematical model to equational form.

(c) Write the initial simplex tableau.

⇐ 3.1 🌐 **56.** Table 4.1.3 gives data for growing cauliflower and head lettuce for an entire growing season in California (entries are in lb/acre).

Table 4.1.3

	Cauliflower	Head Lettuce
Nitrogen	215	173
Phosphate	98	122

SOURCE: U.S. Department of Agriculture, www.usda.gov/nass

Sonja has 210 acres available for cultivating crops. She has purchased a contract for a total of 38,380 pounds of nitrogen and 22,240 pounds of phosphate. She expects to make a profit of $90 per acre on cauliflower and $100 per acre on head lettuce. To determine how many acres of each she should plant to maximize her profit, do the following:

(a) Define the variables and write the mathematical model for this problem.

(b) Use slack variables to adapt the mathematical model to equational form.

(c) Write the initial simplex tableau.

⇐ 3.1 **57.** The Smolko Tent Company makes three types of tents: two-person, four-person, and six-person models. Each tent requires the services, in hours, of three departments, as shown in Table 4.1.4. The fabric cutting, assembly, and packaging departments have available a maximum of 420, 310, and 180 hours each week, respectively.

Table 4.1.4

	Two-person	Four-person	Six-person
Fabric cutting	0.5	0.8	1.1
Assembly	0.6	0.7	0.9
Packaging	0.2	0.3	0.6

The per unit profit is $30, $50, and $80 for the two-, four-, and six-person tents, respectively. To determine how many of each type of tent should be made and sold to maximize weekly profit, do the following:

(a) Define the variables and write the mathematical model for this problem.

(b) Use slack variables to adapt the mathematical model to equational form.

(c) Write the initial simplex tableau.

⇐ 3.1 **58.** Kathy's Secretarial Services has a 1-month intensive training program for receptionist/secretaries for three types of office positions: medical field, small business, and executive secretarial. Each individual receives training in three departments at Kathy's: word processing, telephone and LAN (local area network), and ethics. Table 4.1.5 shows the number of units of training that each type of receptionist/secretary receives.

Table 4.1.5

	Medical Field	Small Business	Executive Secretary
Word processing	0.8	1.1	1.7
Telephone and LAN	0.6	1.3	1.5
Ethics	1.2	0.8	1.3

Each month, word processing, telephone and LAN, and ethics have available a maximum of 32.1 units, 29.7 units and 33 units, respectively. The profit that Kathy's realizes from each individual admitted to medical field, small business, and executive secretary is $1000, $1000, and $1400, respectively. To determine how many individuals should be admitted each month to each program to maximize monthly profit, do the following:

(a) Define the variables and write the mathematical model for this problem.

(b) Use slack variables to adapt the mathematical model to equational form.

(c) Write the initial simplex tableau.

⇐ 3.1 🌐 **59.** The data in Table 4.1.6 give the amount of nitrogen, phosphate, and potash (all in lb/acre) needed to grow asparagus, cauliflower, and celery for an entire growing season in California.

Table 4.1.6

	Asparagus	Cauliflower	Celery
Nitrogen	240	215	336
Phosphate	148	98	171
Potash	42	69	147

SOURCE: U.S. Department of Agriculture, www.usda.gov/nass

Eldrick has 280 acres and can get no more than 65,970 pounds of nitrogen, 31,860 pounds of phosphate, and 21,630 pounds of potash. He expects a profit of $160 per acre for asparagus, $90 per acre for cauliflower, and $80 per acre for celery. To determine how many acres he should allot for each crop to maximize profit, do the following:

(a) Define the variables and write the mathematical model for this problem.

(b) Use slack variables to adapt the mathematical model to equational form.

(c) Write the initial simplex tableau.

🌐**60.** The data in Table 4.1.7 give the amount of nitrogen, phosphate, and potash (all in lb/acre) needed to grow head lettuce, cantaloupe, and bell peppers for an entire growing season in California. Maria has 200 acres and can get no more than 38,430 pounds of nitrogen, 27,845 pounds of phosphate, and 12,985 pounds of potash. She expects a profit of $100 per acre for head lettuce, $90 per acre for

Table 4.1.7

	Head Lettuce	Cantaloupe	Bell Peppers
Nitrogen	173	175	212
Phosphate	122	161	108
Potash	76	39	72

SOURCE: U.S. Department of Agriculture, www.usda.gov/nass

cantaloupe, and $85 per acre for bell peppers. To determine how many acres she should allot for each crop to maximize profit, do the following:

(a) Define the variables and write the mathematical model for this problem.

(b) Use slack variables to adapt the mathematical model to equational form.

(c) Write the initial simplex tableau.

▌ SECTION PROJECT

The biggest change to take place in the field of linear programming solution techniques in the time since George Dantzig developed the simplex method in 1947 was the arrival in 1984 of Karmarkar's algorithm, named after its inventor Narendra Karmarkar. Karmarkar's method often takes significantly less computer time to solve very large-scale linear programming problems.

As we saw in Figure 4.1.2, the simplex method moves from corner point to corner point around the feasible region seeking out an optimal solution. In contrast, Karmarkar's method follows a path of points on the inside of the feasible region.

Although it is likely that the simplex method will continue to be used for many linear programming problems, a new generation of linear programming software built

around Karmarkar's algorithm is already becoming popular. For example, Delta Air Lines became the first commercial airline to use the Karmarkar program (called KORBX), which was developed and sold by AT&T.

For this section project, research and write a short paper on the Karmarkar method. You should include the following in your paper.

(a) The history of the simplex method and George Dantzig and the history of the Karmarkar method and Narendra Karmarkar.

(b) The advantages and/or disadvantages of each method.

(c) Current available computer programs for the simplex method and the Karmarkar method.

(d) An outline of the basic features of each method.

Section 4.2   Simplex Method: Standard Maximum Form Problems

In Section 4.1 we learned how to prepare a standard maximum form linear programming problem for the simplex method. This preparation included the following:

- Introduction of slack variables, one for each constraint
- Using slack variables to convert linear inequality constraints into equational form
- Using the coefficients of the system of linear equations corresponding to the constraints and objective function to write an augmented matrix called the initial simplex tableau

We also reviewed how to pivot about an element in the initial simplex tableau. We saw that pivoting took us from one corner point to another corner point in the feasible region.

At this time we are ready to extend the pivoting process to obtain an optimal solution for the objective function. To see how this is done, let's Flashback to Example 5 from Section 4.1.

Dub, Inc., Revisited

Dub, Inc., is a small manufacturing firm that produces two microwave switches, switch A and switch B. Each switch passes through two departments: assembly and testing. Switch A requires 4 hours of assembly and 1 hour of testing, and switch B requires 3 hours of assembly and 2 hours of testing. The maximum numbers of hours available each week for assembly and testing are 240 hours and 140 hours, respectively. The per unit profit for switch A is $20, whereas the per unit profit of switch B is $30. To determine how many units of each switch should be made and sold each week to maximize profit P, we let x_1 represent the number of units of switch A made each week and x_2 represent the number of units of switch B made each week. The mathematical model is

$$\begin{aligned} \text{Maximize:} \quad & P = 20x_1 + 30x_2 \\ \text{Subject to:} \quad & 4x_1 + 3x_2 \le 240 \\ & x_1 + 2x_2 \le 140 \\ & x_1 \ge 0, x_2 \ge 0 \end{aligned}$$

Since we have two constraints, we need two slack variables, s_1 and s_2. Our constraints in equational form are

$$\begin{aligned} 4x_1 + 3x_2 + s_1 \quad &= 240 \\ x_1 + 2x_2 \quad + s_2 &= 140 \end{aligned}$$

Our objective function is

$$-20x_1 - 30x_2 + P = 0$$

The initial simplex tableau is

$$\begin{array}{ccccc|c} x_1 & x_2 & s_1 & s_2 & P & \\ 4 & 3 & 1 & 0 & 0 & 240 \\ 1 & 2 & 0 & 1 & 0 & 140 \\ \hline -20 & -30 & 0 & 0 & 1 & 0 \end{array}$$

Write the solution corresponding to the initial simplex tableau and interpret the value of each variable.

Flashback Solution

Since s_1, s_2, and P head columns containing exactly one 1 and all other entries are 0, these variables are the **basic variables**. They have the values $s_1 = 240$, $s_2 = 140$, and $P = 0$. Since x_1 and x_2 are the **nonbasic variables**, we set them equal to 0. So the corresponding solution to the initial simplex tableau is

$$x_1 = 0, \quad x_2 = 0, \quad s_1 = 240, \quad s_2 = 140, \quad P = 0$$

The interpretation of each variable is

$x_1 = 0$ (make 0 units of switch A)

$x_2 = 0$ (make 0 units of switch B)

$s_1 = 240$ (there are 240 unused hours in the assembly department)

$s_2 = 140$ (there are 140 unused hours in the testing department)

$P = 0$ (there is a profit of 0 dollars)

Intuitively, this makes sense since, if we make and sell 0 units of switch A and switch B, our profit is $0, and we should still have available the maximum number of hours in each department.

Obviously, a value of zero for Dub, Inc.'s, profit in the Flashback solution is not optimal. Our goal is to find the optimum solution, and we do this by pivoting.

Pivoting and the Simplex Method

After writing the initial simplex tableau, the simplex method requires us to determine on which element in the matrix we perform the pivot operation. To do this we need to determine the **pivot column** and the **pivot row**. To gain some understanding of how we determine the pivot column, consider the objective function in the Flashback, $P = 20x_1 + 30x_2$. Since both coefficients are positive, the value of P can be increased from 0 by increasing x_1 and/or x_2. Since the coefficient of x_2 is larger than the coefficient of x_1, increasing x_2 while holding x_1 constant results in a greater increase in the value of P than increasing x_1 and keeping x_2 constant. Hence we should increase the value of x_2 while keeping x_1 constant. This analysis means that our **pivot column** in the initial simplex tableau is the column headed by x_2. Notice that this column has the most negative of the indicators.

$$\begin{array}{ccccc}
x_1 & x_2 & s_1 & s_2 & P \\
\end{array}$$
$$\left[\begin{array}{ccccc|c}
4 & 3 & 1 & 0 & 0 & 240 \\
1 & 2 & 0 & 1 & 0 & 140 \\
\hline
-20 & -30 & 0 & 0 & 1 & 0
\end{array}\right]$$

Indicators

Recall from Section 4.1 that the indicators are the numbers in the bottom row except for the rightmost 0 and 1. Also recall that the indicators come from rewriting the objective function $P = 20x_1 + 30x_2$ as $-20x_1 - 30x_2 + P = 0$. This suggests the following for determining the pivot column.

▪ Pivot Column

The column with the most negative indicator is the **pivot column**. If there are two such indicators, use the leftmost indicator.

Now that we know how to determine the pivot column, we must determine which row is the pivot row. This will determine precisely on which entry to pivot. Continuing with the problem from the Flashback, we have found the initial

simplex tableau to be

$$
\begin{array}{c}
\begin{array}{ccccc} x_1 & x_2 & s_1 & s_2 & P \end{array} \\
\left[\begin{array}{ccccc|c}
4 & 3 & 1 & 0 & 0 & 240 \\
1 & 2 & 0 & 1 & 0 & 140 \\
\hline
-20 & -30 & 0 & 0 & 1 & 0
\end{array} \right] \\
\uparrow
\end{array}
$$

Pivot column

The initial simplex tableau corresponds to the system

$$
\begin{aligned}
4x_1 + 3x_2 + s_1 &= 240 \\
x_1 + 2x_2 + s_2 &= 140
\end{aligned}
$$

or

$$
\begin{aligned}
s_1 &= 240 - 4x_1 - 3x_2 \\
s_2 &= 140 - x_1 - 2x_2
\end{aligned}
$$

From Your Toolbox

A solution is a basic solution if all values for the variables are nonnegative.

Since pivoting in the column headed x_2 will result in one 1 and all other entries are 0 in this column, the pivot will turn x_2 into a basic variable while x_1 remains nonbasic. Now if x_1 remains nonbasic, we can set $x_1 = 0$. Substituting this into the last two equations gives us

$$
\begin{aligned}
s_1 &= 240 - 3x_2 \\
s_2 &= 140 - 2x_2
\end{aligned}
$$

We know that s_1 and s_2 must be nonnegative in order to be consistent with a basic feasible solution (consult the Toolbox to the left). Since $s_1 \geq 0$ and $s_1 = 240 - 3x_2$, then $240 - 3x_2 \geq 0$. Solving for x_2 gives us

$$
\begin{aligned}
240 - 3x_2 &\geq 0 \\
-3x_2 &\geq -240 \\
x_2 &\leq 80
\end{aligned}
$$

This means that x_2 cannot exceed $\frac{240}{3}$ or 80. Similarly, the equation $s_2 = 140 - 2x_2$ implies that x_2 cannot exceed $\frac{140}{2}$ or 70; in other words, $x_2 \leq 70$. Together, this means that x_2 can be increased by at most 70; which would correspond to the corner point $(0, 70)$. More importantly for our purposes at this time, the quotient $\frac{140}{2} = 70$ tells us the **pivot row**. Observe that the quotients $\frac{240}{3}$ and $\frac{140}{2}$ are simply the elements in the last column (column of constants) divided by the corresponding entry in the pivot column. These observations suggest the following on determining the pivot row.

■ **Pivot Row**

To determine the **pivot row**, divide each number from the rightmost column (column of constants) by the corresponding number from the pivot column. The smallest nonnegative quotient gives the pivot row. If two quotients are equal (and smallest), arbitrarily choose the row nearest the top of the matrix.

▶ **Note:** We do not form a quotient for the bottom row, our indicators. The entry where the pivot column and pivot row intersect is the **pivot element**.

Putting together what has been discussed so far and continuing with the problem from the Flashback yields

$$
\begin{array}{ccccc}
x_1 & x_2 & s_1 & s_2 & P \\
\end{array}
$$

Pivot element →

$$
\left[
\begin{array}{ccccc|c}
4 & 3 & 1 & 0 & 0 & 240 \\
1 & ② & 0 & 1 & 0 & 140 \\
-20 & -30 & 0 & 0 & 1 & 0 \\
\end{array}
\right]
\quad
\begin{array}{l}
\text{Quotients:} \\
240/3 = 80 \\
140/2 = 70 \ \text{Smaller; so} \\
\qquad\qquad \text{pivot row}
\end{array}
$$

Pivot column
(most negative indicator)

Now that we have our pivot element, let's pivot on the circled 2 using our row operations.

$$
\begin{array}{ccccc}
x_1 & x_2 & s_1 & s_2 & P \\
\end{array}
$$

$$
\left[
\begin{array}{ccccc|c}
4 & 3 & 1 & 0 & 0 & 240 \\
1 & ② & 0 & 1 & 0 & 140 \\
-20 & -30 & 0 & 0 & 1 & 0 \\
\end{array}
\right]
\quad \tfrac{1}{2}R_2 \to R_2 \quad
\left[
\begin{array}{ccccc|c}
4 & 3 & 1 & 0 & 0 & 240 \\
\frac{1}{2} & 1 & 0 & \frac{1}{2} & 0 & 70 \\
-20 & -30 & 0 & 0 & 1 & 0 \\
\end{array}
\right]
$$

$$
\begin{array}{c}
-3R_2 + R_1 \to R_1 \\
30R_2 + R_3 \to R_3
\end{array}
\quad
\left[
\begin{array}{ccccc|c}
\frac{5}{2} & 0 & 1 & -\frac{3}{2} & 0 & 30 \\
\frac{1}{2} & 1 & 0 & \frac{1}{2} & 0 & 70 \\
-5 & 0 & 0 & 15 & 1 & 2100 \\
\end{array}
\right]
$$

This last matrix, the result of our first pivot, is called the **second simplex matrix** or **second simplex tableau**.

Example 1   **Interpreting the Second Simplex Tableau**

Interpret the entries in the second simplex tableau from the Flashback.

$$
\begin{array}{ccccc}
x_1 & x_2 & s_1 & s_2 & P \\
\end{array}
$$

$$
\left[
\begin{array}{ccccc|c}
\frac{5}{2} & 0 & 1 & -\frac{3}{2} & 0 & 30 \\
\frac{1}{2} & 1 & 0 & \frac{1}{2} & 0 & 70 \\
-5 & 0 & 0 & 15 & 1 & 2100 \\
\end{array}
\right]
$$

Solution

The basic variables are x_2, s_1, and P and have values $x_2 = 70$, $s_1 = 30$, and $P = 2100$. The nonbasic variables are x_1 and s_2 and have the value of 0. This corresponds to the solution

$$
x_1 = 0, \quad x_2 = 70, \quad s_1 = 30, \quad s_2 = 0, \quad P = 2100
$$

Interpret the Solution: This means that producing and selling 0 units of switch A and 70 units of switch B leaves 30 unused hours in the assembly department and 0 unused hours in the testing department and yields a profit of $2100. ∎

Does the second simplex tableau, which we just interpreted in Example 1, give the optimal solution? The answer is no and is exhibited by the presence of a negative number (-5) in the bottom row of the second simplex tableau. The last row represents the equation

$$-5x_1 + 0x_2 + 0s_1 + 15s_2 + P = 2100$$

which is equivalent to

$$P = 2100 + 5x_1 - 15s_2$$

This last equation tells us that profit is \$2100 if $x_1 = s_2 = 0$. However, if x_1 is a positive number, then the profit will be larger than \$2100. Hence it may be possible to have a larger profit by changing x_1 from 0 to a positive number. We make this change by repeating the pivot process! Continuing with the pivot process as outlined in this section, we have

Pivot element
$$
\begin{array}{ccccc|c}
x_1 & x_2 & s_1 & s_2 & P & \\
\boxed{\tfrac{5}{2}} & 0 & 1 & -\tfrac{3}{2} & 0 & 30 \\
\tfrac{1}{2} & 1 & 0 & \tfrac{1}{2} & 0 & 70 \\
\hline
-5 & 0 & 0 & 15 & 1 & 2100
\end{array}
$$

Quotients:
$30/\tfrac{5}{2} = 12$ Smaller; pivot row
$70/\tfrac{1}{2} = 140$

↑
Pivot column

$$
\begin{array}{ccccc|c}
x_1 & x_2 & s_1 & s_2 & P & \\
\tfrac{5}{2} & 0 & 1 & -\tfrac{3}{2} & 0 & 30 \\
\tfrac{1}{2} & 1 & 0 & \tfrac{1}{2} & 0 & 70 \\
\hline
-5 & 0 & 0 & 15 & 1 & 2100
\end{array}
$$
$\tfrac{2}{5}R_1 \rightarrow R_1$
$$
\begin{array}{ccccc|c}
x_1 & x_2 & s_1 & s_2 & P & \\
1 & 0 & \tfrac{2}{5} & -\tfrac{3}{5} & 0 & 12 \\
\tfrac{1}{2} & 1 & 0 & \tfrac{1}{2} & 0 & 70 \\
\hline
-5 & 0 & 0 & 15 & 1 & 2100
\end{array}
$$

$-\tfrac{1}{2}R_1 + R_2 \rightarrow R_2$
$5R_1 + R_3 \rightarrow R_3$
$$
\begin{array}{ccccc|c}
x_1 & x_2 & s_1 & s_2 & P & \\
1 & 0 & \tfrac{2}{5} & -\tfrac{3}{5} & 0 & 12 \\
0 & 1 & -\tfrac{1}{5} & \tfrac{4}{5} & 0 & 64 \\
\hline
0 & 0 & 2 & 12 & 1 & 2160
\end{array}
$$

The result of the second pivot gives the **third simplex tableau.** Interpreting the entries gives

$$x_1 = 12, \quad x_2 = 64, \quad s_1 = 0, \quad s_2 = 0, \quad P = 2160$$

Notice that the last row of the third simplex tableau represents the following:

$$0x_1 + 0x_2 + 2s_1 + 12s_2 + P = 2160$$
$$0(12) + 0(64) + 2(0) + 12(0) + P = 2160$$
$$P = 2160$$

We have found our optimal (maximum) solution! So each week Dub, Inc., should produce and sell 12 units of switch A and 64 units of switch B to maximize weekly profit at \$2160. Notice that $s_1 = s_2 = 0$ means that all available hours should be used. This result suggests when we stop the pivot process.

The optimal solution has been found when no indicators are negative.

We now summarize what we have done so far in this section. The following steps are used when solving a standard maximum form linear programming problem using the **simplex method.**

■ **Simplex Method for Solving Standard Maximum Form Problems**

1. Write the initial simplex tableau. This involves the following:

   - Determining the objective function
   - Determining all constraints
   - Using slack variables to write each constraint in equational form

2. Determine the pivot element by finding the pivot column and constructing the necessary quotients to determine the pivot row.

3. Use row operations to pivot about the element found in step 2.

4. If the indicators are all nonnegative, this is the final tableau. Proceed to step 5. If the indicators are not all nonnegative, go back to step 2.

5. Read the optimal solution(s) from the final tableau.

Before continuing with the simplex method, we want to compare it to the graphical method in Chapter 3. Figure 4.2.1 shows the feasible region corresponding to the mathematical model in the Flashback. In Chapter 3 we had to evaluate the objective function at **each** corner point to determine the optimal value. See Table 4.2.1.

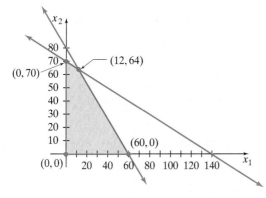

Figure 4.2.1

Table 4.2.1

Corner Point	Value of $P = 20x + 30y$	
$(0, 0)$	$20(0) + 30(0) = 0$	
$(0, 70)$	$20(0) + 30(70) = 2100$	
$(12, 64)$	$20(12) + 30(64) = 2160$	**Maximum**
$(60, 0)$	$20(60) + 30(0) = 1200$	

Notice how the simplex method evaluated the objective function at the corner points $(0, 0)$, $(0, 70)$, and finally $(12, 64)$. There was no need to test the corner point $(60, 0)$ since the optimum value of P was determined before the point was reached. Figure 4.2.2 shows the order in which corner points were evaluated using the simplex method.

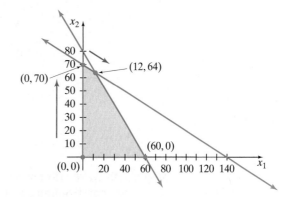

Figure 4.2.2

Example 2 Using the Simplex Method

Use the simplex method to

$$\text{Maximize:} \quad z = 3x_1 + 2x_2$$
$$\text{Subject to:} \quad x_1 + x_2 \leq 5$$
$$2x_1 + x_2 \leq 6$$
$$x_1 \geq 0, x_2 \geq 0$$

Solution

We use the Five-step process:

Step 1: Since the objective function and constraints are given, we need to introduce two slack variables, s_1 and s_2 (one for each constraint), to write the constraints in equational form. This gives

$$x_1 + x_2 + s_1 \quad\quad = 5$$
$$2x_1 + x_2 \quad\quad + s_2 = 6$$

The objective function is written as

$$-3x_1 - 2x_2 + z = 0$$

So our initial simplex tableau is

$$
\begin{array}{ccccc}
x_1 & x_2 & s_1 & s_2 & z
\end{array}
$$
$$
\left[
\begin{array}{ccccc|c}
1 & 1 & 1 & 0 & 0 & 5 \\
2 & 1 & 0 & 1 & 0 & 6 \\
\hline
-3 & -2 & 0 & 0 & 1 & 0
\end{array}
\right]
$$

Step 2: The pivot column, quotients, pivot row, and pivot element are shown next.

$$
\begin{array}{ccccc}
x_1 & x_2 & s_1 & s_2 & z
\end{array} \quad \text{Quotients:}
$$
$$
\text{Pivot element} \quad
\left[
\begin{array}{ccccc|c}
1 & 1 & 1 & 0 & 0 & 5 \\
\text{②} & 1 & 0 & 1 & 0 & 6 \\
\hline
-3 & -2 & 0 & 0 & 1 & 0
\end{array}
\right]
\begin{array}{l}
5/1 = 5 \\
6/2 = 3 \quad \text{Pivot row}
\end{array}
$$
$$
\uparrow
$$
$$
\text{Pivot column}
$$

Step 3: We pivot about the selected 2 as follows:

$$
\begin{array}{ccccc}
x_1 & x_2 & s_1 & s_2 & z
\end{array}
$$
$$
\left[
\begin{array}{ccccc|c}
1 & 1 & 1 & 0 & 0 & 5 \\
\text{②} & 1 & 0 & 1 & 0 & 6 \\
\hline
-3 & -2 & 0 & 0 & 1 & 0
\end{array}
\right]
\frac{1}{2}R_2 \rightarrow R_2
\left[
\begin{array}{ccccc|c}
1 & 1 & 1 & 0 & 0 & 5 \\
1 & \frac{1}{2} & 0 & \frac{1}{2} & 0 & 3 \\
\hline
-3 & -2 & 0 & 0 & 1 & 0
\end{array}
\right]
$$

$$
\begin{array}{ccccc}
x_1 & x_2 & s_1 & s_2 & z
\end{array}
$$
$$
\begin{array}{l}
-R_2 + R_1 \rightarrow R_1 \\
3R_2 + R_3 \rightarrow R_3
\end{array}
\left[
\begin{array}{ccccc|c}
0 & \frac{1}{2} & 1 & -\frac{1}{2} & 0 & 2 \\
1 & \frac{1}{2} & 0 & \frac{1}{2} & 0 & 3 \\
\hline
0 & -\frac{1}{2} & 0 & \frac{3}{2} & 1 & 9
\end{array}
\right]
$$

Step 4: Since we still have negative indicators, we go back to step 2 and repeat. Column 2 is the pivot column, and constructing quotients as before indicates that we choose the $\frac{1}{2}$ in row 1, column 2 as the

pivot element. This gives us

$$\begin{array}{c} \begin{array}{ccccc} x_1 & x_2 & s_1 & s_2 & z \end{array} \\ \left[\begin{array}{ccccc|c} 0 & \boxed{\frac{1}{2}} & 1 & -\frac{1}{2} & 0 & 2 \\ 1 & \frac{1}{2} & 0 & \frac{1}{2} & 0 & 3 \\ \hline 0 & -\frac{1}{2} & 0 & \frac{3}{2} & 1 & 9 \end{array} \right] \end{array} \; 2R_1 \to R_1 \quad \begin{array}{c} \begin{array}{ccccc} x_1 & x_2 & s_1 & s_2 & z \end{array} \\ \left[\begin{array}{ccccc|c} 0 & 1 & 2 & -1 & 0 & 4 \\ 1 & \frac{1}{2} & 0 & \frac{1}{2} & 0 & 3 \\ \hline 0 & -\frac{1}{2} & 0 & \frac{3}{2} & 1 & 9 \end{array} \right] \end{array}$$

$$\begin{array}{c} -\frac{1}{2}R_1 + R_2 \to R_2 \\ \frac{1}{2}R_1 + R_3 \to R_3 \end{array} \quad \begin{array}{c} \begin{array}{ccccc} x_1 & x_2 & s_1 & s_2 & z \end{array} \\ \left[\begin{array}{ccccc|c} 0 & 1 & 2 & -1 & 0 & 4 \\ 1 & 0 & -1 & 1 & 0 & 1 \\ \hline 0 & 0 & 1 & 1 & 1 & 11 \end{array} \right] \end{array}$$

Since the indicators are all nonnegative, this is the final tableau.

Step 5: The solutions are $x_1 = 1$, $x_2 = 4$, $s_1 = 0$, $s_2 = 0$, and $z = 11$. This means that z has a maximum value of 11 when $x_1 = 1$ and $x_2 = 4$ and both slack variables are 0. ∎

✓ **Checkpoint 1**

Now work Exercise 11.

Example 3 **Using the Simplex Method in an Application**

The Gonzalez Office Supplies Company has determined that the profits are $40, $80, and $70 for each Secretarial, Executive, and Presidential chair produced and sold. Each chair requires time on three different machines. These times and the total available time are given in Table 4.2.2, in hours.

Table 4.2.2

	Secretarial	Executive	Presidential	Total Hours Each Week
Hours on machine A	1	1	1	30
Hours on machine B	2	3	2	53
Hours on machine C	1	2	3	47

How many of each type of chair should Gonzalez Office Supplies make each week in order to maximize its profit?

Solution

Step 1: **Understand the Situation:** Since we are asked to determine how many of each type of chair should be made, we let x_1, x_2, and x_3 represent the number of Secretarial, Executive, and Presidential chairs to be made, respectively. The mathematical model is

$$\begin{aligned} \text{Maximize:} \quad & P = 40x_1 + 80x_2 + 70x_3 \\ \text{Subject to:} \quad & x_1 + x_2 + x_3 \le 30 \\ & 2x_1 + 3x_2 + 2x_3 \le 53 \\ & x_1 + 2x_2 + 3x_3 \le 47 \\ & x_1 \ge 0,\; x_2 \ge 0,\; x_3 \ge 0 \end{aligned}$$

Since we have three constraints, we need three slack variables, s_1, s_2, and s_3, in order to write the constraints in equational form. This gives

$$
\begin{aligned}
x_1 + x_2 + x_3 + s_1 \qquad\qquad &= 30 \\
2x_1 + 3x_2 + 2x_3 \qquad + s_2 \qquad &= 53 \\
x_1 + 2x_2 + 3x_3 \qquad\qquad + s_3 &= 47
\end{aligned}
$$

The objective function is written as

$$
-40x_1 - 80x_2 - 70x_3 + P = 0
$$

So the initial simplex tableau is

$$
\begin{array}{ccccccc|c}
x_1 & x_2 & x_3 & s_1 & s_2 & s_3 & P & \\
\hline
1 & 1 & 1 & 1 & 0 & 0 & 0 & 30 \\
2 & 3 & 2 & 0 & 1 & 0 & 0 & 53 \\
1 & 2 & 3 & 0 & 0 & 1 & 0 & 47 \\
\hline
-40 & -80 & -70 & 0 & 0 & 0 & 1 & 0
\end{array}
$$

Step 2: The pivot column, quotients, pivot row, and pivot element are shown next.

Pivot element

$$
\begin{array}{ccccccc|c}
x_1 & x_2 & x_3 & s_1 & s_2 & s_3 & P & \\
\hline
1 & 1 & 1 & 1 & 0 & 0 & 0 & 30 \\
2 & ③ & 2 & 0 & 1 & 0 & 0 & 53 \\
1 & 2 & 3 & 0 & 0 & 1 & 0 & 47 \\
\hline
-40 & -80 & -70 & 0 & 0 & 0 & 1 & 0
\end{array}
$$

Quotients:
$30/1 = 30$
$53/3 = 17\frac{2}{3}$ Pivot row
$47/2 = 23.5$

↑
Pivot column

Step 3: We pivot about the selected 3 as follows:

$$
\begin{array}{ccccccc|c}
x_1 & x_2 & x_3 & s_1 & s_2 & s_3 & P & \\
\hline
1 & 1 & 1 & 1 & 0 & 0 & 0 & 30 \\
2 & ③ & 2 & 0 & 1 & 0 & 0 & 53 \\
1 & 2 & 3 & 0 & 0 & 1 & 0 & 47 \\
\hline
-40 & -80 & -70 & 0 & 0 & 0 & 1 & 0
\end{array}
$$

$\frac{1}{3}R_2 \to R_2$

$$
\begin{array}{ccccccc|c}
x_1 & x_2 & x_3 & s_1 & s_2 & s_3 & P & \\
\hline
1 & 1 & 1 & 1 & 0 & 0 & 0 & 30 \\
\frac{2}{3} & 1 & \frac{2}{3} & 0 & \frac{1}{3} & 0 & 0 & \frac{53}{3} \\
1 & 2 & 2 & 0 & 0 & 1 & 0 & 47 \\
\hline
-40 & -80 & -70 & 0 & 0 & 0 & 1 & 0
\end{array}
$$

$-R_2 + R_1 \to R_1$
$-2R_2 + R_3 \to R_3$
$80R_2 + R_4 \to R_4$

$$
\begin{array}{ccccccc|c}
x_1 & x_2 & x_3 & s_1 & s_2 & s_3 & P & \\
\hline
\frac{1}{3} & 0 & \frac{1}{3} & 1 & -\frac{1}{3} & 0 & 0 & \frac{37}{3} \\
\frac{2}{3} & 1 & \frac{2}{3} & 0 & \frac{1}{3} & 0 & 0 & \frac{53}{3} \\
-\frac{1}{3} & 0 & \frac{5}{3} & 0 & -\frac{2}{3} & 1 & 0 & \frac{35}{3} \\
\hline
\frac{40}{3} & 0 & -\frac{50}{3} & 0 & \frac{80}{3} & 0 & 1 & \frac{4240}{3}
\end{array}
$$

Step 4: Since we do not have all nonnegative indicators, we need to repeat step 2. Column 3 is the pivot column, and constructing quotients as before indicates that we choose the $\frac{5}{3}$ in row 3, column 3 as the pivot element. Continuing the pivot, we begin by performing the row operation $\frac{3}{5} R_3 \to R_3$.

$$
\begin{array}{c}
\begin{array}{ccccccc} x_1 & x_2 & x_3 & s_1 & s_2 & s_3 & P \end{array} \\
\left[\begin{array}{ccccccc|c}
\frac{1}{3} & 0 & \frac{1}{3} & 1 & -\frac{1}{3} & 0 & 0 & \frac{37}{3} \\
\frac{2}{3} & 1 & \frac{2}{3} & 0 & \frac{1}{3} & 0 & 0 & \frac{53}{3} \\
-\frac{1}{3} & 0 & \boxed{\frac{5}{3}} & 0 & -\frac{2}{3} & 1 & 0 & \frac{35}{3} \\
\frac{40}{3} & 0 & -\frac{50}{3} & 0 & \frac{80}{3} & 0 & 1 & \frac{4240}{3}
\end{array} \right]
\end{array}
$$

$$
\begin{array}{cc}
& \begin{array}{ccccccc} x_1 & x_2 & x_3 & s_1 & s_2 & s_3 & P \end{array} \\
\frac{3}{5} R_3 \to R_3 &
\left[\begin{array}{ccccccc|c}
\frac{1}{3} & 0 & \frac{1}{3} & 1 & -\frac{1}{3} & 0 & 0 & \frac{37}{3} \\
\frac{2}{3} & 1 & \frac{2}{3} & 0 & \frac{1}{3} & 0 & 0 & \frac{53}{3} \\
-\frac{1}{5} & 0 & 1 & 0 & -\frac{2}{5} & \frac{3}{5} & 0 & 7 \\
\frac{40}{3} & 0 & -\frac{50}{3} & 0 & \frac{80}{3} & 0 & 1 & \frac{4240}{3}
\end{array} \right]
\end{array}
$$

$$
\begin{array}{cc}
\begin{array}{l} -\frac{1}{3} R_3 + R_1 \to R_1 \\ -\frac{2}{3} R_3 + R_2 \to R_2 \\ \frac{50}{3} R_3 + R_4 \to R_4 \end{array} &
\begin{array}{c}
\begin{array}{ccccccc} x_1 & x_2 & x_3 & s_1 & s_2 & s_3 & P \end{array} \\
\left[\begin{array}{ccccccc|c}
\frac{2}{5} & 0 & 0 & 1 & -\frac{1}{5} & -\frac{1}{5} & 0 & 10 \\
\frac{4}{5} & 1 & 0 & 0 & \frac{3}{5} & -\frac{2}{5} & 0 & 13 \\
-\frac{1}{5} & 0 & 1 & 0 & -\frac{2}{5} & \frac{3}{5} & 0 & 7 \\
10 & 0 & 0 & 0 & 20 & 10 & 1 & 1530
\end{array} \right]
\end{array}
\end{array}
$$

Since all indicators are nonnegative, this is the final tableau.

Step 5: The optimal solution is $x_1 = 0$, $x_2 = 13$, $x_3 = 7$, $s_1 = 10$, $s_2 = 0$, $s_3 = 0$, and $P = 1530$.

Interpret the Solution: This means that to maximize profit Gonzalez Office Supplies should produce 0 Secretarial, 13 Executive, and 7 Presidential chairs each week. This yields a maximum weekly profit of $1530. The value of the slack variable $s_1 = 10$ means that 10 hours of available time on machine A is unused.   ■

 Checkpoint 2

Now work Exercise 35.

We began Section 2.1 with a Flashback to the Gonzalez Office Supplies Company and in Example 7 in Section 2.1 we found that in order to use **all** available time on each machine required Gonzalez Office Supplies to produce 5 Secretarial, 9 Executive, and 8 Presidential chairs each week. Given the profit per chair in Example 3, this production schedule would result in a profit of $40(5) + 80(9) + 70(8) = \$1480$, which is less than the \$1530 we found in Example 3. Example 3 illustrates that through optimal use of the equipment a company can maximize its profit while reducing wear and tear on machines.

Technology Option

Simplex Method on a Calculator

Several computer programs can perform the simplex method. A popular one is LINDO. In Appendix C we supply a calculator program for the TI-83 that will perform the simplex method. This program may be useful, especially when many variables are involved, as in Example 3. If you have a calculator different than a TI-83, consult www.prenhall.com/armstrong for the program for your calculator.

Multiple Solutions and Unbounded Feasible Region

We saw in Chapter 3 that some linear programming problems have an infinite number of solutions. We also saw that if the feasible region is unbounded there may be no optimal (maximum) value. When using the simplex method, here is how to recognize these situations:

- A linear programming problem may have an infinite number of solutions if the bottom row to the left of the vertical dashed line of the final tableau contains a zero in a column that is not a basic variable column.
- A linear programming problem has no solution if the simplex method fails at some point in the process.

An example of the last bullet item is if, when constructing the quotients, we have no nonnegative quotients. In this case, the problem has no solution.

SUMMARY

In this section we continued the process established in Section 4.1. This process of solving linear programming problems is called the **simplex method**. We saw how selecting the **pivot column** led to choosing the correct **pivot row** by constructing quotients. The intersection of the pivot column and the pivot row is the **pivot element**. We then supplied a Five-step process for using the simplex method.

Simplex Method for Solving Standard Maximum Form Problems

1. Write the initial simplex tableau. This involves the following:
   - Determining the objective function
   - Determining all constraints

- Using slack variables to write each constraint in equational form

2. Determine the pivot element by finding the pivot column and constructing the necessary quotients to determine the pivot row.

3. Use row operations to pivot about the element found in step 2.

4. If the indicators are all nonnegative, this is the final tableau. Proceed to step 5. If the indicators are not all nonnegative, go back to step 2.

5. Read the optimal solution(s) from the final tableau.

SECTION 4.2 EXERCISES

In Exercises 1–10, the initial simplex tableau of a linear programming problem is given. Use the simplex method to solve.

1.
$$\begin{array}{ccccc|c}
x_1 & x_2 & s_1 & s_2 & z & \\
\hline
1 & 1 & 1 & 0 & 0 & 5 \\
1 & 2 & 0 & 1 & 0 & 2 \\
\hdashline
-4 & -5 & 0 & 0 & 1 & 10
\end{array}$$

2.
$$\begin{array}{ccccc|c}
x_1 & x_2 & s_1 & s_2 & z & \\
\hline
6 & 2 & 1 & 0 & 0 & 5 \\
11 & 5 & 0 & 1 & 0 & 11 \\
\hdashline
4 & -3 & 0 & 0 & 1 & 15
\end{array}$$

3.
$$\begin{array}{ccccc|c}
x_1 & x_2 & s_1 & s_2 & z & \\
\hline
1 & 4 & 1 & 0 & 0 & 6 \\
2 & 7 & 0 & 1 & 0 & 10 \\
\hdashline
-3 & -7 & 0 & 0 & 1 & 14
\end{array}$$

4.
$$\begin{array}{ccccc|c}
x_1 & x_2 & s_1 & s_2 & z & \\
\hline
1 & 3 & 1 & 0 & 0 & 4 \\
2 & 4 & 0 & 1 & 0 & 7 \\
\hdashline
-6 & -2 & 0 & 0 & 1 & 12
\end{array}$$

5.
$$\begin{array}{cccccc|c}
x_1 & x_2 & s_1 & s_2 & s_3 & z & \\
\hline
-1 & 2 & 1 & 0 & 0 & 0 & 3 \\
3 & 8 & 0 & 1 & 0 & 0 & 6 \\
7 & 5 & 0 & 0 & 1 & 0 & 8 \\
\hdashline
-2 & -7 & 0 & 0 & 0 & 1 & 0
\end{array}$$

6.
$$\begin{array}{cccccc|c}
x_1 & x_2 & s_1 & s_2 & s_3 & z & \\
\hline
2 & 5 & 1 & 0 & 0 & 0 & 3 \\
3 & 6 & 0 & 1 & 0 & 0 & 8 \\
4 & 2 & 0 & 0 & 1 & 0 & 15 \\
\hdashline
-1 & -2 & 0 & 0 & 0 & 1 & 0
\end{array}$$

7.
$$\begin{array}{cccccc|c}
x_1 & x_2 & x_3 & s_1 & s_2 & z & \\
\hline
10 & 4 & 2 & 1 & 0 & 0 & 90 \\
6 & 3 & 4 & 0 & 1 & 0 & 40 \\
\hdashline
-2 & -3 & -8 & 0 & 0 & 1 & 0
\end{array}$$

8.
$$\begin{array}{cccccc|c}
x_1 & x_2 & x_3 & s_1 & s_2 & z & \\
\hline
3 & 6 & 1 & 1 & 0 & 0 & 13 \\
5 & 2 & 2 & 0 & 1 & 0 & 15 \\
\hdashline
-8 & -4 & -10 & 0 & 0 & 1 & 0
\end{array}$$

9.
$$\begin{array}{cccccccc|c}
x_1 & x_2 & x_3 & x_4 & s_1 & s_2 & s_3 & z & \\
1 & 5 & 0 & 0 & 1 & 0 & 0 & 0 & 9 \\
0 & 3 & 0 & 1 & 0 & 1 & 0 & 0 & 13 \\
0 & 2 & 1 & 0 & 0 & 0 & 1 & 0 & 27 \\
\hline
2 & -2 & -1 & 0 & 0 & 0 & 0 & 1 & 80
\end{array}$$

10.
$$\begin{array}{cccccccc|c}
x_1 & x_2 & x_3 & x_4 & s_1 & s_2 & s_3 & z & \\
6 & 0 & 1 & 2 & 1 & 0 & 0 & 0 & 5 \\
2 & 1 & 0 & 10 & 0 & 1 & 0 & 0 & 10 \\
3 & 0 & 2 & 1 & 0 & 0 & 1 & 0 & 8 \\
-10 & -5 & 0 & -2 & 0 & 0 & 0 & 1 & 20
\end{array}$$

For the standard maximum for linear programming problems in Exercises 11–26:

(a) Introduce slack variables and set up the initial simplex tableau.

(b) Use the simplex method to solve.

✓ **11.** Maximize: $z = 2x_1 + x_2$
Subject to: $x_1 + x_2 \le 6$
$x_1 - x_2 \le 2$
$x_1 \ge 0, x_2 \ge 0$

12. Maximize: $z = 3x_1 + x_2$
Subject to: $4x_1 + 5x_2 \le 9$
$2x_1 - x_2 \le 2$
$x_1 \ge 0, x_2 \ge 0$

13. Maximize: $z = 2x_1 + 3x_2$
Subject to: $3x_1 + 2x_2 \le 8$
$x_1 - x_2 \le 7$
$x_1 \ge 0, x_2 \ge 0$

14. Maximize: $z = x_1 + 2x_2$
Subject to: $x_1 + x_2 \le 7$
$2x_1 + 3x_2 \le 16$
$x_1 \ge 0, x_2 \ge 0$

15. Maximize: $z = x_1 + 4x_2$
Subject to: $2x_1 + x_2 \le 32$
$x_1 + x_2 \le 18$
$x_1 + 3x_2 \le 36$
$x_1 \ge 0, x_2 \ge 0$

16. Maximize: $z = 5x_1 + 7x_2$
Subject to: $3x_1 - x_2 \le 8$
$2x_1 + 4x_2 \le 5$
$x_1 + x_2 \le 7$
$x_1 \ge 0, x_2 \ge 0$

17. Maximize: $z = 2x_1 + 2x_2$
Subject to: $3x_1 - 2x_2 \le 0$
$x_1 + x_2 \le 5$
$2x_1 + x_2 \le 6$
$x_1 \ge 0, x_2 \ge 0$

18. Maximize: $z = 2x_1 + 2x_2$
Subject to: $x_1 + x_2 \le 3$
$x_1 - 2x_2 \le 0$
$2x_1 + x_2 \le 2$
$x_1 \ge 0, x_2 \ge 0$

19. Maximize: $z = 5x_1 + x_2$
Subject to: $-x_1 + 2x_2 \le 10$
$2x_1 + x_2 \le 15$
$5x_1 + x_2 \le 8$
$-x_1 + x_2 \le 1$
$x_1 \ge 0, x_2 \ge 0$

20. Maximize: $z = 3x_1 + 5x_2$
Subject to: $-x_1 + 2x_2 \le 10$
$4x_1 + 2x_2 \le 15$
$x_1 + 3x_2 \le 8$
$6x_1 - x_2 \le 4$
$x_1 \ge 0, x_2 \ge 0$

21. Maximize: $z = 3x_1 + 2x_2 + 4x_3$
Subject to: $2x_1 + x_2 + x_3 \le 10$
$5x_1 - x_2 + 2x_3 \le 8$
$x_1 \ge 0, x_2 \ge 0, x_3 \ge 0$

22. Maximize: $z = 2x_1 + x_2 + 3x_3$
Subject to: $3x_1 - x_2 + 2x_3 \le 4$
$2x_1 + x_2 - x_3 \le 7$
$x_1 \ge 0, x_2 \ge 0, x_3 \ge 0$

23. Maximize: $z = 2x_1 + 5x_2 + x_3$
Subject to: $5x_1 + 2x_2 + x_3 \le 20$
$x_1 + 3x_2 + 5x_3 \le 13$
$4x_1 + x_2 + 3x_3 \le 15$
$x_1 \ge 0, x_2 \ge 0, x_3 \ge 0$

24. Maximize: $z = 3x_1 + x_2 + 5x_3$
Subject to: $2x_1 + x_2 + 4x_3 \le 8$
$4x_1 - x_2 + 3x_3 \le 15$
$12x_1 + 2x_2 + 7x_3 \le 12$
$x_1 \ge 0, x_2 \ge 0, x_3 \ge 0$

25. Maximize: $z = 4x_1 + 2x_2 + 3x_3$
Subject to: $2x_1 + 3x_3 \le 8$
$3x_1 + 2x_2 \le 5$
$x_1 \ge 0, x_2 \ge 0, x_3 \ge 0$

26. Maximize: $z = 9x_1 + x_2 + x_3$
Subject to: $4x_1 + x_3 \le 7$
$2x_1 + 3x_2 \le 9$
$x_1 \ge 0, x_2 \ge 0, x_3 \ge 0$

Applications

In the remainder of the exercises, set up and solve each using the simplex method. Interpret solutions as shown in Example 3. Include an interpretation of any nonzero slack variables.

27. The Young Furniture Company produces tables and chairs. Each table requires 2 hours of painting and 4 hours of carpentry work and yields a profit of $70. Each chair requires 1 hour painting and 3 hours of carpentry work and yields a profit of $50. The maximum of painting hours and carpentry hours available each week is 100 and 240, respectively. How many of each should be produced each week to maximize profit?

28. (*continuation of Exercise 27*) Due to weak demand, the Young Furniture Company must lay off some employees. As a result, the maximum number of painting hours and carpentry hours available each week is 50 and 110, respectively. Assuming that all other data remain the same as in Exercise 27, how many tables and chairs should be produced each week to maximize profit?

29. Due to new regulations approved by the Environmental Protection Agency (EPA), the Henderson Chemical Company was required to install a new process to reduce the pollution created when making a particular chemical. The older process releases 4 grams of sulfur and 3 grams of particulates for each gallon of chemical produced. The newer process releases 2 grams of sulfur and 1 gram of particulates for each gallon of chemical produced. The company realizes a profit of $0.40 for each gallon produced under the old process and $0.16 for each gallon produced under the new process. The new EPA regulations do not allow for more than 10,000 grams of sulfur and 7000 grams of particulates to be released daily. How many gallons of this chemical should be produced using the old process to maximize profits and still satisfy the new EPA regulations?

30. (*continuation of Exercise 29*) Barb, the plant manager for Henderson Chemical Company, has learned that the EPA is going to relax the new regulations so that no more than 13,000 grams of sulfur can be released daily. Assuming that all other data remain the same as in Exercise 29, how many gallons of this chemical should be produced using the old process to maximize profits and still satisfy the new EPA regulations?

31. Randy and Rusty have decided to enter the field of personal digital assistants (PDAs). Initially, they plan to design two types of PDAs, the WOW 1000 and the WOW 1500. Because of interest in PDAs, they can sell all that they can produce. However, they need to determine a production rate so as to satisfy various limits with a small production crew. These include a maximum of 150 hours per week for assembly and 100 hours per week for testing. The WOW 1000 requires 3 hours of assembly and 3 hours of testing and the WOW 1500 requires 6 hours of assembly and 3.5 hours of testing. Profit for the WOW 1000 is

estimated at $100 per unit and the profit for the WOW 1500 is estimated at $150 per unit. How many of each type of PDA should be made and sold to maximize weekly profits?

32. (*continuation of Exercise 31*) Randy and Rusty hire some new employees and can, as a result, increase production. By doing so, they now have a maximum of 210 hours per week for assembly and 160 hours per week for testing. Assuming that all other data remain the same as in Exercise 31, how many of each type of PDA should be made and sold to maximize weekly profits?

33. Table 4.2.3 gives data for growing artichokes and asparagus for an entire growing season in California (entries are in lb/acre).

Table 4.2.3

	Artichokes	Asparagus
Nitrogen	197	240
Potash	123	42

SOURCE: U.S. Department of Agriculture, www.usda.gov/nass

Franco has 150 acres available suitable for cultivating crops. He has purchased a contract for a total of 31,880 pounds of nitrogen and 9120 pounds of potash. He expects to make a profit of $140 per acre on artichokes and $160 per acre on asparagus.

(a) How many acres of each should he plant to maximize his profit?

(b) What is the maximum profit?

34. Table 4.2.4 gives data for growing cauliflower and head lettuce for an entire growing season in California (entries are in lb/acre).

Table 4.2.4

	Cauliflower	Head Lettuce
Nitrogen	215	173
Phosphate	98	122

SOURCE: U.S. Department of Agriculture, www.usda.gov/nass

Sonja has 210 acres available for cultivating crops. She has purchased a contract for a total of 38,380 pounds of nitrogen and 22,240 pounds of phosphate. She expects to make a profit of $90 per acre on cauliflower and $100 per acre on head lettuce.

(a) How many acres of each should she plant to maximize her profit?

(b) What is the maximum profit?

35. The Smolko Tent Company makes three types of tents: two-person, four-person and six-person models.

Each tent requires the services, in hours, of three departments as shown in Table 4.2.5. The fabric cutting, assembly, and packaging departments have available a maximum of 420, 310, and 180 hours each week, respectively.

Table 4.2.5

	Two-person	Four-person	Six-person
Fabric cutting	0.5	0.8	1.1
Assembly	0.6	0.7	0.9
Packaging	0.2	0.3	0.6

If the per unit profit is $30, $50, and $80 for the two-person, four-person and six-person tent, respectively, how many of each type tent should be made and sold to maximize weekly profit?

36. Kathy's Secretarial Services has a 1-month intensive training program for receptionist/secretaries for three types of office positions: medical field, small business, and executive secretarial. Each individual receives training in three departments at Kathy's: word processing, telephone and LAN (local area network), and ethics. Table 4.2.6 shows the number of units of training that each type of receptionist/secretary receives.

Table 4.2.6

	Medical Field	Small Business	Executive Secretary
Word processing	0.8	1.1	1.7
Telephone and LAN	0.6	1.3	1.5
Ethics	1.2	0.8	1.3

Each month, word processing, telephone and LAN, and ethics have available a maximum of 32.1 units, 29.7 units, and 33 units, respectively. The profit that Kathy's realizes from each individual admitted to medical field, small business, and executive secretary is $1000, $1000, and $1400, respectively. How many individuals should be admitted each month to each program to maximize monthly profit?

37. The data in Table 4.2.7 give the amount of nitrogen, phosphate, and potash (all in lb/acre) needed to grow asparagus, cauliflower, and celery for an entire growing season in California.

Table 4.2.7

	Asparagus	Cauliflower	Celery
Nitrogen	240	215	336
Phosphate	148	98	171
Potash	42	69	147

SOURCE: U.S. Department of Agriculture, www.usda.gov/nass

If Eldrick has 280 acres and can get no more than 65,970 pounds of nitrogen, 31,860 pounds of phosphate, and 21,630 pounds of potash, and if he expects a profit of $160 per acre for asparagus, $90 per acre for cauliflower, and $80 per acre for celery, how many acres should he allot for each crop to maximize profit?

38. The data in Table 4.2.8 give the amount of nitrogen, phosphate, and potash (all in lb/acre) needed to grow head lettuce, cantaloupe, and bell peppers for an entire growing season in California.

Table 4.2.8

	Head Lettuce	Cantaloupe	Bell Peppers
Nitrogen	173	175	212
Phosphate	122	161	108
Potash	76	39	72

SOURCE: U.S. Department of Agriculture, www.usda.gov/nass

If Maria has 200 acres and can get no more than 38,430 pounds of nitrogen, 27,845 pounds of phosphate, and 12,985 pounds of potash, and if she expects a profit of $100 per acre for head lettuce, $90 per acre for cantaloupe, and $85 per acre for bell peppers, how many acres should she allot for each crop to maximize profit?

39. Table 4.2.9 gives nutritional information for selected Hershey's bites products.

Table 4.2.9

Product	Serving Size	Calories	Protein	Carbohydrates (grams)	Total fat
Reeses cups	7 pieces	90	1	9	6
Cookies 'n' creme	8 pieces	90	2	10	5
York peppermint	9 pieces	90	0	19	1.5

SOURCE: www.hersheys.com

Meagan desires a snack consisting of these three candies. She wants total calories to not exceed 700, total fat to not exceed 36 grams, and total carbohydrates to not exceed 80 grams. How many servings of each should she have if she wants to maximize the total grams of protein?

40. Justin desires a snack consisting of the three candies given in Table 4.2.9. He wants total calories to not exceed 700, total fat to not exceed 30 grams, and total carbohydrates to not exceed 80 grams. How many servings of each should he have if he wants to maximize the total grams of protein?

41. Julia makes and sells three types of bread, Lemon Poppy Seed Glaze, Pumpkin Glaze, and Banana Walnut

Glaze. A loaf of Lemon Poppy Seed requires 1 cup of flour, 3 cups of sugar, and 2 eggs. A loaf of Pumpkin Glaze requires 2 cups of flour, 2 cups of sugar, and 2 eggs. A loaf of Banana Walnut Glaze requires 2 cups of flour, 3 cups of sugar, and 1 egg. Julia makes a profit of $1 per loaf of Lemon Poppy Seed Glaze, $1.50 per loaf of Pumpkin Glaze, and $1.50 per loaf of Banana Walnut Glaze. If Julia has available 24 cups of flour, 45 cups of sugar, and 27 eggs, how many of each type of bread should she make to maximize her profit?

42. Rework Exercise 40 except now assume that Julia has available 30 cups of flour, 56 cups of sugar, and 34 eggs.

■ SECTION PROJECT

Although it is not necessary, the graphing calculator program in Appendix C called SIMPLEX *may be beneficial for this section project.*

Shaver Frames manufactures three types of bicycle frames. The types of frames and the number of minutes that each frame requires on three different machines are given in Table 4.2.10.

Table 4.2.10

	Racing Frame	Hybrid Frame	Mountain Frame
Machine A	2	3	4
Machine B	5	6	5
Machine C	10	5	4

In any given week, machine A has available a maximum of 130 hours; machine B, 90 hours; and machine C, 110 hours. Shaver Frames realizes a profit of $14 for each racing frame, $13.50 for each hybrid frame, and $11.25 for each mountain frame. Shaver Frames desires to make no more than 560 mountain frames and no more than 1200 racing and mountain frames combined. Let x_1 represent the number of racing frames produced, x_2 the number of hybrid frames produced, and x_3 the number of mountain frames produced.

(a) Determine the number of each type of frame that Shaver Frames should produce to maximize profit?

(b) What is the maximum profit?

(c) Interpret the values of all variables, including the slack variables. (*Hint:* Five slack variables are needed and you should convert the hours to minutes.)

Section 4.3 Simplex Method: Standard Minimum Form Problems and Duality

In the first two sections of this chapter we have seen how to use the simplex method to solve linear programming problems in standard maximum form. In this section we learn how to solve linear programming problems in **standard minimum form**. This will quickly lead us to study what is called the **dual**, and we will see how duality plays a role in economics using what is known as **sensitivity analysis**.

Standard Minimum Form

Linear programming problems of the form

$$\text{Minimize:} \quad w = 6y_1 + 7y_2$$
$$\text{Subject to:} \quad 3y_1 + y_2 \geq 2$$
$$y_1 + 2y_2 \geq 0$$
$$y_1 \geq 0, \, y_2 \geq 0$$

are said to be in **standard minimum form**. Notice that we use $y_1, y_2, \ldots$ as decision variables rather than $x_1, x_2, \ldots$. The reason for this will be apparent later.

■ **Standard Minimum Form**

A linear programming problem is said to be in **standard minimum form** if the following are true:

1. The objective function is linear and must be **minimized**.
2. All variables are nonnegative.
3. All constraints, except the nonnegative constraints, are written as a linear expression greater than or equal to a nonnegative constant. Mathematically, this means

$$a_1 y_1 + a_2 y_2 + a_3 y_3 + \cdots + a_n y_n \geq b, \qquad b \geq 0$$

▶ **Note:** The differences between standard maximum form and standard minimum form are that in a standard minimum form problem the objective function is to be **minimized** and the constraints have the $\geq$ symbol (with the constants to the right).

Constructing the Dual

The note following the box defining standard minimum form linear programming problems shows that standard maximum form and standard minimum form problems are somewhat similar. The similarities do not stop there. In fact, we will soon see that the solution of a standard maximum form problem yields a solution to an associated standard minimum form problem, and the solution to a standard minimum form problem gives a solution to an associated standard maximum form problem. We call these associated problems the **dual** of the other. The power of duals is that we can solve standard minimum form problems using the simplex method on its associated standard maximum form problem. But, in order to do this, we must first see how to construct the dual. To illustrate how we form the dual problem, we consider the following minimization problem:

$$\begin{aligned}
\text{Minimize:} \quad & w = 6y_1 + 7y_2 \\
\text{Subject to:} \quad & 3y_1 + y_2 \geq 2 \\
& 4y_1 + 2y_2 \geq 3 \\
& y_1 \geq 0, \, y_2 \geq 0
\end{aligned}$$

In general, the given problem is called the **primal problem**. (The objective function and the inequality constraints associated with the primal problem are called the **primal objective function** and **primal inequalities**, respectively). The first step in forming the dual problem is to rewrite the problem constraints and objective function. We will use this rewritten form to construct an augmented matrix.

$$\begin{array}{cc}
\textit{Rewritten Form} & \textit{Augmented Matrix} \\[4pt]
\begin{aligned}
3y_1 + y_2 &\geq 2 \\
4y_1 + 2y_2 &\geq 3 \\
6y_1 + 7y_2 &= w
\end{aligned}
&
A = \begin{bmatrix} 3 & 1 & | & 2 \\ 4 & 2 & | & 3 \\ 6 & 7 & | & 0 \end{bmatrix}
\end{array}$$

Since we have not introduced slack variables, matrix A is **not** a simplex tableau. The solid vertical and horizontal bars are used to make this **important** distinction.

From Your Toolbox

For a given matrix A, if we interchange the rows and columns, we produce what is called the **transpose of A** and denote it with A^T.

Next we form the transpose of A, denoted A^T. Consult the Toolbox to the left.

$$A = \begin{bmatrix} 3 & 1 & 2 \\ 4 & 2 & 3 \\ 6 & 7 & 0 \end{bmatrix} \qquad A^T = \begin{bmatrix} 3 & 4 & 6 \\ 1 & 2 & 7 \\ 2 & 3 & 0 \end{bmatrix}$$

Now A^T corresponds to a new linear programming problem that is in standard maximum form. Using different variables, x_1 and x_2, so as to avoid confusion, the linear programming problem A^T corresponds to is

$$\begin{aligned} 3x_1 + 4x_2 &\le 6 \\ x_1 + 2x_2 &\le 7 \\ 2x_1 + 3x_2 &= z \end{aligned} \qquad A^T = \begin{bmatrix} 3 & 4 & 6 \\ 1 & 2 & 7 \\ 2 & 3 & 0 \end{bmatrix}$$

To recap, we have:

Primal Problem		*Dual Problem*	
Minimize:	$w = 6y_1 + 7y_2$	Maximize:	$z = 2x_1 + 3x_2$
Subject to:	$3y_1 + y_2 \ge 2$	Subject to:	$3x_1 + 4x_2 \le 6$
	$4y_1 + 2y_2 \ge 3$		$x_1 + 2x_2 \le 7$
	$y_1 \ge 0, y_2 \ge 0$		$x_1 \ge 0, x_2 \ge 0$

We now summarize the procedure for constructing the dual.

Forming the Dual Problem

Given a standard minimum form linear programming problem:

Step 1: Use the coefficients and constants in the primal problem constraints and the objective function to construct an augmented matrix A. Place the coefficients of the objective function in the bottom row.

Step 2: Determine A^T, the transpose of matrix A.

Step 3: Use the rows of A^T to construct a standard maximum form linear programming problem.

Example 1   Constructing the Dual

Form the dual problem for

$$\begin{aligned} \text{Minimize:} \quad & w = 3y_1 + 7y_2 \\ \text{Subject to:} \quad & 4y_1 + y_2 \ge 8 \\ & 5y_1 + 2y_2 \ge 1 \\ & y_1 \ge 0, y_2 \ge 0 \end{aligned}$$

Solution

Step 1: We begin by forming matrix A.

$$A = \begin{bmatrix} 4 & 1 & 8 \\ 5 & 2 & 1 \\ 3 & 7 & 0 \end{bmatrix}$$

Step 2: A^T, the transpose of A, is

$$A^T = \left[\begin{array}{cc|c} 4 & 5 & 3 \\ 1 & 2 & 7 \\ \hline 8 & 1 & 0 \end{array}\right]$$

Step 3: The dual of the given linear programming problem is the standard maximum form problem

$$\begin{aligned} \text{Maximize:} &\quad z = 8x_1 + x_2 \\ \text{Subject to:} &\quad 4x_1 + 5x_2 \le 3 \\ &\quad x_1 + 2x_2 \le 7 \\ &\quad x_1 \ge 0, x_2 \ge 0 \end{aligned}$$

■

✓ **Checkpoint 1**

Now work Exercise 7.

Technology Option

Recall that A^T can be done on a graphing calculator. Figure 4.3.1 shows A on a graphing calculator and Figure 4.3.2 shows A^T.

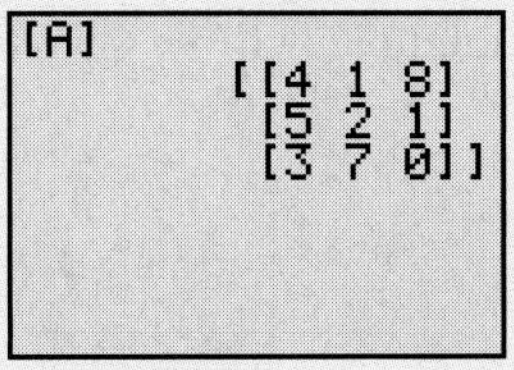

```
[A]
        [[4  1  8]
         [5  2  1]
         [3  7  0]]
```

```
[A] ᵀ
        [[4  5  3]
         [1  2  7]
         [8  1  0]]
```

Figure 4.3.1

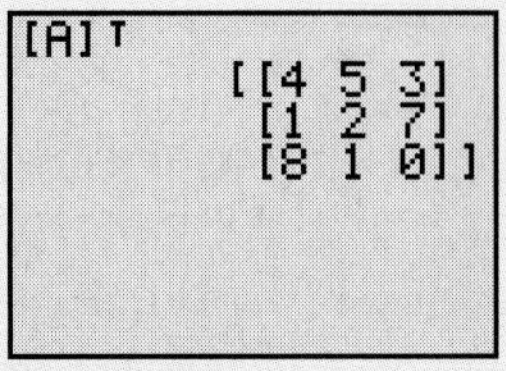

Figure 4.3.2

To review how to perform a matrix transpose on your calculator, consult the online calculator manual at www.prenhall.com/armstrong.

Solutions to Minimization Problems

To gain understanding on why duals are so important, let's consider the primal problem and its dual from the beginning of the section. We solve both in Example 2 using the graphical techniques from Chapter 3

Example 2 Solving Linear Programming Problems Graphically

Solve the following primal and dual problems by graphing the feasible region and evaluating the corner points in the objective function.

	Primal Problem		*Dual Problem*
Minimize:	$w = 6y_1 + 7y_2$	Maximize:	$z = 2x_1 + 3x_2$
Subject to:	$3y_1 + y_2 \ge 2$	Subject to:	$3x_1 + x_2 \le 6$
	$y_1 + 2y_2 \ge 3$		$x_1 + 2x_2 \le 7$
	$y_1 \ge 0, y_2 \ge 0$		$x_1 \ge 0, x_2 \ge 0$

Solution

We begin by graphing the feasible regions and labeling the corner points. See Figure 4.3.3 for the primal problem and Figure 4.3.4 for the dual problem. Notice in Figure 4.3.3 that the axes are labeled y_1 and y_2, whereas in Figure 4.3.4 the axes are labeled x_1 and x_2.

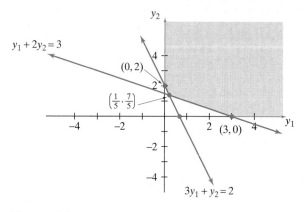

Figure 4.3.3

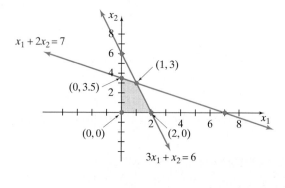

Figure 4.3.4

We now evaluate the objective function at each corner point. See Table 4.3.1 for the primal problem and Table 4.3.2 for the dual problem.

Table 4.3.1 Primal Problem

Corner Point	$w = 6y_1 + 7y_2$	
$(0, 2)$	$6(0) + 7(2) = 14$	
$\left(\frac{1}{5}, \frac{7}{5}\right)$	$6\left(\frac{1}{5}\right) + 7\left(\frac{7}{5}\right) = 11$	**Minimum**
$(3, 0)$	$6(3) + 7(0) = 18$	

Table 4.3.2 Dual Problem

Corner Point	$z = 2x_1 + 3x_2$	
$(0, 0)$	$2(0) + 3(0) = 0$	
$(0, 3.5)$	$2(0) + 3(3.5) = 10.5$	
$(1, 3)$	$2(1) + 3(3) = 11$	**Maximum**
$(2, 0)$	$2(2) + 3(0) = 4$	

For the primal problem we have a minimum value of 11 at the corner point $\left(\frac{1}{5}, \frac{7}{5}\right)$. For the dual problem we have a maximum value of 11 at the corner point $(1, 3)$. ∎

Notice in Example 2 that the two feasible regions are different and the corner points are different, but the optimal values for each objective function are equal. Both have optimal values of 11. This is not coincidental! Example 2 suggests the following theorem, which establishes the connection between the solution of a minimum problem and its dual. The proof of the theorem requires advanced techniques and is beyond the scope of this text.

■ Duality Theorem

The objective function of a minimum linear programming problem has an optimal (minimum) value if and only if the objective function of the associated dual maximum problem has an optimal (maximum) value. Furthermore, the optimal (minimum) value of the minimization problem equals the optimal (maximum) value of the associated dual maximization problem.

The duality theorem tells us that the optimal value of the minimization problem and its dual are equal. As Example 2 illustrated, these optimal values occurred at different corner points. Generally, we cannot determine the variable values that yield the optimal value for the minimum problem by examining the feasible region of the dual problem. But, since the dual is a maximum problem, we can apply the simplex method and find the optimal value and variable values that produce the optimal value to the original primal minimum problem, as we now illustrate in Example 3.

Example 3   **Using the Simplex Method on the Dual**

Use the simplex method to solve the dual problem from Example 2.

Solution

From Your Toolbox

The column with the most negative indicator is the **pivot column**. If there are two such indicators, use the leftmost indicator. To determine the pivot row, divide each number from the rightmost column (column of constants) by the corresponding number from the pivot column. The smallest nonnegative quotient gives the pivot row. If two quotients are equal (and smallest), arbitrarily choose the row nearest the top of the matrix.

Recall that the dual problem in Example 2 is in standard maximum form. After introducing slack variables, the equational form is

$$3x_1 + x_2 + s_1 \qquad\qquad = 6$$
$$x_1 + 2x_2 \qquad + s_2 \qquad = 7$$
$$-2x_1 - 3x_2 \qquad\qquad + z = 0$$

The initial simplex tableau is

$$
\begin{array}{ccccc}
x_1 & x_2 & s_1 & s_2 & z
\end{array}
$$
$$
\left[\begin{array}{ccccc|c}
3 & 1 & 1 & 0 & 0 & 6 \\
1 & 2 & 0 & 1 & 0 & 7 \\
\hline
-2 & -3 & 0 & 0 & 1 & 0
\end{array}\right]
$$

Using the techniques from Section 4.2 (consult the Toolbox to the left), we pivot on the circled 2.

$$
\begin{array}{ccccc}
x_1 & x_2 & s_1 & s_2 & z
\end{array}
$$
$$
\left[\begin{array}{ccccc|c}
3 & 1 & 1 & 0 & 0 & 6 \\
1 & ② & 0 & 1 & 0 & 7 \\
\hline
-2 & -3 & 0 & 0 & 1 & 0
\end{array}\right]
\quad \tfrac{1}{2}R_2 \to R_2 \quad
\left[\begin{array}{ccccc|c}
3 & 1 & 1 & 0 & 0 & 6 \\
\frac{1}{2} & 1 & 0 & \frac{1}{2} & 0 & \frac{7}{2} \\
\hline
-2 & -3 & 0 & 0 & 1 & 0
\end{array}\right]
$$

$$
\begin{array}{ccccc}
x_1 & x_2 & s_1 & s_2 & z
\end{array}
$$
$$
\begin{array}{l}
-R_2 + R_1 \to R_1 \\
3R_2 + R_3 \to R_3
\end{array}
\left[\begin{array}{ccccc|c}
\frac{5}{2} & 0 & 1 & -\frac{1}{2} & 0 & \frac{5}{2} \\
\frac{1}{2} & 1 & 0 & \frac{1}{2} & 0 & \frac{7}{2} \\
\hline
-\frac{1}{2} & 0 & 0 & \frac{3}{2} & 1 & \frac{21}{2}
\end{array}\right]
$$

We now pivot on the circled $\frac{5}{2}$.

$$
\begin{array}{ccccc}
x_1 & x_2 & s_1 & s_2 & z
\end{array}
$$
$$
\left[\begin{array}{ccccc|c}
\frac{5}{2} & 0 & 1 & -\frac{1}{2} & 0 & \frac{5}{2} \\
\frac{1}{2} & 1 & 0 & \frac{1}{2} & 0 & \frac{7}{2} \\
\hline
-\frac{1}{2} & 0 & 0 & \frac{3}{2} & 1 & \frac{21}{2}
\end{array}\right]
\quad \tfrac{2}{5}R_1 \to R_1 \quad
\left[\begin{array}{ccccc|c}
1 & 0 & \frac{2}{5} & -\frac{1}{5} & 0 & 1 \\
\frac{1}{2} & 1 & 0 & \frac{1}{2} & 0 & \frac{7}{2} \\
\hline
-\frac{1}{2} & 0 & 0 & \frac{3}{2} & 1 & \frac{21}{2}
\end{array}\right]
$$

$$
\begin{array}{ccccc}
x_1 & x_2 & s_1 & s_2 & z
\end{array}
$$
$$
\begin{array}{l}
-\tfrac{1}{2}R_1 + R_2 \to R_2 \\
\tfrac{1}{2}R_1 + R_3 \to R_3
\end{array}
\left[\begin{array}{ccccc|c}
1 & 0 & \frac{2}{5} & -\frac{1}{5} & 0 & 1 \\
0 & 1 & -\frac{1}{5} & \frac{3}{5} & 0 & 3 \\
\hline
0 & 0 & \frac{1}{5} & \frac{7}{5} & 1 & 11
\end{array}\right]
$$

As we saw in Example 2, the optimal (maximum) value is $z = 11$ and occurs at $x_1 = 1$, $x_2 = 3$, and $s_1 = s_2 = 0$. ∎

Look carefully at the bottom row in the final tableau in Example 3. In particular, the solution to the **primal minimization problem** in Example 2 is found in the bottom row and in the slack variable columns. As shown in Example 2 and Table 4.3.1, the primal minimization problem has a minimum value of 11 and it occurs at the corner point $(\frac{1}{5}, \frac{7}{5})$, precisely the values in the slack variable columns. This astonishing result is no accident!

An optimal solution to a primal minimum problem can be obtained from the bottom row of the final simplex tableau for the dual maximum problem.

We now have a procedure for solving a minimum problem. It requires applying the simplex method to its dual maximum problem. We summarize this procedure in the following box.

> ■ **Solving Minimization Problems Using Duals**
>
> Given a standard minimum form linear programming problem:
>
> 1. Form the dual, which is a standard maximum form problem.
> 2. Solve the standard maximum form linear programming problem using the simplex method.
> 3. Read the solution to the minimum problem from the bottom row of the final simplex tableau in step 2. Specifically, the solution occurs in the slack variable columns.

▶ **Note:** If the dual maximum problem has no optimal solution, then the primal minimum problem has no solution.

Example 4 **Solving a Minimum Problem**

Solve the following minimum problem by using the dual.

$$\text{Minimize:} \quad w = 3y_1 + 7y_2$$
$$\text{Subject to:} \quad 4y_1 + \ y_2 \geq 8$$
$$5y_1 + 2y_2 \geq 1$$
$$y_1 \geq 0, \, y_2 \geq 0$$

Solution

This is the problem from Example 1, where we found the dual to be

$$\text{Maximize:} \quad z = 8x_1 + x_2$$
$$\text{Subject to:} \quad 4x_1 + 5x_2 \leq 3$$
$$x_1 + 2x_2 \leq 7$$
$$x_1 \geq 0, \, x_2 \geq 0$$

Introducing slack variables and converting to equational form yields

$$4x_1 + 5x_2 + s_1 \qquad\qquad = 3$$
$$x_1 + 2x_2 \qquad + s_2 \qquad = 7$$
$$-8x_1 - \ x_2 \qquad\qquad + z = 0$$

The initial simplex tableau is

$$
\begin{array}{ccccc}
x_1 & x_2 & s_1 & s_2 & z \\
\end{array}
$$
$$
\left[
\begin{array}{ccccc|c}
4 & 5 & 1 & 0 & 0 & 3 \\
1 & 2 & 0 & 1 & 0 & 7 \\
\hline
-8 & -1 & 0 & 0 & 1 & 0 \\
\end{array}
\right]
$$

We pivot on the 4 as follows:

$$
\begin{array}{ccccc}
x_1 & x_2 & s_1 & s_2 & z \\
\end{array}
$$
$$
\left[
\begin{array}{ccccc|c}
④ & 5 & 1 & 0 & 0 & 3 \\
1 & 2 & 0 & 1 & 0 & 7 \\
\hline
-8 & -1 & 0 & 0 & 1 & 0 \\
\end{array}
\right]
\quad \tfrac{1}{4}R_1 \to R_1 \quad
\left[
\begin{array}{ccccc|c}
1 & \frac{5}{4} & \frac{1}{4} & 0 & 0 & \frac{3}{4} \\
1 & 2 & 0 & 1 & 0 & 7 \\
\hline
-8 & -1 & 0 & 0 & 1 & 0 \\
\end{array}
\right]
$$

$$
\begin{array}{cc}
& x_1 \quad x_2 \quad\;\; s_1 \quad\; s_2 \;\; z \\
\begin{array}{c}
-R_1 + R_2 \to R_2 \\
8R_1 + R_3 \to R_3
\end{array}
&
\left[
\begin{array}{ccccc|c}
1 & \frac{5}{4} & \frac{1}{4} & 0 & 0 & \frac{3}{4} \\
0 & \frac{3}{4} & -\frac{1}{4} & 1 & 0 & \frac{25}{4} \\
\hline
0 & 9 & 2 & 0 & 1 & 6 \\
\end{array}
\right]
\end{array}
$$

This is our final tableau. We can read from the bottom row that the solution to the minimum problem is

$$\text{Minimum is } w = 6 \text{ at } y_1 = 2,\ y_2 = 0$$ ■

Interactive Activity

Use the final tableau in Example 4 to read the solution to the dual maximum problem.

Checkpoint 2

Now work Exercise 17.

Applications

Our first application is a Flashback to an example considered in Section 3.4.

 Flashback

Wendy's Revisited

Table 4.3.3 gives nutritional information and current prices for selected items from Wendy's menu.

Table 4.3.3

Item	Protein (g)	Fiber (g)	Price
Breaded Chicken Sandwich	28	2	$2.89
Biggie® fries	7	6	$1.19

SOURCE: Wendy's, www.wendys.com

Joshua desires to eat nothing but Breaded Chicken Sandwiches and Biggie® fries. He wants at least 112 grams of protein and at least 30 grams of fiber daily. If Joshua eats nothing but Breaded Chicken Sandwiches and

Biggie® fries and wishes to meet his minimum nutrient requirements, how much of each food should he eat to minimize daily food costs?

Flashback Solution

Understand the Situation: We let y_1 represent the number of Breaded Chicken Sandwiches and y_2 represent the number of orders of Biggie® fries. The primal problem is

$$\begin{aligned} \text{Minimize:} \quad & C = 2.89y_1 + 1.19y_2 \\ \text{Subject to:} \quad & 28y_1 + 7y_2 \geq 112 \\ & 2y_1 + 6y_2 \geq 30 \\ & y_1 \geq 0,\ y_2 \geq 0 \end{aligned}$$

To get the dual maximum problem, we need the augmented matrix A for the given primal problem. Remember that the objective function goes on the bottom row.

$$A = \begin{bmatrix} 28 & 7 & | & 112 \\ 2 & 6 & | & 30 \\ \hline 2.89 & 1.19 & | & 0 \end{bmatrix}$$

We now determine A^T, the transpose of A, to be

$$A^T = \begin{bmatrix} 28 & 2 & | & 2.89 \\ 7 & 6 & | & 1.19 \\ \hline 112 & 30 & | & 0 \end{bmatrix}$$

This gives the dual maximum problem

$$\begin{aligned} \text{Maximize:} \quad & z = 112x_1 + 30x_2 \\ \text{Subject to:} \quad & 28x_1 + 2x_2 \leq 2.89 \\ & 7x_1 + 6x_2 \leq 1.19 \\ & x_1 \geq 0,\ x_2 \geq 0 \end{aligned}$$

Introducing slack variables and converting to equational form gives us

$$\begin{aligned} 28x_1 + 2x_2 + s_1 \qquad\qquad &= 2.89 \\ 7x_1 + 6x_2 \qquad + s_2 \quad &= 1.19 \\ -112x_1 - 30x_2 \qquad\qquad + z &= 0 \end{aligned}$$

The initial simplex tableau is

$$\begin{array}{ccccc} x_1 & x_2 & s_1 & s_2 & z \end{array}$$
$$\begin{bmatrix} 28 & 2 & 1 & 0 & 0 & | & 2.89 \\ 7 & 6 & 0 & 1 & 0 & | & 1.19 \\ \hline -112 & -30 & 0 & 0 & 1 & | & 0 \end{bmatrix}$$

We leave the pivoting details for you in the Interactive Activity on p. 229. The final simplex tableau is

$$\begin{array}{ccccc} x_1 & x_2 & s_1 & s_2 & z \end{array}$$
$$\begin{bmatrix} 1 & 0 & 0.039 & -0.013 & 0 & | & 0.0971 \\ 0 & 1 & -0.0455 & 0.1818 & 0 & | & 0.085 \\ \hline 0 & 0 & 3 & 4 & 1 & | & 13.43 \end{bmatrix} \quad \text{Entries rounded}$$

We can read from the bottom row that the solution to the primal minimum problem is

$$C = 13.43, \qquad y_1 = 3, \qquad y_2 = 4$$

Interpret the Solution: This means that to minimize daily food costs, while still meeting the daily nutrient requirements, Joshua should eat 3 Breaded Chicken Sandwiches and 4 orders of Biggie® fries. His daily minimum cost is $13.43.

Sensitivity Analysis and Shadow Variables

In some economic applications, the variables in the dual of a linear programming problem measure something useful and are called **shadow variables**. These variables can be used to measure the **sensitivity** of the optimal solution to small changes in the problem requirements.

For example, consider the scenario in the Flashback and Table 4.3.3. If we consider price to be price per unit (that is, price per Breaded Chicken Sandwich or price per Biggie® fries), then x_1 is the value of each gram of protein and x_2 is the value of each gram of fiber. The total value of a Breaded Chicken Sandwich is $28x_1 + 2x_2$, the value of protein plus the value of fiber. This value can be no greater than $2.89, the price of a Breaded Chicken Sandwich. So $28x_1 + 2x_2 \leq 2.89$. Similarly, for Biggie® fries we have $7x_1 + 6x_2 \leq 1.19$.

In the Flashback, Joshua wanted at least 112 grams of protein and at least 30 grams of fiber. It stands to reason that we would want to maximize $112x_1 + 30x_2$, which is the total value of protein and fiber. This means that we want to

$$
\begin{aligned}
\text{Maximize:} \quad & z = 112x_1 + 30x_2 \\
\text{Subject to:} \quad & 28x_1 + 2x_2 \leq 2.89 \\
& 7x_1 + 6x_2 \leq 1.19 \\
& x_1 \geq 0, x_2 \geq 0
\end{aligned}
$$

But this is simply the dual maximum problem that we solved in the Flashback! The final tableau tells us that $x_1 = 0.0971$ and $x_2 = 0.085$. This means that a gram of protein is worth $0.0971 and a gram of fiber is worth $0.085. The maximum nutrient value is $13.43, the same as the minimum daily cost. Observe that the minimum daily cost, $13.43, can be found as follows:

$$
\begin{aligned}
(\$0.0971 \text{ per gram of protein}) \cdot (112 \text{ grams of protein}) &= \$10.8752 \\
+ (\$0.085 \text{ per gram of fiber}) \cdot (30 \text{ grams of fiber}) &= \quad 2.55 \\
&= \$13.4252 \approx \$13.43
\end{aligned}
$$

The numbers $x_1 = 0.0971$ and $x_2 = 0.085$ are called **shadow costs** of the nutrients. These two numbers from the dual (see the final tableau in the Flashback) measure the sensitivity of the optimal solution to changes in the requirements of protein and fiber. Each extra gram of protein costs about $0.0971 and each extra gram of fiber costs about $0.085. For example, if Joshua desires 113 grams of protein and 31 grams of fiber, his new total cost is

$$
\begin{array}{ll}
\$13.43 & (\text{112 grams of protein and 30 grams of fiber}) \\
0.0971 & (\text{extra gram of protein}) \\
+ \quad 0.085 & (\text{extra gram of fiber}) \\
\hline
\$13.6121 \approx \$13.61
\end{array}
$$

SUMMARY

In this section we saw the **standard minimum form** of a linear programming problem. We saw how the **transpose of a matrix** was key in constructing the **dual**. If the **primal** linear programming problem is a standard minimum form, then the dual is a standard maximum form. The primal and the dual have the same optimal value as stated in the Duality Theorem. We supplied a strategy for solving minimization problems using duals. The section concluded with an application and a look at **shadow variables**.

- **Duality Theorem**   The objective function of a minimum linear programming problem has an optimal (minimum) value if and only if the objective function of the associated dual maximum problem has an optimal (maximum)

value. Furthermore, the optimal (minimum) value of the minimization problem equals the optimal (maximum) value of the associated dual maximization problem.

- **Solving Minimum Problems**   Given a standard minimum form linear programming problem:
  1. Form the dual, which is a standard maximum form problem.
  2. Solve the standard maximum form linear programming problem using the simplex method.
  3. Read the solution to the minimum problem from the bottom row of the final simplex tableau in step 2. Specifically, the solution occurs in the slack variable columns.

SECTION 4.3 EXERCISES

In Exercises 1–6 determine the transpose of the given matrix.

1. $\begin{bmatrix} 2 & 1 & 3 \\ 4 & 1 & 2 \end{bmatrix}$

2. $\begin{bmatrix} 3 & 2 & 1 & 0 & 2 \\ 1 & 4 & 0 & 1 & 0 \end{bmatrix}$

3. $\begin{bmatrix} 5 & 7 & 6 & 1 \\ 3 & 2 & 1 & 0 \\ 0 & 1 & 1 & 2 \end{bmatrix}$

4. $\begin{bmatrix} 1 & 3 \\ 2 & 4 \\ 5 & 7 \end{bmatrix}$

5. $\begin{bmatrix} 4 & 1 & 3 \\ 1 & 2 & 7 \\ 8 & 1 & 9 \end{bmatrix}$

6. $\begin{bmatrix} 2 & 1 \\ 3 & 9 \\ 8 & 0 \\ 5 & 4 \end{bmatrix}$

In Exercises 7–12, form the dual problem for the given linear programming problem. See Example 1.

✔ **7.** Minimize: $w = 2y_1 + 3y_2$
   Subject to: $5y_1 + 2y_2 \geq 4$
                   $3y_1 + y_2 \geq 7$
                   $y_1 \geq 0, y_2 \geq 0$

8. Minimize: $w = 7y_1 + 3y_2$
   Subject to: $8y_1 + 10y_2 \geq 15$
                   $y_1 + y_2 \geq 4$
                   $y_1 \geq 0, y_2 \geq 0$

9. Minimize: $w = y_1 + 3y_2 + 4y_3$
   Subject to: $2y_1 + y_2 + 4y_3 \geq 0$
                   $8y_1 + y_2 + y_3 \geq 7$
                   $5y_1 + 2y_2 + 7y_3 \geq 11$
                   $y_1 \geq 0, y_2 \geq 0, y_3 \geq 0$

10. Minimize: $w = 2y_1 + y_2 + 5y_3$
   Subject to: $y_1 + 3y_2 + y_3 \geq 9$
                   $2y_1 + 5y_2 + y_3 \geq 8$
                   $y_1 + 2y_2 + 4y_3 \geq 10$
                   $y_1 \geq 0, y_2 \geq 0, y_3 \geq 0$

11. Minimize: $w = 2y_1 + y_2 + 3y_3$
   Subject to: $5y_1 + y_2 \geq 8$
                   $3y_1 + 8y_3 \geq 7$
                   $y_2 + y_3 \geq 11$
                   $y_1 \geq 0, y_2 \geq 0, y_3 \geq 0$

12. Minimize: $w = 8y_1 + y_2 + 5y_3$
   Subject to: $y_1 + 3y_2 \geq 9$
                   $2y_1 + 4y_3 \geq 8$
                   $3y_2 + y_3 \geq 10$
                   $y_1 \geq 0, y_2 \geq 0, y_3 \geq 0$

In Exercises 13–16, a linear programming problem is given along with the final simplex tableau for the dual problem. Use the final simplex tableau for the dual to:

(a) Find the solution to the primal problem.

(b) Find the solution to the dual problem.

13. Minimize: $w = y_1 + 5y_2$
   Subject to: $y_1 + 2y_2 \geq 6$
                   $3y_1 + y_2 \geq 9$
                   $y_1 \geq 0, y_2 \geq 0$

x_1	x_2	s_1	s_2	z	
1	3	1	0	0	1
0	−5	−2	1	0	3
0	9	6	0	1	6

14. Minimize: $w = 3y_1 + y_2$
   Subject to: $2y_1 + 3y_2 \geq 6$
                   $y_1 + 2y_2 \geq 4$
                   $y_1 \geq 0, y_2 \geq 0$

x_1	x_2	s_1	s_2	z	
0	$-\frac{1}{3}$	1	$-\frac{2}{3}$	0	$\frac{7}{3}$
1	$\frac{2}{3}$	0	$\frac{1}{3}$	0	$\frac{1}{3}$
0	0	0	2	1	2

15. Minimize: $w = 2y_1 + 3y_2$

Subject to: $y_1 + y_2 \geq 12$

$3y_1 + y_2 \geq 9$

$y_1 + 2y_2 \geq 6$

$y_1 \geq 0, y_2 \geq 0$

$$
\begin{array}{cccccc}
x_1 & x_2 & x_3 & s_1 & s_2 & z \\
\left[\begin{array}{cccccc|c}
1 & 3 & 1 & 1 & 0 & 0 & 2 \\
0 & -2 & 1 & -1 & 1 & 0 & 1 \\
\hline
0 & 27 & 6 & 12 & 0 & 1 & 24
\end{array}\right]
\end{array}
$$

16. Minimize: $w = 4y_1 + y_2$

Subject to: $2y_1 + y_2 \geq 8$

$y_1 + 3y_2 \geq 12$

$3y_1 + 2y_2 \geq 6$

$y_1 \geq 0, y_2 \geq 0$

$$
\begin{array}{cccccc}
x_1 & x_2 & x_3 & s_1 & s_2 & z \\
\left[\begin{array}{cccccc|c}
0 & -5 & -1 & 1 & -2 & 0 & 2 \\
1 & 3 & 2 & 0 & 1 & 0 & 1 \\
\hline
0 & 12 & 10 & 0 & 8 & 1 & 8
\end{array}\right]
\end{array}
$$

In Exercises 17–30:

(a) Construct the dual problem.

(b) Use the simplex method on the dual problem to find the solution to the original primal problem.

✓ **17.** Minimize: $w = 7y_1 + 3y_2$

Subject to: $8y_1 + 10y_2 \geq 15$

$y_1 + y_2 \geq 4$

$y_1 \geq 0, y_2 \geq 0$

18. Minimize: $w = 2y_1 + 3y_2$

Subject to: $5y_1 + 2y_2 \geq 4$

$3y_1 + y_2 \geq 7$

$y_1 \geq 0, y_2 \geq 0$

19. Minimize: $w = y_1 + 4y_2$

Subject to: $y_1 + 2y_2 \geq 5$

$y_1 + 3y_2 \geq 6$

$y_1 \geq 0, y_2 \geq 0$

20. Minimize: $w = 9y_1 + 2y_2$

Subject to: $4y_1 + y_2 \geq 13$

$3y_1 + y_2 \geq 12$

$y_1 \geq 0, y_2 \geq 0$

21. Minimize: $w = 2y_1 + y_2$

Subject to: $y_1 + y_2 \geq 8$

$y_1 + 2y_2 \geq 4$

$y_1 \geq 0, y_2 \geq 0$

22. Minimize: $w = 3y_1 + 9y_2$

Subject to: $2y_1 + y_2 \geq 8$

$y_1 + 2y_2 \geq 10$

$y_1 \geq 0, y_2 \geq 0$

23. Minimize: $w = 4y_1 + 5y_2$

Subject to: $2y_1 + 3y_2 \geq 29$

$5y_1 + 4y_2 \geq 55$

$2y_1 + y_2 \geq 18$

$y_1 \geq 0, y_2 \geq 0$

24. Minimize: $w = 5y_1 + 2y_2$

Subject to: $y_1 + 5y_2 \geq 15$

$2y_1 + 4y_2 \geq 24$

$y_1 + y_2 \geq 7$

$y_1 \geq 0, y_2 \geq 0$

25. Minimize: $w = 2y_1 + 3y_2$

Subject to: $15y_1 + y_2 \geq 29$

$y_1 + 6y_2 \geq 20$

$8y_1 + 7y_2 \geq 78$

$y_1 \geq 0, y_2 \geq 0$

26. Minimize: $w = 2y_1 + 7y_2$

Subject to: $y_1 + 4y_2 \geq 40$

$3y_1 + y_2 \geq 36$

$y_1 + y_2 \geq 5$

$y_1 \geq 0, y_2 \geq 0$

27. Minimize: $w = 14y_1 + 8y_2 + 20y_3$

Subject to: $y_1 + y_2 + 3y_3 \geq 6$

$2y_1 + y_2 + y_3 \geq 9$

$y_1 \geq 0, y_2 \geq 0, y_3 \geq 0$

28. Minimize: $w = 5y_1 + 7y_2 + 12y_3$

Subject to: $y_1 + y_2 + 2y_3 \geq 7$

$2y_1 + y_2 + y_3 \geq 4$

$y_1 \geq 0, y_2 \geq 0, y_3 \geq 0$

29. Minimize: $w = 4y_1 + 10y_2 + 9y_3$

Subject to: $y_1 + 3y_2 + y_3 \geq 6$

$y_1 + y_2 + 3y_3 \geq 9$

$3y_1 + y_2 + y_3 \geq 3$

$y_1 \geq 0, y_2 \geq 0, y_3 \geq 0$

30. Minimize: $w = 2y_1 + 3y_2 + 4y_3$

Subject to: $y_1 + 4y_2 + 2y_3 \geq 16$

$y_1 + 3y_2 + 4y_3 \geq 7$

$2y_1 + y_2 + 3y_3 \geq 6$

$y_1 \geq 0, y_2 \geq 0, y_3 \geq 0$

Applications

31. Richie needs to supplement his diet with at least 50 milligrams (mg) of vitamin C and 5 mg of niacin daily. The nutrients are available in two supplement pills, AllDay and Live!. AllDay contains 5 mg of vitamin C and 1 mg of niacin, while Live! contains 10 mg of vitamin C and 2 mg of niacin. Each AllDay pill costs $0.08 and each Live! pill costs $0.09. Let y_1 represent the number of AllDay pills taken daily, and let y_2 represent the number of Live! pills taken daily.

(a) Write a linear programming problem to minimize the cost of adding the nutrients to Richie's diet.

(b) Write the dual of the linear programming problem in part (a).

(c) Use the simplex method on the dual problem in part (b) to solve the original problem in part (a). Your answer should include the number of each pill taken and the daily minimum cost.

32. Roberto needs to supplement his diet with at least 30 mg of vitamin C and 10 mg of niacin daily. The nutrients are available in two supplement pills, GoodDay and Now!. GoodDay contains 3 mg of vitamin C and 2 mg of niacin, while Now! contains 6 mg of vitamin C and 1 mg of niacin. Each GoodDay pill costs $0.06 and each Now! pill costs $0.08. Let y_1 represent the number of GoodDay pills taken daily, and let y_2 represent the number of Now! pills taken daily.

(a) Write a linear programming problem to minimize the cost of adding the nutrients to Roberto's diet.

(b) Write the dual of the linear programming problem in part (a).

(c) Use the simplex method on the dual problem in part (b) to solve the original problem in part (a). Your answer should include the number of each pill taken and the daily minimum cost.

33. Elle, a beagle breeder, can buy a special food, Shine, at $0.30 per pound and a special food, Lustre, at $0.40 per pound. Each pound of Shine has 210 units of protein and 150 units of fat. Each pound of Lustre has 240 units of protein and 380 units of fat. The minimum daily requirements for the beagles (collectively) are 2115 units of protein and 3460 units of fat. Let y_1 represent the number of pounds of Shine used daily and y_2 represent the number of pounds of Lustre used daily.

(a) Write a linear programming problem to minimize Elle's daily food costs while still meeting (or even exceeding) the minimum daily requirements.

(b) Write the dual of the linear programming problem in part (a).

(c) Use the simplex method on the dual problem in part (b) to solve the original problem in part (a). Your answer should include the number of pounds of each mix and the daily minimum cost.

34. (*continuation of Exercise 33*) Ginger, Elle's cousin from Exercise 33, also breeds beagles but in a different state. Ginger can buy Shine at $0.40 per pound and Lustre at $0.50 per pound. Everything else from Exercise 33 remains the same.

(a) Write a linear programming problem to minimize Ginger's daily food costs while still meeting (or even exceeding) the minimum daily requirements.

(b) Write the dual of the linear programming problem in part (a).

(c) Use the simplex method on the dual problem in part (b) to solve the original problem in part (a). Your answer should include the number of pounds of each mix and the daily minimum cost.

35. The AJA Mining Company owns two different mines that produce ore, which is crushed and graded into three classes: high, medium, and low grade. The company has a contract to provide a smelting plant with at least 13 tons of high-grade, 9 tons of medium-grade, and 24 tons of low-grade ore per week. The two mines have different operating characteristics as given in Table 4.3.4.

Table 4.3.4

Mine	Cost per Day (in $1000's)	Production (tons per day)		
		High	**Medium**	**Low**
I	12	3	3	4
II	10	2	1	6

(a) Define the variables and write a linear programming problem to minimize AJA's total operating cost while still meeting (or even exceeding) the smelting plant contract.

(b) Write the dual of the linear programming problem in part (a).

(c) Use the simplex method on the dual problem in part (b) to solve the original problem in part (a). Your answer should include the number of days that each mine should operate and the minimum weekly cost.

36. (*continuation of Exercise 35*) Suppose that the contract with the smelting plant is changed so that the AJA Mining Company needs to provide at least 15 tons of high-grade, 11 tons of medium-grade, and 30 tons of low-grade ore per week. Everything else from Exercise 35 remains the same.

(a) Write a linear programming problem to minimize AJA's total operating cost while still meeting (or even exceeding) the smelting plant contract.

(b) Write the dual of the linear programming problem in part (a).

(c) Use the simplex method on the dual problem in part (b) to solve the original problem in part (a). Your answer should include the number of days that each mine should operate and the minimum weekly cost.

37. Hoofer Oil has two refineries, one in Findlay, Ohio, and one in Ft. Wayne, Indiana. Each refinery can refine the amounts of fuel, in barrels, in a 24-hour period as shown in Table 4.3.5.

Table 4.3.5

Maximum Production of:	Findlay	Ft. Wayne
Low-octane gasoline	800	1000
High-octane gasoline	500	600
Diesel	1000	600

Daily operating costs for the Findlay refinery are $28,000 and daily operating costs for the Ft. Wayne refinery are $33,000. Zachary, the production manager of Hoofer Oil, receives an order for 60,000 barrels of low-octane gasoline, 120,000 barrels of high-octane gasoline, and 150,000 barrels of diesel.

(a) How many days should Hoofer Oil operate each refinery to fulfill the order while minimizing cost?

(b) What is the minimum cost?

38. Shannon Oil has two refineries, one in Lima, Ohio, and one in Detroit, Michigan. Each refinery can refine the amounts of fuel, in barrels, in a 24-hour period as shown in Table 4.3.6.

Table 4.3.6

Maximum Production of:	Lima	Detroit
Low-octane gasoline	900	1100
High-octane gasoline	600	700
Diesel	1100	700

Daily operating costs for the Lima refinery are $35,000 and daily operating costs for the Detroit refinery are $30,000. Joshua, the production manager of Shannon Oil, receives an order for 50,000 barrels of low-octane gasoline, 90,000 barrels of high-octane gasoline, and 130,000 barrels of diesel.

(a) How many days should Shannon Oil operate each refinery to fulfill the order while minimizing cost?

(b) What is the minimum cost?

39. Anita, a dietitian at a local hospital, is to prepare a diet to contain at least 400 units of vitamins C and E, 500 units of minerals (such as zinc and iron), and 1400 calories. She has two foods available, Flakes and Woday. An ounce of Flakes contains 2 units of vitamin C and E, 1 unit of minerals, and 4 calories and costs $0.05. An ounce of Woday contains 1 unit of vitamins C and E, 2 units of minerals, and 4 calories and costs $0.03.

(a) How many ounces of each food should be used to minimize daily food cost?

(b) What is the minimum daily food cost?

40. Lisa, a dietitian at a local hospital, is to prepare two foods to meet certain nutritional requirements. Each pound of Vaca contains 200 units of vitamin C, 80 units of vitamin B_{12}, and 20 units of vitamin E and costs $0.40. Each pound of Carne contains 20 units of vitamin C, 160 units of vitamin B_{12}, and 10 units of vitamin E and costs $0.30. To meet the nutritional needs of the patients, the combination of the two foods must have at least 520 units of vitamin C, 640 units of vitamin B_{12}, and 100 units of vitamin E.

(a) How many pounds of each food should be used to minimize daily food cost?

(b) What is the minimum daily food cost?

41. Table 4.3.7 gives nutritional information in grams for selected items from Taco Bell's menu.

Table 4.3.7

Item	Protein	Fiber	Total Fat
Burrito Supreme®-Beef	17	9	18
Double Burrito Supreme®-Steak	28	3	18

SOURCE: Taco Bell, www.tacobell.com

Warren wants to eat nothing but Burrito Supreme®-Beef and Double Burrito Supreme®-Steak. He wants at least 118 grams of protein and at least 27 grams of fiber daily. If Warren eats nothing but Burrito Supreme®-Beef and Double Burrito Supreme®-Steak and meets his minimum nutrient requirements, answer the following.

(a) How much of each food should Warren eat to minimize his daily total fat intake?

(b) What is this minimum daily total fat intake?

42. Table 4.3.8 gives the nutritional information in grams for selected items from McDonald's menu.

Table 4.3.8

Item	Carbohydrates	Protein	Fat
Cheeseburger	35	16	13
Large French fries	68	8	26
Chocolate shake	60	11	9

SOURCE: McDonald's, www.mcdonalds.com

Diane wishes to eat nothing but cheeseburgers, large French fries, and chocolate shakes. If she wishes to have at least 450 grams of carbohydrates and 80 grams of protein a day, how many of each should she have to minimize her fat intake and still meet her requirements?

Exercises 43–46 pertain to shadow costs. These exercises refer back to previously worked exercises.

43. (a) Determine the shadow costs of the nutrients in Exercise 31.

(b) Use the shadow costs along with the minimum cost answer in Exercise 31(c) to determine the new total cost if Richie changes his nutrient requirements to 51 mg of vitamin C and 6 mg of niacin daily.

44. (a) Determine the shadow costs of the nutrients in Exercise 32.

(b) Use the shadow costs along with the minimum cost answer in Exercise 32(c) to determine the new total cost if Roberto changes his nutrient requirements to 31 mg of vitamin C and 11 mg of niacin daily.

45. (a) Determine the shadow costs of the nutrients in Exercise 33.

(b) Use the shadow costs along with the minimum cost answer in Exercise 33(c) to determine the new total cost if Elle changes the beagles' nutrient requirements to 2116 units of protein and 3461 units of fat daily.

46. (a) Determine the shadow costs of the nutrients in Exercise 34.

(b) Use the shadow costs along with the minimum cost answer in Exercise 33(c) to determine the new total cost if Ginger changes the beagles' nutrient requirements to 2116 units of protein and 3461 units of fat daily.

■ SECTION PROJECT

Ty and Mike are management analysts at a Houston, Texas, laboratory. Table 4.3.9 gives information that Ty and Mike need.

Table 4.3.9			
Experiment	**Biophysicist**	**Biochemist**	**Minimum Needed**
Test 1	8	4	120
Test 2	4	6	115
Test 3	9	4	116

These numbers mean that a biophysicist can complete 8, 4, and 9 of Tests 1, 2, and 3, respectively, per hour. Similarly, a biochemist can complete 4, 6, and 4 of Tests 1, 2, and 3 per

hour. The lab needs a minimum of 120 of Test 1, 115 of Test 2, and 116 of Test 3 done each day. A biophysicist costs $23 per hour, and a biochemist costs $18 per hour.

(a) Define the variables and write a linear programming problem to minimize costs.

(b) Write the dual of the problem in part (a).

(c) Use the simplex method on the dual in part (b) to solve the original problem in part (a). Your answer should include the number of hours for each scientist type and the minimum cost each day.

(d) Determine the solutions to the dual problem in part (b).

(e) Interpret the meaning of the dual and its solution.

(f) Are there any shadow variables in the final simplex tableau in part (a)? Explain.

Section 4.4   Simplex Method: Nonstandard Problems

In Section 4.2, all the linear programming problems that we encountered were of the standard maximum form variety, and in Section 4.3 we saw how the dual was instrumental in solving standard minimum form linear programming problems. In this section we discuss **nonstandard problems**, that is, problems with **mixed constraints** (constraints involving ≤ and ≥). We also discuss another technique for solving minimization problems. This discussion requires the introduction of a new type of variable called a **surplus variable**. We introduce mixed constraints through a Flashback.

Flashback

Dub Inc. Revisited

We were first introduced to Dub, Inc., in Section 3.4 (and revisited in Section 4.2). Recall that Dub, Inc., is a small manufacturing firm that produces two microwave switches, switch A and switch B. Each switch passes through two departments: assembly and testing. Switch A requires 4 hours of assembly and 1 hour of testing, and switch B requires 3 hours of assembly and 2 hours of testing. The maximum numbers of hours available each week for assembly and testing are 240 hours and 140 hours, respectively. The per unit profit for switch A is $20, whereas the per unit profit for switch B is $30. Now suppose that the total number of switches made each week must be at least 10. Graphically determine how many of each switch should be made and sold each week to maximize profit.

Flashback Solution

Understand the Situation: Since we do not know how many of each switch should be produced and sold each week, we let x_1 represent the number of units of switch A made each week and x_2 represent the number of units of switch B made each week. Our mathematical model for this linear programming problem is

$$\text{Maximize:} \quad P = 20x_1 + 30x_2$$
$$\text{Subject to:} \quad 4x_1 + 3x_2 \leq 240$$
$$x_1 + 2x_2 \leq 140$$
$$x_1 + x_2 \geq 10$$
$$x_1 \geq 0, x_2 \geq 0$$

The feasible region is shown in Figure 4.4.1.

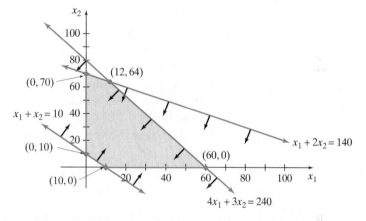

Figure 4.4.1

Table 4.4.1 shows the result of each corner point evaluated in the objective function.

Table 4.4.1

Corner Point	Value of $P = 20x_1 + 30x_2$	
$(10, 0)$	$20(10) + 30(0) = 200$	
$(0, 10)$	$20(0) + 30(10) = 300$	
$(0, 70)$	$20(0) + 30(70) = 2100$	
$(12, 64)$	$20(12) + 30(64) = 2160$	**Maximum**
$(60, 0)$	$20(60) + 30(0) = 1200$	

The optimal (maximum) value in Table 4.4.1 is 2160 and occurs at the corner point $(12, 64)$.

Interpret the Solution: This means that to maximize weekly profit Dub, Inc., should produce and sell 12 units of switch A and 64 units of switch B. This generates a maximum weekly profit of $2160.

Notice that the problem in the Flashback is not a standard maximum form problem since the third inequality, $x_1 + x_2 \geq 10$, does not have the $\leq$ symbol. However, if the total number of switches made is greater than 10 (and it was in the Flashback), then we have some *surplus* switches made. These surplus switches are switches beyond the necessary 10, and we can write

$$(\text{switch A}) + (\text{switch B}) - (\text{surplus switches}) = 10$$

If we let s_3 represent the surplus switches, we can write the above as

$$x_1 + x_2 - s_3 = 10$$

Notice that this equation is simply the third constraint in the Flashback written in equational form. Here s_3 is an example of a **surplus variable**. As with slack variables, surplus variables are nonnegative.

Slack and Surplus Variables

- We **add** a **slack** variable to the left side of $a_1x_1 + a_2x_2 + \cdots + a_nx_n \leq b$ to pick up the slack and create an equation of the form $a_1x_1 + a_2x_2 + \cdots + s_1 = b$.
- We **subtract** a **surplus** variable from the left side of $a_1x_1 + a_2x_2 + \cdots + a_nx_n \geq b$ to take away the excess and create an equation of the form $a_1x_1 + a_2x_2 + \cdots - s_2 = b$.
- Both slack and surplus variables are nonnegative.

Example 1 **Using Slack and Surplus Variables**

Use slack and surplus variables to write the mathematical model in the Flashback in equational form.

Solution

Using s_1 and s_2 for slack variables and s_3 as the surplus variable, we have

$$\begin{aligned}
\text{Maximize:}&\quad P = 20x_1 + 30x_2 \\
\text{Subject to:}&\quad 4x_1 + 3x_2 + s_1 && = 240 \\
&\quad x_1 + 2x_2 && + s_2 && = 140 \\
&\quad x_1 + x_2 && - s_3 = 10
\end{aligned}$$

■

✓ **Checkpoint 1**   Now work Exercise 3.

Let's set up the initial simplex tableau for the system in Example 1. It is

$$
\begin{array}{cccccc}
x_1 & x_2 & s_1 & s_2 & s_3 & P \\
\end{array}
$$

$$
\left[
\begin{array}{ccccccc|c}
4 & 3 & 1 & 0 & 0 & 0 & 240 \\
1 & 2 & 0 & 1 & 0 & 0 & 140 \\
1 & 1 & 0 & 0 & -1 & 0 & 10 \\
\hline
-20 & -30 & 0 & 0 & 0 & 1 & 0
\end{array}
\right]
$$

The row operation $-1R_3 \rightarrow R_3$ gives us

$$
\begin{array}{cccccc}
x_1 & x_2 & s_1 & s_2 & s_3 & P \\
\end{array}
$$

$$
\left[
\begin{array}{cccccc|c}
4 & 3 & 1 & 0 & 0 & 0 & 240 \\
1 & 2 & 0 & 1 & 0 & 0 & 140 \\
-1 & -1 & 0 & 0 & 1 & 0 & -10 \\
\hline
-20 & -30 & 0 & 0 & 0 & 1 & 0
\end{array}
\right]
$$

This tableau gives the solution

$$
x_1 = 0, \quad x_2 = 0, \quad s_1 = 240, \quad s_2 = 140, \quad s_3 = -10, \quad P = 0
$$

From Your Toolbox

A solution is feasible if all variables are nonnegative.

This solution is not **feasible** (consult the Toolbox to the left) since $s_3 = -10$.

Also, observe that the solution $x_1 = 0$ and $x_2 = 0$ corresponds to the point $(0, 0)$. Consulting Figure 4.4.1, we see that the point $(0, 0)$ is not in the feasible region. Why is this observation important? The answer is that the simplex method seeks out an optimal solution by moving from corner point to corner point in the feasible region! Hence, if our initial simplex tableau does not correspond to a feasible solution, we cannot use the simplex method. In order to use the simplex method, we must first get a feasible solution. This requires us to use what is called the **two-phase method**.

The Two-Phase Method

Most mixed constraint problems, such as Example 1, have two phases when solving the simplex method. These two phases are as follows:

- **Phase I: Locating a Corner Point**   Our goal is to locate a feasible solution. To do this we offer the following strategy:
  1. Select a row with a negative entry in the rightmost column.
  2. Pivot on a negative entry in that row.
  3. Repeat this process until our tableau has nonnegative entries in the rightmost column.

- **Phase II: Solving after Phase I**   After using Phase I to obtain a tableau that gives a feasible solution, we then apply the simplex method as usual.

Phase I will take us from the origin to a corner point. Phase II then takes us from corner point to corner point until the optimal value is obtained. In Phase I, if more than one row has a negative entry in the rightmost column, pick a row arbitrarily. To find the negative entry in the row to pivot on, we arbitrarily choose the leftmost one.

Example 2   **Using the Two-Phase Process**

Use the two-phase process on the following tableau to solve the linear programming problem in the Flashback.

$$
\begin{array}{cccccc}
x_1 & x_2 & s_1 & s_2 & s_3 & P \\
\end{array}
$$

$$
\left[
\begin{array}{cccccc|c}
4 & 3 & 1 & 0 & 0 & 0 & 240 \\
1 & 2 & 0 & 1 & 0 & 0 & 140 \\
-1 & -1 & 0 & 0 & 1 & 0 & -10 \\
\hline
-20 & -30 & 0 & 0 & 0 & 1 & 0
\end{array}
\right]
$$

Solution

We begin Phase I by pivoting on the -1 in row 3, column 1. Our first step is to use the row operation $-1R_3 \rightarrow R_3$.

$$
\begin{array}{c}

\end{array}
\begin{array}{cccccc}
x_1 & x_2 & s_1 & s_2 & s_3 & P
\end{array}
$$

$$
\left[
\begin{array}{cccccc|c}
4 & 3 & 1 & 0 & 0 & 0 & 240 \\
1 & 2 & 0 & 1 & 0 & 0 & 140 \\
\boxed{-1} & -1 & 0 & 0 & 1 & 0 & -10 \\
\hdashline
-20 & -30 & 0 & 0 & 0 & 1 & 0
\end{array}
\right]
\quad {-1R_3 \rightarrow R_3} \quad
\left[
\begin{array}{cccccc|c}
x_1 & x_2 & s_1 & s_2 & s_3 & P & \\
4 & 3 & 1 & 0 & 0 & 0 & 240 \\
1 & 2 & 0 & 1 & 0 & 0 & 140 \\
1 & 1 & 0 & 0 & -1 & 0 & 10 \\
\hdashline
-20 & -30 & 0 & 0 & 0 & 1 & 0
\end{array}
\right]
$$

$$
\begin{array}{r}
-4R_3 + R_1 \rightarrow R_1 \\
-R_3 + R_2 \rightarrow R_2 \\
20R_3 + R_4 \rightarrow R_4
\end{array}
\left[
\begin{array}{cccccc|c}
x_1 & x_2 & s_1 & s_2 & s_3 & P & \\
0 & -1 & 1 & 0 & 4 & 0 & 200 \\
0 & 1 & 0 & 1 & 1 & 0 & 130 \\
1 & 1 & 0 & 0 & -1 & 0 & 10 \\
\hdashline
0 & -10 & 0 & 0 & -20 & 1 & 200
\end{array}
\right]
$$

We now have the solution $x_1 = 10$, $x_2 = 0$, $s_1 = 200$, $s_2 = 130$, $s_3 = 0$, and $P = 200$. Since all variables are nonnegative, this is a basic feasible solution and we move to Phase II. [Observe that we are at the corner point $(10, 0)$ in Figure 4.4.1.] We begin Phase II by pivoting on the 4 in row 1, column 5.

$$
\left[
\begin{array}{cccccc|c}
x_1 & x_2 & s_1 & s_2 & s_3 & P & \\
0 & -1 & 1 & 0 & \boxed{4} & 0 & 200 \\
0 & 1 & 0 & 1 & 1 & 0 & 130 \\
1 & 1 & 0 & 0 & -1 & 0 & 10 \\
\hdashline
0 & -10 & 0 & 0 & -20 & 1 & 200
\end{array}
\right]
\quad {\tfrac{1}{4}R_1 \rightarrow R_1} \quad
\left[
\begin{array}{cccccc|c}
x_1 & x_2 & s_1 & s_2 & s_3 & P & \\
0 & -\tfrac{1}{4} & \tfrac{1}{4} & 0 & 1 & 0 & 50 \\
0 & 1 & 0 & 1 & 1 & 0 & 130 \\
1 & 1 & 0 & 0 & -1 & 0 & 10 \\
\hdashline
0 & -10 & 0 & 0 & -20 & 1 & 200
\end{array}
\right]
$$

$$
\begin{array}{r}
-R_1 + R_2 \rightarrow R_2 \\
R_1 + R_3 \rightarrow R_3 \\
20R_1 + R_4 \rightarrow R_4
\end{array}
\left[
\begin{array}{cccccc|c}
x_1 & x_2 & s_1 & s_2 & s_3 & P & \\
0 & -\tfrac{1}{4} & \tfrac{1}{4} & 0 & 1 & 0 & 50 \\
0 & \tfrac{5}{4} & -\tfrac{1}{4} & 1 & 0 & 0 & 80 \\
1 & \tfrac{3}{4} & \tfrac{1}{4} & 0 & 0 & 0 & 60 \\
\hdashline
0 & -15 & 5 & 0 & 0 & 1 & 1200
\end{array}
\right]
$$

Next we pivot on the $\tfrac{5}{4}$ in row 2, column 2.

$$
\left[
\begin{array}{cccccc|c}
x_1 & x_2 & s_1 & s_2 & s_3 & P & \\
0 & -\tfrac{1}{4} & \tfrac{1}{4} & 0 & 1 & 0 & 50 \\
0 & \boxed{\tfrac{5}{4}} & -\tfrac{1}{4} & 1 & 0 & 0 & 80 \\
1 & \tfrac{3}{4} & \tfrac{1}{4} & 0 & 0 & 0 & 60 \\
\hdashline
0 & -15 & 5 & 0 & 0 & 1 & 1200
\end{array}
\right]
\quad {\tfrac{4}{5}R_2 \rightarrow R_2} \quad
\left[
\begin{array}{cccccc|c}
x_1 & x_2 & s_1 & s_2 & s_3 & P & \\
0 & -\tfrac{1}{4} & \tfrac{1}{4} & 0 & 1 & 0 & 50 \\
0 & 1 & -\tfrac{1}{5} & \tfrac{4}{5} & 0 & 0 & 64 \\
1 & \tfrac{3}{4} & \tfrac{1}{4} & 0 & 0 & 0 & 60 \\
\hdashline
0 & -15 & 5 & 0 & 0 & 1 & 1200
\end{array}
\right]
$$

$$
\begin{array}{r}
\tfrac{1}{4}R_2 + R_1 \rightarrow R_1 \\
-\tfrac{3}{4}R_2 + R_3 \rightarrow R_3 \\
15R_2 + R_4 \rightarrow R_4
\end{array}
\left[
\begin{array}{cccccc|c}
x_1 & x_2 & s_1 & s_2 & s_3 & P & \\
0 & 0 & \tfrac{1}{5} & \tfrac{1}{5} & 1 & 0 & 66 \\
0 & 1 & -\tfrac{1}{5} & \tfrac{4}{5} & 0 & 0 & 64 \\
1 & 0 & \tfrac{2}{5} & -\tfrac{3}{5} & 0 & 0 & 12 \\
\hdashline
0 & 0 & 2 & 12 & 0 & 1 & 2160
\end{array}
\right]
$$

We have reached our final tableau. The solution is,

$$
x_1 = 12, \quad x_2 = 64, \quad s_1 = 0, \quad s_2 = 0, \quad s_3 = 66, \quad P = 2160 \qquad \blacksquare
$$

Interactive Activity

Interpret each variable value in the Example 2 solution in the context of the Flashback and Example 1. In particular, what does $s_3 = 66$ mean?

✓ **Checkpoint 2** Now work Exercise 11.

Before we supply a procedure for solving nonstandard problems, let's examine an alternative for minimization. To gain some insight, consider the following objective functions along with the feasible region in Figure 4.4.2.

$$\text{Minimize} \quad w = 3x + y$$
$$\text{Maximize} \quad z = -w = -3x - y$$

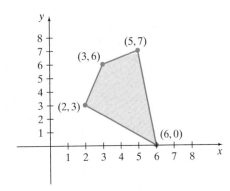

Figure 4.4.2

We evaluate each objective function at the corner points. See Table 4.4.2.

Table 4.2.2

Corner Point	$w = 3x + y$	$z = -3x - y$
(2, 3)	$3(2) + 3 = 9$	$-3(2) - 3 = -9$
(3, 6)	$3(3) + 6 = 15$	$-3(3) - 6 = -15$
(5, 7)	$3(5) + 7 = 22$	$-3(5) - 7 = -22$
(6, 0)	$3(6) + 0 = 18$	$-3(6) - 0 = -18$

Table 4.4.2 indicates that the minimum of w is 9 at the corner point (2, 3). The maximum of z is -9 at the corner point (2, 3). This illustrates the following fact:

The minimum value of an objective function w occurs at the same point as the maximum value of the objective function z, where $z = -w$, provided both have the same constraints.

We illustrate how to use this fact in Example 3.

Example 3   Minimizing With Mixed Constraints

$$\text{Minimize:} \quad w = 2x_1 + 5x_2$$
$$\text{Subject to:} \quad 2x_1 + 3x_2 \leq 5$$
$$4x_1 - x_2 \geq 3$$
$$x_1 \geq 0, x_2 \geq 0$$

Solution

We change this to a maximization problem by letting $z = -w$. We now have

$$\text{Maximize:} \qquad z = -w = -2x_1 - 5x_2$$
$$\text{Subject to:} \qquad 2x_1 + 3x_2 \le 5$$
$$4x_1 - x_2 \ge 3$$
$$x_1 \ge 0,\, x_2 \ge 0$$

Next we add slack variable s_1 and subtract the surplus variable s_2. This gives the equational form

$$2x_1 + 3x_2 + s_1 \qquad\qquad = 5$$
$$4x_1 - x_2 \qquad - s_2 \qquad = 3$$
$$2x_1 + 5x_2 \qquad\qquad + z = 0$$

The initial simplex tableau is

$$
\begin{array}{ccccc}
x_1 & x_2 & s_1 & s_2 & z
\end{array}
\left[\begin{array}{ccccc|c}
2 & 3 & 1 & 0 & 0 & 5 \\
4 & -1 & 0 & -1 & 0 & 3 \\
\hline
2 & 5 & 0 & 0 & 1 & 0
\end{array}\right]
\quad {-1R_2 \to R_2} \quad
\begin{array}{ccccc}
x_1 & x_2 & s_1 & s_2 & z
\end{array}
\left[\begin{array}{ccccc|c}
2 & 3 & 1 & 0 & 0 & 5 \\
-4 & 1 & 0 & 1 & 0 & -3 \\
\hline
2 & 5 & 0 & 0 & 1 & 0
\end{array}\right]
$$

The solution that this represents, $x_1 = x_2 = 0$, $s_1 = 5$, $s_2 = -3$, $z = 0$, is not feasible, so we move into Phase I. We pivot on the -4 in row 2, column 1.

$$
\begin{array}{ccccc}
x_1 & x_2 & s_1 & s_2 & z
\end{array}
\left[\begin{array}{ccccc|c}
2 & 3 & 1 & 0 & 0 & 5 \\
\boxed{-4} & 1 & 0 & 1 & 0 & -3 \\
\hline
2 & 5 & 0 & 0 & 1 & 0
\end{array}\right]
\quad {-\tfrac{1}{4}R_2 \to R_2} \quad
\begin{array}{ccccc}
x_1 & x_2 & s_1 & s_2 & z
\end{array}
\left[\begin{array}{ccccc|c}
2 & 3 & 1 & 0 & 0 & 5 \\
1 & -\tfrac{1}{4} & 0 & -\tfrac{1}{4} & 0 & \tfrac{3}{4} \\
\hline
2 & 5 & 0 & 0 & 1 & 0
\end{array}\right]
$$

$$
\begin{array}{l}
-2R_2 + R_1 \to R_1 \\
-2R_2 + R_3 \to R_3
\end{array}
\quad
\begin{array}{ccccc}
x_1 & x_2 & s_1 & s_2 & z
\end{array}
\left[\begin{array}{ccccc|c}
0 & \tfrac{7}{2} & 1 & \tfrac{1}{2} & 0 & \tfrac{7}{2} \\
1 & -\tfrac{1}{4} & 0 & -\tfrac{1}{4} & 0 & \tfrac{3}{2} \\
\hline
0 & \tfrac{11}{2} & 0 & \tfrac{1}{2} & 1 & -\tfrac{3}{2}
\end{array}\right]
$$

The rightmost column has no negatives and there are no negative indicators, so this is our final tableau. (Note that $z = -\tfrac{3}{2}$. It must be negative since it is the opposite of w, that is, $z = -w$.) To answer the original problem, the solution is $x_1 = \tfrac{3}{4}$ and $x_2 = 0$. Since $z = -w = -\tfrac{3}{2}$, we have the minimum value of $w = \tfrac{3}{2}$.

∎

✓ **Checkpoint 3**

Now work Exercise 17.

To summarize, the following steps are used in solving the nonstandard problems in this section.

■ **Solving Nonstandard Problems**

1. If necessary, convert to a maximization problem.
2. Add slack variables and subtract surplus variables.
3. Write the initial simplex tableau.
4. Employ the Two-Phase Method until an optimal solution is found.

Applications

An important application of linear programming is the minimization of the cost of transporting goods. These types of problems are commonly called **transportation problems**. Example 4 is one of these types of problems.

Example 4 **Minimizing Transportation Costs**

Skinner Bikes sends bicycles from its two plants, Plant A and Plant B, to its bicycle retailers I and II located in Columbus, Ohio. Plant A has a total of 210 bicycles to send and Plant B has 60. Retailer I needs 170 bicycles and retailer II needs 100. Transportation costs per bicycle are \$18 from A to I, \$20 from A to II, \$22 from B to I, and \$16 from B to II. How many bicycles should Skinner Bikes send from each plant to each retailer to minimize transportation costs?

Solution

Understand the Situation: We begin by letting

x_1 represent the number of bicycles sent from A to I

x_2 represent the number of bicycles sent from A to II

x_3 represent the number of bicycles sent from B to I

x_4 represent the number of bicycles sent from B to II

Since Plant A has 210 bicycles to send, this gives us

$$x_1 + x_2 \leq 210$$

Similarly, Plant B has 60 bicycles to send, which means that

$$x_3 + x_4 \leq 60$$

Since retailer I needs 170 bicycles, we have

$$x_1 + x_3 \geq 170$$

Likewise, since retailer II needs 100 bicycles,

$$x_2 + x_4 \geq 100$$

Skinner Bikes wants to minimize transportation costs; hence the objective function is

$$w = 18x_1 + 20x_2 + 22x_3 + 16x_4$$

By letting $z = -w$, we convert the minimization problem to a maximization problem.

Maximize: $z = -w = -18x_1 - 20x_2 - 22x_3 - 16x_4$

Subject to: $x_1 + x_2 \leq 210$

$x_3 + x_4 \leq 60$

$x_1 + x_3 \geq 170$

$x_2 + x_4 \geq 100$

$x_1 \geq 0, x_2 \geq 0, x_3 \geq 0, x_4 \geq 0$

Adding slack variables s_1 and s_2 and subtracting surplus variables s_3 and s_4 gives the equational form

$$
\begin{array}{rcl}
x_1 + x_2 \qquad\qquad + s_1 \qquad\qquad\qquad\qquad &=& 210 \\
x_3 + x_4 \qquad + s_2 \qquad\qquad\qquad &=& 60 \\
x_1 \qquad + x_3 \qquad\qquad\quad - s_3 \qquad\qquad &=& 170 \\
x_2 \qquad + x_4 \qquad\qquad\qquad - s_4 \qquad &=& 100 \\
18x_1 + 20x_2 + 22x_3 + 16x_4 \qquad\qquad\qquad + z &=& 0
\end{array}
$$

The initial simplex tableau is

$$
\begin{array}{ccccccccc}
x_1 & x_2 & x_3 & x_4 & s_1 & s_2 & s_3 & s_4 & z
\end{array}
$$
$$
\left[
\begin{array}{ccccccccc|c}
1 & 1 & 0 & 0 & 1 & 0 & 0 & 0 & 0 & 210 \\
0 & 0 & 1 & 1 & 0 & 1 & 0 & 0 & 0 & 60 \\
1 & 0 & 1 & 0 & 0 & 0 & -1 & 0 & 0 & 170 \\
0 & 1 & 0 & 1 & 0 & 0 & 0 & -1 & 0 & 100 \\
\hline
18 & 20 & 22 & 16 & 0 & 0 & 0 & 0 & 1 & 0
\end{array}
\right]
$$

$$
\begin{array}{ccccccccc}
x_1 & x_2 & x_3 & x_4 & s_1 & s_2 & s_3 & s_4 & z
\end{array}
$$
$$
\begin{array}{c}
\\ \\ -1R_3 \to R_3 \\ -1R_4 \to R_4 \\ \\
\end{array}
\left[
\begin{array}{ccccccccc|c}
1 & 1 & 0 & 0 & 1 & 0 & 0 & 0 & 0 & 210 \\
0 & 0 & 1 & 1 & 0 & 1 & 0 & 0 & 0 & 60 \\
-1 & 0 & -1 & 0 & 0 & 0 & 1 & 0 & 0 & -170 \\
0 & -1 & 0 & -1 & 0 & 0 & 0 & 1 & 0 & -100 \\
\hline
18 & 20 & 22 & 16 & 0 & 0 & 0 & 0 & 1 & 0
\end{array}
\right]
$$

We are now in Phase I. We pivot on the -1 in row 3, column 1. After the pivot we have

$$
\begin{array}{ccccccccc}
x_1 & x_2 & x_3 & x_4 & s_1 & s_2 & s_3 & s_4 & z
\end{array}
$$
$$
\left[
\begin{array}{ccccccccc|c}
0 & 1 & -1 & 0 & 1 & 0 & 1 & 0 & 0 & 40 \\
0 & 0 & 1 & 1 & 0 & 1 & 0 & 0 & 0 & 60 \\
1 & 0 & 1 & 0 & 0 & 0 & -1 & 0 & 0 & 170 \\
0 & -1 & 0 & -1 & 0 & 0 & 0 & 1 & 0 & -100 \\
\hline
0 & 20 & 4 & 16 & 0 & 0 & 18 & 0 & 1 & -3060
\end{array}
\right]
$$

Since $s_4 = -100$, we are still in Phase I. Pivoting on the -1 in row 4, column 2 gives the following:

$$
\begin{array}{ccccccccc}
x_1 & x_2 & x_3 & x_4 & s_1 & s_2 & s_3 & s_4 & z
\end{array}
$$
$$
\left[
\begin{array}{ccccccccc|c}
0 & 0 & -1 & -1 & 1 & 0 & 1 & 1 & 0 & -60 \\
0 & 0 & 1 & 1 & 0 & 1 & 0 & 0 & 0 & 60 \\
1 & 0 & 1 & 0 & 0 & 0 & -1 & 0 & 0 & 170 \\
0 & 1 & 0 & 1 & 0 & 0 & 0 & -1 & 0 & 100 \\
\hline
0 & 0 & 4 & -4 & 0 & 0 & 18 & 20 & 1 & -5060
\end{array}
\right]
$$

We are still in Phase I since $s_1 = -60$. We pivot on the -1 in row 1, column 3 and get

$$
\begin{array}{ccccccccc}
x_1 & x_2 & x_3 & x_4 & s_1 & s_2 & s_3 & s_4 & z
\end{array}
$$
$$
\left[
\begin{array}{ccccccccc|c}
0 & 0 & 1 & 1 & -1 & 0 & -1 & -1 & 0 & 60 \\
0 & 0 & 0 & 0 & 1 & 1 & 1 & 1 & 0 & 0 \\
1 & 0 & 0 & -1 & 1 & 0 & 0 & 1 & 0 & 110 \\
0 & 1 & 0 & 1 & 0 & 0 & 0 & -1 & 0 & 100 \\
\hline
0 & 0 & 0 & -8 & 4 & 0 & 22 & 24 & 1 & -5300
\end{array}
\right]
$$

We have all variables with nonnegative values, so we move to Phase II. (Recall that z must be negative.) Now we pivot on the 1 in row 1, column 4 and get

x_1	x_2	x_3	x_4	s_1	s_2	s_3	s_4	z	
0	0	1	1	−1	0	−1	−1	0	60
0	0	0	0	1	1	1	1	0	0
1	0	1	0	0	0	−1	0	0	170
0	1	−1	0	1	0	1	0	0	40
0	0	8	0	−4	0	14	16	1	−4820

Our next pivot is on the 1 in row 2, column 5. This yields

x_1	x_2	x_3	x_4	s_1	s_2	s_3	s_4	z	
0	0	1	1	0	1	0	0	0	60
0	0	0	0	1	1	1	1	0	0
1	0	1	0	0	0	−1	0	0	170
0	1	−1	0	0	−1	0	−1	0	40
0	0	8	0	0	4	18	20	1	−4820

This is the final tableau. The maximum value of z is -4820, so the minimum value of w is 4820. This occurs when $x_1 = 170$, $x_2 = 40$, $x_3 = 0$, $x_4 = 60$, and $s_1 = s_2 = s_3 = s_4 = 0$.

Interpret the Solution: This means that Plant A should send 170 bicycles to retailer I and 40 bicycles to retailer II, while Plant B should send 0 bicycles to retailer I and 60 bicycles to retailer II. This minimizes the cost at $4820. ∎

SUMMARY

In this section we saw how to solve nonstandard problems, that is, problems with **mixed constraints**. We introduced a new type of variable called a **surplus variable**. The key to solving mixed constraint problems was employing the Two-Phase Method. The section concluded with a classic economics problem called a transportation problem.

- **Two-Phase Method**

  Phase I: Locating a Corner Point   Our goal is to locate a feasible solution. To do this we offer the following strategy: Select a row with a negative entry in the right-

most column and then pivot on a negative entry in that row. We repeat this process until our tableau has non-negative entries in the rightmost column.

  Phase II: Solving after Phase I   After using Phase I to obtain a tableau that gives a feasible solution, we apply the simplex method as usual.

- The minimum value of an objective function w occurs at the same point as the maximum value of the objective function z, where $z = -w$, provided that both have the same constraints.

SECTION 4.4 EXERCISES

In Exercises 1–6, add slack variables and subtract surplus variables to write each system of inequalities in equational form.

1. $3x_1 + 2x_2 \leq 12$
$x_1 + 4x_2 \geq 7$

2. $4x_1 + 2x_2 \leq 8$
$3x_1 + 5x_2 \geq 15$

✓ **3.** $x_1 + 4x_2 + 3x_3 \leq 9$
$2x_1 + x_2 + 5x_3 \geq 10$
$3x_1 + 5x_2 + 2x_3 \leq 8$

4. $5x_1 + 6x_2 + x_3 \leq 20$
$2x_1 + x_2 + 3x_3 \geq 6$
$3x_1 + 4x_2 + x_3 \leq 13$

5. $x_1 + 2x_2 + x_3 \leq 60$
$2x_1 + 3x_3 \geq 18$
$3x_2 + 2x_3 \geq 30$

6. $2x_1 + 3x_2 + 7x_3 \leq 80$
$9x_1 + x_3 \geq 12$
$7x_2 + 4x_3 \geq 20$

In Exercises 7–10, complete the following:

(a) Convert to a maximization problem.

(b) Add slack variables and subtract surplus variables to write in equational form.

(c) Write the initial simplex tableau.

7. Minimize: $w = 8x_1 + 3x_2$
   Subject to: $4x_1 + 2x_2 \leq 8$
                   $x_1 + 4x_2 \geq 7$
                   $x_1 \geq 0, x_2 \geq 0$

8. Minimize: $w = x_1 + 7x_2$
   Subject to: $3x_1 + 2x_2 \leq 12$
                   $5x_1 + 3x_2 \geq 15$
                   $x_1 \geq 0, x_2 \geq 0$

9. Minimize: $w = 2x_1 + x_2 + 3x_3$
   Subject to: $5x_1 + 6x_2 + x_3 \leq 20$
                   $2x_1 + x_2 + 5x_3 \geq 10$
                   $7x_2 + 4x_3 \geq 20$
                   $x_1 \geq 0, x_2 \geq 0, x_3 \geq 0$

10. Minimize: $w = x_1 + 2x_2 + x_3$
   Subject to: $x_1 + 4x_2 + 3x_3 \leq 9$
                   $2x_1 + x_2 + x_3 \geq 7$
                   $3x_2 + 2x_3 \geq 30$
                   $x_1 \geq 0, x_2 \geq 0, x_3 \geq 0$

*In Exercises 11–24, use the method in this section to solve
each linear programming problem.*

✓ **11.** Maximize: $z = 5x_1 + 2x_2$
   Subject to: $x_1 + x_2 \geq 11$
                   $2x_1 + 3x_2 \geq 24$
                   $x_1 + 3x_2 \leq 18$
                   $x_1 \geq 0, x_2 \geq 0$

12. Maximize: $z = 2x_1 + 5x_2$
   Subject to: $x_1 + 2x_2 \leq 18$
                   $2x_1 + x_2 \leq 21$
                   $x_1 + x_2 \geq 10$
                   $x_1 \geq 0, x_2 \geq 0$

13. Maximize: $z = 2x_1 + 3x_2$
   Subject to: $3x_1 + x_2 \leq 15$
                   $x_1 + x_2 \leq 10$
                   $4x_1 + 3x_2 \geq 12$
                   $x_1 \geq 0, x_2 \geq 0$

14. Maximize: $z = 4x_1 + x_2$
   Subject to: $x_1 + 3x_2 \leq 18$
                   $2x_1 + x_2 \leq 12$
                   $3x_1 + 4x_2 \geq 12$
                   $x_1 \geq 0, x_2 \geq 0$

15. Maximize: $z = 2x_1 + x_2 + 3x_3$
   Subject to: $2x_1 + 4x_2 + x_3 \leq 150$
                   $3x_1 + 3x_2 + x_3 \leq 90$
                   $-x_1 + 5x_2 + x_3 \geq 120$
                   $x_1 \geq 0, x_2 \geq 0, x_3 \geq 0$

16. Maximize: $z = x_1 + 2x_2 + 3x_3$
   Subject to: $2x_1 + x_2 + x_3 \leq 30$
                   $x_1 + 2x_2 + 2x_3 \leq 80$
                   $x_1 + 2x_2 - x_3 \geq 10$
                   $x_1 \geq 0, x_2 \geq 0, x_3 \geq 0$

✓ **17.** Minimize: $w = x_1 + 2x_2$
   Subject to: $2x_1 + 3x_2 \leq 12$
                   $3x_1 + x_2 \geq 6$
                   $x_1 \geq 0, x_2 \geq 0$

18. Minimize: $w = 2x_1 + x_2$
   Subject to: $x_1 + 2x_2 \leq 20$
                   $2x_1 + 3x_2 \geq 6$
                   $x_1 \geq 0, x_2 \geq 0$

19. Minimize: $w = 3x_1 + 4x_2$
   Subject to: $x_1 + 2x_2 \leq 18$
                   $x_1 + x_2 \leq 10$
                   $x_1 + 3x_2 \geq 3$
                   $x_1 \geq 0, x_2 \geq 0$

20. Minimize: $w = 4x_1 + x_2$
   Subject to: $3x_1 + x_2 \leq 15$
                   $x_1 + x_2 \leq 12$
                   $2x_1 + 3x_2 \geq 6$
                   $x_1 \geq 0, x_2 \geq 0$

21. Minimize: $w = x_1 + x_2 + 2x_3$
   Subject to: $x_1 - x_2 + x_3 \leq 8$
                   $x_1 + 3x_2 + 2x_3 \geq 10$
                   $-3x_1 + x_2 + 4x_3 \geq 12$
                   $x_1 \geq 0, x_2 \geq 0, x_3 \geq 0$

22. Minimize: $w = x_1 - 2x_2 + x_3$
   Subject to: $x_1 - 2x_2 + 3x_3 \leq 10$
                   $2x_1 + x_2 - 2x_3 \geq 2$
                   $2x_1 + x_2 + 3x_3 \geq 1$
                   $x_1 \geq 0, x_2 \geq 0, x_3 \geq 0$

23. Minimize: $w = 2x_1 - x_2 + 3x_3$
   Subject to: $2x_1 + x_2 + x_3 \geq 2$
                   $x_1 + 3x_2 + x_3 \geq 6$
                   $2x_1 + x_2 + 2x_3 \leq 12$
                   $x_1 \geq 0, x_2 \geq 0, x_3 \geq 0$

24. Minimize: $w = x_1 + 2x_2 + 3x_3$
   Subject to: $x_1 + x_2 + x_3 \leq 20$
                   $2x_1 + x_2 + x_3 \geq 10$
                   $2x_1 + 2x_2 + x_3 \geq 2$
                   $x_1 \geq 0, x_2 \geq 0, x_3 \geq 0$

Applications

25. The Tentsmith Company makes two types of tents: two-person and six-person models. Each tent requires the services, in hours, of two departments as shown in Table 4.4.3.

Table 4.4.3

Department	Labor	
	Two-person	Six-person
Fabric cutting	6	8
Assembly and packaging	2	4

Fabric cutting has a maximum of 580 hours available each week, and assembly and packaging has a maximum of 260 hours available each week. Due to prior contracts, the total number of tents made each week must be at least 20. Each two-person tent yields $40 in profit, and each six-person tent yields $60 in profit.

(a) How many of each type of tent should be made and sold to maximize weekly profit? What is the maximum weekly profit?

(b) Interpret the value of all slack and surplus variables in the final tableau.

26. (*continuation of Exercise 25*) Tanya, the head of research and development at The Tentsmith Company has discovered a way to reduce the amount of time needed for the six-person tent in the fabric cutting department to 7 hours. Everything else from Exercise 25 remains the same.

(a) How many of each type of tent should be made and sold to maximize weekly profit? What is the maximum weekly profit?

(b) Interpret the value of all slack and surplus variables in the final tableau.

27. Denyse and Mike, two recent graduates, have decided to enter the field of microcomputers. Initially, they plan to make two types of microcomputers, Deluxe 1000 and Deluxe 1200. Because of the interest in microcomputers, they can sell all that they can produce. However, they need to determine a production rate so as to satisfy various limits with a small production crew. These limits include a maximum of 150 hours per week for assembly and 100 hours per week for testing. The Deluxe 1000 requires 3 hours of assembly and 3 hours of testing, and the Deluxe 1200 requires 6 hours of assembly and 3.5 hours of testing. For market demonstration purposes, the total number of microcomputers made each week must be at least 8. Profit for the Deluxe 100 is estimated at $300 per unit, and profit for the Deluxe 1200 is estimated at $450 per unit.

(a) How many of each type of microcomputer should be made and sold to maximize weekly profit? What is the maximum weekly profit?

(b) Interpret the value of all slack and surplus variables in the final tableau.

28. (*continuation of Exercise 27*) Due to a favorable response from the market, Denyse and Mike decide to hire more employees and increase production. By doing so, they now have a maximum of 210 hours per week for assembly and 160 hours per week for testing. Everything else from Exercise 27 remains the same.

(a) How many of each type of microcomputer should be made and sold to maximize weekly profit? What is the maximum weekly profit?

(b) Interpret the value of all slack and surplus variables in the final tableau.

29. Table 4.4.4 gives nutritional information in grams and current prices for selected items from Burger King's menu.

Table 4.4.4

Item	Protein	Fiber	Price
Cheeseburger	22	2	$1.19
Small onion rings	3	2	$0.99

SOURCE: Burger King, www.burgerking.com

Mike desires to eat nothing but cheeseburgers and orders of small onion rings. He wants at least 96 grams of protein and at least 26 grams of fiber daily. Also, he wishes to eat at least 1 cheeseburger and at least 1 order of small onion rings. If Mike eats nothing but cheeseburgers and small onion rings and meets his minimum nutrient requirements, answer the following:

(a) How much of each food should Mike eat to minimize his daily food cost?
(b) What is his minimum daily food cost?

30. (*continuation of Exercise 29*) Mike has found out that Burger King is having a promotion during which cheeseburgers sell for $0.49 on Wednesdays. Assuming that all other data in Exercise 29 remain the same, answer the following:

(a) How much of each food should Mike eat to minimize his daily food cost?
(b) What is his minimum daily food cost?

31. Hoofer Oil has two refineries, one in Findlay, Ohio, and one in Ft. Wayne, Indiana. Each refinery can refine the amounts of fuel, in barrels, in a 24-hour period as shown in Table 4.4.5.

Table 4.4.5

Maximum Production of:	Findlay	Ft. Wayne
Low-octane gasoline	800	1000
High-octane gasoline	500	600
Diesel	1000	600

Daily operating costs for the Findlay refinery are $28,000, and daily operating costs for the Ft. Wayne refinery are $33,000. Zachary, the production manager of Hoofer Oil, receives an order for 60,000 barrels of low-octane gasoline, 120,000 barrels of high-octane gasoline, and 150,000 barrels of diesel. Due to budgetary issues, the combined total number of days that the refineries operate cannot exceed 300.

(a) How many days should each refinery operate to minimize cost? What is the minimum cost?

(b) Interpret the value of all slack and surplus variables in the final tableau.

32. Shannon Oil has two refineries, one in Lima, Ohio, and one in Detroit, Michigan. Each refinery can refine the amounts of fuel, in barrels, in a 24-hour period as shown in Table 4.4.6.

Table 4.4.6

Maximum Production of:	Lima	Detroit
Low-octane gasoline	900	1100
High-octane gasoline	600	700
Diesel	1100	700

Daily operating costs for the Lima refinery are $35,000, and daily operating costs for the Detroit refinery are $30,000. Joshua, the production manager of Shannon Oil, receives an order for 50,000 barrels of low-octane gasoline, 90,000 barrels of high-octane gasoline, and 130,000 barrels of diesel. Due to budgetary issues, the combined total number of days that the refineries can operate cannot exceed 200.

(a) How many days should each refinery operate to minimize cost? What is the minimum cost?

(b) Interpret the value of all slack and surplus variables in the final tableau.

33. Anita, a dietitian at a local hospital, is to prepare a diet to contain at least 400 units of vitamins C and E, 500 units of minerals (such as zinc and iron), and 1400 calories. She has two foods available, Goli and Zarobila. An ounce of Goli contains 2 units of vitamin C and E, 1 unit of minerals, and 4 calories and costs $0.05. An ounce of Zarobila contains 1 unit of vitamins C and E, 2 units of minerals, and 4 calories and costs $0.03. Due to restricted availability, the total number of ounces of the two foods is no more than 450 ounces.

(a) How many ounces of each food should be used to minimize daily food cost? What is the minimum daily food cost?

(b) Interpret the value of all slack and surplus variables in the final tableau.

34. Lisa, a dietitian at a local hospital, is to prepare two foods to meet certain nutritional requirements. Each pound of Viega contains 200 units of vitamin C, 80 units of vitamin B_{12}, and 20 units of vitamin E and costs $0.40. Each pound of Zella contains 20 units of vitamin C, 160 units of vitamin B_{12}, and 10 units of vitamin E and costs $0.30. To meet the nutritional needs of the patients, the combination of the two foods must have at least 520 units of vitamin C, 640 units of vitamin B_{12}, and 100 units of vitamin E. Due to restricted availability, the total number of pounds of the two foods is no more than 6 pounds.

(a) How many ounces of each food should be used to minimize daily food cost? What is the minimum daily food cost?

(b) Interpret the value of all slack and surplus variables in the final tableau.

35. A DVD player manufacturer needs to fulfill orders from two retailers in Miami, Florida. The manufacturer has the DVD players stored in two warehouses, warehouse A and warehouse B. Warehouse A has a total of 200 DVD players and warehouse B has 240. Retailer I needs 110 DVD players and retailer II needs 150. Transportation costs per DVD player are $16 from warehouse A to retailer I, $24 from warehouse A to retailer II, $26 from warehouse B to retailer I, and $14 from warehouse B to retailer II. How many DVD players should be sent from each warehouse to each retailer to minimize transportation costs? What is the minimum transportation cost?

36. A manufacturer of computer scanners has two plants, Plant A and Plant B, that can produce up to 400 units and 280 units per month, respectively. The units are shipped to two warehouses, warehouse I and warehouse II. Warehouse I needs at least 360 units each month, and warehouse II needs at least 300 units each month. Transportation costs per scanner are $150 from Plant A to warehouse I, $80 from Plant A to warehouse II, $160 from Plant B to warehouse I, and $60 from Plant B to warehouse II. How many scanners should be sent from each warehouse to each retailer to minimize transportation costs? What is the minimum transportation cost?

■ SECTION PROJECT

Linger Furniture Company manufactures office desks at three locations: Chicago, Detroit, and Fort Lauderdale. The company distributes the desks through regional warehouses located in Phoenix, Boston, and Cleveland. Each month the Chicago plant, Detroit plant, and Fort Lauderdale plant can produce up to 100 units, 300 units, and 300 units, respectively. Each month the Phoenix warehouse, Boston warehouse, and Cleveland warehouse need at least 300 units, 200 units, and 200 units, respectively. The transportation costs per desk are given in Table 4.4.7.

Table 4.4.7

To From	Phoenix	Boston	Cleveland
Chicago	$5	$4	$3
Detroit	$8	$4	$3
Fort Lauderdale	$9	$7	$5

(a) How many desks should be sent from each plant to each warehouse to minimize transportation costs?

(b) What is the minimum transportation cost?

■ Why We Learned It

In this chapter we learned how to solve linear programming problems that have more than two decision variables and more than two constraints. The method discussed to solve these types of linear programming problems is called the simplex method. As in Chapter 3, we saw that many managers in business can use the simplex method to maximize profit and minimize costs or to determine optimum manufacturing times. For example, suppose that a company makes three types of chairs and each chair must spend some time on three different machines. Due to certain restrictions, each machine has a maximum number of hours available each week. Assuming that the company knows the profit that it receives on each chair produced and sold, it would want to determine how many units of each chair to make in a week (given the time constraints of each machine) in order to maximize profit. The simplex method will also show the company how to maximize profit through optimal use of the machines. This means that the company can maximize profit while reducing wear and tear on the machines. We also learned that dietitians can use the simplex method to maximize healthy nutritional elements in a diet, such as vitamins, or minimize elements of a diet, such as total fat intake. Farmers can also use the simplex method to maximize profit, given restrictions of land and the availability of various fertilizers.

■ CHAPTER REVIEW EXERCISES

In Exercises 1 and 2, convert the system of linear inequalities into a system of equations by using slack variables.

1.
$$x_1 + x_2 \le 14$$
$$-2x_1 + 5x_2 \le 6$$

2.
$$2x_1 + x_2 - 5x_3 \le 11$$
$$-3x_1 + 2x_2 - x_3 \le 10$$

In Exercises 3–6, for the given standard maximum form linear programming problem, introduce slack variables and setup the initial simplex tableau.

3. Maximize: $z = x_1 + 3x_2$

Subject to: $x_1 + 2x_2 \le 8$

$x_1 - x_2 \le 4$

$x_1 \ge 0, x_2 \ge 0$

4. Maximize: $z = 3x_1 + 3x_2$

Subject to: $2x_1 - x_2 \le 8$

$x_1 + 3x_2 \le 7$

$5x_1 - 2x_2 \le 4$

$x_1 \ge 0, x_2 \ge 0$

5. Maximize: $z = 2x_1 + 3x_2 + x_3$

Subject to: $x_1 + 2x_2 + 3x_3 \le 14$

$-x_1 + 4x_2 - 2x_3 \le 3$

$x_1 \ge 0, x_2 \ge 0, x_3 \ge 0$

6. Maximize: $z = 3x_1 + 2x_2 + x_3$

Subject to: $x_1 + 2x_2 + 3x_3 \le 9$

$-x_1 + 4x_2 - x_3 \le 4$

$5x_1 - x_2 + 4x_3 \le 12$

$x_1 \ge 0, x_2 \ge 0, x_3 \ge 0$

In Exercises 7 and 8, for each given matrix:

(a) Identify the basic and nonbasic variables.

(b) Determine the solution given by the matrix.

7.

$$\begin{array}{c} \begin{matrix} x_1 & x_2 & s_1 & s_2 & z \end{matrix} \\ \left[\begin{array}{ccccc|c} 0 & -3 & 1 & 5 & 0 & 6 \\ 1 & 2 & 0 & -2 & 0 & 14 \\ \hline 0 & 4 & 0 & 7 & 1 & 5 \end{array}\right] \end{array}$$

8.

$$\begin{array}{c} \begin{matrix} x_1 & x_2 & x_3 & s_1 & s_2 & z \end{matrix} \\ \left[\begin{array}{cccccc|c} 1 & 0 & -2 & 4 & 5 & 0 & 6 \\ 0 & 1 & 3 & 2 & 1 & 0 & 14 \\ \hline 0 & 0 & -1 & 4 & 8 & 1 & 13 \end{array}\right] \end{array}$$

In Exercises 9 and 10:

(a) Pivot about the circled entry.

(b) Read the solution after pivoting.

9.

$$\begin{array}{c} \begin{matrix} x_1 & x_2 & s_1 & s_2 & s_3 & z \end{matrix} \\ \left[\begin{array}{cccccc|c} 2 & 3 & 1 & 0 & 0 & 0 & 3 \\ -1 & ④ & 0 & 1 & 0 & 0 & 4 \\ 1 & -2 & 0 & 0 & 1 & 0 & 12 \\ 3 & 5 & 0 & 0 & 0 & 1 & 2 \end{array}\right] \end{array}$$

$$
\begin{array}{c}
\quad\ x_1\ \ x_2\ \ x_3\ \ x_4\ \ s_1\ \ s_2\ \ s_3\ \ z \\
\textbf{10.}\ \left[\begin{array}{cccccccc|c}
-1 & 1 & 3 & 0 & 0 & 5 & 6 & 0 & 8 \\
2 & 0 & 7 & 1 & 0 & 1 & 2 & 0 & 14 \\
\textcircled{-3} & 0 & 0 & 0 & 1 & 5 & 4 & 0 & 2 \\
\hline
6 & 0 & 3 & 0 & 0 & 4 & -1 & 1 & 4
\end{array}\right]
\end{array}
$$

In Exercises 11–15, perform the setup so that the solution can be determined by the simplex method as illustrated in Section 4.1, Example 5. In Review Exercises 26–30, you will determine optimal solutions for these exercises.

11. Music Notes offers introductory music workshops for drums and guitar. Each student attends an individual music theory session and a music lesson. Drum students attend a 2-hour music theory session and a 4-hour music lesson. Guitar students attend a 4-hour music theory session and a 2-hour music lesson. Each month Music Notes instructors can offer no more than 84 hours of music theory sessions and 78 hours of music lessons. Music Notes makes a profit of $25 for each drum student and $30 for each guitar student. To determine how many of each type of student should be admitted each month to maximize profit, do the following:

(a) Define the variables and write the mathematical model for this problem.

(b) Use slack variables to adapt the mathematical model to equational form.

(c) Write the initial simplex tableau.

12. (*continuation of Exercise 11*) Blues Notes, a local competitor of Music Notes, also offers introductory music workshops for drums and guitar. In addition to an individual music theory session and a music lesson, students also attend an individual music recording session. As with Music Notes, drum students attend a 2-hour music session and a 4-hour music lesson, and guitar students attend a 4-hour music theory session and a 2-hour music lesson. In addition, drum students and guitar students both attend a 2-hour music recording session at Blues Notes. Instructors at Blues Notes each month can offer no more than 64 hours of music theory sessions, 56 hours of music lessons, and 42 hours of music recording. Blues Notes makes a profit of $30 for each drum student and $35 for each guitar student. To determine how many of each type of student should be admitted each month to maximize profit, do the following:

(a) Define the variables and write the mathematical model for this problem.

(b) Use slack variables to adapt the mathematical model to equational form.

(c) Write the initial simplex tableau.

13. Sopris Mountaineering makes two types of sleeping bags: +15 sleeping bag and −15 sleeping bag (+15 and −15 indicate the temperature rating in degrees Fahrenheit). Each bag requires the services of two departments, in hours, as shown in Table R.1.

Table R.1

Department	+15 bag	−15 bag
Fabric cutting	3	6
Assembly and packaging	3	3

Fabric cutting has a maximum of 240 hours available each week, and assembly and packaging has a maximum of 180 hours available each week. Each +15 bag yields $60 in profit and each −15 bag yields $70 in profit. To determine the number of each type of bag that should be made each week to maximize profit:

(a) Define the variables and write the mathematical model for this problem.

(b) Use slack variables to adapt the mathematical model to equational form.

(c) Write the initial simplex tableau.

🌐 **14.** Table R.2 gives data for growing head lettuce and cantaloupe in California for an entire growing season (entries are in lb/acre).

Table R.2

	Head Lettuce	Cantaloupe
Nitrogen	173	175
Phosphate	122	161
Potash	76	39

SOURCE: U.S. Department of Agriculture, www.usda.gov/nass

Amara has 220 acres, and can get no more than 34,700 pounds of nitrogen, 27,845 pounds of phospate, and 13,350 pounds of potash. She expects a profit of $120 per acre for head lettuce and $100 per acre for cantaloupe. To determine how many acres she should allot for each crop to maximize profit, do the following:

(a) Define the variables and write the mathematical model for this problem.

(b) Use slack variables to adapt the mathematical model to equational form.

(c) Write the initial simplex tableau.

🌐 **15.** (*continuation of Exercise 14*) Amara wants to consider the possibility of allocating some acres to growing bell peppers. The amounts of nitrogen, phosphate, and potash needed for the entire growing season are 212 pounds per acre, 108 pounds per acre, and 72 pounds per acre, respectively. Amara expects a profit of $140 per acre for bell peppers. With everything else remaining the same from Exercise 14, to determine how many acres she should allot for each crop to maximize profit, do the following:

(a) Define the variables and write the mathematical model for this problem.

(b) Use slack variables to adapt the mathematical model to equational form.

(c) Write the initial simplex tableau.

In Exercises 16–19, the initial simplex tableau of a linear programming problem is given. Use the simplex method to solve.

16.
$$\begin{array}{ccccc|c}
x_1 & x_2 & s_1 & s_2 & z & \\
4 & 1 & 1 & 0 & 0 & 14 \\
-2 & 1 & 0 & 1 & 0 & 6 \\
\hline
-2 & -3 & 0 & 0 & 1 & 12
\end{array}$$

17.
$$\begin{array}{cccccc|c}
x_1 & x_2 & s_1 & s_2 & s_3 & z & \\
-3 & 2 & 1 & 0 & 0 & 0 & 8 \\
1 & 4 & 0 & 1 & 0 & 0 & 5 \\
-2 & -1 & 0 & 0 & 1 & 0 & 7 \\
\hline
-4 & -5 & 0 & 0 & 0 & 1 & 0
\end{array}$$

18.
$$\begin{array}{cccccc|c}
x_1 & x_2 & x_3 & s_1 & s_2 & z & \\
4 & 5 & 7 & 1 & 0 & 0 & 20 \\
3 & 2 & 1 & 0 & 1 & 0 & 40 \\
\hline
-3 & -4 & -2 & 0 & 0 & 1 & 0
\end{array}$$

19.
$$\begin{array}{cccccccc|c}
x_1 & x_2 & x_3 & x_4 & s_1 & s_2 & s_3 & z & \\
2 & 5 & 0 & 0 & 1 & 0 & 0 & 0 & 5 \\
0 & 1 & 2 & 1 & 0 & 1 & 0 & 0 & 14 \\
2 & 0 & 5 & 0 & 0 & 0 & 1 & 0 & 12 \\
\hline
3 & -2 & -4 & -1 & 0 & 0 & 0 & 1 & 10
\end{array}$$

In Exercises 20–25, for the standard maximum form linear programming problems:

(a) Introduce slack variables and set up the initial simplex tableau.

(b) Use the simplex method to solve.

20. Maximize: $z = 2x_1 + 3x_2$
Subject to: $2x_1 + x_2 \le 7$
$x_1 - x_2 \le 4$
$x_1 \ge 0, x_2 \ge 0$

21. Maximize: $z = 2x_1 + 2x_2$
Subject to: $4x_1 - 2x_2 \le 0$
$x_1 + x_2 \le 7$
$2x_1 + x_2 \le 8$
$x_1 \ge 0, x_2 \ge 0$

22. Maximize: $z = 4x_1 + 3x_2 + 5x_3$
Subject to: $3x_1 + x_2 + x_3 \le 14$
$7x_1 - 2x_2 + 3x_3 \le 4$
$x_1 \ge 0, x_2 \ge 0, x_3 \ge 0$

23. Maximize: $z = 3x_1 + x_2 + x_3$
Subject to: $x_1 - x_2 + x_3 \le 5$
$2x_1 + x_2 + 4x_3 \le 4$
$x_1 \le 0, x_2 \ge 0, x_3 \ge 0$

24. Maximize: $z = 3x_1 + x_2 + 2x_3$
Subject to: $4x_1 + 2x_2 + x_3 \le 10$
$x_1 + 3x_2 + 4x_3 \le 14$
$3x_1 + 2x_2 + 4x_3 \le 12$
$x_1 \ge 0, x_2 \ge 0, x_3 \ge 0$

25. Maximize: $z = 8x_1 + 2x_2$
Subject to: $x_1 + x_2 \le 9$
$2x_1 + 2x_2 \le 4$
$x_1 \ge 0, x_2 \ge 0$

In Exercises 26–30, set up and solve each using the simplex method. Interpret solutions as shown in Section 4.2, Example 3. Include an interpretation of any nonzero slack variables (see Exercises 11–15).

26. Music Notes offers introductory music workshops for drums and guitar. Each student attends an individual music theory session and a music lesson. Drum students attend a 2-hour music theory session and a 4-hour music lesson. Guitar students attend a 4-hour music theory session and a 2-hour music lesson. Each month Music Notes instructors can offer no more than 84 hours of music theory sessions and 78 hours of music lessons. Music Notes makes a profit of $25 for each drum student and $30 for each guitar student. Determine how many of each type of student should be admitted each month to maximize profit.

27. (*continuation of Exercise 26*) Blues Notes, a local competitor of Music Notes, also offers introductory music workshops for drums and guitar. In addition to an individual music theory session and a music lesson, students also attend an individual music recording session. As with Music Notes, drum students attend a 2-hour music theory session and a 4-hour music lesson, and guitar students attend a 4-hour music theory session and a 2-hour music lesson. In addition, drum students and guitar students both attend a 2-hour music recording session at Blues Notes. Instructors at Blues Notes each month can offer no more than 64 hours of music theory sessions, 56 hours of music lessons, and 42 hours of music recording. Blues Notes makes a profit of $30 for each drum student and $35 for each guitar student.

(a) Determine how many of each type of student should be admitted each month to maximize profit.

(b) Compare the results of Exercise 26 and Exercise 27a.

28. Sopris Mountaineering makes two types of sleeping bags: a +15 sleeping bag and a −15 sleeping bag (+15 and −15 indicate the temperature rating in degrees Fahrenheit). Each bag requires the services, in hours, of two departments as shown in Table R.3.

Table R.3		
Department	**+15 bag**	**−15 bag**
Fabric cutting	3	6
Assembly and packaging	3	3

Fabric cutting has a maximum of 240 hours available each week, and assembly and packaging has a maximum of 180 hours available each week. Each +15 bag yields $60 in profit and each −15 bag yields $70 in profit. Determine the number of each type of bag that should be made each week to maximize profit.

🌐 **29.** Table R.4 gives data for growing head lettuce and cantaloupe in California for an entire growing season (entries are in lb/acre).

Table R.4

	Head Lettuce	Cantaloupe
Nitrogen	173	175
Phosphate	122	161
Potash	76	39

SOURCE: U.S. Department of Agriculture, www.usda.gov/nass

Amara has 220 acres and can get no more than 34,700 pounds of nitrogen, 27,845 pounds of phosphate, and 13,350 pounds of potash. She expects a profit for $120 per acre for head lettuce and $100 per acre for cantaloupe. Determine how many acres she should allot for each crop to maximize profit.

🌐 **30.** (*continuation of Exercise 29*) Amara wants to consider the possibility of allocating some acres to growing bell peppers. The amounts of nitrogen, phosphate, and potash needed for the entire growing season are 212 pounds per acre, 108 pounds per acre, and 72 pounds per acre, respectively. Amara expects a profit of $140 per acre for bell peppers.

(a) With everything else remaining the same from Exercise 29, determine how many acres she should allot for each crop to maximize profit.

(b) Compare the results of Exercises 29 and 30a.

In Exercise 31 and 32, determine the transpose of the given matrix.

31. $\begin{bmatrix} 3 & 1 & 0 & 2 \\ 2 & 0 & 3 & 7 \\ 4 & 1 & 8 & 9 \end{bmatrix}$

32. $\begin{bmatrix} 0 & 2 \\ 3 & 4 \\ 1 & 7 \\ 8 & 9 \end{bmatrix}$

In Exercises 33 and 34, form the dual problem for the given linear programming problem.

33. Minimize: $w = 6y_1 + 4y_2$
   Subject to: $4y_1 + 2y_2 \geq 10$
                  $y_1 + 3y_2 \geq 5$
                  $y_1 \geq 0, y_2 \geq 0$

34. Minimize: $w = 3y_1 + 2y_2 + y_3$
   Subject to: $y_1 + y_2 + y_3 \geq 9$
                  $2y_1 + 3y_2 + 8y_3 \geq 10$
                  $y_1 + 5y_2 + y_3 \geq 2$
                  $y_1 \geq 0, y_2 \geq 0, y_3 \geq 0$

In Exercises 35 and 36, a linear programming problem is given along with the final simplex tableau for the dual problem. Use the final simplex tableau for the dual to:

(a) Find the solution to the primal problem.

(b) Find the solution to the dual problem.

35. Minimize: $w = y_1 + 4y_2$
   Subject to: $y_1 + 3y_2 \geq 4$
                  $3y_1 + y_2 \geq 6$
                  $y_1 \geq 0, y_2 \geq 0$

$$
\begin{array}{ccccc}
x_1 & x_2 & s_1 & s_2 & z \\
\end{array}
$$
$$
\left[\begin{array}{ccccc|c}
1 & 3 & 1 & 0 & 0 & 1 \\
0 & -8 & -3 & 1 & 0 & 1 \\
\hline
0 & 6 & 4 & 0 & 1 & 4
\end{array}\right]
$$

36. Minimize: $w = 3y_1 + y_2$
   Subject to: $4y_1 + y_2 \geq 9$
                  $y_1 + 2y_2 \geq 4$
                  $2y_1 + 3y_2 \geq 6$
                  $y_1 \geq 0, y_2 \geq 0$

$$
\begin{array}{cccccc}
x_1 & x_2 & x_3 & s_1 & s_2 & z \\
\end{array}
$$
$$
\left[\begin{array}{cccccc|c}
1 & 0 & \frac{1}{7} & \frac{2}{7} & -\frac{1}{7} & 0 & \frac{5}{7} \\
0 & 1 & \frac{10}{7} & -\frac{1}{7} & \frac{4}{7} & 0 & \frac{1}{7} \\
\hline
0 & 0 & 1 & 2 & 1 & 1 & 7
\end{array}\right]
$$

In Exercises 37–40:

(a) Construct the dual problem.

(b) Use the simplex method in the dual problem to find the solution to the original primal problem.

37. Minimize: $w = 3y_1 + 4y_2$
   Subject to: $4y_1 + 2y_2 \geq 6$
                  $3y_1 + 4y_2 \geq 9$
                  $y_1 \geq 0, y_2 \geq 0$

38. Minimize: $w = 3y_1 + 9y_2$
   Subject to: $y_1 + 5y_2 \geq 24$
                  $5y_1 + y_2 \geq 35$
                  $y_1 + y_2 \geq 8$
                  $y_1 \geq 0, y_2 \geq 0, y_3 \geq 0$

39. Minimize: $w = 16y_1 + 7y_2 + 22y_3$
   Subject to: $y_1 + 2y_2 + 3y_3 \geq 5$
                  $2y_1 + y_2 + y_3 \geq 12$
                  $y_1 \geq 0, y_2 \geq 0, y_3 \geq 0$

40. Minimize: $\quad w = 4y_1 + 5y_2 + 6y_3$

Subject to: $\quad y_1 + 3y_2 + 6y_3 + 18$

$\qquad\qquad y_1 + y_2 + 2y_3 \geq 8$

$\qquad\qquad 3y_1 + y_2 + 4y_3 \geq 5$

$\qquad\qquad y_1 \geq 0, y_2 \geq 0, y_3 \geq 0$

41. Amara needs to supplement her diet with at least 60 milligrams (mg) of vitamin C and 12 mg of niacin. The nutrients are available in two supplement pills, Everyday and Allday. Everyday contains 6 mg of vitamin C and 2 mg of niacin, and Allday contains 6 mg of vitamin C and 1 mg of niacin. Each Everyday pill cost $0.08 and each Allday pill costs $0.07. Let y_1 represent the number of Everyday pills taken daily, and let y_2 represent the number of Allday pills taken daily.

(a) Write a linear programming problem to minimize the cost of adding the nutrients to Amara's diet.

(b) Write the dual of the linear programming problem in part (a).

(c) Use the simplex method on the dual problem in part (b) to solve the original problem in part (a). Your answer should include the numbers of each pill taken and the daily minimum costs.

42. Niel, a dietitian for a sports team, is to prepare an energy snack from two foods, PowerSource and EnergyMax, to meet certain nutritional requirements. The combination of the two foods must have at least 550 grams of carbohydrates, 400 grams of protein, and 200 grams of total fat. The foods have different nutritional characteristics as given in Table R.5.

Table R.5

Food	Cost per Ounce	Carbo-hydrates	Protein	Fat
		Nutrition (grams per ounce)		
PowerSource	$0.10	30	8	2
EnergyMax	$0.12	40	10	4

(a) Define the variables and write a linear programming problem to minimize Niel's cost while still meeting the nutritional requirements.

(b) Write the dual of the linear programming problem in part (a).

(c) Use the simplex method on the dual problem in part (b) to solve the original problem in part (a). Your answer should include the number of ounces of each type of food to use and the minimum cost.

43. Henderson Oil has two refineries, one in Aztec, New Mexico, and one in Buena Vista, Colorado. Each refinery can refine the amounts of fuel, in barrels, in a 24-hour period as shown in Table R.6.

Table R.6

Maximum Production of:	Aztec	Buena Vista
Low-octane gasoline	800	1200
High-octane gasoline	500	600
Diesel	900	600

Daily operating costs for the Aztec refinery are $36,000, and daily operating costs for the Buena Vista refinery are $32,000. Lissa, the production manager at Henderson Oil, receives an order for 120,000 barrels of low-octane gasoline, 100,000 barrels of high-octane gasoline, and 120,000 barrels of diesel.

(a) How many days should Henderson Oil operate each refinery to fulfill (or even exceed) the order while minimizing costs?

(b) What is the minimum cost?

44. Table R.7 gives nutritional information, in grams, for selected items from Wendy's menu.

Table R.7

Item	Carbohydrates	Protein	Fat
Grilled chicken sandwich	44	30	21
Big Bacon Classic®	45	33	31
Large Frosty™	91	14	14

SOURCE: Wendy's, www.wendys.com

(a) Dave wishes to eat nothing but grilled chicken sandwiches, Big Bacon Classics®, and large Frostys™. He wants at least 650 grams of carbohydrates and 130 grams of protein in a day. How many of each item should he have to eat to minimize his fat intake and still meet the nutritional requirements?

(b) What is Dave's minimum daily fat intake?

45. (a) Determine the shadow costs of the nutrients in Exercise 41.

(b) Use shadow costs along with the minimum cost answer in Exercise 41(c) to determine the new cost if Amara changes her nutrient requirements to 61 mg of vitamin C and 13 mg of niacin daily.

In Exercises 46 and 47, add slack variables and subtract surplus variables to write each system of inequalities in equational form.

46. $4x_1 + 3x_2 \leq 9$

$\quad x_1 + 5x_2 \geq 4$

47. $4x_1 + x_2 + 3x_3 \leq 12$

$\quad x_1 + 2x_2 + 5x_3 \geq 9$

$\quad 3x_1 + x_2 + 7x_3 \leq 4$

In Exercises 48 and 49, complete the following:

(a) Convert to a maximization problem.

(b) Add slack variables and subtract surplus variables to write in equational form.

(c) Write the initial simplex tableau.

48. Minimize: $w = x_1 + 4x_2$

Subject to: $2x_1 + 5x_2 \leq 10$

$3x_1 + 4x_2 \geq 9$

$x_1 \geq 0, x_2 \geq 0$

49. Minimize: $w = x_1 + 3x_2 + 8x$

Subject to: $x_1 + 2x_2 + 5x_3 \leq 7$

$x_1 + x_2 + 8x_3 \geq 10$

$4x_2 + 2x_3 \geq 24$

$x_1 \geq 0, x_2 \geq 0, x_3 \geq 0$

In Exercises 50–55, use the method in Section 4.4 to solve each linear programming problem.

50. Maximize: $z = 3x_1 + x_2$

Subject to: $2x_1 + 5x_2 \leq 20$

$4x_1 + 3x_2 \leq 14$

$5x_1 + 6x_2 \geq 10$

$x_1 \geq 0, x_2 \geq 0$

51. Maximize: $z = 4x_1 + x_2$

Subject to: $x_1 + x_2 \geq 10$

$3x_1 + 4x_2 \geq 26$

$x_1 + 4x_2 \leq 16$

$x_1 \geq 0, x_2 \geq 0$

52. Maximize: $z = 2x_1 + 3x_2 + 4x_3$

Subject to: $3x_1 + x_2 + x_3 \leq 25$

$x_1 + 4x_2 + 4x_3 \leq 55$

$x_1 + 3x_2 - x_3 \geq 6$

$x_1 \geq 0, x_2 \geq 0, x_3 \geq 0$

53. Minimize: $w = 3x_1 + 2x_2$

Subject to: $x_1 + 4x_2 \leq 28$

$2x_1 + 5x_2 \geq 10$

$x_1 \geq 0, x_2 \geq 0$

54. Minimize: $w = 6x_1 + 2x_2$

Subject to: $2x_1 + x_2 \leq 18$

$x_1 + x_2 \leq 10$

$3x_1 + 4x_2 \geq 4$

$x_1 \geq 0, x_2 \geq 0$

55. Minimize: $w = 2x_1 + 3x_2 + 4x_3$

Subject to: $x_1 + x_2 + x_3 \leq 30$

$3x_1 + x_2 + x_3 \geq 15$

$4x_1 + 4x_2 + x_3 \geq 5$

$x_1 \geq 0, x_2 \geq 0, x_3 \geq 0$

56. Sopris Mountaineering makes two types of backpacks: 3500-cubic-inch daypacks and 5500-cubic-inch multiday packs. Each backpack requires the services of two departments, in hours, as shown in Table R.8.

Table R.8	Labor	
Department	**Day Pack**	**Multiday Pack**
Fabric cutting	7	10
Assembly and packaging	3	5

Fabric cutting has a maximum of 540 hours available each week, and assembly and packaging has a maximum of 260 hours available each week. Due to prior contracts, the total number of backpacks made each week must be at least 15. Each daypack yields \$20 in profit, and each multiday pack yields \$50 in profit.

(a) How many of each type of backpack should be made and sold to maximize weekly profit? What is the maximum weekly profit?

(b) Interpret the value of all slack and surplus variables in the final tableau.

57. (*continuation of Exercise 56*) Lindsay, the head of research and development at Sopris Mountaineering, has discovered a way to reduce the amount of time needed for the day pack to 6 hours in the fabric-cutting department and 2 hours in the assembly and packaging department, while increasing the profit per day pack to \$35. Everything else from Exercise 56 remains the same.

(a) How many of each type of backpack should be made and sold to maximize weekly profit? What is the maximum weekly profit?

(b) Interpret the value of all slack and surplus variables in the final tableau.

58. Table R.9 gives nutritional information, in grams, and price for selected items from Subway's menu.

Table R.9			
Item	**Protein**	**Fat**	**Price**
6-Inch roasted chicken	25	6	\$3.88
6-Inch Subway Club®	22	5	\$3.55

SOURCE: Subway, www.subway.com

Kris desires to eat nothing but 6-inch roasted chicken sandwiches and 6-inch Subway Clubs. He wants at least 119 grams of protein, but no more than 28 gram of fat. Also, he wishes to eat at least one 6-inch roasted chicken sandwich and at least one 6-inch Subway Club sandwich. If Kris eats nothing but 6-inch roasted chicken sandwiches and 6-inch Subway Club sandwiches and meets his nutritional requirements, answer the following:

(a) How much of each food should Kris eat to minimize his daily food cost?

(b) What is his minimum daily food cost?

🌐 **59.** (*continuation of Exercise 58*) Kris found out that Subway has increased its price of the 6-inch roasted chicken sandwich to $4.00 and increased the price of the 6-inch Subway Club sandwich to $3.70. Assuming that all other data in Exercise 58 remain the same, answer the following:

(a) How much of each food should Kris eat to minimize his daily food cost?

(b) What is his minimum daily food cost?

(c) Compare the results from parts (a) and (b) of Exercises 58 and 59.

60. Henderson Oil has two refineries, one in Aztec, New Mexico, and one in Buena Vista, Colorado. Each refinery can refine the amounts of fuel, in barrels, in a 24-hour period as shown in Table R.10.

Table R.10

Maximum Production of:	Aztec	Buena Vista
Low-octane gasoline	800	1200
High-octane gasoline	500	600
Diesel	900	600

Daily operating costs for the Aztec refinery are $36,000, and daily operating costs for the Buena Vista refinery are $32,000. Lissa, the production manager at Henderson Oil, receives an order for 40,000 barrels of low-octane gasoline, 100,000 barrels of high-octane gasoline, and 120,000 barrels of diesel. Due to budgetary issues, the total number of days that the refineries can operate cannot exceed 250 days.

(a) How many days should each refinery operate to minimize costs while meeting (or even exceeding) the order? What is the minimum cost?

(b) Interpret the value of all slack and surplus variables in the final tableau.

CHAPTER 4 PROJECT

SOURCE: www.thesims.com and *The Sims: Prima's Official Strategy Guide* by Rick Barba

Maxis is a computer game company known for its Sim games. One of the more recent and very popular computer games is called "The Sims." In the game, the player controls people (called Sims) and helps them to get a job and establish relationships and shows them the way to the bathroom. One of the more strategic points of the game is to make the Sims happy. To do this, the player must purchase accessories for the home, such as a couch or a bubble blower. Most objects will affect the Sims' motive(s): those being Hunger, Comfort, Hygiene, Bladder, Energy, Fun, Social, and Room. Some objects do a better job at satisfying needs than others. For instance, would you rather watch T.V. on a 13-inch black and white screen or on a 53-inch HDTV? Unfortunately, the objects that do a better job cost a lot more. The point of this project is to minimize the amount of money spent while satisfying the given needs.

We will consider six objects and only three motives: Fun, Room, and Comfort. Table 1 gives information about these six objects and the needs that they satisfy.

Suppose that a Sim needed at least 16 Fun points, 6 Room points, and 15 Comfort points.

1. Define the six items using x_1 through x_6 as your variables. Write the objective function and the problem constraints.

2. Write the mathematical model in 1 in equational form. Write the initial simplex tableau.

3. Solve the linear program by using the Two-Phase Method in any way that you wish and interpret your results.

4. Referring to the objective function in 1 and the results of 3, why does it make sense that these items were purchased?

5. Referring to 3, for some items more than one is needed. Suppose that we don't want to purchase more than two items. Repeat 1 and 2 and construct new constraints for those items so that no more than two are purchased. Do not solve the corresponding linear programming problem.

Table 1   Items and Corresponding Price, Fun, Room, and Comfort Values

Item	Price § (in Simoleans)	Fun	Room	Comfort
Super Schlooper Bubble Blower	§710	3	0	2
Liebefunkenmann Piano	§5399	4	6	0
Spazmatronic Plasmatronic Go-Go Cage	§1749	7	3	0
WhirlWizard Hot Tub	§6500	2	0	6
Sky Surfer Inflatable Sofa	§190	0	1	3
Dolce Tutti Frutti Sofa	§1450	0	3	9

Chapter 5

Mathematics of Finance

$$FV = PMT \; \frac{(1+i)^n - 1}{i}$$

$$= 82.50 \; \frac{(1+\frac{0.0525}{12})^{360} - 1}{\frac{0.0525}{12}}$$

$$\approx 71{,}924.44$$

The old saying "a penny saved is a penny earned" is more than just a cliché. Consider the fact that in 2000, the average price of a pack of cigarettes in the U.S. was $2.75. Assuming a 30-day month, a pack-a-day smoker spent $82.50 per month on cigarettes. If the smoker quit and invested the monthly cigarette budget of $82.50 in a savings account earning 5.25% interest compounded monthly, the value of the account after 30 years would be greater than $70,000.

What We Know

In Chapters 3 and 4 we studied an important family of problems called linear programming problems. Analysts and managers use this to effectively run and operate their businesses.

Where Do We Go

In this chapter we study interest, another important topic for business people as well as those not necessarily in the management sciences. A good understanding of interest is essential for anyone who wants to manage their own finances or those of other people.

5.1 **Simple Interest**

5.2 **Compound Interest**

5.3 **Future Value of an Annuity**

5.4 **Present Value of an Annuity**

Chapter Review Exercises

Section 5.1 Simple Interest

When we deposit money into a savings account or purchase a certificate of deposit, we have essentially loaned our money to the bank. When we borrow money from the bank to purchase a vehicle or a house, the bank has loaned us money. In either case, the amount of money loaned is called the **principal**. The individual or bank that loans the money receives a fee for lending the money. This fee is called **interest**. See Figure 5.1.1. This interest is classified as either **simple interest** or **compound interest**. We begin our journey into the mathematics of finance with the study of simple interest since the concept of simple interest lays the foundation for much of the material in this chapter.

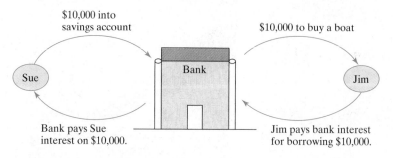

Figure 5.1.1

Simple Interest

From Your Toolbox
Interest rates should be converted to decimals before being used in any formulas. For more review on working with percents, consult the review in Appendix A.1.

When financial institutions advertise **interest rates** on certificates of deposit (CDs) or on loans, it usually means **rate per year**. In general, rates are expressed as percents. Since the word percent means **per hundred**, we can quickly convert from percents to decimals, and vice versa. For our purposes in this chapter, remember the information in the Toolbox to the left.

Simple interest is earned on the amount borrowed and not on any past interest. The **time** that the money is earning interest is given in years. **Simple interest** is simply the product of the principal, rate, and time.

■ **Simple Interest**

To compute **simple interest**, use the formula $I = Prt$, where

$$P = \text{principal (amount invested or borrowed)}$$
$$r = \text{interest rate per year (as a decimal)}$$
$$t = \text{time in years}$$
$$I = \text{interest (earned or paid)}$$

Example 1 Calculating Simple Interest

Lisa borrows $10,000 from her parents for a down payment on a house. She agrees to pay back the loan in 5 years at the rate of 6.5% simple interest. How much interest will Lisa pay?

Solution

Understand the Situation: For this situation, we have that the principal P is $10,000, the interest rate r is 6.5% = 0.065, and the time t is 5 years.

So the interest that Lisa will pay is

$$I = Prt$$
$$= 10{,}000(0.065)(5)$$
$$= \$3250$$

Interpret the Solution: Lisa will pay \$3250 in interest in order to use the \$10,000 for 5 years.   ∎

✓ **Checkpoint 1**

Now work Exercise 1.

When Lisa from Example 1 pays back the loan, she has to pay back the principal, P, plus the interest, I. So the total amount, A, that Lisa pays back to her parents is

$$A = P + I = \$10{,}000 + \$3250 = \$13{,}250$$

For Lisa, the \$10,000 was the amount that she borrowed now, so this amount is also referred to as a **present value**. The \$13,250 she pays back in 5 years is called the **future value**. In other words, \$13,250 is the future value of \$10,000 at a simple interest rate of 6.5% for 5 years.

In general, if a principal P is borrowed at a simple interest rate r, then after t years the borrower owes the lender an amount A, where

$$A = P + I$$
$$= P + Prt$$
$$= P(1 + rt)$$

P is also called the **present value**, and A is also known as the **future value**.

■ **Future Value**

The **future value** A of P dollars at interest rate r (as a decimal) for t years is

$$A = P(1 + rt)$$

Example 2   Calculating Total Amount Due

Juanita borrows \$6500 at 9.9% simple interest for 54 months. How much does Juanita owe at the end of the 54 months?

Solution

Interactive Activity

Using the result of Example 2, what is the future value of \$6500 at an interest rate of 9.9% for 4.5 years?

Understand the Situation: Here P is \$6500 and the simple interest rate, as a decimal, is 0.099. Recall that time t must be in years. So 54 months is $\frac{54}{12} = 4.5$ years. The amount Juanita that owes at the end of 54 months is

$$A = P(1 + rt)$$
$$= 6500[1 + 0.099(4.5)]$$
$$= \$9395.75$$

Interpret the Solution: The total amount due on Juanita's \$6500 loan at 9.9% simple interest for 54 months is \$9395.75.   ∎

The formula for future value, $A = P(1 + rt)$, has four variables. If we have values for any three of these variables, we can solve for the fourth. Examples 3 and 4 illustrate how we can do this.

Example 3 **Determining a Simple Interest Rate**

Joshua borrows $2500 from Zachary and promises to pay back $2630 in 6 months. What simple interest rate will Joshua pay?

Solution

Understand the Situation: We can use the formula for future value with $A = 2630$, $P = 2500$, and $t = \dfrac{6}{12} = 0.5$ and solve for r.

This gives us

$$A = P(1 + rt)$$
$$2630 = 2500[1 + r(0.5)]$$
$$2630 = 2500 + 1250r \qquad \text{Distributive property}$$
$$130 = 1250r$$
$$r = \frac{130}{1250} = 0.104 = 10.4\%$$

Interpret the Solution: This means that Joshua will pay a simple interest rate of 10.4%. ∎

✓ **Checkpoint 2** Now work Exercise 19.

Example 4 **Determining a Present Value**

Shannon desires to earn a simple interest rate of 9% on her investment. How much should she pay, to the nearest cent, for a note that will be worth $7000 in 6 months?

Solution

Understand the Situation: Again we use the formula for future value, $A = P(1 + rt)$. However, here we are interested in finding the principal, P, or the present value. We have $A = 7000$, $r = 0.09$, and $t = \dfrac{6}{12} = 0.5$.

Using these values gives us

$$A = P(1 + rt)$$
$$7000 = P[1 + 0.09(0.5)]$$
$$7000 = 1.045P$$
$$P \approx \$6698.56 \qquad \text{Rounded to the nearest cent}$$

Interpret the Solution: This means that Shannon should pay about $6698.56 for the note. ∎

✓ **Checkpoint 3** Now work Exercise 29.

Table 5.1.1

Time, t, in Years	Amount Owed (future value)
1	$2140
2	2280
4	2560
5	2700
8	3120
10	3400

Future Value and Time

Suppose that Gene borrows $2000 from his Aunt Frances. Frances tells Gene to repay the principal plus 7% simple interest whenever he is able. Using the formula for future value, $A = P(1 + rt)$, Gene computes how much he would owe his aunt if he repaid the loan after various periods of time. Since $P = 2000$ and $r = 0.07$, the formula for future value becomes

$$A = P(1 + rt)$$
$$A = 2000(1 + 0.07t) = 2000 + 140t$$

Table 5.1.1 shows the future values for various values of t.

If we plot the points in Table 5.1.1 and connect them, we get the graph in Figure 5.1.2. Notice in the graph that the future value is a **linear function of time, t**.

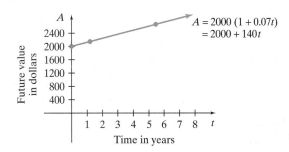

Figure 5.1.2

Technology Option

If we replace A with y and t with x, the equation graphed in Figure 5.1.2 becomes

$$y = 2000 + 140x$$

where y represents the future value and x represents the time in years. We can enter this equation into our graphing calculator and quickly reproduce Table 5.1.1 and Figure 5.1.2. Table 5.1.2 and Figure 5.1.3 show the output.

Table 5.1.2

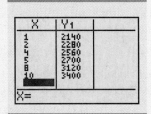
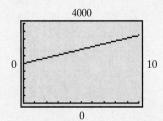

Figure 5.1.3

To learn how to make a table or to use the TABLE command, consult the online graphing calculator manual at www.prenhall.com/armstrong.

Example 5 **Analyzing Future Values as a Function of Time**

Stefani deposits $5000 into a savings account that earns 5.5% simple interest.

(a) Write the future value, A, as a function of time, t.

(b) Make a table to determine A for $t = 1, 3, 5, 7$ and 9.

(c) Graph the equation from part (a). Determine the slope and interpret.

Table 5.1.3

t	$A = 5000 + 275t$
1	$5275
3	5825
5	6375
7	6925
9	7475

Solution

(a) We use the formula for future value, $A = P(1 + rt)$, with $P = 5000$ and $r = 0.055$ to get

$$A = 5000(1 + 0.055t)$$
$$= 5000 + 275t$$

So the future value of Stefani's $5000 at 5.5% simple interest as a function of time is given by $A = 5000 + 275t$.

(b) Table 5.1.3 shows the results for $t = 1, 3, 5, 7$, and 9.

(c) Figure 5.1.4 shows a graph of $A = 5000 + 275t$. The slope of the line is $m = 275$ (consult the Toolbox to the left).

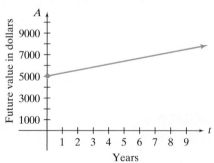

Figure 5.1.4 Graph of $A = 5000 + 275t$

From Your Toolbox

A linear equation in slope–intercept form $y = mx + b$ has slope given by m.

Interpret the Solution: This means that each year Stefani earns $275 in interest on her $5000 deposit.

Doubling Time

Many times we wish to know how long it will take for our money to double. For example, suppose that Stefani in Example 5 wants to know when her deposit of $5000 at 5.5% simple interest has a value of $10,000. Since the future value as a function of time, as found in Example 5(a), is given by

$$A = 5000 + 275t$$

we can algebraically determine the time necessary for her money to double by substituting $10,000 for A and solving for t.

$$10,000 = 5000 + 275t$$
$$5000 = 275t$$
$$t = \frac{5000}{275} \approx 18.18 \text{ years}$$

So it would take about 18.18 years for Stefani's $5000 to double to $10,000 at 5.5% simple interest.

Technology Option

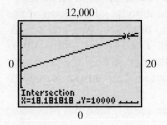

Figure 5.1.5

We can support the algebraic solution for the time required for Stefani's $5000 to double to $10,000 by graphing in the same viewing window $y_1 = 5000 + 275t$ and $y_2 = 10,000$ and using the INTERSECT command. See Figure 5.1.5.

To learn more about using the INTERSECT command on your calculator, consult the online graphing calculator manual at www.prenhall.com/armstrong.

In general, the time required for an amount of money P to double in value to $2P$ at a simple interest rate r can be determined by solving the following equation for t:

$$2P = P(1 + rt)$$

Dividing both sides by P gives us

$$2 = 1 + rt$$

Subtracting 1 from both sides and then dividing by r yields

$$1 = rt$$

$$t = \frac{1}{r}$$

> ### Doubling Time for Simple Interest
>
> The time required for money to double is given by
>
> $$t = \frac{1}{r}$$
>
> where t is time in years and r is the simple interest rate (as a decimal).

Example 6 **Determining a Doubling Time**

How long will it take $8000 to double at 8% simple interest?

Solution

Understand the Situation: The principal (present value) is not important in answering this question. According to the formula, $t = \frac{1}{r}$, we only need to know r to determine the doubling time. Hence we have

$$t = \frac{1}{r}$$

$$= \frac{1}{0.08} = 12.5$$

Interpret the Solution: So at 8% simple interest the doubling time is 12.5 years.

Discounted Loans and Proceeds

Usually, when money is borrowed, it is expected that the amount borrowed (principal) plus interest will have been repaid at the end of the loan term. However, sometimes lenders collect the interest from the amount of the loan at the time that the loan is made. This type of loan is said to be **discounted**, and the interest that is deducted from the loan is the **discount**. The amount that the borrower receives is then called the **proceeds**. When the loan is due, the borrower pays back the amount before the discount was deducted. The amount repaid to the lender is called the **face value** of the loan.

If F represents the amount of the loan (its face value), then the proceeds, p, are simply

$$p = F - D$$

where D is the interest deducted, or the discount. Since the interest is given by

$$D = Frt$$

we have that the proceeds of a discounted loan are

$$p = F - Frt = F(1 - rt)$$

> ■ **Discounted Loans and Proceeds**
>
> The **discount**, D, on a discounted loan of F dollars at annual simple interest rate r over t years is given by
>
> $$D = Frt$$
>
> The **proceeds**, p, of the loan is what the borrower receives and is given by
>
> $$p = F - Frt = F(1 - rt)$$

Example 7   Computing Discount and Proceeds

Franklin signs a note for $20,000 at 10% annual simple interest rate for 8 months. Compute the discount and the proceeds.

Solution

Understand the Situation: We have $F = 20,000, r = 0.10$, and $t = \dfrac{8}{12} = \dfrac{2}{3}$. The discount is

$$D = Frt$$

$$= 20,000(0.10)\left(\frac{2}{3}\right)$$

$$\approx 1333.33 \qquad \text{Rounded to the nearest cent}$$

So the discount is $1333.33. This amount is deducted from the loan amount, F, to give the proceeds.

$$p = F - D$$

$$= 20,000 - 1333.33$$

$$= 18,666.67$$

Interpret the Solution: This means that Franklin receives $18,666.67. However, after 8 months he must repay the face value of $20,000. ■

✓ **Checkpoint 4**   Now work Exercise 65.

SUMMARY

This section focused on discussing **simple interest**. **Future value** and **present value** of money were discussed, as was how to represent future value as a function of time. We analyzed the doubling time for simple interest and concluded the section with a look at **discounted loans** and **proceeds**.

- **Simple Interest:** To compute simple interest, use the formula $I = Prt$, where P = principal (amount invested or borrowed), r = interest rate per year (as a decimal), t = time in years, and I = interest (earned or paid).
- **Future Value:** The **future value** A of P dollars at interest rate r (as a decimal) for t years is $A = P(1 + rt)$.

- **Doubling Time for Simple Interest:** The time required for money to double is given by $t = \dfrac{1}{r}$, where t is time in years and r is the simple interest rate (as a decimal).
- **Discounted Loans and Proceeds:** The **discount**, D, on a discounted loan of F dollars at annual interest rate r over t years is given by $D = Frt$. The **proceeds**, p, of the loan are what the borrower receives and are given by $p = F - Frt = F(1 - rt)$.

SECTION 5.1 EXERCISES

In Exercises 1–8, the amount of money P is borrowed for period of time t at simple interest rate r. Compute the simple interest owed for the use of the money. Round answers to the nearest cent.

✓ **1.** $P = \$1000, r = 6.25\%, t = 3$ years

2. $P = \$2000, r = 7.25\%, t = 2$ years

3. $P = \$10,000, r = 5.5\%, t = 5$ years

4. $P = \$20,000, r = 6.5\%, t = 7$ years

5. $P = \$5000, r = 8.5\%, t = 9$ months

6. $P = \$8000, r = 7.5\%, t = 8$ months

7. $P = \$25,000, r = 8\%, t = 66$ months

8. $P = \$15,000, r = 9\%, t = 78$ months

In Exercises 9–16, the amount of money P is borrowed for period of time t at simple interest rate r. Compute the total amount, A, owed at the end of time t. (Recall that this is the same as future value.) Round answers to the nearest cent.

9. $P = \$2000, r = 7\%, t = 2$ years

10. $P = \$1000, r = 6\%, t = 3$ years

11. $P = \$10,000, r = 6.5\%, t = 5$ years

12. $P = \$25,000, r = 9.5\%, t = 7$ years

13. $P = \$7000, r = 7.5\%, t = 8$ months

14. $P = \$5000, r = 8.5\%, t = 9$ months

15. $P = \$20,000, r = 9.75\%, t = 78$ months

16. $P = \$15,000, r = 8.75\%, t = 66$ months

In Exercises 17–24, the amount of money P loaned for a period of time t along with the simple interest I charged are given. Determine the simple interest rate r of the loan. Round percentages to the nearest hundredth. See Example 3.

17. $P = \$2000, I = \$150, t = 1$ year

18. $P = \$3000, I = \$160, t = 1$ year

✓ **19.** $P = \$5000, I = \$900, t = 2$ years

20. $P = \$5000, I = \$880, t = 2$ years

21. $P = \$10,000, I = \$4280, t = 4.5$ years

22. $P = \$15,000, I = \$6300, t = 4.5$ years

23. $P = \$20,000, I = \$14,200, t = 6$ years, 9 months

24. $P = \$22,000, I = \$13,400, t = 5$ years, 9 months

In Exercises 25–32, determine the present value P given the future value A, time t, and annual simple interest rate r. Round to the nearest cent. See Example 4.

25. $A = \$5000, r = 8\%, t = 2$ years

26. $A = \$7500, r = 7\%, t = 3$ years

27. $A = \$10,000, r = 5.25\%, t = 5$ years

28. $A = \$12,000, r = 6.25\%, t = 6$ years

✓ **29.** $A = \$4000, r = 6.5\%, t = 9$ months

30. $A = \$3000, r = 5.5\%, t = 8$ months

31. $A = \$20,000, r = 7.75\%, t = 78$ months

32. $A = \$25,000, r = 8.25\%, t = 66$ months

In Exercises 33–36, complete the following:

(a) *Write the future value, A, as a function of time t. See Example 5(a).*

(b) *Graph the equation from part (a). Determine the slope and interpret. See Example 5(c).*

33. Sue deposits $8000 into a savings account that earns 6.5% simple interest.

34. Cheryl deposits $5000 into a savings account that earns 5.5% simple interest.

35. Rich deposits $4000 into a savings account that earns 5.25% simple interest.

36. Renee deposits $6000 into a savings account that earns 5.75% simple interest.

In Exercises 37–42, determine how long it will take a deposit to double at the given simple interest rate. Round answers to the nearest hundredth.

37. $r = 10\%$ **38.** $r = 6\%$

39. $r = 8.5\%$ **40.** $r = 6.75\%$

41. $r = 5.75\%$ **42.** $r = 7.25\%$

In Exercises 43–48, determine the discount and the proceeds given the loan amount F, annual interest rate r and time t. See Example 7.

43. $F = \$1000, r = 7\%, t = 8$ months

44. $F = \$1500, r = 8\%, t = 9$ months

45. $F = \$5000, r = 6.5\%, t = 2$ years

46. $F = \$6000, r = 8.5\%, t = 3$ years

47. $F = \$10,000, r = 8.25\%, t = 42$ months

48. $F = \$15,000, r = 7.75\%, t = 66$ months

In Exercises 49–54, determine the loan amount F to produce the given proceeds p at annual interest rate r over time t.

49. $p = \$1000, r = 7\%, t = 8$ months

50. $p = \$1500, r = 8\%, t = 9$ months

51. $p = \$5000, r = 6.5\%, t = 2$ years

52. $p = \$6000, r = 8.5\%, t = 3$ years

53. $p = \$10,000, r = 8.25\%, t = 42$ months

54. $p = \$15,000, r = 7.75\%, t = 66$ months

Applications

55. Shelby borrows $25,000 for 8 years at a simple interest rate of 9%. Determine the interest due on the loan.

56. Lena borrows $22,000 for 7 years at a simple interest rate of 8.5%. Determine the interest due on the loan.

57. Diane borrows $3000 for 8 months at a simple interest rate of 8%. Determine the interest due on the loan.

58. Jeff borrows $5000 for 9 months at a simple interest rate of 9%. Determine the interest due on the loan.

🌐 **59.** In June 2000, the World Bank granted China a loan for $200 million (*Source:* http://suratkabar.com/arsip/

6/2/170615.shtml). Assuming a simple interest rate of 9%, how much interest is due on the loan after 1 year?

🌐 **60.** In November 1996, the World Bank granted the Philippines $281 million in loans (*Source:* www.worldbank.org). Assuming a simple interest rate of 8%, how much interest is due on the loan after 1 year?

61. Ardra borrowed $22,000 from her aunt to start a beauty salon. She repaid her aunt after 2 years. The annual simple interest rate was 8.25%. Determine the total amount she repaid.

62. Maria borrowed $15,000 from her uncle to start an auto parts store. She repaid him after 3 years. The annual simple interest rate was 7.25%. Determine the total amount she repaid.

63. Denyse borrowed $20,000 from her grandfather to start a bakery. She repaid him after 30 months. The annual simple interest rate was 7.25%. Determine the total amount she repaid.

64. Mike borrowed $18,000 from his mother to start a catering business. He repaid her after 42 months. The annual simple interest rate was 8.75%. Determine the total amount he repaid.

✓ **65.** Brian signs a note for $7000 at 8.5% annual simple interest rate for 2 years. Assuming that this is a discounted loan, determine the discount and the proceeds.

66. Sandy signs a note for $6000 at 8.25% annual simple interest rate for 3 years. Assuming that this is a discounted loan, determine the discount and the proceeds.

67. Suppose that Brian in Exercise 65 actually needs $7000, that is, he needs the proceeds of the discounted loan to be $7000. Assuming a 8.5% annual simple interest rate for 2 years, determine the amount that Brian must borrow so that his proceeds are $7000.

68. Suppose that Sandy in Exercise 66 actually needs $6000, that is, she needs the proceeds of the discounted loan to be $6000. Assuming a 8.25% annual simple interest rate for 3 years, determine the amount that Sandy must borrow so that her proceeds are $6000.

69. Melissa wants to buy a $3000 entertainment center in 2 years. How much should she invest now at 6.75% simple interest to have enough for the entertainment center in 2 years?

70. Don wants to buy a $40,000 recreational vehicle in 6 years. How much should he invest now at 7.25% simple interest to have enough for the recreational vehicle in 6 years?

71. In 52 months, Bill wants to build a $20,000 three-season room addition on his house. How much should he invest now at 5.25% simple interest to have enough for the addition in 52 months?

72. In 66 months, Lisa wants to have $8000 worth of remodeling done to her basement. How much should she invest now at 4.75% simple interest to have enough for the remodeling in 66 months?

73. Francine borrows $2500 from Arthur and promises to pay back $2912.50 in 2 years. What interest rate will Francine pay?

74. Brock borrows $1400 from Misty and promises to pay back $1725.50 in 3 years. What interest rate will Brock pay?

75. Lenny borrows $5000 from Robert and promises to pay back $5912.50 in 2.5 years. What interest rate will Lenny pay?

76. Aida borrows $10,000 from Diane and promises to pay back $13,881.25 in 5 years, 9 months. What interest rate will Aida pay?

77. Dylan desires to earn a simple interest rate of 8.5% on his investment. How much should he pay, to the nearest cent, for a note that will be worth $5000 in 3 years?

78. Elle desires to earn a simple interest rate of 7.5% on her investment. How much should she pay, to the nearest cent, for a note that will be worth $8000 in 4 years?

79. As of 8/26/00, BankAtlantic.com offers a 12-month CD (certificate of deposit) at 7.25% given a minimum deposit of $1000 (*Source:* www.bankatlantic.com). Assuming the 7.25% is simple interest, how much, to the nearest cent, should Franco invest to have $2000 after 12 months?

80. As of 8/26/00, BankAtlantic.com offers a 12-month CD (certificate of deposit) at 7.25% given a minimum deposit of $1000 (*Source:* www.bankatlantic.com). Assuming the 7.25% is simple interest, how much, to the nearest cent, should Serena invest to have $3000 after 12 months?

81. Treasury bills (T-bills) can be purchased from the U.S. Treasury Department. At maturity, you are paid the face value of the bill. As of May 17, 2001, you could buy a 13-week T-bill with a face value of $1000 for $990.82 (*Source:* www.publicdebt.treas.gov). What annual simple interest rate does this T-bill earn?

82. Treasury bills (T-bills) can be purchased from the U.S. Treasury Department. At maturity, you are paid the face value of the bill. As of May 17, 2001, you could buy a 26-week T-bill with a face value of $1000 for $981.60 (*Source:* www.publicdebt.treas.gov). What annual simple interest rate does this T-bill earn?

83. If Alicia desires to earn an annual simple interest rate of 9.5% on a 26-week T-bill with a face value of $5000, how much should she pay for the T-bill?

84. If Horace desires to earn an annual simple interest rate of 8.75% on a 26-week T-bill with a face value of $10,000, how much should he pay for the T-bill?

85. Bill is considering a **home equity loan** that has an 8.5% annual simple interest rate. His home has appraised for $191,000 and he still has a balance of $120,000 on his first mortgage.

(a) Compute 80% of the appraisal minus the balance of his first mortgage. This is called the **loan to value** ratio and is the amount that the bank will loan Bill as a home equity loan.

(b) Interest is paid yearly on a home equity loan and is tax deductible. How much will Bill be able to take as a tax deduction in a given year due to interest on the home equity loan?

86. (*continuation of Exercise 85*) Suppose that Bill in Exercise 85 takes the home equity loan as computed in Exercise 85(a).

(a) Assuming the interest rate from Exercise 85, what is the total amount that Bill repays if he has the loan for 6 years?

(b) For the amount in part (a), what would his monthly payment be if he pays the loan off in 6 years with equal monthly payments? (*Hint:* There are 72 months in 6 years.)

SECTION PROJECT

In discounted loans, the actual interest rate can be different from the amount quoted by the bank. The stated rate is called the **nominal rate**, whereas the rate of interest actually paid is called the **effective rate**. For simple interest rate loans, the effective rate and the nominal rates are the same. However, for simple discount notes, the effective rate can be larger than the nominal rate, as this section project illustrates.

Mark Freeman agrees to pay his banker $20,000 in 6 months. The banker subtracts a discount of 8% and gives the rest to Mark.

(a) Compute the discount.
(b) Compute the proceeds.
(c) Using the formula $A = P(1 + rt)$, we can compute the effective rate. Use $A = P(1 + rt)$ with $A = 20,000$ (the loan amount), $P =$ the result from part (b), and $t = 0.5$ (6 months) and solve for r. Convert to a percent and this is the effective rate of interest.
(d) Rework parts (a)–(c) if Mark agrees to pay his banker the $20,000 in 1 year. Assume that the banker subtracts the same discount of 8%.

Section 5.2 Compound Interest

The concept of simple interest and how to compute it presented in Section 5.1 laid the foundation for our next topic, **compound interest**. If the interest earned is added to the original principal so that the next time interest is computed it is based on the original principal plus interest, then this type of interest is said to be **compounded**. In other words,

> **Compound interest** is interest earned on the initial principal plus previously earned interest.

Compounding Periods

An important component in working with situations involving compound interest, is the **compounding period** or simply the **period**. The compounding period is the number of times a year that interest is compounded and is denoted by k. This means that at the end of each compounding period interest is calculated on the initial principal plus all interest earned to that point. Table 5.2.1 gives the most popular compounding periods used by most financial institutions.

Table 5.2.1

Compound Period Name	Number of Compounding Periods per Year, k
Annually	1
Semiannually	2
Quarterly	4
Monthly	12
Weekly	52
Daily	365

When interest is compounded k times per year, we need to determine the **interest per compounding period** or the **periodic rate**. The following shows how to compute the interest per compounding period.

■ **Interest per Compounding Period**

If r is the annual interest rate and k is the number of compounding periods per year, then the interest rate per compounding period, i, is given by

$$i = \frac{r}{k}$$

▶ **Note:** We can think of i as giving the rate of interest for each compounding period.

Example 1 Determining an Interest Rate per Compounding Period

Suppose that a bank pays an annual interest rate of $r = 8\%$. Determine the interest rate per compounding period, i, when the number of compounding periods per year, k, is

    (a) $k = 1$ (compounded annually)

    (b) $k = 4$ (compounded quarterly)

    (c) $k = 12$ (compounded monthly)

Solution

(a) The interest rate per compounding period is

$$i = \frac{r}{k} = \frac{0.08}{1} = 0.08 \quad (\text{or } 8\%)$$

(b) The interest rate per compounding period is

$$i = \frac{r}{k} = \frac{0.08}{4} = 0.02 \quad (\text{or } 2\%)$$

(c) The interest rate per compounding period is

$$i = \frac{r}{k} = \frac{0.08}{12} \approx 0.006667 \quad (\text{or about } 0.6667\%) \qquad \blacksquare$$

> **Interactive Activity**
>
> For the annual interest rate given in Example 1, compute i when the number of compounding periods per year, k, is 2, 52, and 365.

Formula for Compound Interest

To aid us in developing a formula for compound interest, let's review the simple interest formula from Section 5.1 and use it as shown in Example 2.

Example 2 **Computing Compound Interest**

Suppose that a bank pays 5% annual interest compounded quarterly. If we deposit $1000 into a savings account and the quarterly interest earned is left in the account, we can compute how much money is in the account after 1 year. Determine the amount in the account after 1 year.

Solution

Recall from Section 5.1 that the future value A of P dollars at interest rate r (as a decimal) for t years is given by $A = P(1 + rt)$. After 1 quarter (3 months) the amount in the account is

$$A = P(1 + rt)$$

$$= 1000\left[1 + 0.05\left(\frac{3}{12}\right)\right] \qquad \text{Note that } t = \frac{3}{12}.$$

$$= 1000\left[1 + 0.05\left(\frac{1}{4}\right)\right] = \$1012.50$$

To compute the amount after the second quarter, the amount after the first quarter, $1012.50, is now the principal. Hence the amount after the second quarter is

$$A = P(1 + rt)$$

$$= 1012.50\left[1 + 0.05\left(\frac{1}{4}\right)\right]$$

$$\approx \$1025.16 \qquad \text{Rounded to the nearest cent}$$

To compute the amount after the third quarter, the amount after the second quarter, \$1025.16, is now the principal. So the amount after the third quarter is

$$A = P(1 + rt)$$
$$= 1025.16\left[1 + 0.05\left(\frac{1}{4}\right)\right]$$
$$\approx \$1037.97 \quad \text{Rounded to the nearest cent}$$

Finally, to compute the amount after the fourth quarter, the amount after the third quarter, \$1037.97, is now the principal. Hence the amount after the fourth quarter is

$$A = P(1 + rt)$$
$$= 1037.97\left[1 + 0.05\left(\frac{1}{4}\right)\right]$$
$$\approx \$1050.94 \quad \text{Rounded to the nearest cent}$$

Interpret the Solution: This means that after 1 year the total in the savings account is \$1050.94. ∎

We would like to generalize the process in Example 2 and determine a formula for compound interest. Our first observation is that each time we computed the amount at the end of the given quarter in Example 2 we used the formula

$$A = P(1 + rt)$$

Observe that in each calculation the expression in parenthesis $(1 + rt)$ was of the form

$$\left[1 + 0.05\left(\frac{1}{4}\right)\right] = \left(1 + \frac{0.05}{4}\right)$$

Notice that $\dfrac{0.05}{4} = \dfrac{r}{k} = i$, the **interest per compounding period**. So, using Example 2 as a model, in general the amount A_1 after the first compounding period is

$$A_1 = P(1 + i) = P + Pi$$

We notice that the interest earned is given by Pi. Since A_1 takes the place of the principal P for the second compounding period, the amount A_2 after the second compounding period is

$$\begin{aligned} A_2 = A_1 + A_1 i &= A_1(1 + i) \\ &= P(1 + i)(1 + i) \quad \text{Since } A_1 = P(1 + i) \\ &= P(1 + i)^2 \end{aligned}$$

Since A_2 takes the place of the principal P for the third compounding period, the amount A_3 after the third compounding period is

$$\begin{aligned} A_3 = A_2 + A_2 i &= A_2(1 + i) \\ &= P(1 + i)^2(1 + i) \quad \text{Since } A_2 = P(1 + i)^2 \\ &= P(1 + i)^3 \end{aligned}$$

Since A_3 takes the place of the principal P for the fourth compounding period, the amount A_4 after the fourth compounding period is

$$\begin{aligned} A_4 = A_3 + A_3 i &= A_3(1 + i) \\ &= P(1 + i)^3(1 + i) \quad \text{Since } A_3 = P(1 + i)^3 \\ &= P(1 + i)^4 \end{aligned}$$

Interactive Activity

Rework Example 2, except here assume that the bank paid 5% simple interest. In other words, what is the future value of \$1000 at simple interest rate 5% in 1 year?

Generalizing this pattern, after n periods the amount is given by the following:

Amount of Compound Interest

A principal P earning compound interest at a rate of i percent per compounding period has a **compound amount** (or **future value**) after n periods of

$$A = P(1+i)^n$$

where $i = \dfrac{r}{k}$ and $n = kt$. Also,

> P = principal (present value)
> i = interest rate per compounding period
> r = annual interest rate
> k = number of compounding periods in 1 year
> t = time, in years
> n = total number of compounding periods
> A = compound amount (future value) at end of n periods

▶ **Note:** Since $i = \dfrac{r}{k}$ and $n = kt$, an alternative compound interest formula is

$$A = P\left(1 + \frac{r}{k}\right)^{kt}.$$

Example 3 Using the Compound Interest Formula

Suppose that $2000 is invested at an 8% annual interest rate. What is the amount of the investment, rounded to the nearest cent, after 6 years if the 8% annual interest rate is compounded

(a) Annually

(b) Quarterly

(c) Monthly

Solution

For all three parts of the example, $P = 2000$, $r = 0.08$, and $t = 6$.

(a) For annual compounding, we have $i = \dfrac{r}{k} = \dfrac{0.08}{1} = 0.08$ and $n = kt = 1 \cdot 6 = 6$. The amount A is given by

$$A = P(1+i)^n$$
$$= 2000(1 + 0.08)^6 \approx \$3173.75$$

(b) For quarterly compounding, we have $i = \dfrac{r}{k} = \dfrac{0.08}{4} = 0.02$ and $n = kt = 4 \cdot 6 = 24$. The amount A is given by

$$A = P(1+i)^n$$
$$= 2000(1 + 0.02)^{24} \approx \$3216.87$$

(c) For monthly compounding, we have $i = \dfrac{r}{k} = \dfrac{0.08}{12}$ and $n = kt = 12 \cdot 6 = 72$. The amount A is given by

$$A = P(1+i)^n$$
$$= 2000\left(1 + \frac{0.08}{12}\right)^{72} \approx \$3227.00$$

Notice that we did not use a decimal approximation for $i = \dfrac{0.08}{12}$ in the formula. In general, we save any rounding until the final answer. ∎

✓ **Checkpoint 1**   Now work Exercise 7.

Technology Option

Many calculators have built-in financial commands. Figure 5.2.1(a) shows the input for Example 3(c) on a TI-83 using TVM Solver. Figure 5.2.1(b) shows the result (FV) in the TVM Solver window, and Figure 5.2.1(c) shows the result of tvm_FV. (Note that tvm_FV finds the future value.) To learn more about these commands or to see if your calculator has a financial menu, consult the online graphing calculator manual at www.prenhall.com/armstrong.

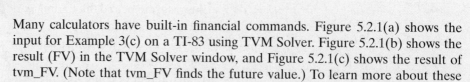

(a)   (b)   (c)

Figure 5.2.1 (a) $PV = -2000$ because it is a deposit. **(b)** Place the cursor after the equals sign for FV and press ALPHA ENTER. An indicator square in the left column designates the solution variable.

As Example 3 indicates, the more compounding periods there are, the greater the future value. Table 5.2.2 shows the results of Example 3 along with other compounding periods.

Table 5.2.2

Annual Rate (%)	Periodic Rate	Compounding Method	Total Number of Periods	Future Value after 6 Years	Interest Earned
8	$\frac{0.08}{1}$	Annually	6	$3173.75	$1173.75
8	$\frac{0.08}{2}$	Semiannually	12	3202.06	1202.06
8	$\frac{0.08}{4}$	Quarterly	24	3216.87	1216.87
8	$\frac{0.08}{12}$	Monthly	72	3227.00	1227.00
8	$\frac{0.08}{52}$	Weekly	312	3230.96	1230.96
8	$\frac{0.08}{365}$	Daily	2190	3231.98	1231.98

It may be surprising that the difference shown in Table 5.2.2 between compounding weekly and compounding daily, $1.02, is relatively small. This small difference is because the interest earned approaches a limit. This limit is when the compounding is done **continuously**. (Continuous compound interest is found by using the formula $A = Pe^{rt}$, where $e \approx 2.718281828\ldots$.) Here, if the interest is compounded continuously, the future value after 6 years of the $2000 principal would be $3232.15, or $0.17 more than daily compounding!

Present Value

Just as we did in Section 5.1, we can determine the present value for some future amount by using the formula $A = P(1 + i)^n$. Example 4 illustrates the algebraic process.

Example 4 **Determining a Present Value**

To the nearest cent, determine how much Brad Bundy should invest now at 5.5% interest compounded monthly to have $10,000 toward the purchase of a minivan in 4 years?

Solution

Understand the Situation: Notice that we are given the future value $A = \$10,000$ for a compound investment and we need to find the present value (or principal) P. We know that $i = \dfrac{r}{k} = \dfrac{0.055}{12}$ and $n = kt = 12(4) = 48$.

This gives us

$$A = P(1 + i)^n$$

$$10,000 = P\left(1 + \frac{0.055}{12}\right)^{48}$$

$$P = \frac{10,000}{\left(1 + \dfrac{0.055}{12}\right)^{48}} \approx \$8029.22$$

Interpret the Solution: This means that Brad should invest $8029.22 now to have $10,000 in 4 years. ∎

✓ **Checkpoint 2**   Now work Exercise 21.

Interactive Activity

Show that a formula for present value can be rewritten as $P = \dfrac{A}{(1+i)^n}$ or as $P = A(1 + i)^{-n}$.

Technology Option

Figure 5.2.2(a) shows the input for Example 4 on a TI-83 using TVM Solver. Figure 5.2.2(b) shows the result (PV) in the TVM Solver window, and Figure 5.2.2(c) shows the result of tvm_PV. (Note that tvm_PV finds the present value.) The negative sign in Figure 5.2.2(b) and (c) means a cash outflow or an investment.

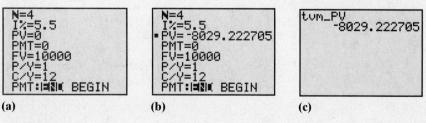

(a)   (b)   (c)

Figure 5.2.2

To learn more about these commands or to see if your calculator has a financial menu, consult the online graphing calculator manual at www.prenhall.com/armstrong.

Future Value and Time

In Section 5.1 we analyzed the future value of simple interest as a function of time. We can do the same thing with compound interest. Suppose that Sara invests \$2000 at 8% interest compounded monthly. Using the formula $A = P(1+i)^n$, where $i = \dfrac{r}{k}$ and $n = kt$, we can determine the amount that Sara has for various periods of time. Since $P = 2000$ and $i = \dfrac{0.08}{12}$, the formula for future value becomes

$$A = P(1+i)^n = P\left(1+\frac{r}{k}\right)^{kt}$$

$$= 2000\left(1+\frac{0.08}{12}\right)^{12t}$$

Table 5.2.3 shows the future value for different values of t.

Table 5.2.3

Time, t, in Years	Future Value, $A = 2000\left(1+\frac{0.08}{12}\right)^{12t}$
1	\$2166.00
2	2345.76
4	2751.33
5	2979.69
8	3784.91
10	4439.28

If we plot the points in Table 5.2.3 and connect them with a smooth curve, we get the graph in Figure 5.2.3.

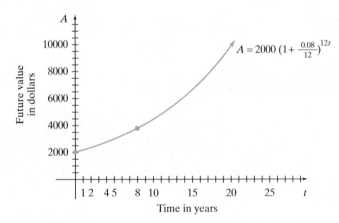

Figure 5.2.3

As we saw in Section 5.1, if the 8% interest was simple interest, we would have Sara's future value given by

$$A = 2000(1+0.08t)$$

Figure 5.2.4 shows the graph of the future value of a compound amount along with the future value for simple interest. With simple interest, the future value $A = 2000(1+0.08t)$ is a linear function of time, t, whereas with compound interest the future value $A = 2000\left(1+\dfrac{0.08}{12}\right)^{12t}$ is an **exponential function of**

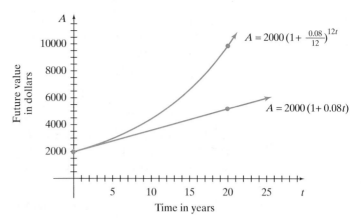

Figure 5.2.4

time, t. Observe that as time increases, the graph for the future value of compound interest starts to increase more rapidly than the graph for the future value of simple interest.

 Technology Option

If we replace t with x, and let y_1 represent the future value for compound interest and y_2 represent the future value for simple interest, the equations in Figure 5.2.4 become

$$\textit{Compound Interest} \qquad\qquad \textit{Simple Interest}$$

$$y_1 = 2000\left(1 + \frac{0.08}{12}\right)^{12x} \qquad y_2 = 2000(1 + 0.08x)$$

We can enter these equations in our graphing calculator and analyze these two functions using the TABLE command as well as graphically. See Table 5.2.4 and Figure 5.2.5(a) and (b).

Table 5.2.4

X	Y1	Y2
1	2166	2160
2	2345.8	2320
4	2751.3	2640
5	2979.7	2800
8	3784.9	3280
10	4439.3	3600
20	9853.6	5200

X=1

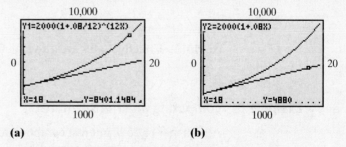

Figure 5.2.5 (a) Future value after 18 years with compound interest. **(b)** Future value after 18 years with simple interest.

Effective Rate

A common problem in financial planning is choosing the best investment from two or more investments. For example, is an investment that pays 10% interest compounded monthly better than one that pays 10.2% interest compounded semiannually? A good way to answer this question is to compare their **effective rates**. An effective rate is the simple interest rate that would give the same amount in 1 year as the compounded rate gives. For example, if $2000 is deposited in a bank that pays 8% interest compounded monthly, then the future value after 1 year is

$$A = P(1 + i)^n$$

$$= 2000 \left(1 + \frac{0.08}{12}\right)^{12} \approx \$2166.00 \qquad \text{Rounded to the nearest cent}$$

Using $A = P(1 + rt)$ from Section 5.1 (future value for simple interest formula), we can determine the simple interest rate by substituting in $A = 2166$, $P = 2000$, and $t = 1$ and solving for r. This gives us

$$A = P(1 + rt)$$

$$2166 = 2000(1 + r)$$

$$\frac{2166}{2000} = 1 + r$$

$$r = \frac{2166}{2000} - 1 = 0.083 = 8.3\%$$

The 8% rate is called the **nominal rate**. The 8.3% rate is the **effective rate** of interest. Notice that this effective rate is a **simple interest rate**. To find a formula relating effective rate, denoted r_{eff}, and nominal rate, r, we simply set the equations for future value for simple interest and future value for compound interest equal to each other. (Note that in both $t = 1$ since it is 1 year.)

$$\begin{array}{c} \text{Future value} \\ \text{(simple interest) in 1 year} \end{array} = \begin{array}{c} \text{Future value} \\ \text{(compound interest) in 1 year} \end{array}$$

$$P(1 + r_{\text{eff}}) = P(1 + i)^k \qquad \text{Since } n = kt \text{ and } t = 1, \text{ we have } n = k.$$

$$1 + r_{\text{eff}} = (1 + i)^k \qquad \text{Divide both sides by } P.$$

$$r_{\text{eff}} = (1 + i)^k - 1$$

■ **Effective Rate**

The **effective rate**, r_{eff}, corresponding to the nominal rate i, where $i = \dfrac{r}{k}$, compounded k times per year, is given by

$$r_{\text{eff}} = (1 + i)^k - 1$$

▶ **Note:** Effective rates are also called **annual percentage yields**, or *APY*.

Example 5   **Computing an Effective Rate**

A bank pays 6.25% interest compounded quarterly. Determine the effective rate rounded to the nearest thousandth.

Solution

We have

$$r_{\text{eff}} = (1 + i)^k - 1$$

$$= \left(1 + \frac{0.0625}{4}\right)^4 - 1$$

$$= (1.015625)^4 - 1$$

$$\approx 0.0639801621$$

$$= 6.398\%$$

So the effective rate is 6.398%.

Interpret the Solution: This means that money invested at 6.398% simple interest earns the same amount of interest in 1 year as money invested at 6.25% interest compounded quarterly. Hence the effective rate of 6.25% interest compounded quarterly is 6.398%. ∎

✓**Checkpoint 3**

Now work Exercise 41.

Example 6 Computing a Nominal Rate

Sam says that he has a CD with an effective rate of 6.5%. What annual nominal rate compounded monthly does this correspond to? Round to the nearest thousandth.

Solution

Understand the Situation: We use the same formula as we did in Example 5. Here we are trying to determine r, the annual nominal rate. Recall that $i = \dfrac{r}{k}$.
Proceeding, we get

$$r_{\text{eff}} = (1 + i)^k - 1$$

$$0.065 = \left(1 + \frac{r}{12}\right)^{12} - 1 \qquad \text{Recall that } i = \frac{r}{k}.$$

$$1.065 = \left(1 + \frac{r}{12}\right)^{12}$$

$$\sqrt[12]{1.065} = 1 + \frac{r}{12}$$

$$\sqrt[12]{1.065} - 1 = \frac{r}{12}$$

$$r = 12(\sqrt[12]{1.065} - 1) \approx 0.0631403313 \quad \text{or} \quad 6.314\%$$

Interpret the Solution: So an annual nominal rate of 6.314% compounded monthly is the same as an effective rate of 6.5%. ∎

Technology Option

Figure 5.2.6 shows how to calculate the effective rate for the situation in Example 5 by using the ▶Eff(command on the TI-83. This command computes the effective rate given the nominal rate and the number of compounding periods in 1 year. Figure 5.2.7 shows the result of using the ▶Nom(command on the TI-83 to answer the situation in Example 6. This command computes the nominal rate given the effective rate and the number of compounding periods in 1 year.

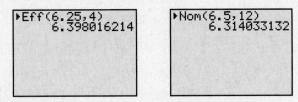

▶Eff(6.25,4)
6.398016214

▶Nom(6.5,12)
6.314033132

Figure 5.2.6   **Figure 5.2.7**

To see if your graphing calculator computes effective rates and nominal rates, consult the online graphing calculator manual at www.prenhall.com/armstrong.

Growth of Money

Many times we wish to know how many compounding periods it will take for a principal to reach a certain dollar amount. We can analyze these questions either graphically or algebraically. Algebraically, we need to use logarithms to solve $A = P(1+i)^n$ for n. (For a review on logarithms, see Appendix B.5.) Proceeding, we get

$$A = P(1+i)^n$$

$$\frac{A}{P} = (1+i)^n$$

$$\ln\left(\frac{A}{P}\right) = \ln(1+i)^n \qquad \text{Take the natural log of both sides.}$$

$$\ln\left(\frac{A}{P}\right) = n\ln(1+i) \qquad \text{Property of logarithms}$$

$$n = \frac{\ln\left(\frac{A}{P}\right)}{\ln(1+i)}$$

Number of Periods for Money to Grow from P to A

The number of compounding periods n for a present value P to reach a future value A with interest rate per compounding period i is given by

$$n = \frac{\ln\left(\frac{A}{P}\right)}{\ln(1+i)}$$

▶ **Note:** The value for n will be rounded up to the nearest integer to guarantee that the value A is reached.

Example 7 **Computing a Growth Time Algebraically**

Determine how long it will take $5000 to grow to at least $8000 if it is invested at 6% interest compounded quarterly.

Solution

Understand the Situation: We begin by using the formula $n = \frac{\ln\left(\frac{A}{P}\right)}{\ln(1+i)}$ and note that $A = 8000$, $P = 5000$, and $i = \frac{0.06}{4}$.
This gives us

$$n = \frac{\ln\left(\frac{8000}{5000}\right)}{\ln\left(1+\frac{0.06}{4}\right)} = \frac{\ln 1.6}{\ln 1.015} \approx 31.56799396$$

Interpret the Solution: This means that it will take 32 quarters, or 8 years, for $5000 to grow to at least $8000 at 6% interest compounded quarterly. ∎

Technology Option

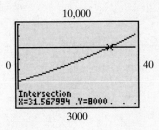

Figure 5.2.8

To analyze Example 7 graphically, we need to graph both sides of the equation

$$A = P(1 + i)^n$$

$$8000 = 5000\left(1 + \frac{0.06}{4}\right)^n$$

We enter $y_1 = 8000$ and $y_2 = 5000\left(1 + \frac{0.06}{4}\right)^x$ into our graphing calculator, graph both in the same viewing window and then use the INTERSECT command. See Figure 5.2.8. Graphically, we reach the same result of 32 quarters for $5000 to grow to $8000.

✓ **Checkpoint 4**   Now work Exercise 37.

SUMMARY

In this section we learned how to compute the future amount when the interest is **compounded**. We derived a formula for compound interest and used it to compute future and present values. We saw that the future value (when interest is compounded) as a function of time is an exponential function. We learned a formula for effective rates, and we finished the section with a formula for computing the number of compounding periods for money to grow from P to A.

SECTION 5.2 EXERCISES

In Exercises 1–10, compute the future value for the given investments at the end of 5 and 8 years. Round answers to the nearest cent.

	Principal or Present Value, P	Annual Interest Rate, r	Compounding Type
1.	$3000	5%	Annually
2.	$5000	6%	Annually
3.	$2000	5.5%	Semiannually
4.	$1500	6.5%	Semiannually
5.	$8000	7.25%	Quarterly
6.	$6000	7.75%	Quarterly
✓ **7.**	$15,000	6.75%	Monthly
8.	$10,000	6.25%	Monthly
9.	$9700	8.125%	Daily
10.	$8200	7.125%	Daily

In Exercises 11–14, use a graphing calculator's built-in financial package to compute the future value for the given investments at the end of 3 and 6 years. Round answers to the nearest cent.

	Principal or Present Value, P	Annual Interest Rate, r	Compounding Type
11.	$8000	5.5%	Quarterly
12.	$5000	6.5%	Monthly
13.	$10,000	5.75%	Monthly
14.	$9000	6.25%	Quarterly

In Exercises 15–24, compute the present value for the given future values. Round answers to the nearest cent.

	Future Value, A	Annual Interest Rate, r	Time, t	Compounding Type
15.	$5000	5%	4	Annually
16.	$4000	6%	6	Annually
17.	$3000	5.5%	5	Semiannually

	Future Value, A	Annual Interest Rate, r	Time, t	Compounding Type
18.	$8000	6.5%	7	Semiannually
19.	$8500	7.25%	6	Quarterly
20.	$6500	7.75%	8	Quarterly
✓ **21.**	$13,000	6.75%	7	Monthly
22.	$9000	6.25%	9	Monthly
23.	$8200	8.125%	10	Daily
24.	$7800	7.125%	15	Daily

In Exercises 25–28, use a graphing calculator's built-in financial package to compute the present value for the given future values. Round answers to the nearest cent.

	Future Value, A	Annual Interest Rate, r	Time, t	Compounding Type
25.	$10,000	6%	6	Quarterly
26.	$8000	5%	5	Monthly
27.	$9000	6.25%	7	Monthly
28.	$13,000	5.75%	8	Quarterly

In Exercises 29–32, match the given description with the correct graph, (a), (b), (c), or (d).

29. Which graph models the future value of $2000 invested in a simple interest rate savings account?

30. Which graph models the future value of $1000 invested in a simple interest rate savings account?

31. Which graph models the future value of $1000 invested in a compound interest rate savings account?

32. Which graph models the future value of $2000 invested in a compound interest rate savings account?

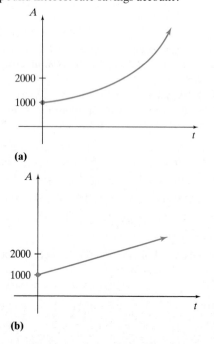

(a)

(b)

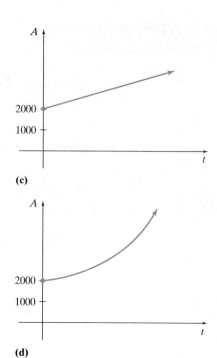

(c)

(d)

In Exercises 33–40, determine how long (that is, the number of compounding periods) it will take the given present value P to grow to the given future value A for the stated interest rate and compounding type. Round the answer up to the nearest integer.

	Present Value, P	Future Value, A	Annual Interest rate, r	Compounding Type
33.	$4000	$5000	6%	Annually
34.	$1000	$1500	5%	Annually
35.	$8000	$13,000	5.5%	Semiannually
36.	$5000	$9000	6.5%	Semiannually
✓ **37.**	$10,000	$16,000	5.25%	Quarterly
38.	$13,000	$23,000	5.75%	Quarterly
39.	$12,000	$20,000	6.75%	Monthly
40.	$11,000	$19,000	6.25%	Monthly

In Exercises 41–48, determine the effective rate corresponding to each of the following nominal rates. Round the percentages to the nearest thousandth.

✓ **41.** 8.5% compounded quarterly

42. 7.5% compounded quarterly

43. 6% compounded monthly

44. 7% compounded monthly

45. 5.25% compounded semiannually

46. 6.25% compounded semiannually

47. 6.75% compounded monthly

48. 5.75% compounded monthly

In Exercises 49–56, use a graphing calculator and its built-in financial package to compute the effective rate.

49. 7.5% compounded quarterly

50. 8.5% compounded quarterly

51. 7% compounded monthly

52. 6% compounded monthly

53. 6.25% compounded semiannually

54. 5.25% compounded semiannually

55. 5.75% compounded monthly

56. 6.75% compounded monthly

In Exercises 57–60, for the given effective rate determine the nominal rate if the compounding is (a) compounded quarterly; or (b) compounded monthly.

57. 6.25%   **58.** 6.75%

59. 5.5%   **60.** 6.5%

In Exercises 61–64, use a graphing calculator and its financial package to determine, for the given effective rate, the nominal rate if the compounding is (a) compounded quarterly; or (b) compounded monthly.

61. 5.75%   **62.** 6.25%

63. 7.5%   **64.** 5.5%

Applications

In Exercises 65–68, use effective rates to determine which is the better investment.

65. 8% compounded monthly or 8.25% compounded annually?

66. 7% compounded monthly or 7.25% compounded quarterly?

67. 8.25% compounded quarterly or 8.125% compounded monthly?

68. 7.875% compounded quarterly or 7.75% compounded monthly?

69. The Consumer Price Index (CPI) is a measure of inflation. The CPI for all urban consumers in the United States during 1999 was 2.7%. If this rate holds true for the next 5 years, what will a $2.29 gallon of milk cost in 5 years if the rate is compounded annually? (*Source:* Bureau of Labor Statistics, http://stats.bls.gov)

70. Assuming the rate and compounding in Exercise 69, what will a $5.99 pound of coffee cost in 5 years?

71. Gene received a $2500 holiday bonus from his employer. He placed the bonus in an account earning 4.25% interest compounded monthly. How much is in his account after 4 years?

72. Kathy won $1000 in a raffle. She placed her winnings in an account earning 4.5% interest compounded quarterly. How much is in her account after 5 years?

73. Susan just purchased a new house costing $135,000. Houses in the area are appreciating at a rate of 3% per year. If this rate holds steady for 7 years, what will the house be worth assuming that this 3% rate is compounded annually?

74. Omar just purchased a new house costing $129,000. Houses in the area are appreciating at a rate of 3.25% per year. If this rate holds steady for 8 years, what will the house be worth assuming that this 3.25% rate is compounded annually?

75. Alec and Lexi compute that they will need $20,000 in 4 years to remodel their kitchen. How much should they invest now at 5.5% interest compounded quarterly to have the $20,000 in 4 years?

76. Lee and Jordan compute that they will need $13,000 in 6 years to build an addition on their house. How much should they invest now at 4.75% interest compounded daily to have the $13,000 in 6 years?

77. Sam and Diane have recently married. They compute that they will need $25,000 in 5 years as a down payment for their first home. How much should they invest now at 8.5% interest compounded monthly to have the $25,000 down payment in 5 years?

78. Brad and Tammy want to buy a new house in 7 years. They estimate that they will need $20,000 as a down payment. How much should they invest now at 7.75% interest compounded quarterly to have the $20,000 down payment in 7 years?

79. Jorge and Jenna have $10,000 to invest toward the purchase of a $16,000 ski boat. How many compounding periods will it take for the $10,000 to grow to at least $16,000 if it is invested at 5.25% compounded quarterly?

80. Doug and Patty have $8000 to invest toward the purchase of a $14,000 fishing boat. How many compounding periods will it take for the $8000 to grow to at least $14,000 if it is invested at 4.8% compounded monthly?

81. As of May 17, 2001, First Internet Bank of Indiana offered a money market account paying 4.50% interest compounded monthly. NetBank offered a money market account paying 4.40% interest compounded daily (*Source:* www.bankrate.com).

(a) Compute the effective rate for the First Internet Bank of Indiana account.

(b) Compute the effective rate for the NetBank account.

(c) Is one account a better investment than the other? Explain.

82. As of August 29, 2000, Key Bank USA offered an interest-bearing checking account paying 5.85% interest compounded daily. Compute the effective rate for this account.

*Exercises 83–86 refer to **zero coupon bonds**. A zero coupon bond is a bond that is sold now at a discount. When it matures at some time in the future, it pays its face value. No interest payments are made.*

83. Paige's grandparents are considering purchasing a $45,000 face value zero coupon bond at her birth to help to finance her college education in 18 years. If money is worth 7.85% compounded annually, how much should they pay for this bond?

84. Repeat Exercise 83, except here assume a $40,000 face value and 8.125% compounded annually.

85. If Beth and Frank pay $7500 for a $40,000 zero coupon bond that matures in 18 years, what is their annual compound rate of return?

86. If Todd and Laurie pay $6235 for a $38,000 zero coupon bond that matures in 17 years, what is their annual compound rate of return?

🌐 **87.** According to an advertisement in the September 2000 issue of *Money* magazine, $10,000 invested in the T. RowePrice Science and Technology Fund on 9/30/87 would have increased in value to $174,440 by 6/30/00. What interest rate compounded annually would account for this increase? (*Hint:* Use $t = 12.75$ years.)

🌐 **88.** According to an advertisement in the August 2000 issue of *Money* magazine, $10,000 invested in the T. RowePrice Mid-Cap Growth Fund on 6/30/92 would have increased in value to $54,518 by 3/31/00. What interest rate compounded annually would account for this increase? (*Hint:* Use $t = 7.75$ years.)

🌐 ▊ **SECTION PROJECT**

In July 2000, the Ohio Lottery started a game called Super Lotto PLUS™. At the time of ticket purchase, you must choose how you would like to receive your winnings (assuming that you win the jackpot!). Your options are yearly equal installments for 30 years or a one-time lump-sum payment. According to the Ohio Lottery Web site (www.ohiolottery.com), a jackpot of $10,000,000 results in either $333,333.33 per year for 30 years or a lump sum of $4,232,055.00. (Both amounts are before taxes.)

(a) Assume that you won this jackpot and took the equal yearly installment plan. If you place the $333,333.33 into a savings account earning 5.5% interest compounded monthly, how much interest would the account earn in 1 month?

(b) How much would the first installment of $333,333.33 grow to in this account in 30 years?

(c) Assume that you won the jackpot and took the lump-sum option. If you place the $4,232,055 into a savings account earning 5.5% interest compounded monthly, how much interest would the account earn in 1 month?

(d) How much would the $4,232,055 grow to in 30 years?

(e) Which option (yearly installment or lump sum) would you take and why?

Section 5.3   Future Value of an Annuity

In Section 5.2 we saw how to find the future value of money when the money was invested at a fixed interest rate that was compounded. Many times, however, people and/or companies do not invest money and then sit back and watch it grow. Instead, they invest smaller amounts at regular intervals. There are many examples of this strategy. Some examples are annual life insurance premiums, loans paid off in monthly installments, and monthly contributions to a 403(b), 401(k), or an IRA account. This section explains the financial tool used by individuals as well as companies to fund large expenses, such as a child's college education or purchasing a large piece of equipment. This tool is called an **annuity**.

Future Value of an Annuity

An **annuity** is simply a sequence of equal periodic payments. The interval between payments, called the **payment period**, can be a year, every 6 months, every quarter, every month, and so on. For example, if you deposit $500 into a savings account at the end of each year and the account earns 8% interest compounded annually, you have created an annuity for which the $500 is the payment and the payment period is every year. Notice that each $500 payment is made at the end of each year; hence the payment period is every year. If payments are made at the end of each period, then the annuity is called an **ordinary annuity**. (In this text we only study ordinary annuities. However, the Section Project in this section does analyze an **annuity due**.) In this section we concern ourselves with computing the amount, or **future value**, of an annuity.

> ■ **Amount or Future Value of an Annuity**
>
> The amount, or **future value**, of an annuity is the sum of all payments made plus all interest earned.

Let's compute the future value of an ordinary annuity and see if it sheds some light on a possible formula. In Example 1, observe how we make use of the compound interest formula from Section 5.2.

Example 1   **Computing the Future Value of an Annuity**

Angela plans on depositing $500 into a savings account at the end of each year for the next 5 years. The bank pays an interest rate of 8% compounded annually. What is the future value of Angela's account at the end of 5 years?

Solution

We first observe that, since Angela is depositing an equal amount ($500) periodically (end of each year) into this account, she is constructing an annuity. Figure 5.3.1 illustrates the payment schedule for Angela.

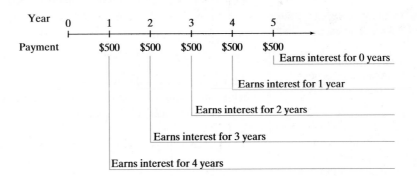

Figure 5.3.1

To determine the future value of Angela's account, we need to use the compound interest formula, $A = P(1 + i)^n$, and determine the future value of each deposit. For example, at the end of the first year Angela deposits $500 and it earns 8% interest compounded annually for 4 years. The future value of this amount, at the end of the fifth year, is given by

$$A = 500 \left(1 + \frac{0.08}{1}\right)^4 = 500(1.08)^4$$

At the end of the second year, Angela deposits $500 and it earns 8% interest compounded annually for 3 years. The future value of this amount, at the end of

the fifth year, is given by

$$A = 500\left(1 + \frac{0.08}{1}\right)^3 = 500(1.08)^3$$

Similarly, the third, fourth, and fifth deposits at the end of years 3, 4, and 5 will earn 8% interest compounded annually for 2, 1, and 0 years, respectively. Their future values are given by

3rd deposit at the end of year 3: $A = 500(1.08)^2$
4th deposit at the end of year 4: $A = 500(1.08)^1$
5th deposit at the end of year 5: $A = 500(1.08)^0 = 500$

So the future value, FV, of Angela's account after 5 deposits, or at the end of 5 years, is

$$FV = 500(1.08)^4 + 500(1.08)^3 + 500(1.08)^2 + 500(1.08)^1 + 500$$

$$= \$2933.30 \quad \text{Rounded to the nearest cent} \qquad ■$$

The second to last line in Example 1 is disguising a useful pattern. If we write this line and factor out the 500, we get

$$FV = 500(1.08)^4 + 500(1.08)^3 + 500(1.08)^2 + 500(1.08)^1 + 500$$

$$= 500[(1.08)^4 + (1.08)^3 + (1.08)^2 + (1.08)^1 + 1]$$

Recognizing this pattern allows us to write an expression for the future value of Angela's account at the end of 8 years, if she continues to deposit $500 at the end of each year. It is

$$FV = 500[(1.08)^7 + (1.08)^6 + (1.08)^5 + (1.08)^4 + (1.08)^3 + (1.08)^2 + (1.08)^1 + 1]$$

Following this pattern, if an amount PMT is deposited into an account at the end of each interest earning period, with i representing the interest rate per period, then the FV (future value) of the annuity at the end of the nth payment period is given by

$$FV = PMT[(1 + i)^{n-1} + (1 + i)^{n-2} + (1 + i)^{n-3} + \cdots + (1 + i) + 1]$$

 From Your Toolbox

The sum of the first n terms of the geometric sequence with common ratio r is $1 + r + r^2 + r^3 + \cdots + r^{n-1} = \dfrac{1 - r^n}{1 - r}$. For a more detailed discussion on geometric sequences and the derivation of the sum formula, see Appendix B.6.

The expression inside the brackets is the sum of a **geometric sequence** with n terms and common ratio $(1 + i)$. A well-known algebraic formula (consult the Toolbox to the left) gives its sum as

$$(1 + i)^{n-1} + (1 + i)^{n-2} + \cdots + (1 + i) + 1 = \frac{1 - (1 + i)^n}{1 - (1 + i)}$$

$$= \frac{1 - (1 + i)^n}{-i} = \frac{(1 + i)^n - 1}{i}$$

Hence the future value, FV, is given by the following:

■ **Future Value of an Ordinary Annuity**

If PMT represents the amount of each payment (or deposit) made at the end of each payment period and i is the interest rate per period, then the future value FV after n payment periods is

$$FV = PMT\frac{(1 + i)^n - 1}{i}$$

▶ **Note:** Some books represent $\dfrac{(1+i)^n - 1}{i}$ with $S_{\overline{n}|i}$, read "S angle n at i."
Before the readily available technology we have today, tables were constructed listing values of $S_{\overline{n}|i}$ for various values of n and i. Thanks to calculators, we can handle many more situations for n and i than any table.

Example 2   Computing a Future Value of an Annuity

Kyle decides to deposit $100 at the end of each month into a savings account earning a guaranteed interest rate of 4.5% compounded monthly. What is the value of this annuity at the end of 10 years?

Solution

Understand the Situation: The periodic payment, PMT, is $100. So $PMT = 100$, $r = 0.045$, and $k = 12$. The interest rate per period, i, is $i = \dfrac{r}{k} = \dfrac{0.045}{12} = 0.00375$. The number of payment periods, n, is $n = kt = 12(10) = 120$.

So the future value of this annuity is given by

$$FV = PMT\frac{(1+i)^n - 1}{i}$$

$$= 100\frac{(1 + 0.00375)^{120} - 1}{0.00375}$$

$$\approx \$15{,}119.81 \qquad \text{Rounded to the nearest cent}$$

Interpret the Solution: This means that after 10 years of depositing $100 at the end of each month into an account that earns 4.5% interest compounded monthly Kyle will have about $15,119.81 in the account. ∎

Interactive Activity

How much of the future value computed in Example 2 is interest?

✓ **Checkpoint 1**

Now work Exercise 5.

Technology Option

```
N=120
I%=4.5
PV=0
PMT=-100
FV=0
P/Y=12
C/Y=12
PMT:END BEGIN
```
(a)

```
N=120
I%=4.5
PV=0
PMT=-100
■FV=15119.80737
P/Y=12
C/Y=12
PMT:END BEGIN
```
(b)

As stated in Section 5.2, some graphing calculators have a special FINANCE menu. To use this feature to solve Example 2, Figure 5.3.2(a) shows what is entered on the TI-83 using the TVM Solver, and Figure 5.3.2(b) shows the result when we solve for FV.

To learn more about the FINANCE menu and TVM Solver, or to learn if your calculator has this feature, consult the online graphing calculator manual at www.prenhall.com/armstrong. In case your graphing calculator does not have a FINANCE menu, we have supplied a program for the graphing calculator in Appendix C that will compute the future value. Consult the online graphing calculator manual at www.prenhall.com/armstrong for the program for your graphing calculator.

Figure 5.3.2 (a) $PMT = -100$ because it is a cash outflow. **(b)** Place the cursor after the equals sign for FV and press ALPHA ENTER. An indicator square in the left column designates the solution variable.

Example 3   Computing a Future Value of an Annuity

The winnings of Super Lotto PLUS™, an Ohio Lottery game, can be taken as a lump-sum distribution or as a 30-year annuity. A jackpot of $22,000,000 has an after-taxes yearly payment (for 30 years) of $502,333.34. (*Source:* www.ohiolottery.com) Assume that Seth wins this jackpot and takes the 30-year annuity option. Assume that he places $250,000 of his yearly payment into an account that earns 6% interest compounded annually. What is the value of this account at the end of 30 years?

Solution

Here we have $PMT = \$250,000$, $i = \dfrac{0.06}{1} = 0.06$, and $n = 30$. So the future value is given by

$$FV = PMT\frac{(1+i)^n - 1}{i}$$
$$= 250,000\frac{(1+0.06)^{30} - 1}{0.06}$$
$$= \$19,764,546.55$$

Observe that Seth would have deposited a total of $250,000(30) = \$7,500,000$, which means that he would earn $\$19,764,546.55 - \$7,500,000 = \$12,264,546.55$ in interest! Also notice that Seth would be able to spend $252,333.34 of his winnings each year.   ∎

Sinking Funds

In many instances, individuals or companies know of an expense that will arise in the future. Examples include college expenses that parents will realize for a newborn baby or the replacement of delivery vehicles for a delivery company. When individuals or companies know of a future expense, many times they make regular payments into a fund to meet the future financial obligation. These funds are special types of annuities called **sinking funds**.

> ■ **Sinking Fund**
>
> In general, any account that is established for the purpose of generating funds to cover a future financial obligation is called a **sinking fund**.

Example 4   Computing the Payment for a Sinking Fund

Craig and Laurie decide to start a college fund for their newborn son Jared. They estimate that they will need $100,000. They decide to deposit money in a bank account that pays 6% interest compounded monthly. How much should they deposit every month to have generated $100,00 after 17 years?

Solution

Understand the Situation: Here we want the future value, FV, to be $100,000. We are trying to find PMT given $FV = 100,000$, $i = \dfrac{0.06}{12} = 0.005$, and $n = 12(17) = 204$.

Interactive Activity

Solve
$$FV = PMT\frac{(1+i)^n - 1}{i}$$
for PMT to get a formula for the sinking fund payment.

Substituting into the formula for future value of an ordinary annuity and solving for *PMT* gives

$$FV = PMT\frac{(1+i)^n - 1}{i}$$

$$100{,}000 = PMT\frac{(1 + 0.005)^{204} - 1}{0.005}$$

Using a calculator, we get

$$PMT \approx \$283.10 \qquad \text{Rounded to the nearest cent}$$

Interpret the Solution: So Craig and Laurie should deposit $283.10 each month into the account. Observe that the total amount that they invest is $283.10(204) = $57,752.40, which means that the interest earned is $100,000 − $57,752.40 = $42,247.60. ∎

Checkpoint 2

Now work Exercise 21.

Technology Option

Figure 5.3.3(a) shows the setup to solve Example 4 using TVM Solver on a TI-83, and Figure 5.3.3(b) shows the result when we solve for PMT.

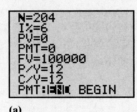

(a)   (b)

Figure 5.3.3

In case your graphing calculator does not have a FINANCE menu, we have supplied a program for the graphing calculator in Appendix C that will compute the payment to fund a sinking fund. Consult the online graphing calculator manual at www.prenhall.com/armstrong for the program for your graphing calculator.

Determining an Interest Rate Using a Graphing Calculator

The last example of this section uses a graphing calculator to determine an interest rate that will achieve an individual's financial goal. This material may be skipped without loss of continuity.

Example 5   Determining an Interest Rate

Carl has started a new job and he decides to deposit $200 each month in an account that pays interest compounded monthly. Carl wishes to have at least

$250,000 after 30 years. What annual interest rate compounded monthly would provide Carl with this amount?

Solution

Here we have $PMT = \$200$, $FV = \$250,000$, and $n = 12(30) = 360$. For Carl, the periodic interest rate i is $i = \dfrac{r}{12}$. We are trying to solve for r, the annual interest rate, in the equation

$$FV = PMT\frac{(1+i)^n - 1}{i}$$

$$250,000 = 200\frac{\left(1 + \dfrac{r}{12}\right)^{360} - 1}{\dfrac{r}{12}}$$

This equation is very difficult to solve algebraically! We opt to solve it graphically by letting x represent the annual interest rate r. We then graph

$$y_1 = 200\frac{\left(1 + \dfrac{x}{12}\right)^{360} - 1}{\dfrac{x}{12}}$$

$$y_2 = 250,000$$

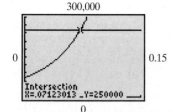

Figure 5.3.4

in the same viewing window and use the INTERSECT command as shown in Figure 5.3.4.

So an annual interest rate, to the nearest hundredth, of 7.13% compounded monthly would provide Carl with $250,000. We rounded to 7.13% to guarantee that at least $250,000 would be accumulated. A rate of 7.12% would not give a sufficient future value. We can check this by comparing the future values at the two rates.

At 7.12%: $\quad FV = PMT\dfrac{(1+i)^n - 1}{i}$

$$= 200\frac{\left(1 + \dfrac{0.0712}{12}\right)^{360} - 1}{\dfrac{0.0712}{12}}$$

$$= \$249,850.89$$

At 7.13%: $\quad FV = PMT\dfrac{(1+i)^n - 1}{i}$

$$= 200\frac{\left(1 + \dfrac{0.0713}{12}\right)^{360} - 1}{\dfrac{0.0713}{12}}$$

$$= \$250,346.20 \qquad \blacksquare$$

We can also use the TVM Solver on the TI-83 to solve Example 5. Figure 5.3.5(a) shows the setup, and Figure 5.3.5(b) shows the result when we solve for I%.

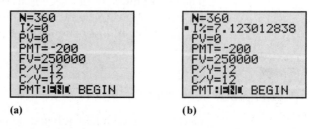

(a)   (b)

Figure 5.3.5

SUMMARY

This section focused on computing the **future value of an ordinary annuity**. The formula that we developed used the compound interest formula from Section 5.2. We used the future value of an ordinary annuity formula to compute future values and to examine **sinking funds**.

- **Future Value of an Ordinary Annuity** If *PMT* represents the amount of each payment (or deposit) made at

the end of each payment period, and i is the interest rate per period, then the future value FV after n payment periods is $FV = PMT\dfrac{(1+i)^n - 1}{i}$.

- **Sinking Fund** In general, any account that is established for the purpose of generating funds to cover a future financial obligation is called a **sinking fund**.

SECTION 5.3 EXERCISES

In Exercises 1–6, determine the amount of each annuity.

1. A deposit of $150 monthly for 10 years at an interest rate of 6% compounded monthly

2. A deposit of $250 monthly for 15 years at an interest rate of 7% compounded monthly

3. A deposit of $2000 annually for 30 years at an interest rate of 7.5% compounded annually

4. A deposit of $1800 annually for 25 years at an interest rate of 6.5% compounded annually

✓ **5.** A deposit of $500 quarterly for 30 years at an interest rate of 6.75% compounded quarterly

6. A deposit of $400 monthly for 25 years at an interest rate of 7.25% compounded quarterly

In Exercises 7–12, determine the amount of each annuity by using the FINANCE menu on your graphing calculator or the ANNFV calculator program found in Appendix C. Consult www.prenhall.com/armstrong for the program for your calculator.

7. A deposit of $250 monthly for 15 years at an interest rate of 7% compounded monthly

8. A deposit of $150 monthly for 10 years at an interest rate of 6% compounded monthly

9. A deposit of $1800 annually for 25 years at an interest rate of 6.5% compounded annually

10. A deposit of $2000 annually for 30 years at an interest rate of 7.5% compounded annually

11. A deposit of $400 quarterly for 25 years at an interest rate of 7.25% compounded quarterly

12. A deposit of $500 quarterly for 30 years at an interest rate of 6.75% compounded quarterly

In Exercise 13–18, determine how much of the amount (or future value) of each annuity is interest.

13. The annuity in Exercise 1

14. The annuity in Exercise 2

15. The annuity in Exercise 3

16. The annuity in Exercise 4

17. The annuity in Exercise 5

18. The annuity in Exercise 6

In Exercises 19–24, determine the periodic payment amount required for each sinking fund.

19. The amount needed is $20,000 in 10 years at 6% interest compounded annually. What is the annual payment?

20. The amount needed is $30,000 in 15 years at 7% interest compounded annually. What is the annual payment?

✓ **21.** The amount needed is $80,000 in 15 years at 6.5% interest compounded monthly. What is the monthly payment?

22. The amount needed is $60,000 in 12 years at 7.5% interest compounded monthly. What is the monthly payment?

23. The amount needed is $30,000 in 8 years at 7.25% interest compounded quarterly. What is the quarterly payment?

24. The amount needed is $40,000 in 7 years at 6.75% interest compounded quarterly. What is the quarterly payment?

In Exercises 25–30, determine the payment required for each sinking fund by using the FINANCE menu on your graphing calculator or the ANNPMT calculator program found in Appendix C. Consult www.prenhall.com/armstrong for the program for your calculator.

25. The amount needed is $30,000 in 15 years at 7% interest compounded annually. What is the annual payment?

26. The amount needed is $20,000 in 10 years at 6% interest compounded annually. What is the annual payment?

27. The amount needed is $60,000 in 12 years at 7.5% interest compounded monthly. What is the monthly payment?

28. The amount needed is $80,000 in 15 years at 6.5% interest compounded monthly. What is the monthly payment?

29. The amount needed is $40,000 in 7 years at 6.75% interest compounded quarterly. What is the quarterly payment?

30. The amount needed is $30,000 in 8 years at 7.25% interest compounded quarterly. What is the quarterly payment?

In Exercises 31–36, refer back to Exercises 19–24 to answer the following. See Example 4.

(a) How much was invested?

(b) How much interest was earned?

31. The sinking fund in Exercise 19

32. The sinking fund in Exercise 20

33. The sinking fund in Exercise 21

34. The sinking fund in Exercise 22

35. The sinking fund in Exercise 23

36. The sinking fund in Exercise 24

In Exercises 37–42, determine the annual interest rate (to the hundredth) that guarantees the given future value of the annuity. See Example 5.

37. The future value is $100,000, where $200 is deposited each month for 20 years into an account paying interest compounded monthly.

38. The future value is $150,000, where $250 is deposited each month for 20 years into an account paying interest compounded monthly.

39. The future value is $125,000, where $2000 is deposited each year for 25 years into an account paying interest compounded annually.

40. The future value is $175,000, where $2000 is deposited each year for 30 years into an account paying interest compounded annually.

41. The future value is $500,000, where $1000 is deposited each quarter for 30 years into an account paying interest compounded quarterly.

42. The future value is $600,000, where $1250 is deposited each quarter for 25 years into an account paying interest compounded quarterly.

Applications

43. Teri deposits $100 per month into a savings account that pays 4.75% interest compounded monthly. Determine the value of this account after 10 years.

44. Steve deposits $80 per month into a savings account that pays 4.25% interest compounded monthly. Determine the value of this account after 15 years.

45. For Teri in Exercise 43, how much of the future value was money that she deposited and how much of the future value was interest earned?

46. For Steve in Exercise 44, how much of the future value was money that he deposited and how much of the future value was interest earned?

47. In 2000, the average price for a pack of cigarettes in the United States was $2.75 (*Source:* www.legislature .state.tn.us/bills/currentga/BILL/SB0192.pdf). Assuming a 30-day month, a typical pack-a-day smoker spends about $82.50 per month on cigarettes. Suppose that a smoker quits smoking and invests the monthly amount spent on cigarettes in a savings account earning 5.25% interest compounded monthly. Determine the value of the account after 30 years.

48. A Senate Office of Research Analysis estimates that California households with gross incomes at or below $15,000 per year with one smoker spend an average of $870 per year on tobacco products (*Source:* www.cbp.org/brief/bb980903.html). If this annual amount were invested in a savings account earning 5.75% interest compounded annually, determine the value of the account after 25 years.

49. From September 1987 through June 2000, the S&P 500 Index had an average annual return of 15.54% (*Source: Money* magazine, September 2000 issue). If Amanda invests $2000 a year in a S&P 500 Index mutual fund (a fund that imitates the S&P 500) for 10 years, how much is her investment worth after 10 years assuming the 15.54% average annual rate?

50. For the 10-year period from 6/30/90 to 6/30/00, the T. RowePrice Science and Technology Fund had an average annual return of 27.43% (*Source: Money* magazine, September 2000 issue). Suppose that Gene invested $2000 a year in this fund during this 10-year period of time. Assuming the 27.43% average annual rate, what is his investment worth after the 10-year period?

51. Jack and Diane are saving money for a down payment on a new house. They decide to deposit $350 per month into an account paying 5.75% interest compounded monthly. Determine the value of their account after 4 years.

52. Hank and Peggy are saving money to purchase a new pickup truck. They decide to deposit $100 per month into an account paying 5.25% interest compounded monthly. Determine the value of their account after 5 years.

53. Maury and Hannah would like to have $30,000 in 6 years in order to purchase a new minivan. What amount should they deposit each month into a savings account paying 5.125% interest compounded monthly?

54. Ted and Kathy would like to have $20,000 in 5 years as a down payment on a new house. What amount should they deposit each month into a savings account paying 5.375% interest compounded monthly?

55. Tamika and Tim have established a sinking fund in order to have $120,000 in 17 years for their twin daughters' college education. How much should they deposit each quarter into a savings account paying 6.125% interest compounded quarterly?

56. Sally and Juan have established a sinking fund in order to have $180,000 in 15 years for their children's college education. How much should they deposit each quarter into a savings account paying 6.5% interest compounded quarterly?

57. For her retirement, Irene wishes to have $200,000 in a 403(b) account 30 years from now. How much should she deposit each month in an account paying 5.75% interest compounded monthly?

58. For his retirement, Dave wishes to have $250,000 in a 403(b) account 25 years from now. How much should he deposit each month in an account paying 5.25% interest compounded monthly?

59. As of October 1999, tuition at 4-year public institutions was increasing at a rate of 3.4% per year (*Source:* www.dailybruin.ucla.edu). For 1998–1999, the average tuition at 4-year public institutions was $3243 (*Source:* www.collegeboard.org).

(a) Assuming the 3.4% rate and the average tuition cost of $3243, how much will tuition at a 4-year public institution be in 15 years? Round to the nearest dollar.

(b) Multiply the answer in part (a) by 4 to get an approximation for 4 years of tuition for a child starting college in 15 years.

(c) If parents wish to establish a sinking fund now to cover 4 years of tuition, how much should they deposit each month in an account paying 5% interest compounded monthly to have the amount in part (b) in 15 years?

60. As of October 1999, tuition at 4-year private schools was increasing at a rate of 4.7% per year (*Source:* www.dailybruin.ucla.edu). From 1998–1999, the average tuition at 4-year private schools was $14,508 (*Source:* www.collegeboard.org).

(a) Assuming the 4.7% rate and the average tuition cost of $14,508, how much will tuition at a 4-year private institution be in 15 years?

(b) Multiply the answer in part (a) by 4 to get an approximation for 4 years of tuition for a child starting college in 15 years.

(c) If parents wish to establish a sinking fund now to cover 4 years of tuition, how much should they deposit each month in an account paying 5% interest compounded monthly to have the amount in part (b) in 15 years?

61. On her 25th birthday, Greg and Chris opened an IRA for their daughter Erika and deposited $2000. On every birthday up to and including her 55th, Erika deposits $2000 into her IRA. How much will be in the account on her 55th birthday if the IRA earns

(a) 7% interest compounded annually?

(b) 9% interest compounded annually?

(c) 11% interest compounded annually?

62. (*continuation of Exercise 61*) Suppose that the $2000 deposit on her 55th birthday is the last that Erika makes. Assume that she leaves the money in the IRA for 10 more years. On her 65th birthday, what is the value of her account if the IRA had earned and continues to earn

(a) 7% interest compounded annually?

(b) 9% interest compounded annually?

(c) 11% interest compounded annually?

63. (a) Debbie deposits $2000 a year into an IRA earning 8% interest compounded annually. She makes her first deposit on her 23rd birthday and her last deposit on her 32nd birthday, a total of 10 deposits. Debbie makes no additional deposits. She leaves the amount accumulated from the 10 deposits in the account, earning 8% interest compounded annually until her 65th birthday. How much is in Debbie's account on her 65th birthday?

(b) Bubba deposits $2000 a year into an IRA earning 8% interest compounded annually. He makes his first deposit on his 33rd birthday and continues to deposit $2000 on every birthday until he is 65, a total of 33 deposits. How much is in his account when he makes his last deposit on his 65th birthday?

(c) Compare the results of parts (a) and (b). In your comparison, determine how much Debbie deposited versus how much Bubba deposited and the amount in each account on their 65th birthday. Are you surprised? Is there an investing moral here?

64. The after-taxes yearly installment for a $10,000,000 Super Lotto PLUS™ jackpot in the Ohio Lottery is $228,333.34 (*Source:* www.ohiolottery.com). This yearly installment is for 30 years and assumes a single winner. Suppose that Kahrin wins such a jackpot. Each year when she receives her annual installment she invests $28,333.34 in an account earning 5.5% interest compounded annually. How much will be in this account in 30 years when she receives her final winning installment?

65. The after-taxes lump-sum payment for a $10,000,000 Super Lotto PLUS™ jackpot in the Ohio Lottery is $2,898,957.68. Suppose that Kyle wins such a jackpot. Kyle deposits his winnings in a savings account paying 5.5% interest compounded monthly.

(a) Kyle quits his job and decides to live off the interest. Each month he has the interest earned in his savings account transferred to his checking account. How much is this monthly amount?

(b) Kyle realizes he can live quite comfortably on $10,000 each month. Of the amount found in part (a), he has $10,000 put into his checking account and the rest invested each month in an account earning 6% interest compounded monthly. How much is this monthly investment?

(c) For the amount invested at 6% interest compounded monthly found in part (b), compute the value of this account after 30 years.

SECTION PROJECT

The formula $FV = PMT\dfrac{(1+i)^n - 1}{i}$ is for ordinary annuities for which payments are made at the *end* of each time period. Annuities in which the payments are made at the *beginning* of each time period are called **annuities due**. The **future value of an annuity due** of n payments of PMT dollars each at the beginning of each period, with i being the interest rate per period, is given by

$$FV = PMT\frac{(1+i)^{n+1} - 1}{i} - PMT$$

(a) Compare and contrast the formulas for ordinary annuities and annuities due.

(b) Rich and Cheryl deposit $200 at the end of each month in an account paying 6.5% interest compounded monthly. This is an ordinary annuity. Compute the value of the account after 30 years.

(c) Suppose that Rich and Cheryl deposit $200 at the beginning of each month in an account paying 6.5% interest compounded monthly. This is an annuity due. Compute the value of the account after 30 years.

(d) How much more money would Rich and Cheryl have with an annuity due versus an ordinary annuity? Consult the results from parts (b) and (c).

Section 5.4   Present Value of an Annuity

In Section 5.2 we studied compound interest and learned the compound interest formula, $A = P(1+i)^n$. Recall that A was called the **future value**. We also learned how to find the **present value**, which is the amount of money needed now to obtain the amount A in the future. A similar idea holds for annuities. For example, suppose that you want to withdraw $500 per month each month for the next 4 years from an account that earns 6.5% compounded monthly. The amount of money initially required in the account to allow this to happen is the sum of the present values of each of these $500 withdrawals. We will return to this scenario in Example 1, but let's first take a Flashback to Example 1 from Section 5.3.

Flashback

Angela's Annuity Revisited

In Example 1 in Section 5.3, we saw that Angela deposits $500 at the end of each year into a savings account that pays an interest rate of 8% compounded annually. The future value of Angela's account at the end of 5 years can be found by using the **future value of an ordinary annuity** formula

$$FV = PMT\frac{(1+i)^n - 1}{i}$$

Using this formula, the future value of Angela's account is

$$FV = 500\frac{(1+0.08)^5 - 1}{0.08} \approx \$2933.30 \qquad \text{Rounded to the nearest cent}$$

How much would Angela have to invest today, in one lump sum, at 8% interest compounded annually to yield the same amount of $2933.30 at the end of 5 years?

Flashback Solution

Understand the Situation: This is nothing more than computing a present value using the compound interest formula discussed in Section 5.2.

Using the compound interest formula,

$$A = P(1+i)^n$$

with $A = 2933.30$, $i = \dfrac{0.08}{1} = 0.08$, and $n = 5$, and solving for P gives us

$$2933.30 = P(1 + 0.08)^5$$
$$2933.30 = P(1.08)^5$$
$$P = \frac{2933.30}{(1.08)^5} \approx \$1996.35 \qquad \text{Rounded to the nearest cent}$$

Interpret the Solution: So Angela would have to deposit \$1996.35 now at 8% interest compounded annually to have \$2933.30 at the end of 5 years.

Present Value of an Annuity

What we did in the Flashback was determine the **present value of an annuity**. If we look back at what we did in the Flashback, we should be able to determine a formula for the present value of an annuity. The future value of Angela's annuity was determined to be \$2933.30 and was found by using

$$FV = PMT\frac{(1+i)^n - 1}{i}$$

The present value of \$1996.35 was found by using the formula

$$A = P(1+i)^n$$

where A was \$2933.30. Since we are trying to find the present value so that $FV = A$, we can simply equate the two equations as follows:

$$P(1+i)^n = PMT\frac{(1+i)^n - 1}{i}$$

To solve for P, we multiply both sides by $(1+i)^{-n}$ and get

$$P = PMT(1+i)^{-n}\frac{(1+i)^n - 1}{i}$$

Distributing $(1+i)^{-n}$ on the right side gives us

$$P = PMT\frac{(1+i)^{-n}(1+i)^n - (1+i)^{-n}}{i} = PMT\frac{1 - (1+i)^{-n}}{i}$$

We have now developed a formula to determine the **present value of an annuity**.

T **From Your Toolbox**

Algebra facts used in developing the formula for present value:

$$b^0 = 1, \qquad b \neq 0$$
$$b^n \cdot b^m = b^{n+m}$$
$$b^n \cdot b^{-n} = b^0 = 1, \qquad b \neq 0$$

■ **Present Value of an Annuity**

The present value PV of an annuity with n payments of PMT dollars made at the end of each payment period with i as the interest rate per period is given by

$$PV = PMT\frac{1 - (1+i)^{-n}}{i}$$

▶ **Note:** Some books represent $\dfrac{1 - (1+i)^{-n}}{i}$ with $a_{\overline{n}|i}$, read "*a* angle *n* at *i*." Before the readily available technology we have today, tables were constructed listing values of $a_{\overline{n}|i}$ for various values of n and i. Thanks to calculators, we can handle many more situations for n and i than can any table.

Example 1   Computing a Present Value of an Annuity

Determine the present value of an annuity that pays $500 per month for 4 years at 6.5% interest compounded monthly. (Note that this is the scenario described prior to the Flashback.)

Solution

Understand the Situation: We use the present value for an annuity formula with $PMT = 500$, $i = \dfrac{0.065}{12}$, and $n = 4(12) = 48$.

This gives us

$$PV = PMT\frac{1 - (1 + i)^{-n}}{i}$$

$$= 500\frac{1 - \left(1 + \dfrac{0.065}{12}\right)^{-48}}{\dfrac{0.065}{12}}$$

$$\approx \$21{,}083.74 \qquad \text{Rounded to the nearest cent}$$

Interpret the Solution: This means that to receive $500 per month each month for 4 years requires that an initial amount of $21,083.74 be placed in the account that earns 6.5% interest compounded monthly. ∎

Interactive Activity

For the situation in Example 1, how much interest has been earned over the 4-year period?

✓ **Checkpoint 1**

Now work Exercise 3.

Technology Option

As we have seen in this chapter, if your calculator has a FINANCE menu, it can be used to solve present value problems such as Example 1. To use these commands to solve Example 1, consult Figure 5.4.1(a), which shows what is entered on the TI-83 using the TVM Solver, and Figure 5.4.1(b), which shows the result when we solve for PV (present value).

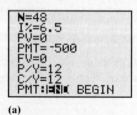

(a)   (b)

Figure 5.4.1

If your graphing calculator does not have a FINANCE menu, we have supplied a program for the graphing calculator that will compute the present value. See Appendix C. Consult the online graphing calculator manual at www.prenhall.com/armstrong for the program for your calculator.

Example 2   **Computing Retirement Income**

Sasha Jones starts her career at the age of 25 and makes equal annual deposits for 35 years into an annuity that earns 6.5% interest compounded annually. She plans on retiring at the age of 60 and would like to withdraw $40,000 annually for 25 years, bringing the account balance to $0. How much should Sasha deposit annually to accumulate enough to provide her with these $40,000 payments?

Solution

Understand the Situation: To answer this question, we need to address both future and present values. As Figure 5.4.2 illustrates, the future value of Sasha's account after 35 years of investing must equal the present value to supply 25 years of yearly $40,000 payments.

We begin by finding the present value required to supply these withdrawals. Using the formula for the present value of an annuity with $PMT = 40,000$, $i = 0.065$, and $n = 25$ gives us

$$PV = PMT\frac{1-(1+i)^{-n}}{i}$$

$$= 40,000\frac{1-(1+0.065)^{-25}}{0.065}$$

$$\approx \$487,915.07 \qquad \text{Rounded to the nearest cent}$$

Figure 5.4.2

Now we need to find the annual deposit that Sasha must make to yield a future value of $487,915.07 in 35 years. Using the formula for future value of an annuity from Section 5.3, with $FV = \$487,915.07$, $i = 0.065$, and $n = 35$, we get

$$FV = PMT\frac{(1+i)^n - 1}{i}$$

$$\$487,915.07 = PMT\frac{(1+0.065)^{35} - 1}{0.065}$$

$$PMT = \$3933.70 \qquad \text{Rounded to the nearest cent}$$

Interpret the Solution: So Sasha needs to deposit $3933.70 annually for 35 years in order to get 25 annual withdrawals of $40,000.

Amortization

One of the most important applications of the present value of an annuity formula occurs in the area of purchasing a home. We introduce this topic in Example 3.

Example 3   **Determining the Cost of a Home**

Frank and Beth have put $30,000 down toward the purchase of a house. To finance the balance of the purchase price of the house, they agree to pay a bank $950 per month for 30 years at 8% annual interest charged monthly to pay off the mortgage (loan) on the house. (Mortgage rates are given as **annual percentage**

rates, *APR*, but it is understood that the monthly rate is $\dfrac{1}{12}$ of the quoted annual rate and that the interest is compounded monthly.) Determine how much the house originally cost and how much interest was paid.

Solution

Understand the Situation: This scenario is the same as Example 1. In other words, we are finding the present value of an annuity of $950 per month at 8% for 30 years.

Using the present value of an annuity formula with $PMT = 950$, $i = \dfrac{0.08}{12}$, and $n = 30(12) = 360$ gives us

$$PV = PMT \frac{1 - (1+i)^{-n}}{i}$$

$$= 950 \frac{1 - \left(1 + \dfrac{0.08}{12}\right)^{-360}}{\dfrac{0.08}{12}}$$

$$\approx \$129{,}469.32 \qquad \text{Rounded to the nearest cent}$$

So the original cost of the house is

$$\$129{,}469.32 + \$30{,}000 = \$159{,}469.32$$

Since Beth and Frank are paying $950 each month for 30 years (360 months), they pay a total of

$$950(360) = \$342{,}000$$

for the house. So the total interest paid is

$$\text{Total paid} - \text{Present value} = \text{Total interest paid}$$
$$\$342{,}000 - \$129{,}469.32 = \$212{,}530.68$$

This shockingly large amount of interest is typical of what happens with a long mortgage. ∎

There are two different ways we can look at the loan in Example 3.

1. If Frank and Beth pays $950 per month for 30 years (360 months) with an interest rate of 8% compounded monthly, then the house is theirs.
2. The loan is the same as an annuity that the bank bought from Frank and Beth. The bank paid Frank and Beth $129,469.32 (a present value) for an annuity that then paid the bank $950 per month for 360 months at 8% interest compounded monthly.

No matter how we look at it, the loan that Frank and Beth had in Example 3 had a fixed rate of interest and the monthly payments of $950 included both principal and interest. In general, a loan is said to be **amortized** if both principal and interest are paid by a sequence of equal payments over equal time periods. Many times, we are interested in computing the equal payments that amortize the loan. If we solve the present value of an annuity formula for *PMT*, we obtain the following formula:

■ **Payments for an Amortized Loan**

A loan of *PV* dollars at interest rate *i* per period is **amortized** in *n* equal payments of *PMT* dollars, where *PMT* is given by

$$PMT = PV \frac{i}{1 - (1+i)^{-n}}$$

Example 4 **Determining a Monthly House Payment**

As of Thursday 5/17/01, Ohio Savings Bank was offering a 30-year fixed-rate mortgage of 7.25% (*Source: The Plain Dealer*). Juan and Andrea Martinez purchase a house for $140,000. After putting 20% down as a down payment, they finance the balance with Ohio Savings Bank.

(a) Determine the size of the down payment and the amount financed.

(b) Determine the monthly payment needed to amortize the loan.

Solution

(a) The down payment is 20% of $140,000. This dollar amount is

$$0.20(140,000) = \$28,000$$

So with a down payment of $28,000, Juan and Andrea need to finance

$$\$140,000 - \$28,000 = \$112,000$$

(b) The monthly payment needed to amortize the $112,000 loan is determined by using the payments for an amortized loan formula. Here $PV = 112,000$, $i = \dfrac{0.0725}{12}$, and the number of monthly payments is $n = 12(30) = 360$. This gives us

$$PMT = PV \frac{i}{1-(1+i)^{-n}}$$

$$= 112,000 \frac{\dfrac{0.0725}{12}}{1-\left(1+\dfrac{0.0725}{12}\right)^{-360}}$$

$$\approx \$764.04 \qquad \text{Rounded to the nearest cent}$$

Interpret the Solution: This means that monthly payments of $764.04 are needed to amortize the loan. ∎

Interactive Activity

Find the total interest paid when the loan in Example 4 is amortized over 30 years. Now determine the monthly payment if the loan in Example 4 is to be amortized in 15 years and compare to the monthly payment in Example 4. Find the total interest paid when the loan in Example 4 is amortized over 15 years. Comment on your results.

Technology Option

Figure 5.4.3(a) and (b) shows the result of Example 4b using the FINANCE menu on a graphing calculator.

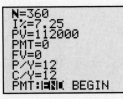

(a)   (b)

Figure 5.4.3

✓ **Checkpoint 2**   Now work Exercise 45.

Amortization Schedules

When amortizing a loan such as the 30-year mortgage in Example 4, the situation may arise where at some point in time we wish to pay off the remainder of the debt in one lump-sum payment. Every time a home with an outstanding mortgage is sold, such a situation arises. To understand what happens in a situation like this, we need a closer analysis of the amortization process. In Example 5 we offer a simple example that allows us to analyze the effect that each payment has on an amortized loan.

Example 5   **Constructing an Amortization Schedule**

Manny borrows $1000 that he agrees to repay in 12 equal monthly payments at an interest rate of 9%. How much of each monthly payment goes toward reducing the unpaid balance (or principal) of the loan and how much goes toward interest?

Solution

First we observe that an interest rate of 9% has a periodic rate of $i = \dfrac{0.09}{12} = 0.0075$ per month. Next we use the formula for payments of an amortized loan to determine the monthly payment. With $PV = 1000$, $i = 0.0075$, and $n = 12$, this gives

$$PMT = PV \frac{i}{1 - (1+i)^{-n}}$$

$$= 1000 \frac{0.0075}{1 - (1 + 0.0075)^{-12}}$$

$$\approx \$87.45 \qquad \text{Rounded to the nearest cent}$$

Now, at the end of the first month, interest is due on the entire $1000 since this is the unpaid balance. This interest due is

$$\$1000(0.0075) = \$7.50$$

The monthly payment of $87.45 is split into two parts: payment of the interest and reduction of the unpaid balance (or principal). This gives

Monthly payment		Interest due		Reduction of unpaid balance
$87.45	=	$7.50	+	$79.95

So the unpaid balance for the next month (month 2) is

Prior unpaid balance		Reduction of unpaid balance		New unpaid balance
$1000	−	$79.95	=	$920.05

The interest due at the end of the second month is based on the $920.05 and is computed to be

$$\$920.05(0.0075) \approx \$6.90 \qquad \text{Rounded to the nearest cent}$$

So the monthly payment of $87.45 is split as follows:

Monthly payment		Interest due		Reduction of unpaid balance
$87.45	=	$6.90	+	$80.55

Hence the unpaid balance for the next month (month 3) is

Prior unpaid balance		Reduction of unpaid balance		New unpaid balance
$920.05	−	$80.05	=	$839.50

We continue this process until all payments have been made and the unpaid balance is $0. Table 5.4.1 shows the entire process month by month and is known as an **amortization schedule**.

Table 5.4.1

Payment Number	Amount of Payment	Interest for Period	Reduction of Unpaid Balance	Unpaid Balance (principal)
0				$1000
1	$87.45	$7.50	$79.95	$920.05
2	87.45	6.90	80.55	839.50
3	87.45	6.30	81.15	758.35
4	87.45	5.69	81.76	676.59
5	87.45	5.07	82.38	594.21
6	87.45	4.46	82.99	511.22
7	87.45	3.83	83.62	427.60
8	87.45	3.21	84.24	343.36
9	87.45	2.58	84.87	258.49
10	87.45	1.94	85.51	172.98
11	87.45	1.30	86.15	86.83
12	87.48	0.65	86.83	0

Observe that the last payment needs to be increased by $0.03 in order to reduce the unpaid balance to $0. This minor change is due to rounding off in the computations and is typical of an amortized loan. ■

Example 6 Finding an Unpaid Loan Balance

In Example 4, we saw that Juan and Andrea Martinez financed a balance of $112,000 at 7.25% over 30 years to purchase a house. The monthly payments needed to amortize this loan were computed to be $764.04. After 10 years they wish to sell their house and purchase a new one. What is the unpaid balance on the mortgage after 10 years?

Solution

Understand the Situation: One way to answer this question is by constructing an amortization schedule like the one in Table 5.4.1. However, this table would be 120 lines long! Luckily, there is an easier method to answer this question. After 10 years (or 120 payments) the unpaid balance is equivalent to the amount of the loan that can be paid off with the remaining 240 payments (or 20 years). As we stated earlier in this section, the bank views mortgages as an annuity bought from individuals. Hence, for Juan and Andrea, the unpaid balance of their loan with 240 remaining payments is the same as the present value of an annuity with monthly payments of $764.04 for 20 years.

Using the present value of an annuity formula with $PMT = 764.04$, $n = 240$, and $i = \dfrac{0.0725}{12}$, we get

$$PV = PMT\dfrac{1-(1+i)^{-n}}{i}$$

$$= 764.04\dfrac{1-\left(1+\dfrac{0.0725}{12}\right)^{-240}}{\dfrac{0.0725}{12}}$$

$$\approx \$96{,}667.92 \quad \text{Rounded to the nearest cent}$$

Interpret the Solution: This means that after 10 years of making monthly payments of \$764.04 Juan and Andrea still owe \$96,667.92 on their \$112,000 mortgage. ∎

The scenario in Example 6 illustrates the following for computing an unpaid balance of an amortized loan.

Unpaid Balance of an Amortized Loan

The unpaid balance of an amortized loan is equal to the present value of the remaining (unpaid) payments.

✓ Checkpoint 3

Now work Exercise 49.

The unpaid loan balance in Example 6 may seem too large an amount to owe after making monthly payments for 10 years. But if we compute the amount of interest due on the first payment, we would get

$$\$112{,}000\dfrac{0.0725}{12} \approx \$676.67 \quad \text{Rounded to the nearest cent}$$

So the first monthly payment of \$764.04 was divided as

Monthly payment		Interest due		Reduction of unpaid balance
\$764.04	=	\$676.67	+	\$87.37

Only \$87.37 was used to reduce the unpaid balance with the first payment. Again, it is typical in long-term amortization loans to start out with small reductions in the unpaid balance.

SUMMARY

This section concluded our discussion of the mathematics of finance. Here we learned how to compute the present value of an annuity and applied this concept to retirement planning and loans. We also supplied a formula to compute payments for an amortized loan. We discussed how to construct an amortization schedule, a task ideally suited for a spreadsheet. The section concluded with an example illustrating how to compute an unpaid balance on a loan.

• **Present Value of an Annuity:** The present value PV of an annuity with n payments of PMT dollars made at the end of each payment period, with i as the interest rate per period, is $PV = PMT\dfrac{1-(1+i)^{-n}}{i}$.

• **Payments for an Amortized Loan:** A loan of PV dollars at interest rate i per period is **amortized** in n equal payments of PMT dollars, where PMT is given by $PMT = PV\dfrac{i}{1-(1+i)^{-n}}$.

SECTION 5.4 EXERCISES

In Exercises 1–6, determine the present value of each annuity.

1. Payments of $800 monthly for 8 years at an interest rate of 6% compounded monthly

2. Payments of $600 monthly for 15 years at an interest rate of 7% compounded monthly

✓ **3.** Payments of $900 monthly for 20 years at an interest rate of 7.5% compounded monthly

4. Payments of $700 monthly for 25 years at an interest rate of 6.5% compounded monthly

5. Payments of $2000 yearly for 30 years at an interest rate of 8.25% compounded annually

6. Payments of $1500 yearly for 20 years at an interest rate of 7.25% compounded annually

In Exercises 7–12, determine the present value of each annuity by using the FINANCE menu on your graphing calculator or the ANNPV calculator program found in Appendix C. Consult www.prenhall.com/armstrong *for the program for your graphing calculator.*

7. Payments of $600 monthly for 15 years at an interest rate of 7% compounded monthly

8. Payments of $800 monthly for 8 years at an interest rate of 6% compounded monthly

9. Payments of $700 monthly for 25 years at an interest rate of 6.5% compounded monthly

10. Payments of $900 monthly for 20 years at an interest rate of 7.5% compounded monthly

11. Payments of $1500 yearly for 20 years at an interest rate of 7.25% compounded annually

12. Payments of $2000 yearly for 30 years at an interest rate of 8.25% compounded annually

In Exercises 13–18, find the payment required to amortize each loan.

13. $25,000 at 7.5% compounded monthly for 10 years

14. $20,000 at 8.5% compounded monthly for 5 years

15. $100,000 at 7.75% compounded monthly for 30 years

16. $100,000 at 7.75% compounded monthly for 20 years

17. $13,000 at 6.25% compounded quarterly for 5 years

18. $20,000 at 7.25% compounded quarterly for 10 years

In Exercises 19–24, find the payment required to amortize each loan by using the FINANCE menu on your graphing calculator or the AMORTPAY calculator program found in Appendix C. Consult www.prenhall.com/armstrong *for the program for your graphing calculator.*

19. $20,000 at 8.5% compounded monthly for 5 years

20. $25,000 at 7.5% compounded monthly for 10 years

21. $100,000 at 7.75% compounded monthly for 20 years

22. $100,000 at 7.75% compounded monthly for 30 years

23. $20,000 at 7.25% compounded quarterly for 10 years

24. $13,000 at 6.25% compounded quarterly for 5 years

In Exercises 25–30, refer back to Exercises 13–18 and answer the following:

(a) Determine the total amount paid over the life of the loan.

(b) Determine how much interest was paid over the life of the loan.

25. The loan in Exercise 13

26. The loan in Exercise 14

27. The loan in Exercise 15

28. The loan in Exercise 16

29. The loan in Exercise 17

30. The loan in Exercise 18

Applications

31. Teri deposits $100 per month in a savings account that pays 4.75% interest compounded monthly for 10 years. Find the lump sum that Teri needs to deposit now to yield the same amount as the annuity after 10 years. (*Hint:* See Exercise 43, Section 5.3, for the future value of Teri's annuity.)

32. Steve deposits $80 per month in a savings account that pays 4.25% compounded monthly for 15 years. Find the lump sum that Steve needs to deposit now to yield the same amount as the annuity after 15 years. (*Hint:* See Exercise 44, Section 5.3, for the future value of Steve's annuity.)

33. Al and Tootie purchased a powerboat costing $34,000. They paid 20% down and amortized the rest with equal monthly payments over a 10-year period. They agree to pay 10% annual interest compounded monthly.

(a) Determine the monthly payment.

(b) Determine the total amount of interest paid over the 10 years.

34. Rework parts (a) and (b) from Exercise 33 keeping everything the same, except here compute for a 5-year period.

35. Irene deposits $200 each month into a retirement account earning 6.5% interest compounded monthly for 30 years. After the 30 years, she retires and withdraws equal monthly payments from the account for 25 years, at which time the account balance is $0. What monthly payment will Irene receive for the last 25 years?

36. Nick deposits $2400 each year into a retirement account earning 6.5% interest compounded annually for

30 years. After the 30 years, he retires and withdraws equal yearly payments from the account for 25 years at which time the account balance is $0. What yearly payments will Nick receive for the last 25 years?

37. (*continuation of Exercises 35 and 36*)

(a) How much did Irene in Exercise 35 deposit over the 30 years?

(b) How much did Nick in Exercise 36 deposit over the 30 years?

(c) How much do Irene and Nick receive on an annual basis once they retire and begin withdrawing money? Why aren't their annual amounts equal?

38. Rework Exercise 36, except here assume that Nick can earn 7% interest compounded annually.

39. Malika Agwani starts her career at the age of 25 and makes equal monthly deposits for 35 years into an annuity that earns 6.25% compounded monthly. She plans on retiring at the age of 60 and would like to withdraw $3000 monthly for 25 years, bringing the account balance to $0. How much should Malika deposit monthly to accumulate enough to provide her with these $3000 monthly payments?

40. Kahuna Smolko starts his career at the age of 30 and makes equal monthly deposits for 30 years into an annuity that earns 6.25% interest compounded monthly. He plans on retiring at the age of 60 and would like to withdraw $3000 monthly for 25 years, bringing the account balance to $0. How much should Kahuna deposit monthly to accumulate enough to provide him with these $3000 monthly payments?

41. In the Ohio Super Lotto PLUS™ lottery game, the smallest jackpot is $4 million. If there is one winner of this jackpot, he or she can take 30 annual payments (before taxes) of $133,333.33 or a lump sum that is equivalent to the present-day cash value (*Source:* www.ohiolottery.com). This present-day cash value is determined by the interest rate at which the money could be invested.

(a) Determine the present value of a $4 million lottery jackpot distributed in 30 annual payments (before taxes) of $133,333.33 using an interest rate of 6% compounded annually.

(b) Determine the present value of a $4 million lottery jackpot distributed in 30 annual payments (before taxes) of $133,333.33 using an interest rate of 7% compounded annually.

42. In the Lotto Virginia lottery game, a jackpot of $10 million has 26 annual payments (before taxes) of about $384,620 (*Source:* www.valottery.com). A lump sum that is equivalent to the present-day cash value can be determined by using the interest rate at which the money could be invested.

(a) Determine the present value of a $10 million lottery jackpot distributed in 26 annual payments (before taxes) of $384,620 using an interest rate of 6% compounded annually.

(b) Determine the present value of a $10 million lottery jackpot distributed in 26 annual payments (before taxes) of $384,620 using an interest rate of 7% compounded annually.

43. Sue and Tim state at a dinner party that they paid $20,000 down toward a new house. They also state that they have a 30-year mortgage for which they pay $870 per month. If interest is 8% compounded monthly, what did the house that Sue and Tim purchased originally cost?

44. Laurie and Craig have purchased a new home. They paid $25,000 down and have a 15-year mortgage for which they pay $980 per month. If interest is 8.25% compounded monthly, what did the house that Laurie and Craig purchased originally cost?

45. As of Thursday 5/17/01, Firstar Bank was offering a 30-year fixed-rate mortgage of 7.38% (*Source: The Plain Dealer*). George and Debbie Ashton purchase a house for $180,000. After putting 20% down as a down payment, they finance the balance with Firstar Bank.

(a) Determine the size of the down payment and the amount financed.

(b) Determine the monthly payment needed to amortize the loan.

46. (*continuation of Exercise 45*) As of Thursday 5/17/01, Firstar Bank was offering a 15-year fixed-rate mortgage of 7.00% (*Source: The Plain Dealer*). Repeat Exercise 45, except here assume that George and Debbie Ashton take out a 15-year mortgage.

47. (*continuation of Exercises 45 and 46*)

(a) Find the total interest that George and Debbie Ashton will pay if they take out a 30-year mortgage.

(b) Find the total interest that George and Debbie Ashton will pay if they take out a 15-year mortgage.

48. As of Thursday 5/17/01, Dollar Bank was offering a 30-year fixed-rate mortgage of 7.25% (*Source: The Plain Dealer*). Mike and Leslie Williams have taken out a $150,000 30-year mortgage with Dollar Bank.

(a) Determine the monthly payment needed to amortize the loan.

(b) Determine the total interest paid on this 30-year mortgage.

49. For the Ashton family in Exercise 45, determine the unpaid balance on their mortgage after

(a) 10 years **(b)** 15 years **(c)** 20 years

50. For the Williams family in Exercise 48, determine the unpaid balance on their mortgage after

(a) 10 years **(b)** 15 years **(c)** 20 years

Exercises 51 and 52 use the following:

It can be shown that the unpaid balance on a 30-year mortgage after x payments can be approximated by the function

$$y = PMT \left[\frac{1 - (1 + i)^{-(360-x)}}{i} \right]$$

where PMT is the monthly payment and i is the periodic rate.

51. (a) For the Ashton family in Exercise 45, write the function that approximates the unpaid balance after x payments.

(b) In the viewing window [0, 360] by [0, 150,000], graph the function from part (a).

(c) Use the graph to approximate the unpaid balance after 10 years ($x = 120$), 15 years ($x = 180$), and 20 years ($x = 240$). Compare the results with Exercise 49.

52. (a) For the Williams family in Exercise 48, write the function that approximates the unpaid balance after x payments.

(b) In the viewing window [0, 360] by [0, 160,000], graph the function from part (a).

(c) Use the graph to approximate the unpaid balance after 10 years ($x = 120$), 15 years ($x = 180$), and 20 years ($x = 240$). Compare the results with Exercise 50.

53. A loan of $10,000 with interest of 9% compounded monthly is repaid with equal monthly payments over 1 year. Construct the amortization schedule for this loan.

54. A loan of $4800 with interest of 12% compounded monthly is repaid with equal monthly payments over 2 years. Construct the amortization schedule for this loan.

SECTION PROJECT

Bill and Lisa wish to take out a **home equity loan** to finance their sons' college education. A local bank offers them a home equity loan called an **80% LTV ratio**. (LTV means loan-to-value and is typical of many home equity loans.) To compute the amount available as a loan, Bill and Lisa need to know the current value of their house and the unpaid balance on their mortgage. The amount available in an 80% LTV home equity loan is then computed to be

(80% of home's value) − (balance on mortgage)

Bill and Lisa purchased their home for $160,000 and put 20% down. They took out a 30-year fixed-rate mortgage at 6.75%.

(a) Determine the size of the down payment.

(b) Determine the amount financed.

(c) Determine the monthly payment needed to amortize the loan.

(d) Many things determine the current value of a home. With all items factored in, assume that the home appreciates in value at a rate of 2.75% per year. Determine the value of the home in 18 years.

(e) Determine the unpaid balance on the mortgage after 18 years.

(f) Determine the maximum amount that Bill and Lisa can receive in a home equity loan assuming an 80% LTV ratio loan.

▪ Why We Learned It

In this chapter we learned the basics of the mathematics of finance. To make good financial decisions, companies and individuals need to have an understanding of finance. For example, suppose that a company knows that in 5 years it will need to replace a piece of machinery that costs $1 million. Instead of coming up with the $1 million all at once in 5 years, the company could make fixed monthly deposits into an account with the goal that the future value of the account will be $1 million in 5 years. Or the company could simply make a lump-sum deposit now that would grow in value to $1 million in 5 years. The first situation requires an understanding of future value of an annuity, while the second situation uses knowledge of zero coupon bonds. Individuals can also use annuities to plan for retirement. For example, suppose that an individual wants to withdraw $1000 per month each month for 25 years to supplement his or her retirement income. If this individual is just beginning his or her career and has 30 years of investing, he or she would need to determine how much money is needed in the account in 30 years to finance $1000 monthly withdraws for 25 years. This situation would require a knowledge of future value as well as present value of an annuity.

CHAPTER REVIEW EXERCISES

In Exercises 1 and 2, the amount of money P borrowed for period of time t at simple interest rate r is given.

(a) Compute the simple interest owed for the use of the money.

(b) Compute the total amount, A, owed at the end of time t.

   1. $P = \$12,000, r = 8.25\%, t = 3$ years

   2. $P = \$8000, r = 7.75\%, t = 9$ months

In Exercises 3 and 4, the amount of money P loaned for period of time t along with the simple interest I charged is given. Determine the simple interest rate r of the loan. Round percentages to the nearest hundredth.

   3. $P = \$5000, I = \$750, t = 2$ years

   4. $P = \$18,000, I = \$6600, t = 5$ years, 9 months

In Exercises 5 and 6, determine the present value P given the future value A, time t, and annual simple interest rate r. Round to the nearest cent.

   5. $A = \$8500, r = 8\%, t = 4$ years

   6. $A = \$17,000, r = 7.25\%, t = 75$ months

   7. Yoshii deposits $6000 into a savings account that earns 6.25% simple interest.

(a) Write the future value A as a function of time t.

(b) Graph the equation from part (a).

(c) Determine the slope and interpret.

In Exercises 8 and 9, determine how long it will take a deposit to double at the given simple interest rate. Round answers to the nearest hundredth.

   8. $r = 7.75\%$

   9. $r = 6.25\%$

In Exercises 10 and 11, determine the discount and the proceeds given the loan amount F, annual interest rate r, and time t.

   10. $F = \$1800, r = 7\%, t = 9$ months

   11. $F = \$12,000, r = 8.75\%, t = 4$ years

In Exercises 12 and 13, determine the loan amount F to produce the given proceeds p at annual interest rate r over time t. Round to the nearest cent.

   12. $P = \$4000, r = 6\%, t = 2$ years

   13. $P = \$18,000, r = 7.25\%, t = 66$ months

   14. Troy borrows $4000 for 9 months at a simple interest rate of 7.5%. Determine the interest due on the loan.

   15. Nicole borrowed $16,000 from her father to start an accounting firm. She repaid him after 54 months. The annual simple interest rate was 6.75%. Determine the total amount that she repaid.

   16. Rebecca wants to buy a $10,000 sailboat in 3 years. How much should she invest now, to the nearest cent, at 7.25% simple interest to have enough for the sailboat in 3 years?

   17. Quinn borrows $1500 from Melissa and promises to pay back $1800 in 2 years. What simple interest rate will Quinn pay?

   18. If Mark desires to earn an annual simple interest rate of 8.5% on a 26-week T-bill with a face value of $15,000, how much should he pay for the T-bill?

In Exercises 19 and 20, compute the future value for the given principal or present value P, annual interest rate r, and given compounding type at the end of 4 and 9 years. Round answers to the nearest cent.

   19. $P = \$7000, r = 6.75\%$, compounded semiannually

   20. $P = \$11,000, r = 7.25\%$, compounded monthly

In Exercises 21 and 22, use a graphing calculator's built-in financial package to compute the future value for the given principal or present value P, annual interest rate r, and given compounding type at the end of 2 and 5 years. Round answers to the nearest cent.

   21. $P = \$9000, r = 5.75\%$, compounded annually

   22. $P = \$12,000, r = 6.5\%$, compounded quarterly

In Exercises 23 and 24, compute the present value for the given future value A, annual interest rate r, time t, and given compounding type. Round answers to the nearest cent.

   23. $A = \$7500, r = 6.75\%, t = 6$ years, compounded monthly

   24. $A = \$4500, r = 6.5\%, t = 4$ years, compounded daily

In Exercises 25 and 26, use a graphing calculator's built-in financial package to compute the present value for the given future value A, annual interest rate r, time t, and given compounding type. Round answers to the nearest cent.

   25. $A = \$12,000, r = 6\%, t = 5$ years, compounded semiannually

   26. $A = \$8500, r = 7.25\%, t = 3$ years, compounded quarterly

In Exercises 27 and 28, determine how long (that is, is the number of compounding periods) it will take the given present value P to grow to the given future value A for the stated annual interest rate r and compounding type. Round the answer to the nearest integer.

   27. $P = \$4000, A = \$7500, r = 6.5\%$, compounded annually

   28. $P = \$10,000, A = \$12,000, r = 6.75\%$, compounded quarterly

In Exercises 29 and 30, determine the effective rate corresponding to each of the following nominal rates. Round the percentages to the nearest thousandth.

   29. 7.75% compounded quarterly

   30. 6.5% compounded semiannually

In Exercises 31 and 32, for the given effective rate determine the nominal rate if the compounding is (a) quarterly; or (b) monthly.

31. 5.75%   **32.** 7.25%

🌐 **33.** The Consumer Price Index (CPI) is a measure of inflation. The CPI for all urban consumers in 2000 was 3.4%. If this rate holds true for the next 3 years, what will a $1.69 half-gallon of orange juice cost in 3 years if the rate is compounded annually (*Source:* Bureau of Labor Statistics, http://stats.bls.gov).

34. Kris and Rachel purchased a new condominium costing $159,000. Condominiums in the area are appreciating at a rate of 2.25% per year. If this rate holds steady for 6 years, what will the condominium be worth, assuming that this 2.25% rate is compounded annually? Round to the nearest thousand.

35. Mike wants to buy a new boat in 2 years. He estimates that he will need $8000 for the boat. How much should he invest now, to the nearest cent, at 6.75% interest compounded quarterly to have the $8000 in 2 years?

36. Paul wishes to invest some money in a money market account. He finds a bank that has an account paying 4.60% interest compounded monthly and another bank with an account paying 4.38% compounding daily.

(a) Compute the effective rate for the account paying 4.60% interest compounded monthly to the nearest hundredth.

(b) Compute the effective rate for the account paying 4.38% interest compounded daily to the nearest hundredth.

(c) Is one account a better investment than the other? Explain.

37. Beth's parents are considering purchasing a $50,000 face value zero coupon bond at her birth to help to finance her college education in 17 years. If money is worth 7.65% compounded annually, how much, to the nearest cent, should they pay for this bond?

In Exercises 38 and 39, determine the amount of each annuity.

38. A deposit of $200 monthly for 10 years at an interest rate of 6.25% compounded monthly

39. A deposit of $3000 annually for 30 years at an interest rate of 6.75% compounded annually

📟 **40.** Use the FINANCE menu on your graphing calculator or the ANNFV calculator program found in Appendix C to find the amount of a deposit of $300 monthly for 20 years at an interest rate of 7.25% compounded monthly. (Consult www.prenhall.com/armstrong for the program for your calculator.)

📟 **41.** Determine how much of the amount (or future value) of the annuity in Exercise 39 is interest.

In Exercises 42 and 43, determine the periodic payment amount required for each sinking fund.

42. The amount needed is $20,000 in 15 years at 6.75% interest compounded monthly. What is the monthly payment?

43. The amount needed is $50,000 in 10 years at 7.25% interest compounded quarterly. What is the quarterly payment?

📟 **44.** Use the FINANCE menu on your graphing calculator or the ANNPMT calculator program found in

Appendix C to determine the annual payment required for a sinking fund for which the amount needed is $20,000 in 8 years at 6% interest compounded annually. (Consult www.prenhall.com/armstrong for the program for your calculator.)

45. Refer back to Exercise 42 to answer the following:

(a) How much was invested?

(b) How much interest was earned?

In Exercises 46 and 47, determine the annual interest rate (to the nearest hundredth) that guarantees the future value of the annuity.

📟 **46.** The future value is $130,000; $300 is deposited each month for 20 years into an account paying interest compounded monthly.

📟 **47.** The future value is $500,000; $2000 is deposited each quarter for 30 years into an account paying interest compounded quarterly.

48. Kerri deposits $150 per month into a savings account that pays 4.5% interest compounded monthly.

(a) Determine the value of this account after 5 years.

(b) How much of the future value was money that Kerri deposited and how much of the future value was interest earned?

49. Meagan estimates that she spends $150 per month on cigarettes. She decides that she will quit and invest the monthly amount spent on cigarettes in an account earning 6.25% interest compounded monthly. At the end of 10 years, she will donate the interest earned on her account to her favorite charity. How much money can the charity expect to get in 10 years?

50. Jen and Mide would like to have $25,000 in 5 years as a down payment on a new house. What amount should they deposit each month into a savings account paying 6.375% interest compounded quarterly?

51. For his retirement, Kris wishes to have $225,000 in a 403(b) account 25 years from now. How much should he deposit each month in an account paying 6.25% interest compounded monthly?

52. (a) Grace deposits $3000 a year into an IRA earning 7% interest compounded annually. She makes her first deposit on her 25th birthday and her last deposit on her 35th birthday, a total of 11 deposits. Grace makes no additional deposits. She leaves the amount accumulated from the 11 deposits in the account, earning 7% interest compounded annually until her 65th birthday. How much is in Grace's account on her 65th birthday?

(b) Ted, Grace's friend, encouraged Grace to deposit $1500 into the same IRA account from part (a) each year beginning on her 25th birthday and ending with a deposit on her 65th birthday, a total of 41 deposits. How much would be in Grace's account when she makes her last deposit on her 65th birthday if she had taken Ted's advice?

(c) Compare the results of parts (a) and (b). In your comparison, determine how much Grace actually deposited and how much she would have deposited had she taken Ted's advice.

In Exercises 53 and 54, determine the present value of each annuity.

53. Payments of $750 monthly for 20 years at an interest rate of 7.25% compounded monthly

54. Payments of $1000 yearly for 15 years at an interest rate of 8.25% compounded annually

In Exercises 55 and 56, determine the present value of each annuity by using the FINANCE menu on your graphing calculator or the ANNPV calculator program found in Appendix C. (Consult www.prenhall.com/armstrong *for the program for your graphing calculator.)*

55. Payments of $800 monthly for 6 years at an interest rate of 7.5% compounded monthly

56. Payments of $2500 yearly for 30 years at an interest rate of 8.75% compounded annually

In Exercises 57 and 58, find the payment required to amortize each loan.

57. $30,000 at 6.75% compounded monthly for 15 years

58. $25,000 at 8.25% compounded quarterly for 20 years

In Exercises 59 and 60, find the payment required to amortize each loan by using the FINANCE menu on your graphing calculator or the AMORTPAY calculator program found in Appendix C. (Consult: www.prenhall.com/armstrong *for the program for your graphing calculator.)*

59. $75,000 at 7.25% compounded monthly for 30 years

60. $18,000 at 8.5% compounded quarterly for 15 years

61. Refer back to Exercise 57 to answer the following:

(a) Determine the total amount paid over the life of the loan.

(b) Determine how much interest was paid over the life of the loan.

62. Lena deposits $200 per month in a savings account that pays 4.5% interest compounded monthly for 8 years. Find the lump sum that Lena needs to deposit now to yield the same amount as the annuity after 8 years.

63. Al and Peggy purchased 15 acres of land costing $40,000. They paid 25% down and amortized the rest with equal monthly payments over a 15-year period. They agree to pay 8.25% annual interest compounded monthly.

(a) Determine the monthly payment.

(b) Determine the total amount of interest paid over the 15 years.

64. Lindsay deposits $240 each month into a retirement account earning 6.75% interest compounded monthly for 25 years. After 25 years, she retires and withdraws equal monthly payments from the account for 20 years, at which time the account balance is $0.

(a) What monthly payment will Lindsay receive for the last 20 years?

(b) How much did Lindsay deposit over the 25 years?

65. (*continuation of Exercise 64*) Suppose that Lindsay deposits $400 each month into a retirement account earning 6.75% interest compounded monthly for 15 years. She then leaves the amount accumulated after 15 years in the account, still earning 6.75% interest compounded monthly, for another 10 years without making any further deposits. She then withdraws equal monthly payments from the account for 20 years, at which time the account balance is $0.

(a) What monthly payment will Lindsay receive for the last 20 years?

(b) Compare your results from Exercise 64(a) and 65(a). In your comparison include the total amount that Lindsay deposits in each case.

66. Amy and Mike have purchased a new home. They paid $20,000 down and have a 30-year mortgage for which they pay $725 per month; the interest is 7.25% compounded monthly.

(a) What did the house that Amy and Mike purchased originally cost?

(b) Determine the total interest paid on this 30-year mortgage.

(c) Suppose that Amy and Mike decide to sell their home after 15 years. Determine the unpaid balance on their mortgage.

67. A loan of $6000 with interest of 8% compounded monthly is repaid with equal monthly payments over 1 year. Construct the amortization schedule for this loan.

In Exercises 68–71, do the following:

(a) Determine if the problem is a simple interest, compound interest, future value of an annuity, or present value of an annuity problem.

(b) Solve the problem.

68. Spencer's parents have established a sinking fund in order to have $150,000 in 17 years for his college education. How much should they deposit each quarter into a savings account paying 5.75% interest compounded quarterly?

69. Ryan borrows $3000 from Kris and promises to pay back $3400 in 3 years. What interest rate, to the nearest hundredth, will Ryan pay?

70. Lindsay and Mark purchase a house for $130,000. After putting 20% down, they finance the balance with a 30-year fixed-rate mortgage of 7.45% and make equal monthly payments to pay off the balance. If Lindsay and Mark decide to sell their home in 10 years, what is their unpaid balance?

71. How much should Ron invest now at 8.25% interest compounded quarterly to have $15,000 in 5 years to purchase a new boat?

CHAPTER 5 PROJECT

In this project we analyze credit cards and debt. We will learn how the finance charge is calculated, how to find the minimum payment due, and the importance of paying off the debt as soon as possible.

Kahuna Smolko just graduated from Old State University and wishes to reward himself with a new television. Recently, he received a Big Bucks credit card in the mail that offered a 12.99% fixed APR (annual percentage rate). He used the card to purchase a 65-inch wide screen HDTV from Sony for $10,000.00 (including sales tax) on the first of the month (*Source:* Sony Style Electronics: www.sonystyle.com). After this purchase, Kahuna cut up his credit card. He decides to pay only the minimum payment due for each bill. The credit card company receives his payments on the 16th day of the month.

To compute the finance charge, many credit card companies use a process called a *two-cycle average daily balance* since it results in a higher interest payment by the consumer. For simplicity, we will assume 30-day months and 360 days in a year.

On the first bill, the balance will be $10,000.00. Big Bucks computes the minimum payment due as 2.5% of the balance. Thus, the minimum payment due will be $10,000.00 \times 0.025 = 250.00$. Normally, the minimum payment is truncated (meaning if the decimal were actually $250.43, $250 will still be the minimum payment). Thus the remaining balance will be $10,000.00 - 250.00 = 9750.00$ (remember, this payment is received by Big Bucks on the 16th). We will assume that Kahuna must pay at least $25.00 per month.

Since this is a two-cycle average daily balance, Kahuna will be charged interest for the previous month *plus* the current month (on a one-cycle method, only the current month is charged interest). To calculate the finance charge for the current bill, we use the formula

$$\left(\begin{array}{c}\text{Finance charge of}\\ \text{previous month}\end{array}\right) + \underbrace{\left(\begin{array}{c}\text{Average daily balance}\\ \text{of current month}\end{array}\right) \times \dfrac{\text{APR}}{360} \times 30}_{\text{Finance charge for the current month}}$$

As mentioned, the previous month's average daily balance (ADB) was $10,000 (since no payment was made in the first month). To compute the ADB for the second month, we use

$$(\underbrace{10,000}_{\substack{\text{Balance before}\\ \text{payment made}}} \times \underbrace{15}_{\substack{\text{Number of days}\\ \text{balance was held}}} + \underbrace{9750}_{\substack{\text{Balance after}\\ \text{payment made}}}$$

$$\times \underbrace{15}_{\substack{\text{Number of}\\ \text{days remaining}\\ \text{in month}}}) \div \underbrace{30}_{\substack{\text{Number of}\\ \text{days in billing cycle}}} = 9875$$

The finance charge for the first month is $10,000 \times \dfrac{0.1299}{360} \times 30 = 108.25$. The finance charge for the second month is $9875 \times \dfrac{0.1299}{360} \times 30 = 106.896875$. The sum of the finance charges will be $108.25 + 106.896875 = 215.146875 \approx 215.14$. Thus, the new balance will be

$$\underbrace{9750}_{\substack{\text{Balance after}\\ \text{payment made}}} + \underbrace{215.14}_{\substack{\text{Finance charge}\\ \text{for current bill}}} = 9965.14$$

The new monthly payment will be 2.5% of this new balance, or $0.025 \times 9965.14 = 249.00$ after truncating.

1. Calculate the next two cycles. Be sure to find the new balance after the last minimum monthly payment was made, the new ADB, the new finance charge, and the new minimum monthly payment.

2. Visit www.prenhall.com/armstrong and input the program called Credit into your calculator. Check your answers to question 1 with this program.

3. Using the calculator program, determine the number of cycles necessary to pay off the credit card completely. (As you may have guessed, it will take many cycles.) Also, determine the total amount paid.

4. Suppose that Kahuna pays $300 a month instead of the given minimum monthly balance. Compute the first three cycles by hand and compare your results to question 1.

5. Visit www.prenhall.com/armstrong and input the program called Credit2 into your calculator. Check your answers to question 4 with this program. Use the program to determine the number of cycles necessary to pay off the credit card completely and the total amount paid. Compare total amount paid here with the total amount paid in question 3.

6

Sets and the Fundamentals of Probability

Odds to Probability
$P(A) = \dfrac{a}{a+b}$

Probability of Tiger Woods Winning the 2001 British Open
$P(A) = \dfrac{a}{a+b} = \dfrac{3}{3+2} = \dfrac{3}{5}$

We frequently hear about the odds of something occurring. But how do odds relate to probability? In 2001, golf pro Tiger Woods was "hot" and the odds of his winning the British Open were 3 to 2. Using the formula for converting odds to probability, we can translate odds into the probability of a given event occurring. In this case, the probability of Tiger Woods winning the Open was 3/5. Although the probability of a win was high, Tiger actually suffered a stunning defeat. Probability is a valuable tool for predicting likely outcomes, but it does not ensure that a given event will occur.

What We Know

So far in this text, we have seen many applications involving matrices and linear programming, as well as the mathematics of finance. We have seen how an understanding of these topics can be important for individuals in many diverse areas.

Where Do We Go

In this chapter we begin our journey into the study of probability. Areas as diverse as insurance policy analysis, weather forecasting, medical research, and financial planning utilize concepts of probability. Since probability is different from the mathematics of the first part of this text, we ease into its study with a section dedicated to problem-solving techniques in probability.

Problem Solving: Probability and Statistics

The idea of chance is a part of our lives everyday. Many of the decisions that we make are influenced, in part, by chance. When we use the word *chance* we mean *the likelihood that an event will occur*. For example, an individual may decide whether or not to take an umbrella with him or her to work based on the chance that it will rain that day. In this scenario, the **event** in question is "it will rain that day." Another example is taking a chance that guessing C as an answer for a question on a multiple-choice test is the correct answer. Or we may take a chance on whether the balding tires on our car will blow out this summer. All these events and many others are based on **chance or probability**. You probably have heard about games of chance, that is, games that involve probability, such as a state lottery, a card game, or slot machines. However, probability is a necessary tool in areas such as insurance policy analysis, weather forecasting, medical research, and financial planning. In general terms, we can say that probability is the branch of mathematics that produces a number that expresses the likelihood of occurrence of a specific event.

The challenge that many students face when beginning the study of probability is the use of language in probability applications. This means that you must **read the problems closely** and **think carefully** about what you are asked to find. An author takes a great deal of care to select just the right words for a probability problem. You need to read the problem slowly and carefully, and you may need to read it more than once so that you understand what the author is saying. Then you must think about the problem in a careful, step-by-step fashion to figure out what is needed and how you can achieve the correct results. In this section, we will introduce some of the key concepts that are necessary to solve the application problems commonly encountered in the study of probability. We will also demonstrate how to read closely and think carefully when confronted with an application. The ideas in this section can be applied to the remainder of Chapter 6 and also in Chapter 8.

Thinking about Populations and Samples

We begin with two terms fundamental to the study of probability: population and sample. We will revisit these definitions later in the text.

■ **Population and Sample**
- A **population** is the collection of **all** outcomes, responses, measurements, or counts that are of interest.
- A **sample** is a portion (or subset) of the population.

When we study statistics, the companion topic to probability, we take a sample at random and use the results to make estimates of the population as a whole. By at random, we mean that every item in the population has an equal chance of being included in the sample. We call this branch of statistics **inferential statistics**. In the study of **probability**, we do the reverse. That is, we use the information about a population to make inference about the possible nature of the sample. This contrast is illustrated in Figure 6.1.1(a) and (b).

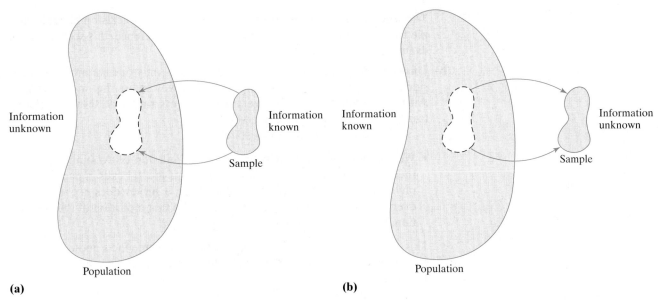

(a)   (b)

Figure 6.1.1 (a) Inferential statistics involves making conclusions about a population based on information gathered in a sample. **(b)** Probability involves making statements about a sample based on information that is known about a population.

Example 1   Distinguishing between Statistical Inference and Probability

Classify each of the following as an example of probability or of statistical inference.

(a) Past records show that about 15% of the population of Austinburg use the emergency 911 number each year. In a random sample of 50 Austinburg citizens, the chances are really high that at least one used the 911 emergency number last year.

(b) A survey shows that in the town of Austinburg, about 1 out of every 14 residents say that they use public transportation at least once per week.

Solution

(a) **Understand the Situation:** The first task when solving a probability problem is to read the problem slowly and repeatedly until you recognize exactly what the writer is talking about. Make sure that you understand exactly what the writer says. In this case, there are two statements. In the first statement, we know that 15% of Austinburg citizens use 911. In the second statement, we are told that if we take a random sample of 50 citizens the chances are really high that one has called 911 during the past year.

Now let's think about these two statements.

- The statement "15% of Austinburg citizens use 911" is a statement about a population (the people of Austinburg).

- The statement "In a random sample of 50 Austinburg citizens, there is a good chance that at least one used the 911 emergency number last year" is a statement about a sample (portion) of the population. We can make this statement because we know that 15% of the people (or 15 out of 100) used 911 last year.

The second statement shows that we are drawing a conclusion about a sample from information known about a population. Hence, this scenario is an example of probability.

(b) **Understand the Situation:** In this case there is only one statement to consider. In the statement, we are told that a survey shows that in the town of Austinburg about 1 out of every 14 residents says that they use public transportation at least once per week.

Now let's think about this statement.

• We are not given any facts about the population of Austinburg. Here, there are no past records, as was the case for the 911 calls in part (a), simply a survey. Under normal circumstances, a survey does not imply that every resident was contacted. It does imply that a portion of the population, that is, a sample, was contacted.

This statement indicates that we are estimating some facts about a population based on some sample results. Hence, this scenario is an example of statistical inference. ■

✓ **Checkpoint 1**   Now work Exercise 3.

Thinking about Probability Experiments

In order to collect information that can be used in situations involving probability, we conduct **probability experiments**. A probability experiment can be as simple as flipping a coin or as complex as a medical research study.

> ■ **Probability Experiment**
>
> A **probability experiment** is a process that leads to well-defined results, called **outcomes**. All possible outcomes are predictable, but any given outcome cannot be predicted with certainty.

Notice from the definition that a probability experiment must have two conditions satisfied. The first condition is that the outcomes must be unambiguous. In other words, the result of each individual outcome must be able to be easily placed into a specific category. The second condition is that all possible outcomes are predictable, but a specific outcome cannot be predicted with certainty. For example, when flipping a coin we know that all possible outcomes are heads or tails, yet we cannot predict with certainty if the flip will yield heads or tails.

Example 2   **Determining If a Process Is a Probability Experiment**

Determine whether each of the following processes represents a probability experiment.

(a) 100 people were called on the telephone and asked, "How was your day?"

(b) 100 people were called on the telephone and asked, "How was your day?
1. Good   2. Fair   3. Bad   4. None of your business   5. No comment"

(c) A bag contains three marbles painted with the numbers 1, 2, and 3. A marble is drawn from the bag, its number is recorded, and it is then returned to the bag. Another marble is drawn and its number is recorded.

Solution

(a) Understand the Situation: To ask someone a general question such as "How was your day?" could elicit many different kinds of responses. This is typical of an open-ended question. Since open-ended questions can lead to many different kinds of responses, this indicates that the outcomes (responses) are not well defined. So this process is not a probability experiment.

(b) Understand the Situation: At first glance, this seems the same as part (a). But when we read the scenario carefully, we see that it is different. Specifically, people surveyed are asked to pick one of five possible responses. This means that we do not have an open-ended question! Each outcome is well defined (all possible outcomes are Good, Fair, Bad, None of your business, or No comment). We know that the outcome of a single telephone call is going to be recorded as one of these five responses, but the outcome of each telephone call is not predictable with certainty. We have no way of knowing exactly how an individual will respond to the question. Since the outcomes are well defined, but any given outcome is not predictable with certainty, this process is an example of a probability experiment.

(c) Understand the Situation: In this case, each individual outcome is well defined (all possible outcomes are the whole numbers 1, 2, or 3). We know that the outcome of the first draw is going to be a 1, 2, or 3. We know that the outcome of the second draw is also going to be a 1, 2, or 3. However, the outcome of each draw is not predictable with certainty. This means that the process is a probability experiment. ■

Thinking about Ordered and Unordered Samples

When studying the concepts of probability, we commonly have a need to count the number of different outcomes that we can have for a probability experiment. In Example 2(c), we can see that there are nine possible outcomes. These nine possible outcomes are (1, 1), (1, 2), (1, 3), (2, 1), (2, 2), (2, 3), (3, 1), (3, 2), and (3, 3). But there are times when counting the number of possible outcomes can be more complicated. If we want to know the number of ways that a 12-member board of trustees can choose a president, vice-president, and CEO, for example, we need additional counting techniques. One of the keys that enables us to count these possible outcomes depends on whether order is important. By order being important, we mean that:

- The order of arrangement makes a difference, or
- There is a ranking of the members of the arrangement.

For example, let's say that 12 horses are entered in this year's Jasper Derby. If we want to count the number of ways that the horses can finish first, second, and third, then the order of finish is important. In other words, there is a ranking of the first three finishers. We will see how to compute the actual number of outcomes for a probability experiment like this one in Section 6.3. For now, we are going to focus on determining whether order is important.

Example 3   **Determining If Order Is Important**

For each of the following, determine whether order is important.

(a) There are 24 members in the New Strawsburg Junior High School Computer Club. When computing how many ways the club can elect a president, vice-president, secretary, and treasurer, is order important?

(b) The trauma unit at Gothicheath Hospital has 44 staff members. When computing how many ways a committee of three can be formed to investigate salary discrepancies, is order important?

Solution

(a) **Understand the Situation:** The names that we associate with the president, vice-president, secretary, and treasurer make a difference. This indicates that order is important.

(b) **Understand the Situation:** The hospital staff wants to form a committee of **any** three from among the possible 44 staff members. Notice that no ranking of committee members is involved. This means that, in this case, order is not important.   ∎

✔ **Checkpoint 2**

Now work Exercise 21.

Thinking about Simultaneous Events

Many times in the study of probability we want to know if it is possible for two things to happen at the same time. These things can be regarded as a collection of outcomes of a probability experiment. We call this collection of outcomes **events**. (We will discuss events in more depth in Section 6.4.)

For example, if a six-sided die is tossed in a probability experiment, then an event could be that the outcome of the roll is an odd number. The outcomes associated with this event are the numbers 1, 3, and 5. Another event could be that the outcome of the roll is an even number. The outcomes associated with this event are the numbers 2, 4, and 6. Since a single roll of a die cannot produce **both** an odd number **and** an even number, these events cannot happen at the same time.

By contrast, consider the event that a randomly chosen student from Berliner College is registered for a chemistry class this semester, and the other event is that the randomly chosen student from Berliner College is also registered for a Spanish class this semester. It is possible for a student to sign up for a chemistry class and for a Spanish class during the same semester. Hence, this is an example of two events that can occur simultaneously.

Before proceeding, let's pause for a word of caution. When we ask questions such as if two events can happen at the same time, we are asking the question to be considered under normal circumstances.

A common error made by many individuals who are just beginning to study probability is trying to read too much information into the questions in an attempt to find some exception.

So when posed questions in the study of probability, such as if two events can happen at the same time, we mean for the question to be considered under everyday, usual, commonsense circumstances. For example, when flipping a coin, the only reasonable outcomes are heads or tails. Some may say that the coin could land on its edge, but this is not a usual or common sense result.

Example 4   ## Determining If Events Can Happen at the Same Time

Determine if the following events can happen at the same time.

(a) The event that Sandy gets a speeding ticket and the event that Sandy gets a ticket for not wearing a seat belt.

(b) The event that, during the same at bat, a softball player, Ariel, hits a single and the event that Ariel hits a triple.

(c) The event that during a single roll of a six-sided die Don rolls a 2 and the event that Don rolls a 4.

Solution

(a) Understand the Situation: To help us determine what is happening in this case, we will list the two events.

- Event 1: Sandy gets a speeding ticket.
- Event 2: Sandy gets a ticket for not wearing her seat belt.

It appears that these events, receiving a speeding ticket and receiving a ticket for not wearing a seat belt, can happen at the same time.

(b) Understand the Situation: As an aid to help us determine what is happening here, we again will list the two events.

- Event 1: During an at bat, Ariel hits a single.
- Event 2: During the same at bat, Ariel hits a triple.

During the same at bat, a softball player cannot hit a single **and** a triple. This means that these events cannot happen at the same time.

(c) Understand the Situation: As in parts (a) and (b), we list the two events.

- Event 1: During the roll of a die, Don rolls a 2.
- Event 2: During the same roll, Don rolls a 4.

During the same roll of a die, an individual cannot roll a 2 **and** a 4. This means that these events cannot happen at the same time.   ∎

✓ **Checkpoint 3**

Now work Exercise 29.

Thinking about Whether One Event Influences Another Event

Many times, a crucial piece of information that we want to know about two successive events is whether the result of one event causes the next event to be more or less likely to occur. For example, let's say that one event is that Alan takes an aspirin and the next event is that Alan's headache is relieved. For most people, taking an aspirin increases the likelihood that the headache will be relieved. This indicates that the first event (taking an aspirin) does influence the second event (headache is relieved).

Now suppose that Michelle oversleeps because she did not set her alarm clock and then tries to start her car only to discover that its battery is dead. We know that oversleeping does not make it more or less likely for her car to have a dead battery. So the outcome of the first event (oversleeping) does not influence the second event (the car having a dead battery).

Once again, we want to point out that we should not read too much into the event(s) that we are given. Even though some may say that Michelle may be suffering from bad karma or perhaps the alarm clock could have somehow drained the car battery or that not setting the alarm clock indicates irresponsibility, hence she may also be irresponsible of car maintenance, we can conclude that under normal, usual circumstances, one event (oversleeping) does not influence the other event (dead car battery).

Example 5 Determining If One Event Influences the Other

Determine if the outcome of one event influences the outcome of the other event.

(a) The event of receiving 9 inches of snow on Sunday night and the event that the local school district cancels classes on Monday morning.

(b) The event that the first toss of a coin is heads and the event that the second toss of the coin is tails.

Solution

(a) Understand the Situation: As we did in Example 4, we will list the two events separately as an aid to determine what is happening here.

- Event 1: Receiving 9 inches of snow Sunday night.
- Event 2: Local school district cancels classes Monday morning.

Since getting a large snowfall makes it more likely for school to be canceled, we conclude that the outcome of the first event (receiving 9 inches of snow Sunday night) does influence the second event (local school district cancels classes Monday morning).

(b) Understand the Situation: Again, we begin by listing the two events.

- Event 1: First toss of a coin is heads.
- Event 2: Second toss of a coin is tails.

Tossing heads with the first toss does not alter the likelihood of tossing tails on the second toss. So, in this case, the outcome of the first event (first toss of a coin is heads) does not influence the second event (second toss of a coin is tails). ∎

✓ **Checkpoint 4**

Now work Exercise 53.

Solving Applications in Probability

We conclude this section by offering some general tips on solving applications in the area of probability.

■ **Problem Solving Tips for Probability**

1. Read the application problem closely and think carefully.
2. For the question that is being asked, ask yourself whether or not the following apply:

   (a) Is order important?

   (b) Can the events occur simultaneously?

   (c) Can the outcome of one event influence the likelihood of another event?

3. Do not conjecture about a problem too much. Answer the question as it would occur under ordinary circumstances.
4. Answer the question being posed using a complete sentence.

SUMMARY

In this section, we explored the unique use of language when studying probability. We found that many of the applications addressed one of three key questions. The first question asks, "Is order important?", meaning does the order of arrangement make a difference, or is there a ranking of the members of the arrangement. The second central question that we examined is "Can events occur simultaneously?", meaning can the outcome of a probability experiment satisfy two events at once. The third of the equations that we addressed is "Can the outcome of one event influence the likelihood of another event?" The ability to answer these questions will make the voyage through the probability and statistics part of this text less arduous.

SECTION 6.1 EXERCISES

In Exercises 1–8, classify each of the following as an example of probability or statistical inference.

$\stackrel{\Rightarrow}{8.3}$ 🌐 **1.** A 1992 report on prisoners and education showed that 49% of the prisoners did not have a high school diploma. In a sample of 10 prisoners selected at random, 6 do not have a high school diploma. (*Source:* National Center for Education Statistics, www.nces.ed.gov.)

$\stackrel{\Rightarrow}{8.3}$ 🌐 **2.** A 1994 report revealed that 32.6% of all births in the United States were to unmarried women. In a random sample of 30 women who gave birth in 1994, nine of them were unmarried. (*Source:* Center for Disease Control, www.cdc.gov.)

✓ 🌐 **3.** To determine the attitudes of Americans concerning the political leadership in Russia, a June 2001 Gallup poll asked a random sample of 1011 adults 18 years and older if they are more likely to consider Russian President Vladimir Putin friendly or unfriendly to the United States; 435 of them considered him friendly to the United States. (*Source:* The Gallup Organization, www.gallup.com.)

🌐 **4.** A randomly selected sample of 1061 American adults 18 years and older were asked, "Could the entire moon program have been an elaborate deception staged to fool the public?" 944 of them did not believe that the U.S. government staged or faked the *Apollo* moon landing. (*Source:* The Gallup Organization, www.gallup.com.)

🌐 **5.** A May 2001 survey determined that 76% of all Internet users check their e-mail daily. In a random sample of 35 Internet users, 25 said that they check their e-mail daily. (*Source:* UCLA Center for Communication Policy, www.ccp.ucla.edu.)

🌐 **6.** In a report on the adult public, 53.1% viewed American children negatively. In a sample of 20 adults selected at random, 14 viewed children negatively. (*Source:* www.publicagenda.org.)

$\stackrel{\Rightarrow}{8.3}$ 🌐 **7.** A report on minorities and transportation found that 59% of African Americans in New York City had no automobiles. In a random sample of 15 African Americans from New York City, 9 of them did not have an automobile. (*Source:* American Highway Users Alliance, www.highways.org.)

$\stackrel{\Rightarrow}{8.3}$ 🌐 **8.** In 1997, 10.8% of female smokers smoked cigars. In a sample of 20 female smokers selected at random, two of them smoked cigars. (*Source:* American Cancer Society, www.cancer.org.)

In Exercises 9–12, determine whether each of the following processes represents a probability experiment.

6.4 **9.** A box contains 5 identical tickets labeled 1 through 5. In a process, a ticket is drawn, its number is recorded, and it is returned to the box. Then a second draw is taken. The sum of the two numbers is recorded.

6.4 **10.** The suggestions card at Pappy's restaurant asks customers to rate the service as fair, good, or excellent and then to rate the quality of the food on a scale from 1 to 5, with 1 being superior and 5 being lousy.

11. In a mail survey, 400 people were asked, "What is your least favorite song?"

12. A random sample of 1050 people was asked, "If you could change your name, what would you change it to?"

In Exercises 13 and 14, consult the customer satisfaction card supplied.

The Store	Excellent	Above Average	Average	Below Average	Poor
Overall shopping experience	○	○	○	○	○
The Merchandise					
Selection	○	○	○	○	○
In stock	○	○	○	○	○
Value	○	○	○	○	○
The Sales Personnel					
Sales people available	○	○	○	○	○
Sales people knowledgeable	○	○	○	○	○
The Checkout Area					
Speed	○	○	○	○	○
Friendliness	○	○	○	○	○
The Service Desk					
Courtesy	○	○	○	○	○
Helpfulness	○	○	○	○	○

Customer Type   Do-It-Yourselfer ○   Contractor / Commercial Customer ○

Additional Comments: _____

13. Which item(s) represents a probability experiment. Explain.

14. Which item(s) does not represent a probability experiment. Explain.

In Exercises 15–24, determine whether order is important.

6.3 **15.** The board of directors at the SafeSide Company consists of 12 members. When computing how many different ways a chief executive officer, a vice-president and a chief financial officer can be selected, is order important?

6.3 **16.** The Chess Club at Dyeston High School has 15 members. When computing how many different ways a president, a vice-president, and a treasurer can be selected, is order important?

6.3 **17.** The Miss Farrago County Fair queen contest has 19 entrants. When computing how many different ways a winner, first runner-up, and second runner-up can be selected, is order important?

6.3 **18.** Traditionally, 33 cars start the Indianapolis 500. When computing how many different ways the drivers can place first, second, and third, is order important?

6.3 **19.** In Crawfords County, 6 judges are running for 3 county judge positions. When computing how many different ways the 3 county judges can be elected, is order important?

6.3 **20.** The central office of the Higgens Corporation has to choose 3 of its 8 regional offices for an audit. When computing how many different ways the regional offices can be chosen, is order important?

6.3 **21.** A group of 20 professors in an Archeology department want to choose a research team of 5 to study a new archeological site. When computing how many different ways the research team can be formed, is order important?

⇒ **22.** For a literature class, a student has to read 4 books
6.3 from a list of 21. When computing how many different ways
the 4 books can be chosen, is order important?

⇒ 🌐 **23.** In the Maine State lottery called *Megabucks*,
6.3 6 balls numbered 1 to 42 are drawn from a hopper. When
computing how many different ways the 6 numbers can be
chosen, is order important? (*Source:* Maine State Lottery
Commission, www.mainelottery.com.)

⇒ 🌐 **24.** Jelly Belly manufactures 40 different flavors of
6.3 jelly beans. If a person goes to a Jelly Belly store, he can get
a sampler cup of 8 jelly beans, in which the 8 jelly beans are
all different flavors. When computing how many different
sampler cups are possible, is order important? (*Source:*
Jelly Belly, www.jellybelly.com.)

*In Exercises 25–34, determine if the following events can
happen at the same time.*

⇒ **25.** A person in a mall is selected at random. The event
6.5 that this the person has brown hair and the event that this
person has blue eyes.

⇒ **26.** A person in a mall is selected at random. The event
6.5 that the person has a college degree and the event that the
person is employed.

⇒ **27.** In a state that requires voters to register for a single
6.5 party, a voter is selected at random for an exit poll. The
event that the person is a registered Republican and the
event that the person is a registered Democrat.

⇒ **28.** In a state that uses electronic ballots, a voter is se-
6.5 lected at random for an exit poll. The event that the person
voted for issue 1 on the ballot and the event that the person
voted against issue 1 on the ballot.

✓ ⇒ **29.** A student at Brockston College is selected at ran-
6.5 dom. The event that the student is a biology major and the
event that the student is a senior.

⇒ **30.** A student at Brockston College is selected at ran-
6.5 dom. The event that the student receives financial aid and
the event that the student is on the Dean's list.

⇒ **31.** A box contains 30 identical tickets numbered 1 to 30.
6.5 One ticket is drawn at random. The event that the ticket is
labeled with an even number and the event that the ticket is
labeled with an odd number.

⇒ **32.** A box contains 30 identical tickets numbered 1 to 30.
6.5 One ticket is drawn at random. The event that the ticket is
labeled with a prime number and the event that the ticket is
labeled with a composite number.

⇒ **33.** Two respondents from a nationwide survey are se-
6.5 lected at random for a follow-up interview. The event that
the two respondents are brothers and the event that the two
respondents are sisters.

⇒ **34.** Two respondents from a nationwide survey are se-
6.5 lected at random for a follow-up interview. The event that

the two respondents are brothers and the event that the two
respondents are not related.

*In Exercises 35–40, an event is given. Write another event so
that together the events cannot happen at the same time.*

35. If a coin is flipped, the event that the outcome is heads.

36. If a card is drawn from a shuffled, standard 52-card
deck of cards, the event that the card is red.

37. If a person wakes up from a nap, the event that it will
be dark outside.

38. The event that a randomly selected student will live in a
coed dormitory.

39. The event that a randomly selected person has a hearing aid.

40. The event that a randomly selected student taking history passes a history test.

*In Exercises 41–46, an event is given. Write another event so
that together the events can happen at the same time.*

41. In an automobile, the event that a person is arrested for
fleeing from the police.

42. At an airport, the event that the flight from Chicago to
Washington is delayed due to bad weather in Chicago.

43. During the fall semester, the event that a randomly
selected student receives a failing grade in biology.

44. During a meeting of the school's chess club, the event
that Thom is elected club vice-president.

45. When drawing a card from a deck, the event that a
black card is drawn.

46. When drawing a card from a deck, the event that a
numbered (nonface) card is drawn.

*In Exercises 47–56, determine if the outcome of one event
influences the outcome of the other event.*

⇒ **47.** The event that a flash food occurs at Windsor Park
6.6 on Tuesday afternoon and the event that the Tuesday
evening company picnic at Windsor Park is postponed.

⇒ **48.** The event that Rebecca completes her political
6.6 science course reading assignment every day and the event
that Rebecca passes her political science course.

⇒ **49.** The event that Kahrin's car battery is dead and the
6.6 event that Kahrin's oven light is burned out.

⇒ **50.** The event that the telephone in Rich's office does
6.6 not work and the event that the vending machine in Rich's
daughter's high school is empty.

⇒ **51.** The event that Clifford gets a raise and a promotion
6.6 and the event that Clifford purchases a new car.

⇒ **52.** The event that Lana is exposed to ultraviolet radia-
6.6 tion for three decades and the event that Lana develops
skin cancer.

✓ 6.6 **53.** A box contains 5 tickets numbered 1 to 5. In an experiment, a ticket is drawn, its number is recorded, and it is replaced. Then a second ticket is drawn. The event that the first ticket is a 3 and the event that the second ticket is a 3.

⇒ 6.6 **54.** A bag contains 5 green and 4 purple marbles. In an experiment, a marble is drawn, its color is recorded, and it is replaced. Then a second marble is drawn. The event that the first marble is purple and the event that the second marble is purple.

⇒ 6.6 **55.** The event that a person has brown eyes and the event that the person's son has brown eyes.

⇒ 6.6 **56.** The event that Jim has received seven speeding tickets and the event that Jim has high car insurance premiums.

In Exercises 57–62, an event is given. Write another event so that together the outcome of one event does not influence the outcome of the other event.

57. The event that it will be sunny on Wednesday.

58. The event that the next president will be a Republican.

59. A child pulls out two cookies from a jar containing five chocolate chip and five oatmeal cookies. The event that the first cookie taken is an oatmeal cookie.

60. If 2 cards are drawn from a shuffled, standard 52-card deck of cards, the event that the first card drawn is an ace.

61. If a die is tossed, the event that the outcome is an even number.

62. Ten identical balls numbered 0 through 9 are placed in a hopper for the Mini Lotto game. If three numbered balls are drawn for the Mini Lotto game, the event that the winning numbers the first week of August are 3–8–9.

In Exercises 63–68, an event is given. Write another event so that together the outcome of one event does influence the outcome of the other event.

63. If a person goes outside to sunbathe, the event that the person does not put on sunscreen.

64. In Olympic skating, the event that Tara Lapinski wins the gold medal in freestyle skating.

65. In a graduation ceremony, the event that Tammy gets the award for valedictorian.

66. At an ice cream parlor, the event that Ben eats four banana splits.

67. For a political science exam, the event that Ron did not go to class or complete the assigned reading to prepare for the exam.

68. At a stop sign down the street from her house, the event that Chandra does not come to a complete stop at the intersection.

SECTION PROJECT

In this section, we have seen that events can have certain properties, such as if the outcome of one event influences another or if two events can occur at the same time. Another important property to know is if the list of events includes every possible outcome of an experiment. If we toss a coin, for example, we can say that the possible events associated with the experiment are to flip a head and to flip a tail. On any given flip, we know that both events cannot occur at the same time. We also know that on any given flip we must either flip a head or flip a tail. So the list of events, flip a head and flip a tail, includes every possible outcome of the experiment.

A standard deck of 52 cards consists of four suits: hearts, diamonds, clubs, and spades. Each suit, in turn, consists of 13 cards: Ace, two, three, . . . , ten, jack, queen, and king. We draw a card at random from the deck. Let's say that one event is that it is a diamond and another event is that it is a heart. Even though these events cannot occur at the same time, the list of these two events does not include every possible outcome. Knowing that two events cannot occur at the same time does not necessarily mean that every possible outcome is accounted for.

Let's test our understanding of these ideas by completing the table by writing *yes* or *no* for each situation.

Card Draws	Can the two events occur at the same time?	Will the two events list all possible outcomes?
(a) Draw a spade and draw a club		
(b) Draw a face card and a numbered card		
(c) Draw a king and a 2		
(d) Draw a spade and a nonspade		
(e) Draw a 9 and a heart		
(f) Draw a black card and a club		

Section 6.2   Sets and Set Operations

We often categorize things in life so that we can make sense of them. We categorize cars by their make, model, color, list of options, and so forth. Creating collections of objects helps us to find order and meaning. When we categorize objects, we are creating **sets**. In this section, we will learn how to write sets of objects, perform operations on sets, and explore some properties of sets. Later, we will apply many of these ideas when we study probability.

Set Notation

A **set** is a well-defined collection of objects. For example, the months January, February, May, and July represent the set of all months of the year that end in the letter y. On the other hand, a list of the five best female tennis players of all time is not a set since it is *not* well defined. This is because naming the five best female tennis players is a subjective matter; different people could have different lists.

Capital letters, such as A, B, C, and so on, are usually used to name sets. The objects in a set are its **elements**. The elements of a set are separated by commas and enclosed in **braces**, { }. For example, the set of all even numbers from 1 to 10, inclusive, can be represented by

$$A = \{2, 4, 6, 8, 10\}$$

Notice that if $B = \{4, 8, 10, 6, 2\}$ then the sets A and B have exactly the same elements, just written in a different order. We call the sets A and B **equal**, and write $A = B$. To indicate that 4 is an element of A, we write

$$4 \in \{2, 4, 6, 8, 10\} \quad \text{or equivalently} \quad 4 \in A$$

The symbol $\in$ means "belongs to" or "is an element of." To indicate that 5 is not an element of A, we write

$$5 \notin \{2, 4, 6, 8, 10\} \quad \text{or equivalently} \quad 5 \notin A$$

Some sets have no elements. For example, the set of U.S. coins in which each coin is worth exactly 7 cents has no elements. We call a set with no elements the **empty set** and designate it with the symbol $\emptyset$ or { }.

There are generally two ways to express the elements in a set. The first is the **listing method** where we simply list all the elements in a set. However, at times this method is not practical to use. Let's say that set C consists of the even numbers between 1 and 1000, inclusive. Instead of listing all 500 numbers in the set, we could use **set-builder notation** and write

$$C = \{x \mid x \text{ is an even number between 1 and 1000, inclusive}\}$$

This is read, "C is the set of all elements x such that x is an even number between 1 and 1000, inclusive." Note that x is the name of the typical element in C and that the vertical bar is read "such that."

Example 1   Writing Sets

Use the listing method to rewrite each of the following sets:

(a) $A = \{x \mid x \text{ is a U.S. coin less than a dollar in value}\}$

(b) $B = \{x \mid x \text{ is an even number and } 10 \le x \le 20\}$

Solution

(a) Here we write the names of the desired coins in set notation.

$$A = \{\text{penny, nickel, dime, quarter, half-dollar}\}$$

(b) For B, the designated numbers are listed and separated by commas.
$$B = \{10, 12, 14, 16, 18, 20\}$$

✓ **Checkpoint 1**

Now work Exercise 5.

At times the elements of one set are contained within another set. For example, if we let $A = \{x | x$ is a whole number and $1 \le x \le 30\}$ and $B = \{5, 10, 15, 20\}$, then we observe that every element of B is also an element of A. We call B a **subset** of A, denoted by $B \subseteq A$. In fact, there are elements of A that are not elements of B, so we call B a **proper subset** of A and write $B \subset A$.

■ **Subset and Proper Subset**

- If every element of set B is also an element of set A, then we say that B is a **subset** of A, denoted by $B \subseteq A$.
- If set B is a subset of set A and there is at least one element in A that is not in B, then we say that B is a **proper subset** of A, denoted by $B \subset A$.

▷ **Note:** From the definition of subset, we observe that for any set A and any set B the following are true:

1. The empty set is a subset of every set. That is $\emptyset \subseteq A$ and $\emptyset \subseteq B$.
2. Every set is a subset of itself. So, for example, $A \subseteq A$ and $B \subseteq B$.
3. If $A \subseteq B$ and $A \ne B$, then $A \subset B$.
4. If $A \subseteq B$ and $B \subseteq A$, then $A = B$.
5. If $A \subset B$, then $A \subseteq B$. However, the converse (if $A \subseteq B$, then $A \subset B$) is not true in general.

Example 2 Listing Subsets

List all possible subsets for the set $A = \{3, 4, 5\}$.

Solution

We know from the note just prior to this example that the empty set and A itself are subsets. So the list of all possible subsets is

$$\emptyset, \{3, 4, 5\}, \{3\}, \{4\}, \{5\}, \{3, 4\}, \{3, 5\}, \{4, 5\}$$

■

Set Operations

In arithmetic, we have four basic **operations** of numbers: addition, subtraction, multiplication, and division with the operation symbols $+$, $-$, $\times$, and $\div$. In sets, we also have operations. One such operation is the **intersection** of sets. Consider the sets $A = \{1, 2, 3, 4, 5, 6\}$ and $B = \{2, 4, 6, 8, 10\}$. Notice that A and B have the elements 2, 4, and 6 in common. We say that the intersection of A and B, denoted by $A \cap B$, is the set $\{2, 4, 6\}$. Visually, we can use **Venn diagrams** to construct graphical representations of set operations. A Venn diagram starts with a

Interactive Activity

Notice that in Example 2 a set with three elements has eight subsets. Now list all the subsets for the sets {3, 4} and {3, 4, 5, 6}. Do you see a pattern? Complete the following statement. *A set of n distinct elements has _____ subsets.*

universal set U that contains all the elements under consideration. Let's say that the universal set for A and B is $U = \{x|x$ is a whole number and $1 \leq x \leq 15\}$. Then the Venn diagram for $A \cap B$ is as shown in Figure 6.2.1.

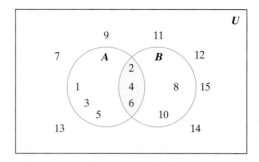

Figure 6.2.1 The shaded region shows the intersection of sets A and B, that is, the elements in both A and B.

Intersection of Sets

For sets A and B, the set of all elements that belong to both A and B is the **intersection** of A and B, denoted $A \cap B$. Mathematically, this is written as

$$A \cap B = \{x|x \in A \textbf{ and } x \in B\}$$

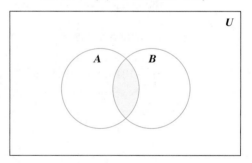

Shaded region shows $A \cap B$.

Example 3   Determining the Intersection of Sets

Let $A = \{1, 5, 7, 9\}$, $B = \{3, 4, 5, 6, 7\}$, and $C = \{2, 3, 4\}$ with $U = \{x|x$ is a whole number and $1 \leq x \leq 10\}$. Write the desired set and draw a shaded Venn diagram for each of the following:

(a) $A \cap B$ **(b)** $A \cap C$

Solution

(a) For $A \cap B$, we notice that A and B only have the elements 5 and 7 in common. This means that $A \cap B = \{5, 7\}$. This is shown in the Venn diagram in Figure 6.2.2.

(b) Here we see that sets A and C have no elements in common. This means that the intersection of A and C is the empty set; that is, $A \cap C = \emptyset$. Notice in the Venn diagram in Figure 6.2.3 that the two circles representing the sets do not overlap.

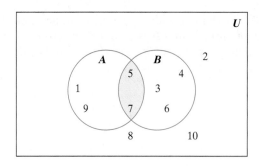

Figure 6.2.2 Shaded region shows $A \cap B$.   **Figure 6.2.3**   ■

In part (b) of Example 3, we saw that $A \cap C = \emptyset$. We call sets that have no elements in common **disjoint sets**.

■ **Disjoint Sets**

> If $A \cap B = \emptyset$, then the sets A and B are **disjoint**.

Sometimes we want to know what makes up the merging of two or more sets. Let's say that we have the sets $A = \{1, 2, 3, 4, 5, 6\}$ and $B = \{2, 4, 6, 8, 10\}$. The merging of A and B is the set $\{1, 2, 3, 4, 5, 6, 8, 10\}$. Mathematically, we call the merging of A and B the **union** of A and B.

■ **Union of Sets**

> For sets A and B, the set of all elements that belong to at least one of A and B is the **union** of A and B, denoted by $A \cup B$. Mathematically, this is written as
>
> $$A \cup B = \{x \mid x \in A \text{ or } x \in B \text{ or } x \in A \cap B\}$$

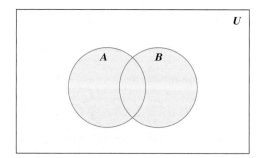

> Shaded region shows $A \cup B$.

▶ **Note:** If an element is a member of set A **or** is a member of set B **or both**, it is in the union of A and B.

Example 4 **Determining the Union of Sets**

Let $A = \{1, 3, 5, 7, 9\}$, $B = \{3, 4, 5, 6, 7\}$, and $C = \{1, 2, 3, 4\}$ with $U = \{x \mid x$ is a whole number and $1 \le x \le 10\}$. Write the desired set and draw a shaded Venn diagram for each of the following:

(a) $A \cup B$ **(b)** $(A \cup B) \cap C$

Solution

(a) Here we take the elements that are in set A or in set B or in both to get $A \cup B = \{1, 3, 4, 5, 6, 7, 9\}$. See Figure 6.2.4.

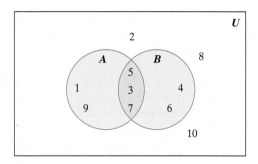

Figure 6.2.4 Shaded region shows $A \cup B$.

(b) When simplifying arithmetic expressions, we know from the order of operations that we must perform the operations in the parentheses first. The same applies to set operations. So the expression $(A \cup B) \cap C$ means that we must perform $A \cup B$ first. We then take this result and intersect with set C. In part (a) we found $A \cup B$. So here we simply take the result of part (a) and intersect with set C. This gives us

$$(A \cup B) \cap C = \{1, 3, 4, 5, 6, 7, 9\} \cap \{1, 2, 3, 4\} = \{1, 3, 4\}$$

This result of $(A \cup B) \cap C$ is shown in the Venn diagram in Figure 6.2.5.

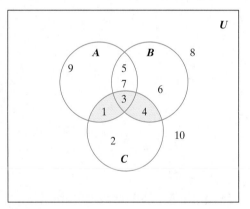

Figure 6.2.5   ∎

✔ **Checkpoint 2**   Now work Exercise 57.

Example 4(b) illustrates that we complete set operations that are in parentheses first, just as we do with arithmetic and algebraic expressions. In fact, the properties associated with set operations are very similar to those in arithmetic and algebra.

▨ Properties of Set Operations

Let U be a universal set. If A, B, and C are subsets of U, then we have the following:

1. Commutative Property for Union of Sets $A \cup B = B \cup A$
2. Commutative Property for Intersection of Sets $A \cap B = B \cap A$

3. Associative Property for Union of Sets $(A \cup B) \cup C = A \cup (B \cup C)$
4. Associative Property for Intersection of Sets $(A \cap B) \cap C = A \cap (B \cap C)$
5. Distributive Properties $$A \cup (B \cap C) = (A \cup B) \cap (A \cup C)$$ $$A \cap (B \cup C) = (A \cap B) \cup (A \cap C)$$

Many times, we are interested in simply knowing what elements are **not** part of a particular set. Let's say we have $U = \{x | x$ is an employee at Farberg Hospital$\}$ and $A = \{x | x$ is a female employee at Farberg Hospital$\}$; then the set of elements *not* in A is $\{x | x$ is a male employee at Farberg Hospital$\}$. We call this the **complement** of A.

■ **Complement of a Set**

If U is a universal set and A is a subset of U, then the **complement** of set A, denoted by A^C, is the set of all elements not in set A. Mathematically, this is written as

$$A^C = \{x | x \notin A\}$$

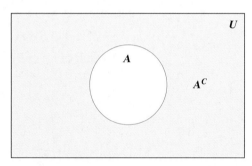

Shaded area represents A^C.

Example 5   **Determining the Complement of a Set**

Let $A = \{1, 5, 7, 9\}$, $B = \{3, 4, 5, 6, 7\}$, and $C = \{2, 3, 4\}$ with $U = \{x | x$ is a whole number and $1 \leq x \leq 10\}$. Write the desired set and draw a shaded Venn diagram for each of the following:

(a) A^C **(b)** $(A \cap B)^C$ **(c)** $A^C \cup B^C$

Solution

(a) For A^C, we want the elements that are not in A, so we have $A^C = \{2, 3, 4, 6, 8, 10\}$. See Figure 6.2.6.

(b) From Example 3(a), we know that $A \cap B = \{5, 7\}$. The complement of this set is $(A \cap B)^C = \{1, 2, 3, 4, 6, 8, 9, 10\}$. The Venn diagram for $(A \cap B)^C$ is in Figure 6.2.7. In words, this is the set of all elements of U not in the intersection of A and B.

(c) From part (a), we know that $A^C = \{2, 3, 4, 6, 8, 10\}$, and we can see that the set of elements not in B is $B^C = \{1, 2, 8, 9, 10\}$. If we take the union of these

Interactive Activity

Using the sets given in Example 5, determine $(A \cup B)^C$ and then determine $A^C \cap B^C$. Compare the results. What do you notice?

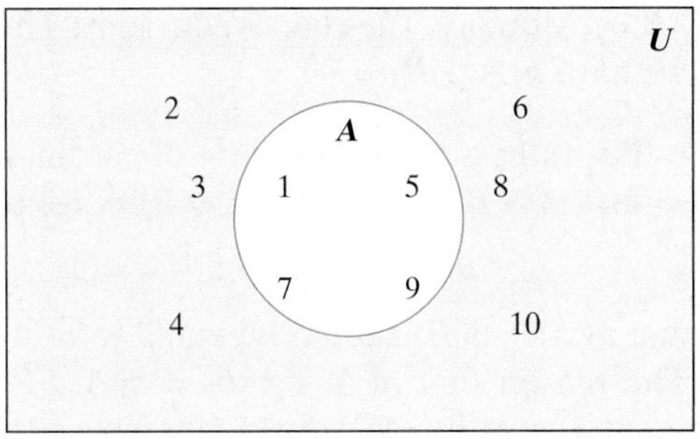

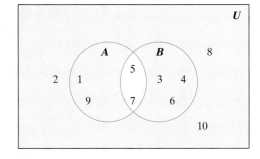

Figure 6.2.6 Shaded region represents A^C.

Figure 6.2.7 Shaded region represents $(A \cap B)^C$.

sets we get

$$A^C \cup B^C = \{2, 3, 4, 6, 8, 10\} \cup \{1, 2, 8, 9, 10\} = \{1, 2, 3, 4, 6, 8, 9, 10\}$$

$A^C \cup B^C$ is shown graphically in Figure 6.2.8.

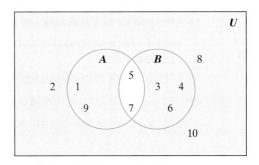

Figure 6.2.8 Shaded region represents $A^C \cup B^C$. ■

Cardinality of Sets

Often our primary interest in a set is to know the number of elements in the set. We call the number of elements in a set A its **cardinality** and denote it with $n(A)$. For example, if $B = \{25, 26, 27, 30\}$, then $n(B) = 4$, since B has four elements. Similarly, if $A = \{15, 20, 25, 30, 35\}$, then $n(A) = 5$.

Example 6   **Determining Cardinality**

Let $A = \{$Monday, Tuesday, Wednesday, Thursday, Friday$\}$ and $B = \{$Friday, Saturday, Sunday, Monday$\}$. Determine the following:

(a) $A \cap B$

(b) $n(A \cap B)$

(c) $A \cup B$

(d) $n(A \cup B)$

Solution

(a) For $A \cap B$, we see that A and B only have the elements Monday and Friday in common. This means that $A \cap B = \{$Monday, Friday$\}$.

(b) Since $A \cap B = \{$Monday, Friday$\}$, we have $n(A \cap B) = 2$.

(c) Here we take the elements that are in set A or in set B or in both to get $A \cup B = \{$Monday, Tuesday, Wednesday, Thursday, Friday, Saturday, Sunday$\}$.

(d) Since $A \cup B = \{$Monday, Tuesday, Wednesday, Thursday, Friday, Saturday, Sunday$\}$, we have $n(A \cup B) = 7$. ∎

Example 6 illustrates a useful property of cardinality of sets. Observe that $n(A) = 5$, $n(B) = 4$, $n(A \cap B) = 2$, and $n(A \cup B) = 7$. From arithmetic, we have

$$n(A) + n(B) = 5 + 4 = 9$$

It might seem that $n(A) + n(B)$ should be equal to $n(A \cup B)$, but we know that $n(A \cup B) = 7$. The reason that $n(A) + n(B) \neq n(A \cup B)$ relies on the fact that when we summed $n(A) + n(B) = 5 + 4 = 9$, there were two elements common to both A and B. These two elements in $A \cap B$ are being counted twice when we compute $n(A) + n(B)$. So to compute $n(A \cup B)$, it appears that $n(A \cup B)$ is simply the sum of $n(A)$ and $n(B)$ minus $n(A \cap B)$. In other words,

$$n(A \cup B) = n(A) + n(B) - n(A \cap B)$$

This serves as motivation for the first of the following properties of cardinality of sets.

Properties of Cardinality of Sets

Let U be a universal set. If A and B are subsets of U, then we have the following:

1. If A and B are sets with elements in common, then
$$n(A \cup B) = n(A) + n(B) - n(A \cap B)$$

2. If A and B are disjoint sets, then
$$n(A \cup B) = n(A) + n(B)$$

3. $n(A^C) = n(U) - n(A)$

4. $n(\emptyset) = 0$

▷ **Note:** Property 2 is a special case of property 1. If A and B are disjoint sets, then $n(A \cap B) = 0$ since $n(\emptyset) = 0$. Thus

$$\begin{aligned} n(A \cup B) &= n(A) + n(B) - n(A \cap B) \\ &= n(A) + n(B) - 0 \\ &= n(A) + n(B) \end{aligned}$$

These properties of the cardinality of sets can be used in applications as shown in Example 7.

Example 7 Determining the Cardinality of a Set

The numbers of public elementary and secondary school teachers in the states of Texas and New York during 1998 are shown in Table 6.2.1.

Table 6.2.1

Grade Level: State	Elementary	Secondary	Total
Texas	132,500	122,200	254,700
New York	98,800	98,400	197,200
Total	231,300	220,600	451,900

SOURCE: U.S. Department of Education, www.ed.gov

We let A represent the set of elementary school teachers from the states of Texas and New York, B represent the set of secondary school teachers from the states of Texas and New York, C represent the set of elementary and secondary school teachers from Texas, and D represents the set of elementary and secondary school teachers from New York. With these definitions of sets A, B, C, and D, Table 6.2.1 can be written as shown in Table 6.2.2.

Table 6.2.2

Grade Level: State	Elementary (A)	Secondary (B)	Total
Texas (C)	132,500	122,200	254,700
New York (D)	98,800	98,400	197,200
Total	231,300	220,600	451,900

SOURCE: U.S. Department of Education, www.ed.gov

Determine the following and interpret each.

(a) $n(A)$　　　(b) $n(A \cap C)$　　　(c) $n(A \cup C)$

Solution

(a) **Understand the Situation:** Here we are being asked to determine the number of elements in the set of elementary school teachers from the states of Texas and New York.

　　If we look down the column represented by Elementary (A), we see a total of 231,300. So $n(A) = 231,300$.

Interpret the Solution: This means that during 1998 there were about 231,300 elementary school teachers in the states of Texas and New York.

(b) **Understand the Situation:** We are being asked to find the number of elements common to the set of elementary school teachers from the states of Texas and New York *and* the set of elementary and secondary school teachers from Texas.

　　If we look down the column represented by Elementary (A) and across the row represented by Texas (C), we see a value of 132,500, meaning that $n(A \cap C) = 132,500$.

Interpret the Solution: This means that during 1998 there were about 132,500 elementary school teachers in the state of Texas.

(c) First note that if we look across the row represented by Texas (C) we see that $n(C) = 245,700$. So, using the solutions from part (a) and part (b) along with the first property of cardinality of sets, we get

$$n(A \cup C) = n(A) + n(C) - n(A \cap C)$$
$$= 231,300 + 245,700 - 132,500 = 344,500$$

Interpret the Solution: This means that in 1998 there were about 344,500 who were either an elementary school teacher from Texas or New York or a public school teacher in Texas. ∎

✓ **Checkpoint 3**　　Now work Exercise 71.

SUMMARY

In this section, we introduced **sets**, which are well-defined collections of objects. We discussed the two methods, **listing** and **set-builder**, for writing sets. We then discussed some of the operations associated with sets.

- The set of all elements that belong to both A and B is the **intersection** of A and B, denoted $A \cap B$. That is, $A \cap B = \{x | x \in A \text{ and } x \in B\}$.
- The set of all elements that belong to at least one of A and B is the **union** of A and B, denoted by $A \cup B$. That is, $A \cup B = \{x | x \in A \text{ or } x \in B \text{ or } x \in A \cap B\}$.

- The **complement** of set A, denoted by A^C, is the set of all elements not in A.
- If A and B are any sets, then $n(A \cup B) = n(A) + n(B) - n(A \cap B)$.
- If A and B are disjoint sets, then $n(A \cup B) = n(A) + n(B)$.

SECTION 6.2 EXERCISES

In Exercises 1–6, rewrite the set using the listing method.

1. $A = \{x | x \text{ is a whole number and } 1 \leq x \leq 7\}$

2. $B = \{x | x \text{ is a whole number and } 2 \leq x \leq 10\}$

3. $C = \{x | x \text{ is an even number and } 2 \leq x \leq 25\}$

4. $D = \{x | x \text{ is an odd number and } 2 \leq x \leq 25\}$

✔ **5.** $E = \{x | x \text{ is a month that ends with the letters b-e-r}\}$

6. $F = \{x | x \text{ is a day of the week that starts with the letter S}\}$

In Exercises 7–12, rewrite the set using set-builder notation.

7. $A = \{2, 3, 4, 5, 6\}$

8. $B = \{12, 13, 14, 15, 16\}$

9. $C = \{5, 10, 15, 20, 25, 30\}$

10. $D = \{20, 30, 40, 50, 60, 70, 80\}$

11. $E = \{\text{Tuesday, Thursday}\}$

12. $F = \{\text{January, June, July}\}$

In Exercises 13–18, write all possible subsets for the given set.

13. $\{2\}$

14. $\{5\}$

15. $\{\text{Honda, Ford}\}$

16. $\{5, 7\}$

17. $\{8, 9, 10\}$

18. $\{\text{knitting, needlepoint, quilting}\}$

In Exercises 19–24, fill in the blanks with $\subset, \subseteq, \supset,$ or $\supseteq$ so that each statement is true. More than one answer may be possible.

19. $\{1, 2, 3\}$_____$\{2, 3\}$

20. $\{10, 20, 30\}$_____$\{10, 20\}$

21. $\{a, v, r\}$_____$\{r, a, v\}$

22. $\{\text{Fred, Ruth, Ralph}\}$_____$\{\text{Ralph, Fred, Ruth}\}$

23. $\emptyset$_____$\emptyset$

24. For any set A, A_____A.

In Exercises 25–46, let $U = \{x | x \text{ is a whole number and } 1 \leq x \leq 15\}$, $A = \{1, 2, 3, 4, 5\}$, $B = \{2, 4, 6, 8, 10\}$, and $C = \{11, 12, 13, 14, 15\}$. Write each of the following sets using the listing method.

25. $A \cap B$ **26.** $A \cap C$

27. $B \cap C$ **28.** $B \cap U$

29. $A \cup B$ **30.** $A \cup C$

31. $B \cup C$ **32.** $B \cup U$

33. A^C **34.** C^C

35. $B \cap C^C$ **36.** $B^C \cap C$

37. U^C **38.** $\emptyset^C$

39. $(A \cap C)^C$ **40.** $(B \cap C)^C$

41. $A^C \cup C$ **42.** $B^C \cup C$

43. $A \cup \emptyset$ **44.** $C \cup \emptyset$

45. $\emptyset \cap B^C$ **46.** $\emptyset \cap A^C$

In Exercises 47–56, use the Venn diagram to write the desired set.

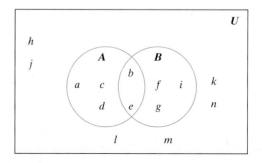

47. A

48. B

49. U

50. $A \cup B$

51. $A \cap B$

52. B^C

53. A^C

54. $(A \cap B)^C$

55. $(A \cup B)^C$

56. $A \cap B^C$

In Exercises 57–66, let $U = \{1, 2, 3, 4, 5, 6, 7, 8, 9\}$, $A = \{2, 4, 6, 8\}$, $B = \{4, 5, 6\}$, and $C = \{2, 3, 4, 5, 6\}$. Write a set and draw a shaded Venn diagram for each of the following:

✓ **57.** $A \cup (B \cap C)$

58. $A \cap (B \cup C)$

59. $A \cap (B \cup C)^C$

60. $A^C \cap (B \cup C)$

61. $(A \cup B) \cap (A \cup C)$

62. $(A \cap B) \cup (A \cap C)$

63. $A^C \cup B^C$

64. $A^C \cup C^C$

65. $(A \cap B)^C$

66. $(A \cap C)^C$

🌐 *For Exercises 67–72, the values of arms transfers, in millions of dollars, from the United States and the United Kingdom to India and Pakistan during 1996 are as shown in the table.*

Recipient/ Supplier	United States (A)	United Kingdom (B)	Total
India (C)	110	0	$n(C) = 110$
Pakistan (D)	140	70	$n(D) = 210$
Total	$n(A) = 250$	$n(B) = 70$	$n(U) = 320$

SOURCE: U.S. Statistical Abstract, www.census.gov/statab/www

Here A represents the set of arms supplied by the United States, B represents the set of the arms supplied by the United

Kingdom, C represents the set of the arms received by India, and D represents the set of the arms received by Pakistan. Determine the following and interpret each.

67. $n(A)$

68. $n(C)$

69. $n(B \cap D)$

70. $n(A \cap C)$

✓ **71.** $n(B \cup D)$

72. $n(A \cup C)$

🌐 *For Exercises 73–78, the number of American households in 1997 in the midwestern and the southern United States, measured in millions, whose household incomes are below and above $35,000 are as shown in the table.*

Region/ Income	Under $35,000 ($A$)	$35,000 and over ($B$)	Total
Midwest (C)	35.3	13.1	48.4
South (D)	18.5	18.0	36.5
Total	53.8	31.1	84.9

SOURCE: U.S. Statistical Abstract, www.census.gov/statab/www

Here A represents the set of households in the regions with incomes under $35,000, B represents the set of households in the regions with incomes of $35,000 and over, C represents the set of households in the Midwest, and D represents the set of households in the South. Determine the following and interpret each.

73. $n(A^C)$

74. $n(C^C)$

75. $n(A \cup D)$

76. $n(A \cup C)$

77. $n(A^C \cup D)$

78. $n(A \cup C^C)$

▌ SECTION PROJECT

In Example 6, we illustrated, for two sets A and B that are not disjoint, that $n(A \cup B) = n(A) + n(B) - n(A \cap B)$. For this section project, consider the following sets:

$A = \{1, 2, 3, 4, 5\}, \quad B = \{2, 4, 6, 8\}, \quad C = \{1, 3, 5, 7, 9, 11\}$

(a) Determine $n(A)$.

(b) Determine $n(B)$.

(c) Determine $n(C)$.

(d) Determine $n(A \cap B)$.

(e) Determine $n(A \cap C)$.

(f) Determine $n(B \cap C)$.

(g) Determine $n(A \cap B \cap C)$.

(h) Verify that

$$n(A \cup B \cup C) = n(A) + n(B) + n(C) - n(A \cap B)$$
$$- n(A \cap C) - n(B \cap C) + n(A \cap B \cap C)$$

Explain why this formula is correct in general.

Section 6.3   Principles of Counting

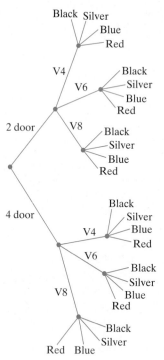

Figure 6.3.1

In the previous section we learned that cardinality was an important idea in the discussion of sets. Now we want to generalize our discussion of counting objects. In this section we will discuss the techniques of counting objects under different conditions and assumptions. The study of counting techniques is called **combinatorics**.

Fundamental Principles

Let's say that a person goes to a new car lot to order a new sport utility vehicle called the Trailmaster. The SUV comes in a two-door and a four-door model. There are three types of engines: V4, V6, and V8. Additionally, the SUV comes in the colors black, silver, red, and blue. To see in how many distinct ways a person can order a Trailmaster, we can draw a **tree diagram** to visually show how many ways the vehicle can be ordered. The person has to make choices in three steps. Choosing the number of doors is the first step, choosing the engine type is the second step, and selection of a color is the third step. We will draw **branches** on the tree diagram for the choices in each step. The tree diagram for ordering the Trailmaster is in Figure 6.3.1. We can see from counting the branches on the right side of the tree diagram that there are 24 ways to order a Trailmaster.

Example 1   Constructing a Tree Diagram

A survey given by a local pollster has three questions.

> Question 1: Are you a smoker?
>
> Question 2: Do you consider yourself to be a liberal, moderate, or conservative?
>
> Question 3: Do you support or oppose capital punishment?

Draw a tree diagram to represent all possible answers in the survey. How many ways could a respondent answer all three questions?

Solution

Understand the Situation: An individual who completes this survey has to respond to three questions. Each question requires the individual to make a choice. The tree diagram for all the choices in the survey is in Figure 6.3.2.

Interpret the Solution: If we count the number of branches on the right side of the tree diagram, we see that there are a total of 12 ways to respond to the three questions.   ■

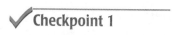

Figure 6.3.2

✓ **Checkpoint 1**

Now work Exercise 3.

Let's inspect Example 1 a little more closely. The first question, "Are you a smoker?", can be answered in 2 ways, the second question, "Do you consider yourself to be liberal, moderate or conservative?", can be answered in 3 ways, and the third question, "Do you support or oppose capital punishment?", can be answered in 2 ways. Notice that if we multiply these values together, we get

Possible answers for Question 1		Possible answers for Question 2		Possible answers for Question 3	
2	·	3	·	2	= 12 ways

This is the same result that we determined by using the tree diagram in Figure 6.3.2! This illustrates one of the fundamental counting techniques, the **Multiplication Principle**.

> ■ **Multiplication Principle**
>
> If a choice consists of k steps, of which the first can be made in n_1 ways, the second in n_2 ways, ..., and the last step can be made in n_k ways, then the number of different ways that the total choice can be made is given by
>
> $$n_1 \cdot n_2 \cdot n_3 \cdots \cdot n_k$$

Even though tree diagrams give us a good visual perspective of counting objects, it is not practical for many applications. An application of the Multiplication Principle is given in Example 2. Notice in this example that drawing a tree diagram would be very difficult.

Example 2   **Applying the Multiplication Principle**

A certain state's automobile license plates are embossed with three letters followed by four digits. Assuming that there are no restrictions on numbers or letters, how many different license plates can be made?

Solution

Understand the Situation: In this case, we have to make choices in seven steps. There is a choice for each of the three letters and there is a choice for each of the four digits.

Since there are 26 letters in the alphabet and there are 10 digits, the total number of ways to make license plates in this state is given by

Choices for:	Letter 1		Letter 2		Letter 3		Digit 1		Digit 2		Digit 3		Digit 4
	26	·	26	·	26	·	10	·	10	·	10	·	10

$$= 26^3 \cdot 10^4 = 175{,}760{,}000$$

Interpret the Solution: This means that in this particular state a total of 175,760,000 different license plates can be made if there are no restrictions on numbers or letters. ■

Interactive Activity

Let's say in Example 2 that a law was passed so that no letter could be repeated on a license plate. Under the new law, how many different license plates could be made?

Permutations

Let's explore another counting technique. Suppose that there are 4 people who need to sit for a photograph and the photographer must arrange them in a row. How many ways can the photographer do this? To find out, we can again use the multiplication principle. The first choice can be made in 4 ways since all the people are waiting to be seated. The second choice can be made in 3 ways since the first person chosen is already sitting for the photo. The third choice can be made in 2 ways and the fourth choice can be made in only 1 way, since there is only one person left to be seated. So the total number of ways is

Choice 1   Choice 2   Choice 3   Choice 4
$$4 \quad \cdot \quad 3 \quad \cdot \quad 2 \quad \cdot \quad 1 \quad = 12 \text{ ways}$$

So there are 12 ways to seat the 4 people for the photograph. For simplicity, special products such as $4 \cdot 3 \cdot 2 \cdot 1$ are given the special name **factorial**. Factorials appear frequently in combinatorics.

Factorial Notation

If n is a natural number, then n **factorial,** denoted by $n!$, is given by

$$n! = n(n-1)(n-2) \cdots \cdot 3 \cdot 2 \cdot 1$$

As a special case, we define $0! = 1$.

▶ **Note:** We call n the first factor of $n!$, $n-1$ is called the second factor of $n!$, and so forth.

Example 3   Using Factorial Notation

In the Jasper Derby, 12 horses start the race.

(a) In how many ways can the horses finish first, second, and third?
(b) Rewrite the answer in part (a) using factorial notation.

Solution

(a) Of the 12 horses, there are 3 places to de determined. For the first-place horse, there are 12 possibilities. For the second-place horse there are 11 possibilities (the first-place horse is already determined). Finally, for the third-place horse there are 10 possibilities. So the total number of possibilities that these 12 horses can finish first, second, or third is

Choice 1   Choice 2   Choice 3
$$12 \quad \cdot \quad 11 \quad \cdot \quad 10 \quad = 1320$$

So the horses can finish first, second, and third in 1320 different ways.

(b) We know by the definition of n factorial that

$$12! = 12 \cdot 11 \cdot 10 \cdot 9 \cdot 8 \cdot 7 \cdot 6 \cdot 5 \cdot 4 \cdot 3 \cdot 2 \cdot 1$$

Since we need only the first three factors of $12!$, we can rewrite $12!$ as

$$12! = (12 \cdot 11 \cdot 10) \cdot (9 \cdot 8 \cdot 7 \cdot 6 \cdot 5 \cdot 4 \cdot 3 \cdot 2 \cdot 1)$$
$$= (12 \cdot 11 \cdot 10) \cdot 9!$$

To get the solution in part (a), we can divide $12!$ by $9!$ to get

$$\frac{12!}{9!} = \frac{(12 \cdot 11 \cdot 10) \cdot 9!}{9!} = 12 \cdot 11 \cdot 10 = 1320$$

So we see that by using factorial notation the solution to part (a) can be written as $\dfrac{12!}{9!}$. ∎

Two aspects to the scenario in Example 3 are critical. First, we see that the **order is important**. Recall from Section 6.1, by order being important we mean that

- The order of arrangement makes a difference, or
- There is a ranking of the members of the arrangement.

Next we see that there is **no repetition**, meaning that, for example, a horse cannot finish both first and third. These conditions are critical components of the counting technique called **permutation**.

■ **Formula for Permutation**

The number of ways in which r objects can be selected in a specific order from among n objects is given by the **permutation** $_nP_r$ (read "n select r"), where

$$_nP_r = \frac{n!}{(n-r)!}$$

▶ **Note:** Permutations have two important characteristics:

1. Order is important
2. There is no repetition.

Let's take a Flashback to an example first presented in Section 6.1 to practice our newest counting technique.

Interactive Activity

For any value of n, use the definition of factorial to compute $_nP_n$ and $_nP_0$. What special properties do these permutations have and how can they be interpreted? Also, compare the two quantities and see if there is a relation between them.

Flashback

Computer Club Revisited

There are 24 members in the New Strawsburg Junior High School Computer club. How many ways can the club pick a president, a vice-president, a secretary, and a treasurer?

Flashback Solution

In Section 6.1 we saw that order is important and there is no repetition. The names that we associate with the president, vice-president, secretary, and treasurer make a difference. So the total number of ways that the 4 club officers can be chosen is a 24 select 4 permutation, which is calculated by

$$_{24}P_4 = \frac{24!}{(24-4)!} = \frac{24!}{20!} = 24 \cdot 23 \cdot 22 \cdot 21 = 255{,}024$$

Interpret the Solution: So the Computer club officers can be selected in 255,024 different ways.

Combinations

Often when we count objects, the order is *not* important. For these types of applications, we use a counting technique called the **combination**.

> ■ **Combination**
>
> Unordered selections (order not important) of distinct objects are called **combinations**. An unordered selection using r of the n distinct objects is given by the combination $_nC_r$ (read "n choose r"). $_nC_r$ is defined to be the number of ways of choosing r distinct objects from a set of n distinct objects without regard to the order of the selection.

▶ **Note:** 1. For the combination, order is **not** important and there is no repetition.

2. Some textbooks express the combination $_nC_r$ by writing $\binom{n}{r}$ or $C_{n,r}$.

Before we supply a formula to determine $_nC_r$, let's do an example to fully understand the difference between when order is important and when order is not important, as well as to gain some insight into the formula.

Example 4   **Arranging Letters**

Consider the letters w, x, y, and z. Determine in how many ways we can choose three without repeating any letter where

(a) Order is important

(b) Order is not important

Solution

(a) If order is important, the total number of ways that 3 letters letters can be chosen is a 4 select 3 permutation calculated by

$$_4P_3 = \frac{4!}{(4-3)!} = 24$$

All 24 possible selections are shown in the right-hand column of the tree diagram in Figure 6.3.3.

(b) When order is not important, only 4 of the 24 selections determined in part (a) are listed. They are

$$wxy \quad wxz \quad wyz \quad xyz$$

Notice that when selecting the letters w, x, and y different results are obtained depending on whether order is important. When order is not important, it does not matter how we arrange the letters. However, when order is important, it does matter how the letters are arranged. This can be summarized as follows:

Order Is Not Important	Order Is Important
wxy	$wxy, \; wyx, \; xwy, \; xyw, \; ywx, \; yxw$

Notice that the three letters wxy in the "order is not important" list allow for six rearrangements in the "order is important" list. ■

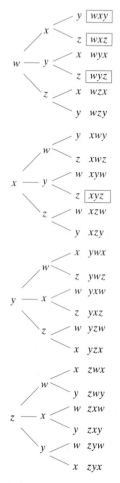

Figure 6.3.3

Observe in Example 4 that the number of possible selections when order is important, 24, is six times the number of possibilities when order is not important, 4. The reason for this is that each unordered selection consisted of three letters. As part (b) of Example 4 illustrated, these three letters allow for 6 (or 3!) rearrangements. We can use this observation to obtain a formula for $_nC_r$. In general, **each** unordered selection of r objects gives rise to $r!$ ordered selections. But the number of ordered selections is given by the permutation $_nP_r$. So the number of ordered selections, $_nP_r$, is $r!$ times the number of unordered selections. Since the number of unordered selections is given by $_nC_r$, we have

$$_nP_r = r! \cdot {}_nC_r$$

Since $_nP_r = \dfrac{n!}{(n-r)!}$, we can substitute this and solve for $_nC_r$ and get

$$_nP_r = r! \cdot {}_nC_r$$

$$\frac{n!}{(n-r)!} = r! \cdot {}_nC_r \qquad \text{Substitute.}$$

$$_nC_r = \frac{n!}{r!(n-r)!} \qquad \text{Divide both sides by } r!.$$

Formula for Combination

The number of different selections of r objects chosen from n distinct objects when no object is repeated and order is not important is given by the **combination** $_nC_r$, where $_nC_r$ is computed by

$$_nC_r = \frac{n!}{r!(n-r)!}$$

Example 5   Computing Combinations

A hospital has 44 staff members. In how many ways can a committee of 3 be formed to investigate salary discrepancies?

Solution

Understand the Situation: Here we have $n = 44$ distinct objects. The order is not important since there is no ranking of committee members. Also notice that there is no repetition.

So the number of ways that the committee can be formed is the number of 3 combinations of 44 objects, which is calculated by

$$_{44}C_3 = \frac{44!}{3!(44-3)!} = \frac{44!}{3! \cdot 41!} = 13{,}224$$

Interpret the Solution: This means that there are 13,224 possible committees of size 3 chosen from the 44 available staff members.   ■

✓ **Checkpoint 2**   Now work Exercise 65.

Technology Option

Calculators can perform permutations and combinations. Figure 6.3.4 shows the result of the Flashback and Figure 6.3.5 shows the result of Example 5.

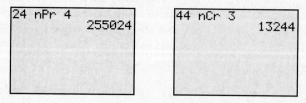

| Figure 6.3.4 | Figure 6.3.5 |

To learn how to compute permutations and combinations on your calculator, consult the online graphing calculator manual at www.prenhall.com/armstrong.

There are times when the combination and the multiplication principle are used together. Let's say that a city council consists of 6 Democrats and 5 Republicans. The council wants to form a committee of 4 to investigate possible campaign finance irregularities. If the committee is to consist of equal representation from each political party, then we have two tasks ahead of us. Specifically, we need to choose 2 Democrats and 2 Republicans. The number of ways that we can choose the 2 Democrats and the number of ways that we can choose the 2 Republicans are $_6C_2$ and $_5C_2$, respectively. These choices are illustrated in Figure 6.3.6.

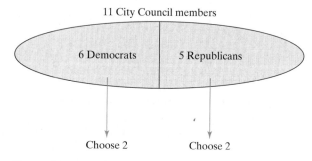

Figure 6.3.6

So the number of ways to form the committee is given by

$$_6C_2 \cdot _5C_2 = 15 \cdot 10 = 150$$

Hence, by the Multiplication Principle, there are 150 different ways to form the committee.

Example 6 **Combining the Multiplication Principle and the Combination**

A shipment of 20 suits arrives at the Businessman's Discount Warehouse. The shipment contains 7 suits with flaws in the fabric.

(a) Determine $_{20}C_5$ and interpret.

(b) In how many ways can 5 suits be chosen from the shipment so that exactly 2 of them have flaws?

Solution

(a) **Understand the Situation:** We are being asked to determine the number of ways of choosing 5 suits, in any order, from the shipment of 20 suits.

The number of ways to choose 5 suits in any order from a shipment of 20 is given by

$$_{20}C_5 = \frac{20!}{5!(20-5)!} = \frac{20!}{5! \cdot 15!} = 15{,}504$$

Interpret the Solution: So there are 15,504 ways that 5 suits can be chosen from the shipment of 20 suits.

(b) **Understand the Situation:** Notice that if we want to choose 5 suits so that exactly 2 of them have flaws this means that the other 3 suits have no flaws. Since 7 of the 20 suits have flaws, we want to choose 2 of the 7 with flaws and choose 3 of the 13 without flaws. Figure 6.3.7 shows the graphical representation of this part of the example.

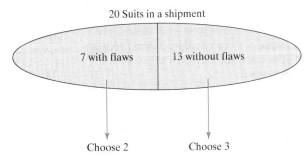

20 Suits in a shipment

7 with flaws   13 without flaws

Choose 2   Choose 3

Figure 6.3.7

So the desired number of possibilities is given by

$$_7C_2 \cdot {}_{13}C_3 = \frac{7!}{2!(7-2)!} \cdot \frac{13!}{3!(13-3)!} = \frac{7!}{2!5!} \cdot \frac{13!}{3!10!} = 21 \cdot 286 = 6006$$

Interpret the Solution: This means that there are 6006 ways to choose 5 suits from the shipment so that 2 suits have flaws and 3 suits do not have flaws. ■

SUMMARY

In this section, we learned several counting techniques, such as the **Multiplication Principle**, the **permutation**, and the **combination**.

- **Multiplication Principle**   If a choice consists of k steps, of which the first can be made in n_1 ways, the second in n_2 ways, . . . , and the last step can be made in n_k ways, then the number of different ways that the total choice can be made is given by $n_1 \cdot n_2 \cdot n_3 \cdots \cdots n_k$.

- **n factorial** $n! = n(n-1)(n-2) \cdots \cdots 3 \cdot 2 \cdot 1$
- **Number of Permutations** $_nP_r = \dfrac{n!}{(n-r)!}$. Order is important and there is no repetition.
- **Number of Combinations** $_nC_r = \dfrac{n!}{r!(n-r)!}$. Order is not important and there is no repetition.

SECTION 6.3 EXERCISES

1. A two-question phone survey asks a person the following questions:

Question 1: Have you donated to the local Fraternal Order of Police?

Question 2: Do you feel safe or unsafe in your home at night?

(a) Draw a tree diagram to represent all possible answers in the survey.

(b) In how many ways could a respondent answer both questions?

2. A two-question survey by the Find-It Company asks a person the following questions:

Question 1: Have you used an Internet search engine in the last week?

Question 2: Have you made an online purchase in the last month?

(a) Draw a tree diagram to represent all possible answers in the survey.

(b) In how many ways could a respondent answer both questions?

✓ **3.** A person wants a new bookshelf and finds that there are two color choices: white and oak. There is also a size choice of three, four, or five shelves.

(a) Draw a tree diagram to represent all possible bookshelves.

(b) In how many ways could a person create a bookshelf?

4. A customer who orders a computer from the Z-Dron Company has the choice of three microprocessors: budget, faster, and fastest. The customer also has a choice of a 17-inch or a 19-inch monitor.

(a) Draw a tree diagram to represent all possible computer systems.

(b) In how many ways could a customer create a computer system?

5. The new XZR-200 motorcycle comes in a choice of four colors: black, silver, green, and magenta. There is also an option for a luggage rack.

(a) Draw a tree diagram to represent all possible motorcycle packages.

(b) In how many ways could a customer create a motorcycle package?

6. A ceiling fan can come from one of four factories: East, South, North, and Midwest. Each factory includes the option of a brass or silver fixture.

(a) Draw a tree diagram to represent all possible ceiling fans.

(b) In how many ways could a customer create a ceiling fan?

In Exercises 7–20, use the Multiplication Principle to determine the following:

7. A new concert hall features 12 classical and 16 rock concerts in its premiere season. In how many different ways can a person construct a package deal as a present that consists of a ticket for a classical concert and a ticket for a rock concert?

8. The Blossom Blouse Company has 8 warehouses and 24 retail outlets. In how many different ways can a person order a blouse from the warehouse and another blouse from a retail outlet?

9. A football squad has 21 linemen, 8 receivers, and 9 running backs. A most-improved player award will be given to each position. In how many ways can a most-improved

player award be given to a lineman, a receiver, and a running back?

10. The SpectraSpectacle Company makes its Laservision sunglasses with 9 different colored lenses, 6 different colored ear pieces, and 4 different colored nose pieces. In how many different ways can a person order a pair of Laservision sunglasses?

11. The menu at the Top Class restaurant has a choice of 4 appetizers, 6 entrees, 6 beverages, and 5 desserts. In how many different ways can a customer order an appetizer, entree, beverage, and dessert at the restaurant?

12. A child has 6 T-shirts, 7 pairs of pants, 4 pairs of shoes, and 10 pairs of socks. Assuming that the clothes are color coordinated, in how many different ways can the child pick a T-shirt, pants, shoes, and socks to wear?

13. A textbook committee has to choose from among 10 prealgebra, 13 college algebra, 5 precalculus, and 7 calculus books. In how many ways can the committee chose a textbook from each subject area?

14. The call letters of a radio station consists of 4 letters. The first letter must be a K or a W. If the last 3 letters can be repeated, how many different call letters can be made?

15. The Deanol sports car comes in a choice of 5 colors and 3 engine sizes. There is also an air conditioner and a turbocharger option. In how many different ways can the car be ordered?

16. An identification card used at the Linger Corporation consists of two digits from 0 to 4 and then 3 letters. If the digits and letters can be repeated, in how many different ways can an identification card be made?

17. The lock on the Takeaway brief case has 5 digits. If the numbers cannot be repeated, how many different possibilities are there for the lock?

18. The password to gain entrance to the Techway Company is a sequence of 5 letters from A to K. If a letter can be repeated, how many different passwords can be made?

19. A 25-question multiple-choice test in engineering has choice of A, B, C, or D for each question. In how many different ways can a student answer all the questions?

20. A history quiz consists of 8 true–false questions. In how many different ways can a student answer all the questions?

In Exercises 21–28, evaluate the expression without using a calculator.

21. $7!$ **22.** $10!$

23. $_5P_3$ **24.** $_7P_2$

25. $_{10}P_4$ **26.** $_9P_5$

27. $_8P_0$ **28.** $_8P_8$

In Exercises 29–34, use a calculator to determine the following:

29. $_{50}P_{20}$

30. $_{30}P_{15}$

31. $_{40}P_{15}$

32. $_{45}P_{28}$

33. $_{90}P_{30}$

34. $_{50}P_{41}$

35. The board of directors at the SafeSide Company consists of 12 members. In how many different ways can a chief executive officer, a vice-president and a chief financial officer be selected?

36. The Chess Club at Dyeston High School has 15 members. In how many different ways can a president, vice-president, and treasurer be selected?

37. The Miss Farrago County Fair queen contest has 19 entrants. In how many different ways can a winner, first runner-up, and second runner-up be selected?

38. Traditionally, 33 cars start the Indianapolis 500. In how many different ways can the drivers place first, second, and third?

39. Food critic Hero Maramoto has eaten at 25 restaurants this year. To introduce the top 4 restaurants in his newspaper column, how many choices does he have?

40. In a customer preference survey, a person is asked to try 16 different brands of applesauce and rank the first 5 in order of preference. How many different rankings are possible?

41. Cabin cruisers can communicate with each other by lofting 5 from among 12 flags up a mast. If the message is determined by the order in which the flags are raised, how many different messages are possible?

42. A researcher has found that 13 drugs are effective in treating a certain ailment. She conjectures that the sequencing of the administration of the drugs can affect their effectiveness. In how many different ways can 6 of the drugs be administered?

43. The starting lineup for the Jasper Comets softball team consists of 9 members. In how many ways can the manager arrange the batting order?

44. The 1921 class of Stringsbrug High School consists of 7 surviving members. In how many ways can they be arranged in 7 chairs, in a row, for a class photo?

In Exercises 45–56, evaluate the expression without using a calculator.

45. $_{5}C_{3}$

46. $_{6}C_{2}$

47. $_{10}C_{6}$

48. $_{9}C_{5}$

49. $_{12}C_{5}$

50. $_{11}C_{4}$

51. $_{7}C_{1}$

52. $_{9}C_{1}$

53. $_{13}C_{0}$

54. $_{17}C_{0}$

55. $_{12}C_{12}$

56. $_{14}C_{14}$

In Exercises 57–62, use a calculator to determine the following:

57. $_{50}C_{20}$

58. $_{30}C_{15}$

59. $_{40}C_{15}$

60. $_{45}C_{28}$

61. $_{90}C_{30}$

62. $_{50}C_{41}$

63. In Crawfords County, 6 judges are running for 3 county judge positions. In how many different ways can the 3 county judges be elected?

64. The central office of the Higgens Corporation has to choose 3 of its 8 regional offices for an audit. In how many different ways can the regional offices be chosen?

65. A group of 20 professors in an Archeology department wants to choose a research team of 5 to study a new archeological site. In how many different ways can the research team be formed?

66. For a literature class, a student has to read 4 books from a list of 21. In how many different ways can the 4 books be chosen?

67. There are 29 companies bidding for the construction of a new building. In how many different ways can 5 of the companies be chosen for hire?

68. How many different 5-card hands can be dealt from a standard deck of 52 cards?

69. In the Maine State lottery called *Megabucks*, 6 balls numbered 1 to 42 are drawn from a hopper. In how many different ways can the 6 numbers be chosen? (*Source:* www.mainelottery.com.)

70. In the Washington State lottery called Lotto, 6 balls numbered 1 to 49 are drawn from a hopper. In how many different ways can the 6 numbers be chosen? (*Source:* www.wa.gov/lot/home.htm.)

71. Jelly Belly manufactures 40 different flavors of jelly beans. If a person goes to a Jelly Belly store, they can get a sampler cup of 8 jelly beans in which the 8 jelly beans are all different flavors. How many different sampler cups are possible? (*Source:* www.jellybelly.com.)

72. Sweet Dreams has 34 different flavors of ice cream. One of its ice cream cones is a triple dip of 3 different flavors for which the order of the 3 different flavors does not matter. How many different kinds of triple-dip ice cream cone can be made?

73. A city council has 5 Democrats and 7 Republicans. In how many different ways can a subcommittee be formed with 2 Democrats and 4 Republicans?

74. The Euratal Company has 4 women and 8 men on its board of directors. In how many different ways can a gender-equality subcommittee of 4 be formed so that the numbers of men and women are equal?

75. In a shipment of 30 car stereos, 8 are known to be defective. In how many ways can a sample of 10 be chosen so that 2 are defective and 8 are not defective?

76. A box of 50 pieces of taffy has 20 that are pink and 30 that are yellow. In how many ways can a handful of 8 pieces of taffy be pulled out so that 2 are pink and 6 are yellow?

77. A group of 20 environmentalists is made up of 6 Republicans, 8 Democrats, and 6 members of the Green Party. In how many different ways can a public relations committee of 7 be formed with 2 Republicans, 2 Democrats, and 3 Green Party members?

▌ SECTION PROJECT

(a) Compute $_{10}C_3$ and $_{10}C_7$. Compare the solutions.

(b) Prove that for any whole numbers n and r, where $0 \leq r \leq n$, $_nC_r = {_nC_{n-r}}$. (*Hint:* Use the factorial definition to write the left and right sides. Simplify the right side and compare it to the left side.)

(c) Compute the sum $_{15}C_3 + {_{15}C_4}$. Compare the solution to $_{16}C_4$.

(d) Prove that for any whole numbers n and r, where $0 \leq r \leq n$, $_nC_r + {_nC_{r+1}} = {_{n+1}C_{r+1}}$. Use the hint given in part (b).

Section 6.4   Introduction to Probability

Probability is a topic that many people experience in their everyday lives, yet know little about. When a weather forecaster states that there is an 80% chance of rain tomorrow or when a physician asserts that a certain surgical procedure has a 97% chance of success, they are often stating a probability. **Probability** can be thought of as the branch of mathematics that produces a number that expresses the likelihood of occurrence of a specific event. The study of probability is called **probability theory**. In this section, we will introduce the terminology and notation of probability theory and examine three fundamental ways in which probabilities can be computed.

Probability Experiments

To illustrate some of the terminology of probability theory, let's start by examining an activity that involves probability. Recall from Section 6.1 that a **probability experiment** is a process that leads to well-defined results called **outcomes**. All possible outcomes are predictable, but any given outcome cannot be predicted with certainty. Let's take a Flashback to an example from Section 6.1 to refamiliarize ourselves with probability experiments.

 Flashback

Marbles in a Bag Revisited

A bag contains three marbles that are painted with the numbers 1, 2, and 3. A marble is drawn randomly from the bag, its number is recorded, and it is then returned to the bag. Another marble is drawn randomly and its number is recorded. Construct a **tree diagram** showing all possible outcomes of this experiment.

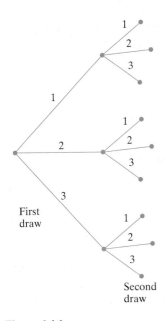

First draw

Second draw

Figure 6.4.1

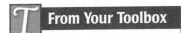
Flashback Solution

First, since we can observe each pair of numbers that are drawn, this is an example of a probability experiment. Each result of an experiment or a **trial**, such as drawing the marbles and recording the numbers, is called an outcome. We can display all the outcomes of this experiment several ways. One way is to construct a tree diagram as shown in Figure 6.4.1. We can see from the right side of the tree diagram that there are a total of nine possible outcomes for the experiment.

As stated at the end of the Flashback solution, there are nine possible outcomes for the experiment. These nine outcomes are **equally likely**, since the marbles are identical and the ones drawn are selected randomly. By **equally likely**, we mean that each outcome has an equal chance of occurring. We can write these nine possible equally likely outcomes for the experiment in the Flashback using **set notation** as shown in set S.

$$S = \{(1, 1), (1, 2), (1, 3), (2, 1), (2, 2), (2, 3), (3, 1), (3, 2), (3, 3)\}$$

Set S is a special set in probability called a **sample space**.

■ **Sample Space**

For a probability experiment, the **sample space**, denoted by S, is the set of all possible outcomes.

Many times we are interested in the number of elements in a sample space. Recall that in Section 6.2 we called this the **cardinality** of a set. Consult the Toolbox to the left. For our experiment of drawing the marbles from the bag, the size of the sample space, denoted by $n(S)$, is $n(S) = 9$.

Example 1   Identifying the Sample Space of a Probability Experiment

Consider a box that contains 4 identical tickets. Suppose that a ticket is labeled with a star, a ticket is labeled with a triangle, a ticket is labeled with a square, and a ticket is labeled with a circle. Also consider a bag that contains 2 marbles identical

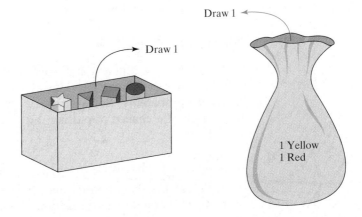

in every way except color. One is a yellow and the other is red. In an experiment, one random draw is made from the box and one random draw is made from the bag and the outcome is recorded. Write the sample space S and determine $n(S)$.

Solution

The draws from the box can be represented by the symbols ★, △, ■, and ○. For the bag, let Y indicate that the yellow marble is drawn and R indicate that the red marble is drawn. This gives us the sample space

$$S = \{(★, Y), (★, R), (△, Y), (△, R), (■, Y), (■, R), (○, Y), (○, R)\}$$

By counting the outcomes, we see that $n(S) = 8$. ■

✓ **Checkpoint 1**

Now work Exercise 3.

Note in Example 1 that we really did not have to list the whole sample space to determine $n(S)$. We could have also used the **Multiplication Principle**, which was discussed in Section 6.3. Since the draw from the box can be made in 4 ways and the draw from the bag can be made in 2 ways, we get

Choice 1   *Choice 2*
Draw from box   Draw from bag
$$4 \qquad \cdot \qquad 2 \qquad = 8 = n(S)$$

Now let's return to the experiment of drawing a marble numbered 1, 2, or 3 from the bag, replacing it, and drawing again. Recall that in this experiment the set of all possible outcomes (that is, the sample space) is

$$S = \{(1, 1), (1, 2), (1, 3), (2, 1), (2, 2), (2, 3), (3, 1), (3, 2), (3, 3)\}$$

Suppose that we want to list the outcomes that are associated when a 2 is drawn. If we label this set of outcomes with A, we get

$$A = \{(1, 2), (2, 1), (2, 2), (2, 3), (3, 2)\}$$

We call A an **event**. In words, we can say that A represents the event that at least one 2 is drawn. Notice that $A \subset S$.

■ **Event**

An **event** is simply a subset of the sample space.

▶ **Note:** We usually label events with capital letters such as A, B, C, and D or B_1, B_2, and so on.

Types of Probability

There are several fundamental ways to compute probability. The first way is one that almost everyone has done at some point in his or her life and is called **subjective probability**. For example, when a student frowns and says, "I don't have much of a chance of passing this test," the student is stating a subjective probability.

■ **Subjective Probability**

A probability that is based on estimates, educated guesses, intuition, emotion, or other indirect information is called a **subjective probability**.

Although subjective probability can be useful in sorting out certain decisions that a person can make, it has little or no mathematical basis. So we now turn to another type of probability that uses many of the mathematical tools that we have already studied. This type is called **classical probability**.

■ **Classical Probability**

Assume that S is a sample space that has $n(S)$ outcomes, where each outcome is equally likely. If A is an event associated with S, then the probability of event A occurring, denoted by $P(A)$, is given by

$$\begin{array}{l}\text{Probability of} \\ \text{event } A \text{ occurring}\end{array} = P(A) = \frac{n(A)}{n(S)} = \frac{\text{number of outcomes of event } A}{\text{total number of possible outcomes}}$$

▶ **Note:** 1. Recall that by equally likely we mean that every element in the sample space has an equal chance of being selected.
2. Probabilities can be expressed as either fractions, decimals, or percents.

Example 2 **Computing Classical Probability**

In Example 1, we were given the sample space

$$S = \{(\bigstar, Y), (\bigstar, R), (\triangle, Y), (\triangle, R), (\blacksquare, Y), (\blacksquare, R), (\bigcirc, Y), (\bigcirc, R)\}$$

Let A represent the event that a ticket with a $\bigstar$ is drawn and let B represent the event that a red marble is drawn.

(a) Determine $P(A)$ and interpret.
(b) Determine $P(B)$ and interpret.

Solution

(a) **Understand the Situation:** When we are asked to determine $P(A)$, we are being asked to determine the probability of A, the event that a ticket with a $\bigstar$ is drawn. To do this, we need to know the number of outcomes associated with event A as well as the total number of equally likely outcomes, that is, the cardinality of the sample space.

First, recall from Example 1 that the total number of possible outcomes is given by $n(S) = 8$. The outcomes associated with event A are $(\bigstar, Y), (\bigstar, R)$. Therefore,

$$A = \{(\bigstar, Y), (\bigstar, R)\}$$

So, using classical probability, we get

$$P(A) = \frac{\text{number of ways to draw a } \bigstar}{\text{total number of possible outcomes}} = \frac{n(A)}{n(S)} = \frac{2}{8} = \frac{1}{4}$$

Interpret the Solution: This means that the probability of having an outcome that includes a ticket with a $\bigstar$ is $\frac{1}{4}$.

(b) **Understand the Situation:** When we are asked to determine $P(B)$, we are being asked to determine the probability of B, the event that a red marble is drawn. To do this, we need to know the number of outcomes associated with event B as well as the total number of equally likely outcomes, that is, the cardinality of the sample space.

The outcomes associated with event B are elements of the following set:

$$B = \{(\star, R), (\triangle, R), (\blacksquare, R), (\bigcirc, R)\}$$

So the probability of event B is

$$P(B) = \frac{\text{number of ways to draw a red marble}}{\text{total number of possible outcomes}} = \frac{n(B)}{n(S)} = \frac{4}{8} = \frac{1}{2}$$

Interpret the Solution: This means that the probability of having an outcome that includes a red marble is $\frac{1}{2}$.

✓ **Checkpoint 2**

Now work Exercise 13.

When computing classical probability, we can also use some of the counting techniques discussed in Section 6.3. Example 3 illustrates how to use the multiplication principle.

Example 3 **Computing Classical Probability**

In an experiment, Conor tosses a six-sided die and records the number. He tosses it again and records the number from the second toss. Let A represent the event that the sum of the two tosses is an 8. Determine $P(A)$ and interpret.

Solution

Understand the Situation: When we are asked to determine $P(A)$, we are being asked to determine the probability of A, the event that the sum of the two tosses is 8. To do this, we need to know the number of outcomes associated with event A as well as the total number of equally likely outcomes, that is, the cardinality of the sample space.

We begin by determining the total number of possible outcomes. Since any of the numbers 1, 2, 3, 4, 5, or 6 is possible on the first toss as well as the second toss, we can employ the multiplication principle to determine the total number of possible outcomes. This gives

Table 6.4.1

First Toss	Second Toss	Sum
2	6	8
3	5	8
4	4	8
5	3	8
6	2	8

$$\underset{6}{\underset{\text{Possibilities for first toss}}{}} \cdot \underset{6}{\underset{\text{Possibilities for second toss}}{}} = 36$$

So the sample space has 36 equally likely outcomes. This means that $n(S) = 36$. Now the number of ways to have the sum of the two tosses equal 8 is shown in Table 6.4.1, which indicates that there are 5 ways to sum the results of the two tosses of the die and get a sum of 8. So the probability of event A is

$$P(A) = \frac{\text{number of ways to sum the tosses to get 8}}{\text{total number of possible outcomes}} = \frac{n(A)}{n(S)} = \frac{5}{36} \approx 0.139$$

Interpret the Solution: This means that the probability of having the sum of the two tosses of a six-sided die equal 8 is $\frac{5}{36} \approx 0.139$. That is, there is a 13.9% chance that the sum of the two tosses is 8.

Permutations and combinations, two counting techniques discussed in Section 6.3, can also be used in computing classical probability. The use of combinations to compute classical probability is shown in Example 4.

Example 4 **Computing Classical Probability with Combinations**

The Garforth County Board of Trustees is made up of 4 Republicans and 5 Democrats. The Board has agreed to form a subcommittee of 4 to investigate the impact of commercial development in the county. The 4 will be chosen at random.

(a) In how many ways can a subcommittee of 4 be formed?

(b) In how many ways can a subcommittee be formed that has 2 Republicans and 2 Democrats?

(c) Let A represent the event that the subcommittee is formed randomly with 2 Republicans and 2 Democrats. Determine $P(A)$ and interpret.

Solution

(a) **Understand the Situation:** Here we have to choose 4 trustees from among a total of 9 trustees. Since order is not important and there is no repetition, the total number of ways that the subcommittee can be formed is the number of 4 combinations of 9 objects.

The total number of ways that the subcommittee can be formed is given by

$$_9C_4 = \frac{9!}{4!(9-4)!} = \frac{9 \cdot 8 \cdot 7 \cdot 6}{4 \cdot 3 \cdot 2 \cdot 1} = 126$$

So the subcommittee of 4 can be formed in 126 ways.

(b) Using the techniques introduced in Section 6.3 and consulting Figure 6.4.2, we see that the number of ways to form the subcommittee with 2 Republicans and 2 Democrats is given by

$$\begin{array}{ccc} \textit{Choice 1} & & \textit{Choice 2} \\ \text{Republicans} & & \text{Democrats} \\ _4C_2 & \cdot & _5C_2 \\ 6 & \cdot & 10 & = 60 \end{array}$$

where $_4C_2 = \frac{4!}{2!2!} = 6$ and $_5C_2 = \frac{5!}{3!2!} = 10$. So there are 60 ways to form a subcommittee of 4 with 2 Republicans and 2 Democrats.

Interactive Activity

For Example 4, determine the probability that the subcommittee is randomly formed with all members being Republicans.

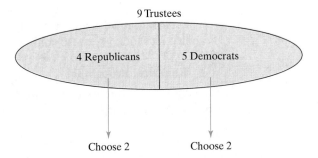

9 Trustees — 4 Republicans | 5 Democrats — Choose 2 — Choose 2

Figure 6.4.2

(c) Using classical probability along with the results of parts (a) and (b), we get

$$P(A) = \frac{\text{number of ways to form a subcommittee with 2 Republicans and 2 Democrats}}{\text{total number of ways to form a subcommittee of 4}} = \frac{{}_4C_2 \cdot {}_5C_2}{{}_9C_4} = \frac{60}{126} \approx 0.476$$

Interpret the Solution: This means that there is about a 47.6% probability that the subcommittee is formed with 2 Republicans and 2 Democrats. ∎

✓ **Checkpoint 3**

Now work Exercise 25.

The third type of probability, called **empirical probability**, makes use of frequency distributions of qualitative data. Frequency distributions and qualitative data will be studied in more depth in Chapter 7. For our purposes at this time, think of a **frequency**, denoted by f, as telling us how often something occurs or how many of a certain item there are. A **frequency distribution** is a table that summarizes the frequencies. Think of **qualitative data** as being data that are organized by categories. (Some even call qualitative data categorical data.) See Table 6.4.2, which gives the distribution of the U.S. population by race for 1998.

Table 6.4.2

Race	Population in millions, f
White	195.44
Black	32.72
American Indian, Eskimo	2.00
Asian	9.90
Hispanic	30.25

SOURCE: U.S. Census Bureau, www.census.gov

In Table 6.4.2, race is the qualitative data and is shown in the left column. The frequency for each racial category is given in the right column. If we sum the numbers (frequencies) in the right column, we get

$$195.44 + 32.72 + 2.00 + 9.90 + 30.25 = 270.31$$

Since these numbers are in millions, the sum represents 270.31 million. This number represents the population of the United States in 1998. Mathematically, we use the symbol $\sum$ (Greek capital letter sigma) to communicate sum. Hence, a more compact way to represent summing the numbers in the right column of Table 6.4.2 is by writing

$$\sum f = 270.31$$

Now that we have addressed some terminology common to empirical probability, let's look at how to compute it.

■ **Empirical Probability**

Given a frequency distribution of qualitative data, if A is an event associated with a qualitative data item, then the probability of event A is given by

$$P(A) = \frac{\text{frequency of event } A}{\text{sum of all frequencies}} = \frac{f}{\sum f}$$

Example 5   **Computing Empirical Probability**

The distribution of the U.S. population by race in 1998 is shown in Table 6.4.2. If a person in the United States is selected at random for a survey, determine the probability that the person is Hispanic.

Solution

Understand the Situation: We need to determine the U.S. population in 1998 and the total number of the U.S. Hispanic population in 1998.

To determine the U.S. population, we must add the frequencies in the table to get

$$\sum f = 195.44 + 32.72 + 2.00 + 9.90 + 30.25 = 270.31$$

So in 1998 the U.S. population was about 270.31 million. Table 6.4.2 also indicates that there were 30.25 million U.S. Hispanics in 1998. Now we let A represent the event that the randomly selected person is Hispanic. To determine $P(A)$ using empirical probability, we compute

$$P(A) = \frac{\text{frequency of Hispanics}}{\text{sum of frequencies}} = \frac{30.25}{270.31} \approx 0.112$$

Interpret the Solution: This means that in 1998 the probability that a randomly selected person in the United States is Hispanic is about 0.112 or about 11.2%.

✓ Checkpoint 4

Now work Exercise 33.

So far in this section we have seen many examples that include the phrase **selected at random**. We will use this phrase often in this chapter. When we use this phrase, we mean that all elements in the sample space are equally likely to be selected. Now let's turn our attention to the basic rules of probability.

◼ **Basic Rules of Probability**

For sample space S, assume that $A \subseteq S$ is an event and let $P(A)$ represent the probability of A.

1. The probability of an event has a value between 0 and 1, inclusive. In other words, for any event A, $0 \le P(A) \le 1$. Equivalently, using percentages, $0\% \le P(A) \le 100\%$.
2. If event A is certain to occur, then $P(A) = 1$. If event A is certain not to occur, then $P(A) = 0$.

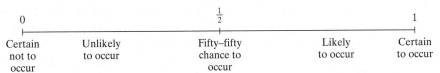

3. The event that A does *not* occur is the **complement of A**, denoted by A^C. If $P(A) = a$, then $P(A^C) = 1 - a$. That is, $P(A^C) = 1 - P(A)$.

In Example 6 we begin to apply some of these rules of probabilities.

Example 6   Using Probability Rules

Table 6.4.3

Marital Status	Percentage (%)
Never married	25
Married	55
Widowed	10
Divorced	10

SOURCE: U.S. Census Bureau, www.census.gov

The marital status of U.S. females 15 years or older in 2000 is given by the distribution in Table 6.4.3. For a study, a U.S. female 15 years or older was selected at random. Let W represent the event that she is widowed. Determine $P(W^C)$ and interpret.

Solution

Understand the Situation: Since W represents the event that a female 15 years or older is widowed, W^C (the complement of W) is the event that the female is *not* widowed.

We can see from reading Table 6.4.3 that $P(W) = 10\%$ or 0.10. From the basic rules of probability, we know that

$$P(W^C) = 1 - P(W) = 1 - 0.10 = 0.90$$

Interpret the Solution: This means that in 2000 the probability that a 15 year old or older female selected at random in the United States was *not* widowed is about 0.90 or 90%.

SUMMARY

In this section, we learned that there are three types of probabilities: **subjective**, **classical**, and **empirical**. We also learned some terminology concerning probability theory and the basic rules of probability.

- For a probability experiment, the **sample space**, denoted by S, is the set of all possible outcomes.
- For a probability experiment, an **event** is simply a subset of the sample space.
- **Subjective Probability**   A probability that is based on estimates, educated guesses, intuition, emotion, or other information.
- **Classical Probability**

$$\begin{aligned}\text{Probability of}\atop\text{event } A \text{ occurring} &= P(A) = \frac{n(A)}{n(S)}\\ &= \frac{\text{number of outcomes of event } A}{\text{total number of possible outcomes}}\end{aligned}$$

- **Empirical Probability**

$$\begin{aligned}\text{Probability of}\atop\text{event } A \text{ occurring} &= P(A)\\ &= \frac{\text{frequency of event } A}{\text{sum of all frequencies}}\\ &= \frac{f}{\sum f}\end{aligned}$$

- **Basic Rules of Probability**
  1. For any event A, $0 \leq P(A) \leq 1$.
  2. If event A is certain to occur, then $P(A) = 1$. If event A is certain not to occur, then $P(A) = 0$.
  3. The event that A does *not* occur is the **complement of A**, denoted by A^C. If $P(A) = a$, then $P(A^C) = 1 - a$. That is, $P(A^C) = 1 - P(A)$.

SECTION 6.4 EXERCISES

In Exercises 1–10, write the sample space S using set notation and then determine n(S).

1. A box contains 9 identical tickets labeled 1 through 9. In an experiment, a ticket is drawn at random and its number is recorded.

2. A jar contains 11 identical marbles labeled 1 through 11. In an experiment, a marble is drawn at random and its number is recorded.

6.1 ✓ **3.** A box contains 5 identical tickets labeled 1 through 5. In an experiment, a ticket is drawn at random,

its number is recorded, and it is returned to the box. Then a second ticket is drawn at random. The sum of the two numbers is recorded.

4. A box contains 5 identical tickets labeled 1 through 5. In an experiment, a ticket is drawn at random, its number is recorded, and it is returned to the box. Then a second ticket is drawn at random. The difference of the first minus the second number is recorded.

5. A pollster asks a respondent to classify her political ideology as conservative, moderate, or liberal.

6. The Holster Company asks 500 users of their eyeliner if their eye color is blue, brown, green, or hazel.

7. A warranty card for a new camping tent asks two questions. The first is to rate the quality of the tent as good, excellent, or superior, and the second asks to rate how waterproof the tent is on a scale from 1 to 4, with 1 being extremely waterproof and 4 being not very waterproof at all.

6.1 **8.** The suggestions card at Pappy's restaurant asks customers to rate the service as fair, good, or excellent and then to rate the quality of the food on a scale from 1 to 5, with 1 being superior and 5 being lousy.

9. A pollster asks the question "Do you think that the quality of television programming is better or worse than it was 15 years ago?" The pollster then asks, "What do you listen to the most, AM or FM radio?"

10. A pollster asks the question "Is your personal financial situation better, worse, or about the same as compared to 5 years ago?" The pollster then asks, "Have you ever declared bankruptcy?"

In Exercises 11–20, use classical probability to compute $P(A)$ and $P(B)$ for the indicated events A and B.

11. For the experiment in Exercise 1, a ticket is drawn at random and its number is recorded. Let A represent the event that the outcome is an odd number and B represent the event that the outcome is an even number.

12. For the experiment in Exercise 2, a marble is drawn at random and its number is recorded. Let A represent the event that the outcome is an odd number and B represent the event that the outcome is an even number.

13. For the experiment in Exercise 3, a ticket is drawn at random, its number is recorded, and it is returned to the box. Then a second ticket is drawn randomly. The sum of the two numbers is recorded. Let A represent the event that the sum is even and B represent the event that the sum is odd.

14. For the experiment in Exercise 4, a ticket is drawn randomly, its number is recorded, and it is returned to the box. Then a second ticket is drawn at random. The difference of the first minus the second number is recorded. Let A represent the event that the difference is positive and B represent the event that the difference is negative. Explain why $P(A) + P(B) \neq 1$.

15. A bag contains 10 identical marbles painted with the numbers 1 through 10. In an experiment, 2 marbles are drawn in succession and randomly, and the first marble is not replaced before the second draw. The numbers are recorded. Let A represent the event that the numbers are the same and B represent the event that a 2 is drawn.

16. A box contains 10 tickets numbered 1 through 10. In an experiment, 2 tickets are drawn in succession and randomly and the first ticket is not replaced before the second draw. The numbers are recorded. Let A represent the event that a

3 is drawn and B represent the event that the numbers add up to 5.

17. For the experiment in Exercise 7, a warranty card is drawn at random. Let A represent the event that the tent quality is rated as excellent and B represent the event that the tent's waterproof rating is a 2.

18. For the experiment in Exercise 8, a suggestions card is drawn at random. Let A represent the event that the service is rated as good and B represent the event that the food quality is rated as a 1.

19. A box contains 4 identical tickets numbered 1 through 4. In an experiment, a ticket is drawn at random, its number recorded, the ticket is returned to the box, and a second ticket is drawn randomly. Let A represent the event that the numbers are the same and B represent the event that a 4 is drawn.

20. A bag contains 4 identical marbles with 1 painted red, 1 painted yellow, 1 painted green, and 1 painted blue. In an experiment, a marble is drawn randomly, its color is recorded, the marble is returned to the bag, and a second marble is drawn at random. Let A represent the event that the marbles have the same color and B represent the event that a blue marble is drawn.

21. A cookie jar contains 15 identical chocolate chip cookies and 15 identical oatmeal raisin cookies. Kim takes two cookies out of the jar without looking at what kind they are. Let S be the sample space.

(a) Determine $n(S)$.

(b) If A represents the event that both cookies that Kim takes are oatmeal raisin, determine $n(A)$.

(c) Use classical probability to determine $P(A)$ and interpret.

22. Mrs. Perez mixes 10 plain and 10 peanut M&M's in a sack for Chris's school lunch. After finishing the rest of his lunch, Chris turns over the sack and two M&M's fall out. Let S be the sample space.

(a) Determine $n(S)$.

(b) If A represents the event that both M&M's are plain, determine $n(A)$.

(c) Use classical probability to determine $P(A)$ and interpret.

23. A box contains 30 identical marbles of which 15 are painted red and 15 are painted blue. Cole reaches into the box and pulls out 3 marbles in a single draw at random.

(a) Determine $n(S)$.

(b) If A represents the event that all three marbles are painted blue, determine $n(A)$.

(c) Use classical probability to determine $P(A)$ and interpret.

24. A box contains 20 identical marbles of which 10 are painted with an **X** and 10 are painted with a **Y**. Lexi reaches into the box and pulls out 4 marbles in a single draw at random.

(a) Determine $n(S)$.

(b) If A represents the event that all four marbles are labeled with an X, determine $n(A)$.

(c) Use classical probability to determine $P(A)$ and interpret.

✔ **25.** The city council in Staten City is comprised of 5 Democrats and 7 Republicans.

(a) In how many different ways can a subcommittee of 6 council persons be formed?

(b) In how many different ways can a subcommittee be formed with 2 Democrats and 4 Republicans?

(c) If A represents the event that a subcommittee is formed randomly with 2 Democrats and 4 Republicans, determine $P(A)$ and interpret.

26. The Euratal Company has 4 women and 8 men on its board of directors.

(a) In how many different ways can a gender-equality subcommittee of 4 be formed?

(b) In how many different ways can a gender-equality subcommittee of 4 be formed so that the number of men and women are equal?

(c) If A represents the event that a gender-equality subcommittee of 4 is formed at random so that the number of men and women are equal, determine $P(A)$ and interpret.

27. In a shipment of 30 car stereos, 8 are known to be defective.

(a) In how many different ways can a sample of 10 car stereos be chosen?

(b) In how many different ways can a sample of 10 be chosen so that 2 are defective and 8 are not defective?

(c) If A represents the event that in a sample of 10 car stereos 2 are defective and 8 are not defective, determine $P(A)$ and interpret.

28. In a box of 50 pieces of taffy, 20 pieces are pink and 30 pieces are yellow. If A represents the event that a handful of 8 pieces of taffy is pulled out at random so that 2 are pink and 6 are yellow, determine $P(A)$ and interpret.

29. A group of 20 environmentalists is made up of 6 Republicans, 8 Democrats, and 6 members of the Green Party. If A represents the event that a public relations committee of 7 is formed randomly with 2 Republicans, 2 Democrats, and 3 Green Party members, determine $P(A)$ and interpret.

🌐 **30.** In the Washington State lottery game called Lotto, 6 balls numbered 1 to 49 are drawn from a hopper. (*Source:* www.wa.gov/lot/home.htm.) If A represents the event that a person buys one ticket and wins, determine $P(A)$ and interpret. Write the solution as a proper fraction.

In Exercises 31–40, use empirical probability to answer the following questions.

🌐 **31.** The number and type of U.S. periodicals published in 1980 is shown in the table. A journalism professor selects a periodical from 1980 at random for a paper on reporting quality.

Type	Number
Weekly	1716
Semimonthly	645
Monthly	3985
Bimonthly	1114
Quarterly	1444

SOURCE: U.S. Census Bureau, www.census.gov

(a) Determine the probability that the periodical was a quarterly.

(b) Determine the probability that the periodical was a weekly.

🌐 **32.** The number and type of U.S. periodicals published in 1998 is shown in the table. A journalism professor selects a periodical from 1998 at random for a paper on reporting quality.

Type	Number
Weekly	364
Semimonthly	156
Monthly	3363
Bimonthly	2168
Quarterly	3309

SOURCE: U.S. Census Bureau, www.census.gov

(a) Determine the probability that the periodical was a quarterly.

(b) Determine the probability that the periodical was a weekly.

🌐 ✔ **33.** The type of robbery crimes, in thousands, reported in the United States in 1996 is shown in the table. A criminal justice professor selects a reported robbery crime from 1996 at random for a case study.

Type of Crime	Crimes (in thousands)
Street	274
Commercial	72
Gas station	13
Convenience store	32
Residence	57
Bank	11

SOURCE: U.S. Bureau of Crime Statistics, www.ojp.osdoj.gov

(a) Determine the probability that the crime involved a gas station.

(b) Determine the probability that the crime involved a bank.

34. The type of weapon, in thousands, used to commit crimes in 1995 is shown in the table. A criminal justice professor selects a reported robbery crime from 1995 at random for a case study.

Weapon	Number (in thousands)
Firearm	218
Knife	48
Other dangerous weapon	62
Strongarm	207

SOURCE: U.S. Bureau of Crime Statistics, www.ojp.osdoj.gov

(a) Determine the probability that the crime involved a firearm.

(b) Determine the probability that the crime involved a knife.

35. The number of U. S. enlisted military personnel, in thousands, in 1997 is shown in the table. In 1997 an enlisted military personnel was selected at random for a survey.

Branch of Service	Thousands Enlisted
Army	162.3
Navy	93.3
Marines	50.0
Air Force	74.5

SOURCE: U.S. Census Bureau, www.census.gov

(a) Determine the probability that the person was in the Army.

(b) Determine the probability that the person was in the Navy.

36. The number of recipients, in thousands, of master's degrees in science-related areas during 1995 and 1996 is shown in the table.

Degree Field	Master's Recipients (in thousands)
Computer science	18.2
Mathematics	7.9
Life science	15.3
Physical science	9.7
Social science	25.1

SOURCE: U.S. Department of Education, www.ed.gov

(a) If a master's degree recipient from 1995 or 1996 is selected at random, determine the probability that the degree was in the social science field.

(b) If a master's degree recipient from 1995 or 1996 is selected at random, determine the probability that the degree was in the life science fields.

37. The number of scientists, in thousands, employed in the United States in 1996 is shown in the table. In 1996 a scientist was selected at random for a survey.

Type of Scientist	Number (in thousands)
Social scientist	263.5
Physical scientist	206.7
Life scientist	180.0
Mathematical scientist	15.6

SOURCE: U.S. Census Bureau, www.census.gov

(a) Determine the probability that the scientist is a life scientist.

(b) Determine the probability that the scientist is not a life scientist.

38. The number of scientists, in thousands, employed by the U.S. government in 1996 is shown in the table. In 1996 a U.S. government scientist was selected at random for a survey.

Type of Scientist	Number (in thousands)
Social scientist	74.7
Physical scientist	71.2
Life scientist	47.2
Mathematical scientist	4.8

SOURCE: U.S. Census Bureau, www.census.gov

(a) Determine the probability that the scientist is a life scientist.

(b) If a U.S. government scientist from 1996 is selected at random, determine the probability that the scientist is not a life scientist.

39. The number of U.S. houses sold at various prices, in dollars, in 1998 is shown in the table. For an upcoming story, a national magazine selects at random a house that sold in 1998.

Sales Price	Number of Houses
Under $100,000	140,000
$100,000 to $119,999	112,000
$120,000 to $149,999	183,000
$150,000 to $199,999	208,000
$200,000 and over	248,000

SOURCE: U.S. Department of Housing and Urban Development, www.hud.gov

(a) Determine the probability that it sold for anywhere between $120,000 to $149,999.

(b) Determine the probability that it sold for $120,000 or more.

40. The percentage of U.S. houses built in 1999 of various sizes is shown in the table. For an upcoming story, a national magazine selects at random a house built in 1999.

Area (ft²)	Houses (%)
Under 1200	7
1200 to 1599	19
1600 to 1999	22
2000 to 2399	18
2400 and over	34

SOURCE: U.S. Department of Housing and Urban Development, www.hud.gov

(a) Determine the probability that the house had an area of 1600 to 1999 square feet.

(b) Determine the probability that the house had an area of 1600 square feet or more.

In Exercises 41–46, classify the following as an example of classical, empirical, *or* subjective *probability.*

41. Samantha claims that the probability that the telephone rings just when she is getting out of the shower seems close to 80%.

42. A girl tells a boy at the local high school, "There is zero chance of me going on a date with you!"

43. A survey shows that 45% of all Americans watched one NASCAR race in 1999. (*Source:* www.harrisinteractive.com.)

44. A recent poll shows that 81% of Americans feel that the choice of a vice-president has no influence on who they vote for president of the United States. (*Source:* www.harrisinteractive.com.)

45. The probability of winning a drawing when 500 tickets have been sold is $\frac{1}{500}$.

46. The probability of tossing a quarter and having the flip be a tail is 50%.

SECTION PROJECT

A traditional example of classical probability is computing what is commonly known as the *Birthday Problem*. Suppose that a group of r randomly selected people are together in a room. What is the probability that at least two of them have the same birthday? Assume that by the same birthday we mean month and day (and not necessarily the year) and we ignore the effect of leap years. To find out, let A represent the event that at least two of the r people in the room have the same birthday. Determining $P(A)$ directly can be a difficult task, but what we can do is compute $P(A^C)$, where A^C represents the event that no two people in the room have the same birthday (that is, all people in the room have different birthdays).

(a) Write an equation to show how $P(A)$ and $P(A^C)$ are related.

(b) To determine $P(A^C)$ using classical probability, we need to compute

$$P(A^C) = \frac{\text{number of ways that } r \text{ randomly selected people}}{\text{the number of all possible sets of birthdays for } r} \\ \frac{\text{can have different birthdays}}{\text{randomly selected people}}$$

For the denominator, we know that the first person can have a birthday 365 ways, the second person can have a birthday 365 ways, the third person can have a birthday 365 ways, and so on. Use the multiplication principle to determine the denominator of $P(A^C)$.

(c) For the numerator of $P(A^C)$, we need the number of ways that r randomly selected people can have different birthdays. This means that the first person can have a birthday 365 ways, the second person 364 ways, the third 363 ways, and so on. Note that the order is important and there is no repetition. Write an expression for the numerator of $P(A^C)$. Then write a fraction that gives a formula to compute $P(A^C)$.

(d) Complete the table below to determine the probability that at least two people in a room of r randomly selected people have the same birthday.

Number of People, r	Probability $P(A)$
5	
20	
30	
40	
50	
60	

(e) Using the table in part (d) as a guide, determine how many randomly selected people should be in a room so that there is at least a 50% chance that at least two people in the room have the same birthday.

Section 6.5   Computing Probabilities Using the Addition Rule

In the study of probability theory, we frequently need to determine a probability involving two or more events. For example, suppose that you roll a six-sided die and want to know the probability that on a single roll the result is a 1 **or** a 6. Or suppose that a medical researcher wants to know the probability that a newborn has low birth weight **or** whether the mother takes prenatal vitamins. These two situations illustrate what are called **compound events**. In this section we will learn a rule called the **Addition Rule**, which we can use to compute the probability of certain kinds of compound events. We will also examine the notion of **mutually exclusive events** and learn how the probabilities of such events are determined using the Addition Rule.

Mutually Exclusive Events

In Section 6.1 we looked at whether or not two events could happen at the same time. For example, in Section 6.1, Example 4(a), we saw that the event that Sandy gets a speeding ticket and the event that Sandy gets a ticket for not wearing a seat belt could happen at the same time. Whereas in Section 6.1, Example 4(b), we saw that the event that, during the same at bat, a softball player, Ariel, hits a single and the event that Ariel hits a triple were two events that could not happen at the same time. Events that have no outcomes in common, such as Ariel hitting a single and a triple in the same at bat, are what we call **mutually exclusive events**.

■ **Mutually Exclusive Events**

Two events A and B are **mutually exclusive** if A and B cannot occur at the same time. In other words, $A \cap B = \emptyset$.

▶ **Note:** Recall in Section 6.1 that we offered the following warning. A common error that many who are just beginning to study probability make is trying to read too much information into the questions in an attempt to find some exception.

The Venn diagrams in Figure 6.5.1 contrast mutually exclusive and non-mutually exclusive events.

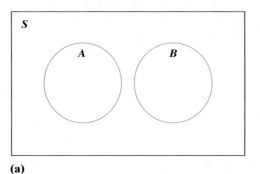

(a)

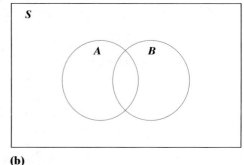

(b)

Figure 6.5.1 **(a)** A and B are mutually exclusive events. **(b)** A and B are not mutually exclusive events.

▶ **Note:** In Section 6.2, we defined **disjoint sets** as sets with no elements in common. Mutually exclusive events have no outcomes in common, so they are disjoint subsets of the sample space.

Example 1 **Determining If Events Are Mutually Exclusive**

For the following, determine if the events A and B are mutually exclusive.

(a) A blood donor is selected at random. Let A represent the event that the donor has type A blood and B represent the event that the donor has type B blood.

(b) A student from Harperston Business College is selected at random. Let A represent the event that the student is an accounting major and B represent the event that the student is a female.

Solution

(a) Since the person cannot have two different blood types, we see that events A and B are mutually exclusive.

(b) In this case, we know that there must be students who are accounting majors, since it is a business college. It also seems reasonable that some accounting majors are female. Since these events can happen at the same time, events A and B are not mutually exclusive. ∎

✓ **Checkpoint 1**

Now work Exercise 7.

Table 6.5.1

Branch of Service	Enlistment
Army (A)	162,300
Navy (B)	93,300
Marines (C)	50,000
Air Force (D)	74,500

SOURCE: U.S. Department of Defense, www.defenselink.mil

Now suppose that we want to determine the probability that, given two events, **either** of the events occur. To see how we compute this type of probability, consider Table 6.5.1, which shows the distribution of enlisted U.S. military personnel in 1997. Notice that the events A, B, C, and D are assigned to the events that an enlisted person is in the Army, Navy, Marines, or Air Force, respectively.

First notice that if we add the enlistments in each of the branches we get $\sum f = 380,100$. Suppose that a researcher selects an enlisted person at random. To determine the probability that the enlisted person is in the Navy *or* the Air Force, we begin by determining the total number that is enlisted in the Navy or Air Force. This sum is $93,300 + 74,500 = 167,800$. We have already determined that the total number of enlisted personnel is 380,100. So the probability that the randomly selected enlisted person is in the Navy **or** Air Force is

$$P(\text{Navy or Air Force}) = P(B \cup D)$$

$$= \frac{\text{number enlisted in the Navy or Air Force}}{\text{total enlisted military personnel}}$$

$$= \frac{167,800}{380,100} \approx 0.441$$

This means that in 1997 the probability that a randomly selected enlisted person is a member of the Navy or the Air Force is about 44.1%.

Notice that the probability that the enlisted person is in the Navy is given by $P(B) = \dfrac{93,300}{380,100}$, while the probability that the enlisted person is in the Air

Force is given by $P(D) = \dfrac{74,500}{380,100}$. If we sum these probabilities, we get

$$P(B) + P(D) = \frac{93,300}{380,100} + \frac{74,500}{380,100} = \frac{167,800}{380,100} = P(B \cup D)$$

So it appears for this situation that $P(B \cup D) = P(B) + P(D)$. This is not a coincidence! Since events B and D are mutually exclusive, these calculations illustrate the Addition Rule for Mutually Exclusive Events.

■ **Addition Rule for Mutually Exclusive Events**

If A and B are mutually exclusive events (this means that only one of them can occur or, equivalently, $A \cap B = \emptyset$), then the probability that either A occurs **or** B occurs is given by

$$P(A \textbf{ or } B) = P(A \cup B) = P(A) + P(B)$$

▶ **Note:** 1. In Section 6.2 we discussed the cardinality of sets and showed that, if sets A and B are disjoint, then $n(A \cup B) = n(A) + n(B)$.

2. If A and B are mutually exclusive events, then $P(A \cap B) = 0$ since $A \cap B = \emptyset$.

Example 2   **Applying the Addition Rule for Mutually Exclusive Events**

Table 6.5.2 shows the number of successful space launches for all countries that launched during 1999. (Success is defined as attainment of Earth orbit or Earth escape.) Notice that the events A, B, C, D, E, and F are assigned to the events that a launch was from Russia, United States, ESA, China, India, or Ukraine, respectively. A space exploration researcher selects a 1999 successful space launch at random. Determine $P(B \cup C)$ and interpret.

Table 6.5.2

Country	Number of Successful Launches
Russia (A)	26
United States (B)	30
ESA (European Space Agency) (C)	10
China (D)	4
India (E)	1
Ukraine (F)	2

SOURCE: Statistical Abstract of the United States, www.census.gov/statab/www/

Interactive Activity

For the experiment in Example 2 determine $P(D \cup E)$.

Solution

Understand the Situation: We begin by noting that events B and C are mutually exclusive. The same space launch cannot take place from the United States (event B) and the ESA (event C) at the same time. This means that we can use the Addition Rule for Mutually Exclusive Events. Summing the numbers in the right-hand column of Table 6.5.2 determines that in 1999 a total of 73 successful space launches took place. This tells us that $P(B) = \dfrac{30}{73}$ and $P(C) = \dfrac{10}{73}$.

Since we know that events B and C are mutually exclusive, we have

$$P(B \cup C) = P(B) + P(C)$$
$$= \frac{30}{73} + \frac{10}{73} = \frac{40}{73}$$

Interpret the Solution: In 1999 the probability that a randomly selected successful space launch came either from the United States or the ESA is $\frac{40}{73}$. ∎

✓ **Checkpoint 2**

Now work Exercise 23.

The General Addition Rule

We have seen how to compute $P(A \cup B)$ if A and B are mutually exclusive events. The next step is to see how we compute $P(A \cup B)$ if A and B are **not** mutually exclusive. For example, suppose that you roll a six-sided die and are interested in the probability of rolling a number that is less than 3 or rolling an even number. In Figure 6.5.2 we can see that these events are not mutually exclusive, since 2 is an outcome of both events.

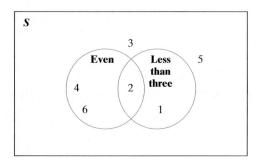

Figure 6.5.2

We know that there are 6 possible outcomes on a single roll of a die. Figure 6.5.2 illustrates that 4 of these 6 are less than 3 or even. This means that

$$P(\text{less than 3 or even}) = \frac{4}{6} = \frac{2}{3}$$

Notice that it appears that the probability of rolling a number that is less than three or an even number can be expressed as

$$P(\text{less than 3 or even}) = P(\text{less than 3}) + P(\text{even}) - P(\text{less than 3 and even})$$
$$= \frac{2}{6} + \frac{3}{6} - \frac{1}{6} = \frac{4}{6} = \frac{2}{3} \approx 0.667$$

To generalize what we just did, we need to consult the Toolbox below to recall a useful rule from Section 6.2, as well as a type of probability that we explored in Section 6.4.

𝒯 **From Your Toolbox**

- If A and B are any sets, then $n(A \cup B) = n(A) + n(B) - n(A \cap B)$.
- **Classical Probability**

$$\begin{array}{c} \text{Probability of} \\ \text{event } A \text{ occurring} \end{array} = P(A) = \frac{n(A)}{n(S)} = \frac{\text{number of outcomes of event } A}{\text{total number of possible outcomes}}$$

We can let A and B represent two events where the numbers of outcomes for these events are given by $n(A)$ and $n(B)$, respectively. For a sample space S, we can use classical probability to get

$$P(A \cup B) = \frac{\text{number of ways } A \text{ or } B \text{ can occur}}{\text{total number of possible outcomes}} = \frac{n(A) + n(B) - n(A \cap B)}{n(S)}$$

$$= \frac{n(A)}{n(S)} + \frac{n(B)}{n(S)} - \frac{n(A \cap B)}{n(S)} = P(A) + P(B) - P(A \cap B)$$

We call this rule the General Addition Rule.

■ **General Addition Rule**

For any events A and B, the probability that at least one of A and B occurs (that is, A occurs or B occurs or both A and B occur) is given by

$$P(A \cup B) = P(A) + P(B) - P(A \cap B)$$

▶ **Note:** Notice that the Addition Rule for Mutually Exclusive Events is just a special case of the General Addition Rule. If A and B are mutually exclusive, then $P(A \cap B) = 0$.

Venn diagrams are commonly used to visualize applications of the General Addition Rule. These diagrams can either show the number of outcomes associated with given events or can display the probabilities assigned to these events. This is illustrated in Example 3.

Example 3 **Applying the General Addition Rule**

The curator of a science museum surveys 1100 patrons at random and finds that 800 saw the space probe exhibit, 500 attended the 3D movie, and 300 saw both the exhibit and the movie. Let A represent the event that a patron saw the space probe exhibit and B represent the event that he or she saw the 3D movie.

(a) Draw a Venn diagram to represent the sample space and the events A and B.

(b) Determine $P(A \cup B)$ and interpret.

(c) Determine $P(A \cup B)^C$ and interpret.

Solution

(a) The Venn diagram in Figure 6.5.3 shows the number of outcomes associated with events A, B, $A \cap B$ and sample space S. Let's make a few observations

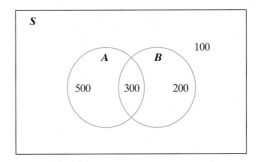

Figure 6.5.3

about the Venn diagram. Notice that if we sum all of the outcomes in the circle associated with event A we get $500 + 300 = 800$, and if we sum the outcomes in the circle associated with event B, we get $300 + 200 = 500$. Also notice that if we sum all of the values in the diagram we get $500 + 300 + 200 + 100 = 1100$, which is the size of the sample space.

(b) **Understand the Situation:** First note that

$$P(A) = \frac{\text{number surveyed who saw the space probe exhibit}}{\text{total number surveyed}} = \frac{800}{1100} = \frac{8}{11} \quad \text{and}$$

$$P(B) = \frac{\text{number surveyed who saw the 3D movie}}{\text{total number surveyed}} = \frac{500}{1100} = \frac{5}{11}.$$

Since 300 patrons saw both the exhibit and the movie, we know that

$$P(A \cap B) = \frac{300}{1100} = \frac{3}{11}.$$

So the General Addition Rule yields

$$P(A \cup B) = P(A) + P(B) - P(A \cap B) = \frac{8}{11} + \frac{5}{11} - \frac{3}{11} = \frac{10}{11}$$

Interpret the Solution: This means that if a surveyed patron is selected at random the probability that she or he saw either the space probe exhibit, the 3D movie, or both is $\frac{10}{11}$.

(c) From the definition of complementary events, we know that

$$P(A \cup B)^C = 1 - P(A \cup B)$$

So, using the solution from part (b), we get

$$P(A \cup B)^C = 1 - P(A \cup B) = 1 - \frac{10}{11} = \frac{1}{11}$$

Interpret the Solution: This means that if a surveyed patron is selected at random the probability that she or he attended neither the space probe exhibit nor the 3D movie is $\frac{1}{11}$. ∎

There is a useful observation embedded within Example 3. If we take each number shown in Figure 6.5.3 and divide it by 1100 (the size of the sample space), we get what is called a **probability Venn diagram**. This diagram is shown in Figure 6.5.4. Note that the sum of all the values in the probability Venn diagram add up to 1.

The General Addition Rule can also be used to compute probabilities based on data that are in tabular form. In practice, many probabilities involving two or more events are computed using this method.

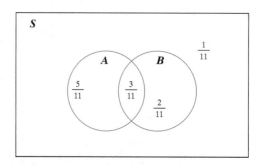

Figure 6.5.4 Probability Venn Diagram.

Example 4 Using the Addition Rule with Tabular Data

The source of payment for patients discharged from U.S. hospitals during 1996 for two age groups is shown in Table 6.5.3. The number of discharges is given in thousands. For the labeled events A, B, C, and D, determine the following probabilities and interpret.

(a) $P(B \cap D)$ **(b)** $P(A \cup C)$

Table 6.5.3

Age Group	Private (A)	Government (B)	Total
Less than 44 years old (C)	6,090	6,442	12,532
45 years or older (D)	4,636	13,337	17,973
Total	10,726	19,779	30,505

SOURCE: U.S. National Center for Health Statistics, www.cdc.gov/nchs

Solution

(a) Understand the Situation: First notice that event B is the event that the source of payment for a discharged patient in 1996 was the government. Event D is the event that a discharged patient was 45 years old or older. So when we compute $P(B \cap D)$, we are computing the probability that a randomly selected patient discharged from a hospital in 1996 was 45 years old or older and the source of payment was the government.

By inspecting Table 6.5.3, we see that if we follow down the event B column and across the event D row they intersect at the cell with the value 13,337 thousand. Since the total number of patients discharged in 1996 was 30,505 thousand, we have

$$P(B \cap D) = \frac{\text{value where the event } B \text{ column and event } D \text{ row intersect}}{\text{total patients discharged}}$$

$$= \frac{13,337}{30,505} \approx 0.44$$

Interpret the Solution: This means that if we select a discharged patient at random the probability that he or she had government health insurance **and** was 45 year or older is about 0.44 or about 44%.

(b) Understand the Situation: Since event A is the event that the source of payment was private and event C is the event that the discharged patient was less than 44 years old, we are being asked to determine the probability that a randomly selected patient has private health insurance, is 44 years old or younger, or both. Table 6.5.3 indicates that

$$P(A) = \frac{\text{total in event } A \text{ column}}{\text{total patients discharged}} = \frac{10,726}{30,505},$$

$$P(C) = \frac{\text{total in event } C \text{ row}}{\text{total patients discharged}} = \frac{12,532}{30,505},$$

and where event A column and event C row intersect gives us

$$P(A \cap C) = \frac{6090}{30,505}.$$

The General Addition Rule gives us

$$P(A \cup C) = P(A) + P(C) - P(A \cap C)$$

$$= \frac{\text{total in event } A \text{ column}}{\text{total patients discharged}} + \frac{\text{total in event } C \text{ row}}{\text{total patients discharged}}$$

$$- \frac{\text{value where the event } A \text{ column and event } C \text{ row intersect}}{\text{total patients discharged}}$$

$$= \frac{10{,}726}{30{,}505} + \frac{12{,}532}{30{,}505} - \frac{6090}{30{,}505} = \frac{17{,}168}{30{,}505} \approx 0.56$$

Interpret the Solution: This means that if we select a discharged patient at random the probability that he or she either has private health insurance, is 44 years or younger, or both is about 0.56 or about 56%. ∎

Checkpoint 3

Now work Exercise 35.

Odds

We conclude this section with a brief discussion of **odds**. Sometimes, when we know the probability of an event A, we speak of the odds for (or against) the event A, rather than the probability of event A. This is particularly true in cases involving betting on sporting events. For example, we may hear that "The odds of Tiger Woods winning the British Open are 3 to 2." We will return to this example later, but first we introduce a definition that ties together the concept of odds and probabilities.

Probability to Odds

If $P(A)$ is the probability of an event A occurring, then

1. The **odds in favor of A** are "$P(A)$ to $1 - P(A)$" or, equivalently, "$P(A)$ to $P(A^C)$." This is represented by

$$\frac{P(A)}{1 - P(A)} = \frac{P(A)}{P(A^C)}, \qquad P(A) \neq 1$$

2. The **odds against A** are "$1 - P(A)$ to $P(A)$" or, equivalently, "$P(A^C)$ to $P(A)$." This is represented by

$$\frac{1 - P(A)}{P(A)} = \frac{P(A^C)}{P(A)}, \qquad P(A) \neq 0$$

▶ **Note:** When possible, odds are expressed as ratios of whole numbers.

Example 5 Converting from Probability to Odds

Dillon rolls a six-sided die once. The probability of rolling a 2 is $\frac{1}{6}$. Determine the odds in favor of rolling a 2.

Solution

Understand the Situation: We let A represent the event that Dillon rolls a 2. This means that we are given $P(\text{rolling a 2}) = P(A) = \frac{1}{6}$. This also gives us

Interactive Activity

For the situation in Example 5, determine the odds against rolling a 2.

$P(A^C) = \dfrac{5}{6}$. The odds in favor of A (rolling a 2) are $\dfrac{1}{6}$ to $\dfrac{5}{6}$ or, equivalently,

$$\frac{P(A)}{1 - P(A)} = \frac{\dfrac{1}{6}}{1 - \dfrac{1}{6}} = \frac{\dfrac{1}{6}}{\dfrac{5}{6}} = \frac{1}{5}$$

Interpret the Solution: So the odds in favor of rolling a 2 are $\dfrac{1}{5}$. We could read this as "Odds in favor of rolling a 2 are 1 to 5." ∎

If the situation in Example 5 were in a gaming situation where money was wagered, we would call it a **fair game** if the following occurred:

- If Dillon bet \$1 on a 2 turning up, he would lose his \$1 if any number other than a 2 turns up.
- If Dillon bet \$1 on a 2 turning up, he would win \$5 (and have his bet of \$1 returned) if a 2 turns up.

Generally, if the odds in favor of an event A are "a to b", the **game is fair** if a bet of \$$a$ is lost if event A does not occur, but a win of \$$b$ (as well as returning the bet of \$$a$) is realized if event A does occur.

We have seen how to convert probability to odds. Now suppose that we are given the odds for an event, but wish to determine the probability of the event. If the odds in favor of event A occurring are "a to b," using part 1 of the Probability to Odds formula box allows us to write

$$\frac{a}{b} = \frac{P(A)}{1 - P(A)}$$

Solving for $P(A)$ (we leave the details for Exercise 75), yields the following:

> ■ **Odds to Probability**
>
> If the odds for event A are a to b, then the probability of event A occurring is
>
> $$P(A) = \frac{a}{a + b}$$

Example 6   **Converting Odds to Probability**

As of July 9, 2001, the odds of Tiger Woods winning the 2001 British Open are 3 to 2. Express these odds as a probability. (*Source:* www.vegasinsider.com/u/futures.)

Solution

Understand the Situation: Let A be the event that Tiger Woods wins the British Open. The odds in favor of A are given as 3 to 2. So we have $a = 3$ and $b = 2$.

Using the formula for converting odds to probability gives us

$$P(A) = \frac{a}{a + b} = \frac{3}{3 + 2} = \frac{3}{5}$$

Interpret the Solution: If the odds of Tiger Woods winning the British Open are 3 to 2, then the probability that he will win the British Open is $\dfrac{3}{5}$. ∎

✓ **Checkpoint 4**   Now work Exercise 69.

SUMMARY

In this section we learned how to compute compound events of the form $P(A \cup B)$. We saw that the rule that we used to compute this probability depended on the events being mutually exclusive. We also learned the relationship between probability and odds.

- **Mutually Exclusive Events** Two events A and B are **mutually exclusive** if A and B cannot occur at the same time; that is, $A \cap B = \emptyset$.
- **Addition Rule for Mutually Exclusive Events** $P(A \cup B) = P(A) + P(B)$
- **General Addition Rule** $P(A \cup B) = P(A) + P(B) - P(A \cap B)$
- If $P(A)$ is the probability of an event A occurring, then

1. The **odds in favor of A** are "$P(A)$ to $1 - P(A)$" or, equivalently, "$P(A)$ to $P(A^C)$." This is represented by

$$\frac{P(A)}{1 - P(A)} = \frac{P(A)}{P(A^C)}, \qquad P(A) \neq 1$$

2. The **odds against A** are "$1 - P(A)$ to $P(A)$" or, equivalently, "$P(A^C)$ to $P(A)$." This is represented by

$$\frac{1 - P(A)}{P(A)} = \frac{P(A^C)}{P(A)}, \qquad P(A) \neq 0$$

- **Odds to Probability** If the odds for event A are a to b, then the probability of event A occurring is $P(A) = \dfrac{a}{a + b}$.

SECTION 6.5 EXERCISES

In Exercises 1–10, determine if the events A and B are mutually exclusive.

⇐6.1 **1.** A person in a mall is selected at random. A represents the event that the person has brown hair and B represents the event that the person has blue eyes.

⇐6.1 **2.** A person in a mall is selected at random. A represents the event that the person has a college degree and B represents the event that the person is employed.

⇐6.1 **3.** In a state that requires voters to register for a single party, a voter is selected at random for an exit poll. The event that the person is a registered Republican and the event that the person is a registered Democrat.

⇐6.1 **4.** In a state that uses electronic ballots, a voter is selected at random for an exit poll. The event that the person voted for issue 1 on the ballot and the event that the person voted against issue 1 on the ballot.

⇐6.1 **5.** A student at Brockston College is selected at random. A represents the event that the student is a biology major and B represents the event that the student is a senior.

⇐6.1 **6.** A student at Brockston College is selected at random. A represents the event that the student receives financial aid and B represents the event that the student is on the dean's list.

⇐6.1 ✓ **7.** A box contains 30 identical tickets numbered 1 to 30. One ticket is drawn at random. A represents the event that the ticket is labeled with an even number and B represents the event that the ticket is labeled with an odd number.

⇐6.1 **8.** A box contains 30 identical tickets numbered 1 to 30. One ticket is drawn at random. A represents the event that the ticket is labeled with a prime number and B represents the event that the ticket is labeled with a composite number.

⇐6.1 **9.** Two respondents from a nationwide survey are selected at random for a follow-up interview. A represents the event that the two respondents are brothers and B represents the event that the two respondents are sisters.

⇐6.1 **10.** Two respondents from a nationwide survey are selected at random for a follow-up interview. A represents the event that the two respondents are brothers and B represents the event that the two respondents are not related.

In Exercises 11–14, consider a box that contains 10 identical tickets numbered 1 to 10. If a draw is made at random, determine the following probabilities.

11. A ticket labeled with a 1 or a 6 is drawn.

12. A ticket labeled with a 2 or a 3 is drawn.

13. A ticket labeled with a 5 or a 12 is drawn.

14. A ticket labeled with a 4 or a 14 is drawn.

In Exercises 15–18, consider a bag that contains 6 red, 5 blue, and 4 yellow marbles of identical shapes. If a draw is made at random, determine the following probabilities.

15. A red or a blue marble is drawn.

16. A red or a yellow marble is drawn.

17. A black or a blue marble is drawn.

18. A red or a white marble is drawn.

For Exercises 19–24, consider a survey of all 500 workers at the Canal Junction water treatment plant. Let A represent the event that the worker took a 3-week vacation in the last year and B represent the event that the worker did more than 5 weeks of overtime in the last year. The results are shown in the Venn diagram. Assuming that a worker is selected at random, use the diagram to answer the following:

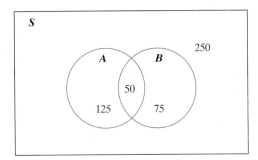

19. Determine $P(A)$ and interpret.

20. Determine $P(B)$ and interpret.

21. Determine the probability that the worker took a 3-week vacation in the last year and did more than 5 weeks of overtime in the last year.

22. Determine the probability that the worker did not take a 3-week vacation in the last year and did not do more than 5 weeks of overtime in the last year.

✓ **23.** Determine $P(A \cup B)$ and interpret.

24. Determine $P(A \cup B)^C$ and interpret.

In Exercises 25–30, consider a survey of a group of seniors, selected at random, at Clayborne College who were asked about their attitudes concerning academic probation. Let A represent the event that the student had been placed on academic probation for at least one semester and B represent the event that the person had been placed on the dean's list for at least one semester. The results are shown in the probability Venn diagram. Assuming that a senior in the survey group is selected at random, use the diagram to answer the following:

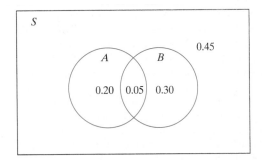

25. Determine $P(B)$ and interpret.

26. Determine $P(A)$ and interpret.

27. Determine $P(A \cap B)^C$ and interpret.

28. Determine $P(A \cap B)$ and interpret.

29. Determine the probability that the student had been placed on academic probation for at least one semester

or had been placed on the dean's list for at least one semester.

30. Determine the probability that the student had not been placed on academic probation for at least one semester or had not been placed on the dean's list for at least one semester.

🌐 For Exercises 31–40 consider the following. The number of doctorates conferred in the social sciences (anthropology, sociology, economics, and political science) and the physical sciences (astronomy, physics, and chemistry) during 1998 is shown in the following table. The data are classified as being a U.S. citizen or a foreign-born citizen.

	Social Sciences (A)	Physical Sciences (B)	Total
U.S. citizen (C)	2353	2415	4768
Foreign-born citizen (D)	760	1111	1871
Total	3113	3526	6639

SOURCE: U.S. Census Bureau, www.census.gov

For the events labeled in the table as A, B, C, and D, answer the following:

31. Determine $P(A)$ and interpret.

32. Determine $P(B)$ and interpret.

33. Determine $P(A^C)$ and interpret.

34. Determine $P(B^C)$ and interpret.

✓ **35.** Determine the probability that the doctorate is in social sciences and is conferred on a U.S. citizen.

36. Determine the probability that the doctorate is in physical sciences and is conferred on a foreign-born citizen.

37. Determine $P(A \cap B)$ and interpret.

38. Determine $P(C \cap D)$ and interpret.

39. Determine the probability that the doctorate is in social sciences or is conferred on a foreign-born citizen.

40. Determine the probability that the doctorate is in physical sciences or is conferred on a U.S. citizen.

🌐 For Exercises 41–50, consider the following. In 1996, the percentage of American adults who exercised for 30 minutes or more at least 5 times per week is shown in the table. The percentages are classified by gender.

	Did Not Exercise (A)	Exercised Regularly (B)	Total (%)
Men (C)	22.0	25.6	47.6
Women (D)	25.6	26.8	52.4
Total	47.6	52.4	100

SOURCE: National Center for Chronic Disease Prevention, www.cdc.gov/nccdphp

For the events labeled in the table as A, B, C, and D, answer the following:

41. Determine $P(B)$ and interpret.

42. Determine $P(A)$ and interpret.

43. Determine $P(B^C)$ and interpret.

44. Determine $P(A^C)$ and interpret.

45. Determine the probability that the person exercised regularly and is female.

46. Determine the probability that the person did not exercise regularly and is male.

47. Determine $P(C \cap D)$ and interpret.

48. Determine $P(A \cap B)$ and interpret.

49. Determine the probability that the person exercised regularly or is male.

50. Determine the probability that the person did not exercise regularly or is female.

51. The probability of event A occurring is 0.8.

(a) Determine the odds in favor of A occurring.

(b) Determine the odds against A occurring.

52. The probability of event A occurring is 0.2.

(a) Determine the odds in favor of A occurring.

(b) Determine the odds against A occurring.

53. The probability of event A not occurring is 0.7.

(a) Determine the odds in favor of A occurring.

(b) Determine the odds against A occurring.

54. The probability of event A not occurring is 0.6.

(a) Determine the odds in favor of A occurring.

(b) Determine the odds against A occurring.

55. The odds in favor of event A occurring are 5 to 3. Determine the probability of A occurring.

56. The odds in favor of event A occurring are 7 to 5. Determine the probability of A occurring.

57. The odds against event A occurring are 2 to 5. Determine the probability of A not occurring.

58. The odds against event A occurring are 4 to 7. Determine the probability of A not occurring.

In Exercises 59–62, determine the odds in favor of obtaining the specified event.

59. A tail in a single toss of a coin

60. An even number in a single roll of a die

61. At least 1 tail when a coin is tossed 2 times

62. At least 1 tail when a coin is tossed 3 times

In Exercises 63–66, determine the odds against the specified event.

63. A head in a single toss of a coin

64. A number divisible by 3 in a single roll of a die

65. Two tails when a coin is tossed 2 times

66. An even number in a single roll of a die

67. Latasha hears that the probability of rain tomorrow is 30%.

(a) Determine the odds that it will rain tomorrow.

(b) Determine the odds that it will not rain tomorrow.

68. Frank, a carpet sales representative, believes that the odds are 9 to 4 that he will close a deal of carpeting the offices of a certain company. Determine the probability that Frank will make the sale.

69. As of April 23, 2001, the odds in favor of the Dallas Cowboys winning the 2002 Super Bowl were listed as 1 to 75. As of April 23, 2001, determine the probability that the Dallas Cowboys will win the 2002 Super Bowl. (*Source:* www.vegasinsider.com/u/futures.)

70. As of April 23, 2001, the odds in favor of the St. Louis Rams winning the 2002 Super Bowl were listed as 1 to 4. As of April 23, 2001, determine the probability that the St. Louis Rams will win the 2002 Super Bowl. (*Source:* www.vegasinsider.com/u/futures.)

71. As of July 2, 2001, the odds in favor of the Los Angeles Dodgers winning the 2001 World Series were listed as 1 to 25. As of July 2, 2001, determine the probability that the Los Angeles Dodgers will win the 2001 World Series. (*Source:* www.vegasinsider.com/u/futures.)

72. As of July 2, 2001, the odds in favor of the Cleveland Indians winning the 2001 World Series were listed as 1 to 6. As of July 2, 2001, determine the probability that the Cleveland Indians will win the 2001 World Series. (*Source:* www.vegasinsider.com/u/futures.)

73. As of July 2, 2001, the odds in favor of the Los Angeles Lakers winning the 2002 NBA Championship were listed as 5 to 7. As of July 2, 2001, determine the probability that the Los Angeles Lakers will win the 2002 NBA Championship. (*Source:* www.vegasinsider.com/u/futures.)

74. As of July 2, 2001, the odds in favor of the Charlotte Hornets winning the 2002 NBA Championship were listed as 5 to 7. As of July 2, 2001, determine the probability that the Charlotte Hornets will win the 2002 NBA Championship. (*Source:* www.vegasinsider.com/u/futures.)

75. If the odds in favor of event A occurring are a to b, show that

$$P(A) = \frac{a}{a+b}$$

Hint: Start with the equation $\dfrac{a}{b} = \dfrac{P(A)}{1 - P(A)}$ and solve for $P(A)$.

SECTION PROJECT

For the following probability Venn diagram, $P(A) = 0.45$, $P(B) = 0.45$, and $P(C) = 0.40$. Determine the values of w, x, y, and z.

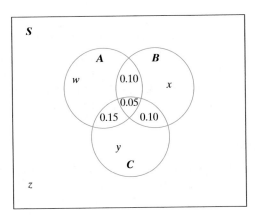

Section 6.6

Computing Probabilities Using the Multiplication Rule

In the previous section we used the Addition Rule to determine the probability that **either** of two events would occur. To compute this probability, we first had to know if the events were mutually exclusive. In this section we will use a new rule called the **Multiplication Rule**, which determines the probability, given two events, that **both** events occur. To compute this probability, we often have to determine if the events have a condition called **independence**. We will also examine a new type of probability called **conditional probability**.

Independent Events

Recall that in Section 6.1 we saw that often the occurrence of an event has no particular effect on the occurrence of some other event. For example, in Section 6.1 Example 5(b), we analyzed the event that the first toss of a coin is heads and the event that the second toss of the coin is tails. The likelihood of tossing heads or tails is the same, so tossing heads with the first toss does not alter the likelihood of tossing tails on the second toss. So, in this case, the outcome of the first event (first toss of a coin is heads) does not influence the second event (second toss of the coin is tails). We call events of this type **independent events**.

> ■ **Independent Events**
>
> Two events A and B are **independent** if the occurrence (or nonoccurrence) of one event does not affect the other's chance of occurring. If A and B are not independent, we say that they are **dependent events**.

▷ **Note:** We again offer a warning first presented in Section 6.1. A common error that many who are just beginning to study probability make is trying to read too much information into the questions in an attempt to find some exception.

Example 1 Determining If Events Are Independent

For each of the following, determine if the events A and B are independent.

(a) A represents the event that Austin practices golf 6 hours each day for 10 years and B represents the event that Austin becomes a professional golfer.

(b) Each week, 5 numbers from 1 to 40 are randomly selected from a bin in the Ultra-Lotto game. A represents the event that the five winning numbers in the first Ultra-Lotto held this year are 7, 10, 19, 29, and 33, and B represents the event that the five winning numbers in the second Ultra-Lotto held this year are 7, 10, 19, 29, and 33.

Solution

(a) **Understand the Situation:** If Austin practices golf 6 hours each day for 10 years, his chances of becoming a professional golfer are increased. So in this case, events A and B are dependent.

(b) **Understand the Situation:** Since all the numbers drawn in a lottery are placed back into the bin for the next drawing, the occurrence or nonoccurrence of A does not change the probability of B. Thus, A and B are independent. ∎

✓ **Checkpoint 1**

Now work Exercise 3.

Now let's see how independence can be used to find probabilities of two or more events occurring simultaneously. Suppose that a box contains 3 identical tickets colored red, blue, and green. In an experiment, a ticket is drawn at random, its color is recorded, and it is returned to the box. Then a second ticket is drawn at random from the box. If we let R, B, and G represent that a red, blue, or green ticket is drawn respectively, we can make a tree diagram for the experiment as shown in Figure 6.6.1.

If we write the sample space for this experiment, we get

$$S = \{RR, RB, RG, BR, BB, BG, GR, GB, GG\}$$

Suppose that we want to find the probability that we get a blue ticket on the first draw *and* a green ticket on the second draw. To do this we let

A represent the event that the blue ticket is drawn on the first draw

B represent the event that the green ticket is drawn on the second draw

The only associated outcome of this experiment is BG. So, using classical probability, we get

$$P(\text{blue first } \textbf{and} \text{ green second}) = P(A \cap B)$$

$$= \frac{\text{number of ways to draw a}}{\text{blue ticket and then a green ticket}} \Big/ \text{total number of possible outcomes}$$

$$= \frac{n(A \cap B)}{n(S)} = \frac{1}{9}$$

Also notice that, on any given draw, the probability that a blue ticket is drawn is $P(A) = \dfrac{1}{3}$ and the probability that a green ticket is drawn is $P(B) = \dfrac{1}{3}$. Notice

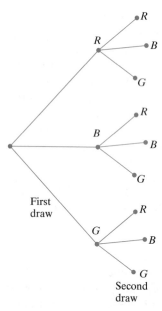

First draw

Second draw

Figure 6.6.1

that if we multiply these two probabilities together we get

$$P(A) \cdot P(B) = \frac{1}{3} \cdot \frac{1}{3} = \frac{1}{9} = P(A \cap B)$$

It is no coincidence that these probabilities are the same. This illustrates the **Multiplication Rule for Independent Events**.

▩ **Multiplication Rule for Independent Events**

If two events A and B are independent, then the probability of both A and B occurring is given by

$$P(A \textbf{ and } B) = P(A \cap B) = P(A) \cdot P(B)$$

▶ **Note:** Make sure that you understand the difference between *independent events* and *mutually exclusive events*. Independent events pertain to how the occurrence of one event affects the occurrence of another event. Mutually exclusive events pertain to whether events can occur at the same time.

Example 2   **Applying the Multiplication Rule for Independent Events**

In a July 2000 Harris poll, it was found that 47% of the American public had independently read or heard something about the missile defense system that the Pentagon was developing. If two Americans are selected at random, determine the probability that both had heard about the missile defense system. (*Source:* www.harrisinteractive.com.)

Solution

Interactive Activity

Recompute Example 2, except here determine the probability that neither of the two randomly selected Americans had read or heard anything about the missile defense system.

Understand the Situation: Let A represent the event that the first randomly selected American had heard about the missile defense system and B represent the event that the second randomly selected American had heard about the missile defense system. The events A and B are independent since whether the first person has heard about the system does not affect whether the second person has heard about the system.

Since events A and B are independent, we can use the Multiplication Rule for Independent Events. This gives us

$$P(A \cap B) = P(A) \cdot P(B) = (0.47)(0.47) \approx 0.22$$

Interpretation: This means that if two Americans are selected at random there is about a 22% probability that both of them have read or heard something about the missile defense system. ▪

Conditional Probability

We have seen that events are independent if the occurrence of one event does not affect the occurrence of the other. Our next step is to determine the probability $P(A \cap B)$ if the events are dependent. If the events are dependent, we must compute a new type of probability called **conditional probability** denoted by $P(B \,|\, A)$. We read $P(B \,|\, A)$ as "the probability of B given A." This means that we compute the probability of B given that A has occurred.

To see how conditional probability works, let's consider a bag that contains 3 red, 4 white and 3 green marbles. In an experiment, a marble is drawn at random, its color recorded, and then a second marble is drawn **without replacing** the first. Let A represent the event that a white marble is drawn on the first draw and B represent the event that a green marble is drawn on the second draw. Suppose that we want to determine $P(B \mid A)$. This is the probability that we draw a green marble on the second draw, given that the first draw produced a white marble. If the first marble drawn is white, then there are 9 marbles left in the bag and 3 of those are green. So for the second draw this gives us

$$P(B \mid A) = \dfrac{\text{number of green marbles left in the bag after a white marble is drawn}}{\text{number of marbles in the bag after the first draw}} = \dfrac{3}{9} = \dfrac{1}{3}$$

Frequently, sample spaces are often much larger than the number of marbles that were in the bag for the experiment that we just considered. Hence, computing conditional probabilities using the method just outlined is not always practical. Fortunately, there is a formula that we can use to compute conditional probabilities.

> ### Conditional Probability
>
> If A and B are events with $P(A) \neq 0$, then the **conditional probability** that B occurs, given that A has occurred, is given by
>
> $$P(B \mid A) = \dfrac{P(A \cap B)}{P(A)}$$

We can think of the conditional probability $P(B \mid A)$ as the probability that event B occurs if the sample space is restricted to the outcomes associated with event A. This idea is illustrated visually in Figure 6.6.2.

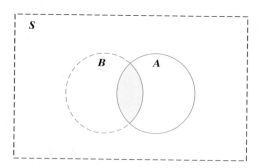

Figure 6.6.2

Example 3 Determining a Conditional Probability Using a Venn Diagram

In a random sample of 700 business people, 350 said that they had an airline flight delayed in the last month, 200 said that they had an airline flight canceled in the last month, and 75 said that they had both a flight delay and a cancellation in the last month. Let D represent the event that a businessperson had a flight delayed and C represent the event that the businessperson had a flight canceled.

(a) Draw a Venn diagram to represent the business people sampled with events D and C. Compute $P(D \cap C)$ and interpret.

(b) Determine $P(C \mid D)$ and interpret.

Solution

(a) The Venn diagram for the sample is shown in Figure 6.6.3. To compute $P(D \cap C)$, we see that there are 75 business people who had both a cancellation and a flight delay and there were 700 business people surveyed. This gives us

$$P(D \cap C) = \frac{\text{business people who had}}{\text{total number surveyed}} = \frac{75}{700} = \frac{3}{28} \approx 0.107$$

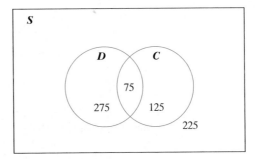

Figure 6.6.3

Interpret the Solution: This means that if one of the business people is selected at random, the probability that the person had **both** a cancellation **and** a flight delay in the last month is about 10.7%.

(b) First note that $P(D) = \dfrac{350}{700} = \dfrac{1}{2}$. Using the formula for computing conditional probability, we get

$$P(C \mid D) = \frac{P(D \cap C)}{P(D)} = \frac{\frac{75}{700}}{\frac{350}{700}} = \frac{75}{350} = \frac{3}{14} \approx 0.214$$

Interpret the Solution: This means that the probability that a randomly chosen businessperson had a flight canceled, given that he or she had a delayed flight, is about 21.4%.   ■

✓ **Checkpoint 2**

Now work Exercise 37.

We need to make some observations based on Example 3. First note that

$$P(D \mid C) = \frac{P(C \cap D)}{P(C)} = \frac{\frac{75}{700}}{\frac{200}{700}} = \frac{75}{200} = \frac{3}{8}$$

which is different from the result in part (b). This is evidence that conditional probability is not commutative, meaning that generally, for events A and B,

$$P(B \mid A) \neq P(A \mid B)$$

Also note that, when we computed $P(C \mid D)$ in part (b) of Example 3, we got the fraction $\frac{75}{350}$. This implies that we can think of conditional probability as a way of restricting the sample space. For $P(C \mid D)$ in Example 3, we can say that we determined the probability that a businessperson had a canceled flight as we restrict the sample space just to those who had delayed flights. This is displayed graphically in Figure 6.6.4.

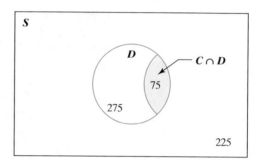

Figure 6.6.4 $P(C \mid D) = \frac{75}{350} = \frac{3}{14}$

General Multiplication Rule

Using the formula for conditional probability, we can derive a general formula for computing $P(A \cap B)$.

$$P(B \mid A) = \frac{P(A \cap B)}{P(A)} \qquad \text{Formula for conditional probability}$$

$$P(A) \cdot P(B \mid A) = P(A \cap B) \qquad \text{Multiply both sides by } P(A), P(A) \neq 0.$$

This gives us the **General Multiplication Rule**.

■ **General Multiplication Rule**

For any events A and B, the probability that A and B **both** occur is given by

$$P(A \cap B) = P(A) \cdot P(B \mid A)$$

Recall that the Multiplication Rule for Independent Events A and B is given by

$$P(A \cap B) = P(A) \cdot P(B)$$

By comparing the General Multiplication Rule to the Multiplication Rule for Independent Events A and B, we can see that if A and B are independent events then $P(B \mid A) = P(B)$. This means that the probability of B given A equals the probability of B. Another way of looking at it is that the fact that the event A has occurred does not affect the probability of event B. This gives us the following alternative for independent events.

■ **B Is Independent of A**

Let A and B be two events of a sample space S, where $P(B) > 0$. The event B is independent of A if and only if

$$P(B \mid A) = P(B)$$

3 Red
5 Blue

Example 4   Applying the General Multiplication Rule

A bag contains 3 red and 5 blue marbles. In an experiment, a marble is drawn at random, its color is recorded, and then a second marble is drawn at random and *without* replacing the first. Determine the probability that first a red and then a blue marble are drawn.

Solution

Understand the Situation: Let R_1 represent the event that first a red marble is drawn. Let B_2 represent the event that the second draw is a blue marble.

By the General Multiplication Rule, we have

$$P(R_1 \cap B_2) = \begin{array}{c}\text{probability that the}\\ \text{first marble is red}\end{array} \cdot \begin{array}{c}\text{probability that the second marble is}\\ \text{blue, given that the first was red}\end{array}$$

$$= P(R_1) \cdot P(B_2 \mid R_1) = \frac{3}{8} \cdot \frac{5}{7} = \frac{15}{56}$$

Interpret the Solution: So the probability that first a red and then a blue marble is drawn is $\frac{15}{56}$.   ∎

Interactive Activity

In Example 4, we computed $P(R_1 \cap B_2) = \frac{15}{56}$. Compute $P(B_1 \cap R_2)$ and compare to the solution of Example 4.

We can also use tabular data to determine conditional probabilities and probabilities for other compound events. When using data from tables, it is important to remember that we get a compound fraction when computing a conditional probability. This is illustrated in Example 5.

Example 5   Computing Conditional Probability Using a Table

In 1987, the number of men and women enrolled in 2- and 4-year colleges and universities, measured in thousands, is shown in Table 6.6.1.

Table 6.6.1

	2-Year (*A*)	4-Year (*B*)	Total
Male (M)	1522	3356	4,878
Female (F)	2127	3299	5,426
Total	3649	6655	10,304

SOURCE: U.S. Census Bureau, www.census.gov

A 2- or 4-year college student is selected at random. For the labeled events *A, B, M,* and *F,* determine the following probabilities and interpret.

(a) Determine the probability that a college student selected at random is female.

(b) Given that the randomly selected student is female, determine the probability that she attends a 4-year college.

Solution

(a) **Understand the Situation:** Here we are trying to determine $P(F)$.

If we follow across the row associated with event F, we get a total of 5426. So dividing by the total number of college enrollees (in thousands), we get

$$P(F) = \frac{5426}{10{,}304} \approx 0.527$$

Interpret the Solution: This means that if a 2- or 4-year college student is selected at random there is about a 52.7% chance that the enrollee is female.

(b) **Understand the Situation:** We are really being asked to determine the probability that the randomly chosen student attends a 4-year college, *given* that she is a female. Mathematically, this means we are trying to compute

$$P(B \mid F)$$

The formula for conditional probability gives us

$$P(B \mid F) = \frac{P(F \cap B)}{P(F)} = \frac{\dfrac{3299}{10{,}304}}{\dfrac{5426}{10{,}304}} = \frac{3299}{5426} \approx 0.608$$

Interpret the Solution: This means that if a college student is selected at random the probability that the student attends a 4-year college, given that he or she is a female, is about 60.8%. ∎

✓ **Checkpoint 3**

Now work Exercise 53.

SUMMARY

In this section we studied how to use the multiplication rule to determine $P(A \cap B)$. We saw that the formula we used depended on whether the events were independent.

- Two events A and B are **independent** if the occurrence (or nonoccurrence) of one event does not affect the other's chance of occurring.
- **Multiplication Rule for Independent Events** $P(A \cap B) = P(A) \cdot P(B)$

- **Conditional Probability** $P(B \mid A) = \dfrac{P(A \cap B)}{P(A)}$, $P(A) \neq 0$
- **General Multiplication Rule** $P(A \cap B) = P(A) \cdot P(B \mid A)$

SECTION 6.6 EXERCISES

In Exercises 1–10, determine if the events A and B are independent or dependent. Explain your reasoning.

⇐ 6.1 **1.** *A* represents the event that a flash food occurs at Windsor Park on Tuesday afternoon and *B* represents the event that the Tuesday evening company picnic at Windsor Park is postponed.

⇐ 6.1 **2.** *A* represents the event that Rebecca completes her political science course reading assignment every day and *B* represents the event that Rebecca passes her political science course.

⇐ 6.1 ✓ **3.** *A* represents the event that Kahrin's car battery is dead and *B* represents the event that Kahrin's oven light is burned out.

⇐ 6.1 **4.** *A* represents the event that the telephone in Rich's office does not work and *B* represents the event

that the vending machine in Rich's daughter's high school is empty.

⇐ 6.1 **5.** *A* represents the event that Clifford gets a raise and a promotion and *B* represents the event that Clifford purchases a new car.

⇐ 6.1 **6.** *A* represents the event that Lana is exposed to ultraviolet radiation for three decades and *B* represents the event that Lana develops skin cancer.

⇐ 6.1 **7.** A box contains 5 tickets numbered 1 to 5. In an experiment, a ticket is drawn at random, its number is recorded, and it is replaced. Then a second ticket is drawn randomly. *A* represents the event that the first ticket is a 3 and *B* represents the event that the second ticket is a 3.

⇐ 6.1 **8.** A bag contains 5 green and 4 purple marbles. In an experiment, a marble is drawn at random, its color is recorded, and it is replaced. Then a second marble is drawn randomly. *A* represents the event that the first marble is purple and *B* represents the event that the second marble is purple.

⇐ 6.1 **9.** *A* represents the event that Jim has received seven speeding tickets and *B* represents the event that Jim has high car insurance premiums.

⇐ 6.1 **10.** *A* represents the event that a person has brown eyes and *B* represents the event that the person's son has brown eyes.

11. A box contains 3 tickets numbered 1 to 3. In an experiment, a ticket is drawn at random, its number is recorded, and it is returned to the box. Then a second ticket is drawn randomly. Determine the probability that both tickets are labeled with the number 2.

12. A box contains 2 tickets numbered 1 and 2. In an experiment, a ticket is drawn at random, its number is recorded, and it is returned to the box. Then a second ticket is drawn randomly. Determine the probability that both tickets are labeled with the number 1.

13. A bag contains 5 green and 4 blue marbles. In an experiment, a marble is drawn at random, its color is recorded, and then it is returned to the bag. Then a second marble is drawn randomly. Determine the probability that two green marbles are drawn.

14. A bag contains 3 red and 8 yellow marbles. In an experiment, a marble is drawn at random, its color is recorded, and then it is returned to the bag. Then a second marble is drawn randomly. Determine the probability that two yellow marbles are drawn.

🌐 **15.** A June 2000 Gallup poll showed that 14% of American adults said independently that the decline of ethics, morals, and family values was the most important problem facing the country. A sociologist selects two Americans at random for an interview. Let *A* represent the event that the first person believes that the decline of ethics, morals, and family values is the most important problem facing the country and *B* represent the event that the second person believes that the decline of ethics, morals, and family values is the most important problem facing the country. (*Source:* www.gallup.com.)

(a) Are events *A* and *B* independent? Explain.
(b) Determine $P(A \cap B)$ and interpret.
(c) Determine $P(A^C)$ and $P(B^C)$ and interpret each.
(d) Determine $P(A^C \cap B^C)$ and interpret.

🌐 **16.** A February 1999 Gallup poll showed that 34% of adult Americans said independently of each other that AIDS was the most urgent health problem facing the United States. A medical researcher selects two Americans at random for a report. Let *A* represent the event that the first person believes that AIDS is the most urgent health problem facing the United States and *B* represent the

event that the second person believes that AIDS is the most urgent health problem facing the United States. (*Source:* www.gallup.com.)

(a) Are events *A* and *B* independent? Explain.
(b) Determine $P(A \cap B)$ and interpret.
(c) Determine $P(A^C)$ and $P(B^C)$ and interpret each.
(d) Determine $P(A^C \cap B^C)$ and interpret.

🌐 **17.** A March 2000 Gallup poll showed that 25% of adult Americans reported that their doctors diagnosed a condition or suggested a treatment that turned out to be wrong. If an HMO selects four adult Americans at random, determine the probability that all four had their doctors make such an error. (*Source:* www.gallup.com.)

🌐 **18.** An April 2000 Gallup poll showed that 46% of American adults said independently that the federal government should be more involved in public education than it currently is. If an educational political action committee selects four adult Americans at random, determine the probability that all four believe the federal government should be more involved in public education. (*Source:* www.gallup.com.)

🌐 **19.** In 1996 the U.S. Census Bureau reported that 35 million Americans owned personal computers. Among the 9.9 million Americans who used their computers more than 16 hours per week, 4.7 million said that they used their computer for personal use only, 2.1 million said that they used their computer for business use only, and 3.1 million said that they used their computer for both personal and business use. (*Source:* U.S. Census Bureau, www.census.gov.)

(a) A representative from a computer manufacturer selects at random a computer owner who uses his or her computer for more than 16 hours per week. Let *A* represent the event that the computer was for personal use and *B* represent the event that the computer was for business use. Draw a Venn diagram to represent the computer owners who use their computer for more than 16 hours per week with events *A* and *B*.
(b) Compute $P(A \cap B)$ and interpret.
(c) Compute $P(B \mid A)$ and interpret.

In Exercises 20–23, a Venn diagram is given. Determine $P(A \cap B)$ and $P(B \mid A)$.

20.

21.

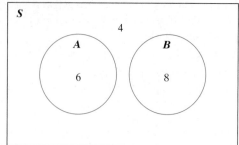

22.

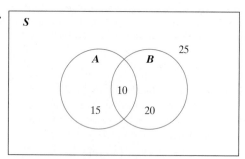

23.

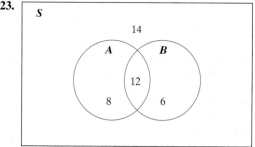

In Exercise 24–27, a probability Venn diagram is given. Determine $P(A \cap B)$ and $P(B \mid A)$.

24.

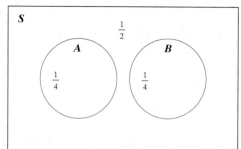

25.

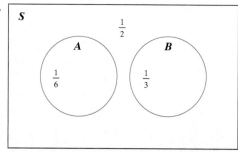

26.

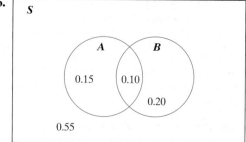

27.

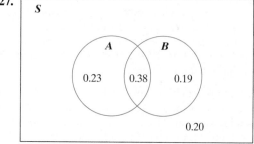

In Exercises 28–31, a table of data is given. Determine $P(A \cap C)$ and $P(C \mid A)$.

28.

	A	B	Total
C	10	20	30
D	15	5	20
Total	25	25	50

29.

	A	B	Total
C	40	60	100
D	55	45	100
Total	95	105	200

30.

	A	B	Total
C	26	33	59
D	18	48	66
Total	44	81	125

31.

	A	B	Total
C	87	52	139
D	61	50	111
Total	148	102	250

In Exercises 32–35, a table of probabilities is given. Determine $P(B \cap D)$ and $P(D \mid B)$.

32.

	A	B	Total
C	$\frac{1}{4}$	$\frac{1}{4}$	$\frac{1}{2}$
D	$\frac{1}{8}$	$\frac{3}{8}$	$\frac{1}{2}$
Total	$\frac{3}{8}$	$\frac{5}{8}$	1

33.

	A	*B*	**Total**
C	$\frac{1}{10}$	$\frac{3}{10}$	$\frac{2}{5}$
D	$\frac{1}{5}$	$\frac{2}{5}$	$\frac{3}{5}$
Total	$\frac{3}{10}$	$\frac{7}{10}$	1

34.

	A	*B*	**Total**
C	0.22	0.15	0.37
D	0.42	0.21	0.63
Total	0.64	0.36	1.00

35.

	A	*B*	**Total**
C	0.17	0.24	0.41
D	0.20	0.39	0.59
Total	0.37	0.63	1.00

36. A bag contains 5 yellow and 6 green marbles. In an experiment, a marble is drawn at random, its color is recorded, and a second marble is drawn randomly without replacing the first.

(a) Determine the probability that the first marble is green and the second marble is yellow.

(b) Determine the probability that both marbles drawn are green.

✓ **37.** A bag contains 6 red and 4 blue marbles. In an experiment, a marble is drawn at random, its color is recorded, and a second marble is drawn randomly without replacing the first.

(a) Determine the probability that the first marble is red and the second marble is blue.

(b) Determine the probability that both marbles drawn are blue.

38. A box contains 10 tickets with 4 of them labeled with a △ and 6 tickets labeled with a ◯. In an experiment, 3 tickets are drawn consecutively without replacement and at random.

(a) Determine the probability that all three tickets are labeled with a △.

(b) Determine the probability that the first ticket is labeled with a △ and the second and third tickets are labeled with a ◯.

39. A box contains 10 tickets with 7 tickets labeled with a ★ and 3 tickets labeled with a ◆. In an experiment, 3 tickets are drawn consecutively without replacement and at random.

(a) Determine the probability that all three tickets are labeled with a ◆.

(b) Determine the probability that the first ticket is labeled with a ◆ and the second and third tickets are labeled with a ★.

🌐 *For Exercises 40–47, consider the following: The number of doctorates conferred in the social sciences (anthropology, sociology, economics, and political science) and the physical sciences (astronomy, physics, and chemistry) during 1998 is shown in the following table. The data are classified as being a U.S. citizen or a foreign-born citizen.*

	Social Sciences (*A*)	Physical Sciences (*B*)	**Total**
U.S. citizen (C)	2353	2415	4768
Foreign-born citizen (D)	760	1111	1871
Total	3113	3526	6639

SOURCE: U.S. Census Bureau, www.census.gov

For the events labeled in the table as *A*, *B*, *C*, and *D* answer the following:

40. Determine $P(A \cap B)$ and interpret.

41. Determine $P(C \cap D)$ and interpret.

42. Determine the probability that the doctorate is in social sciences and is conferred on a U.S. citizen.

43. Determine the probability that the doctorate is in physical sciences and is conferred on a foreign-born citizen.

44. Determine $P(C \mid A)$ and interpret.

45. Determine $P(D \mid B)$ and interpret.

46. Determine the probability that the doctorate is in physical sciences, given that the recipient is a U.S. citizen.

47. Determine the probability that the doctorate is in social sciences, given that the recipient is a foreign-born citizen.

🌐 *For Exercises 48–55, consider the following: In 1996, the percentage of American adults who exercised for 30 minutes or more at least 5 times per week is shown in the table. The percentages are classified by gender.*

	Did Not Exercise (*A*)	Exercised Regularly (*B*)	**Total**
Men (C)	22.0%	25.6%	47.6%
Women (D)	25.6%	26.8%	52.4%
Total	47.6%	52.4%	100%

SOURCE: National Center for Chronic Disease Prevention, www.cdc.gov/nccdphp

For the events labeled in the table as A, B, C, and D answer the following.

48. Determine $P(C \cap D)$ and interpret.

49. Determine $P(A \cap B)$ and interpret.

50. Determine $P(D \cap B)$ and interpret.

51. Determine $P(A \cap C)$ and interpret.

52. Determine the probability that the person is a female given that the person exercises regularly.

✓ **53.** Determine the probability that the person is a male given that the person does not exercise regularly.

54. Determine $P(A \mid D)$ and interpret.

55. Determine $P(B \mid C)$ and interpret.

SECTION PROJECT

We can use the Multiplication Rule for Independent Events, along with the complement of an event, to determine the probability that **at least one** outcome occurs. For example, let's say that a student takes a six-question multiple-choice test, with each question having choices A, B, C, and D. The student guesses on each question. If we let A represent the event that the student gets *at least one* question correct, then the complement of the event, A^C, represents the event that the student gets *no* questions correct. Since the trials are independent and the probability of getting a question wrong is $\frac{3}{4}$, we get

$$P(A^C) = \frac{3}{4} \cdot \frac{3}{4} \cdot \frac{3}{4} \cdot \frac{3}{4} \cdot \frac{3}{4} \cdot \frac{3}{4} = \frac{3^6}{4^6} \approx 0.178$$

This means the probability of getting at least one question correct is

$$P(A) = 1 - P(A^C) = 1 - 0.178 = 0.822$$

Thus, the probability that the student gets at least one question correct is about 82.2%. Use this technique to answer the following:

In an August 1999 Harris poll, 24% of all Americans knew that Trent Lott was the U.S. Senate Majority Leader. If eight Americans are chosen at random, determine the probability that at least one knows that Lott was the U.S. Senate Majority Leader. (*Source:* www.harrisinteractive.com.)

Section 6.7 Bayes' Theorem and Its Applications

In the previous section we saw that conditional probability is an important tool for computing probabilities by the Multiplication Rule. Now we want to expand our discussion of conditional probability. First, we want to see if there is a relationship between the probabilities $P(B \mid A)$ and $P(A \mid B)$. This will lead us to discover a powerful formula called **Bayes' Theorem**, which can be used to compute certain conditional probabilities.

The Relationship between $P(B \mid A)$ and $P(A \mid B)$

In Section 6.6 we showed that, generally, $P(B \mid A) \neq P(A \mid B)$. To see how the probabilities $P(B \mid A)$ and $P(A \mid B)$ are related, however, we must first understand the difference between these two expressions. This difference is illustrated in Example 1.

Example 1 Computing Conditional Probabilities

In a January 1999 Harris poll, a random sample of American adults were asked the question "By the year 2020, do you think the quality of life in your community is more likely to get better or worse?" The percentages of the results, separated by gender, are shown in Table 6.7.1.

Table 6.7.1

	Male (A)	Female (B)	Total
Better (C)	31%	30%	61%
Worse (D)	19%	20%	39%
Total	50%	50%	100%

SOURCE: www.harrisinteractive.com

Let's select an American adult at random. For the labeled events A, B, C, and D, determine the following probabilities and interpret.

(a) $P(B \mid C)$ **(b)** $P(C \mid B)$

Solution

(a) **Understand the Situation:** We are being asked to determine the probability that we select a female (event B), given that the response to the question is "better" (event C).

Using the definition of conditional probability, we need to compute $P(B \mid C) = \dfrac{P(C \cap B)}{P(C)}$. Since the values in the table are in percentages, we can write them in decimal form to get

$$P(B \mid C) = \frac{P(C \cap B)}{P(C)} = \frac{0.30}{0.61} \approx 0.492$$

Interpret the Solution: This means that if we select an American adult at random the probability is about 49.2% that the person is female, given that the adult's opinion is that the community's quality of life will get better.

(b) **Understand the Situation:** In this case, we are being asked to compute the probability that the response to the question is "better" (event C) given that the selected person is a female (event B).

Using the definition of conditional probability, we have

$$P(C \mid B) = \frac{P(B \cap C)}{P(B)} = \frac{0.30}{0.50} = 0.60$$

Interpret the Solution: This means that if we select an American adult at random the probability is 60% that the person thinks his or her community's quality of life will get better, given that the selected individual is female.   ∎

✓ **Checkpoint 1**   Now work Exercise 3.

Example 1 illustrates that, if we are given a complete table of probabilities, computing a particular conditional probability is not difficult. But there are times when we do not have a complete table of values. To see how $P(B \mid A)$ and $P(A \mid B)$ are related, let's turn to our definition of conditional probability. First, we know that $P(B \mid A) = \dfrac{P(A \cap B)}{P(A)}$, so after multiplying both sides by $P(A)$ [note that $P(A) \neq 0$], we get

$$P(A) \cdot P(B \mid A) = P(A \cap B)$$

We also know that $P(A \mid B) = \dfrac{P(B \cap A)}{P(B)}$, which, after multiplying both sides by $P(B)$ [note that $P(B) \neq 0$], yields

$$P(B) \cdot P(A \mid B) = P(B \cap A)$$

Since we know that, for any events A and B, $P(B \cap A) = P(A \cap B)$, we can set the two equations equal to each other to get

$$P(A) \cdot P(B \mid A) = P(B) \cdot P(A \mid B)$$

If we divide both sides of the equation by $P(A)$, we get

$$P(B \mid A) = \frac{P(B) \cdot P(A \mid B)}{P(A)}$$

We now have the following relationship between $P(A \mid B)$ and $P(B \mid A)$.

■ **Relationship between $P(A \mid B)$ and $P(B \mid A)$**

The conditional probabilities $P(A \mid B)$ and $P(B \mid A)$ are related by

$$P(B \mid A) = \frac{P(B) \cdot P(A \mid B)}{P(A)}$$

where $P(A) \neq 0$.

The relationship between events involving conditional probabilities are often expressed in the form of tree diagrams like the one shown in Figure 6.7.1. Notice that the tree diagram has two **branches**. There is one branch for B and another for its complement B^C. We will soon see that we use B^C to guarantee that all elements of the sample space are accounted for.

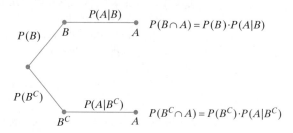

Figure 6.7.1

Let's say that we want to compute $P(A)$. To do this, we must follow the first branch to compute $P(B \cap A)$ **and** we must follow the second branch to compute $P(B^C \cap A)$. From the General Multiplication Rule, the probability associated with the top branch, $P(B \cap A)$, is $P(B) \cdot P(A \mid B)$, and the probability associated with the second branch, $P(B^C \cap A)$, is $P(B^C) \cdot P(A \mid B^C)$. Since B and B^C, along with their associated branches, are mutually exclusive, we can use the Addition Rule for Mutually Exclusive Events to get

$$P(A) = P((B \cap A) \cup (B^C \cap A))$$
$$= P(B \cap A) + P(B^C \cap A)$$
$$= P(B) \cdot P(A \mid B) + P(B^C) \cdot P(A \mid B^C)$$

Informally, we can say that **$P(A)$ is equal to the sum of the branches**. Let's see how these formulas can be used in an application.

Example 2 Determining $P(B \mid A)$ If $P(A \mid B)$ Is Given

In a March 2000 survey, a Harris poll queried 1149 adults; 43% of the adults had disabilities and the remainder did not have disabilities. The survey found that 30% of the respondents who had disabilities accessed the Internet from their home and 48% of the respondents without disabilities accessed the Internet from their home. Let B represent the event that a respondent had disabilities and A represent the event that the respondent accessed the Internet from home. (*Source:* www.harrisinteractive.com.)

(a) Construct a tree diagram for the events A, B, and B^C.

(b) Determine $P(A)$ and interpret.

(c) Determine $P(B \mid A)$ and interpret.

Solution

$P(A|B) = 0.30$
$P(B) = 0.43$ B A
$P(B^C) = 0.57$ $P(A|B^C) = 0.48$
B^C A

Figure 6.7.2

(a) First notice that we are given that $P(B) = 0.43$ and $P(B^C) = 1 - 0.43 = 0.57$, where B^C represents the event that a respondent does not have disabilities. We are also given that $P(A \mid B) = 0.30$ and $P(A \mid B^C) = 0.48$. These events and probabilities are shown in the tree diagram in Figure 6.7.2.

(b) To calculate $P(A)$, we must compute the sum of the branches to get

$$P(A) = P(B) \cdot P(A \mid B) + P(B^C) \cdot P(A \mid B^C)$$
$$= (0.43)(0.30) + (0.57)(0.48) \approx 0.403$$

Interpret the Solution: This means that if a respondent is selected at random there is about a 40.3% probability that she or he accesses the Internet from home.

(c) To determine $P(B \mid A)$, we need to use the relationship between $P(A \mid B)$ and $P(B \mid A)$ formula.

$$P(B \mid A) = \frac{\text{first branch}}{\text{sum of branches}} = \frac{P(B) \cdot P(A \mid B)}{P(A)}$$
$$= \frac{P(B) \cdot P(A \mid B)}{P(B) \cdot P(A \mid B) + P(B^C) \cdot P(A \mid B^C)}$$
$$= \frac{(0.43)(0.30)}{(0.43)(0.30) + (0.57)(0.48)} \approx 0.320$$

Interpret the Solution: This means that if a respondent is selected at random the probability that she or he has a disability, given that they access the Internet from home, is about 32%. ∎

Interactive Activity

Use the method shown in part (c) of Example 2 to compute and interpret $P(B^C|A)$.

✓ **Checkpoint 2**

Now work Exercise 9.

The formula that we have derived and used in Example 2 is a version of a theorem called **Bayes' Theorem**, which is named after Reverend Thomas Bayes who, during the 18th century, studied calculations much like those computed in Example 2. Let's now summarize the formula that we have seen thus far.

▨ **Bayes' Theorem for Two Branches**

For events A and B with complement B^C,

• The tree diagram for the events A, B and B^C is as follows:

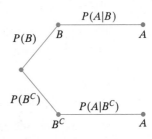

- $P(A) = $ sum of branches $= P(B) \cdot P(A \mid B) + P(B^C) \cdot P(A \mid B^C)$
- $P(B \mid A) = \dfrac{\text{first branch}}{\text{sum of branches}} = \dfrac{P(B) \cdot P(A \mid B)}{P(B) \cdot P(A \mid B) + P(B^C) \cdot P(A \mid B^C)}$
- $P(B^C \mid A) = \dfrac{\text{second branch}}{\text{sum of branches}} = \dfrac{P(B^C) \cdot P(A \mid B^C)}{P(B) \cdot P(A \mid B) + P(B^C) \cdot P(A \mid B^C)}$

In Example 3, we apply Bayes' Theorem in a factory setting.

Example 3   Computing Probabilities Using Bayes' Theorem

Suppose that $\dfrac{7}{10}$ of the computer keyboards made by the TouchRight Company come from factory A, while factory B produces the rest. Quality control records show that $\dfrac{1}{5}$ of the keyboards from factory A do not pass inspection and $\dfrac{1}{3}$ of the keyboards from factory B do not pass inspection. Determine the probability that if a randomly selected keyboard does not pass inspection, the keyboard came from factory B.

Solution

Understand the Situation: To apply Bayes' Theorem, we need to first assign events. We let

    A represent the event that the keyboard came from factory A

    A^C represent the event that the keyboard came from factory B

    D represent the event that the keyboard did not pass inspection

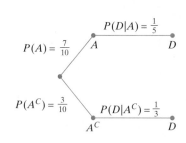

Figure 6.7.3

We have been given the probabilities $P(A) = \dfrac{7}{10}$, $P(A^C) = 1 - \dfrac{7}{10} = \dfrac{3}{10}$, $P(D \mid A) = \dfrac{1}{5}$, and $P(D \mid A^C) = \dfrac{1}{3}$. These probabilities are shown in the tree diagram in Figure 6.7.3.

To determine the probability that if a randomly selected keyboard does not pass inspection the keyboard came from factory B, we need to compute $P(A^C \mid D)$. This gives us

$$P(A^C \mid D) = \frac{\text{second branch}}{\text{sum of branches}} = \frac{P(A^C) \cdot P(D \mid A^C)}{P(A) \cdot P(D \mid A) + P(A^C) \cdot P(D \mid A^C)}$$

$$= \frac{\dfrac{3}{10} \cdot \dfrac{1}{3}}{\dfrac{7}{10} \cdot \dfrac{1}{5} + \dfrac{3}{10} \cdot \dfrac{1}{3}} = \frac{\dfrac{1}{10}}{\dfrac{7}{50} + \dfrac{1}{10}} = \frac{\dfrac{1}{10}}{\dfrac{6}{25}} = \frac{5}{12}$$

Interpret the Solution: This means that there is a $\dfrac{5}{12}$ probability that a randomly selected keyboard that did not pass inspection was manufactured at factory B.

The Generalized Bayes' Theorem

So far we have seen how to apply the Bayes' Theorem when we have mutually exclusive events B and B^C, which results in two branches in a tree diagram. In some applications, however, there can be many branches in the tree diagram. In order to use Bayes' Theorem in these cases, we need to have all the events take up the whole sample space. We call events of this type **collectively exhaustive**.

■ Collectively Exhaustive Events

Events $B_1, B_2, B_3, \ldots, B_k$ are **collectively exhaustive** if $B_1 \cup B_2 \cup B_3 \cup \cdots \cup B_k = S$, where S is the sample space.

▶ **Note:** 1. In terms of their probabilities, we can see that the events $B_1, B_2, B_3, \ldots, B_k$ are collectively exhaustive if $P(B_1 \cup B_2 \cup B_3 \cup \cdots \cup B_k) = 1$ and they are mutually exclusive.
2. The events $B_1, B_2, B_3, \ldots, B_k$ are mutually exclusive if any two of them are mutually exclusive.

We can now state the Generalized Bayes' Theorem. This generalization of the theorem was developed by the mathematician Pierre LaPlace in 1814, several years after Bayes' death.

■ Generalized Bayes' Theorem

For mutually exclusive and collectively exhaustive events $B_1, B_2, B_3, \ldots, B_k$ and for any event A with $i = 1, 2, 3, \ldots, k$, we have the following:

• The tree diagram for the events is as follows:

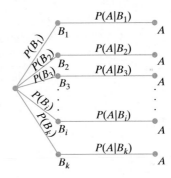

• $P(A) = P(B_1) \cdot P(A \mid B_1) + P(B_2) \cdot P(A \mid B_2) + \cdots + P(B_k) \cdot P(A \mid B_k)$
  $=$ sum of branches
• For any value i where $1 \le i \le k$, we have

$$P(B_i \mid A) = \frac{P(B_i) \cdot P(A \mid B_i)}{P(B_1) \cdot P(A \mid B_1) + P(B_2) \cdot P(A \mid B_2) + \cdots + P(B_k) \cdot P(A \mid B_k)}$$

$$= \frac{i\text{th branch}}{\text{sum of branches}}$$

Now let's see how we can apply the Generalized Bayes' Theorem.

Example 4   Applying the Generalized Bayes' Theorem

In a September 1999 Harris poll, a random sample of 1009 Americans were asked about the aftermath of the fire at the Branch Dividian compound in

Waco, Texas. In the sample, 43% of the respondents identified themselves as Republicans, 36% as Democrats, and 21% as Independents. They were asked, "Do you think Attorney General Janet Reno should resign because of her handling of what happened in Waco?" The results showed that 33% of the Republicans, 10% of the Democrats, and 25% of the Independents said that she should resign. Let B_1, B_2, and B_3 represent the event that a randomly selected respondent was a Republican, Democrat, or Independent, respectively, and let A represent the event that the respondent said that Reno should resign. (*Source:* www.harrisinteractive.com.)

(a) Construct a tree diagram of the events.

(b) Compute and interpret $P(A)$.

(c) Compute and interpret $P(B_3 \mid A)$.

Solution

(a) **Understand the Situation:** Let's begin by organizing our information in tabular form as shown in Table 6.7.2. We can use the table to construct the tree diagram in Figure 6.7.4, where the percentages have been converted to decimal form.

Table 6.7.2

Party Affiliation	Percentage of Sample	Percentage Favoring Resignation
Republican	43%	33%
Democrat	36%	10%
Independent	21%	25%

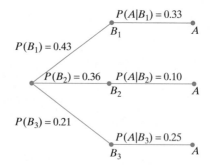

Figure 6.7.4

(b) To determine $P(A)$, we use the Generalized Bayes' Theorem and compute the sum of the three branches to get

$$P(A) = (B_1) \cdot P(A \mid B_1) + P(B_2) \cdot P(A \mid B_2) + P(B_3) \cdot P(A \mid B_3)$$
$$= (0.43)(0.33) + (0.36)(0.10) + (0.21)(0.25) \approx 0.230$$

Interpret the Solution: This means that if we select a respondent at random there is about a 23% probability that she or he feels that Attorney General Reno should resign.

(c) To determine $P(B_3 \mid A)$ we need to compute

$$P(B_3 \mid A) = \frac{\text{third branch}}{\text{sum of branches}}$$

$$= \frac{P(B_3) \cdot P(A \mid B_3)}{P(B_1) \cdot P(A \mid B_1) + P(B_2) \cdot P(A \mid B_2) + P(B_3) \cdot P(A \mid B_3)}$$

$$= \frac{(0.21)(0.25)}{(0.43)(0.33) + (0.36)(0.10) + (0.21)(0.25)} \approx 0.228$$

Interpret the Solution: This means that if we select a respondent at random the probability that the respondent is an Independent, given that she or he favored Reno's resignation, is about 23%.   ∎

SUMMARY

In this section we learned about the association between the probabilities $P(B \mid A)$ and $P(A \mid B)$. We then generalized this relationship with a theorem called Bayes' Theorem.

- **Bayes' Theorem for Two Branches**   For events A, B, and B^C:

  1. $P(A) = $ sum of branches
     $= P(B) \cdot P(A \mid B) + P(B^C) \cdot P(A \mid B^C)$

  2. $P(B \mid A) = \dfrac{\text{first branch}}{\text{sum of branches}}$

     $= \dfrac{P(B) \cdot P(A \mid B)}{P(B) \cdot P(A \mid B) + P(B^C) \cdot P(A \mid B^C)}$

  3. $P(B^C \mid A) = \dfrac{\text{second branch}}{\text{sum of branches}}$

     $= \dfrac{P(B^C) \cdot P(A \mid B^C)}{P(B) \cdot P(A \mid B) + P(B^C) \cdot P(A \mid B^C)}$

- The events $B_1, B_2, B_3, \ldots, B_k$ are mutually exclusive if any two of them are mutually exclusive.

- Events $B_1, B_2, B_3, \ldots, B_k$ are **collectively exhaustive** if $B_1 \cup B_2 \cup B_3 \cup \cdots \cup B_k = S$

- **Generalized Bayes' Theorem**   For mutually exclusive and collectively exhaustive events $B_1, B_2, B_3, \ldots, B_k$ and for any event A:

  1. $P(A) = P(B_1) \cdot P(A \mid B_1) + P(B_2) \cdot P(A \mid B_2) +$
     $\cdots + P(B_k) \cdot P(A \mid B_k)$

  2. For any value i, where $1 \le i \le k$, we have

     $P(B_i \mid A)$

     $= \dfrac{P(B_i) \cdot P(A \mid B_i)}{\begin{array}{c} P(B_1) \cdot P(A \mid B_1) + P(B_2) \cdot P(A \mid B_2) + \\ \cdots + P(B_k) \cdot P(A \mid B_k) \end{array}}$

     $= \dfrac{i\text{th branch}}{\text{sum of branches}}$

SECTION 6.7 EXERCISES

In Exercises 1–4, a table of data is given. Compute the values of $P(D \mid B)$ and $P(B \mid D)$.

1.

	A	B	Total
C	10	20	30
D	15	5	20
Total	25	25	50

2.

	A	B	Total
C	40	60	100
D	55	45	100
Total	95	105	200

✓ **3.**

	A	B	Total
C	26	33	59
D	18	48	66
Total	44	81	125

4.

	A	B	Total
C	87	52	139
D	61	50	111
Total	148	102	250

In Exercises 5–8, a table of probabilities is given. Compute the values of P(C | A) and P(A | C).

5.

	A	B	Total
C	$\frac{1}{4}$	$\frac{1}{4}$	$\frac{1}{2}$
D	$\frac{1}{8}$	$\frac{3}{8}$	$\frac{1}{2}$
Total	$\frac{3}{8}$	$\frac{5}{8}$	1

6.

	A	B	Total
C	$\frac{1}{10}$	$\frac{3}{10}$	$\frac{2}{5}$
D	$\frac{1}{5}$	$\frac{2}{5}$	$\frac{3}{5}$
Total	$\frac{3}{10}$	$\frac{7}{10}$	1

7.

	A	B	Total
C	0.22	0.15	0.37
D	0.42	0.21	0.63
Total	0.64	0.36	1.00

8.

	A	B	Total
C	0.17	0.24	0.41
D	0.20	0.39	0.59
Total	0.37	0.63	1.00

9. The number of doctorates conferred in the social sciences (anthropology, sociology, economics, and political science) and the physical sciences (astronomy, physics, and chemistry) during 1998 is shown in the following table. The data are classified as being a U.S. citizen or a foreign-born citizen.

	Social Sciences (A)	Physical Sciences (B)	Total
U.S. citizen (C)	2353	2415	4768
Foreign-born citizen (D)	760	1111	1871
Total	3113	3526	6639

SOURCE: U.S. Census Bureau, www.census.gov

For the events labeled in the table as A, B, C, and D, compute and interpret the following probabilities:

(a) $P(D | B)$

(b) $P(B | D)$

10. In 1996, the percentage of American adults who exercised for 30 minutes or more at least 5 times per week is shown in the table. The percentages are classified by gender.

	Did Not Exercise (A)	Exercised Regularly (B)	Total
Men (C)	22.0%	25.6%	47.6%
Women (D)	25.6%	26.8%	52.4%
Total	47.6%	52.4%	100%

SOURCE: National Center for Chronic Disease Prevention, www.cdc.gov/nccdphp

For the events labeled in the table as A, B, C, and D, compute and interpret the following probabilities:

(a) $P(C | A)$

(b) $P(A | C)$

In Exercises 11–16, a tree diagram is given. Determine $P(A)$, $P(B | A)$, and $P(B^C | A)$.

11.

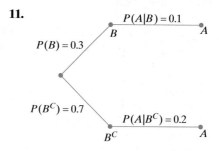

12.

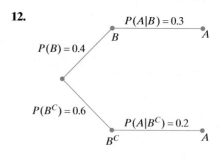

13.

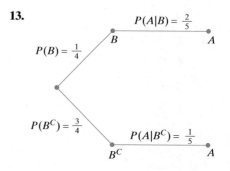

14.

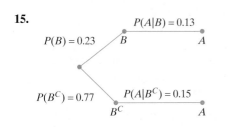

15.

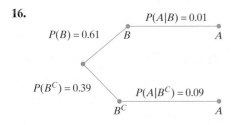

16.

$P(B) = 0.61$ $P(A|B) = 0.01$
B A

$P(B^C) = 0.39$ $P(A|B^C) = 0.09$
B^C A

17. The Garforth Company runs a day and a night shift at its factory. Of the 500 workers, 300 work the day shift and the remainder work the night shift. When asked whether they would like to have a paid holiday for St. Patrick's Day or for Columbus Day next year, 150 of the workers on the day shift favored St. Patrick's Day and 150 of the workers on the night shift favored St. Patrick's Day. A union leader selects a worker at random. Let B represent the event that the worker is on the day shift, B^C represent the event that the worker is on the night shift, and A represent the event that the worker favored a paid holiday for St. Patrick's Day.

(a) Determine $P(A)$ and interpret.

(b) Determine $P(B \mid A)$ and interpret.

18. The Clanford Mk V device is made at two factories. The northern plant manufactures 40% of the devices and the southern plant manufacturers the remainder of the devices. Records show that 2% of the devices from the northern plant are defective and 6% of the devices from the southern plant are defective. A quality control engineer selects a Clanford Mk V device at random. Let B present the event that the device came from the northern plant, B^C represent the event that the device came from the southern plant, and A represent the event that the device is defective.

(a) Determine $P(A)$ and interpret.

(b) Determine $P(B \mid A)$ and interpret.

19. At Ketterson College, 66% of the student body is underclass persons and the remainder is upperclass persons. In a survey, 33% of the underclass persons favored switching from quarters to semesters and 12% of the upperclass persons favored switching from quarters to semesters. A prospective student visiting the campus selects a Ketterson College student at random. Let B represent the event that the selected student is an underclass person, B^C represent the event that the selected student is an upperclass person, and A represent the event that the selected student favored switching from quarters to semesters.

(a) Determine $P(A)$ and interpret.

(b) Determine $P(B \mid A)$ and interpret.

20. In a May 2000 Gallup poll, 54% of the respondents said that they had money invested in the stock market and the remaining respondents did not. They were asked the question, "If you had $1000 to spend, do you think investing it in the stock market would be a good idea." Of the stockholders, 81% said that it was a good idea and 45% of the nonstockholders said that it was a good idea. A national money magazine selects a respondent at random. Let B represent the event that the respondent was a stockholder and let A represent the event that he or she thought that investing $1000 in the stock market would be a good idea. (*Source:* www.gallup.com.)

(a) Determine $P(A)$ and interpret.

(b) Determine $P(B \mid A)$ and interpret.

(c) Determine $P(B^C \mid A)$ and interpret.

21. In a March 2000 Gallup poll, 26% of the respondents called themselves fans of professional golf and the remaining respondents did not. When asked who they felt was the greatest active golfer, 81% of the golf fans replied that Tiger Woods was the greatest, while 64% of those who said they were not a fan said that Tiger Woods was the greatest. A national golf magazine selects a respondent at random. Let B represent the event that the respondent was a golf fan and let A represent the event that he or she felt that Tiger Woods was the greatest active golfer. (*Source:* www.gallup.com.)

(a) Determine $P(A)$ and interpret.

(b) Determine the probability that the respondent is a golf fan given that he or she felt that Tiger Woods is the greatest active golfer.

(c) Determine $P(B^C \mid A)$ and interpret.

22. In a November 1999 Gallup poll, 45% of the respondents said that they were baseball fans and the remaining respondents said that they were not. Former Major League player Pete Rose is ineligible for baseball's Hall of Fame due to charges that he bet on Major League baseball games. When asked if Rose should be eligible for the Hall of Fame, 62% of the baseball fans said that he should be eligible, while 57% of those who were not baseball fans said that he should be eligible. Select a respondent at random and let B represent the event that the respondent was a baseball fan and A represent the event that the respondent said that Rose should be eligible for the Hall of Fame. (*Source:* www.gallup.com.)

(a) Determine $P(A)$ and interpret.

(b) Determine the probability that the respondent is a baseball fan given that the respondent felt that Rose should be eligible for the Hall of Fame.

(c) Determine $P(B^C \mid A)$ and interpret.

23. In a November 1999 Gallup poll, 45% of the respondents said that they were baseball fans and the remaining respondents said that they were not. Former Major League player Shoeless Joe Jackson is ineligible for baseball's Hall of Fame due to charges that he took money from gamblers in exchange for fixing the 1919 World Series. When asked if Jackson should be eligible for the Hall of Fame, 44% of the baseball fans said that he should be eligible, while 33% of those who were not baseball fans said that he should be eligible. A national sports magazine selects a respondent at random. Let B represent the event that the respondent was a baseball fan and let A represent the event that the respondent said that Jackson should be eligible for the Hall of Fame. (*Source:* www.gallup.com.)

(a) Determine $P(A)$ and interpret.

(b) Determine the probability that the respondent is a baseball fan, given that the respondent felt that Jackson should be eligible for the Hall of Fame.

(c) Determine $P(B^C \mid A)$ and interpret.

24. In a July 1999 Gallup poll, half of the respondents called themselves a regular viewer of the Cable News Network (CNN) and the other half said that they were not regular viewers. When asked if they were interested in the results of polls about political campaigns and elections, 50% of the CNN regular viewers said that they were very interested, while 39% of those who were not regular CNN viewers said that they were very interested. A reporter selects a respondent at random. Let B represent the event that the respondent was a regular CNN viewer and let A represent the event that the respondent said that she or he was very interested in the results of polls about political campaigns and elections. (*Source:* www.gallup.com.)

(a) Determine $P(A)$ and interpret.

(b) Determine the probability that the respondent is a regular CNN viewer, given that the respondent said that she or he is very interested in the results of polls about political campaigns and elections.

(c) Determine $P(B^C \mid A)$ and interpret.

25. In a February 2000 poll, a group of adults were asked if they ate organic foods on a regular basis and if they were willing to spend higher prices for organic foods. The results are summarized in the following table:

Eat Organic Foods Regularly	Percent	Willing to Pay Higher Prices for Organic Food
Yes	46	65%
No	54	34%

SOURCE: www.abcnews.go.com

Select a respondent at random and let B represent the event that the respondent is a regular eater of organic foods and A represent the event that the respondent is willing to pay higher process for organic foods.

(a) Determine $P(A)$ and interpret.

(b) Determine the probability that the respondent is a regular eater of organic foods, given that the respondent is willing to pay higher prices for organic foods.

(c) Determine $P(B^C \mid A)$ and interpret.

26. In a September 2000 poll, a group of adults were asked if they were gun owners and if the National Rifle Association (NRA) had too much influence on the issue of gun control. The results are summarized in the following table:

Gun Owner	Percent	Too Much NRA Influence
Yes	29	31%
No	71	55%

SOURCE: www.abcnews.go.com

Select a respondent at random and let B represent the event that the respondent was a gun owner and A represent the event that the respondent believed that the NRA had too much influence on the issue of gun control.

(a) Determine $P(A)$ and interpret.

(b) Determine the probability that the respondent is a gun owner, given that the respondent believes that the NRA has too much influence on the issue of gun control.

(c) Determine $P(B^C \mid A)$ and interpret.

27. In a fall 1998 survey, a sample of adults was asked if they were either parents or teachers and if they felt that kids whose parents are not involved in school sometimes get shortchanged. The results are summarized in the following table:

Group	Percent	Kids Get Shortchanged
Parents	55	72%
Teachers	45	48%

SOURCE: www.publicagenda.org

Select a respondent at random and let B_1 represent the event that the respondent is a parent, B_2 represent the event that the respondent is a teacher, and A represent the event that the respondent felt that kids whose parents are not involved in school sometimes get shortchanged.

(a) Determine $P(A)$ and interpret.

(b) Determine the probability that the respondent is a parent, given that the respondent felt that kids whose parents are not involved in school sometimes get shortchanged.

(c) Determine $P(B_2 \mid A)$ and interpret.

In Exercises 28–33, a tree diagram is given. Determine if the events are collectively exhaustive.

28.

$$P(B_1) = \tfrac{1}{3}$$

$$P(A|B_1) = \tfrac{1}{8}$$

$$P(B_2) = \tfrac{2}{3}$$

$$P(A|B_2) = \tfrac{1}{4}$$

29.

$$P(B_1) = 0.71 \qquad P(A|B_1) = 0.05$$

$$P(B_2) = 0.29 \qquad P(A|B_2) = 0.20$$

30.

$$P(B_1) = \tfrac{5}{8} \qquad P(A|B_1) = \tfrac{3}{8}$$

$$P(B_2) = \tfrac{1}{4} \qquad P(A|B_2) = \tfrac{1}{8}$$

31.

$$P(B_1) = 0.25 \qquad P(A|B_1) = 0.12$$

$$P(B_2) = 0.70 \qquad P(A|B_2) = 0.19$$

32.

$$P(B_1) = 0.35 \qquad P(A|B_1) = 0.15$$

$$P(B_2) = 0.30 \qquad P(A|B_2) = 0.20$$

$$P(B_3) = 0.35 \qquad P(A|B_3) = 0.10$$

33.

$$P(B_1) = 0.25 \qquad P(A|B_1) = 0.30$$

$$P(B_2) = 0.40 \qquad P(A|B_2) = 0.20$$

$$P(B_3) = 0.35 \qquad P(A|B_3) = 0.15$$

In Exercises 34–39, a tree diagram is given. Determine $P(A)$, $P(B_1 \mid A)$, and $P(B_3 \mid A)$.

34.

$$P(B_1) = 0.3 \qquad P(A|B_1) = 0.2$$

$$P(B_2) = 0.2 \qquad P(A|B_2) = 0.1$$

$$P(B_3) = 0.5 \qquad P(A|B_3) = 0.3$$

35.

$$P(B_1) = 0.2 \qquad P(A|B_1) = 0.4$$

$$P(B_2) = 0.4 \qquad P(A|B_2) = 0.3$$

$$P(B_3) = 0.4 \qquad P(A|B_3) = 0.1$$

36.

$$P(B_1) = \tfrac{1}{4} \qquad P(A|B_1) = \tfrac{1}{8}$$

$$P(B_2) = \tfrac{1}{2} \qquad P(A|B_2) = \tfrac{1}{4}$$

$$P(B_3) = \tfrac{1}{4} \qquad P(A|B_3) = \tfrac{3}{8}$$

37.

$$P(B_1) = \tfrac{1}{3} \qquad P(A|B_1) = \tfrac{1}{4}$$

$$P(B_2) = \tfrac{1}{6} \qquad P(A|B_2) = \tfrac{1}{8}$$

$$P(B_3) = \tfrac{1}{2} \qquad P(A|B_3) = \tfrac{1}{10}$$

38.

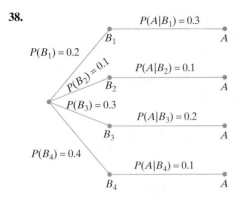

$P(B_1) = 0.2$, $P(A|B_1) = 0.3$, B_1, A

$P(B_2) = 0.1$, $P(A|B_2) = 0.1$, B_2, A

$P(B_3) = 0.3$, $P(A|B_3) = 0.2$, B_3, A

$P(B_4) = 0.4$, $P(A|B_4) = 0.1$, B_4, A

39.

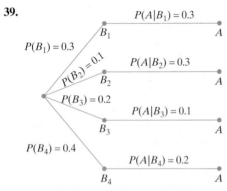

$P(B_1) = 0.3$, $P(A|B_1) = 0.3$, B_1, A

$P(B_2) = 0.1$, $P(A|B_2) = 0.3$, B_2, A

$P(B_3) = 0.2$, $P(A|B_3) = 0.1$, B_3, A

$P(B_4) = 0.4$, $P(A|B_4) = 0.2$, B_4, A

40. In Binfort County, the local election board keeps records of the number of registered voters from each party for school board elections. The voter turnout for the last school board election is shown in the table.

Party Affiliation	Percent of Electorate	Percent Turnout
Republican	35	20
Democrat	40	25
Independent	25	10

A newspaper reporter selects a registered voter at random. Let B_1, B_2, and B_3 represent the event that the voter is a Republican, Democrat, or Independent, respectively, and let A represent the event that the voter turned out to vote for the last school board election.

(a) Construct a tree diagram.

(b) Determine $P(A)$ and interpret.

(c) Determine the probability that the voter is a Republican, given that the voter turned out to vote for the last school board election.

(d) Determine $P(B_2 \mid A)$ and interpret.

41. The Juniper Company makes light fixtures at three factories. From past records they know the proportion of output from each factory and the percentage of defects that come from each factory. This information is shown in the table.

Factory	Percent Output	Percent Defects
I	40	1
II	25	2
III	35	4

A quality control engineer selects a Juniper Company light fixture at random. Let B_1, B_2, and B_3 represent the events that the light fixture comes from factory I, factory II, or factory III, respectively, and let A represent the event that the light fixture is defective.

(a) Construct a tree diagram.

(b) Determine $P(A)$ and interpret.

(c) Determine the probability that the light fixture came from factory II, given that the light fixture is defective.

(d) Determine $P(B_3 \mid A)$ and interpret.

42. In the United States in 1998, the income of parents and the percentage of children under 18 living with their father only is shown in the table.

Income	Percent of Population	Percent Living with Father Only
Under $25,000 ($B_1$)	17	6
$25,000 to $40,000 ($B_2$)	30	7
Over $40,000 ($B_3$)	53	3

SOURCE: U.S. Census Bureau, www.census.gov

A researcher selects an American child under 18 at random. Let B_1, B_2, and B_3 represent the events labeled in the table and A represent the event that a child is living alone with his or her father.

(a) Construct a tree diagram.

(b) Determine $P(A)$ and interpret.

(c) Determine the probability that parental income is under $25,000, given that the child is living alone with his or her father.

(d) Determine $P(B_3 \mid A)$ and interpret.

43. In the United States in 1995, the percentage of women from 15 to 44 years of age and the percentage of women who had never been married or cohabited is shown in the table.

Age	Percent of Study	Percent Never Married or Cohabited
15 to 24 (B_1)	30	67
25 to 34 (B_2)	34	16
35 to 44 (B_3)	36	6

SOURCE: U.S. Census Bureau, www.census.gov

A medical researcher selects an American woman aged 15 to 44 at random. Let B_1, B_2, and B_3 represent the events labeled in the table and A represent the event that a woman had never been married or cohabited.

(a) Construct a tree diagram.

(b) Determine $P(A)$ and interpret.

(c) Determine the probability the woman is from 25 to 34 years of age, given that the woman has never been married or cohabited.

(d) Determine $P(B_3 \mid A)$ and interpret.

 ## SECTION PROJECT

Some applications of the Bayes Theorem can be somewhat daunting. For example, the following table shows the percent of remarriages of women in the United States for several age groups.

Age	Percent of Population	Percent remarried
Under 20 (B_1)	21.1	1.7
20 to 24 (B_2)	37.1	15.3
25 to 29 (B_3)	18.7	26.7
30 to 34 (B_4)	9.3	20.6
35 to 44 (B_5)	7.8	20.8
45 to 64 (B_6)	5	14.3
65 and over (B_7)	1	2.9

SOURCE: U.S. Census Bureau, www.census.gov

A medical researcher selects an American woman at random. Let B_1, B_2, B_3, B_4, B_5, B_6, and B_7 represent the events labeled in the table, and let A represent the event that the woman has been remarried. Use the BAYES calculator program in Appendix C in answering parts (b) through (e).

(a) Construct a tree diagram.

(b) Determine $P(A)$ and interpret.

(c) Determine $P(B_4 \mid A)$ and interpret.

(d) Determine $P(B_5 \mid A)$ and interpret.

(e) Determine $P(B_7 \mid A)$ and interpret.

■ Why We Learned It

In this chapter, we learned the fundamentals of counting and the fundamentals of probability. State lottery officials use the counting technique combination to help determine rules for lottery games. We saw that many researchers use probability to help understand the nature of a sample if some information about its population is known. For example, researchers use probability theory to study diverse areas as stock market fluctuations, information theory, coding theory, and weather forecasting just to name a few. Quality control managers use probability to help determine if a given item has a defect. These are just a few of the ways in which probability can be used.

■ CHAPTER REVIEW EXERCISES

In Exercises 1 and 2, classify each of the following as an example of probability or statistical inference.

1. An April 2000 Gallup poll asked a random sample of 1004 adults if they had voluntarily recycled newspapers, glass, aluminum, motor oil, or other items in the past year; 90% said that they had voluntarily recycled these items in the past year. (*Source:* www.publicagenda.org.)

2. In a 1998 report on the American teen-age public, 72% of American teen-agers said that they can always trust their parents to be there for them when needed. In a

sample of 30 American teen-agers selected at random, 23 said that they could trust their parents to be there when needed. (*Source:* www.publicagenda.org.)

In Exercises 3–5, determine whether each of the following processes represents a probability experiment.

3. A teacher evaluation asks students to rate their teacher's energy level on a scale from 1 to 5, with 1 being lethargic and 5 being tireless.

4. A random sample of 2010 people were asked, "If you could live anywhere in the world, where would it be?"

5. A survey asked respondents to fill in their political party affiliation as Republican, Democratic, or Independent.

In Exercises 6–8, determine whether order is important.

6. There are 24 students who attend a field trip to a science museum. When computing in how many ways the teacher can select a pilot, copilot, and navigator for a flight simulator, is order important?

7. A music festival has 8 different musicians playing. When computing in how many different ways the musicians can be first, second, and third to take the stage, is order important?

8. From a group of 8 backpackers, 2 are to be chosen to guide the group on their backpacking trip in Escalante, Utah. When computing in how many different ways the team of backpackers can choose their leaders, is order important?

In Exercises 9–11, determine if the following events can happen at the same time.

9. A respondent from a nationwide survey is selected at random for an interview. The event that the respondent was born in Michigan and New Hampshire.

10. A resident of Colorado Springs, Colorado, is selected at random. The event that the resident is a student and has hiked Pikes Peak.

11. A spectator at a soccer match is selected at random. The event that the person is seated in section A and the event that the person is seated in section H.

In Exercise 12, an event is given.

(a) Write another event so that together the events cannot happen at the same time.

(b) Write another event so that together the events can happen at the same time.

12. The event that a student is a senior.

In Exercises 13 and 14, determine if the outcome of one event influences the outcome of the other event.

13. A box contains 5 blue chips and 4 red chips. In an experiment, a chip is drawn at random, its color is recorded, and it is replaced. Then a second chip is drawn randomly. The event that the first chip is blue and that the second chip is blue.

14. The event that Rebecca's car breaks down and the event that Rebecca rides her bike to work the next day.

In Exercise 15, an event is given.

(a) Write another event so that together the outcome of one event does not influence the outcome of another event.

(b) Write another event so that together the outcome of one event influences the outcome of the other event.

15. On a science exam, the event that Ron has not prepared.

In Exercises 16 and 17, rewrite the set using the listing method.

16. $A = \{x \mid x$ is a prime number and $1 \le x \le 20\}$

17. $B = \{x \mid x$ is a state that starts with the word "New"$\}$

In Exercises 18 and 19, rewrite the set using set-builder notation.

18. $A = \{2, 4, 6, 8, 10\}$

19. $B = \{$Alabama, Alaska, Arizona, Arkansas$\}$

In Exercises 20 and 21, write all possible subsets for the given set.

20. $\{10, 12\}$

21. $\{$McKinley, Pikes Peak, Washington$\}$

22. Let $A = \{x \mid x$ is a Review Exercise that is an example of probability and $1 \le x \le 2\}$ and $B = \{x \mid x$ is a Review Exercise that is an example of statistical inference and $1 \le x \le 2\}$.

(a) Rewrite sets A and B using the listing method.

(b) Find $A \cap B$.

(c) Are A and B disjoint? Explain.

In Exercises 23–25, let $U = \{x \mid x$ is a whole number and $16 \le x \le 30\}$, $A = \{16, 17, 18, 19, 20\}$, $B = \{17, 19, 21, 23, 25, 27, 29\}$, and $C = \{26, 27, 28, 29, 30\}$. Write each of the following sets using the listing method.

23. $A \cap B$

24. $A \cap C^C$

25. $(A \cup C)^C$

In Exercises 26–28, use the Venn diagram to write the desired set.

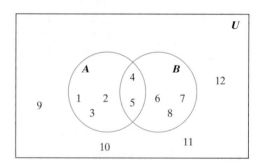

26. B

27. $A^C \cap B$

28. $(A \cap B)^C$

29. Let $U = \{1, 2, 3, 4, 5, 6\}$, $A = \{1, 3, 5\}$, $B = \{1, 2, 3\}$, and $C = \{2, 3, 4, 5\}$. Write a set and draw a shaded Venn diagram for $A^C \cap (B \cup C)$.

In Exercises 30 and 31, the number of students enrolled in public and private schools, in thousands, in the United States in grades 9 to 12 and college in 1997 is shown in the table.

Type of School	Grades 9–12 (A)	College (B)	Total
Public (C)	13,054	11,146	24,200
Private (D)	1,308	3,199	4,507
Total	14,362	14,345	28,707

SOURCE: U.S. National Center for Education Statistics, www.nces.ed.gov

Here A represents the set of students in grades 9 to 12 at public and private schools, B represents the set of students in college at public and private schools, C represents the set of students in grades 9 to 12 and college at public schools, and D represents the number of students in grades 9 to 12 and college at private schools. Determine the following and interpret each.

 30. $n(D^C)$ **31.** $n(A \cup C)$

32. A three-question phone survey of registered voters asks the following questions:

Question 1: Do you worry about the loss of open spaces and natural habitats for wildlife?

Question 2: Do you favor or oppose President Bush's decision to drill in the Arctic National Wildlife Refuge?

Question 3: Are you registered Republican, Democrat, or Independent?

Draw a tree diagram to represent all possible answers in the survey. How many ways could a respondent answer all three questions?

In Exercises 33–35, determine the following:

33. A new sports pack feature offered through a ticket agency allows you to construct a package deal that includes one ticket for a professional soccer game and one ticket for a professional football game. If there are 33 games in the soccer season and 16 games in a football season, how many ways can a person construct a package deal?

34. A password consists of 4 letters and 3 digits. If letters and digits can be repeated, in how many different ways can a password be created?

35. A science quiz consists of ten true–false questions. In how many different ways can a student answer all the questions?

In Exercises 36 and 37, evaluate the expression without using a calculator.

36. 6!   **37.** $_6P_2$

In Exercises 38 and 39, use a calculator to determine the following:

38. $_{35}P_{15}$ **39.** $_{60}P_{10}$

In Exercises 40 and 41, evaluate the expression without using a calculator.

40. $_6C_3$ **41.** $_8C_4$

In Exercises 42 and 43, use a calculator to determine the following:

42. $_{40}C_{20}$ **43.** $_{60}C_{40}$

44. There are 24 students who attend a field trip to a science museum. In how many different ways can the teacher select a pilot, copilot, and navigator for a flight simulator?

45. From a group of 8 backpackers, 2 are to be chosen to guide the group on their backpacking trip in Escalante, Utah. In how many different ways can the team of backpackers choose their leaders?

46. A music festival has 8 different musicians playing. In how many different ways can the musicians be first, second, and third to take the stage?

47. In a shipment of 40 CD players, 6 are known to be defective. In how many ways can a sample of 8 be chosen so that 3 are defective and 5 are not defective?

48. A city council has 6 Democrats, 5 Republicans, and 3 Independents. In how many different ways can a subcommittee be formed with 3 Democrats, 2 Republicans, and 1 Independent?

In Exercises 49–51, write the sample space S using set notation and then determine n(S).

49. A box contains 12 cards labeled A through L. In an experiment, a card is drawn out and its letter is recorded.

50. A box contains 1 yellow, 1 red, and 1 blue marble. A bag contains a ticket labeled with a 1 and a ticket labeled with a 3. In an experiment, one draw is made from the box and one draw is made from the bag and the outcome is recorded.

51. A pollster asks the question, "Have you ever received a student loan?" The pollster then asks, "Did you attend a public or private school?"

In Exercises 52–55, use classical probability to compute P(A) and P(B) for the indicated events A and B.

52. For the experiment in Exercise 50, one draw is made from the box and one draw is made from the bag and the outcome is recorded. Let A represent the event that a yellow marble is drawn and B represent the event that a ticket labeled with a 3 is drawn.

53. In an experiment, Rebecca performs a pH test on a sample of drinking water and records the result as acidic, neutral, or basic. She then performs a nitrate test on the same sample of water and records the result as normal or excessive. Let A represent the event that the water is acidic and B represent the event that the water is neutral and the nitrate level is normal.

54. For the experiment in Exercise 51, a pollster asks two questions and records the results. Let A represent the event that the respondent answered "yes" to question 1 and "private" to question 2 and B represent the event that the respondent answered "no" to question 1.

55. In an experiment, Chad tosses a six-sided die and records the number. He tosses it again and records the number from the second toss. Let A represent the event that the numbers on each toss are the same and B represent the event that the sum of the two tosses is a 6.

56. A fruit basket contains 10 oranges and 10 apples. Kris takes two pieces of fruit from the basket without looking at what kind they are.

(a) Determine $n(S)$.

(b) If A represents the event that both pieces of fruit are apples, determine $n(A)$.

(c) Use classical probability to determine $P(A)$ and interpret.

57. The city council in Platt City is comprised of 6 Democrats and 8 Republicans.

(a) In how many different ways can a subcommittee of 5 council persons be formed?

(b) In how many different ways can a subcommittee be formed with 3 Democrats and 2 Republicans?

(c) If A represents the event that a subcommittee is formed with 3 Democrats and 2 Republicans, determine $P(A)$ and interpret.

58. In a shipment of 40 pairs of waders, 6 are known to leak.

(a) In how many different ways can a sample of 8 pairs of waders be chosen?

(b) In how many different ways can a sample of 8 be chosen so that 3 pairs of waders leak and 5 do not leak?

(c) If A represents the event that in a sample of 8 pairs of waders 3 leak and 5 do not leak, determine $P(A)$ and interpret.

🌐 **59.** The number of workers 16 years old or older in the United States and their means of transportation to work in 2000 is shown in the table.

Transportation	Number (in millions)
Car, truck, or van (drove alone)	97.2
Car, truck, or van (carpooled)	14.3
Public transportation	6.6
Walked	3.4
Other means	1.8
Worked at home	4.1

SOURCE: U.S. Census Bureau, www.census.gov

In 2000, a working individual 16 years old or older was selected at random.

(a) Determine the probability that a working individual 16 years old or older drives alone to work in a car, truck, or van.

(b) Determine the probability that a working individual 16 years old or older does not drive alone to work in a car, truck, or van.

In Exercises 60 and 61, classify the following as an example of classical, empirical, *or* subjective *probability.*

60. A local weatherman uses the fact that it has rained 35 days out of 50 days that have shown the same weather patterns as today to state that there is a 70% chance of rain today.

61. The probability of rolling a 4 on one toss of a die is $\frac{1}{6}$.

In Exercises 62 and 63, determine if the events A and B are mutually exclusive.

62. A respondent from a nationwide survey is selected at random for an interview. A represents the event that the respondent was born in Michigan and B represents the event that the respondent was born in New Hampshire.

63. A resident of Colorado Springs, Colorado, is selected at random. A represents the event that the resident is a student and B represents the event that the resident has hiked Pikes Peak.

64. A box contains 12 tickets numbered 1 to 12. If a draw is made at random, determine the probability that the ticket is labeled with a 1 or an even number.

65. A bowl of cereal contains 15 red, 8 green, 14 yellow, and 10 orange pieces. If a piece of cereal is selected at random from the bowl, determine the probability that a red or orange piece is drawn.

For Exercises 66–71, consider a survey of a group of 400 students at Osceola College. Let A represent the event that the student has season football tickets and B represent the event that the student is an upperclassperson. The results are shown in the Venn diagram. Assuming that a student in the survey group is selected at random, use the diagram to answer the following:

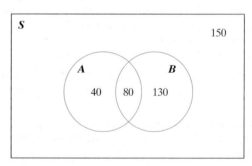

66. Determine $P(A)$ and interpret.

67. Determine $P(B)$ and interpret.

68. Determine $P(A \cup B)$ and interpret.

69. Determine $P(A \cap B)^C$ and interpret.

70. Determine the probability that the student has season football tickets and is an upperclassperson.

71. Determine the probability that the student does not have season football tickets and is not an upperclassperson.

🌐 *For Exercises 72–74, consider the following. In 1998, the percentage of persons living in counties in California or Texas that do not meet U.S. Environmental Protection*

Agency air quality standards is shown in the following table.

	Meeting Standards, % (A)	Not Meeting Standards, % (B)	Total (%)
California (C)	6.7	55.7	62.4
Texas (D)	18.8	18.8	37.6
Total	25.5	74.5	100

SOURCE: National Center of Chronic Disease Prevention, www.cdc.gov, U.S. Census Bureau, www.census.gov

72. Determine $P(A)$ and interpret.

73. Determine the probability that a randomly selected person from California or Texas is living in California and in a county that is not meeting the U.S. Environmental Protection Agency air quality standards.

74. Determine the probability that a randomly selected person from California or Texas is living in Texas or in a county that is not meeting U.S. Environmental Protection Agency air quality standards.

75. The probability of event A occurring is 0.4.

(a) Determine the odds in favor of A occurring.

(b) Determine the odds against A occurring.

76. The odds in favor of event A occurring are 7 to 2. Determine the probability of A occurring.

77. Determine the odds in favor of a prime number on a single roll of a ten-sided die containing the whole numbers 1 to 10.

78. Determine the odds against a number divisible by 3 on a single roll of a ten-sided die containing the whole numbers 1 to 10.

79. As of July 11, 2001, the odds in favor of the Detroit Red Wings winning the 2002 Stanley Cup are listed as 1 to 5. Determine the probability that the Detroit Red Wings will win the 2002 Stanley Cup. (*Source:* www.vegasinsider.com/u/futures.)

In Exercises 80 and 81, determine if events A and B are independent or dependent. Explain your reasoning.

80. A box contains 5 blue chips and 4 red chips. In an experiment, a chip is drawn at random, its color is recorded, and it is replaced. Then a second chip is drawn randomly. Let A represent the event that the first chip is blue and B represent the event that the second chip is blue.

81. A is the event that Rebecca's car breaks down and B is the event that Rebecca rides her bike to work the next day.

82. A March 2001 Gallup poll showed that 18% of adult Americans fear flying on an airplane. If an airline

chooses 4 adult Americans at random to receive a free promotional flight, what is the probability that all 4 fear flying? (*Source:* www.publicagenda.org.)

83. In a sample of 25 students, 14 said that they studied at least 3 hours for their Biology exam, 18 said that they received a grade of 80% or higher, and 12 said that they studied at least 3 hours and received a grade of 80% or higher. Let A represent the event that a student studied at least 3 hours and B represent the event that a student received a grade of 80% or higher.

(a) Draw a Venn diagram to represent the students sampled with events A and B.

(b) Compute $P(A \cap B)$ and interpret.

(c) Determine $P(B \mid A)$ and interpret.

84. Use the probability Venn diagram to determine $P(A \cap B)$ and $P(B \mid A)$.

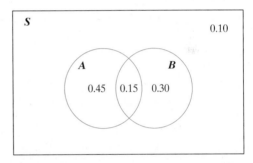

85. Use the table of data to determine $P(A \cap C)$ and $P(C \mid A)$.

	A	B	Total
C	15	20	35
D	25	45	70
Total	40	65	105

86. Use the table of probabilities to determine $P(B \cap D)$ and $P(D \mid B)$.

	A	B	Total
C	0.34	0.18	0.52
D	0.28	0.20	0.48
Total	0.62	0.38	1.00

87. A bag contains 6 yellow and 8 blue marbles. In an experiment, a marble is drawn at random, its color is recorded, and a second marble is drawn randomly without replacing the first.

(a) Determine the probability that the first marble is blue and the second marble is yellow.

(b) Determine the probability that both marbles are yellow.

🌐 *For Exercises 88–92, consider the following: In 1998, the percentage of persons living in counties in California or Texas that do not meet U.S. Environmental Protection Agency air quality standards is shown in the following table:*

	Meeting Standards, % (A)	Not Meeting Standards, % (B)	Total (%)
California (C)	6.7	55.7	62.4
Texas (D)	18.8	18.8	37.6
Total	25.5	74.5	100

SOURCE: National Center of Chronic Disease Prevention, www.cdc.gov, U.S. Census Bureau, www.census.gov

88. Determine $P(A \cap B)$ and interpret.

89. Determine $P(C \cap D)$ and interpret.

90. Determine $P(B \mid C)$ and interpret.

91. Determine the probability that a randomly selected person lives in a county not meeting U.S. Environmental Protection Agency air quality standards, given that they are from Texas.

92. Determine $P(C \mid B)$.

93. Use the table of data to compute the values of $P(D \mid B)$ and $P(B \mid D)$.

	A	B	Total
C	20	30	50
D	15	10	25
Total	35	40	75

94. Use the table of probabilities to compute the values of $P(C \mid A)$ and $P(A \mid C)$.

	A	B	Total
C	$\frac{1}{3}$	$\frac{1}{3}$	$\frac{2}{3}$
D	$\frac{1}{4}$	$\frac{1}{12}$	$\frac{1}{3}$
Total	$\frac{7}{12}$	$\frac{5}{12}$	1

95. A pollster conducts a phone survey of 500 adult men and 516 adult women. The pollster asks each person if he or she is afraid of heights. The results are shown in the table.

	Afraid (A)	Not Afraid (B)	Total
Men (C)	312	188	500
Women (D)	289	227	516
Total	601	415	1016

For the events labeled in the table as A, B, C, and D, compute and interpret the following probabilities.

(a) $P(C \mid A)$ (b) $P(A \mid C)$

In Exercises 96 and 97, a tree diagram is given. Determine $P(A)$, $P(B \mid A)$ and $P(B^C \mid A)$.

96.

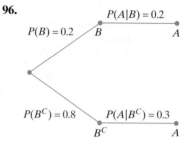

97.

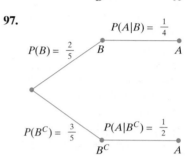

98. At Osceola College, 58% of the student body is underclasspeople and the remainder is upperclasspeople. In a survey, 60% of the underclasspeople favored the construction of a new parking structure and 30% of the upperclasspeople favored the construction of a new parking structure. A prospective student visiting the campus selects an Osceola College student at random. Let B represent the event that the selected student is an underclassperson, B^C represent the event that the selected student is an upperclassperson, and A represent the event that the selected student favored the construction of a new parking garage.

(a) Determine $P(A)$ and interpret.

(b) $P(B \mid A)$ and interpret.

(c) Determine $P(B^C \mid A)$ and interpret.

🌐 **99.** In a March 1999 poll, a random sample of American adults were asked if they were aware that American and British air forces were continuing to attack Iraqi air defenses and radar stations in no-fly zones. This sample group was then asked if they support the attacks. (For those who answered "no" to the first question, they were asked if they support the attacks now that they are aware of them.) The results are summarized in the following table:

Aware of Attacks	Percent	Support Attacks
Yes	69%	61%
No	31%	48%

SOURCE: www.harrisinteractive.com

Suppose that a respondent was selected at random. Let B represent the event that the respondent is aware of the attacks and A represent the event that the respondent supports the attacks.

(a) Determine $P(A)$ and interpret.

(b) Determine the probability that the respondent is aware of the attacks, given that the respondent supports the attacks.

(c) Determine $P(B^C \mid A)$ and interpret.

In Exercises 100 and 101, a tree diagram is given. Determine if the events are collectively exhaustive.

100.

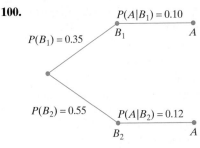

$P(B_1) = 0.35$

$P(A \mid B_1) = 0.10$ B_1 A

$P(B_2) = 0.55$

$P(A \mid B_2) = 0.12$ B_2 A

101.

$P(A \mid B_1) = \frac{1}{8}$

$P(B_1) = \frac{1}{4}$ B_1 A

$P(B_2) = \frac{3}{4}$ $P(A \mid B_2) = \frac{3}{8}$ B_2 A

In Exercises 102 and 103 a tree diagram is given. Determine $P(A)$, $P(B_1 \mid A)$, and $P(B_3 \mid A)$.

102.

$P(B_1) = \frac{1}{4}$ B_1 $P(A \mid B_1) = \frac{1}{8}$ A

$P(B_2) = \frac{1}{8}$ $P(A \mid B_2) = \frac{3}{16}$ B_2 A

$P(B_3) = \frac{5}{8}$ $P(A \mid B_3) = \frac{1}{4}$ B_3 A

103.

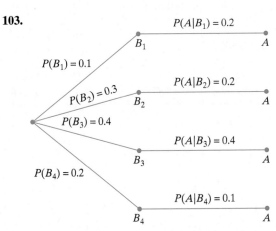

$P(B_1) = 0.1$

$P(A \mid B_1) = 0.2$ B_1 A

$P(B_2) = 0.3$ $P(A \mid B_2) = 0.2$ B_2 A

$P(B_3) = 0.4$ $P(A \mid B_3) = 0.4$ B_3 A

$P(B_4) = 0.2$ $P(A \mid B_4) = 0.1$ B_4 A

104. At Monadrock High School, the student council handed out surveys to determine if students favored adding lacrosse to the athletic program. The percent of surveys returned from each class is shown in the table.

Class	Student Body (%)	Surveys Returned (%)
Freshman	30	75
Sophomore	25	50
Junior	20	40
Senior	25	10

A student working for the school newspaper selects a student at random. Let B_1, B_2, B_3, and B_4 represent the event that the student is a freshman, sophomore, junior, or senior, respectively, and let A represent the event that the student returned the survey.

(a) Construct a tree diagram.

(b) Determine $P(A)$ and interpret.

(c) Determine the probability that the voter is a freshman, given that the voter returned the survey.

(d) Determine $P(B_3 \mid A)$ and interpret.

105. Sopris Mountaineering Company makes ski bindings at three factories. From past records, they know the proportion of output from each factory and the percentage of defects that come from each factory. The information is shown in the table.

Factory	Output (%)	Defects (%)
I	30	2
II	25	1
III	45	3

A quality control engineer selects a ski binding at random. Let B_1, B_2, and B_3 represent the event that the binding comes from factory I, factory II, or factory III, respectively, and let A represent the event that the binding has a defect.

(a) Construct a tree diagram.

(b) Determine $P(A)$ and interpret.

(c) Determine the probability that the ski binding came from factory II, given that the ski binding is defective.

(d) Determine $P(B_3 \mid A)$ and interpret.

CHAPTER 6 PROJECT

SOURCE: University of Minnesota: www.umn.edu

In this project, we will consider the enrollment of the Spring 2001 Semester at the University of Minnesota. We will use these data to compute different types of probabilities, such as conditional probabilities. Consider the Tables 1 and 2 below.

For a case study on university life, suppose that a national education magazine selects a student at random.

1. Using Table 1, let M stand for medical student, G stand for graduate student, U stand for undergraduate.

(a) Find the probability that the student is a medical student. Find the probability that the student is an undergraduate Medical student.

(b) Determine the probability that the student is between the ages of 21 and 24, given that the student is a medical student.

(c) Determine the probability that the student is a medical student, given that the student is between the ages of 21 and 24.

(d) Find $P(M \mid G)$ and interpret.

(e) Find $P(G \mid M)$ and interpret.

(f) Compare the results from parts (a) and (b) and also parts (d) and (e). Why do the answers differ?

2. Using Table 2, let W stand for white, AA stand for African American, and TC stand for twin cities campus.

(a) Find the probability that the student attends the Duluth campus.

(b) Find the probability that the student goes to the Morris campus, given that the student is American Indian.

(c) Find $P(W \cup AA)$ and interpret.

(d) Find $P[(W \cup AA)^C]$ and interpret.

(e) Find the probability that the student attends the Crookston campus or the Twin Cities campus, given that she or he is an International student. Assume that a student cannot attend both colleges.

3. There are a total of 1448 white medical students at the Twin Cities campus. Using Tables 1 and 2, find the probability that the student is a white medical student, given that he or she attends the Twin Cities campus. Let WMS stand for white medical student.

Table 1   Number of Students Attending the University of Minnesota's Twin Cities Campus

Age	Undergraduate Under 21	21–24	25–34	Other	Graduate Under 21	21–24	25–34	Other	Professional Under 21	21–24	25–34	Other	Nondegree Under 21	21–24	25–34	Other	Total
Medical	6	64	30	14	0	95	174	41	0	195	537	45	0	3	519	182	1,897
Total	11,873	9918	2476	1023	5	1809	5840	2333	8	930	1417	185	1035	959	1967	1616	43,394

Table 2   Number of Students Attending the University of Minnesota Classified by Race

	African American	American Indian	Asian/Pacific	Chicano/Latino	White	International	N/A	Total
Crookston	40	20	19	30	1,736	32	440	2,317
Duluth	72	110	149	69	7,659	153	291	8,503
Morris	96	102	49	21	1,444	8	54	1,774
Twin Cities	1432	273	2917	755	31,890	3245	2882	43,394
Total	1640	505	3134	875	42,729	3438	3667	55,988

7 Graphical Data Description

Type of asset	Value of asset in billions of dollars
Real estate	823
Machinery	89
Livestock	62
Financial assets	55
Crops	30

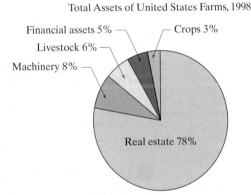

Total Assets of United States Farms, 1998

Financial assets 5% — Crops 3%
Livestock 6% —
Machinery 8% —
Real estate 78%

It is often said that a picture is worth a thousand words. This adage is never more apparent than when trying to make sense of a set of data. Consider the five categories of total assets of farms in the U.S. in 1995, as listed in the table. Although the data shows that a farms' greatest asset is the land itself, the statistical graph, or pie chart, makes this information perfectly clear in a quick visual summary.

What We Know

In the previous chapter we learned how to construct sets and how to find the number of elements in a set. This led us to some counting principles, which then led us to an introductory study of probability.

Where Do We Go

In this chapter we will study ways to communicate important information on sets of data, including the representation of data in graphical form and interpreting data that are presented in graphical form. We will also study the different ways of measuring the centrality of a data set and the overall dispersion of the data.

Section 7.1   Graphing Qualitative Data

Often in newspapers and magazines and on television, we see graphs and charts that summarize data. These graphs are used because they provide a quick, easy, and effective way to summarize raw data. In this section we will explore the properties of these graphs and study how they can be constructed.

Types of Data

We often hear statements such as "In 1998, 12.7% of the U.S. population was at least 65 years old," or "From 1980 to 1988, there were 2.62 residents per U.S. household." These statements are made after analyzing **data**. Data are made up of information coming from responses to surveys, from making observations, and from taking measurements. A collection of data is called a **data set**. Data sets come from two sources, called populations and samples. Recall from Section 6.1 that we defined **population** and **sample** as follows:

■ **Population and Sample**

  • A **population** is the collection of *all* outcomes, responses, measurements, or counts that are of interest.

  • A **sample** is a portion (or subset) of the population.

Example 1   Identifying a Population and a Sample

In an April 29, 2000, poll conducted by CNN/USA Today and the Gallup Organization, 1003 adult Americans were asked if they would favor or oppose the registration of all handguns; 76% of the adults favored registration. Identify the population and sample. Estimate the number in the data set who favor registration and the number who oppose registration (*Source:* www.gallup.com).

Solution

The **population** consists of the responses, or possible responses, of all American adults. The **sample** is made up of the 1003 American adults who participated in the survey. Since 76% of 1003 is about 762, the data set consists of the 762 responses favoring registration and the 241 opposing responses.   ■

✓ **Checkpoint 1**   Now work Exercise 11.

Data can be classified in various ways. For now, we will classify data as either **qualitative** or **quantitative**.

■ **Qualitative and Quantitative Data**

  • Data that can be placed into distinct categories according to some characteristic, attribute, or nonnumerical label are called **qualitative data**. Since this type of data can be placed into distinct categories, they are sometimes referred to as **categorical data**.

  • Data that are numerical in nature and can be ordered or ranked are called **quantitative data**.

Example 2

Classifying Data as Qualitative or Quantitative

The average gasoline price, in dollars per gallon, in Phoenix, Arizona, on June 23, 2000, for several types of gasoline is shown in Table 7.1.1. Which data are qualitative data and which are quantitative data? Explain.

Table 7.1.1

Type	Price in Dollars per Gallon
Regular	1.42
Premium	1.56
Diesel	1.55

SOURCE: www.gallup.com

Solution

The data in this table can be separated into two sets. One set consists of the type of gasoline and the other set consists of the average per gallon price. The gasoline types are nonnumerical and are separated into **categories**, so these are **qualitative** data. The average prices are numerical and can be ordered, so they make up the **quantitative** data. ∎

We will call a table like Table 7.1.1 a **data table**. The remainder of this section will analyze the graphical techniques describing qualitative data.

Bar Graphs

One of the most widely used graphs for qualitative data is the **bar graph**. This type of graph is popular because bar graphs are easy to make and easy to interpret. Consider the bar graph in Figure 7.1.1, which represents the top five U.S. cities visited, in millions, by overseas travelers in 1997. From the graph, we see that New York City had the most overseas travelers at about 5 million. Los Angeles had just under 4 million overseas visitors. Note that the cities are listed in descending order by the number of overseas travelers. The horizontal axis lists the qualitative data values, that is, the cities. The vertical bars show the number of overseas travelers. The number of overseas travelers visiting these U.S. cities

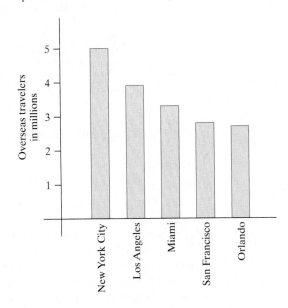

Top 5 United States Cities Visited by Overseas Travelers, 1997

Figure 7.1.1

SOURCE: U.S. Census Bureau, www.census.gov

is called the **frequency**. The bars on the graph are all of equal width, and there is an equal space between each bar. We now list some of the features of the bar graph.

■ **Properties of Bar Graphs**

1. The horizontal axis contains the qualitative data values.
2. The vertical axis is labeled with the frequencies.
3. The data are arranged from largest to smallest in value as the graph is read from left to right.
4. The height of the bars is determined by the frequency.
5. The bars are of equal width and have an equal space between them.
6. The bar graph includes labels and units in the axes and a descriptive title.

▶ **Note:** Sometimes the axes in items 1 and 2 are interchanged. However, in this text we will use the outline above unless otherwise indicated.

Example 3 **Constructing a Bar Graph**

The data table in Table 7.1.2 shows the amount spent, in billions of dollars, in the United States in 1997 on various recreation services.

Table 7.1.2

Type of Recreation Service	Amount Spent (in billions of dollars)
Physical fitness facilities	5.7
Public golf courses	4.3
Video arcades	3.7
Amusement parks	7.3
Sports clubs	7.7

SOURCE: U.S. Census Bureau, www.census.gov

Construct a bar graph for the data.

Solution

We begin by ranking the recreation services in descending order by the amount spent. The types of recreation services are labeled on the horizontal axis since they are the qualitative values. The vertical axis is labeled with the amount spent in billions of dollars. See Figure 7.1.2.

Now we construct the vertical bars with equal width and with equal spacing between them. The height of the bars is determined by the amounts spent, which are the frequencies. The completed bar graph is in Figure 7.1.3. Observe that, when we start with the leftmost bar and count to the right, physical fitness facilities rank third in total amount spent in 1997.

Amount Spent on Recreation Services, 1997

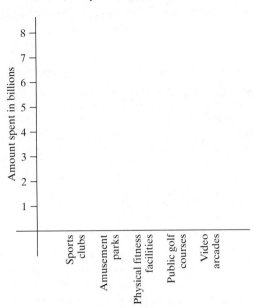

Amount Spent on Recreation Services, 1997

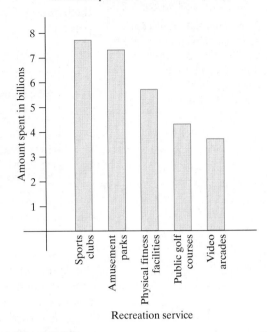

Figure 7.1.2   **Figure 7.1.3**

Technology Option

Bar graphs can be made using computer spreadsheet programs such as Excel or statistical packages such as Minitab. Figure 7.1.4 shows the bar graph obtained for the data given in Example 3 using Excel.

Amount Spent on Recreation Services, 1998

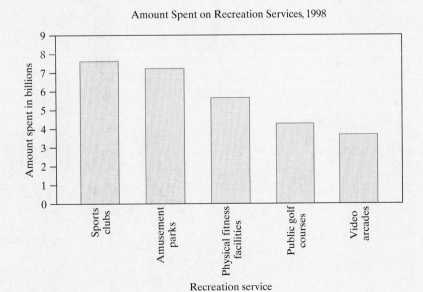

Figure 7.1.4

Bar graphs can also be used to display two sets of data. Table 7.1.3 shows the highest educational attainment of U.S. men and women in 1998, measured by percent.

Table 7.1.3

Educational Attainment	Percent Male	Percent Female
No high school diploma	17.2	17.1
High school graduate	32.3	35.2
Some college	17.1	17.3
Associate degree	6.9	8.0
Bachelor's degree	17.1	15.8
Advanced college degree	9.4	6.6

SOURCE: The U.S. Statistical Abstract, www.census.gov/statab/www/

We can compare the highest educational attainment by gender if we draw dual adjacent bars for each educational attainment level. When we have two separate sets of data, ordering them as in Example 3 is difficult, if not impossible. In this case we will organize the data by the listed educational attainment. The bar graph is shown in Figure 7.1.5. Note that the **legend** in the upper-right corner of the graph differentiates the educational attainment of men and of women.

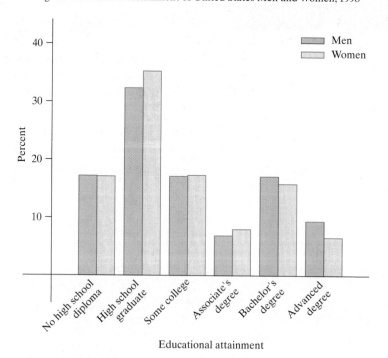

Highest Educational Attainment of United States Men and Women, 1998

Figure 7.1.5

Pie Charts

One type of graph that is used extensively in descriptive statistics is the **pie chart**. This chart is ideally suited to show how a whole quantity is divided into several parts. The pie chart is made up of a circle divided into various slices called **sectors**. The size of each sector is determined by the percentage that each qualitative data value is of the whole. Let's take a look at an example of a pie chart. Table 7.1.4 shows the marital status of U.S. females 15 years or older in 2000. The corresponding pie chart for this data table is shown in Figure 7.1.6.

Marital Status of United States Females 15 Years or Older, 2000

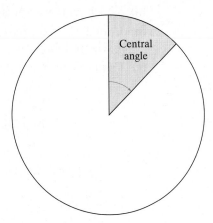

Table 7.1.4

Marital Status	Percentage of Females
Never married	25
Married	55
Widowed	10
Divorced	10

SOURCE: U.S. Census Bureau, www.census.gov

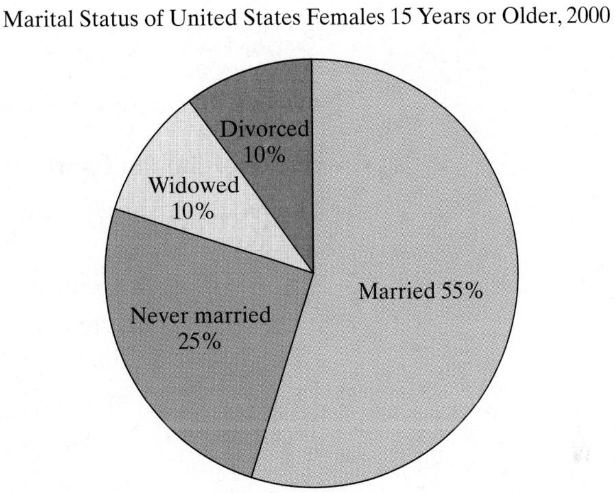

Figure 7.1.6

The graph clearly shows that the majority of females age 15 or older were married in 2000, while a relatively small percentage were divorced. This illustrates how effectively and quickly we can access data from a pie chart as compared to a numerical table.

When constructing a pie chart by hand, one key is to determine how large each sector will be. To determine its size, we need to find the measure, in degrees, of its **central angle**. The central angle is shown in Figure 7.1.7.

Figure 7.1.7

If we are given percentages as was done in Table 7.1.4, we can multiply the percentage, converted to decimal form, by 360°, the number of degrees in a circle. The measure of each central angle of the pie chart in Figure 7.1.6 is shown in Table 7.1.5. The data have been ordered from the highest to the lowest percentage.

Table 7.1.5

Marital Status	Percentage of Females	Degrees
Married	55	$0.55(360°) = 198°$
Never married	25	$0.25(360°) = 90°$
Widowed	10	$0.10(360°) = 36°$
Divorced	10	$0.10(360°) = 36°$

Now let's take a look at some of the properties of the pie chart.

 Properties of the Pie Chart

1. The pie chart shows the part of the whole for each qualitative (categorical) data item.
2. The size of each sector in the pie chart corresponds to its frequency.
3. Each sector is labeled and includes its frequency or a percentage. Every pie chart has a title.
4. The sectors are arranged with the largest in the upper-right portion of the circle and, then clockwise in descending order.

Our main focus with pie charts will be reading and interpreting what they tell us. This is illustrated in Example 4.

Example 4 Interpreting a Pie Chart

The total assets of U.S. farms in 1998, divided into five categories, are shown in the pie chart in Figure 7.1.8. Note that, since the labels are too large to fit inside the smaller sectors, we use **leader lines** to connect the label with its sector (*Source:* U.S. Statistical Abstract, www.census.gov/statab/www).

(a) Which type of asset is the largest and which is the smallest?

(b) If total assets for these five categories were $1059 billion, estimate the assets for livestock.

(c) Estimate the assets that were not real estate.

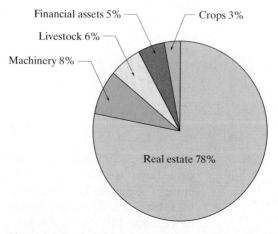

Total Assets of United States Farms, 1998

Figure 7.1.8

Solution

(a) Since real estate is the largest sector and has the highest percentage, it is the largest asset. On the other hand, crops is the smallest sector and has the lowest percentage, which means that it is the smallest asset.

(b) From Figure 7.1.8 we know that livestock is about 6% of the total assets. Since the total assets for these five categories in 1998 totaled $1059 billion, the assets for livestock are estimated by

$$0.06(1059) = 63.54$$

So we can estimate that the assets for livestock in 1998 were about $63.54 billion.

(c) Figure 7.1.8 indicates that about 78% of the total assets were in real estate. This means that 100% − 78% = 22% was in the other four categories. Our estimate for the assets not in real estate is found by

$$0.22(1059) = 232.98$$

So we can estimate that the assets not in real estate in 1998 were about $232.98 billion.   ■

Technology Option

Pie charts can be made using computer spreadsheets such as Excel or statistical packages such as Minitab. Figure 7.1.9 shows the pie chart for Example 4 using Excel.

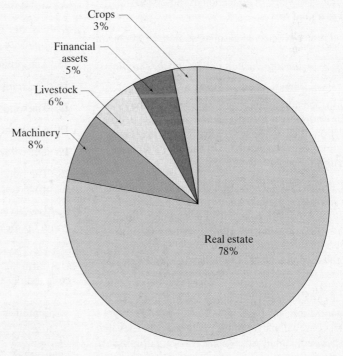

Total Assets of United States Farms, 1998

Crops 3%
Financial assets 5%
Livestock 6%
Machinery 8%
Real estate 78%

Figure 7.1.9

✓ **Checkpoint 2**   Now work Exercise 37.

SUMMARY

In this section, we studied two types of commonly used graphs, the **bar graph** and the **pie chart**. We explored the properties, construction techniques, and how to interpret these graphs.

- **Data** are made up of information coming from responses to surveys, from making observations, and from taking measurements.

- A **population** is the collection of *all* outcomes, responses, measurements, or counts that are of interest.

- A **sample** is a portion (or subset) of the population.
- **Qualitative data** are data that can be placed into distinct categories according to some characteristic, attribute, or nonnumerical label. These are sometimes called **categorical data**.
- **Quantitative data** are data that are numerical in nature and can be ordered or ranked.

SECTION 7.1 EXERCISES

In Exercises 1–8, determine whether the data set is a population or a sample. Explain your reasoning.

1. The age of every member of the U.S. Senate.

2. The annual salary of every employee at a certain corporation.

3. The party affiliation of every third voter exiting a polling booth.

4. A survey of 50 workers at a factory that employs 3000.

5. The annual income of all homeowners in California.

6. The number of children of all families who own home computers.

7. The number of years of education of those who responded to a mail survey.

8. The fuel prices at selected gasoline stations in Michigan.

In Exercises 9–14, identify the population and the sample.

9. In a Spring 2000 survey conducted by the Gallup Organization, 568 adult Americans were asked if they would favor or oppose a proposal that would allow people to put a portion of their Social Security payroll taxes into personal retirement accounts; 65% of the adults favored the proposal (*Source:* www.gallup.com).

10. In a June 1999 survey conducted by NBC and the *Wall Street Journal*, 491 adult Americans were asked, "In terms of setting Medicare policy dealing with prescription drug benefits, who do you trust more: President Clinton or companies that make prescription drugs?" Fifty-seven percent of the respondents said that they trusted President Clinton more (*Source:* NBC/*Wall Street Journal*).

11. In an October 1998 survey conducted by the Gallup Organization, 467 baseball fans were asked, "Who is the greatest baseball player in the game today?" Eleven percent responded Sammy Sosa (*Source:* www.gallup.com).

12. In a March 2000 survey, 205 golf fans were asked by the Gallup Organization, if they could play any golf course in the world, which one would it be? Thirty-seven percent of the respondents said Pebble Beach (*Source:* www.gallup.com).

13. CNN conducted an on-line, interactive quick vote on May 20, 1999. They asked 19,026 respondents, "A study found that bioengineered corn can harm butterflies; should such crops be put on hold pending more study?" Seventy-five percent said there should be a hold on genetically engineered crops until further study had been conducted (*Source:* www.cnn.com).

14. In a summer 2000 *USA Today* online survey, 1909 readers were asked, "Which activist, leader, or politician has done the most to raise the visibility of women?" to which 34.6% responded by saying Mother Teresa (*Source:* www.usatoday.com).

In Exercises 15–20, determine whether the data are qualitative or quantitative.

15. The gender of the members of a book reading club.

16. The religious preference of the citizens of a small town.

17. The number of student desks in a classroom.

18. The length of bass caught in Lake Superior.

19. The colors of blazers in a ladies clothing store.

20. The ranking of service (excellent, good, poor) at a suburban restaurant.

In Exercises 21–30, a table of data is given. Construct a bar graph for the data.

21. The total cash receipts, in billions of dollars, from U.S. farm marketings for four of the top farm livestock commodities in 1998 are shown in the table.

Commodity	Receipts (in billions of dollars)
Cattle	33.7
Hogs	9.4
Chickens	15.1
Turkeys	2.7

SOURCE: U.S. Statistical Abstract, www.census.gov/statab/www

22. The total cash receipts, in billions of dollars, from U.S. farm marketings for four of the top farm crop commodities in 1998 are shown in the table.

Commodity	Receipts (in billions of dollars)
Corn	17.1
Soybeans	15.4
Wheat	7.0
Cotton	6.0

SOURCE: U.S. Statistical Abstract, www.census.gov/statab/www

23. The number of establishments, in thousands, for selected U.S. service industries in 1992 is shown in the table.

Establishment	Number (in thousands)
Social services	140.8
Personal services	197.1
Business services	306.6
Automotive repair	172.0
Health services	465.3

SOURCE: U.S. Statistical Abstract, www.census.gov/statab/www/

24. The number of establishments, in thousands, for selected U.S. service industries in 1997 is shown in the table.

Establishment	Number (in thousands)
Social services	162.3
Personal services	204.6
Business services	398.2
Automotive repair	192.3
Health services	498.9

SOURCE: U.S. Statistical Abstract, www.census.gov/statab/www/

25. The number of units, in millions, of various recorded audio media shipped in the United States in 1990 is shown in the table.

Media Type	Number Shipped (in millions)
Compact disks	286.5
Cassettes	442.2
Cassette singles	87.4
Vinyl albums	39.9

SOURCE: U.S. Census Bureau, www.census.gov

26. The number of units, in millions, of various recorded audio media shipped in the United States in 1997 is shown in the table.

Media Type	Number Shipped (in millions)
Compact disks	847.0
Cassettes	158.5
Cassette singles	26.4
Vinyl Albums	8.8

SOURCE: U.S. Census Bureau, www.census.gov

27. The number of titles of new books imported to the United States in various subjects during 1990 is shown in the table.

Subject	New Titles Imported
Education	234
History	329
Fiction	241
Medicine	588
Technology	546

SOURCE: U.S. Census Bureau, www.census.gov

28. The number of titles of new books imported to the United States in various subjects during 1997 is shown in the table.

Subject	New Titles Imported
Education	190
Fiction	273
History	512
Medicine	706
Technology	501

SOURCE: U.S. Census Bureau, www.census.gov

29. The number of foreign students, in thousands, who came to study in the United States in 1995 from various countries is shown in the table.

Country	Foreign Students (in thousands)
Canada	22.7
Mexico	9.0
Japan	45.2
South Korea	33.6
Spain	5.1
United Kingdom	7.8
Turkey	6.7

SOURCE: U.S. Department of Education, www.ed.gov

30. The number of foreign students, in thousands, who went to study in the United Kingdom in 1995 from various countries is shown in the table.

Country	Foreign Students (in thousands)
Canada	1.7
France	8.9
Japan	2.6
Germany	9.5
Greece	10.3
United States	6.2
Ireland	9.8

SOURCE: U.S. Department of Education, www.ed.gov

In Exercises 31–36, answer the questions associated with the given bar graph.

31. The number of cigarettes, in billions, exported by various countries in 1998 is shown in the bar graph on the next page (*Source:* U.S. Census Bureau, www.census.gov).

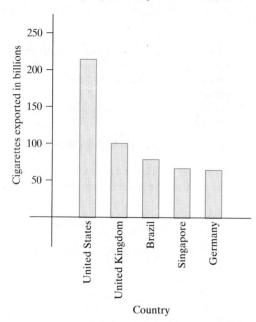

Top Cigarette Export Countries, 1998

(a) Which countries exceeded 100 billion cigarettes exported in 1998?

(b) What countries had exports that were about the same amount?

🌐 **32.** The number of cigarettes, in billions, imported by various countries in 1998 is shown in the bar graph (*Source:* U.S. Census Bureau, www.census.gov).

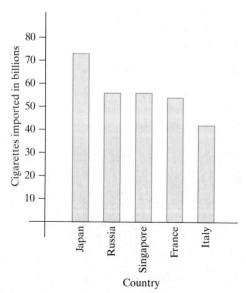

Top Cigarette Import Countries, 1998

(a) Which countries exceeded 50 billion cigarettes imported in 1998?

(b) What countries had imports that were about the same amount?

🌐 **33.** The number of monitoring stations in the U.S. for various forms of air pollution in 1998 is shown in the bar

graph (*Source:* U.S. Environmental Protection Agency, www.epa.gov).

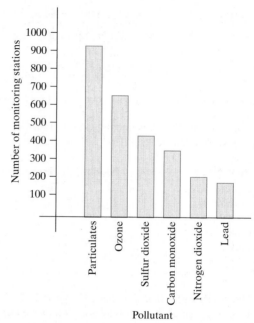

Number of Monitoring Stations for various Pollutants, 1998

(a) Which pollutants have less than 500 monitoring stations?

(b) About how many more monitoring stations are there for particulates than there are for lead?

🌐 **34.** The amount of U.S. municipal waste generated, in millions of tons, by various items in 1997 is shown in the bar graph (*Source:* U.S. Census Bureau, www.census.gov).

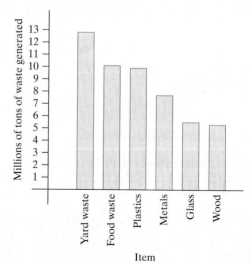

Amount of Municipal Waste Generated by Specific Items, 1997

(a) About how many millions of tons of yard waste was generated in 1997?

(b) About how many millions of tons of waste was generated by food and yard wastes combined?

35. The number of U.S. households, in thousands, who went on a leisure vacation of various types in 1995 is shown in the bar graph (*Source:* U.S. Census Bureau, www.census.gov).

Type of Leisure Trip Taken, 1995

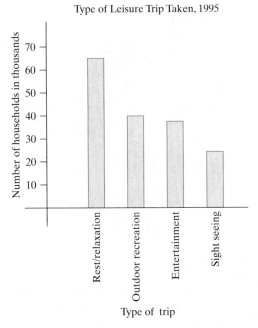

(a) About how many households took trips for rest and relaxation in 1995?

(b) About how many more rest and relaxation trips were taken in 1995 than sightseeing trips?

36. The number of traffic fatalities involving a large truck in 1997 from various states in the United States is shown in the bar graph (*Source:* U.S. Department of Transportation, www.dot.gov).

Number of Fatalities Involving a Large Truck, 1997

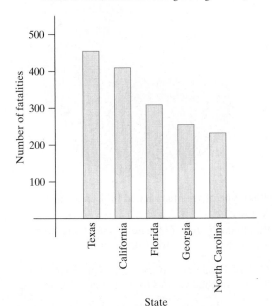

(a) Which states had more than 300 traffic fatalities involving a truck?

(b) Would the data from the bar graph be suitable for a pie chart? Explain.

In Exercises 37–46, answer the questions associated with the given pie chart.

37. The percent of the amount of debt from various sources held by American families in 1995 is shown in the pie chart (*Source:* U.S. Census Bureau, www.census.gov).

Debt Held by American Families, 1995

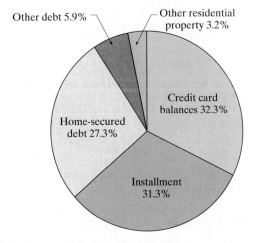

(a) What percentage of family debt is from home-secured debt and installment?

(b) What percentage of family debt is from credit-card balances and installment?

38. The percent of the amount of debt from various sources held by American families in 1998 is shown in the pie chart (*Source:* U.S. Census Bureau, www.census.gov).

Debt Held by American Families, 1998

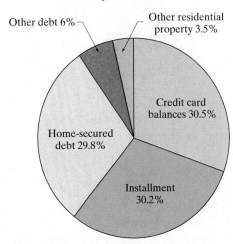

(a) What percentage of family debt is from home-secured debt and installment?

(b) What was the third largest source of family debt in 1995?

39. The types of motorized transportation used by the public in France during 1996 are shown in the pie chart. Numbers are in thousands (*Source:* U.S. Census Bureau, www.census.gov).

Types of Motorized Transportation in France, 1996

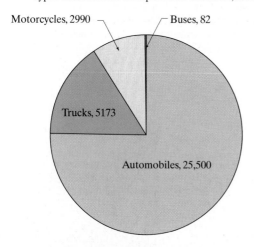

(a) What type of motorized transportation was used the least in France in 1996?

(b) What percentage of all transportation vehicles were motorcycles? Round to the nearest tenth.

40. The types of motorized transportation used by the public in the United Kingdom during 1996 are shown in the pie chart. Numbers are in thousands (*Source:* U.S. Census Bureau, www.census.gov).

Types of Motorized Transportation in the United Kingdom, 1996

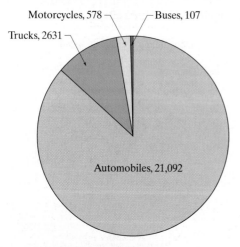

(a) What type of motorized transportation was used the least in the United Kingdom in 1996?

(b) What percentage of all motorized transportation were trucks? Round to nearest tenth.

41. The number of scientists, in thousands, employed in the United States in 1996 is shown in the pie chart (*Source:* U.S. Census Bureau, www.census.gov).

Scientists Employed in the United States, 1996

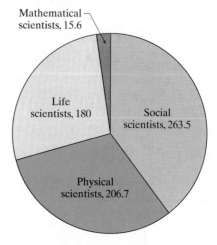

(a) What type of scientist ranked third in terms of employment?

(b) What percentage of all scientists employed in the United States were social scientists? Round to nearest tenth.

42. The number of scientists, in thousands, employed by the U.S. government in 1996 is shown in the pie chart (*Source:* U.S. Census Bureau, www.census.gov).

Scientists Employed by the United States Government, 1996

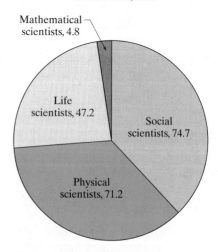

(a) What type of scientist ranked third in terms of employment?

(b) What percentage of U.S. government scientists were social scientists? Round to nearest tenth.

43. The number of hours per person per year that Americans watched various forms of television in 1992 is shown in the pie chart (*Source:* U.S. Census Bureau, www.census.gov).

Television Viewing (hours per person per year) in the United States, 1992

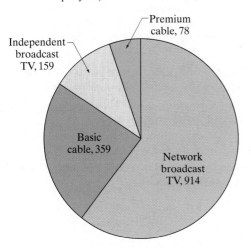

(a) What form of television was least watched in terms of hours per person per year?

(b) What percentage of television viewing was basic cable? Round to nearest tenth.

44. The number of hours per person per year that Americans watched various forms of television in 2000 is shown in the pie chart (*Source:* U.S. Census Bureau, www.census.gov).

Television Viewing (hours per person per year) in the United States, 2000

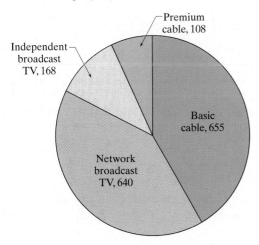

(a) What form of television was least watched in terms of hours per person per year?

(b) What percentage of television viewing was basic cable? Round to nearest tenth.

45. The types of robbery crimes, in thousands, reported in the United States in 1996 are shown in the pie chart (*Source:* U.S. Bureau of Crime Statistics, www.ojp.usdoj.gov/bjs).

Reported Robbery Crimes in the United States, 1996

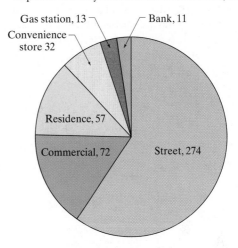

(a) About how many gas station and convenience store crimes were reported in 1996?

(b) What percentage of reported robbery crimes were residence robbery crimes? Round to nearest tenth.

46. The types of weapon used, in thousands, to commit crimes in 1995 in the United States are shown in the pie chart (*Source:* U.S. Bureau of Crime Statistics, www.ojp.usdoj.gov/bjs).

Weapons Used in Crimes in the United States, 1995

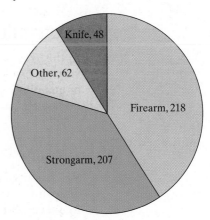

(a) About how many crimes in 1995 involved firearms and knives?

(b) What percentage of crimes involved firearms? Round to nearest tenth.

SECTION PROJECT

There are times when a **stacked bar graph** shows not only the total frequency, but also the frequency for a certain portion of the whole group. The stacked bar graph that follows shows the total births in 1993 for five of the 20 largest U.S. metropolitan areas. Births to teen mothers are also displayed. The frequencies are shown at the top of each bar for accuracy (*Source:* U.S. National Center for Health Statistics, www.cdc.gov/nchs).

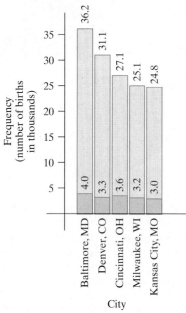

Use the graph to answer the following.

(a) Determine the percentage of births of teen mothers in Cincinnati. Use the formula

$$\frac{\text{number of births to teen mothers}}{\text{total births}} \cdot 100\%.$$

(b) Determine the percentage of births to teen mothers in Baltimore. Use the formula shown in part (a).

(c) Of all the cities displayed, which city had the highest percentage of births to teen mothers?

(d) Of all the cities displayed, which city had the lowest percentage of births to teen mothers?

(e) Determine the total percentage of births to teen mothers in these five cities.

Section 7.2   Graphing Quantitative Data

In the previous section we saw how we could use bar graphs and pie charts to visually display qualitative data. But much of the data that we study are numeric or **quantitative data**. In this section we will examine the techniques of organizing, displaying, and interpreting this type of data.

The Frequency Distribution

When a set of quantitative data in raw and ungrouped form has many values, it is difficult to see trends and patterns. We can organize the data by the use of the **frequency distribution**.

■ **Frequency Distribution**

A table that categorizes raw data into **classes** is called a **frequency distribution**. The number of data values that falls into each class is called the class **frequency**, denoted by f.

By raw data, we mean data that are given in their original form. Let's examine a frequency distribution to see how it summarizes data. Table 7.2.1 displays a frequency distribution for the number of community hospitals in the 50 states of the United States in 1990.

Table 7.2.1

Class (number of hospitals)	Frequency, f (number of states)
8 to 62	19
63 to 117	13
118 to 172	10
173 to 227	4
228 to 282	2
283 to 337	0
338 to 392	0
393 to 447	2

SOURCE: U.S. Department of Health and Human Services, www.hhs.gov

We can see that there are eight classes in this frequency distribution. The first class has 8 to 62 hospitals, the second class from 63 to 117 hospitals, and so on. The frequency of the first class is 19. This means that 19 states in the United States had from 8 to 62 hospitals in 1990. We can also see from inspecting Table 7.2.1 that in 1990 no states had from 338 to 392 community hospitals.

The leftmost number in each class is called the **lower class limit**. For example, in Table 7.2.1 the lower limit of the first class is 8 and the lower limit of the third class is 118. The rightmost number in each class is called the **upper class limit**. In Table 7.2.1, 62 and 172 are the upper limits for the first and third classes, respectively. Notice that if we take the difference between any two lower class limits we get the same number.

$$63 - 8 = 55, \quad 118 - 63 = 55, \quad 173 - 118 = 55, \quad 228 - 173 = 55, \quad \text{and so on}$$

This value is called the **class width**. The class width of the frequency distribution given in Table 7.2.1 is 55. If we add up all the frequencies, we get 50 as expected. In Chapter 6 we used the capital Greek letter $\sum$ (sigma) to signify "to sum." Hence, we can denote the sum of the frequencies by writing $\sum f = 50$. We call this sum the **sample size** and denote it with the letter n. In other words, $\sum f = n$. Note that the **population** is the set of *all* hospitals in the 50 states, not only the community hospitals.

One useful type of frequency is the **relative frequency**. This tells us what portion or percent falls into each class. To compute the relative frequency of a class, we take the frequency f and divide it by the sample size n.

■ **Relative Frequency**

The **relative frequency** of a class is the portion of observations that falls into that class and is given by

$$\text{relative frequency} = \frac{\text{frequency of a class}}{\text{sum of frequencies}} = \frac{f}{n} = \frac{f}{\sum f}$$

▶ **Note:** Relative frequencies can be expressed as either whole number percents or decimals rounded to the hundredths place.

Example 1 | **Computing Relative Frequency**

For the frequency distribution in Table 7.2.1, include a third column that gives the relative frequency of each class. Write the frequencies in decimal form.

Solution

We have already seen that, for the frequency distribution in Table 7.2.1, $n = 50$. For the first class, the relative frequency is

$$\frac{\text{frequency of first class}}{\text{sum of frequencies}} = \frac{f}{n} = \frac{19}{50} = 0.38$$

Continuing in this manner, we get the relative frequencies shown in Table 7.2.2.

Table 7.2.2

Class (number of hospitals)	Frequency, f (number of states)	Relative Frequency
8 to 62	19	0.38
63 to 117	13	0.26
118 to 172	10	0.20
173 to 227	4	0.08
228 to 282	2	0.04
283 to 337	0	0.00
338 to 392	0	0.00
393 to 447	2	0.04

Interactive Activity

Sum the relative frequencies in Table 7.2.2. How do you explain the value of the sum? Will the sum of relative frequencies always be this number? Explain.

Constructing Frequency Distributions

Now that we have seen some of the features of the frequency distribution, we can use the guidelines that follow to construct a frequency distribution from raw data.

■ Constructing a Frequency Distribution

Step 1: Decide on the number of classes that you would like the frequency distribution to have. In general, there should be between 5 and 15 classes. (Many times, the number of classes to be used is given.)

Step 2: Determine the class width, denoted by CW. This is given by

$$CW = \frac{\text{largest data value} - \text{smallest data value}}{\text{number of classes}}$$

As a convention, we round CW up to the next whole number.

Step 3: Set the class limits. The smallest data value is the lower limit of the first class. The other lower class limits are determined by adding the class width to the lower limit of the preceding class. The upper limits are one unit less than the lower limit of the next class.

Step 4: Tally the data to determine the frequency of each class. Sorting the data in ascending order can simplify this task.

▶ **Note:** When setting the class limits, we should keep in mind the following:

1. Each data item can fall into only one class. That is, the classes cannot overlap.

2. Each class should have the same width. (An exception could occur if there are outliers.)
3. Every data item must fall into a class. If this does not occur, classes must be added or the class width must be increased.

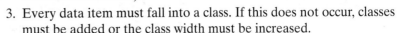

Example 2 Constructing a Frequency Distribution

The average annual expenditures in dollars on vehicle purchases in 28 U.S. metropolitan areas during 1995 are shown in Table 7.2.3.

Table 7.2.3

1225	1755	1815	1926	1967	2240	2270
2339	2401	2454	2640	2656	2684	2828
2833	2931	3108	3184	3235	3241	3286
3339	3521	3521	3598	4225	5136	5814

SOURCE: U.S. Statistical Abstract, www.census.gov/statab/www/

Construct a frequency distribution using six classes.

Solution

For this example, we will follow the four steps in the frequency distribution guidelines.

Step 1: Here we are given that we should have six classes.

Step 2: Since the largest data value is 5814 and the smallest data value is 1225, we determine that the approximate class width is

$$\frac{5814 - 1225}{6} \approx 764.83$$

Rounding up to the next whole number, we get $CW = 765$.

Step 3: The smallest data value is 1225, so this gives us the lower limit of the first class. The lower limits of the successive classes are found by adding the class width to the lower limit of the previous class. This means that the lower limit of the second class is $1225 + 765 = 1990$, the lower limit of the third class is $1990 + 765 = 2755$, and so on. For the upper class limits, we know that the upper limit of the first class is 1989, which is one less than the lower limit of the second class. The upper limit of the second class is 2754, which is one less than the upper limit of the third class, and so on. The classes and the class limits are given in Table 7.2.4.

Table 7.2.4

Class Number	Lower Limit	Upper Limit
1	1225	1990 − 1 = 1989
2	1225 + 765 = 1990	2754
3	2755	3519
4	3520	4284
5	4285	5049
6	5050	5814

Step 4: Now we tally the number of data items that fall into each class. This is the class frequency. For example, since there are five data values between 1225 and 1989 inclusive, the frequency of the first class is 5. The completed frequency distribution is shown in Table 7.2.5.

Table 7.2.5

Class (expenditure on vehicle purchases)	Frequency, f (number of metro areas)
1225 to 1989	5
1990 to 2754	8
2755 to 3519	9
3520 to 4284	4
4285 to 5049	0
5050 to 5814	2

✓ **Checkpoint 1**

Now work Exercise 5.

The Histogram

One of the most popular graphical data representations of a frequency distribution is the **histogram**. To plot histograms, we first need to compute the **class boundaries**.

■ **Class Boundary**

For the given class limits of a frequency distribution, we have the following:

1. The **lower class boundary** is the lower class limit minus one-half unit. That is,

$$\text{lower boundary} = \text{lower limit} - 0.5$$

2. The **upper class boundary** is the upper class limit plus one-half unit. That is,

$$\text{upper boundary} = \text{upper limit} + 0.5$$

In Table 7.2.6, we add a column of class boundaries to the frequency distribution of the state community hospitals given in Table 7.2.1. Notice in the table

Table 7.2.6

Class (number of hospitals)	Class Boundaries	Frequency, f (number of states)
8 to 62	7.5 to 62.5	19
63 to 117	62.5 to 117.5	13
118 to 172	117.5 to 172.5	10
173 to 227	172.5 to 227.5	4
228 to 282	227.5 to 282.5	2
283 to 337	282.5 to 337.5	0
338 to 392	337.5 to 392.5	0
393 to 447	392.5 to 447.5	2

that the upper boundary of a class is equal to the lower boundary of the next class. We can see from the properties of the histogram that follow that this is no coincidence. We want the rectangles in a histogram to have no gaps between them, as was seen with the bar graph.

■ **Properties of the Histogram**

1. The horizontal axis is labeled with the class boundaries.
2. The vertical axis is labeled with the frequency.
3. Adjacent bars are constructed with the width determined by the class boundaries and the height determined by the class frequency.
4. The histogram can also be constructed using the relative frequencies.
5. Every histogram is given a title.

Example 3 **Constructing a Histogram**

Interactive Activity

Add a column to the frequency distribution in Table 7.2.7 for the relative frequency of each class. Then construct a histogram using the relative frequencies as the label on the vertical axis. How does the shape of this histogram compare to the histogram in Figure 7.2.1?

Construct a histogram for the frequency distribution in Example 2.

Solution

First we need to add a column containing the class boundaries to the frequency distribution. This is shown in Table 7.2.7.
     Figure 7.2.1 shows the histogram for the data.

Table 7.2.7

Expenditure on Vehicle Purchases	Class Boundaries	Frequency, f
1225 to 1989	1224.5 to 1989.5	5
1990 to 2754	1989.5 to 2754.5	8
2755 to 3519	2754.5 to 3519.5	9
3520 to 4284	3519.5 to 4284.5	4
4285 to 5049	4284.5 to 5049.5	0
5050 to 5814	5049.5 to 5814.5	2

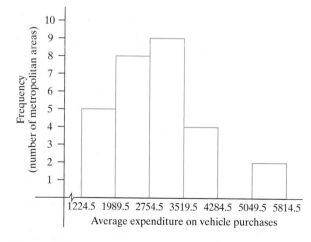

Figure 7.2.1

✓ **Checkpoint 2**   Now work Exercise 15.

Technology Option

Histograms can also be made using a graphing calculator. Figure 7.2.2(a) shows the histogram for Example 3 made on a graphing calculator, and Figure 7.2.2(b) shows the histogram for Example 3 made in Excel.

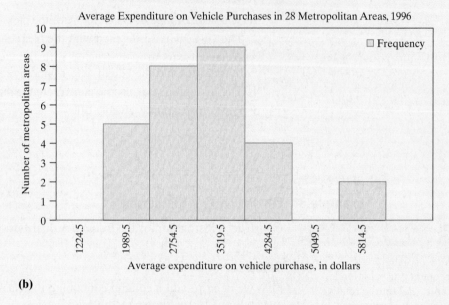

(a)

(b)

Figure 7.2.2

For more on making histograms on your calculator, consult the online graphing calculator manual at www.prenhall.com/armstrong.

The Frequency Polygon

Another way to graph a frequency distribution is to construct a **frequency polygon**. This type of graph uses the **midpoints** of the classes on the horizontal axis.

■ **Class Midpoint**

Given a frequency distribution, the **class midpoint**, denoted by m, is given by

$$m = \frac{\text{lower class limit} + \text{upper class limit}}{2}$$

The frequency polygon is usually, but not always, associated with the classes being some measure of time, with the units being minutes, hours, or years.

■ **Properties of the Frequency Polygon**

1. The horizontal axis is labeled with the class midpoints.
2. The vertical axis is labeled with the frequency or the relative frequency.
3. Points are plotted using the midpoints and the frequencies.
4. The points are connected with line segments.

5. Extra *tie-down* points are added to the beginning and end of the graph. The tie-down points have a frequency of zero. The left tie-down point is the first class midpoint minus the class width. The right tie-down point is the last class midpoint plus the class width.

Example 4 Constructing a Frequency Polygon

The number of Americans, measured in millions, aged 5 to 74 who were covered by private or government health insurance in 1997 is shown in the frequency distribution in Table 7.2.8.

Table 7.2.8

Class (age group)	Frequency, f (number covered in millions)
5 to 14	38
15 to 24	35
25 to 34	40
35 to 44	45
45 to 54	34
55 to 64	22
65 to 74	32

SOURCE: U.S. Department of Health and Human Services, www.hhs.gov

(a) Compute the class width.

(b) Add a column to the frequency distribution that contains the class midpoints.

(c) Construct a frequency polygon for the distribution.

Solution

(a) The difference between the lower class limits of the first two classes is $15 - 5 = 10$, which is also true for each successive class. So the class width is $CW = 10$.

(b) The midpoint of the first class is $\dfrac{5 + 14}{2} = 9.5$. Continuing in this fashion, we get the distribution shown in Table 7.2.9. Notice that the difference between any two consecutive class midpoints is also the class width.

Table 7.2.9

Class (age group)	Class Midpoint, m	Frequency, f (number covered in millions)
5 to 14	9.5	38
15 to 24	$\frac{15+24}{2} = 19.5$	35
25 to 34	29.5	40
35 to 44	39.5	45
45 to 54	49.5	34
55 to 64	59.5	22
65 to 74	69.5	32

(c) Before we construct the frequency polygon, we determine the tie-down points. To get the left tie-down point, we subtract the class width from the first class midpoint to get $9.5 - 10 = -0.5$. The right tie-down point is found by adding the class width to the last class limit, which gives us $69.5 + 10 = 79.5$. The completed frequency polygon is shown in Figure 7.2.3. Note that, even though it may not make sense to have a negative value on the horizontal axis, the idea is to include tie-down points so that a closed polygon is formed.

Americans Covered by Private or Government Health Insurance, 1997

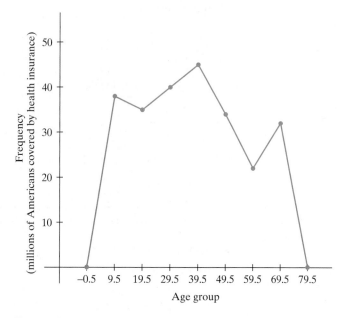

Figure 7.2.3   ■

✓ **Checkpoint 3**

Now work Exercise 27.

The Ogive

Many times we want to know how many data values are either below or above a certain quantitative value or even between two given values. This can be done by constructing an **ogive** (pronounced ójive). An ogive is a graph of the **cumulative frequencies** of each class.

■ **Cumulative Frequency**

The **cumulative frequency** of a class is the frequency of that class added to the frequencies of the previous classes.

We can see from the definition that the cumulative frequency of the first class is just that class's frequency. The cumulative frequency of the last class in a frequency distribution is the same as the sample size n.

■ **Properties of the Ogive**

1. The horizontal axis contains the upper class boundaries.
2. The vertical axis contains the cumulative frequencies.
3. Points are plotted with the upper class boundaries and the cumulative frequencies. The points are connected with line segments.
4. A leftmost point is plotted with a cumulative frequency of zero. The point is plotted on the horizontal axis and is given by the first upper class boundary minus the class width.

Example 5   Constructing an Ogive

For the frequency distribution in Example 3, complete the following:

(a) Compute the cumulative frequency of each class.

(b) Construct an ogive for the distribution.

Solution

(a) The cumulative frequency of the first class is 5 since there are no classes that come before it. The cumulative frequency of the second class is $5 + 8 = 13$. Continuing in this fashion, we get the distribution shown in Table 7.2.10.

Table 7.2.10

Class	Class Boundaries	Frequency, f	Cumulative Frequency
1225 to 1989	1224.5 to 1989.5	5	5
1990 to 2754	1989.5 to 2754.5	8	$5 + 8 = 13$
2755 to 3519	2754.5 to 3519.5	9	$13 + 9 = 22$
3520 to 4284	3519.5 to 4284.5	4	$22 + 4 = 26$
4285 to 5049	4284.5 to 5049.5	0	$26 + 0 = 26$
5050 to 5814	5049.5 to 5814.5	2	$26 + 2 = 28$

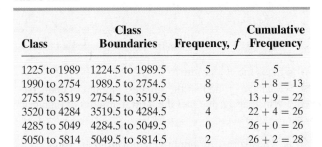

Average Expenditure on Vehicle Purchases

Figure 7.2.4

(b) Since the class width is 765, we see that the left tie-down point is $1989.5 - 765 = 1224.5$. Note that this is also the lower boundary for the first class. The ogive is shown in Figure 7.2.4. Among other observations, we can see from the graph that about 20 of the U.S. metropolitan areas had average vehicle purchases of about $3500 or less.   ■

✓ **Checkpoint 4**   Now work Exercise 35.

SUMMARY

In this section, we saw that we could organize a set of quantitative data by constructing a **frequency distribution**. From this distribution, we could make the **histogram**, **frequency polygon**, and the **ogive**.

- The **relative frequency** of a class is given by

$$(\text{relative frequency}) = \frac{\text{frequency of a class}}{\text{sum of frequencies}} = \frac{f}{n} = \frac{f}{\sum f}.$$

- The **class width**, denoted by CW, is given by

$$CW = \frac{\text{largest data value} - \text{smallest data values}}{\text{number of classes}},$$

where CW is rounded up to the next whole number.

- The **lower class boundary** is the lower class limit minus one-half unit.
- The **upper class boundary** is the upper class limit plus one-half unit.
- The **class midpoint**, denoted by m, is given by

$$m = \frac{\text{lower class limit} + \text{upper class limit}}{2}.$$

- The **cumulative frequency** of a class is the frequency of that class added to the frequencies of the previous classes.

SECTION 7.2 EXERCISES

In Exercises 1–10, a frequency distribution is given. Answer the following.

1. The amount of square footage for single-family homes, in thousands, in the United States during 1997 is shown in the frequency distribution.

Class (square feet)	Frequency (in thousands)
501 to 999	9,764
1000 to 1499	17,030
1500 to 1999	15,007
2000 to 2499	10,680
2500 to 2999	5,944
3000 to 3499	3,282
3500 to 3999	2,188

SOURCE: U.S. Census Bureau, www.census.gov

(a) What is the class width?

(b) How many houses had square footage from 2000 to 2499 square feet?

2. The amount of square footage for *vacant* single family homes, in thousands, in the United States during 1997 is shown in the frequency distribution.

Class (square feet)	Frequency (in thousands)
501 to 999	799
1000 to 1499	1130
1500 to 1999	741
2000 to 2499	373
2500 to 2999	238
3000 to 3499	127
3500 to 3999	85

SOURCE: U.S. Census Bureau, www.census.gov

(a) What is the class width?

(b) How many vacant houses had square footage from 2000 to 2499 square feet?

3. The average amount of financial debt, in thousands of dollars, held by U.S. families during 1995 is classified by age group in the frequency distribution.

Class (age group)	Frequency (in thousands of dollars)
25 to 34	15.0
35 to 44	37.6
45 to 54	40.7
55 to 64	21.5
65 to 74	7.6
75 to 84	1.9

SOURCE: U.S. Census Bureau, www.census.gov

(a) What is the class width?

(b) Which age group had the smallest amount of financial debt? Which age group had the largest amount of financial debt?

4. The average amount of *credit-card* debt, rounded to the nearest hundred dollars, held by U.S. families during 1995 is classified by age group in the frequency distribution.

Class (age group)	Frequency (in dollars)
25 to 34	1300
35 to 44	1900
45 to 54	2000
55 to 64	1300
65 to 74	800
75 to 84	400

SOURCE: U.S. Census Bureau, www.census.gov

(a) What is the class width?

(b) Which age group had the smallest amount of credit-card debt? Which age group had the largest amount of credit-card debt?

5. The number of farms, in thousands, owned by Americans during 1992 is classified by age group in the following frequency distribution.

Class (age group)	Frequency (in thousands of farms)
15 to 24	28
25 to 34	179
35 to 44	382
45 to 54	429
55 to 64	430
65 to 74	478

SOURCE: U.S. Department of Agriculture, www.usda.gov

(a) How many farms were owned by people who were 55 years or older?

(b) How many farms were owned by people who were 25 to 54 years old?

6. The number of farms, in thousands, owned by Americans during 1997 is classified by age group in the following frequency distribution.

Class (age group)	Frequency (in thousands of farms)
15 to 24	21
25 to 34	128
35 to 44	371
45 to 54	467
55 to 64	427
65 to 74	497

SOURCE: U.S. Department of Agriculture, www.usda.gov

(a) How many farms were owned by people who were 55 years or older?

(b) How many farms were owned by people who were 25 to 54 years old?

7. The number of Americans, in millions, aged 15 to 74 covered by private or government health insurance by age group in 1997 is shown in the frequency distribution.

Class (age group)	Frequency (in millions)
15 to 24	32.3
25 to 34	39.4
35 to 44	44.5
45 to 54	34.1
55 to 64	22.3
65 to 74	32.1

SOURCE: U.S. Department of Health and Human Services, www.hhs.gov

(a) According to the frequency distribution, how many Americans aged 15 to 74 were covered by private or government health insurance in 1997?

(b) Which age group had the most Americans covered by private or government health insurance in 1997?

8. The number of Americans, in millions, covered by private or government health insurance classified by household income in 1997 is shown in the frequency distribution.

Class (income in dollars)	Frequency (in millions)
1 to 25,000	72.2
25,001 to 50,000	80.4
50,001 to 75,000	56.2
75,001 to 100,000	60.3

SOURCE: U.S. Department of Health and Human Services, www.hhs.gov

(a) According to the frequency distribution, how many Americans with household incomes of $100,000 or less were covered by private or government health insurance in 1997?

(b) How many Americans with household incomes of $50,000 or less were covered by private or government health insurance in 1997?

(c) How many Americans with household incomes of $50,001 or more were covered by private or government health insurance in 1997?

9. The costs for homeowners, in thousands, in the northeastern United States whose monthly housing costs were under $1000 are shown in the frequency distribution.

Class (cost in dollars)	Frequency (in thousands of homeowners)
200 to 299	1923
300 to 399	1493
400 to 499	1339
500 to 599	967
600 to 699	849
700 to 799	765
800 to 899	482
900 to 999	722

SOURCE: U.S. Census Bureau, www.census.gov

(a) How many northeastern homeowners had monthly costs of less than $500?

(b) How many northeastern homeowners had monthly costs of $300 or more, but less than $700?

10. The costs for homeowners, in thousands, in the midwestern United States whose monthly housing costs were under $1000 are shown in the frequency distribution.

Class (cost in dollars)	Frequency (in thousands of homeowners)
200 to 299	4912
300 to 399	2186
400 to 499	1430
500 to 599	1379
600 to 699	1241
700 to 799	1078
800 to 899	700
900 to 999	1052

SOURCE: U.S. Census Bureau, www.census.gov

(a) How many midwestern homeowners had monthly costs of less than $500?

(b) How many midwestern homeowners had monthly costs of $300 or more, but less than $700?

In Exercises 11–16, complete the following:

(a) Determine the class boundaries.

(b) Construct a histogram.

11. Use the frequency distribution in Exercise 1.

12. Use the frequency distribution in Exercise 2.

13. Use the frequency distribution in Exercise 3.

14. Use the frequency distribution in Exercise 4.

✓ **15.** Use the frequency distribution in Exercise 5.

16. Use the frequency distribution in Exercise 6.

In Exercises 17–20, complete the following:

(a) Determine the class boundaries.

(b) Determine the relative frequency for each class. Round to the nearest hundredth.

(c) Construct a relative frequency histogram.

17. Use the frequency distribution in Exercise 7.

18. Use the frequency distribution in Exercise 8.

19. Use the frequency distribution in Exercise 9.

20. Use the frequency distribution in Exercise 10.

For the data in Exercises 21–26, complete the following:

(a) Write a frequency distribution. See Example 2.

(b) Use your calculator to construct a histogram.

21. The amount in dollars, spent on new clothes at the Happening Zone for a sample of 30 preteens is shown in the table. Use five classes.

48	345	127	33	48	260
189	388	190	286	370	100
247	255	400	130	189	263
0	144	188	266	241	188
378	330	249	287	274	91

22. The time, in minutes, spent watching the local news during one day for a sample of 28 citizens of Lewisburg is shown in the table. Use five classes.

45	120	25	10	35	29	39
15	51	14	38	210	180	185
125	60	35	40	81	36	110
85	66	90	41	50	35	15

23. The area, measured in square miles, of selected major bodies of water in the Gulf Coast area of the United States is shown in the table. Use six classes.

813	733	631	616	511
310	271	253	245	236
212	189	151	146	129
122	118	112	112	104
101	99	94	93	89
85	83	79	75	

SOURCE: U.S. Department of the Interior, www.doi.gov

24. The area, measured in square miles, of selected major bodies of water in Alaska is shown in the table. Use six classes.

1559	1382	1199	1022	792	791
791	702	640	463	447	436
433	403	393	348	345	324
324	322	310	310	301	295
293	241	229	225	210	206

SOURCE: U.S. Department of the Interior, www.doi.gov

25. The length, in miles, of the 25 longest rivers in the United States is shown in the table. Use five classes.

2540	2340	1980	1900	1900
1460	1450	1420	1310	1290
1280	1240	1040	990	926
906	886	862	800	774
743	724	692	659	649

SOURCE: U.S. Department of the Interior, www.doi.gov

26. The number of visitors, in thousands, by overseas travelers to the top 35 traveled states in the United States in 1996 is shown in the table. Use six classes.

22,658	6004	5710	4804	3059	2062	1292	1156	1178	974
897	634	657	702	612	566	363	385	498	408
340	363	385	498	408	340	363	249	227	272
272	227	181	159	159					

SOURCE: U.S. Census Bureau, www.census.gov

In Exercises 27–34, a frequency distribution is given. Complete the following:

(a) Determine the midpoint of each class.

(b) Construct a frequency polygon.

✓ **27.** The ages of the employees at the NewFinish Corporation are shown in the frequency distribution.

Class (age)	Frequency (in employees)
20 to 29	15
30 to 39	50
40 to 49	56
50 to 59	30
60 to 69	18

28. The pretest scores of 24 entering freshmen at Festivelle High School are shown in the frequency distribution.

Class (test score)	Frequency (in students)
96 to 102	2
103 to 109	3
110 to 116	3
117 to 123	4
124 to 130	5
131 to 137	1
138 to 144	4
145 to 151	1
152 to 158	1

29. The weights, in pounds, of 100 sixth-grade students at P.H. Johnson elementary school are shown in the frequency distribution.

Class (pounds)	Frequency (in students)
59 to 61	4
62 to 64	8
65 to 67	12
68 to 70	13
71 to 73	21
74 to 76	15
77 to 79	12
80 to 82	9
83 to 85	4
86 to 88	2

30. The total fines, in dollars, for outstanding parking tickets in the village of Kettersville are shown in the frequency distribution.

Class (dollars)	Frequency
10 to 24	5
25 to 39	8
40 to 54	9
55 to 69	15
70 to 84	5
85 to 99	2

31. Use the frequency distribution in Exercise 1.

32. Use the frequency distribution in Exercise 2.

33. Use the frequency distribution in Exercise 3.

34. Use the frequency distribution in Exercise 4.

In Exercises 35–40, complete the following:

(a) Determine the cumulative frequency for each class.

(b) Construct an ogive.

✓ **35.** Use the frequency distribution in Exercise 27.

36. Use the frequency distribution in Exercise 28.

37. Use the frequency distribution in Exercise 29.

38. Use the frequency distribution in Exercise 30.

39. Use the frequency distribution in Exercise 6.

40. Use the frequency distribution in Exercise 7.

🌐 *For Exercises 41–44, use the histogram for U.S. mothers aged 15 to 39 of newborns in 1990 to answer the following (Source: U.S. Census Bureau,* www.census.gov*).*

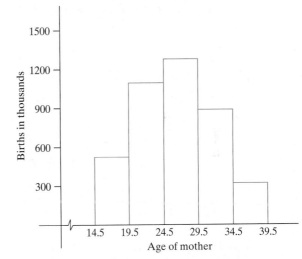

Age of Mother to Newborns, 1990

41. What age group had the most newborns?

42. What age group had the least newborns?

43. What is the class width?

44. What are the class *limits* of the third class?

🌐 *For Exercises 45–48, use the histogram (see on the next page) for U.S. mothers aged 15 to 44 of newborns in 1998 to answer the following (Source: U.S. Census Bureau,* www.census.gov*).*

45. What age group had the most newborns?

46. What age group had the least newborns?

47. What is the class width?

48. What are the class *limits* of the third class?

Age of Mother to Newborns, 1998

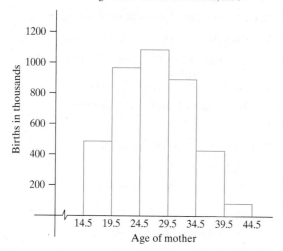

For Exercises 49–52, use the frequency polygon for the amount spent on child care and education to raise a child for husband–wife families whose incomes are between $36,000 and $60,600 in 1998 to answer the following (Source: U.S. Department of Health and Human Services, www.hhs.gov).

Annual Cost of Childcare and Education, 1998

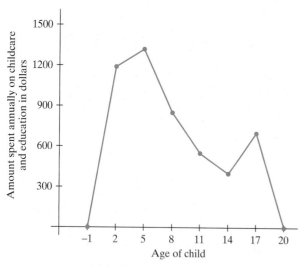

49. What is the class *width*?

50. What are the *limits* of the fourth class?

51. In which class do parents spend over $1300 annually on child care and education?

52. In which class do parents spend about $400 annually on child care and education?

For Exercises 53–56, use the frequency polygon for the number of deaths, in thousands, caused by cancer in the United States in 1997 to answer the following (Source: U.S. National Center for Health Statistics, www.cdc.gov/nchs).

53. What is the class *width*?

54. What are the class *limits* of the third class?

Deaths Caused by Cancer, 1997

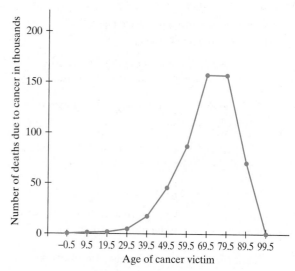

55. In which class(es) is the number of deaths caused by cancer close to 160,000?

56. In which class is the number of deaths caused by cancer close to 70,000?

For Exercises 57–60, use the ogive for the number of U.S. drivers, in thousands, involved in alcohol-related fatal crashes in 1998 to answer the following (Source: U.S. Department of Transportation, www.dot.gov).

Alcohol Involvement in Fatal Auto Crashes, 1998

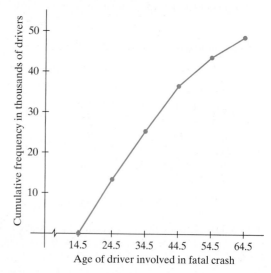

57. What are the class *limits* of the first class?

58. What is the class *width*?

59. About how many drivers aged 34.5 years and below were involved with alcohol-related fatal crashes in 1998?

60. From inspecting the ogive, which class has the greatest frequency? Explain.

For Exercises 61–64, use the ogive for the number of U.S. deaths, in thousands, caused by accidental and adverse effects in 1997 (Source: U.S. National Center for Health Statistics, www.cdc.gov/nchs).

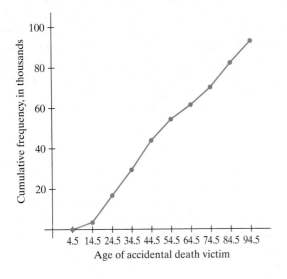

Deaths Caused by Accidental and Adverse Effects, 1997

61. What are the class *limits* of the first class?

62. What is the class *width*?

63. About how many individuals aged 54.5 years and below died from accidental and adverse effects in 1997?

64. From inspecting the ogive, which class has the greatest frequency? Explain.

SECTION PROJECT

Another quick way to obtain a graph similar to a histogram is to use a **stemplot** (sometimes called a stem and leaf plot) to plot the data. On the stemplot, the data are separated by a **stem** (the leftmost digits) and a **leaf** (the rightmost digit) by a vertical bar. For example, consider the following data for the test results for 20 students taking a philosophy final exam.

| 22 | 40 | 51 | 52 | 62 | 62 | 68 | 69 | 69 | 74 |
| 78 | 79 | 80 | 82 | 85 | 86 | 88 | 90 | 94 | 99 |

The stemplot for the data is given in Figure 7.2.5.

The average fuel consumption, in miles per gallon, for U.S. automobiles for selected years from 1970 to 1996 is shown in the table (*Source:* U.S. Department of Transportation, www.dot.gov).

| 13.5 | 14.0 | 16.0 | 16.4 | 16.8 | 17.0 | 17.3 | 17.5 | 17.4 | 18.0 |
| 18.7 | 18.9 | 20.2 | 21.1 | 20.9 | 20.5 | 20.7 | 21.1 | 21.3 | 21.4 |

(a) Construct a stemplot for the average fuel consumption for automobiles. Use the digits to the left of the decimal

Stem	Leaves
2	2
3	
4	0
5	1 2
6	2 2 8 9 9
7	4 8 9
8	0 2 5 6 8
9	0 4 9

Figure 7.2.5

point for the stems and the digit to the right of the decimal point for the leaves.

(b) Which row has the greatest number of leaves? How would you interpret this?

(c) Are there any rows that have no leaves? What does this mean?

(d) If a histogram of the fuel consumption were constructed, how would its shape compare to that of the stemplot?

Section 7.3 Measures of Centrality

Many times we are given a set of data to inspect, but there are so many numbers, or the numbers are so diverse, that it is difficult to analyze the data. To make sense of sets of data, we often want to know what mathematical characteristics data items possess. These characteristics are often referred to as **measures of centrality**, because they give us the value that represents the center or core of the data. When we hear statements such as "The average salary of American workers in 1997 was $29,261" or "The median sales price of a home in the United States in 1996 was $115,800," we are hearing measures of centrality. In this section we will see how these measures are computed and interpreted, the properties of these measures, and how to approximate a measure of centrality even if we do not have all the data at our disposal.

The Mean

One of the most frequently used measures in statistics is the **mean**, which is commonly called the **average**. We can determine the mean of a sample of n data items by determining its sum and dividing the sum by the sample size. To express the sum of the data items, we will use sigma notation and write $\sum x$, which we read as "the sum of all x-values."

■ **Mean**

- For a **sample** of n data items, the **sample mean**, denoted by $\bar{x}$ (read "x-bar") is given by

$$\bar{x} = \frac{\sum x}{n} = \frac{\text{sum of data}}{\text{sample size}}$$

- If we have a **population** of N data items, then the **population mean**, denoted by μ (the Greek lowercase letter mu), is given by

$$\mu = \frac{\sum x}{N} = \frac{\text{sum of data}}{\text{population size}}$$

> **From Your Toolbox**
>
> A **population** is the set of all elements under consideration. A **sample** is a subset of the population.

▶ **Note:** Observe that we use n for the sample size and N for population size.

We can see from the definition of the mean and the note following the definition that we write $\bar{x}$ if we have a sample and μ if the data are regarded as a population. Since one is divided by n and the other by N, we must use two different notations.

Example 1 Determining a Mean

The seven most popular recreational activities of Americans in 1997 are shown in Table 7.3.1. Determine the mean and interpret.

Table 7.3.1

Activity	Participation (in millions)
Exercise walking	76.3
Swimming	59.5
Exercising with equipment	47.9
Camping	46.6
Bicycle riding	45.1
Bowling	44.8
Fishing	39.0

SOURCE: U.S. Census Bureau, www.census.gov

Solution

Understand the Situation: Notice that the data in Table 7.3.1 are for the seven most popular recreational activities. The table does not list all the recreational activities. This means that the data in Table 7.3.1 are a sample of all recreational activities. As a result, we are to compute a sample mean, $\bar{x}$.

Here we have a sample with size $n = 7$. The formula for the sample mean yields

$$\bar{x} = \frac{\sum x}{n} = \frac{76.3 + 59.5 + 47.9 + 46.6 + 45.1 + 44.8 + 39.0}{7} = \frac{359.2}{7} \approx 51.3$$

Interpret the Solution: Since inevitably some people participate in more than one activity, this does not mean that in 1997 an average of 51.3 million Americans participated in the top seven recreational activities. What it does mean is that in 1997 an average of 51.3 million participations were made in the top seven recreational activities by Americans. Some of the participations may have been made by the same people. ∎

✓ **Checkpoint 1**

Now work Exercise 7.

Technology Option

All graphing calculators compute the mean for a given set of data, but the way that the data are entered and used for statistical computations varies from model to model. In Figure 7.3.1(a), we have entered the data from Example 1 into a list. In Figure 7.3.1(b), we have used the 1-Variable Stats menu. Notice in Figure 7.3.1(b)

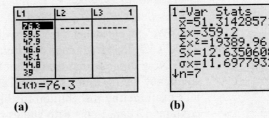

(a)   (b)

Figure 7.3.1

that $\bar{x} \approx 51.3$, the same result achieved in Example 1. Other items of interest in Figure 7.3.1(b) are $\sum x = 359.2$ and $n = 7$. Both of these were also determined in Example 1.

To see how your calculator computes the mean, consult the online graphing calculator manual at www.prenhall.com/armstrong.

We learned in Section 7.2 that when samples of data are relatively large we can categorize the data into classes and summarize them in a **frequency distribution**. However, if the data have already been summarized in a frequency distribution, it is still possible to estimate the mean. Before we illustrate how this is done, let's consider another example of computing a mean.

Example 2 **Determining a Mean**

The length in miles, of the 25 longest U.S. rivers is shown in Table 7.3.2. Determine the mean and interpret.

Table 7.3.2

2540	2340	1980	1900	1900
1460	1450	1420	1310	1290
1280	1240	1040	990	926
906	886	862	800	774
743	724	692	659	649

SOURCE: U.S. Department of the Interior, www.doi.gov

Solution

Understand the Situation: Table 7.3.2 shows the lengths of the 25 longest U.S. rivers. The table does not contain all U.S. rivers, which indicates that we have a sample. As a result, we need to compute a sample mean, $\bar{x}$.

Here we have a sample with size $n = 25$. The formula for the sample mean yields

$$\bar{x} = \frac{\sum x}{n} = \frac{2540 + 2340 + 1980 + \cdots + 692 + 659 + 649}{25} = \frac{30,761}{25} = 1230.44$$

Interpret the Solution: This means that the average length of the 25 longest U.S. rivers is about 1230 miles. ∎

Table 7.3.3

Class	Frequency, f
649 to 1027	12
1028 to 1406	5
1407 to 1785	3
1786 to 2164	3
2165 to 2543	2

We can construct a frequency distribution, using five classes, for the data in Table 7.3.2. This frequency distribution is shown in Table 7.3.3.

Now suppose that we did not have the actual data and were only given the frequency distribution shown in Table 7.3.3. To see how we can approximate $\bar{x}$, let's examine the distribution further. We know, for example, that the first class contains 12 data items with values ranging anywhere from 649 to 1027. If we do not know what these 12 values are, the best guess would be the value in the middle of the class. In Section 7.2, we called this value the **class midpoint**, x_m. The midpoint of each class is shown in Table 7.3.4.

Table 7.3.4

Class	Midpoint, x_m	Frequency, f
649 to 1027	$\frac{649 + 1027}{2} = 838$	12
1028 to 1406	$\frac{1028 + 1406}{2} = 1217$	5
1407 to 1785	1596	3
1786 to 2164	1975	3
2165 to 2543	2354	2

From Your Toolbox

The **midpoint** of a class is determined by computing

$$\frac{\text{lower limit} + \text{upper limit}}{2} = x_m.$$

To approximate $\bar{x}$, we first multiply the class midpoint by the class frequency. This operation gives us an approximation for the sum of the values in a class. Then we add the products and divide by the sum of the frequencies to determine our estimate. For the distribution in Table 7.3.4, we get

$$\bar{x} \approx \frac{838(12) + 1217(5) + 1596(3) + 1975(3) + 2354(2)}{25} = 1262.48$$

This means, using the frequency distribution, that the mean of the 25 longest U.S. rivers is about 1262 miles. Notice that this value differs from the actual mean found in Example 2 by only about 32 miles.

■ **Mean from a Frequency Distribution**

Given a frequency distribution, an approximation for the **mean** of the summarized data is given by

$$\bar{x} \approx \frac{\sum x_m \cdot f}{n}$$

where

x_m is the midpoint of each class
f is the class frequency
n is the sample size, where $n = \sum f$

The Median

Table 7.3.5

2	1	12	3	2	3	2

Even though the mean is the most commonly used measure of centrality, it often does not give us enough information about the data. To see how this can happen, let's suppose that a group of seven students from Vanderburg Junior High School are participants in the summer *Got the Need to Read* program. The number of books read by each participant is shown in Table 7.3.5.

The mean of the data is $\bar{x} = \dfrac{\sum x}{n} = \dfrac{2 + 1 + 12 + 3 + 2 + 3 + 2}{7} = \dfrac{25}{7} \approx 3.6.$

This number is somewhat misleading. Notice that the mean is about 3.6, yet there are six values less than 3.6 and only one greater than 3.6. The reason that the mean is so large is because one motivated student read 12 books; nine more than any other student in the group. We call this value of 12 an **outlier** because it "lies outside" all the other data items. Since the mean uses every data element when it is calculated, it is susceptible to being distorted by outliers. When outliers are present, and in other cases as well, we may find another measure of centrality called the **median** useful.

Interactive Activity

We saw that if $n = 7$ then the median was the fourth item of the ordered data. Which item in the ordered data would be the median if $n = 11$ or $n = 23$? Write a general expression telling which item in the ordered data would be the median for a sample of size n, where n is an odd number.

■ **Median**

For a set of data, the **median**, denoted by M, is the middle value of the set of data after the data have been ordered from smallest to largest.

▶ **Note:** For a set of data, we can think of the median as being similar to the "median" in a divided highway.

If we order the data from the summer reading program from smallest to largest, we get

$$1, \quad 2, \quad 2, \quad 2, \quad 3, \quad 3, \quad 12$$

Since there are $n = 7$ data items, the median is the fourth value in the order, meaning that $M = 2$.

Computing the median when we have an odd number of data items is fairly easy, but what happens when we have an even number of data? Example 3 illustrates how to handle this situation.

Example 3 **Determining the Median**

The total number of votes cast, in millions, in U.S. presidential elections from 1960 to 1996 is shown in Table 7.3.6. Determine the median M.

Table 7.3.6

Year	Total Votes Cast (in millions)
1960	68.8
1964	70.6
1968	73.2
1972	77.7
1976	81.6
1980	86.5
1984	92.7
1988	91.6
1992	104.4
1996	96.3

SOURCE: U.S. Census Bureau, www.census.gov

Solution

First we must rank the data from smallest to largest. The ordered data are

$$68.8, 70.6, 73.2, 77.7, 81.6, 86.5, 91.6, 92.7, 96.3, 104.4$$

Since there are $n = 10$ data items, an even number of data items, no single data item is the middle. In this situation we must find the mean (average) of the fifth and sixth values.

This yields

$$M = \frac{81.6 + 86.5}{2} = 84.05$$

Interpret the Solution: This means that between 1960 and 1996 in half of the presidential elections less than 84.05 million Americans voted and in the other half more than 84.05 million Americans voted. ■

Technology Option

Graphing calculators with statistical capabilities can compute the median of a set of entered data. Scientific calculators generally do not compute the median. For Figures 7.3.2(a) and (b), we have entered the data in Example 3 into a list. We then used the 1-Variable Stats menu. In Figure 7.3.2(a), can you determine what the mean for the set of data is? Figure 7.3.2(b) shows us that Med = 84.05, the same result attained in Example 3. It also tells us that $n = 10$.

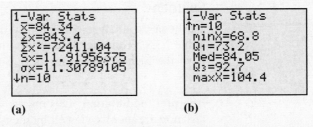

(a)   (b)

Figure 7.3.2

To see how your calculator computes the median, consult the online graphing calculator manual at www.prenhall.com/armstrong.

Since the median is computed by determining the position of an item in a set of data, rather than by the data values themselves, the median is not affected by outliers. Because it is more resistant to extreme values, the median is sometimes called a **resistant** measure of centrality.

The Mode

So far, we have seen how to determine measures of centrality for quantitative data. If the data are qualitative (categorical), however, the mean and median cannot be determined. The **mode** is a measure of centrality that sometimes is useful in analyzing qualitative data.

▪ Mode

For a set of data, either quantitative or qualitative, the **mode**, denoted by M_0, is the data item or data items that occur most frequently.

▶ **Note:** 1. If no data items are repeated, then the set of data has no mode.
2. If more than one data item has the highest frequency, then each of these data items is a mode.

Example 4   Determining the Mode

ActionNews 16 conducts a focus group to determine which newscast, morning, noon, or evening, these viewers watch most frequently. The results are shown in Table 7.3.7.

Table 7.3.7

evening	morning	evening	noon
morning	evening	morning	morning
evening	noon	morning	evening

Determine the mode, M_0, and interpret.

Table 7.3.8

Morning	5
Noon	2
Evening	5

Solution

We begin to analyze the data by constructing a frequency table as shown in Table 7.3.8. From the table we see that both the morning and evening newscasts have the highest frequency of 5. So both morning and evening newscasts are modes.

Interpret the Solution: This means that the morning and the evening newscasts are most frequently viewed by members of this focus group. Since we have *two* modes in this data set, we would say that it is **bimodal**. ∎

 **Checkpoint 2**

Now work Exercise 29.

Technology Option

Most graphing calculators do not explicitly compute the mode of a set of entered quantitative data. The mode can be determined by first sorting the quantitative data in ascending order and then inspecting the data for the item with the highest frequency. For example, the quantitative data 3, 1, 4, 1, 5, 9, 2 are stored in a list as shown in Figure 7.3.3(a). In Figure 7.3.3(b), they are sorted in ascending order and we can see that the mode is 1.

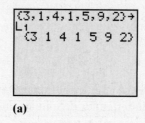 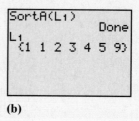

(a)   (b)

Figure 7.3.3

To see how data are sorted on your calculator, consult the online graphing calculator manual at www.prenhall.com/armstrong.

The Midrange

There are times when we are not given the complete set of data, but just the extreme values of a set of data. For example, the local weather report usually gives the low and high temperatures for the day, but no other temperatures. In situations like these, we can still determine a measure of centrality by computing the **midrange** of the data. The type of measure of centrality that the midrange gives us is similar to the mean (or average).

◼ **Midrange**

For a set of data, the **midrange**, denoted by MR, is given by

$$MR = \frac{(\text{minimum data value}) + (\text{maximum data value})}{2}$$

Example 5   **Determining the Midrange**

When researching the total market value of major league baseball franchises in the United States and Canada in 1996, it was found that the New York Yankees had the highest value at $241 million and the Minnesota Twins had the lowest value at $77 million (*Source:* www.infoplease.com). Determine *MR*.

Solution

For these data we have a minimum value of 77 and a maximum value of 241, so the midrange is

$$MR = \frac{77 + 241}{2} = \frac{318}{2} = 159$$

This means that the midrange of the major league baseball franchises in the United States and Canada in 1996 was about $159 million. ∎

Technology Option

Generally, calculators do not have built-in capabilities to compute the midrange. However, many compute the minimum and maximum values from which the midrange can be found. To see how to get the minimum and maximum values for a set of data on your calculator, consult the online graphing calculator manual at www.prenhall.com/armstrong.

When looking at the market values of **every** major league baseball franchise in 1996, it can be shown that the median franchise market value is $M = \$128$ million and the mean is $\mu = \$134$ million. The reason that the midrange is so much different from the median or mean is because the New York Yankees' market value was significantly more than the rest of the baseball teams in 1996. This illustrates that, even though the midrange can be used to determine a measure of centrality when little information is given, it is very susceptible to outliers.

Skewness

Frequency polygons and histograms can have many shapes. Some graphs have an important feature called **skewness**. Let's say that in a frequency distribution the higher frequencies occur in the beginning, or lower classes. Then we say the graph is **skewed right**. On the other hand, if the higher frequencies occur in the ending, or higher classes, then the graph is **skewed left**. If the higher frequencies occur in the middle classes, we call the graph **symmetric**. Examples of these cases are shown in Figure 7.3.4(a)–(c).

The graph in Figure 7.3.4(a) is skewed right because the data *tails off* or *falls off* to the right. We see that the skewness is determined by the direction of the tail, and not where the data are clustered. The severity of the skewness is often determined by how much the mean and median differ in value.

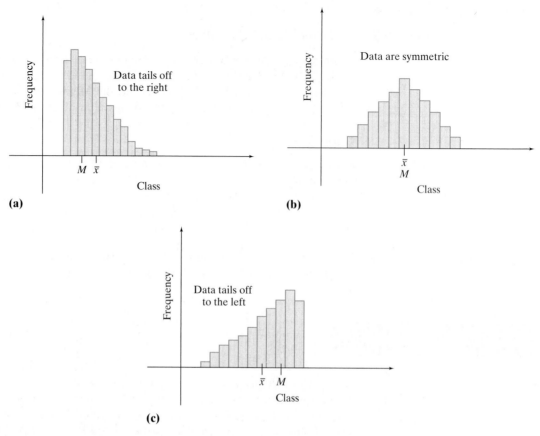

Figure 7.3.4 **(a)** Histogram is skewed right. **(b)** Histogram is symmetric. **(c)** Histogram is skewed left.

Example 6   Determining the Skewness of a Distribution

To decide if a new parking garage is needed, managers at the William Welsh Office Tower survey the employees to see how many had at least one parking ticket in the last 3 months. The distribution of those who had at least one ticket in the last 3 months is shown in Table 7.3.9.

(a) Determine the mean number of tickets given to those who received tickets.

(b) Construct a histogram of the number of tickets given in the 3-month period to those who received tickets.

(c) Classify the histogram as skewed left, skewed right, or symmetric.

Table 7.3.9

Number of Tickets	Frequency, f
1	23
2	35
3	20
4	8
5	4
6	1

Solution

(a) First note that the size of the sample is $n = \sum f = 23 + 35 + \cdots + 1 = 91$. To find the mean, we can use the formula $\bar{x} \approx \dfrac{\sum x_m \cdot f}{n}$, where x_m represents the number of tickets.

$$\bar{x} \approx \frac{(1)23 + (2)35 + (3)20 + (4)8 + (5)4 + 6(1)}{91} = \frac{211}{91} \approx 2.32$$

So, among those receiving parking tickets in the 3-month period, the mean number of tickets received was about 2.32.

(b) Here we have six classes, yet there are no class boundaries. We can construct the histogram by drawing the rectangles directly over the values on the horizontal axis, as shown in Figure 7.3.5.

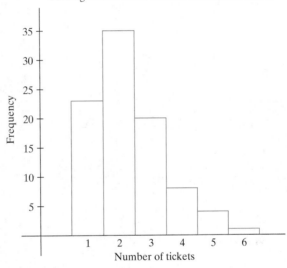

Parking Tickets at the William Welsh Office Tower

Figure 7.3.5

(c) We can see in Figure 7.3.5 that the histogram tails off to the right. This tells us that the histogram is skewed right. ∎

SUMMARY

- For a sample of n data items, the **sample mean** is given by $\bar{x} = \dfrac{\sum x}{n} = \dfrac{\text{sum of data}}{\text{sample size}}$.
- Given a frequency distribution, an approximation for the **mean** of the summarized data is given by $\bar{x} \approx \dfrac{\sum x_m \cdot f}{n}$, where x_m is the midpoint of each class and f is the class frequency.
- For a set of data, the **median**, denoted by M, is the middle value of the set of data after the data have been ordered from smallest to largest.

- For a set of data, either quantitative or qualitative, the **mode**, denoted by M_0, is the data item (or items) that occurs most frequently.
- For a set of data, the **midrange**, denoted by MR, is given by

$$MR = \frac{\left(\begin{array}{c}\text{minimum data}\\\text{value}\end{array}\right) + \left(\begin{array}{c}\text{maximum data}\\\text{value}\end{array}\right)}{2}.$$

SECTION 7.3 EXERCISES

In Exercises 1–10, a set of data is given. Determine the
sample mean $\bar{x}$ and the mode M_0.

1.

6	3	6	5	9	8

2.

9	8	4	5	8	1

3.

4	8	6	1	1
2	3	4	7	0

4.

5	1	4	0	3
4	0	7	8	2

5.

3	17	18	14	9	7
16	2	15	4	19	12

6.

21	15	25	2	13	8
22	17	18	9	24	7

✓ 7.

14.3	15.8	17.4	18.3	19.1
15.8	11.2	9.0	11.2	19.5

8.

7.1	6.8	4.9	5.2	6.3
4.9	7.0	6.3	7.1	5.5

9.

0.25	0.59	0.63	0.58	0.21
0.11	0.78	0.65	0.93	0.27

10.

0.55	0.73	0.18	0.69	0.71
0.83	0.63	0.78	0.68	0.03

In Exercises 11–16, determine the population mean μ for the
given population data.

11.

6	2	4	8	9
2	5	2	11	8

12.

4	7	8	1	25
5	6	2	3	14

13.

135	120	148	165	123
140	156	187	203	162
154	186	155	139	147

14.

248	291	254	211	214
258	265	247	251	283
281	265	263	267	288

15.

50.5	20.1	38.7	71.2
67.7	59.1	39.8	47.5
82.6	41.4	61.1	58.6
60.5	48.8	61.7	55.3

16.

7.7	8.3	6.9	1.1
6.8	9.3	7.1	8.5
6.3	5.7	4.9	3.8
1.3	7.6	8.5	4.1

In Exercises 17–26, a set of data is given. Determine the
median M and the midrange MR.

17.

6	3	5	11	4

18.

5	11	2	9	0

19.

3	9	8	4	5	5

20.

9	8	4	7	6	9

21.

18	16	17	14	21
11	29	25	27	29

22.

55	68	51	63	54
59	64	63	63	57

23.

0.36	0.25	0.81	0.93	0.66
0.31	0.58	0.33	0.47	0.44
0.58	0.36	0.09	0.87	0.25

24.

1.7	5.0	3.6	8.4	2.2
7.3	3.9	4.5	8.6	3.1
4.7	7.5	6.8	7.7	8.6

25.

168	173	189	201
159	256	547	209
256	163	211	177
198	253	261	256

26.

530	691	598	603
552	641	538	614
540	390	548	626
584	572	629	540

In Exercises 27–30, determine the mode for the qualitative
data.

27.

Medium	Small	Medium
Small	Medium	Small
Medium	Large	Medium

28.

Good	Fair	Poor
Fair	Poor	Good
Poor	Fair	Poor

✓ **29.**

Blue	Brown	Brown	Black
Blue	Blue	Brown	Blue
Black	Black	Brown	Black
Green	Brown	Green	Blue

30.

Maple	Oak	Birch	Oak
Oak	Oak	Pine	Maple
Oak	Maple	Maple	Oak
Birch	Maple	Birch	Maple

In Exercises 31–38, a frequency distribution is given. Estimate the mean $\bar{x}$.

31.

Class	Frequency, f
0 to 4	4
5 to 9	7
10 to 14	3

32.

Class	Frequency, f
0 to 9	4
10 to 19	10
20 to 19	5

33.

Midpoint, x_m	Frequency, f
15	5
25	8
35	8
45	5

34.

Midpoint, x_m	Frequency, f
30	2
40	6
50	6
60	2

35.

Class	Frequency, f
2.1 to 2.5	1
2.6 to 3.0	1
3.1 to 3.5	6
3.6 to 4.0	7
4.1 to 4.5	10

36.

Class	Frequency, f
42.5 to 45.4	15
45.5 to 48.4	12
48.5 to 51.4	5
51.5 to 54.4	1
54.5 to 57.4	1

37.

Midpoint, x_m	Frequency, f
25	6
50	8
75	12
100	22
125	10
150	9

38.

Midpoint, x_m	Frequency, f
13	3
18	5
23	9
28	10
33	8
38	6

In Exercises 39–46, classify the distribution as skewed right, skewed left, or symmetric.

39.

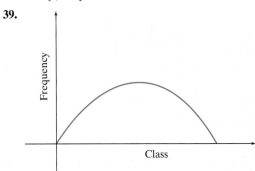

40.

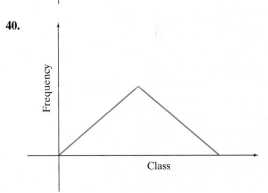

41.

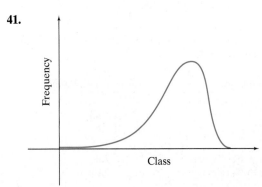

42.

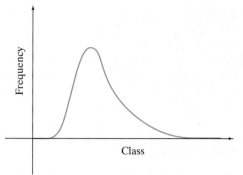

43.

Class	Frequency
100 to 109	100
110 to 119	150
120 to 129	300
130 to 139	850
140 to 149	1025
150 to 159	925

44.

Class	Frequency
0.20 to 0.29	7
0.30 to 0.39	18
0.40 to 0.49	26
0.50 to 0.59	26
0.60 to 0.69	19
0.70 to 0.79	6

45.

Class	Frequency
0.10 to 0.19	3
0.20 to 0.29	14
0.30 to 0.39	19
0.40 to 0.49	19
0.50 to 0.59	13
0.60 to 0.69	2

46.

Class	Frequency
10 to 19	50
20 to 29	60
30 to 39	90
40 to 49	300
50 to 59	635
60 to 69	440

 47. The top (most visited) six Web sites in May 2000 are shown in the table.

Web Site	Visitors (in millions)
AOL Network	59.2
Microsoft	49.3
Yahoo	48.9
Lycos	32.5
Excite@Home	28.8
Go Network	23.1

SOURCE: www.infoplease.com

(a) Determine $\bar{x}$ and interpret.
(b) Determine M and interpret.

 48. The top (most visited) news, information, and entertainment Web sites in May 2000 are shown in the table.

Web Site	Visitors (in millions)
ZDnet	9.8
CNET	9.6
About.com	9.4
MSNBC.com	8.9
Weather.com	7.6
Disney Online	6.5

SOURCE: www.infoplease.com

(a) Determine $\bar{x}$ and interpret.
(b) Determine M and interpret.

 49. The nine largest oil reserves by country in 1999 are shown in the table.

Country	Reserves (in billions of barrels)
Saudi Arabia	259
Iraq	113
Kuwait	93
Adu Dhabi	93
Iran	90
Venezuela	73
Russia	49
Mexico	39
Libya	30

SOURCE: ogj.pennnet.com

(a) Determine $\bar{x}$ and interpret.
(b) Determine M_0 and interpret.

 50. The nine largest oil reserves by country in 2000 are shown in the table.

Country	Reserves (in billions of barrels)
Saudi Arabia	261
Iraq	113
Kuwait	94
Adu Dhabi	92
Iran	90
Venezuela	73
Russia	49
Libya	29
Mexico	29

SOURCE: ogj.pennnet.com

(a) Determine $\bar{x}$ and interpret.
(b) Determine M_0 and interpret.

 51. The top 15 countries in personal computer usage in 1998 are shown in the table.

Country	PC's in Use (in millions)
United States	129.0
Japan	32.8
Germany	21.1
United Kingdom	18.3
France	15.4
Canada	11.8
Italy	10.6
China	8.3
Australia	7.7
South Korea	6.7
Spain	5.7
Russia	5.6
Brazil	5.2
Netherlands	5.1
Mexico	4.6

SOURCE: www.c-i-a.com

(a) Determine M and interpret.

(b) Determine MR and interpret.

52. The top 15 countries in Internet usage in 1998 are shown in the table.

Country	Weekly Internet Users (in millions)
United States	76.0
Japan	9.8
United Kingdom	8.1
Germany	7.1
Canada	6.5
Australia	4.4
France	2.8
Sweden	2.6
Italy	2.1
Taiwan	2.1
South Korea	2.0
Spain	2.0
Netherlands	2.0
China	1.6
Finland	1.6

SOURCE: www.c-i-a.com

(a) Determine M_0 and interpret.

(b) Determine MR and interpret.

53. The number of emergency room visits in the United States during 1997, classified by age from 5 to 84, is shown in the frequency distribution.

Age	Number of visits (in millions)
5 to 14	15.7
15 to 24	14.4
25 to 34	14.8
35 to 44	14.6
45 to 54	12.1
55 to 64	9.1
65 to 74	6.2
75 to 84	8.6

SOURCE: U.S. National Center for Health Statistics, www.cdc.gov/nchs

(a) Determine $n = \sum f$ and interpret.

(b) Estimate the mean of the frequency distribution and interpret.

54. The ages of U.S. prisoners sentenced to death in 1980 are shown in the table.

Age	Number of Prisoners
15 to 24	184
25 to 34	334
35 to 44	144
45 to 54	42
55 to 64	10

SOURCE: U.S. Census Bureau, www.census.gov

(a) Estimate the mean of the frequency distribution and interpret.

(b) Determine if the distribution of the data is generally skewed right, skewed left, or symmetric.

55. The ages of U.S. prisoners sentenced to death in 1997 are shown in the table.

Age	Number of Prisoners
15 to 24	289
25 to 34	1075
35 to 44	1050
45 to 54	768
55 to 64	153

SOURCE: U.S. Census Bureau, www.census.gov

(a) Estimate the mean of the frequency distribution and interpret.

(b) Determine if the distribution of the data is generally skewed right, skewed left, or symmetric.

SECTION PROJECT

Let's examine some of the properties of the measures of centrality. Historically, the traditional holiday bonuses given to six employees at the Happy Healthcare Company are

| 250 | 450 | 600 | 400 | 650 | 600 |

(a) Determine the mean, median, and mode of the data.

(b) Now suppose that the company CEO decided to give an additional $300 in bonuses to each employee last year. Determine the mean, median, and mode of this new set of data.

(c) Compare the measures from parts (a) and (b). What effect does adding the same number to a set of data have on the mean, median, and mode?

(d) Now suppose that unexpected company profits result in the CEO doubling the traditional bonuses this year.

Determine the mean, median, and mode of this new set of data.

(e) Compare the measures from parts (a) and (d). What effect does multiplying every number in a set of data by a constant have on the mean, median, and mode?

Section 7.4   Measures of Dispersion

In the previous section we saw that the measures of centrality can help us to analyze a set of data. But there is another piece of information that is useful when summarizing data numerically. We also want to know how the data are spread out. For example, if we survey the ages of patrons at a fast-food restaurant, the ages can vary greatly, because anyone from infants to senior citizens could frequent the restaurant. By contrast, if we survey the ages of guests at a senior citizen party, the ages would be relatively the same and not very spread out. In this section we will study **measures of dispersion**, which are numerical values associated with how much the data in a set are spread out.

The Range

The simplest measure of dispersion to compute is the **range**. The range is the difference between the minimum and maximum values in a set of data. The advantage of using the range is that it is easy to compute. The disadvantage is that it, much like the midrange, is very susceptible to outliers, because only the minimum and maximum data values are used. The range can be useful in comparing the dispersions of two sets of data.

> ■ **Range**
>
> For a set of data, the **range**, denoted by R, is given by
>
> $$R = (\text{maximum data value}) - (\text{minimum data value})$$

Example 1   Comparing the Ranges of Data

In Example 5 in Section 7.3, we saw that in 1996 the New York Yankees had the highest franchise market value at $241 million and the Minnesota Twins had the lowest value at $77 million. During the same year in the National Football League, the Dallas Cowboys had the highest franchise market value at $320 million and the Indianapolis Colts had the lowest value at $170 million. Compare the ranges of the two sets of data (*Source:* www.infoplease.com).

Solution

For the major league baseball franchises, the range is

$$R = 241 - 77 = 164$$

and for the National Football League franchises, the range is

$$R = 320 - 170 = 150$$

So it appears from inspecting the ranges that the spread of the market value of the major league baseball franchises is greater than that of the National Football League franchises.   ∎

✓ **Checkpoint 1**   Now work Exercise 5.

The Standard Deviation

A drawback of using the range to determine how much the data are spread out is that it is always affected by outliers. The **standard deviation** is a more accurate measure of dispersion. In general terms, the standard deviation gives an "average" distance that each data item lies away from the mean. To see how the standard deviation is computed, consider the data in Table 7.4.1, which gives the number, in thousands, of Persian Gulf War veterans from each state in the western region of the United States. By defining the western region to be the six states shown in Table 7.4.1, these data can be considered population data.

Table 7.4.1

State	Gulf War Veterans (in thousands)
California	172
Oregon	28
Washington	46
Idaho	11
Arizona	30
Nevada	9

SOURCE: U.S. Census Bureau, www.census.gov

The population mean of the data is

$$\mu = \frac{\sum x}{N} = \frac{176 + 28 + 46 + 11 + 30 + 9}{6} = \frac{296}{6} \approx 49.33$$

Now let's plot μ along with each data item and its distance from the mean. See Figure 7.4.1. These distances are shown numerically in Table 7.4.2. Since these distances are obtained by subtracting the mean from the data value, we denote this difference by writing $x - \mu$.

We can see that the value of $x - \mu$ is positive when the data item is greater than the mean and $x - \mu$ is negative when the data item is less than the mean. Since

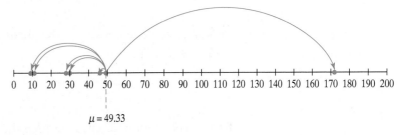

$\mu = 49.33$

Figure 7.4.1

Table 7.4.2

State	x	$x - \mu$
California	172	$172 - 49.33 = 122.67$
Oregon	28	$28 - 49.33 = -21.33$
Washington	46	-3.33
Idaho	11	-38.33
Arizona	30	-19.33
Nevada	9	-40.33

we are trying to determine the "average" distance that each data item lies away from the mean, and distances are always nonnegative, let's square each $x - \mu$ value so that none of them will be negative. This squaring is shown in Table 7.4.3.

Table 7.4.3

State	x	$x - \mu$	$(x - \mu)^2$
California	172	$172 - 49.33 = 122.67$	$(122.67)^2 = 15{,}047.9289$
Oregon	28	$28 - 49.33 = -21.33$	$(-21.33)^2 = 454.9689$
Washington	46	-3.33	11.0889
Idaho	11	-38.33	1469.1889
Arizona	30	-19.33	373.6489
Nevada	9	-40.33	1626.5089

If we compute the sum of these squares, we get $\sum(x - \mu)^2 = 18{,}983.3334$. Dividing this sum by the population size gives us $\dfrac{\sum(x - \mu)^2}{N} = \dfrac{18{,}983.3334}{6} \approx$ 3163.89. Now to "undo" the squaring of the distance from the mean, we take the the square root of this value to get $\sqrt{\dfrac{\sum(x - \mu)^2}{N}} = \sqrt{3163.89} \approx 56.25$. This value tells us that the number of Persian Gulf veterans in the six western states varied on average by about 56.25 thousand from the mean. We have just seen how the **population standard deviation** is computed. Since we primarily use sample data in our computations, let's define the formula for the **sample standard deviation**.

■ **Standard Deviation**

- Given a sample of n data items, the **sample standard deviation**, denoted by s, is given by

$$s = \sqrt{\frac{\sum(x - \bar{x})^2}{n - 1}}$$

- For a population of size N, the **population standard deviation**, denoted by σ, is given by

$$\sigma = \sqrt{\frac{\sum(x - \mu)^2}{N}}$$

▶ **Note:** The main difference between the sample standard deviation and the population standard deviation is the denominator. In the sample standard deviation, we divide by $n - 1$; in the population standard deviation, we divide by N. Since we often use the sample standard deviation to estimate the population standard deviation, for some technical mathematical reasons that are beyond the scope of this text, dividing by $n - 1$ rather than N produces a more accurate estimate.

Example 2   ## Determining the Sample Standard Deviation

The 1990 population size of seven randomly selected cities in the Boston metropolitan area is shown in Table 7.4.4. Use the formula for sample standard deviation to determine s.

Table 7.4.4

City	Population (in thousands)	City	Population (in thousands)
Brockton	236	Manchester	174
Fitchburg	138	New Bedford	176
Lawrence	353	Worcester	478
Lowell	281		

SOURCE: U.S. Census Bureau, www.census.gov

Solution

Let's start by computing the mean.

$$\bar{x} = \frac{236 + 138 + 353 + 281 + 174 + 176 + 478}{7} = \frac{1836}{7} \approx 262.29$$

Using the formula for sample standard deviation, we get

$$s = \sqrt{\frac{(236 - 262.29)^2 + (138 - 262.29)^2 + \cdots + (176 - 262.29)^2 + (478 - 262.29)^2}{7 - 1}}$$

$$= \sqrt{\frac{86,489.4287}{6}} \approx 120.06$$

Interpret the Solution: This means that the populations of the seven cities varied on average by about 120,060 from the mean population of 262,290 people. ∎

Interactive Activity

Recompute the standard deviation in Example 2 as if the data were a population. Which value is larger, s or σ? For any given set of data, will this always be true? Explain.

Technology Option

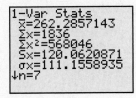

Figure 7.4.2

Almost all scientific and graphing calculators compute the standard deviation of a set of data. On many graphing calculators, the sample standard deviation is denoted by Sx and the population standard deviation is denoted by σx. For Figure 7.4.2, the data in Example 2 have been entered in a list and the 1-Variable Stats menu has been utilized. Notice that $Sx \approx 120.06$, the same value that we obtained in Example 2. Also, notice that Figure 7.4.2 contains other items that we found in Example 2.

To see how your calculator computes the standard deviation, consult the online graphing calculator manual at www.prenhall.com/armstrong.

Standard Deviation from a Frequency Distribution

In Section 7.3 we learned that we can approximate the mean of a set of values summarized in a frequency distribution. We can also approximate the standard deviation of such summarized data. To see how this is done, let's return to the frequency distribution of the 25 longest U.S. rivers, in miles, that we used in Section 7.3. This distribution is given in Table 7.4.5.

Table 7.4.5

Class	Midpoint, x_m	Frequency, f
649 to 1027	838	12
1028 to 1406	1217	5
1407 to 1785	1596	3
1786 to 2164	1975	3
2165 to 2543	2354	2

To approximate the standard deviation, we must first compute how much each class midpoint varies from the mean. Recall that in Section 7.3 we *approximated* the mean $\bar{x}$ to be about 1262. But we must also account for the class frequencies. Knowing that the sum of the frequencies is $n = \sum f = 25$, we get

$$s \approx \sqrt{\frac{\begin{array}{c}(838 - 1262)^2 \cdot 12 + (1217 - 1262)^2 \cdot 5 + (1596 - 1262)^2 \cdot 3 + \\ \cdots + (2354 - 1262)^2 \cdot 2\end{array}}{25 - 1}} = \sqrt{\frac{6,412,140}{24}}$$

$$\approx 516.89$$

So, using the frequency distribution for the 25 longest U.S. rivers, an approximation for the mean is 1262 miles and an approximation for the standard deviation is about 517 miles. We now generalize how to approximate the standard deviation from a frequency distribution.

■ **Standard Deviation from a Frequency Distribution**

Given a frequency distribution, an approximation for the **sample standard deviation** of the summarized data is given by

$$s = \sqrt{\frac{\sum(x_m - \bar{x})^2 \cdot f}{n - 1}}$$

where

x_m is the midpoint of each class
f is the frequency of each class
n is the sample size
$\bar{x}$ is the sample mean

Example 3 **Estimating the Mean and Standard Deviation from a Frequency Distribution**

The number of U.S. males living alone aged 15 to 84 in 1998 is given in Table 7.4.6. Estimate $\bar{x}$ and s and interpret each.

Table 7.4.6

Age Group	Number Living Alone (in millions)
15 to 24	0.72
25 to 34	2.22
35 to 44	2.56
45 to 54	1.90
55 to 64	1.26
65 to 74	1.11
75 to 84	1.23

SOURCE: U.S. Census Bureau, www.census.gov

Table 7.4.7

Midpoint, x_m	Frequency, f
$\frac{15+24}{2} = 19.5$	0.72
29.5	2.22
39.5	2.56
49.5	1.90
59.5	1.26
69.5	1.11
79.5	1.23

Solution

First, we must determine the midpoint of each class as shown in Table 7.4.7. Using the formula $\bar{x} = \dfrac{\sum x_m \cdot f}{n}$, where $n = \sum f = 0.72 + 2.22 + 2.56 + \cdots + 1.23 = 11$, we get

$$\bar{x} = \frac{19.5(0.72) + 29.5(2.22) + 39.5(2.56) + \cdots + 79.5(1.23)}{11}$$

$$= \frac{14.04 + 65.49 + 101.12 + \cdots + 97.785}{11} = \frac{524.6}{11} \approx 47.69$$

Now, to get the standard deviation, we must compute

$$s = \sqrt{\frac{\sum (x_m - \bar{x})^2 \cdot f}{n-1}}$$

$$= \sqrt{\frac{\begin{array}{c}(19.5 - 47.69)^2 \cdot 0.72 + (29.5 - 47.69)^2 \cdot 2.22 + (39.5 - 47.69)^2 \cdot 2.56 + \\ \cdots + (79.5 - 47.69)^2 \cdot 1.23\end{array}}{11 - 1}}$$

$$\approx 18.53$$

Interpret the Solution: According to the frequency distribution, the mean age of U.S. males living alone in 1998 was about 47.69 years with a standard deviation of about 18.53 years. This means that the age of U.S. males living alone in 1998 varied on average by about 18.53 years from the mean of 47.69 years. ∎

Technology Option

The program MEANSD in Appendix C approximates the sample mean and sample standard deviation of a frequency distribution, where the class midpoints are entered in L_1 and the class frequencies are entered in L_2. The program is written for the Texas Instruments TI-83. The input and output from executing the MEANSD calculator program for the data in Example 3 are shown in Figure 7.4.3.

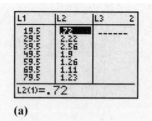
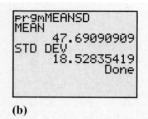

(a)   (b)

Figure 7.4.3

To see the program listing for your calculator, consult the textbook's companion Web site at www.prenhall.com/armstrong.

 Checkpoint 2

Now work Exercise 31.

Pearson Skewness Coefficient

Our final topic in this section revisits the concept of skewness that was introduced in Section 7.3. We revisit this concept through a Flashback.

Flashback

Parking Tickets Revisited

Table 7.4.8

Number of Tickets	Frequency, f
1	23
2	35
3	20
4	8
5	4
6	1

In Example 6 in Section 7.3, we saw that in an effort to decide if a new parking garage is needed the managers at the William Welsh Office Tower surveyed employees to see how many had at least one parking ticket in the last 3 months. The distribution of those who had at least one ticket in the last 3 months is shown in Table 7.4.8 (to the left), and a histogram for the data is given in Figure 7.4.4.

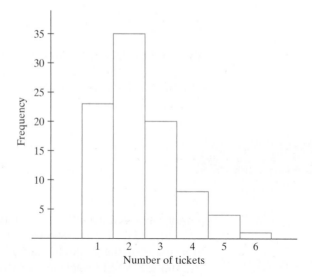

Parking Tickets at the William Welsh Office Tower

Figure 7.4.4

As Figure 7.4.4 indicates, the histogram is skewed right since the histogram tails off or falls off to the right. In Example 6 in Section 7.3, we determined that the mean number of tickets received in the 3-month period is $\bar{x} \approx 2.32$. Determine the standard deviation s.

Flashback Solution

We know that the formula for standard deviation is

$$s = \sqrt{\frac{\sum(x - \bar{x})^2}{n - 1}}$$

Table 7.4.8 indicates that we have $n = \sum f = 91$ data items. In using the formula for standard deviation, the quantity $(x - \bar{x})^2 = (1 - 2.32)^2$ would occur 23 times, since Table 7.4.8 shows the frequency of one ticket is 23. Similarly, $(x - \bar{x})^2 = (2 - 2.32)^2$ would occur 35 times, $(x - \bar{x})^2 = (3 - 2.32)^2$ would occur 20 times, and so on. Rather than list all these occurrences, we can use multiplication and simplify our work as follows:

$$s = \sqrt{\frac{\begin{array}{c}(1 - 2.32)^2 \cdot 23 + (2 - 2.32)^2 \cdot 35 + (3 - 2.32)^2 \cdot 20 + (4 - 2.32)^2 \cdot 8 \\ + (5 - 2.32)^2 \cdot 4 + (6 - 2.32)^2 \cdot 1\end{array}}{91 - 1}}$$

$$= \sqrt{\frac{117.7584}{90}}$$

$$\approx 1.14$$

So the standard deviation for the number of parking tickets is 1.14.

It can be shown that the median, M, for the number of parking tickets in the Flashback is $M = 2$. There exists a numerical value using M, $\bar{x}$, and s that tells us if the distribution of the data is skewed left, skewed right, or symmetric. This numerical value is called the **Pearson Skewness Coefficient**.

■ **Pearson Skewness Coefficient**

A measure to determine the skewness of a distribution is called the **Pearson Skewness Coefficient**, denoted SK. The formula for SK is

$$SK = \frac{3(\bar{x} - M)}{s}$$

where $\bar{x}$ is the mean, M is the median, and s is the standard deviation. The values for SK usually range from -3 to 3.

- If $SK < 0$, then the distribution is skewed left.
- If $SK > 0$, then the distribution is skewed right.
- If $SK \approx 0$, then the distribution is symmetrical.

Example 4 Determining the Pearson Skewness Coefficient

For the data in the Flashback, determine the Pearson Skewness Coefficient given that $\bar{x} \approx 2.32$, $M = 2$, and $s \approx 1.14$.

Solution

We compute the Pearson Skewness Coefficient to be

$$SK = \frac{3(\bar{x} - M)}{s}$$

$$= \frac{3(2.32 - 2)}{1.14}$$

$$\approx 0.84$$

Since $SK \approx 0.84$, that is, $SK > 0$, the distribution is skewed right. This can be visually verified by consulting Figure 7.4.4. ∎

✓ **Checkpoint 3**

Now work Exercise 43.

Final Word on Standard Deviation

The standard deviation can give us more information about a set of data than already discussed. For example, let's draw a curve through the midpoints of the tops of the rectangles for the histogram in Figure 7.4.5. Assume that the histogram is for a relatively large frequency distribution. If the resulting curve is roughly bell shaped, then it can be shown that about 68% of the data lies within 1 standard deviation of the mean. In other words, 68% of the data lies in the interval $(\bar{x} - s, \bar{x} + s)$. About 95% of the data lies within 2 standard deviations of the mean, that is, about 95% of the data lies in the interval $(\bar{x} - 2s, \bar{x} + 2s)$. We will have more to say about bell-shaped curves and this way of looking at standard deviation in Chapter 8.

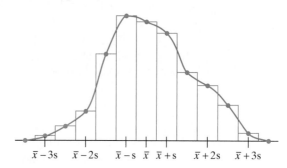

Figure 7.4.5

SUMMARY

- The **range**, denoted by R, is given by $R = \left(\begin{array}{c}\text{maximum data}\\\text{value}\end{array}\right) - \left(\begin{array}{c}\text{minimum data}\\\text{value}\end{array}\right)$.

- The **sample standard deviation**, denoted by s, is given by $s = \sqrt{\dfrac{\sum(x - \bar{x})^2}{n - 1}}$.

- The **population standard deviation**, denoted by σ, is given by $\sigma = \sqrt{\dfrac{\sum(x - \mu)^2}{N}}$.

- The approximation for the **sample standard deviation**

for a given frequency distribution is given by $s \approx \sqrt{\dfrac{\sum(x_m - \bar{x})^2 \cdot f}{n - 1}}$.

- The **Pearson Skewness Coefficient**, denoted SK, is given by $SK = \dfrac{3(\bar{x} - M)}{s}$, where $\bar{x}$ is the mean, M is the median, and s is the standard deviation. The values for SK usually range from -3 to 3. If $SK < 0$, then the distribution is skewed left. If $SK > 0$, then the distribution is skewed right. If $SK \approx 0$, then the distribution is symmetrical.

SECTION 7.4 EXERCISES

In Exercises 1–10, a set of data is given. Determine the range R.

1.

6	3	6	5	9	8

2.

9	8	4	5	8	1

3.

4	8	6	1	1
2	3	4	7	0

4.

5	1	4	0	3
4	0	7	8	2

✔ **5.**

3	17	18	14	9	7
16	2	15	4	19	12

6.

21	15	25	2	13	8
22	17	18	9	24	7

7.

14.3	15.8	17.4	18.3	19.1
15.8	11.2	9.0	11.2	19.5

8.

7.1	6.8	4.9	5.2	6.3
4.9	7.0	6.3	7.1	5.5

9.

0.25	0.59	0.63	0.58	0.21
0.11	0.78	0.65	0.93	0.27

10.

0.55	0.73	0.18	0.69	0.71
0.83	0.63	0.78	0.68	0.03

In Exercises 11–16, determine the population standard deviation for the given population data. Use the formula

$$\sigma = \sqrt{\frac{\sum (x - \mu)^2}{N}}.$$

11.

6	2	4	8	9
2	5	2	11	8

12.

4	7	8	1	25
5	6	2	3	14

13.

135	120	148	165	123
140	156	187	203	162
154	186	155	139	147

14.

248	291	254	211	214
258	265	247	251	283
281	265	263	267	288

15.

50.5	20.1	38.7	71.2
67.7	59.1	39.8	47.5
82.6	41.4	61.1	58.6
60.5	48.8	61.7	55.3

16.

7.7	8.3	6.9	1.1
6.8	9.3	7.1	8.5
6.3	5.7	4.9	3.8
1.3	7.6	8.5	4.1

In Exercises 17–26, a set of data is given. Determine the sample standard deviation s.

17.

6	3	5	11	4

18.

5	11	2	9	0

19.

3	9	8	4	5	5

20.

9	8	4	7	6	9

21.

18	16	17	14	21
11	29	25	27	29

22.

55	68	51	63	54
59	64	63	63	57

23.

0.36	0.25	0.81	0.93	0.66
0.31	0.58	0.33	0.47	0.44
0.58	0.36	0.09	0.87	0.25

24.

1.7	5.0	3.6	8.4	2.2
7.3	3.9	4.5	8.6	3.1
4.7	7.5	6.8	7.7	8.6

25.

168	173	189	201
159	256	547	209
256	163	211	177
198	253	261	256

26.

530	691	598	603
552	641	538	614
540	390	548	626
584	572	629	540

In Exercises 27–34, a frequency distribution is given. Estimate the standard deviation using the formula

$$s = \sqrt{\frac{\sum (x_m - \bar{x})^2 \cdot f}{n - 1}}.$$

27.

Class	Frequency, f
0 to 4	4
5 to 9	7
10 to 14	3

28.

Class	Frequency, f
0 to 9	4
10 to 19	10
20 to 29	5

29.

Midpoint, x_m	Frequency, f
15	5
25	8
35	8
45	5

30.

Midpoint, x_m	Frequency, f
30	2
40	6
50	6
60	2

✔ **31.**

Class	Frequency, f
2.1 to 2.5	1
2.6 to 3.0	1
3.1 to 3.5	6
3.6 to 4.0	7
4.1 to 4.5	10

32.

Class	Frequency, f
42.5 to 45.4	15
45.5 to 48.4	12
48.5 to 51.4	5
51.5 to 54.4	1
54.5 to 57.4	1

33.

Midpoint, x_m	Frequency, f
25	6
50	8
75	12
100	22
125	10
150	9

34.

Midpoint, x_m	Frequency, f
13	3
18	5
23	9
28	10
33	8
38	6

35. The area of selected major bodies of water in the Pacific Coast region of the United States is shown in the table.

Body of Water	Area (in square miles)
Puget Sound, WA	808
San Francisco Bay, CA	264
Willapa Bay, WA	125
Hood Canal, WA	117

SOURCE: U.S. Census Bureau, www.census.gov

(a) Determine the range and interpret.

(b) Determine the sample standard deviation and interpret.

36. The area of selected major bodies of water in the Atlantic Coast region of the United States is shown in the table.

Body of Water	Area (in square miles)
Chesapeake Bay, MD	2747
Pamlico Sound, NC	1622
Long Island Sound, NY	914
Delaware Bay, DE	614
Cape Cod, MA	598
Albemarle Sound, NC	492

SOURCE: U.S. Census Bureau, www.census.gov

(a) Determine the range and interpret.

(b) Determine the sample standard deviation and interpret.

37. The ages of U.S. prisoners sentenced to death in 1980 are shown in the table. Estimate the standard deviation of the frequency distribution and interpret.

Age	Number of Prisoners
15 to 24	184
25 to 34	334
35 to 44	144
45 to 54	42
55 to 64	10

SOURCE: U.S. Census Bureau, www.census.gov

38. The ages of U.S. prisoners sentenced to death in 1997 are shown in the table. Estimate the standard deviation of the frequency distribution and interpret.

Age	Number of Prisoners
15 to 24	289
25 to 34	1075
35 to 44	1050
45 to 54	768
55 to 64	153

SOURCE: U.S. Census Bureau, www.census.gov

39. The number of emergency room visits in the United States during 1997, classified by age from 5 to 84, is shown in the frequency distribution. Estimate the standard deviation of the frequency distribution and interpret.

Age	Number of visits (in millions)
5 to 14	15.7
15 to 24	14.4
25 to 34	14.8
35 to 44	14.6
45 to 54	12.1
55 to 64	9.1
65 to 74	6.2
75 to 84	8.6

SOURCE: U.S. National Center for Health Statistics, www.cdc.gov/nchs

In Exercises 40–45, determine the Pearson Skewness Coefficient and classify the distribution as skewed right, skewed left, or symmetric.

40.

Summary	Statistics
$\bar{x}$	35.7
M	36
s	0.93

41.

Summary	Statistics
$\bar{x}$	103.5
M	108
s	2.1

42. A sample of data has a sample mean $\bar{x} = 48$ with a median $M = 48$ and standard deviation of $s = 4.01$.

✓ **43.** A sample of data has a sample mean $\bar{x} = 7.5$ with a median $M = 7.5$ and standard deviation of $s = 1.03$.

44. The ages of U.S. females in 1990 had a mean of 36.1 years, with a median age of 34.0 and standard deviation of 22.9 years (*Source:* U.S. Census Bureau, www.census.gov).

45. The ages of U.S. males in 1990 had a mean of 33.4 years, with a median age of 31.6 and standard deviation of 21.6 years (*Source:* U.S. Census Bureau, www.census.gov).

SECTION PROJECT

Let's examine some of the properties of the measures of variation. Suppose that the traditional holiday bonuses given to six employees at the Happy Healthcare Company are

250	450	600	400	650	600

(a) Determine the range and sample standard deviation of the data.

(b) Now suppose that the company CEO decided to give an additional $300 in bonuses this year to the six employees. Determine the range and sample standard deviation of this new set of data.

(c) Compare the measures from parts (a) and (b). What effect does adding the same number to a set of data have on the range and sample standard deviation?

(d) Now suppose that in an effort to have a consistent bonus policy the company CEO gives each of the six employees a $500 bonus. Determine the range and sample standard deviation of this new set of data.

(e) What can we say about the value of the range and sample standard deviation of a set of data that is comprised of a single numerical value?

■ Why We Learned It

In this chapter, we learned how to analyze data sets both numerically and graphically. We looked at pie charts and saw that the ability to read a pie chart can give much useful information. For example, a social scientist given a pie chart listing the number of hours per person per year watching various forms of television can sum the numbers given in the pie chart to determine the total hours per person per year watching television. Once this is done, the scientist can also determine the percentages watching the various forms of television. We also saw how the histogram, bar graph, frequency polygon, and ogive can visually represent a data set. The latter part of this chapter was dedicated to measures of central tendency, along with ways to measure how spread out the data are. For example, if a criminal justice professor has the ages of all U.S. prisoners sentenced to death, he or she may be interested in the mean (average) age of this group of prisoners.

To measure how spread out the ages are, the professor could then compute the standard deviation. The professor could also compute the median of the ages and the mode. If the professor constructed a histogram of the data, he or she may be interested in whether the histogram is skewed. This can be analyzed visually or numerically by using the Pearson Skewness Coefficient.

CHAPTER REVIEW EXERCISES

In Exercises 1 and 2, identify the population and the sample.

1. In a March 2001 survey conducted by the Gallup Organization, 1060 adult Americans were asked if the quality of the environment as a whole is getting better or getting worse; 36% of adults responded that the overall quality of the environment is getting better (*Source:* www.gallup.com).

2. In a November 2000 survey, 501 college football fans were asked by the Gallup Organization what team they considered to be the number 1 team in the country; 28% responded that the number 1 team was Oklahoma (*Source:* www.gallup.com).

In Exercises 3–5, determine whether the data are qualitative or quantitative.

3. The number of students per class in each course offered at a university

4. The type of fish caught during a fishing tournament

5. The movies playing at the local cinema

In Exercises 6–8, a table of data is given. Construct a bar graph for the data.

6. The total numbers of females, in thousands, participating in various NCAA sports during the 1997–1998 academic year are shown in the table.

Sport	Number of Females Participating (in thousands)
Basketball	13.8
Soccer	16.0
Cross country	10.5
Tennis	8.1

SOURCE: U.S. Statistical Abstract, www.census.gov/statab/www

7. The total numbers of pounds, in millions, of various domestic fish and shellfish caught in 1998 are shown in the table.

Species	Number of Pounds (in millions)
Lobsters	79.6
Halibut	73.2
Atlantic cod	24.5
Sablefish	43.5
Sea scallops	13.1

SOURCE: U.S. Statistical Abstract, www.census.gov/statab/www

8. The numbers of adults, in millions, attending various sporting events one or more times a month in 1998 are shown in the table.

Sporting Event	Number of Adults Attending (in millions)
Auto racing	3.6
Baseball	8.6
College football	5.0
Ice hockey	2.9
Soccer	3.4
Golf	2.2

SOURCE: U.S. Statistical Abstract, www.census.gov/statab/www

In Exercises 9–11, answer the questions associated with the given bar graph.

9. The average yearly incomes, in thousands, of 25- to 34-year-olds by highest degree earned for 1998 is shown in the bar graph (*Source:* U.S. Statistical Abstract, www.census.gov/statab/www).

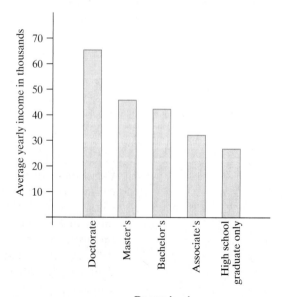

Average Yearly Income of 25-to-34-Year Olds by Highest Degree Earned, 1998

(a) Which degrees were held by 25- to 34-year-olds who earned at least $40,000 in 1998?

(b) About how much more was the average yearly income of a 25- to 34-year-old in 1998 holding a doctorate degree than someone in this age group who was only a high school graduate?

10. The numbers of people, in millions, using home computers in 1997 by various age groups is shown in the bar graph (*Source:* U.S. Statistical Abstract, www.census.gov/statab/www).

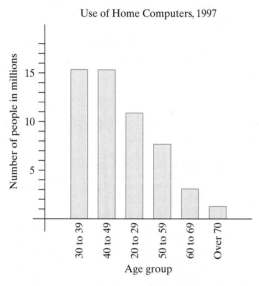

Use of Home Computers, 1997

(a) Which age groups had about the same number of users?

(b) Which age groups had under 6 million users in 1997?

11. The numbers of people 5 years old and over, in thousands, speaking various languages other than English at home in 1990 are shown in the bar graph (*Source:* U.S. Statistical Abstract, www.census.gov/statab/www).

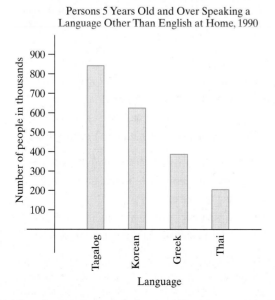

Persons 5 Years Old and Over Speaking a Language Other Than English at Home, 1990

(a) About how many people 5 years old and over spoke Thai at home in 1990?

(b) About how many more people 5 years old and over spoke Tagalog than Thai at home in 1990?

In Exercises 12–14, answer the questions associated with the given pie chart.

12. The percent of toxic chemical releases in 1998 by industry is shown in the pie chart (*Source:* U.S. Statistical Abstract, www.census.gov/statab/www).

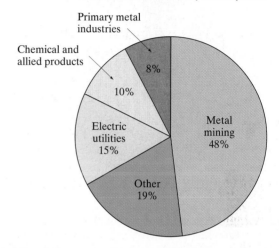

Toxic Chemical Releases by Industry, 1998

(a) What industry was the largest source of toxic chemical releases in 1998?

(b) If 7,307,000,000 pounds of toxic chemicals was released by industries in 1998, how many pounds were released by electric utilities?

13. Juvenile arrests, in thousands, for drug possession in 1998 are shown in the pie chart (*Source:* U.S. Statistical Abstract, www.census.gov/statab/www).

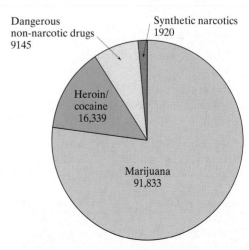

Juvenile Arrest for Drug Possession, 1998

(a) What type of possession ranked third in terms of arrests?

(b) If there were 119,237 thousand arrests in 1998 for drug possession, what percentage of those were for possession of marijuana? Round to the nearest tenth.

14. The number of lottery ticket sales, in billions of dollars, by various types of games in 1999 is shown in the pie chart (*Source:* U.S. Statistical Abstract, www.census.gov/statab/www).

Lottery Ticket Sales, 1999

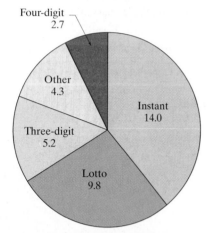

(a) Which type of game had the lowest number of ticket sales?

(b) If total ticket sales in 1999 were $36 billion, what percent of sales was from instant tickets and 3-digit tickets? Round to the nearest tenth.

In Exercises 15 and 16, answer the questions associated with the given frequency distribution.

15. The total number of adults, in millions, who attended at least one movie in 1996 is classified by age group in the frequency distribution.

Class (age group)	Frequency (in millions)
25 to 34	31.7
35 to 44	33.1
45 to 54	21.9
55 to 64	9.6
65 to 74	7.4

SOURCE: U.S. Statistical Abstract, www.census.gov/statab/www

(a) What is the class width?

(b) How many adults age 45 to 54 attended at least one movie in 1996?

16. The number of adults, in millions, who had read the newspaper in the week prior to a spring 2000 survey is classified by household income in the frequency distribution.

Class (income in dollars)	Frequency (in millions)
10,000 to 19,999	16.7
20,000 to 29,999	26.1
30,000 to 39,999	18.6
40,000 to 49,999	17.3

SOURCE: U.S. Statistical Abstract, www.census.gov/statab/www

(a) How many adults whose income was between $10,000 and $19,999 in spring 2000 read the newspaper at least once in the week prior to the survey?

(b) If there were 24,055,000 adults whose income was between $30,000 and $39,999 in spring 2000, what percentage of them read the newspaper at least once in the week prior to the survey? Round to the nearest tenth.

17. Use Exercise 15 to complete the following.

(a) Determine the class boundaries.

(b) Construct a frequency histogram.

18. Use Exercise 16 to complete the following.

(a) Determine the class boundaries.

(b) Determine the relative frequency for each class.

(c) Construct a relative frequency histogram.

For the data in Exercises 19 and 20, complete the following:

(a) Write a frequency distribution.

(b) Use your calculator to construct a histogram.

19. The time, in minutes, spent exercising during one day for a sample of 30 adults is shown in the table. Use seven classes.

0	15	25	0	45	30
60	120	15	35	40	45
45	30	30	0	0	60
90	0	120	0	45	60
15	30	20	40	80	0

20. The amount, in dollars, spent on recreation per month for a sample of 18 teen-agers is shown in the table. Use five classes.

40	60	15	10	5	30
97	25	20	18	30	12
80	20	40	42	37	26

In Exercises 21 and 22, complete the following:

(a) Determine the midpoint of each class.

(b) Construct a frequency polygon.

21. Use the frequency distribution in Exercise 15.

22. Use the frequency distribution in Exercise 16.

In Exercises 23 and 24, complete the following:

(a) Determine the cumulative frequency for each class.

(b) Construct an ogive.

23. Use the frequency distribution in Exercise 15.

24. Use the frequency distribution in Exercise 16.

For Exercises 25–27, use the frequency polygon for death rates from cancer per 100,000 females in the given age groups for ages 25 to 74 in 1998 (Source: U.S. Statistical Abstract, www.census.gov/statab/www).

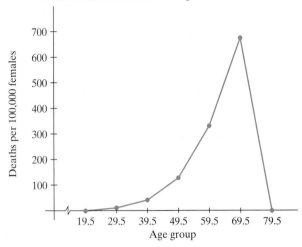

Death Rates from Cancer for Ages 25 to 74, 1998

25. What is the class *width*?

26. What are the *limits* of the third class?

27. In which class are there about 330 deaths per 100,000 females?

🌐 *For Exercises 28–30, use the ogive for the number of deaths reported due to injuries at work for persons age 15 to 64 in 1999 (Source: U.S. National Center for Health Statistics, www.cdc.gov/nchs).*

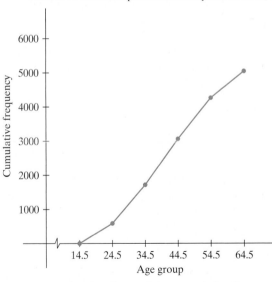

Number of Deaths Reported Due to Injuries at Work, 1999

28. What are the class *limits* of the second class?

29. What is the class *width*?

30. About how many individuals age 15 to 44.5 years old died from work-related injuries in 1999?

In Exercises 31 and 32, a sample of data is given. Determine the sample mean $\bar{x}$ and the mode M_0.

31.

5	8	9	2	2
3	7	6	1	8

32.

0.22	0.53	0.77	0.64	0.21
0.14	0.08	0.27	0.89	0.47

In Exercises 33 and 34, determine the population mean μ for the given population data.

33.

122	128	134	188	133
118	121	126	150	145
115	119	141	143	112

34.

2.6	3.4	1.1	3.3
4.1	5.6	5.3	3.4
1.8	9.8	1.9	2.7

In Exercises 35 and 36, a sample of data is given. Determine the median M and the midrange MR.

35.

10	13	27	29	19
11	31	18	21	15

36.

320	325	415	584
608	409	353	305
223	287	619	387
298	521	409	508

In Exercises 37 and 38, determine the mode for the qualitative data.

37.

First	First	Third
Second	Second	First
Third	Second	First

38.

Biology	Math	Music	Physics
Physics	Biology	Music	Music
Physics	Math	Math	Physics
Math	Math	Biology	Biology

In Exercises 39 and 40, a frequency distribution is given. Estimate the mean $\bar{x}$.

39.

Class	Frequency, f
0 to 4	5
5 to 9	6
10 to 14	4
15 to 19	3

40.

Midpoint, x_m	Frequency, f
20	2
30	5
40	4
50	3
60	8

In Exercises 41 and 42, classify the distribution as skewed right, skewed left, or symmetric.

41.

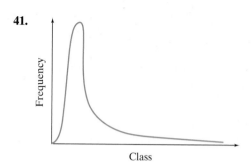

42.

Class	Frequency
0.10 to 0.14	2
0.15 to 0.19	3
0.20 to 0.24	4
0.25 to 0.29	10
0.30 to 0.34	12

43. The top (highest grossing) six summer films of all time are shown in the table.

Film	Domestic gross (in millions)
Star Wars	461.0
Star Wars: Episode One— The Phantom Menace	431.1
E.T.	399.8
Jurassic Park	357.1
Forrest Gump	329.7
The Lion King	312.9

SOURCE: www.infoplease.com

(a) Determine $\bar{x}$ and interpret.
(b) Determine M and interpret.

44. The top nine hockey teams to make the most appearances in the Stanley Cup finals are given in the table along with their total number of appearances.

Team	Appearances
Montreal Canadiens	32
Toronto Maple Leafs	21
Detroit Red Wings	21
Boston Bruins	17
New York Rangers	10
Chicago Blackhawks	10
Philadelphia Flyers	7
Edmonton Oilers	6
New York Islanders	5

SOURCE: www.infoplease.com

(a) Determine $\bar{x}$ and interpret.
(b) Determine M_0 and interpret.
(c) Determine MR and interpret.

45. The total number of arrests in 1999 of 10- to 18-year-olds is given in the table.

Age	Number of Arrests
10 to 12	117,679
13 to 15	662,847
16 to 18	1,244,361

SOURCE: www.infoplease.com

(a) Estimate the mean of the frequency distribution and interpret.
(b) Determine if the distribution of the data is generally skewed right, skewed left, or symmetric.

In Exercises 46–48, a sample of data is given. Determine the range R.

46.

3	2	7	12	5

47.

4	18	25	27	38	20
10	12	42	16	5	31

48.

0.10	0.57	0.32	0.19	0.89	0.63
0.22	0.64	0.05	0.67	0.74	0.44

In Exercises 49–51, determine the population standard deviation for the given population data. Use the formula

$$\sigma = \sqrt{\frac{\sum(x - \mu)^2}{N}}.$$

49.

3	12	10	17	6
5	8	12	13	14

50.

138	94	87	126	312
220	119	184	155	175
198	123	146	158	202

51.

3.4	2.8	5.1	3.2
2.7	8.9	9.9	1.6
3.4	5.7	1.8	9.5
3.8	7.2	7.8	9.4

In Exercises 52–54, a sample of data is given. Determine the sample standard deviation s.

52.

2	8	3	1	9

53.

74	71	70	68	77
62	65	70	75	76

54.

3.8	4.7	2.6	8.9	5.6
2.1	1.5	3.7	7.4	2.3
4.1	1.8	1.7	2.3	6.6

In Exercises 55 and 56, a frequency distribution is given. Estimate the standard deviation using the formula

$$s = \sqrt{\frac{\sum (x_m - \bar{x})^2 \cdot f}{n - 1}}.$$

55.

Class	Frequency, f
13 to 17	2
18 to 22	6
23 to 27	8
28 to 32	3

56.

Midpoint, x_m	Frequency, f
10	12
20	15
30	8
40	7
50	20
60	16

57. The 1999 population size of seven randomly selected counties in New Hampshire is shown in the table.

County	Population (in thousands)
Belknap	54
Coos	33
Grafton	79
Hillsbourough	367
Merrimack	130
Stafford	111
Sullivan	40

SOURCE: U.S. Census Bureau, www.census.gov

(a) Determine the range and interpret.

(b) Determine the sample standard deviation and interpret.

58. The total number of arrests in 1999 of 10- to 18-year-olds is given in the table. Estimate the standard deviation of the frequency distribution and interpret.

Age	Number of Arrests
10 to 12	117,679
13 to 15	662,847
16 to 18	1,244,361

SOURCE: www.infoplease.com

In Exercises 59 and 60, determine the Pearson Skewness Coefficient and classify the distribution as skewed right, skewed left, or symmetric.

59.

Summary	Statistics
$\bar{x}$	42.8
M	38
s	0.74

60. A sample of data has a sample mean $\bar{x} = 32$ with a median $M = 32$ and standard deviation of $s = 3.98$.

CHAPTER 7 PROJECT

SOURCE: www.nielsenmedia.com and 7metasearch.com

One application of the standard deviation is Chebyshev's Theorem. The theorem states that for any set of data, population or sample, *at least* $p = 1 - \dfrac{1}{k^2}$ of the data lie within k standard deviations of the mean, where $k > 1$. For a sample of data, this means that at least $p\%$ of the data lies between $\bar{x} - ks$ and $\bar{x} + ks$.

Consider the following table of a sample of the number of unique visitors to random Internet sites during September 2001.

8169	7903	7255	6517	7556	7350	8295
6100	6618	6968	6216	5054	4907	6275
6188	6660	6510	6009	6111	6076	6639
6387	6401	5866	6405	5229	5729	6069
5264	5502	5316	5176	5092	5796	6048
5530	6104	5439	5379	6261	5943	5414
5722	5428	5551	5971	6069	5432	5764
5939						

1. Determine the sample mean and the sample standard deviation. Round to hundredths place.

2. According to Chebyshev's Theorem, at least 75% of the data lies within k standard deviations of the mean where $1 - \dfrac{1}{k^2} = \dfrac{3}{4}$ (that is 75%). Solve this equation to determine the proper k value.

3. Determine the values $n_1 = \bar{x} - ks$ and $n_2 = \bar{x} + ks$ using the $\bar{x}$- and s-values found in step 1 along with the k-value found in step 2.

4. What number represents 75% of the data in the table? That is, what is 75% of the sample size?

5. Tally the data to determine the number of unique visitors to random Internet sites during September 2001 that are between n_1 and n_2.

6. How do you account for the difference in the solutions between steps 4 and 5?

7. Repeat steps 2 through 6 for at least 88.9% $\left(\text{that is,} \dfrac{8}{9}\right)$ of the data.

Probability Distributions

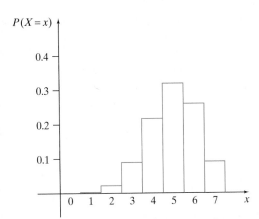

$$P(X=x) = {}_7C_x(0.71)^x(0.29)^{7-x}$$

$$P(X=5) = {}_7C_5(0.71)^5(0.29)^2$$

$$\approx 0.319$$

The Internet has opened a world of information to anyone who can go online. This is especially true for people seeking medical information. An August 2000 Harris poll found that 71% of Americans who are online use the Internet to access health and medical information. The probability that any number of 7 randomly selected online Americans use the Internet to access this information can be shown using a probability histogram. The binomial distribution gives the probability that 5 of 7 randomly selected Americans go online to access health and medical information. This probability is about 0.319 or 31.9%.

What We Know

In Chapter 6 we learned the fundamentals of probability, and in Chapter 7 we learned about describing data graphically. We also learned about measures of central tendency and measures of dispersion.

Where Do We Go

In this chapter we will study random variables as a precursor to the study of probability distributions. This allows us to group outcomes that share similar attributes and to compute probabilities. We will also revisit the measure of central tendency and the measure of dispersion.

Section 8.1   Discrete Random Variables

Thus far in our study of probability we have learned how to determine the probability of individual events. Now we want to find how to determine the probabilities of many events with specific common characteristics. To do this we use **random variables**. We will also examine numerical tables of values, sometimes called **probability distributions**, that can be used to determine probabilities.

Discrete Random Variables

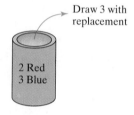

Draw 3 with replacement

Figure 8.1.1

Let's begin with a typical example from Chapter 6 and see how we can use the results already known to assign a random variable to the experiment and to construct a probability distribution. Suppose that a jar contains 2 red and 3 blue marbles. In an experiment, we pull out a marble and record its color and then return it to the jar. We repeat this two more times until we have made three draws (Figure 8.1.1). The sample space for the experiment is

$$S = \{RRR, RRB, RBR, BRR, RBB, BRB, BBR, BBB\}$$

To compute the probabilities for these outcomes, we can use the classical probability formula from Chapter 6. (Consult the Toolbox below.)

From Your Toolbox

Classical Probability: Assume that S is a sample space that has $n(S)$ outcomes for which each outcome is equally likely. If A is an event associated with S, then the probability of event A occurring, denoted by $P(A)$, is given by

$$\text{probability of event } A \text{ occurring} = P(A) = \frac{n(A)}{n(S)} = \frac{\text{number of outcomes of event } A}{\text{total number of possible outcomes}}$$

We notice that on any given draw the probability of drawing a red marble, denoted by $P(R)$, can be found by dividing the number of red marbles in the jar by the total number of marbles in the jar. Hence, for any given single draw,

$$P(R) = \frac{\text{number of red marbles in the jar}}{\text{total number of marbles in the jar}} = \frac{2}{5}$$

The probability of drawing a blue marble on any single draw is

From Your Toolbox

Two events A and B are **independent** if the outcome of one event does not affect the other's chance of occurring. If two events A and B are independent, then the probability of both A and B occurring is given by $P(A \text{ and } B) = P(A \cap B) = P(A) \cdot P(B)$.

$$P(B) = \frac{\text{number of blue marbles in the jar}}{\text{total number of marbles in the jar}} = \frac{3}{5}$$

Since the marble is replaced before the next draw, we know this means that the trials are independent. (Consult the Toolbox to the left.) So, to determine the probability of drawing three marbles, we can apply the multiplication rule for independent events. (Consult the Toolbox to the left.) If we compute this for all eight possible outcomes, we get the results shown in Table 8.1.1.

We can see that some of the probability values in Table 8.1.1 are the same. Let's find a way to group them. If we let the variable X represent the number of

Table 8.1.1

Outcome	P (outcome)
RRR	$\frac{2}{5} \cdot \frac{2}{5} \cdot \frac{2}{5} = \frac{8}{125} = 0.064 = 6.4\%$
RRB	$\frac{2}{5} \cdot \frac{2}{5} \cdot \frac{3}{5} = \frac{12}{125} = 0.096 = 9.6\%$
RBR	$\frac{2}{5} \cdot \frac{3}{5} \cdot \frac{2}{5} = \frac{12}{125} = 0.096 = 9.6\%$
BRR	$\frac{3}{5} \cdot \frac{2}{5} \cdot \frac{2}{5} = \frac{12}{125} = 0.096 = 9.6\%$
BBR	$\frac{3}{5} \cdot \frac{3}{5} \cdot \frac{2}{5} = \frac{18}{125} = 0.144 = 14.4\%$
BRB	$\frac{3}{5} \cdot \frac{2}{5} \cdot \frac{3}{5} = \frac{18}{125} = 0.144 = 14.4\%$
RBB	$\frac{2}{5} \cdot \frac{3}{5} \cdot \frac{3}{5} = \frac{18}{125} = 0.144 = 14.4\%$
BBB	$\frac{3}{5} \cdot \frac{3}{5} \cdot \frac{3}{5} = \frac{27}{125} = 0.216 = 21.6\%$

blue marbles drawn, we see that the possible values of X can be $x = 0$, $x = 1$, $x = 2$, and $x = 3$, because anywhere from zero to three blue marbles can be drawn. If we associate the eight outcomes with the values of X, we get Table 8.1.2.

Table 8.1.2

x-value	$x = 0$	$x = 1$	$x = 2$	$x = 3$
Associated outcomes	RRR	RRB, RBR, BRR	BBR, BRB, RBB	BBB

If we assign probabilities to these x-values, we get the table of values shown in Table 8.1.3. Notice that for the values of $x = 1$ and $x = 2$ we must multiply 0.096 and 0.144, respectively, by 3, because they are associated with three different outcomes.

Table 8.1.3

$X = x$	$P(X = x)$
$X = 0$	$P(X = 0) = 0.064$
$X = 1$	$P(X = 1) = 3(0.096) = 0.288$
$X = 2$	$P(X = 2) = 3(0.144) = 0.432$
$X = 3$	$P(X = 3) = 0.216$

We want Table 8.1.3 to illustrate that X assigns a probability to each value of x. We call X a **random variable**. Since there is a finite number of possible values of X for this experiment, we specifically call X a **discrete random variable**.

■ **Discrete Random Variable**
- A **random variable**, X, is a rule that assigns a real number to each outcome of a probability experiment.
- A random variable is **discrete** if its set of values is finite or countably infinite.

▶ **Note:** 1. X represents the random variable for the experiment, whereas x represents the possible value for the random variable.

2. By *countably infinite*, we mean that all the possible values of the random variable can be arranged so that one is called the first, one the second, one the third, and so on. That is, they can be listed in some order.

Since Table 8.1.3 is a table of a random variable and its possible values and their corresponding probabilities, some call this type of table a **probability distribution**. (We will use *table of values* and *probability distribution* interchangeably in this text.) Notice that in our probability distribution all the probability values are between 0 and 1 and the sum of the probabilities is 1. That is,

$$\sum_{x=0}^{3} P(X=x) = P(X=0) + P(X=1) + P(X=2) + P(X=3)$$

$$= 0.064 + 0.288 + 0.432 + 0.216 = 1$$

Note that in the expression $\sum_{x=0}^{3} P(X=x)$ we use an **index of summation** below and above the $\sum$ symbol to show where we start and stop adding probabilities. Now let's list the two conditions that define a discrete probability distribution.

■ **Discrete Probability Distribution**

A distribution of a discrete random variable X is a **discrete probability distribution** if:

1. $0 \le P(X=x) \le 1$ That is, all the probability values are between 0 and 1, inclusive.

2. $\sum_{\substack{\text{all } x \\ \text{values}}} P(X=x) = 1$ The sum of the probabilities equals 1.

Sometimes we are given a function that is used to determine the probabilities for given values of the random variable. We determine these probabilities by simply evaluating the function. Such functions are called **probability mass functions**. A probability mass function will help us to compute the probabilities for given values of X.

■ **Probability Mass Function**

A **probability mass function** is a discrete probability distribution for which the probabilities are computed using a function.

▶ **Note:** A probability mass function refers to a discrete random variable.

For a function to be a probability mass function, it must meet the two criteria for a discrete probability distribution. Example 1 illustrates how to determine if a function is a probability mass function.

Example 1   **Determining If a Function Is a Probability Mass Function**

For the function

$$P(X = x) = \begin{cases} \frac{1}{10}x, & x = 1, 2, 3, 4 \\ 0, & \text{otherwise} \end{cases}$$

complete the following:

(a) Write a table of values for $P(X = x)$.

(b) Determine if $P(X = x)$ is a probability mass function.

Table 8.1.4

$X = x$	$P(X = x)$
$X = 1$	$P(X = 1) = \frac{1}{10} = 0.1$
$X = 2$	$P(X = 2) = \frac{1}{5} = 0.2$
$X = 3$	$P(X = 3) = \frac{3}{10} = 0.3$
$X = 4$	$P(X = 4) = \frac{2}{5} = 0.4$

Solution

(a) To make a table of values for $P(X = x)$, we substitute the values of x and evaluate as shown in Table 8.1.4. The bottom part of $P(X = x)$ that reads "0 otherwise" means that, if any values other than $x = 1, 2, 3, 4$ are used, then $P(X = x) = 0$.

(b) To determine if $P(X = x)$ is a probability mass function, we must see if it meets the two criteria outlined in the definition of discrete probability distribution.

1. From Table 8.1.4, we see that the probability values for $P(X = x)$ are $0.1, 0.2, 0.3$, and 0.4. So all the probability values are between 0 and 1.

2. The sum of the probabilities is

$$\sum_{x=1}^{4} P(X = x) = P(X = 1) + P(X = 2) + P(X = 3) + P(X = 4)$$

$$= 0.1 + 0.2 + 0.3 + 0.4 = 1$$

Since $P(X = x)$ meets the two criteria, it is a probability mass function. ∎

✓ Checkpoint 1

Now work Exercise 15.

In addition to making a numerical table of probabilities, we can also make a graph of the probability values. In Chapter 7 we learned how to construct histograms. Here we will construct what we call **probability histograms**. The probability histogram for $P(X = x)$ in Example 1 is shown in Figure 8.1.2. In a

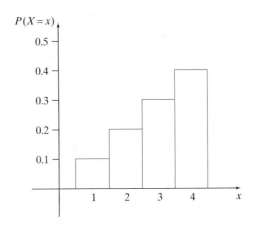

Figure 8.1.2 Probability histogram for Table 8.1.4.

probability histogram, the length of the base of each rectangle is 1, and x is the midpoint of the base. The height of each rectangle is equal to the probability value for the specified value of x.

Example 2   **Determining the Value That Makes $P(X = x)$ a Probability Mass Function**

For the function

$$P(X = x) = \begin{cases} \frac{1}{k}x, & x = 2, 3, 4, 5 \\ 0, & \text{otherwise} \end{cases}$$

determine the value of k so that $P(X = x)$ is a probability mass function.

Solution

Using the definition of probability mass function, we need a value of k such that $\sum_{x=2}^{5} P(X = x) = 1$. The sum gives

$$\sum_{x=2}^{5} \frac{1}{k}x = \frac{1}{k} \cdot 2 + \frac{1}{k} \cdot 3 + \frac{1}{k} \cdot 4 + \frac{1}{k} \cdot 5 = \frac{1}{k}(2 + 3 + 4 + 5) = \frac{1}{k} \cdot 14$$

Setting this sum equal to 1 yields the equation $\frac{1}{k} \cdot 14 = 1$. Solving for k, we see that $14 = k$. This gives the probability mass function

$$P(X = x) = \begin{cases} \frac{1}{14}x, & x = 2, 3, 4, 5 \\ 0, & \text{otherwise} \end{cases}$$

A quick check show that the values of $P(X = x)$ are between 0 and 1 and that $\sum_{x=2}^{5} \frac{1}{14}x = 1$. This verifies that $P(X = x)$ is a probability mass function. ■

To simplify the writing of a probability mass function, we will use the following agreement throughout the remainder of this chapter.

> **We will only write a probability mass function for values of x for which it is positive and assume that $P(X = x) = 0$ where it is not positive. For example, this means that we can write the solution to Example 2 as**

$$P(X = x) = \frac{1}{14}x, \qquad x = 2, 3, 4, 5$$

Probability between Two Values

Let's return to the probability experiment that started this section. Recall that a jar contains 2 red and 3 blue marbles. We pull out a marble and record its color and then return it to the jar. We repeat this two more times until we have made three draws. Suppose that we want to compute the probability that we draw either 2 or 3 blue marbles from the jar. To do this, we can simply add the probability values for $P(X = 2)$ and $P(X = 3)$. Recall that the values for $P(X = 2)$ and $P(X = 3)$ are in Table 8.1.3. Symbolically, this is

$$P(2 \leq X \leq 3) = \sum_{x=2}^{3} P(X = x) = P(X = 2) + P(X = 3)$$

$$= 0.432 + 0.216 = 0.648$$

This means that, as the experiment is repeated, the probability that we draw either 2 or 3 blue marbles from the jar is 0.648 or 64.8%. We can use the probability histogram in Figure 8.1.3 to get this same result. To do this, we simply add the areas of the shaded rectangles that correspond to $P(X = 2)$ and $P(X = 3)$. (Recall that the base of each rectangle is 1 and the height is simply the probability value.)

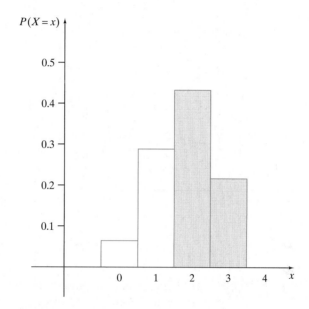

Figure 8.1.3 The sum of the areas of the two shaded rectangles gives $P(2 \leq X \leq 3)$.

■ **Computing the Probability between Two Values**

For a probability mass function $P(X = x)$, $P(a \leq X \leq b) = \sum_{x=a}^{b} P(X = x)$.

Example 3 **Computing the Probability between Two Values**

In a probability experiment, a coin is tossed 4 times and the result is recorded. Let X represent the total number of heads recorded. Figure 8.1.4 shows all 16 possible outcomes along with the value of X for each outcome.

(a) Construct a probability histogram of the distribution.

(b) Determine $P(1 \leq X \leq 3)$ and interpret.

```
                          HTTH
                          HTHT
              HTTT    THTH    HHHT
              THTT    HHTT    HHTH
              TTHT    THHT    HTHH
    TTTT    TTTH    TTHH    THHH    HHHH
    X = 0    X = 1    X = 2    X = 3    X = 4
```

Figure 8.1.4

Solution

(a) **Understand the Situation:** To aid us in drawing the probability histogram, we begin by making a table of values for $P(X = x)$. See Table 8.1.5.

Using these values, we construct the probability histogram shown in Figure 8.1.5.

Table 8.1.5

$X = x$	$P(X = x)$
$X = 0$	$P(X = 0) = \frac{1}{16} = 0.0625$
$X = 1$	$P(X = 1) = \frac{4}{16} = 0.25$
$X = 2$	$P(X = 2) = \frac{6}{16} = 0.375$
$X = 3$	$P(X = 3) = \frac{4}{16} = 0.25$
$X = 4$	$P(X = 4) = \frac{1}{16} = 0.0625$

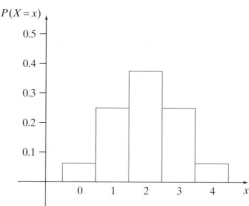

Figure 8.1.5 Probability histogram for the number of heads in four tosses of a coin.

Interactive Activity

Compute $P(X = 0)$ for the situation in Example 3. Also, compute $1 - P(X = 0)$ and interpret what this means.

(b) **Understand the Situation:** When asked to determine $P(1 \leq X \leq 3)$, we are being asked to determine the probability that 4 tosses of a coin will result in 1, 2, or 3 heads.

We compute this probability to be

$$P(1 \leq X \leq 3) = P(X = 1) + P(X = 2) + P(X = 3)$$
$$= 0.25 + 0.375 + 0.25$$
$$= 0.875$$

Interpret the Solution: When a coin is tossed 4 times, the probability that 1, 2, or 3 heads occurs is 0.875, or 87.5%. ∎

✓ **Checkpoint 2**

Now work Exercise 33.

Notice that the result of Example 3b corresponds to the sum of the areas of the shaded rectangles for the probability histogram as shown in Figure 8.1.6.

Probability Distributions from Empirical Probabilities

Our final topic in this introductory section to Chapter 8 revisits empirical probabilities. To recall what empirical probabilities are, consult the Toolbox on p. 469.

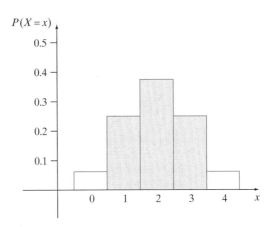

Figure 8.1.6 The sum of the areas of the three shaded rectangles equals $P(1 \leq X \leq 3)$.

 From Your Toolbox

Empirical Probability: Given a frequency distribution of qualitative data, if A is an event associated with a qualitative data item, then the probability of event A is given by $P(A)$

$$= \frac{\text{frequency of event } A}{\text{sum of all frequencies}}$$

$$= \frac{f}{\sum f}.$$

Now let's examine an application in which we use empirical probabilities to construct a table of values (probability distribution).

Example 4   Constructing a Probability Distribution Using Relative Frequencies

Between 1993 and 1997, the number of two-bedroom rental units, in thousands, under construction at various monthly rent levels, in dollars, in the United States is shown in Table 8.1.6.

Table 8.1.6

Monthly Rent (rounded to the nearest hundred)	Number of Units (in thousands)
100	62
200	377
300	205
400	233
500	187
600	198
700	236

SOURCE: U.S. Census Bureau, www.census.gov

(a) Construct a table of values for the data.

(b) If a housing unit under construction is chosen at random for a housing study, determine the probability that its rent will be $500 or more per month.

Solution

(a) **Understand the Situation:** To construct a table of values, we must first find the sum of the frequencies, f, in the second column of Table 8.1.6. The total number of units is

$$\sum f = 62 + 377 + 205 + 233 + 187 + 198 + 236 = 1498$$

Now, if we let X be a discrete random variable representing the monthly rent, in dollars, we can construct the table of values for $P(X = x)$.

The table of values for $P(X = x)$ is shown in Table 8.1.7. Note that the decimals for probability values are rounded to the nearest thousandth. As a result of this rounding, the decimal values may not sum to exactly 1.

Table 8.1.7

$X = x$	$P(X = x)$
$X = 100$	$P(X = 100) = \frac{62}{1498} \approx 0.041$
$X = 200$	$P(X = 200) = \frac{377}{1498} \approx 0.252$
$X = 300$	$P(X = 300) = \frac{205}{1498} \approx 0.137$
$X = 400$	$P(X = 400) = \frac{233}{1498} \approx 0.156$
$X = 500$	$P(X = 500) = \frac{187}{1498} \approx 0.125$
$X = 600$	$P(X = 600) = \frac{198}{1498} \approx 0.132$
$X = 700$	$P(X = 700) = \frac{236}{1498} \approx 0.158$

Interactive Activity

Use the result of $P(X \geq 500)$ from Example 4b to determine $1 - P(X \geq 500)$. Explain what this result represents.

(b) Understand the Situation: Since X is a random variable representing monthly rent, we want to determine $P(500$ or more$)$, which symbolically is $P(X \geq 500)$.

Continuing, we get

$$P(X \geq 500) = P(500) + P(600) + P(700)$$
$$\approx 0.125 + 0.132 + 0.158 = 0.415$$

Interpret the Solution: The probability that a randomly selected housing unit under construction will have a monthly rent of $500 or more is 0.415, or 41.5%.

∎

✓ **Checkpoint 3**   Now work Exercise 47.

SUMMARY

- A **random variable** is a rule that assigns a real number to each outcome of a probability experiment.
- A random variable is **discrete** if its set of values is finite or countably infinite.
- A function P for a discrete random variable X is a **probability mass function** if $0 \leq P(X = x) \leq 1$ and

$$\sum_{\substack{\text{all } x \\ \text{values}}} P(X = x) = 1.$$

- For a probability mass function $P(X = x)$,

$$P(a \leq X \leq b) = \sum_{x=a}^{b} P(X = x).$$

SECTION 8.1 EXERCISES

In Exercises 1–10, determine if the table meets the two criteria of a probability distribution.

1.

$X = x$	$P(X = x)$
$X = 0$	$P(X = 0) = \frac{1}{2}$
$X = 1$	$P(X = 1) = \frac{1}{4}$
$X = 2$	$P(X = 2) = \frac{1}{4}$

2.

$X = x$	$P(X = x)$
$X = 1$	$P(X = 1) = \frac{1}{3}$
$X = 2$	$P(X = 2) = \frac{2}{3}$
$X = 3$	$P(X = 3) = \frac{1}{3}$

3.

$X = x$	$P(X = x)$
$X = 0$	$P(X = 0) = \frac{1}{10}$
$X = 1$	$P(X = 1) = \frac{1}{20}$
$X = 2$	$P(X = 2) = \frac{1}{5}$
$X = 3$	$P(X = 3) = \frac{2}{5}$
$X = 4$	$P(X = 4) = \frac{1}{4}$

4.

$X = x$	$P(X = x)$
$X = 1$	$P(X = 1) = \frac{3}{32}$
$X = 2$	$P(X = 2) = \frac{3}{16}$
$X = 3$	$P(X = 3) = \frac{5}{32}$
$X = 4$	$P(X = 4) = \frac{3}{8}$
$X = 5$	$P(X = 5) = \frac{3}{16}$

5.

$X = x$	$P(X = x)$
$X = -2$	$P(X = -2) = 0.03$
$X = -1$	$P(X = -1) = 0.13$
$X = 0$	$P(X = 0) = 0.40$
$X = 1$	$P(X = 1) = 0.35$
$X = 2$	$P(X = 2) = 0.09$

6.

$X = x$	$P(X = x)$
$X = -1$	$P(X = -1) = 0.174$
$X = 0$	$P(X = 0) = 0.111$
$X = 2$	$P(X = 2) = 0.260$
$X = 4$	$P(X = 4) = 0.300$
$X = 6$	$P(X = 6) = 0.155$

7.

$X = x$	$P(X = x)$
$X = 10$	$P(X = 10) = -0.1$
$X = 20$	$P(X = 20) = 0.3$
$X = 30$	$P(X = 30) = 0.6$
$X = 40$	$P(X = 40) = 0.2$

8.

$X = x$	$P(X = x)$
$X = 5$	$P(X = 5) = 0$
$X = 10$	$P(X = 10) = 1.2$
$X = 25$	$P(X = 25) = 0.1$
$X = 35$	$P(X = 35) = -0.3$

9.

$X = x$	$P(X = x)$
$X = 8$	$P(X = 8) = \frac{1}{5}$
$X = 9$	$P(X = 9) = \frac{1}{5}$
$X = 10$	$P(X = 10) = \frac{1}{10}$
$X = 11$	$P(X = 11) = \frac{3}{20}$
$X = 12$	$P(X = 12) = \frac{7}{20}$

10.

$X = x$	$P(X = x)$
$X = 0$	$P(X = 0) = \frac{11}{20}$
$X = 1$	$P(X = 1) = \frac{1}{4}$
$X = 2$	$P(X = 2) = \frac{1}{20}$
$X = 3$	$P(X = 3) = \frac{1}{10}$
$X = 4$	$P(X = 4) = \frac{1}{20}$

In Exercises 11–20, a formula for $P(X = x)$ is given. Complete the following:

(a) Write a table of values for $P(X = x)$.

(b) Determine if the table meets the two criteria of a probability distribution. Remember that $P(X = x) = 0$ when the x-values are not defined.

11. $P(X = x) = \dfrac{1}{6}x, \quad x = 0, 1, 2, 3$

12. $P(X = x) = 0.01x, \quad x = 10, 20, 30, 40$

13. $P(X = x) = \dfrac{1}{4}, \quad x = 1, 2, 3, 4$

14. $P(X = x) = \dfrac{1}{5}, \quad x = 0, 1, 2, 3, 4$

✓ **15.** $P(X = x) = \dfrac{1}{15}x^2, \quad x = 0, 1, 2, 3, 4$

16. $P(X = x) = \dfrac{1}{18}x^3, \quad x = 0, 1, 2, 3, 4$

17. $P(X = x) = 0.05x, \quad x = 2, 3, 4, 6$

18. $P(X = x) = \dfrac{1}{55}x, \quad x = 5, 10, 15, 20$

19. $P(X = x) = \dfrac{x + 2}{10}, \quad x = 0, 1, 2, 3$

20. $P(X = x) = \dfrac{x - 4}{14}, \quad x = 6, 7, 8, 9$

In Exercises 21–26, determine the values of k so that $P(X = x)$ is a probability mass function.

21. $P(X = x) = \dfrac{1}{k}x, \quad x = 0, 1, 2, 3$

22. $P(X = x) = \dfrac{2}{k}x, \quad x = 1, 2, 3, 4$

23. $P(X = x) = kx, \quad x = 5, 10, 15, 20$

24. $P(X = x) = kx, \quad x = 3, 6, 9, 12, 15$

25. $P(X = x) = \dfrac{1}{k}x^2, \quad x = 0, 1, 2, 4$

26. $P(X = x) = \dfrac{1}{k}x^3, \quad x = 0, 1, 2, 3, 5$

In Exercises 27–32, a table of frequencies is given. Construct the corresponding probability distribution with $X = x$ in the left column and $P(X = x)$ in the right column. See Example 4.

27.

Value	Frequency
1	5
2	5
3	10

28.

Value	Frequency
0	6
1	10
2	14

29.

Value	Frequency
1	10
2	20
3	15
4	5
5	25

30.

Value	Frequency
0	55
10	45
20	15
30	10

31.

Value	Frequency
0	150
1	600
2	400
3	200

32.

Value	Frequency
0	1500
10	0
20	1750
30	2000

Applications

✓ **33.** The Videoshack has determined that the number of videos that a customer rents on any given visit can be represented by a discrete random variable X with the following probability distribution.

$X = x$	$P(X = x)$
$X = 0$	$P(X = 0) = 0.05$
$X = 1$	$P(X = 1) = 0.50$
$X = 2$	$P(X = 2) = 0.30$
$X = 3$	$P(X = 3) = 0.15$

(a) Construct a probability histogram of the distribution.
(b) Determine $P(1 \le X \le 2)$ and interpret.

34. The MoccaMania coffee shop has found that its number of daily customers can be represented by a discrete random variable X with the following probability distribution.

$X = x$	$P(X = x)$
$X = 45$	$P(X = 45) = 0.10$
$X = 46$	$P(X = 46) = 0.30$
$X = 47$	$P(X = 47) = 0.25$
$X = 48$	$P(X = 48) = 0.20$
$X = 49$	$P(X = 49) = 0.05$
$X = 50$	$P(X = 50) = 0.10$

(a) Construct a probability histogram of the distribution.
(b) Determine $P(46 \le X \le 49)$ and interpret.

35. Milford College hires the next four staff employees without regard to gender. The personnel office knows that the pool of applicants is large and there is an equal chance of each applicant being hired. Records show that the number of women hired can be represented by a discrete random variable X with the following probability distribution.

$X = x$	$P(X = x)$
$X = 0$	$P(X = 0) = 0.1$
$X = 1$	$P(X = 1) = 0.2$
$X = 2$	$P(X = 2) = 0.4$
$X = 3$	$P(X = 3) = 0.2$
$X = 4$	$P(X = 4) = 0.1$

(a) Construct a probability histogram of the distribution.
(b) Determine $P(2 \le X \le 4)$ and interpret.

36. When four different households are surveyed on Tuesday night, the National Rating Service Company determines that the discrete random variable X represents the number of households who are watching the new drama "Alien from Within."

$X = x$	$P(X = x)$
$X = 0$	$P(X = 0) = 0.2$
$X = 1$	$P(X = 1) = 0.3$
$X = 2$	$P(X = 2) = 0.3$
$X = 3$	$P(X = 3) = 0.1$
$X = 4$	$P(X = 4) = 0.1$

(a) Construct a probability histogram of the distribution.
(b) Determine $P(0 \le X \le 2)$ and interpret.

37. The Longberg Elementary School PTA has determined that the number of commercials per show aired during afterschool television shows can be represented by a discrete random variable X with the following probability distribution.

$X = x$	$P(X = x)$
$X = 0$	$P(X = 0) = 0.20$
$X = 1$	$P(X = 1) = 0.15$
$X = 2$	$P(X = 2) = 0.25$
$X = 3$	$P(X = 3) = 0.20$
$X = 4$	$P(X = 4) = 0.05$
$X = 5$	$P(X = 5) = 0.10$
$X = 6$	$P(X = 6) = 0.05$

(a) Construct a probability histogram of the distribution.
(b) Determine $P(X \le 3)$ and interpret.

38. The North-Am racing series has found that the number of yellow flag caution laps, rounded to the nearest 10, during any given race can be represented by a discrete random variable X with the following probability distribution.

$X = x$	$P(X = x)$
$X = 0$	$P(X = 0) = 0.20$
$X = 10$	$P(X = 10) = 0.25$
$X = 20$	$P(X = 20) = 0.38$
$X = 30$	$P(X = 30) = 0.10$
$X = 40$	$P(X = 40) = 0.07$

(a) Construct a probability histogram of the distribution.

(b) Determine $P(X \geq 10)$ and interpret.

39. Parmiter College has found that the number of students who drop out of introductory psychology can be represented by a discrete random variable X with the following probability distribution.

$X = x$	$P(X = x)$
$X = 12$	$P(X = 12) = 0.09$
$X = 13$	$P(X = 13) = 0.18$
$X = 14$	$P(X = 14) = 0.38$
$X = 15$	$P(X = 15) = 0.20$
$X = 16$	$P(X = 16) = 0.15$

(a) Determine the probability that at most 14 students will drop out of introductory psychology during any given quarter.

(b) According to the distribution, what is the least likely number of students to drop out of introductory psychology during any given quarter?

40. The Supreme Pie pizzeria determines that its number of late-night pizza deliveries on any given Wednesday can be represented by a discrete random variable X with the following probability distribution.

$X = x$	$P(X = x)$
$X = 4$	$P(X = 4) = 0.32$
$X = 5$	$P(X = 5) = 0.11$
$X = 6$	$P(X = 6) = 0.40$
$X = 7$	$P(X = 7) = 0.12$
$X = 8$	$P(X = 8) = 0.05$

(a) Determine the probability that at least 5 late-night deliveries will be made on any given Wednesday night.

(b) According to the distribution, what is the least likely number of late-night pizza deliveries on any given Wednesday?

41. Officials of Stone Temple City have determined that the size of family, including homeowner and all relatives, can be represented by a discrete random variable X. The probability distribution for family size is shown in the following table. A city official chooses a family at random to complete a questionnaire about civil planning.

$X = x$	$P(X = x)$
$X = 2$	$P(X = 2) = 0.15$
$X = 3$	$P(X = 3) = 0.23$
$X = 4$	$P(X = 4) = 0.21$
$X = 5$	$P(X = 5) = 0.15$
$X = 6$	$P(X = 6) = 0.14$
$X = 7$	$P(X = 7) = 0.12$

(a) Determine the probability that the size of the family will be at least 4.

(b) According to the distribution, what is the least likely size that the chosen family will be?

42. Researchers at the Hobart Sleep Institute have determined that the number of hours of sleep (rounded to the nearest hour) that the subjects of a large study get during a sleep study can be represented by a random variable X that has the probability distribution shown in the following table. A sleep study subject is chosen at random to answer follow-up questions.

$X = x$	$P(X = x)$
$X = 4$	$P(X = 4) = 0.11$
$X = 5$	$P(X = 5) = 0.15$
$X = 6$	$P(X = 6) = 0.20$
$X = 7$	$P(X = 7) = 0.33$
$X = 8$	$P(X = 8) = 0.12$
$X = 9$	$P(X = 9) = 0.09$

(a) Determine the probability that the subject slept at least 6 hours.

(b) According to the distribution, what is the least likely number of hours that a subject slept during the study?

43. The enrollment at Hagerty College is given in the following frequency table.

Academic Rank	Enrollment
Freshmen	7800
Sophomores	5500
Juniors	3200
Seniors	2500

(a) Let X be a discrete random variable representing the number of years of college completed by a randomly selected student from Hagerty College, meaning that $x = 0, 1, 2, 3$. Construct a probability distribution for the data.

(b) Compute $P(X = 3)$ and interpret.

44. The number of defective blue jeans discovered after inspection by the Saxsworth Company is shown in the following table:

Number of Inspections	Number of Defects
1	1500
2	800
3	200
4	50

(a) Let X be a discrete random variable where X represents the number of inspections completed. Construct a probability distribution for the data.

(b) Compute $P(X = 2)$ and interpret.

45. The number of passengers who miss the overnight flight from Tokyo to Honolulu, as indicated by the annual records of Pacific Rim Airlines, are shown in the table.

Passengers Who Miss the Flight	Number of Flights
0	53
1	127
2	280
3	128
4	77

(a) Let X be a discrete random variable representing the number of passengers who miss the overnight flight. Construct a probability distribution for the data.

(b) Compute $P(X = 3)$ and interpret.

46. A marketing research firm determines that if groups of 5 youths are chosen the number (of the 5) who own the new Dream Boy hand-held video game are as shown in the table.

Number of Youths	Number Who Own Dream Boy
0	233
1	207
2	170
3	118
4	82
5	30

(a) Let X be a discrete random variable representing the number of youths who own the Dream Boy. Construct a probability distribution for the data.

(b) Compute $P(X = 2)$ and interpret.

✓ 🌐 **47.** The number of drivers aged 16 to 24 who were in automobile accidents in 1996 is shown in the table.

Age	Number of Accidents (in millions)
16	1.6
17	2.2
18	2.4
19	2.7
20	2.8
21	2.9
22	2.9
23	3.1
24	3.6

SOURCE: National Safety Council, www.nsc.org

(a) Let X be a discrete random variable representing the age of a driver in an accident. Construct a probability distribution for the data.

(b) Compute $P(X = 20)$ and interpret.

(c) Compute $P(X \leq 20)$ and interpret.

🌐 **48.** The distribution of weights of American males aged 70 to 79 from 1988 to 1994 is shown in the table.

Weight (in pounds)	Percentage	Weight (in pounds)	Percentage
100–109	0.2	200–209	9.5
110–119	0.2	210–219	5.2
120–129	1.3	220–229	5.4
130–139	4.1	230–239	2.5
140–149	7.4	240–249	1.0
150–159	10.1	250–259	0.7
160–169	11.9	260–269	1.1
170–179	13.2	270–279	0.9
180–189	11.1	280–289	0.4
190–199	13.6	290 and above	0.2

SOURCE: U.S. Center for Health Statistics, www.cdc.gov/nchs

(a) Let X be a discrete random variable representing the weight of an American male aged 70 to 79 from 1988 to 1994. Construct a probability distribution for the data.

(b) Compute $P(X < 160)$ and interpret.

(c) Compute $P(X \geq 200)$ and interpret.

🌐 **49.** In 1997, the number of school activities in which at least one parent of a child in grades K–5 participated is shown in the following table:

Activities Attended	Percentage
0	3.5
1	7.4
2	19.7
3	32.0
4 or more	37.4

SOURCE: U.S. Census Bureau, www.census.gov

Let X be a discrete random variable representing the number of school activities that at least one parent attended in 1997.

(a) Compute $P(X = 2)$ and interpret.

(b) Compute $P(X \geq 2)$ and interpret.

🌐 **50.** In 1997, the number of school activities in which at least one parent of a child in grades 6–8 participated is shown in the following table:

Activities Attended	Percentage
0	7.3
1	11.7
2	25.8
3	32.2
4 or more	23.0

SOURCE: U.S. Census Bureau, www.census.gov

Let X be a discrete random variable representing the number of school activities that at least one parent attended in 1997.

(a) Compute $P(X = 3)$ and interpret.

(b) Compute $P(X \geq 3)$ and interpret.

◼ SECTION PROJECT

Suppose that a mathematics department has nine women and five men as full-time faculty members.

(a) In how many ways can a committee of three women and one man be selected from the department?

(b) In how many ways can a committee of four be selected from the department.

(c) Using the results from (a) and (b), determine the probability that a committee of four from the department will consist of three women and one man.

The result of part (c) illustrates a special probability distribution called the **hypergeometric distribution**. The hypergeometric distribution is a distribution of a random variable that has two outcomes when selection is done without replacement. The distribution is described as follows: Given a population with only two types (males and females, for example) such that there are a items of one kind and b items of another kind and $a + b$ equals the total population, the

probability $P(X = x)$ of selecting without replacement a sample of size n with X items of type a and $n - X$ items of type b is $P(X) = \dfrac{(_a C_X) \cdot (_b C_{n-X})}{_{(a+b)} C_n}$.

(d) For the mathematics department with nine women and five men in (a)–(c), if we were to use the hypergeometric distribution to determine the probability that a committee of four consists of three women and one man, determine the values of a, b, X, and n that would be used in the hypergeometric formula. Use the hypergeometric formula to verify your answer to part (c).

(e) Use the hypergeometric distribution to determine the following: A recent housing study by a major homeowners insurance company found that exactly four out of nine houses on a certain street were underinsured. If five houses are selected from the nine houses, determine the probability that two are underinsured.

Section 8.2   Expected Value and Standard Deviation of a Discrete Random Variable

Given a particular probability distribution or a probability mass function, we often want to know what the *average* outcome would be if we repeat the experiment over and over many times. Since we are working with a probability distribution, this average is computed in a different way from computing the mean of a set of data.

Expected Value

Table 8.2.1

$X = x$	$P(X = x)$
$X = 1$	$P(X = 1) = \frac{1}{6}$
$X = 2$	$P(X = 2) = \frac{1}{6}$
$X = 3$	$P(X = 3) = \frac{1}{6}$
$X = 4$	$P(X = 4) = \frac{1}{6}$
$X = 5$	$P(X = 5) = \frac{1}{6}$
$X = 6$	$P(X = 6) = \frac{1}{6}$

To discover how an average of a probability distribution can be computed, let's start with a familiar experiment of tossing a six-sided die. Suppose that we roll a die 4 times and the results of the rolls are 2, 3, 6, and 1. Then the average of the rolls is $\dfrac{4 + 3 + 6 + 2}{4} = \dfrac{15}{4} = 3.75$. It is important to note that the average of the 4 rolls is not a value that occurs in a single roll. It is not possible to roll a 3.75!

If we wanted to roll a die, say, 300 times, the average of the rolls is computed in a similar fashion. To determine the average, we can sum the 300 outcomes and divide the result by 300. To see how we can approximate this sum, we let X represent the outcome on any roll of the die, where the possible values of X are $x = 1, 2, 3, 4, 5, 6$. Since the outcomes of the rolls are equally likely, we get the values of the probability distribution $P(X = x)$ shown in Table 8.2.1.

Let's examine the next to last row of the table. Since $P(X = 5) = \frac{1}{6}$, this tells us that if we roll a die 300 times we expect that about $\frac{1}{6}$ of the 300 rolls will produce a 5. So, in 300 rolls the number of times that we expect to see a 5 is found by $\frac{1}{6} \cdot 300 = 50$. Hence, we expect to roll a 5 about 50 times if we toss the die a total of 300 times. Notice that Table 8.2.1 indicates that we would expect to roll **any** number $\frac{1}{6}$ of the time. This means that in 300 rolls we expect to see each number occur 50 times. We summarize this information in Table 8.2.2.

Table 8.2.2

$X = x$	$P(X = x)$	Number of Times $X = x$ Is Expected
$X = 1$	$P(X = 1) = \frac{1}{6}$	$\frac{1}{6} \cdot 300 = 50$
$X = 2$	$P(X = 2) = \frac{1}{6}$	$\frac{1}{6} \cdot 300 = 50$
$X = 3$	$P(X = 3) = \frac{1}{6}$	$\frac{1}{6} \cdot 300 = 50$
$X = 4$	$P(X = 4) = \frac{1}{6}$	$\frac{1}{6} \cdot 300 = 50$
$X = 5$	$P(X = 5) = \frac{1}{6}$	$\frac{1}{6} \cdot 300 = 50$
$X = 6$	$P(X = 6) = \frac{1}{6}$	$\frac{1}{6} \cdot 300 = 50$

To determine the expected sum of the 300 rolls, we do the following:

$$\text{sum of rolls} = 1 \cdot \left(\begin{array}{c} \text{number of} \\ \text{1's expected} \end{array} \right) + 2 \cdot \left(\begin{array}{c} \text{number of} \\ \text{2's expected} \end{array} \right)$$

$$+ 3 \cdot \left(\begin{array}{c} \text{number of} \\ \text{3's expected} \end{array} \right) + \cdots + 6 \cdot \left(\begin{array}{c} \text{number of} \\ \text{6's expected} \end{array} \right)$$

$$= 1 \cdot 50 + 2 \cdot 50 + 3 \cdot 50 + 4 \cdot 50 + 5 \cdot 50 + 6 \cdot 50$$

$$= 1050$$

Dividing this sum by the number of rolls, 300, gives us our expected average. This is

$$\text{average of rolls} = \frac{\text{sum of rolls}}{\text{number of rolls}} = \frac{1050}{300} = 3.5$$

Notice that if we write this average in the form of a fraction we get

$$\text{average of rolls} = \frac{\text{sum of rolls}}{\text{number of rolls}}$$

$$= \frac{1 \cdot \left(\begin{array}{c} \text{number of} \\ \text{1's expected} \end{array} \right) + 2 \cdot \left(\begin{array}{c} \text{number of} \\ \text{2's expected} \end{array} \right) + 3 \cdot \left(\begin{array}{c} \text{number of} \\ \text{3's expected} \end{array} \right) + \cdots + 6 \cdot \left(\begin{array}{c} \text{number of} \\ \text{6's expected} \end{array} \right)}{300}$$

$$= \frac{1 \cdot \left[\left(\frac{1}{6} \right)(300) \right] + 2 \cdot \left[\left(\frac{1}{6} \right)(300) \right] + 3 \cdot \left[\left(\frac{1}{6} \right)(300) \right] + \cdots + 6 \cdot \left[\left(\frac{1}{6} \right)(300) \right]}{300}$$

$$= \frac{300 \left[1 \cdot \left(\frac{1}{6} \right) + 2 \cdot \left(\frac{1}{6} \right) + 3 \cdot \left(\frac{1}{6} \right) + \cdots + 6 \cdot \left(\frac{1}{6} \right) \right]}{300}$$

$$= 1 \cdot \left(\frac{1}{6} \right) + 2 \cdot \left(\frac{1}{6} \right) + 3 \cdot \left(\frac{1}{6} \right) + \cdots + 6 \cdot \left(\frac{1}{6} \right)$$

$$= \frac{21}{6} = 3.5$$

The value 3.5 tells us that, if we repeat the experiment over and over many times, the average of the outcomes is expected to be about 3.5. Again, we note that the average is not a value that occurs in a single roll. What we have just determined is the **mean** or average value of a random variable X. Often we call the mean of a random variable X the **expected value** of X. (The expected value can also be thought of as a probability weighted average, since each outcome is *weighted* by its probability.)

■ **Expected Value of a Discrete Random Variable**

For a discrete random variable X having values $x_1, x_2, \ldots, x_n$, the **expected value**, **$E(X)$**, (or **mean**, **μ_X**) of the distribution is

$$E(X) = \mu_X = x_1 P(X = x_1) + x_2 P(X = x_2) + \cdots + x_n P(X = x_n)$$

▶ **Note:** The formula in the preceding box is for the case of a finite number of random variable values. In the case that X is countably infinite, the general definition is $\mu_X = \sum\limits_{\substack{\text{all } x \\ \text{values}}} x \cdot P(X = x)$, assuming that the sum exists.

Before we leave the experiment of tossing a die, let's make a final observation. When we compute the expected value as a fraction, notice that the 300 for the number of rolls was factored from the numerator and then canceled with the 300 in the denominator. This tells us that the actual number of trials does not matter, because the expected value represents the **long-run average** of the outcomes. In Example 1 we determine the expected value for another probability distribution.

Example 1 **Determining an Expected Value**

In Example 3 of Section 8.1, we conducted a probability experiment of tossing a coin 4 times where X represented the total number of heads recorded. The table of values for $P(X = x)$ is shown in Table 8.2.3. Determine the expected value and interpret.

Table 8.2.3

$X = x$	$P(X = x)$
$X = 0$	$P(X = 0) = \frac{1}{16}$
$X = 1$	$P(X = 1) = \frac{4}{16}$
$X = 2$	$P(X = 2) = \frac{6}{16}$
$X = 3$	$P(X = 3) = \frac{4}{16}$
$X = 4$	$P(X = 5) = \frac{1}{16}$

Solution

Using the formula for expected value, we get

$$E(X) = \mu_X = 0 \cdot P(X = 0) + 1 \cdot P(X = 1) + 2 \cdot P(X = 2) + 3 \cdot P(X = 3)$$
$$+ 4 \cdot P(X = 4)$$

$$= 0 \cdot \frac{1}{16} + 1 \cdot \frac{4}{16} + 2 \cdot \frac{6}{16} + 3 \cdot \frac{4}{16} + 4 \cdot \frac{1}{16}$$

$$= 0 + \frac{4}{16} + \frac{12}{16} + \frac{12}{16} + \frac{4}{16} = \frac{32}{16} = 2$$

Interpret the Solution: This means that if the experiment of tossing a coin 4 times is repeated over time we can expect on average that 2 of the 4 tosses will be heads. ■

If we make a probability histogram for the distribution in Table 8.2.3 and look at where the expected value, $E(X) = 2$, found in Example 1 occurs, we can obtain a geometric meaning of expected value. The expected value is similar to a fulcrum (a balancing point) that would perfectly "balance" the probability histogram. See Figure 8.2.1.

Example 1 had proper fractions for all the probability values. In applications, many of the probabilities are given as decimals. Example 2 shows this type of application.

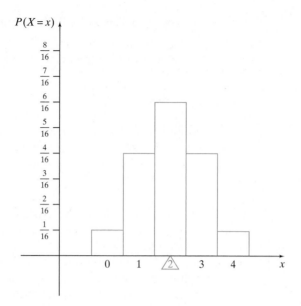

Figure 8.2.1

Example 2 Applying an Expected Value

Between 1993 and 1997, the number of two-bedroom rental units, in thousands, under construction at various monthly rent levels X, in dollars, in the United States can be represented by the probability distribution in Table 8.2.4.

Table 8.2.4

$X = x$	$P(X = x)$
$X = 100$	$P(X = 100) \approx 0.041$
$X = 200$	$P(X = 200) \approx 0.252$
$X = 300$	$P(X = 300) \approx 0.137$
$X = 400$	$P(X = 400) \approx 0.156$
$X = 500$	$P(X = 500) \approx 0.125$
$X = 600$	$P(X = 600) \approx 0.132$
$X = 700$	$P(X = 700) \approx 0.158$

SOURCE: U.S. Census Bureau, www.census.gov

Compute the expected value and interpret.

Solution

Using the formula for expected value, we get

$$E(X) = \mu_X = 100(0.041) + 200(0.252) + 300(0.137) + 400(0.156)$$
$$+ 500(0.125) + 600(0.132) + 700(0.158)$$
$$= 410.30$$

Interpret the Solution: This means that, if many two-bedroom housing units were selected randomly between 1993 and 1997 in the United States, the average of the monthly rent of all selected rentals would have been about $410.30. ■

Checkpoint 1

Now work Exercise 41.

Example 3   **Applying an Expected Value**

The Ashton Wrestling Club is having a fund-raising raffle. One thousand tickets have been sold for $5 each. There is 1 first-place prize of $1000, 3 second-place prizes of $500 each, 5 third-place prizes of $200 each, and 20 consolation prizes of $50 each. Let X represent the net winnings, that is, the winnings minus the cost of purchasing a ticket. Determine $E(X)$ and interpret.

Table 8.2.5

Type of Ticket	Net Winnings
Losing	$0 - 5 = -5$
Consolation	$50 - 5 = 45$
Third place	$200 - 5 = 195$
Second place	$500 - 5 = 495$
First place	$1000 - 5 = 995$

Solution

Understand the Situation: We need to determine the possible values of X, that is, the value of a losing ticket, a first-place ticket, a second-place ticket, a third-place ticket, and a consolation prize ticket. After we determine the probability of having each type of ticket, we can then construct a table of values for $P(X = x)$. We can then use the table to determine $E(X)$.

The net winnings of any ticket, which give us the values of X, is simply

$$\text{winnings} - \text{cost of a ticket} = \text{net winnings}$$

We summarize the net winnings of each type of ticket in Table 8.2.5.

Since there is 1 first-place prize, the probability of having a first-place ticket is

$$P(\text{first place}) = \frac{1}{1000}$$

We know that there are 3 second-place tickets, which means the probability of having a second-place ticket is

$$P(\text{second place}) = \frac{3}{1000}$$

Continuing in this manner allows us to make the table of values for $P(X = x)$ shown in Table 8.2.6, from which we can compute $E(X)$ to be

$$E(X) = -5(0.971) + 45(0.02) + 195(0.005) + 495(0.003) + 995(0.001)$$
$$= -0.5$$

Table 8.2.6

$X = x$	$P(X = x)$
$X = -5$	$P(X = -5) = \frac{971}{1000} = 0.971$
$X = 45$	$P(X = 45) = \frac{20}{1000} = 0.02$
$X = 195$	$P(X = 195) = \frac{5}{1000} = 0.005$
$X = 495$	$P(X = 495) = \frac{3}{1000} = 0.003$
$X = 995$	$P(X = 995) = \frac{1}{1000} = 0.001$

Interpret the Solution: The expected value gives the long-run average loss for an owner of one ticket. In other words, if someone participated in such a raffle by purchasing just one ticket each time, in the long run this individual can expect to lose $0.50 per raffle on average.   ∎

Standard Deviation

In addition to knowing the expected value, or mean, of a random variable X, we also want to know how spread out or *dispersed* the probability distribution is.

Consider the two distributions whose probability histograms are given in Figure 8.2.2. Figure 8.2.2(a) shows a probability distribution that is fairly concentrated about its expected value (mean), whereas Figure 8.2.2(b) is more dispersed or spread about the expected value (mean).

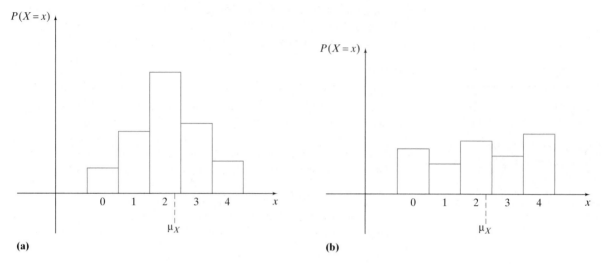

(a)   **(b)**

Figure 8.2.2

We can give a numerical value to this dispersion by computing the **standard deviation** of the probability distribution. The standard deviation for a probability distribution is denoted by the Greek lowercase letter σ (pronounced *sigma*). We can think of the standard deviation as the average distance that each random variable value lies away from the mean, when accounting for probability.

■ **Standard Deviation of a Discrete Random Variable**

For a discrete random variable X having values $x_1, x_2, \ldots, x_n$, the **standard deviation** is

$$\sigma_X = \sqrt{\sum_{i=1}^{n} x_i^2 P(x = x_i) - \mu_X^2}$$

▶ **Note:** The formula in the preceding box is for the case of a finite number of random variable values. In the case when X is countably infinite, the general definition is

$$\sigma_X = \sqrt{\sum_{\substack{\text{all } x \\ \text{values}}} x^2 \cdot P(X = x) - \mu_X^2}$$

Using the formula to compute the standard deviation can appear to be somewhat perplexing. To simplify computations, we can compute the standard deviation in three steps:

Step 1: Compute

$$\sum_{i=1}^{n} x_i^2 P(X = x_i) = x_1^2 P(X = x_1) + x_2^2 P(X = x_2) + \cdots + x_n^2 P(X = x_n)$$

Step 2: Subtract the square of the mean, expected value, from the result in step 1 to get

$$\sum_{i=1}^{n} x_i^2 P(X = x_i) - \mu_X^2$$

This quantity is called the **variance** and is denoted by σ_X^2.

Step 3: Take the square root of the result in step 2 to get

$$\sigma_X = \sqrt{\sum_{i=1}^{n} x_i^2 P(X = x_i) - \mu_X^2}$$

Example 4 illustrates how these three steps are computed.

Example 4   **Computing the Standard Deviation of a Discrete Random Variable**

For the probability distribution of monthly rents in Example 2, compute the variance and standard deviation.

Solution

We will follow the three steps in computing the standard deviation.

Step 1: $\displaystyle\sum_{i=1}^{7} x_i^2 P(X = x_i) \approx 100^2(0.041) + 200^2(0.252) + 300^2(0.137)$

$$+ 400^2(0.156) + 500^2(0.125) + 600^2(0.132)$$
$$+ 700^2(0.158) = 203{,}970$$

Step 2: Now we square the expected value $E(X)$, or mean μ_X, that we found in Example 2 and subtract it from the value found in step 1.

$$\sum_{i=1}^{7} x_i^2 P(X = x_i) - \mu_X^2 = 203{,}970 - (410.30)^2 = 35{,}623.91$$

So the variance is $\sigma_X^2 = 35{,}623.91$.

Step 3: To determine the standard deviation, we simply take the square root of the variance. This gives us

$$\sigma_X = \sqrt{\sigma_X^2}$$
$$= \sqrt{35{,}623.91} \approx 188.74$$

Interpret the Solution: So the standard deviation of the monthly rents is about $188.74. This means that the monthly rents varied on average by about $188.74 from the mean of $410.30. ∎

✓ **Checkpoint 2**

Now work Exercise 31.

Discrete Probability Distribution Models

At times we may know some specific properties of a probability experiment. In some cases we can use a predetermined function called a **probability distribution**

model to determine $P(X = x)$. For example, suppose that eight tickets labeled with the numbers 1 through 8 are placed into a box. In an experiment, one ticket is drawn out randomly. If we let X represent the outcome of any draw, then the possible values of X are $x = 1, 2, 3, 4, 5, 6, 7, 8$. Since each outcome is equally likely, the probability of getting any of these outcomes is $\frac{1}{8}$. This is shown in the probability histogram in Figure 8.2.3.

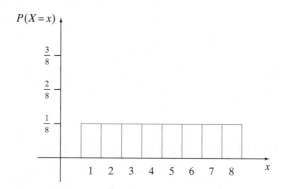

Figure 8.2.3

The probability mass function for X is

$$P(X = x) = \frac{1}{8}, \qquad x = 1, 2, 3, 4, 5, 6, 7, 8$$

Since the rectangles in the probability histogram are evenly or *uniformly* dispersed and each rectangle has the same height, this type of distribution is called the **discrete uniform distribution**.

Discrete Uniform Distribution

Let X be a discrete random variable having values $x = 1, 2, \ldots, N$. If all the probabilities associated with X are equal, then X is a **discrete uniform random variable** with probability mass function

$$P(X = x) = \begin{cases} \frac{1}{N}, & x = 1, 2, \ldots, N \\ 0, & \text{otherwise} \end{cases}$$

Example 5   Determining Probabilities with the Discrete Uniform Distribution

A new radio station in Evansburg called Red Hot Radio airs a promotion for their Red Hot Chili Cookoff contest. The station manager knows that the number of times that the promotion airs for the contest is a random variable uniformly distributed between 1 and 5 times per hour, for any given hour.

(a) Determine the probability mass function $P(X = x)$ for the distribution. Sketch a probability histogram for the distribution.

(b) Determine $P(X > 3)$ and interpret.

Solution

(a) Here X is a discrete uniform random variable for which the possible values of X are $x = 1, 2, 3, 4, 5$. So, knowing that $N = 5$, the probability mass function for the distribution is

$$P(X = x) = \frac{1}{5}, \qquad x = 1, 2, 3, 4, 5$$

Figure 8.2.4 shows the probability histogram for the distribution.

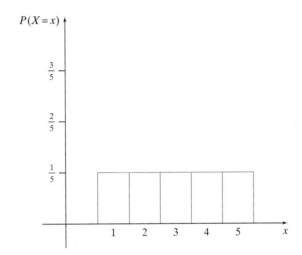

Figure 8.2.4

(b) **Understand the Situation:** To compute $P(X > 3)$, we need to add the probability values that are associated with X being greater than 3. This means that we can have $X = 4$ or $X = 5$, or $P(X \geq 4)$.

So we get

$$P(X > 3) = P(X \geq 4) = P(X = 4) + P(X = 5)$$
$$= \frac{1}{5} + \frac{1}{5} = \frac{2}{5}$$

Interpret the Solution: This means that, during any given hour, there is a $\frac{2}{5}$ or 40% chance that the promotion will be aired more than 3 times. ∎

✓ **Checkpoint 3**

Now work Exercise 57 (a) and (d).

Observe in Example 5(b) that $P(X > 3) = P(X \geq 4)$ can be represented graphically by summing the areas of the shaded rectangles for the probability histogram shown in Figure 8.2.5 on p. 484.

The mean, or expected value, and standard deviation can be computed for the discrete uniform distribution using the techniques outlined in this section. But discrete probability distribution models, including the discrete uniform distribution, also have shortcut formulas that can be used to determine the mean (expected value) and standard deviation. The derivation of these formulas requires more advanced mathematics.

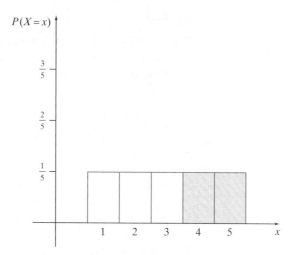

Figure 8.2.5 The sum of the areas of the shaded rectangles equals $P(X \geq 4)$.

■ **Mean and Standard Deviation of the Discrete Uniform Distribution**

For discrete uniform distributions, the **mean**, or **expected value**, and **standard deviation** are given by the formulas

$$E(X) = \mu_X = \frac{N+1}{2} \quad \text{and} \quad \sigma_X = \sqrt{\frac{(N-1)(N+1)}{12}}$$

Example 6 **Determining the Mean and Standard Deviation for the Discrete Uniform Distribution**

In Example 5 we determined that the probability mass function for the number of times that the Red Hot Chili Cookoff airs per hour is given by

$$P(X = x) = \frac{1}{5}, \qquad x = 1, 2, 3, 4, 5$$

(a) Determine the mean using the formula

$$E(X) = \mu_X = x_1 P(X = x_1) + x_2 P(X = x_2) + \cdots + x_n P(X = x_n)$$

and verify by using the shortcut formula $\mu_X = \dfrac{N+1}{2}$.

(b) Determine the standard deviation, σ_X, of the distribution using the shortcut formula.

Solution

(a) Using the definition of expected value, we get

$$\mu_X = 1 \cdot \frac{1}{5} + 2 \cdot \frac{1}{5} + 3 \cdot \frac{1}{5} + 4 \cdot \frac{1}{5} + 5 \cdot \frac{1}{5}$$

$$= \frac{1}{5} + \frac{2}{5} + \frac{3}{5} + \frac{4}{5} + \frac{5}{5} = \frac{15}{5} = 3$$

With the shortcut formula using $N = 5$, we get

$$\mu_X = \frac{N+1}{2} = \frac{5+1}{2} = \frac{6}{2} = 3$$

Interpret the Solution: Using either formula, we see that the promotion airs, on average, 3 times an hour.

(b) Using the shortcut formula, we get

$$\sigma_X = \sqrt{\frac{(N-1)(N+1)}{12}} = \sqrt{\frac{(5-1)(5+1)}{12}} = \sqrt{\frac{(4)(6)}{12}} = \sqrt{\frac{24}{12}}$$

$$= \sqrt{2} \approx 1.41$$

Interpret the Solution: So the standard deviation of the distribution is about 1.41. This means that the number of times that the promotion airs varied on average by about 1.41 times per hour from the mean of 3 times per hour. ∎

SUMMARY

In this section we learned how to determine the long-run average of a discrete probability distribution. We also learned how to compute the standard deviation of these distributions.

- For a discrete random variable, the **expected value** or **mean** is

$$E(X) = \mu_X = x_1 P(X = x_1) + x_2 P(X = x_2) + \cdots$$
$$+ x_n P(X = x_n)$$

- For a discrete random variable, the **standard deviation** is

$$\sigma_X = \sqrt{\sum_{i=1}^{n} x_i^2 P(x) - \mu^2}$$

- If all the probabilities associated with X are equal, then X is a **discrete uniform random variable** with probability mass function

$$P(X = x) = \begin{cases} \frac{1}{N}, & x = 1, 2, \ldots, N \\ 0, & \text{otherwise} \end{cases}$$

- For the discrete uniform distribution, the **mean (expected value)** and **standard deviation** are given by the formulas

$$\mu_X = \frac{N+1}{2} \quad \text{and} \quad \sigma_X = \sqrt{\frac{(N-1)(N+1)}{12}}$$

SECTION 8.2 EXERCISES

In Exercises 1–14, a table of values for the probability distribution $P(X = x)$ is given. Determine the expected value $E(X)$ of the random variable X having the given discrete probability distribution.

1.

$X = x$	$P(X = x)$
$X = 0$	$P(X = 0) = \frac{1}{2}$
$X = 1$	$P(X = 1) = \frac{1}{4}$
$X = 2$	$P(X = 2) = \frac{1}{4}$

2.

$X = x$	$P(X = x)$
$X = 1$	$P(X = 1) = \frac{1}{3}$
$X = 2$	$P(X = 2) = \frac{1}{3}$
$X = 3$	$P(X = 3) = \frac{1}{3}$

3.

$X = x$	$P(X = x)$
$X = 0$	$P(X = 0) = \frac{1}{10}$
$X = 1$	$P(X = 1) = \frac{1}{20}$
$X = 2$	$P(X = 2) = \frac{1}{5}$
$X = 3$	$P(X = 3) = \frac{2}{5}$
$X = 4$	$P(X = 4) = \frac{1}{4}$

4.

$X = x$	$P(X = x)$
$X = 1$	$P(X = 1) = \frac{3}{32}$
$X = 2$	$P(X = 2) = \frac{3}{16}$
$X = 3$	$P(X = 3) = \frac{5}{32}$
$X = 4$	$P(X = 4) = \frac{3}{8}$
$X = 5$	$P(X = 5) = \frac{3}{16}$

5.

X = x	P(X = x)
X = -2	P(X = -2) = 0.03
X = -1	P(X = -1) = 0.13
X = 0	P(X = 0) = 0.40
X = 1	P(X = 1) = 0.35
X = 2	P(X = 2) = 0.09

6.

X = x	P(X = x)
X = -1	0.174
X = 0	0.111
X = 2	0.260
X = 4	0.300
X = 6	0.155

7.

X = x	P(X = x)
X = 10	P(X = 10) = 0.1
X = 20	P(X = 20) = 0.1
X = 30	P(X = 30) = 0.6
X = 40	P(X = 40) = 0.2

8.

X = x	P(X = x)
X = 5	P(X = 5) = 0.3
X = 10	P(X = 10) = 0.2
X = 25	P(X = 25) = 0.1
X = 35	P(X = 35) = 0.4

9.

X = x	P(X = x)
X = 8	$P(X = 8) = \frac{1}{5}$
X = 9	$P(X = 9) = \frac{1}{5}$
X = 10	$P(X = 10) = \frac{1}{10}$
X = 11	$P(X = 11) = \frac{3}{20}$
X = 12	$P(X = 12) = \frac{7}{20}$

10.

X = x	P(X = x)
X = 0	$P(X = 0) = \frac{11}{20}$
X = 1	$P(X = 1) = \frac{1}{4}$
X = 2	$P(X = 2) = \frac{1}{20}$
X = 3	$P(X = 3) = \frac{1}{10}$
X = 4	$P(X = 4) = \frac{1}{20}$

11.

X = x	P(X = x)
X = 0	P(X = 0) = 0.5
X = 1	P(X = 1) = 0.2
X = 2	P(X = 2) = 0.1
X = 3	P(X = 3) = 0.2

12.

X = x	P(X = x)
X = 0	P(X = 0) = 0.2
X = 2	P(X = 2) = 0.4
X = 3	P(X = 3) = 0.1
X = 4	P(X = 4) = 0.3

13.

X = x	P(X = x)
X = 0	P(X = 0) = 0.05
X = 2	P(X = 2) = 0.05
X = 3	P(X = 3) = 0.10
X = 4	P(X = 4) = 0.20
X = 5	P(X = 5) = 0.60

14.

X = x	P(X = x)
X = 0	P(X = 0) = 0.15
X = 1	P(X = 1) = 0.15
X = 2	P(X = 2) = 0.25
X = 3	P(X = 3) = 0.10
X = 4	P(X = 4) = 0.35

In Exercises 15–26, a probability mass function $P(X = x)$ is given. Determine the expected value.

15. $P(X = x) = \frac{1}{6}x, \quad x = 0, 1, 2, 3$

16. $P(X = x) = 0.01x, \quad x = 10, 20, 30, 40$

17. $P(X = x) = \frac{1}{8}x, \quad x = 0, 1, 2, 5$

18. $P(X = x) = \frac{1}{100}x, \quad x = 10, 30, 60$

19. $P(X = x) = \frac{1}{4}, \quad x = 1, 2, 3, 4$

20. $P(X = x) = \frac{1}{5}, \quad x = 0, 1, 2, 3, 4$

21. $P(X = x) = \frac{1}{14}x^2, \quad x = 1, 2, 3$

22. $P(X = x) = \frac{1}{36}x^3, \quad x = 1, 2, 3$

23. $P(X = x) = 0.05x, \quad x = 2, 3, 4, 5, 6$

24. $P(X = x) = \frac{1}{50}x, \quad x = 5, 10, 15, 20$

25. $P(X = x) = \frac{x + 2}{10}, \quad x = -1, 0, 1, 2$

26. $P(X = x) = \frac{x - 4}{14}, \quad x = 6, 7, 8, 9$

In Exercises 27–40, determine the standard deviation, σ_X, for the following:

27. For the discrete probability distribution in Exercise 1

28. For the discrete probability distribution in Exercise 2

29. For the discrete probability distribution in Exercise 3

30. For the discrete probability distribution in Exercise 4

✔ **31.** For the discrete probability distribution in Exercise 7

32. For the discrete probability distribution in Exercise 8

33. For the probability mass function in Exercise 15

34. For the probability mass function in Exercise 16

35. For the probability mass function in Exercise 19

36. For the probability mass function in Exercise 20

37. For the probability mass function in Exercise 21

38. For the probability mass function in Exercise 22

39. For the probability mass function in Exercise 25

40. For the probability mass function in Exercise 26

Applications

✓ **41.** From prior records the owner of the Video4U has determined that the number of videos that a customer rents on any given visit can be represented by a discrete random variable X with the following probability distribution.

$X = x$	$P(X = x)$
$X = 0$	$P(X = 0) = 0.05$
$X = 1$	$P(X = 1) = 0.50$
$X = 2$	$P(X = 2) = 0.30$
$X = 3$	$P(X = 3) = 0.15$

Determine the expected value, $E(X)$, and interpret.

42. From prior records the manager of Javah! coffee shop has determined that its total number of daily customers can be represented by a discrete random variable X with the following probability distribution.

$X = x$	$P(X = x)$
$X = 45$	$P(X = 45) = 0.10$
$X = 46$	$P(X = 46) = 0.30$
$X = 47$	$P(X = 47) = 0.25$
$X = 48$	$P(X = 48) = 0.20$
$X = 49$	$P(X = 49) = 0.05$
$X = 50$	$P(X = 50) = 0.10$

Determine the expected value, $E(X)$, and interpret.

43. The Franklinton Group, a consortium of educators, has determined that the number of commercials per show aired during afterschool television shows on a certain network can be represented by a discrete random variable X with the following probability distribution.

$X = x$	$P(X = x)$
$X = 0$	$P(X = 0) = 0.20$
$X = 1$	$P(X = 1) = 0.15$
$X = 2$	$P(X = 2) = 0.25$
$X = 3$	$P(X = 3) = 0.20$
$X = 4$	$P(X = 4) = 0.05$
$X = 5$	$P(X = 5) = 0.10$
$X = 6$	$P(X = 6) = 0.05$

(a) Determine the expected value and interpret.

(b) What is the most frequently occurring number of commercials on any given show? (That is, what x-value yields the greatest $P(X = x)$-value?) This value is called the **mode** of the distribution.

44. The North-Am racing series has determined that the number of yellow flag caution laps, rounded to the nearest 10, during any given race can be represented by a discrete random variable X having the following probability distribution.

$X = x$	$P(X = x)$
$X = 0$	$P(X = 0) = 0.20$
$X = 10$	$P(X = 10) = 0.25$
$X = 20$	$P(X = 20) = 0.38$
$X = 30$	$P(X = 30) = 0.10$
$X = 40$	$P(X = 40) = 0.07$

(a) Determine the expected value and interpret.

(b) What is the most frequently occurring number of yellow flag caution laps run during any given race? (That is, what x-value yields the greatest $P(X = x)$-value?) This value is called the **mode** of the distribution.

45. Parmiter College has determined that the number of students who drop out of introductory psychology can be represented by a discrete random variable X having the following probability distribution.

$X = x$	$P(X = x)$
$X = 12$	$P(X = 12) = 0.09$
$X = 13$	$P(X = 13) = 0.18$
$X = 14$	$P(X = 14) = 0.38$
$X = 15$	$P(X = 15) = 0.20$
$X = 16$	$P(X = 16) = 0.15$

According to the distribution, how many students can we expect to drop out of introductory psychology?

46. Linger's Pizzeria management team has determined that the number of late-night pizza deliveries on any given Wednesday can be represented by a discrete random variable X having the following probability distribution.

$X = x$	$P(X = x)$
$X = 4$	$P(X = 4) = 0.32$
$X = 5$	$P(X = 5) = 0.11$
$X = 6$	$P(X = 6) = 0.40$
$X = 7$	$P(X = 7) = 0.12$
$X = 8$	$P(X = 8) = 0.05$

According to the distribution, how many late-night pizzas can the pizzeria expect to deliver on any given Wednesday night?

47. The Sedgewick Baseball Club is having a fund-raising raffle. Twenty thousand raffle tickets have been sold for $10 each. Tickets are drawn at random and cash prizes are awarded as follows: 1 prize of $10,000; 2 prizes of $5000; 10 prizes of $1000, and 20 prizes of $500. Determine the expected value of this raffle if 1 ticket is purchased.

48. The Library Society is having a fund-raising raffle. Fifteen thousand raffle tickets have been sold for $5 each. Tickets are drawn at random and cash prizes are awarded as follows: 1 prize of $5000, 3 prizes of $1000, 5 prizes of $750, and 10 prizes of $300. Determine the expected value of this raffle if 1 ticket is purchased.

49. The number of defective springs discovered after inspection by the Springsworth Company is shown in the following table.

Number of Inspections	Number of Defects
1	1500
2	800
3	200
4	50

(a) Determine the expected value and interpret.

(b) Determine the standard deviation of the distribution and interpret.

50. The number of drivers aged 16 to 24 who were in automobile accidents in 1996 is shown in the table.

Age	Number of Accidents (in millions)
16	1.6
17	2.2
18	2.4
19	2.7
20	2.8
21	2.9
22	2.9
23	3.1
24	3.6

SOURCE: National Safety Council, www.nsc.org

(a) Determine the expected value and interpret.

(b) Determine the standard deviation of the distribution and interpret.

51. The distribution of weights of American males aged 70 to 79 from 1988 to 1994 is shown in the table.

Weight (in pounds)	Percentage	Weight (in pounds)	Percentage
100–109	0.2	200–209	9.5
110–119	0.2	210–219	5.2
120–129	1.3	220–229	5.4
130–139	4.1	230–239	2.5
140–149	7.4	240–249	1.0
150–159	10.1	250–259	0.7
160–169	11.9	260–269	1.1
170–179	13.2	270–279	0.9
180–189	11.1	280–289	0.4
190–199	13.6	290 and over	0.2

SOURCE: U.S. Center for Health Statistics, www.cdc.gov/nchs

(a) According to the distribution, what can we expect the average weight of an American male aged 70 to 79 from 1988 to 1994 to be?

(b) Determine the standard deviation of the distribution and interpret.

52. John Spears wishes to purchase a 7-year term-life insurance policy that pays the beneficiary $100,000 in the event that his death occurs during the next 7 years. From inspecting life insurance tables, John determines that the probability that he will live another 7 years is about 0.97. To calculate the minimum premium (payment) for this insurance, determine when the insurance company's profit is zero.

53. Chris Thomas purchased a $100,000 1-year term-life insurance policy for $276. Assuming that the probability that she will live another year is 0.981, determine the company's expected gain.

54. Chenush Gas is a natural gas drilling company. After testing and analysis by geologists, Chenush Gas is considering drilling on two different sites. The geologists have estimated that site I will net $50 million if successful (probability 0.3) and lose $5 million if not successful (probability 0.7). Site II is estimated to net $80 million if successful (probability 0.2) and lose $6 million if not successful (probability 0.8).

(a) Determine the expected return for each site.

(b) Based on part (a), which site would you advise the company to select.

(c) Which site has less risk? (This corresponds to which has a smaller variance.)

55. Chenush Oil is an oil drilling company. After testing and analysis by geologists, Chenush Oil is considering drilling on two different sites. The geologists have estimated that site I will net $40 million if successful (probability 0.25) and lose $4 million if not successful (probability 0.75). Site II is estimated to net $70 million if successful (probability 0.12) and lose $5 million if not successful (probability 0.88).

(a) Determine the expected return for each site.

(b) Based on part (a), which site would you advise the company to select.

(c) Which site has less risk? (This corresponds to which has a smaller variance.)

56. The number of times that KTTV radio gives the time of day each hour is a random variable uniformly distributed between 1 and 5, inclusive. Let X be a discrete uniform random variable representing the number of times that the time of day is given each hour.

(a) Write a probability mass function for X.

(b) Determine μ_X and interpret.

(c) Compute σ_X for the distribution and interpret.

(d) Determine the probability that the time of day will be given two or more times in a given hour.

✓ **57.** The Kronix Company knows that the number of times that its old printing machine breaks down during a shift is a random variable uniformly distributed between 1 and 10 times inclusive. Let X be a discrete uniform random variable representing the number of times that the printing machine breaks down during a shift.

✓ **(a)** Write a probability mass function for X.

(b) Determine μ_X and interpret.

(c) Compute σ_X for the distribution and interpret.

✓ **(d)** Determine the probability that during any given shift the printing machine will break down three or fewer times.

58. Manufacturers of Mushy Slushy drinks know that the machine that dispenses the Slushys keeps the drink mixture at a temperature (rounded to the nearest degree) that is a random variable uniformly distributed between 1 and 6 degrees Celsius, inclusive. Let X be a discrete uniform random variable representing the temperature in degrees Celsius.

(a) Write a probability mass function for X.

(b) Determine μ_X and interpret.

(c) Compute σ_X for the distribution and interpret.

(d) Determine the probability that if the Mushy Slushy drink dispenser is checked the temperature will be at most 3 degrees Celsius.

59. The Rockford Company manufactures plastic automobile parts and uses the Rockford hardness scale to determine the hardness of the plastic pieces manufactured. The scale and the manufacturing process can be represented by a discrete uniform random variable from 1 to 25. Let X be a discrete uniform random variable representing the hardness on the Rockford hardness scale.

(a) Write a probability mass function for X.

(b) Determine μ_X and interpret.

(c) Compute σ_X for the distribution and interpret.

(d) Determine the probability that if a plastic piece is selected at random the piece will have a Rockford hardness scale value of more than 20.

SECTION PROJECT

Let's explore some of the properties of the expected value and standard deviation of discrete random variables. Consider the following three probability distributions:

Distribution A

$X = x$	$P(X = x)$
$X = 10$	$P(X = 10) = 0.1$
$X = 20$	$P(X = 20) = 0.4$
$X = 30$	$P(X = 30) = 0.3$
$X = 40$	$P(X = 40) = 0.2$

Distribution B

$X = x$	$P(X = x)$
$X = 15$	$P(X = 15) = 0.1$
$X = 25$	$P(X = 25) = 0.4$
$X = 35$	$P(X = 35) = 0.3$
$X = 45$	$P(X = 45) = 0.2$

Distribution C

$X = x$	$P(X = x)$
$X = 90$	$P(X = 90) = 0.1$
$X = 180$	$P(X = 180) = 0.4$
$X = 270$	$P(X = 270) = 0.3$
$X = 360$	$P(X = 360) = 0.2$

(a) How are the x-values between Distribution A and Distribution B related? That is, how can we get the x-values for Distribution B using the x-values from Distribution A?

(b) How are the x-values between Distribution A and Distribution C related? That is, how can we get the x-values for Distribution C using the x-values from Distribution A?

(c) Compute the mean μ_X of each distribution. How is the mean of Distribution A related to the mean of

Distribution B? How is the mean of Distribution A related to the mean of Distribution C?

(d) Compute the standard deviation σ_X of each distribution. How is the standard deviation of Distribution A related to the standard deviation of Distribution B? How is the standard deviation of Distribution A related to the standard deviation of Distribution C?

Section 8.3 The Binomial Distribution

Many of the types of probability that we study have exactly two possible outcomes. For example, a person whose television viewing habits are being monitored can either watch or not watch a specific program. A student taking a multiple-choice test can get a question either correct or incorrect. A college football team can either win or lose a particular game. In this section we will study the **binomial distribution**, which is used to determine probabilities of events concerning repeated experiments that have only two possible outcomes in each experiment. We will also explore the properties of the binomial distribution, including the mean and standard deviation.

The Binomial Distribution

Generally, the trials associated with probability experiments that have exactly two outcomes, like the ones we mentioned earlier, are called **Bernoulli trials**. Named after the Swiss mathematician Jakob Bernoulli, these experiments are regarded as having an outcome of **success** (watching a particular television show, getting a test question correct, or winning a football game) or **failure**. If we combine a series of independent Bernoulli trials, we get a **binomial experiment**.

> ■ **Binomial Experiment**
>
> A **binomial experiment** is a probability experiment that satisfies the following four assumptions:
>
> 1. Each trial has exactly two possible outcomes: success or failure.
> 2. The probability of success for each trial is the same. We call this fixed probability of success p.
> 3. The experiment has a finite number of trials, denoted by n.
> 4. Each trial is independent of the others.

Example 1 **Identifying a Binomial Experiment**

A 1995 report revealed that 31.4% of African Americans were smokers (*Source:* www.ymn.org/newstat). Suppose that a sample of 20 African Americans is selected at random. Determine if this meets the assumptions of a binomial experiment and, if so, identify the values of p and n.

Solution

Understand the Situation: Let's check the four assumptions for a binomial experiment.

1. There are two possible outcomes. These are defined as
   Success: member of the sample is a smoker
   Failure: member of the sample is not a smoker
2. The probability of success for each trial is the same, with the probability of success written as $p = 0.314$ in decimal form.
3. The experiment has a fixed number of trials with $n = 20$.
4. Since the fact that one member of the sample being a smoker does not affect whether the other members of the sample smoke, the trials are independent. Since all four assumptions are satisfied, we have a binomial experiment. ■

 Checkpoint 1

Now work Exercise 11.

Example 1 shows a peculiar aspect of binomial experiments. We defined a success as identifying a person from the sample as a smoker. Even though smoking is not regarded as a "success," since we are trying to determine probabilities concerning smokers in the sample, this is how we would define a success. We must be aware that success and failure are merely mathematical conventions to characterize binomial random variables. Sometimes what in real life is clearly a failure is, for mathematical convention, defined to be a success.

The possible results of binomial experiments and probabilities associated with them give us a type of discrete probability distribution called the **binomial distribution**. This important distribution has its own probability mass function, based on the values of n and p.

> ■ **Binomial Distribution**
>
> In a binomial experiment, the probability of having x successes in n trials is given by the probability mass function.
>
> $$P(X = x) = {}_nC_x \cdot p^x(1 - p)^{n-x}, \qquad x = 0, 1, 2, \ldots, n$$
>
> where
>
> > n is the number of trials
> > p is the fixed probability of success
> > x is the number of successes in n trials

▶ **Note:** 1. The binomial distribution is a special type of discrete distribution. This means that the information presented in Sections 8.1 and 8.2 can be used here.

2. Some textbooks write the binomial probability mass function as $P(X = x) = {}_nC_x \cdot p^x \cdot q^{n-x}$, where $q = 1 - p$ represents the probability of failure.

Now let's see how the binomial distribution can be used to calculate probabilities in some classic probability experiments.

Example 2   Determining a Binomial Probability

Draw 3 with replacement

1 Blue
1 Green

Figure 8.3.1

A bag contains 1 blue and 1 green marble. In an experiment, a marble is drawn at random, its color is recorded, and it is returned to the bag. Two more draws are done in the same way, for a total of three draws (see Figure 8.3.1).

(a) Use classical probability to compute the probability that in 3 draws exactly 2 will be green marbles.

(b) Use the binomial distribution to determine the probability that in 3 draws exactly 2 will be green marbles.

Solution

(a) The sample space for the experiment has 8 elements:

$$S = \{BBB, BBG, BGB, BGG, GBG, GGB, GBB, GGG\}$$

The outcomes associated with drawing exactly 2 green marbles are BGG, GBG, GGB. So, using classical probability, we get

$$P(2 \text{ green marbles}) = \frac{\text{number of ways to get 2 green marbles}}{\text{total number of ways to draw 3 marbles}} = \frac{3}{8}$$

(b) Understand the Situation: We first note that we have two possible outcomes: drawing a blue marble or drawing a green marble. Since we want to determine the probability that in 3 draws exactly 2 will be green marbles, let's define success as drawing a green marble. The probability of success for each trial is

$$p = \frac{\text{number of green marbles in the bag}}{\text{total number of marbles in the bag}} = \frac{1}{2}$$

Since the number of draws is 3, this means that there are a fixed number of trials, $n = 3$. We replace the marble before the next draw, which means that the trials are independent. So we have a binomial experiment, with $p = \frac{1}{2}$ and $n = 3$, and the probability mass function is

$$P(X = x) = {}_nC_x \cdot p^x(1-p)^{n-x}, \qquad x = 0, 1, 2, \ldots, n$$

$$= {}_3C_x \left(\frac{1}{2}\right)^x \left(1-\frac{1}{2}\right)^{3-x}, \qquad x = 0, 1, 2, 3$$

where x represents the number of green marbles drawn in 3 trials.

To determine the probability of drawing 2 green marbles, we need to compute $P(X = x)$ for $x = 2$. Consulting the Toolbox to the left for evaluating ${}_3C_2$, we get

$$P(X = 2) = {}_3C_2 \left(\frac{1}{2}\right)^2 \left(1-\frac{1}{2}\right)^{3-2}$$

$$= \frac{3!}{2!(3-2)!} \left(\frac{1}{2}\right)^2 \left(\frac{1}{2}\right)^1 = 3 \cdot \frac{1}{4} \cdot \frac{1}{2} = \frac{3}{8}$$

Interpret the Solution: The probability of drawing 2 green marbles is $\frac{3}{8}$. Notice that this is the same result obtained in part (a). ∎

One way that we use the binomial distribution is to examine the number of successes, x, in a **simple random sample** of size n. By a simple random sample of size n, we mean that the sample selected from the population has n elements and every set of n elements has an equal chance to be in the sample actually selected. Let's see how we can use the binomial distribution to approximate probabilities of a simple random sample.

From Your Toolbox

$${}_nC_r = \frac{n!}{r!(n-r)!}$$

Interactive Activity

Repeat Example 2 if there are 5 draws from the bag.

Example 3 Applying the Binomial Distribution

An August 2000 Harris poll found that 71% of Americans who are on-line use the Internet to access health and medical information (*Source:* www.harrisinteractive.com). A simple random sample of 7 Americans who are on-line is selected.

(a) Make a table of values for $P(X = x)$. Construct a probability histogram for the distribution.

(b) Determine $P(X = 5)$ and interpret.

(c) Determine $P(X \geq 5)$ and interpret.

Solution

(a) Understand the Situation: To make a table of values, we define success as the selected on-line American uses the Internet to access health and medical information. Since 71% of all on-line Americans polled use the Internet to access health and medical information, we have a probability of success

$p = 0.71$. There are a fixed number of trials $n = 7$, so the probability mass function for the distribution is

$$P(X = x) = {}_nC_x \cdot p^x(1-p)^{n-x}, \qquad x = 0, 1, 2, \ldots, n$$
$$= {}_7C_x(0.71)^x(1-0.71)^{7-x}$$
$$= {}_7C_x(0.71)^x(0.29)^{7-x}, \qquad x = 0, 1, 2, \ldots, 7$$

Using this formula, we get the table of values for $P(X = x)$ shown in Table 8.3.1. The probabilities are rounded to three decimal places.

Table 8.3.1

$X = x$	$P(X = x)$
$X = 0$	$P(X = 0) = {}_7C_0(0.71)^0(0.29)^7 \approx 0.000$
$X = 1$	$P(X = 1) = {}_7C_1(0.71)^1(0.29)^6 \approx 0.003$
$X = 2$	$P(X = 2) = {}_7C_2(0.71)^2(0.29)^5 \approx 0.022$
$X = 3$	$P(X = 3) = {}_7C_3(0.71)^3(0.29)^4 \approx 0.089$
$X = 4$	$P(X = 4) = {}_7C_4(0.71)^4(0.29)^3 \approx 0.217$
$X = 5$	$P(X = 5) = {}_7C_5(0.71)^5(0.29)^2 \approx 0.319$
$X = 6$	$P(X = 6) = {}_7C_6(0.71)^6(0.29)^1 \approx 0.260$
$X = 7$	$P(X = 7) = {}_7C_7(0.71)^7(0.29)^0 \approx 0.091$

Using the values in Table 8.3.1, we get the probability histogram shown in Figure 8.3.2.

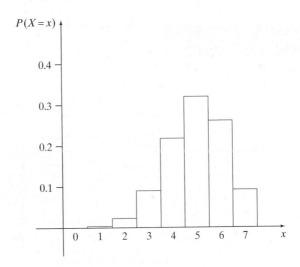

Figure 8.3.2

(b) Using Table 8.3.1, we determine that $P(X = 5) \approx 0.319$.

Interpret the Solution: This means that if 7 Americans who are on-line are selected at random, the probability that 5 of them use the Internet to access health and medical information is about 31.9%.

(c) To compute $P(X \geq 5)$, we have

$$P(X \geq 5) = P(X = 5) + P(X = 6) + P(X = 7)$$
$$\approx 0.319 + 0.260 + 0.091 = 0.670$$

Interpret the Solution: This means that if we select 7 Americans who are on-line at random, there is about a 67% chance that 5 or more of them use the Internet to access health and medical information. Notice that this corresponds to summing the areas of the shaded rectangles of the probability histogram, as shown in Figure 8.3.3.

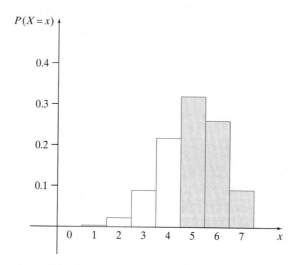

Figure 8.3.3

 Checkpoint 2

Now work Exercise 45.

Technology Option

The output from using the BINOMPDF command on a graphing calculator to compute, say, $P(X = 5)$ in part (a) of Example 3 is shown in Figure 8.3.4(a).

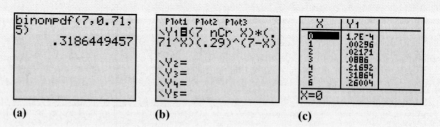

(a)   (b)   (c)

Figure 8.3.4

To quickly reproduce the table of values in Table 8.3.1, we can enter the probability mass function $P(X = x)$ in our $y =$ editor [see Figure 8.3.4(b)] and use the TABLE command [see Figure 8.3.4(c)].

To learn about these commands for your calculator, consult the on-line graphing calculator manual at www.prenhall.com/armstrong.

Two phrases frequently used in probability applications are "at most" and "at least." Example 4 demonstrates how these phrases are used in applications.

Example 4   Using the Phrases "at most" and "at least"

A May 2000 CBS News/NY Times poll asked "Will Social Security have money for you?" Thirty percent of American adults aged 45 years and younger responded yes (*Source:* www.CBSnews.com). Suppose that a random sample of 9 American adults 45 years and younger is selected at random.

(a) Determine the probability that at least 5 respond yes to the Social Security question.

(b) Determine the probability that at most 5 respond yes to the Social Security question.

Solution

Understand the Situation: To aid us in both parts of this example, we construct a table of values for $P(X = x)$. Here we have a binomial random variable X with $n = 9$ and $p = 0.30$. This means that

$$P(X = x) = {}_9C_x \cdot (0.30)^x (1 - 0.30)^{9-x}, \qquad x = 0, 1, 2, \ldots, 9$$
$$= {}_9C_x \cdot (0.30)^x (0.70)^{9-x}, \qquad x = 0, 1, 2, \ldots, 9$$

See Table 8.3.2.

Table 8.3.2

$X = x$	$P(X = x)$
$X = 0$	$P(X = 0) = {}_9C_0 \cdot (0.30)^0 (1 - 0.30)^{9-0} \approx 0.040$
$X = 1$	$P(X = 1) = {}_9C_1 \cdot (0.30)^1 (1 - 0.30)^{9-1} \approx 0.156$
$X = 2$	$P(X = 2) \approx 0.267$
$X = 3$	$P(X = 3) \approx 0.267$
$X = 4$	$P(X = 4) \approx 0.172$
$X = 5$	$P(X = 5) \approx 0.074$
$X = 6$	$P(X = 6) \approx 0.021$
$X = 7$	$P(X = 7) \approx 0.004$
$X = 8$	$P(X = 8) \approx 0.000$
$X = 9$	$P(X = 9) \approx 0.000$

(a) To say "at least 5" means that we could have 5 or more respond yes to the question. So we compute

$$P(\text{at least } 5) = P(X \geq 5)$$
$$= P(X = 5) + P(X = 6) + P(X = 7) + P(X = 8) + P(X = 9)$$
$$\approx 0.074 + 0.021 + 0.004 + 0.000 + 0.000 = 0.099$$

Interpret the Solution: This means that there is about a 9.9% chance that at least 5 will respond yes to the question.

(b) To say "at most 5" means that we could have 5 or less respond yes to the question. So we compute

$$P(\text{at most } 5) = P(X \leq 5)$$
$$= P(X = 0) + P(X = 1) + P(X = 2) + P(X = 3)$$
$$+ P(X = 4) + P(X = 5)$$
$$\approx 0.040 + 0.156 + 0.267 + 0.267 + 0.172 + 0.074 = 0.976$$

Interactive Activity

Why is it that the probabilities found in parts (a) and (b) of Example 4 are not complements?

Interpret the Solution: This means that there is about a 97.6% chance that at most 5 will respond yes to the question. ∎

✔ **Checkpoint 3**

Now work Exercise 49.

Mean and Standard Deviation of the Binomial Distribution

Probability distribution models such as the binomial distribution have specific formulas for the mean and standard deviation. To discover what the formulas are, let's first revisit an example presented in Section 8.1.

Flashback

Coin Toss Revisited

In Section 8.1, Example 3, we examined a probability experiment in which a coin is tossed 4 times and the result is recorded. If we let X represent the total number of heads recorded, show that the experiment can be considered a binomial experiment and determine its probability mass function $P(X = x)$.

Flashback Solution

To show that the experiment can be considered a binomial experiment, we check the four assumptions for a binomial experiment given at the beginning of this section.

1. There are two possible outcomes that we define as
   Success: toss is a head
   Failure: toss is a tail
2. The probability of success for each trial is the same, with the probability of success written as $p = 0.5$.
3. The experiment has a fixed number of trials, with $n = 4$.
4. Since the result of one toss does not affect any other toss, the trials are independent.

Since all four assumptions are met, we have a binomial experiment. We have determined that $p = 0.5$ and $n = 4$, so the probability mass function for the binomial distribution is given by

$$P(X = x) = {_4C_x} \cdot (0.5)^x (1 - 0.5)^{4-x}, \qquad x = 0, 1, 2, 3, 4$$
$$= {_4C_x} \cdot (0.5)^x (0.5)^{4-x}, \qquad x = 0, 1, 2, 3, 4$$

Table 8.3.3

$X = x$	$P(X = x)$
$X = 0$	$P(X = 0) = 0.0625$
$X = 1$	$P(X = 1) = 0.25$
$X = 2$	$P(X = 2) = 0.375$
$X = 3$	$P(X = 3) = 0.25$
$X = 4$	$P(X = 4) = 0.0625$

In the Flashback, we determined that X is a binomial random variable with $n = 4$ and $p = 0.5$. Using the probability mass function,

$$P(X = x) = {_4C_x} \cdot (0.5)^x (0.5)^{4-x}, \qquad x = 0, 1, 2, 3, 4$$

we can make a table of values for $P(X = x)$ as shown in Table 8.3.3.

Using the formula from Section 8.2 to compute the mean (consult the Toolbox on p. 497), we get

$$\mu_X = E(X) = 0(0.0625) + 1(0.25) + 2(0.375) + 3(0.25) + 4(0.0625)$$
$$= 0 + 0.25 + 0.75 + 0.75 + 0.25 = 2$$

Observe that this is the same as computing $np = 4(0.5) = 2$. This is no coincidence! This suggests that for the binomial distribution the mean of the distribution is $\mu_X = np$.

From Your Toolbox

For any discrete probability distribution, the mean (expected value) is given by

$$E(X) = \mu_X = x_1 P(X = x_1) + x_2 P(X = x_2) + \cdots + x_n P(X = x_n)$$

with standard deviation

$$\sigma_X = \sqrt{\sum_{i=1}^{n} x_i^2 P(X = x_i) - [E(X)]^2}$$

Now let's compute the standard deviation for the same distribution. Using the formula from Section 8.2 to compute the standard deviation (consult the Toolbox above) and following our three steps outlined in Section 8.2, we start by computing the value of $\sum x^2 P(X = x)$.

$$\sum x^2 P(X = x) = 0^2(0.0625) + 1^2(0.25) + 2^2(0.375) + 3^2(0.25) + 4^2(0.0625) = 5$$

Subtracting the square of the mean gives us

$$\sum x^2 P(X = x) - \mu_X^2 = 5 - (2)^2 = 1$$

Taking the square root of this result gives us the standard deviation.

$$\sigma_X = \sqrt{\sum x^2 P(X) - \mu^2} = \sqrt{1} = 1$$

Observe that this is the same as computing

$$\sigma_X = \sqrt{np(1 - p)} = \sqrt{4 \cdot (0.5)(1 - 0.5)} = \sqrt{4 \cdot (0.5) \cdot (0.5)} = \sqrt{1} = 1$$

There is no coincidence here either! We have just illustrated that the mean and standard deviation of the binomial distribution are based on the values of n and p.

Mean and Standard Deviation of the Binomial Random Variable

Let X be a binomial random variable with number of trials n and probability of success p. Then the **mean**, or expected value, of the distribution is

$$E(X) = \mu_X = np$$

with **standard deviation** given by

$$\sigma_X = \sqrt{np(1 - p)}$$

Example 5   Determining the Mean and Standard Deviation

In Example 1 we saw that 31.4% of African Americans were smokers and a sample of 20 African Americans is selected at random. Compute μ_X and σ_X and interpret each.

Solution

Understand the Situation: Here we have a binomial random variable with $n = 20$ and $p = 0.314$.

So the mean is

$$E(X) = \mu_X = np = 20(0.314) = 6.28$$

with a standard deviation

$$\sigma_X = \sqrt{np(1-p)} = \sqrt{20(0.314)(1-0.314)} = \sqrt{20(0.314)(0.686)} \approx 2.08$$

Interpret the Solution: This means that, as we repeat the experiment over time, taking many samples of 20, in the long run we would expect about 6.28 of those in the sample to be smokers. The number of smokers varied, on average, by about 2.08 from the mean of 6.28. ∎

✓ **Checkpoint 4**   Now work Exercise 53.

SUMMARY

- A **binomial experiment** is a probability experiment such that each trial has exactly two possible outcomes: success or failure. The probability of success for each trial is the same and is denoted by p. The experiment has a finite number of trials, denoted by n, and each trial is independent.

- In a binomial experiment the probability of having x successes in n trials is given by the binomial **prob-**ability mass function: $P(X = x) = {_nC_x} \cdot p^x(1-p)^{n-x}$, $x = 0, 1, 2, \ldots, n$.

- The **mean**, or expected value, of the binomial distribution is $\mu_X = np$.

- The **standard deviation** of the binomial distribution is $\sigma_X = \sqrt{np(1-p)}$.

SECTION 8.3 EXERCISES

In Exercises 1–10, evaluate the binomial probability mass function $P(X = x) = {_nC_x} \cdot p^x(1-p)^{n-x}$ for the given values of n, p, and x.

1. $n = 4$, $p = \dfrac{1}{2}$, $x = 2$

2. $n = 3$, $p = \dfrac{1}{3}$, $x = 2$

3. $n = 5$, $p = 0.2$, $x = 0$

4. $n = 4$, $p = 0.3$, $x = 0$

5. $n = 8$, $p = \dfrac{2}{3}$, $x = 6$

6. $n = 7$, $p = \dfrac{3}{5}$, $x = 5$

7. $n = 10$, $p = 0.08$, $x = 5$

8. $n = 10$, $p = 0.11$, $x = 6$

9. $n = 20$, $p = 0.50$, $x = 10$

10. $n = 12$, $p = 0.6$, $x = 6$

In Exercises 11–20, determine if the assumptions of a binomial experiment are reasonably satisfied. If so, identify the values for p and n.

✓ **11.** A bag contains 4 blue and 4 white marbles. A marble is drawn at random, its color is recorded and it is returned to the bag. This is repeated for a total of 30 draws.

12. Fifty people are independently surveyed at a grocery store to see if they like the store brand of baked beans.

13. In a survey, 200 people selected at random and independently are asked which brand of motor oil they use.

14. In a public opinion poll, respondents are asked in which state they would most like to retire.

15. Ninety percent of the students at Plymouth High School pass the mandatory competency test. A random sample of 12 students is asked if they passed the test.

16. Twenty-three percent of the citizens in Klossington feel that the local police department is too tough on crime. A random sample of 50 Klossington citizens is asked if the local police department is too tough on crime.

17. A new brand of breakfast cereal called Hearty Flakes is liked by 11% of the market. A telemarketer conducts a phone survey and calls consumers until a consumer says that he or she likes Hearty Flakes.

18. Fifteen percent of the shoppers at the Manson Mall buy an item at the food court. A mall employee stands at the door and asks people if they intend to go to the food court until she gets an affirmative response.

19. One out of every five of the households in Classton call 911 during any given year. 150 Classton households randomly selected are asked in a direct mailing if they called 911 within the last year.

20. Two out of every seven children in Sherman Elementary School wear glasses. In a mail questionnaire, a local health organization asks a random sample of 15 students if they wear glasses.

In Exercises 21–26, a binomial probability mass function is given. Complete the following.

(a) Make a table of values for $P(X = x)$.

(b) Construct a probability histogram for the distribution found in part (a).

21. $P(X = x) = {}_3C_x \cdot \left(\dfrac{1}{2}\right)^x \left(1 - \dfrac{1}{2}\right)^{3-x}, \quad x = 0, 1, 2, 3$

22. $P(X = x) = {}_2C_x \cdot \left(\dfrac{1}{4}\right)^x \left(1 - \dfrac{1}{4}\right)^{2-x}, \quad x = 0, 1, 2$

23. $P(X = x) = {}_5C_x \cdot (0.3)^x (1 - 0.3)^{5-x}, \quad x = 0, 1, 2, 3, 4, 5$

24. $P(X = x) = {}_4C_x \cdot (0.6)^x (1 - 0.6)^{4-x}, \quad x = 0, 1, 2, 3, 4$

25. $P(X = x) = {}_5C_x \cdot (0.7)^x (1 - 0.7)^{5-x}, \quad x = 0, 1, 2, 3, 4, 5$

26. $P(X = x) = {}_4C_x \cdot (0.4)^x (1 - 0.4)^{4-x}, \quad x = 0, 1, 2, 3, 4$

In Exercises 27–32, match the probability statement with the appropriately shaded probability histogram.

(a)

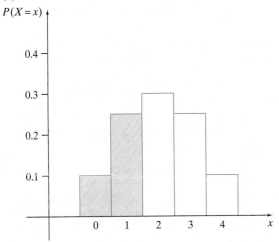

(b)

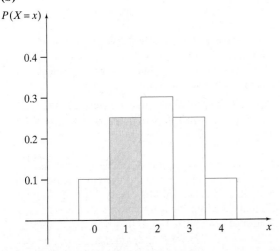

(c)

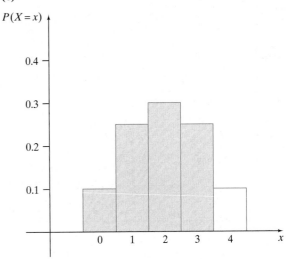

(d)

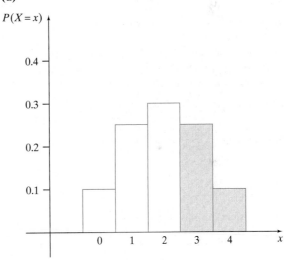

(e)

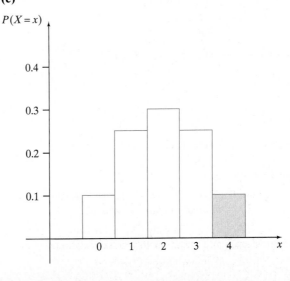

(f)

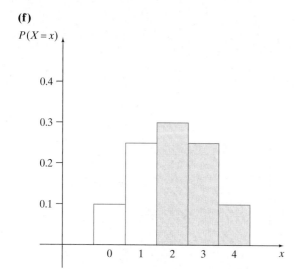

27. $P(X = 1)$ **28.** $P(X = 4)$

29. $P(X \le 3)$ **30.** $P(X \le 1)$

31. $P(X \ge 2)$ **32.** $P(X \ge 3)$

In Exercises 33–40, assume that X represents a binomial random variable. Determine the mean $E(X) = \mu_X$ and standard deviation σ_X of the distribution.

33. $n = 12,\ p = \dfrac{1}{2}$

34. $n = 18,\ p = \dfrac{1}{3}$

35. $n = 20,\ p = 0.25$

36. $n = 16,\ p = 0.3$

37. $P(X = x) = {}_6C_x \cdot \left(\dfrac{1}{4}\right)^x \left(1 - \dfrac{1}{4}\right)^{6-x}$,

  $x = 0, 1, 2, 3, 4, 5, 6$

38. $P(X = x) = {}_8C_x \cdot \left(\dfrac{1}{10}\right)^x \left(1 - \dfrac{1}{10}\right)^{8-x}$,

  $x = 0, 1, 2, 3, \ldots, 7, 8$

39. $P(X = x) = {}_{18}C_x \cdot \left(\dfrac{1}{8}\right)^x \left(1 - \dfrac{1}{8}\right)^{18-x}$,

  $x = 0, 1, 2, 3, \ldots, 17, 18$

40. $P(X = x) = {}_{22}C_x \cdot \left(\dfrac{1}{10}\right)^x \left(1 - \dfrac{1}{10}\right)^{22-x}$,

  $x = 0, 1, 2, 3, \ldots, 21, 22$

Applications

In Exercises 41–58, assume that all applications can be expressed by a binomial random variable X.

41. The Vector Tech Company gives a preemployment screening exam and finds that 40% of those who take the exam pass. A group of 7 prospective employees takes the exam.

(a) Verify that the assumptions of a binomial experiment are reasonably satisfied.
(b) Evaluate $P(X = 5)$ and interpret.

42. A survey at Harperston College reveals that 70% of the students sold their philosophy book back to the bookstore at the end of the course. The bookstore selects a group of 6 philosophy students at random.

(a) Verify that the assumptions of a binomial experiment are reasonably satisfied.
(b) Evaluate $P(X = 3)$ and interpret.

43. The Direct Satisfaction Company knows from experience that 40% of the people who enter their sweepstakes also order a magazine. A new national magazine selects a group of 12 sweepstakes entrants at random.

(a) Determine $P(X = 4)$ and interpret.
(b) Determine $P(4 \le X \le 5)$ and interpret.

44. In the town of Litherville, 30% of the households say that they are satisfied with the garbage service. The mayor's office selects 10 houses at random.

(a) Determine $P(X = 3)$ and interpret.
(b) Determine $P(3 \le X \le 4)$ and interpret.

45. When asked, "Who do you think was the greatest American president?" 23% of the Republicans responded Abraham Lincoln (*Source:* www.CBSNews.com). A sample of 15 Republicans is selected at random.

(a) Determine $P(X = 3)$ and interpret.
(b) Determine $P(X \le 3)$ and interpret.

46. When asked, "Who do you think was the greatest American president?" 26% of the Democrats responded John F. Kennedy (*Source:* www.CBSNews.com). A sample of 12 Democrats is selected at random.

(a) Determine $P(X = 4)$ and interpret.
(b) Determine $P(X \le 4)$ and interpret.

47. In January 1999, a study showed that 75% of all American households had cable TV (*Source:* www.infoplease.com). A satellite TV company selects 8 American households at random.

(a) Determine the probability that 7 have cable TV.
(b) Determine the probability that at most 5 houses have cable TV.

48. A 1997 mayoral report on curfews stated that 88% of the cities with curfews felt that they made the city safer (*Source:* U.S. Conference of Mayors, www.usmayors.org). Suppose that a Justice Department official selects 15 cities with curfews at random.

(a) Determine the probability that 12 mayors felt that the curfew made the city safer.

(b) Determine the probability that at least 12 mayors felt that the curfew made the city safer.

49. In February 2000, 2.8% of Colorado's labor force was unemployed (*Source:* U.S. Bureau of Labor Statistics, www.bls.gov). A national money magazine selects 50 people from Colorado's labor force at random.

(a) Determine the probability that 1 was unemployed.

(b) Determine the probability that at most 1 was unemployed.

50. In February 2000, 2.2% of Iowa's labor force was unemployed (*Source:* U.S. Bureau of Labor Statistics, www.bls.gov). A national money magazine selects 60 people from Iowa's labor force at random.

(a) Determine the probability that 1 was unemployed.

(b) Determine the probability that at least 1 was unemployed.

51. In 1997, 31.2% of male smokers smoked cigars (*Source:* American Cancer Society, www.cancer.org). A cigar importer selects 13 male smokers at random.

(a) Determine the probability that 3 were cigar smokers.

(b) Determine the probability that at most 3 were cigar smokers.

52. In 1997, 10.8% of female smokers smoked cigars (*Source:* American Cancer Society, www.cancer.org). A cigar importer selects 20 female smokers at random.

(a) Determine the probability that 2 were cigar smokers.

(b) Determine the probability that at most 2 were cigar smokers.

53. A report on minorities and transportation found that 59% of African Americans in New York City had no automobiles (*Source:* American Highway Users Alliance, www.highways.org). A poll selects 10 African Americans from New York City at random.

(a) Construct a probability histogram for the distribution of New York City African Americans with no automobiles.

(b) Compute μ_X and σ_X for the distribution and interpret each.

54. A report on minorities and transportation found that 40% of African Americans in Baltimore had no

automobiles (*Source:* American Highway Users Alliance, www.highways.org). A poll selects 10 African Americans from Baltimore at random.

(a) Construct a probability histogram for the distribution of Baltimore African Americans with no automobiles.

(b) Compute μ_X and σ_X for the distribution and interpret each.

55. A 1992 survey on prisoners and education showed that 49% of the prisoners did not have a high school diploma (*Source:* National Center for Education Statistics, www.nces.ed.gov). A social worker selects 20 prisoners at random.

(a) How many are expected not to have a high school diploma?

(b) Construct a table of values for $P(X = x)$ and determine the most likely number of prisoners that did not have a high school diploma. That is, find the value x that corresponds to the greatest probability.

56. A 1992 survey on prisoners and education showed that 36% of the parents of prisoners did not have a high school diploma (*Source:* National Center for Education Statistics, www.nces.ed.gov). A social worker selects parents of 10 prisoners at random.

(a) How many are expected not to have a high school diploma?

(b) Construct a table of values for $P(X = x)$ and determine the most likely number of parents of prisoners that did not have a high school diploma. That is, find the value x that corresponds to the greatest probability.

57. A 1994 report revealed that 32.6% of all U.S. births were to unmarried women (*Source:* Centers for Disease Control, www.cdc.gov). A parenting magazine selects 30 women who gave birth in 1994 at random.

(a) How many are expected to be unmarried?

(b) What is the probability that at most 10 of the women were unmarried?

58. In a survey of adults, 53.1% viewed American children negatively (*Source:* www.publicagenda.org). A parenting magazine selects 20 adults at random.

(a) How many are expected to view American children negatively?

(b) What is the probability that at most 10 view American children negatively?

▌ SECTION PROJECT

Consider a binomial random variable X with $n = 8$ and $p = \dfrac{1}{2}$.

(a) Construct a table of values for $P(X = x)$. Do you see a pattern in the probabilities? Explain.

(b) Draw a probability histogram for the distribution found in part (a). Classify the probability histogram as skewed left, skewed right, or symmetric.

(c) For the equation $P(x) = P(k - x)$, find the value of k that satisfies this equation for this distribution.

Section 8.4   The Normal Distribution

Thus far we have examined probability distributions that correspond to discrete random variables. The possible x-values of the random variable X up until now have been a finite subset of the whole numbers $0, 1, 2, \ldots, n$ or have been countably infinite. But let's say that we have a probability experiment in which we test the life length, in hours, of a new rechargeable battery. If we were interested in the probability that a battery lasted more than 5.5 hours, for example, we would have fractions or irrational numbers for the random variable values. In other words, the possible values of x are in the interval $[5.5, \infty)$, which in no way can be ordered. This means that the possible values of x are uncountable. To accommodate this situation, we need a different kind of random variable, called a **continuous random variable**. In this section we introduce continuous random variables. This will allow us to examine a continuous probability distribution called the **normal distribution**. This is the best known and most used of all probability distribution models. We will also see how we can use the normal distribution in applications.

Normal Random Variables

In Section 8.3 we saw that the binomial random variable had a corresponding probability mass function that was used to compute probabilities for known values of x. If the random variable is continuous, the probabilities are found using a function called a **probability density function**.

■ **Continuous Random Variable**

- If a random variable X can assume values corresponding to any of the points contained in one or more intervals of real numbers, then X is a **continuous random variable**.
- The probability function associated with a continuous random variable is called a **probability density function**.
- The graph of a probability density function is called a **density curve**.

▶ **Note:** In Section 8.1 we noted that a probability mass function was associated with a discrete random variable. Here we note that a **probability density function** is associated with a **continuous random variable**.

The **normal distribution** is a particular type of continuous probability distribution with a particular type of probability density function. It is sometimes called the Gaussian distribution after one of its developers, German mathematician Karl Frederick Gauss.

■ **Normal Random Variable**

If X is a continuous random variable with probability density function

$$f(x) = \frac{1}{\sigma\sqrt{2\pi}}e^{-\frac{1}{2}\left(\frac{x-\mu}{\sigma}\right)^2}$$

then X is a **normal random variable**. The possible values of X are the set of real numbers.

▶ **Note:** In this section we will use μ for μ_X and σ for σ_X for simplicity.

We will not be directly computing with this rather complicated looking formula, but instead will often refer to its graph, or **density curve**. This normal density curve, shown in Figure 8.4.1, is sometimes called the **bell curve** (or **normal curve**), because of its distinctive shape.

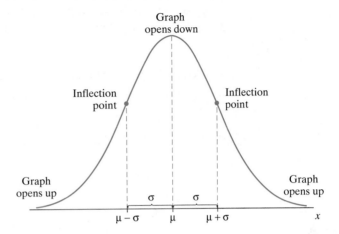

Figure 8.4.1

Figure 8.4.1 shows that the two important values that define the shape of the normal density curve are μ and σ. We can also see from the graph in Figure 8.4.1 that if we move the distance of 1 standard deviation to the left of the mean we get to the point on the graph where it switches from opening up (or bowing up) to opening down. This point is called an **inflection point**. If we move the distance of 1 standard deviation to the right of the mean, we get to the point on the curve where the graph switches from opening down to opening up again (another inflection point).

To compute probabilities with the normal distribution, we will transform the normal distribution into a simpler distribution called the **standard normal distribution**. First, let's define the **standard normal random variable**.

■ **Standard Normal Random Variable**

If Z is a continuous random variable with probability density function

$$f(Z) = \frac{1}{\sqrt{2\pi}} e^{-\frac{1}{2}z^2}$$

then Z is called a **standard normal random variable**. The possible values of Z are the set of real numbers.

▶ **Note:** A standard normal random variable is a normal random variable with mean $\mu = 0$ and standard deviation $\sigma = 1$.

To compute probabilities with normal distributions, we first use the following rule to transform the normal distribution to a standard normal distribution.

Let X be a normal random variable with mean μ and standard deviation σ; then

$$Z = \frac{X - \mu}{\sigma}$$

is a standard normal random variable.

▶ **Note:** We will refer to the value $z = \dfrac{x - \mu}{\sigma}$ as a **z-score**.

The standard normal distribution allows us to determine probabilities associated with the normal distribution all in the same way. Since we will be using the z-scores and the standard normal curve for all our normal distribution applications, let's look at some of the properties of the standard normal distribution.

■ **Properties of the Standard Normal Distribution**

1. The graph of the standard normal curve is shown in Figure 8.4.2.

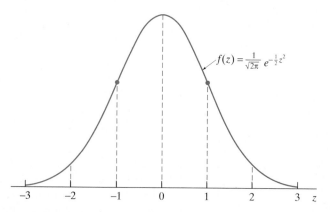

$$f(z) = \frac{1}{\sqrt{2\pi}}\, e^{-\frac{1}{2}z^2}$$

Figure 8.4.2

2. The total area under the standard normal curve is 1. That is, $P(-\infty < Z < \infty) = 1$.
3. The curve is bell shaped.
4. The mean, median, and mode are equal and centered in the middle of the graph.
5. The graph is symmetric about the mean. This means that the shape of the curve is the same on both sides of a vertical line through the center of the distribution, that is, at $z = 0$. This line divides the area under the curve into two equal parts.
6. The maximum occurs at $\mu = 0$.
7. About 68% of the area under the curve is within 1 standard deviation of the mean $\mu = 0$. That is, about 68% of the area under the curve is in the interval $(-1, 1)$. About 95% of the area under the curve is within 2 standard deviations of the mean $\mu = 0$, or in the interval $(-2, 2)$. About 99.7% of the area under the curve is within 3 standard deviations of the mean $\mu = 0$, or in the interval $(-3, 3)$.

The key to using the standard normal distribution is the computation of the z-scores. The z-score tells us how many standard deviations a value lies away from the mean. For example, a z-score of $z = 1$ means that the data value is 1 standard deviation greater than the mean μ. A z-score of $z = -1$ means that the data value is 1 standard deviation less than the mean μ. Let's see how z-scores can be interpreted in a familiar setting.

Example 1   **Interpreting *z*-scores**

The results of Rachel's midterms for Chemistry and English classes are shown in Table 8.4.1, along with the class mean and standard deviation. Assume that the distribution of scores in each class is normal. Compute Rachel's *z*-score for each class and interpret.

Table 8.4.1

Midterm Results	Chemistry	English
Midterm score, *x*	72	70
Mean, μ	68	76
Standard deviation, σ	3.2	8.0

Solution

Let's call the *z*-score for the Chemistry midterm z_1. Using the *z*-score formula, we get

$$z_1 = \frac{x - \mu}{\sigma} = \frac{72 - 68}{3.2} = 1.25$$

Interpret the Solution: In Chemistry, her score was 1.25 standard deviations greater than the mean.

The *z*-score for the English midterm, which we will call z_2, is

$$z_2 = \frac{70 - 76}{8.0} = -0.75$$

Interpret the Solution: In English, her score was 0.75 standard deviation less than the mean. This shows that Rachel's grade in Chemistry was much better than her grade in English *relative* to other students.

The *z*-scores are plotted on the standard normal curve in Figure 8.4.3.

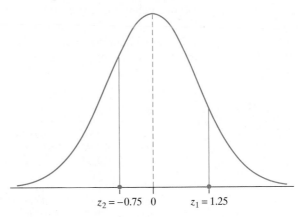

$z_2 = -0.75$ 0 $z_1 = 1.25$

Figure 8.4.3

Interactive Activity

In Example 1 suppose that, Rachel's English midterm was graded incorrectly and she really earned a score of 81. Recompute her *z*-score for the English midterm. With this new score, relative to the rest of the students, on which midterm did she do better.

✓ **Checkpoint 1**

Now work Exercise 3.

Finding Areas under the Standard Normal Curve

When using the binomial distribution in Section 8.3, we could add up the area of the rectangles in the probability histogram to determine probabilities. For the normal distribution we must compute the amount of area under the standard normal curve. Since we have a continuous random variable, determining the area requires techniques of calculus. However, tables have been created using calculus techniques, and we can use these tables to determine these areas. A table used to calculate these areas, called the Standard Normal Table, is in Appendix D. This table gives the area from $z = 0$ to some given z-score. See Figure 8.4.4.

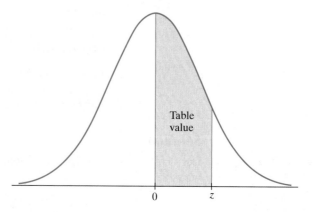

Figure 8.4.4

A table value is all that is needed to compute a probability. To understand how to use this table, let's say we want to compute $P(0 < Z < 1.24)$. In this case we need the table value at $z = 1.24$. To get this value, we look down the left column until we find 1.2 (the z-score accurate to the tenths place). Then we look across the row until we find 0.04 at the top of the page (the hundredths place digit of the z-score). From the table, we get $P(0 < Z < 1.24) = $ (value at $z = 1.24$) $= 0.3925$. See Figure 8.4.5. The area that $z = 1.24$ corresponds to is shown on the standard normal curve in Figure 8.4.6. This means that the probability that the z-score value is between 0 and 1.24 is about 39.25%. Note that we said that the "z-score value is between 0 and 1.24," but did not specify if the values of 0 and 1.24 were included or not. This is because lines have length and not width, so the area between these two z-scores does not change whether we include these

z	.00	.01	.02	.03	.04	.05
0.0	0.0000	0.0040	0.0080	0.0120	0.0160	0.0199
0.1	0.0398	0.0438	0.0478	0.0517	0.0557	0.0596
0.2	0.0793	0.0832	0.0871	0.0910	0.0948	0.0987
0.3	0.1179	0.1217	0.1255	0.1293	0.1331	0.1368
0.4	0.1554	0.1591	0.1628	0.1664	0.1700	0.1736
0.5	0.1915	0.1950	0.1985	0.2019	0.2054	0.2088
0.6	0.2257	0.2291	0.2324	0.2357	0.2389	0.2422
0.7	0.2580	0.2611	0.2642	0.2673	0.2704	0.2734
0.8	0.2881	0.2910	0.2939	0.2967	0.2995	0.3023
0.9	0.3159	0.3186	0.3212	0.3238	0.3264	0.3289
1.0	0.3413	0.3438	0.3461	0.3485	0.3508	0.3531
1.1	0.3643	0.3665	0.3686	0.3708	0.3729	0.3749
1.2	0.3849	0.3869	0.3888	0.3907	0.3925	0.3944
1.3	0.4032	0.4049	0.4066	0.4082	0.4099	0.4115

Figure 8.4.5

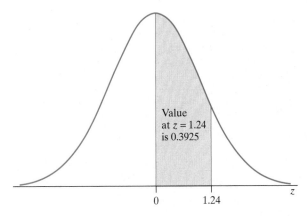

Figure 8.4.6 The value at $z = 1.24$ gives the area of the shaded region.

endpoints or not. So, for continuous random variables, including the normal random variable, we can use the following notation rule:

> ■ **Notation for Endpoints of Continuous Random Variables**
>
> For real numbers a and b and a continuous random variable X, such as a normally distributed random variable, we have the following:
>
> 1. $P(a < X < b) = P(a \le X \le b)$
> 2. $P(X < a) = P(X \le a)$
> 3. $P(X > b) = P(X \ge b)$

In Example 2 we illustrate how to use the Standard Normal Table to compute other probabilities, as well as use some properties of the standard normal distribution.

Example 2   Computing Areas under the Standard Normal Curve

Determine the following areas under the standard normal curve.

(a) $P(Z \ge 1.68)$ (b) $P(-1.31 < Z < 0.50)$

Solution

(a) From Property 2 of Properties of the Standard Normal Distribution, we know that the total area under the standard normal curve is 1. Using this fact along with Property 5 (graph is symmetric about the mean, $z = 0$), we know that $P(Z \ge 0) = 0.5$. This is shown in Figure 8.4.7 on p. 508.

To determine $P(Z \ge 1.68)$, we can simply take away the area for $P(0 < Z < 1.68)$ from $P(Z \ge 0)$. See Figure 8.4.8 on p. 508.

The table value at $z = 1.68$ is 0.4535 (see Standard Normal Table in Appendix D). So we have

$$P(Z \ge 1.68) = P(Z \ge 0) - P(0 < Z < 1.68)$$

$$= 0.50 - (\text{value at } z = 1.68) = 0.50 - 0.4535$$

$$= 0.0465$$

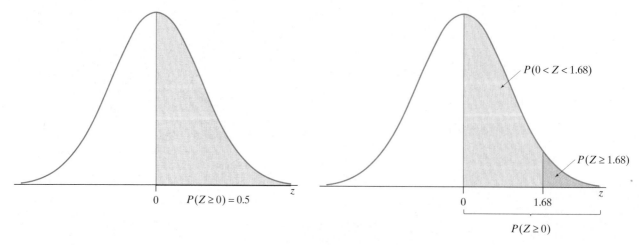

Figure 8.4.7

Figure 8.4.8 $P(Z \geq 1.68) = P(Z \geq 0) - P(0 < Z < 1.68)$.

(b) To compute $P(-1.31 < Z < 0.50)$ corresponds to finding the area shown in Figure 8.4.9.

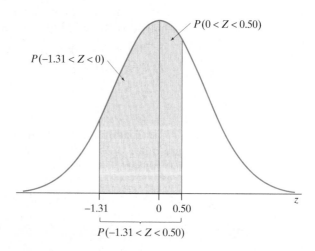

Figure 8.4.9

Notice in Figure 8.4.9 that

$$P(-1.31 < Z < 0.50) = P(-1.31 < Z < 0) + P(0 < Z < 0.50)$$

$P(0 < Z < 0.50)$ is immediately found by consulting the Standard Normal Table in Appendix D. Since the value at $z = 0.50$ is 0.1915, we have

$$P(0 < Z < 0.50) = 0.1915$$

To determine $P(-1.31 < Z < 0)$, we again utilize Property 5 (graph is symmetric about the mean, $z = 0$). The area shaded in Figure 8.4.10 corresponds to $P(0 < Z < 1.31)$ and is identical to $P(-1.31 < Z < 0)$, thanks to symmetry. This means that the table values for $z = -1.31$ and $z = 1.31$ are the same.

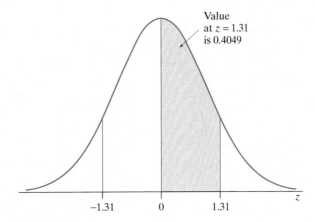

Figure 8.4.10

We can now compute

$$P(-1.31 < Z < 0.50) = P(-1.31 < Z < 0) + P(0 < Z < 0.50)$$

$$= 0.4049 + 0.1915$$

$$= 0.5964$$

Observe that in computing $P(-1.31 < Z < 0.50)$ one z-score was negative and the other was positive. What we ended up doing was simply adding the table value for $z = 1.31$ and $z = 0.50$.   ■

In Table 8.4.2 on p. 510 we summarize how to determine areas under the standard normal curve for different z-values. We will use Table 8.4.2 frequently.

Applications of the Normal Distribution

The normal distribution can be used on a wide variety of applications. The only requirement is that the distribution have the distinctive bell-shaped curve. We can assume that in this section all the applications come from data that are normally distributed. To solve applications of the normal distribution, we will use the following six-step process.

■ **Solving Normal Distribution Applications**

1. After reading the application problem, identify the mean μ and the standard deviation σ.
2. Write the desired probability in terms of the normal random variable X.
3. Convert the given x-values to standard z-scores using the formula
$$z = \frac{x - \mu}{\sigma}.$$
4. Sketch the standard normal curve and shade the desired area.
5. Use Table 8.4.2 and the Standard Normal Table in Appendix D to determine the probability.
6. Interpret the solution.

Table 8.4.2

Description of Area	Graph	Computation of Area
1. Between 0 and any value		$P(a < Z < 0) = $ (value at $z = a$)
		$P(0 < Z < b) = $ (value at $z = b$)
2. A left or right tail		$P(Z < a) = 0.50 - $ (value at $z = a$)
		$P(Z > b) = 0.50 - $ (value at $z = b$)
3. Between any two values where the z-scores have the same sign		$P(a < Z < b) = $ (value at $z = a$) $-$ (value at $z = b$)
		$P(a < Z < b) = $ (value at $z = b$) $-$ (value at $z = a$)
4. Between any two values where the z-scores have opposite sign		$P(a < Z < b) = $ (value at $z = a$) $+$ (value at $z = b$)
5. Less than any z-score to the right of the mean		$P(Z < b) = 0.50 + $ (value at $z = b$)
6. Greater than any z-score to the left of the mean		$P(Z > a) = 0.50 + $ (value at $z = a$)

Now let's try this procedure in an application. **For all applications in this section, assume that the quantity discussed can be represented by a normal random variable**. When computing z-scores, always round to the nearest hundredths place so that you can use the Standard Normal Table in Appendix D.

Example 3   Applying the Normal Distribution

The amount of floor area in a new, privately owned, single-family house built in 1970 had an average of 1500 square feet with a standard deviation of 683 square feet (*Source:* U.S. Census Bureau, www.census.gov). For a special featured article, a national home improvement magazine selects a house built in 1970 at random.

(a) Determine the probability that the amount of floor space in the house is less than 1200 square feet.

(b) Determine the probability that the amount of floor space in the house is between 1600 and 2000 square feet.

Solution

For part (a) of this example, we will list the six steps.

(a)

Step 1: We let X be a normal random variable. We are given the mean $\mu = 1500$ and $\sigma = 683$.

Step 2: Here we need the probability

$$P(\text{house has less than 1200 square feet}) = P(X < 1200)$$

Step 3: Converting this value to a z-score gives us

$$P(X < 1200) = P\left(Z < \frac{1200 - 1500}{683}\right) \approx P(Z < -0.44)$$

Step 4: This region is shown in Figure 8.4.11.

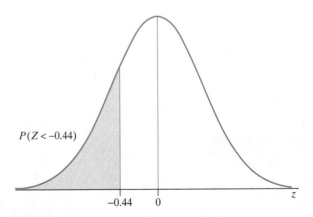

$P(Z < -0.44)$

$-0.44 \quad 0$

z

Figure 8.4.11

Step 5: Table 8.4.2 indicates that to determine $P(Z < -0.44)$ we do the following:

$$P(Z < -0.44) = 0.50 - (\text{value at } z = 0.44)$$

The Standard Normal Table in Appendix D gives us the value at $z = 0.44$. Continuing, we get

$$P(Z < -0.44) = 0.50 - (\text{value at } z = 0.44) = 0.50 - 0.1700 = 0.330$$

Step 6: This means that there is about a 33% probability that a randomly chosen house built in 1970 will have a floor area of less than 1200 square foot.

(b) Now we need to find the probability that the square footage is between 1600 and 2000. In terms of our random variable, this is $P(1600 < X < 2000)$. Converting to a z-score gives us

$$P(1600 < X < 2000) = P\left(\frac{1600 - 1500}{683} < Z < \frac{2000 - 1500}{683}\right)$$
$$\approx P(0.15 < Z < 0.73)$$

The area corresponding to this probability is shown in Figure 8.4.12. To find this area, we use Table 8.4.2 and the Standard Normal Table for the values of $z = 0.73$ and $z = 0.15$ to get

$$P(0.15 < Z < 0.73) = (\text{value at } z = 0.73) - (\text{value at } z = 0.15)$$
$$= 0.2673 - 0.0596 = 0.2077$$

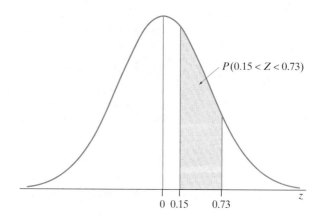

Figure 8.4.12

This means that there is about a 21% chance that a randomly chosen house built in 1970 will have floor area between 1600 and 2000 square feet. ∎

✓ **Checkpoint 2**   Now work Exercise 45.

Technology Option

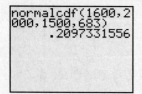

Figure 8.4.13

The solution to part (b) of Example 3, using the NORMALCDF command on a graphing calculator, is shown in Figure 8.4.13. The difference in the answers is due to the rounding of the z-score to the hundredths place.

To see how your calculator computes normal probabilities, consult the on-line calculator manual at www.prenhall.com/armstrong.

Let's try another application in order to practice computing probabilities using the standard normal distribution. In this application we want an area to the right of a given z-score and also two z-scores that *straddle* the mean.

Example 4   **Applying the Normal Distribution**

The average age of a farmer in 1992 was 51.9 years with a standard deviation of 8.7 years (*Source:* U.S. Census Bureau, www.census.gov). An agriculture economics professor selects a farmer from 1992 at random for a case study paper.

(a) Determine the probability that the farmer's age exceeded 40 years.

(b) Determine the probability that the farmer's age was between 45 and 55 years.

Solution

(a) We are given the mean $\mu = 51.9$ and the standard deviation $\sigma = 8.7$. We seek the probability that a farmer is older than 40. So we want $P(X > 40)$. Converting this value to a z-score gives

$$P(X > 40) = P\left(Z > \frac{40 - 51.9}{8.7}\right) \approx P(Z > -1.37)$$

This area is shown in Figure 8.4.14.

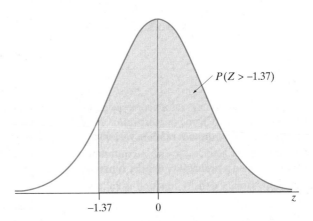

Figure 8.4.14

Using Table 8.4.2 and the Standard Normal Table in Appendix D for $z = 1.37$ gives us

$$P(Z > -1.37) = (\text{value at } z = 1.37) + 0.50 = 0.4147 + 0.50 = 0.9147$$

Interpret the Solution: This means that if a farmer is randomly chosen there is about a 91% chance that the farmer is older than 40.

(b) Here we want the probability that the farmer is between 45 and 55 years of age. So we need to compute

$$P(45 < X < 55) = P\left(\frac{45 - 51.9}{8.7} < Z < \frac{55 - 51.9}{8.7}\right) \approx P(-0.79 < Z < 0.36)$$

This area is shown in Figure 8.4.15.

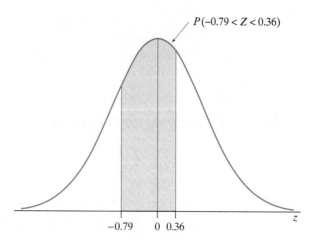

$P(-0.79 < Z < 0.36)$

−0.79   0 0.36

z

Figure 8.4.15

Using Table 8.4.2 and the Standard Normal Table in Appendix D for $z = 0.79$ and $z = 0.36$ gives us

$$P(-0.79 < Z < 0.36) = \text{(value at } z = 0.79) + \text{(value at } z = 0.36)$$
$$= 0.2852 + 0.1406 = 0.4258$$

Interpret the Solution: This means that if a farmer is chosen a random there is about a 43% chance that the farmer is between 45 and 55 years of age. ■

✓ **Checkpoint 3**   Now work Exercise 49.

SUMMARY

- If a random variable X can assume values corresponding to any of the points contained in one or more intervals of real numbers, then X is a **continuous random variable**.
- The probability function associated with a continuous random variable is called a **probability density function**.
- The graph of a probability density function is called a **density curve**.
- If X is a continuous random variable with probability density function

$$f(x) = \frac{1}{\sigma\sqrt{2\pi}} e^{-\frac{1}{2}\left(\frac{x-\mu}{\sigma}\right)^2}$$

then X is a **normal random variable**. The possible values of x are the set of real numbers.
- If X is a normal random variable with mean μ and standard deviation σ, then we define Z as a **standard normal random variable**, where $Z = \dfrac{X - \mu}{\sigma}$.
- The value $z = \dfrac{x - \mu}{\sigma}$ is called a **z-score**.
- The mean of the standard normal distribution is $\mu = 0$ with a standard deviation of $\sigma = 1$.

SECTION 8.4 EXERCISES

Assume that each of the exercises in this section can be represented by a normal random variable X.

1. The mean and standard deviation of the gas mileages for the Jupiter and the Atlantis mid-sized cars are shown in the table.

	Jupiter	**Atlantis**
μ	23.5	26.0
σ	3.5	5.1

(a) In a test, a Jupiter gets 27 mpg, while an Atlantis gets 29.1 mpg. Compute the z-score for each car's gas mileage.

(b) Graph the z-scores from part (a) on the standard normal curve. Relative to the rest of the cars of the same make, which car got better mileage?

2. The mean and standard deviation of the wages of those who work on first and second shift at the Homburg Manufacturing Plant are shown in the table.

	First Shift	Second Shift
μ	22.0	19.5
σ	1.1	1.9

(a) A randomly chosen worker from the first shift makes $23.87 per hour, while a randomly chosen worker from the second shift makes $23.11. Compute the z-score for each of the worker's wages.

(b) Graph the z-scores from part (a) on the standard normal curve. Relative to the rest of the workers in their respective shifts, which shift worker made higher wages?

✓ **3.** The mean and standard deviation of the entrance exam scores for the Sunny Day Prep School for the 1999–2000 and 2000–2001 academic years are shown in the table.

	1999–2000	2000–2001
μ	153	142
σ	10.6	8.7

(a) A randomly chosen student who took the test in 1999–2000 scored 137, while a randomly chosen student who took the test in 2000–2001 scored 131. Compute the z-score for each student's exam.

(b) Graph the z-scores from part (a) on the standard normal curve. Relative to the rest of the students taking the test that year, which student did better?

4. The mean and standard deviation of the annual snowfall in the neighboring communities of Nashbrook and Hampton are shown in the table.

	Nashbrook	Hampton
μ	36	41
σ	1.5	0.8

(a) Last year Nashbrook received 35.1 inches of snow, while Hampton received 40.6 inches of snow. Compute the z-score for each community's snowfall.

(b) Graph the z-scores from part (a) on the standard normal curve. Relative to the rest of the years, which community received less snow?

5. The mean and standard deviation of the number of miles traveled each month by the sales divisions in the MegaStar and HemelGraph companies are shown in the table.

	MegaStar	HemelGraph
μ	1570	1645
σ	53	42

(a) Last month the sales division of MegaStar traveled 1702.5 miles, while the sales division of HemelGraph traveled 1741.6 miles. Compute the z-score for each company's mileages.

(b) Graph the z-scores from part (a) on the standard normal curve. Relative to the rest of the months, which company's sales division traveled less?

6. The mean and standard deviation of the number of bags lost each week by AirCom and BlueSky airlines are shown in the table.

	AirCom	BlueSky
μ	120	78
σ	2.8	2.7

(a) Last week Aircom lost 114 bags and BlueSky lost 74 bags. Compute the z-score for each airline's lost bags.

(b) Graph the z-scores from part (a) on the standard normal curve. Relative to the rest of the weeks, which company lost fewer bags?

In Exercises 7–16, find the shaded area using Table 8.4.2 and the Standard Normal Table in Appendix D.

7.

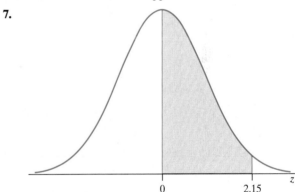

8.

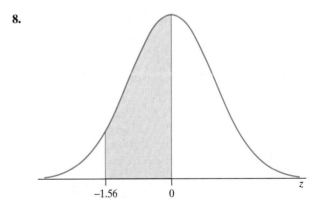

9.

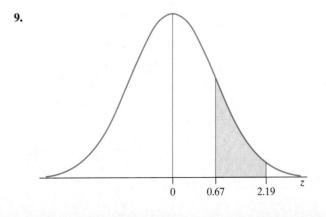

10.

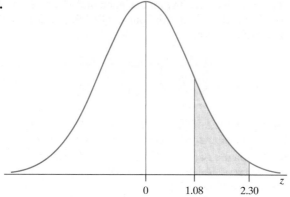

11.

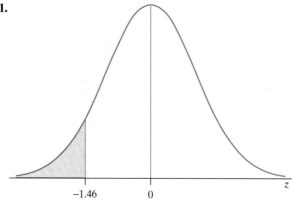

12.

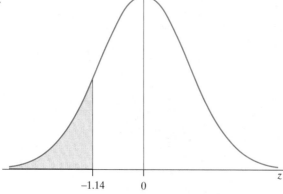

13.

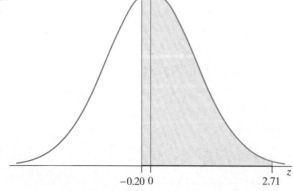

14.

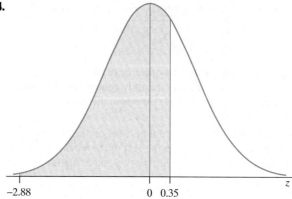

15.

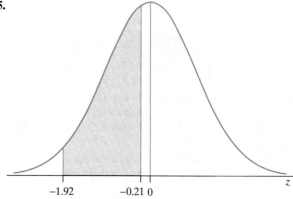

16.

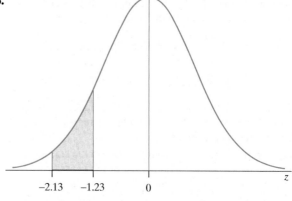

In Exercises 17–26, find the area between the given z-scores using Table 8.4.2 and the Standard Normal Table in Appendix D.

17. Between $z = 0$ and $z = 1.50$

18. Between $z = 0$ and $z = 1.70$

19. Between $z = 0$ and $z = -0.65$

20. Between $z = 0$ and $z = -0.85$

21. Between $z = -1.23$ and $z = 0.61$

22. Between $z = -0.75$ and $z = 1.45$

23. Between $z = 0.53$ and $z = 2.11$

24. Between $z = 0.67$ and $z = 2.04$

25. Between $z = -1.25$ and $z = -0.19$

26. Between $z = -1.98$ and $z = -0.51$

In Exercises 27–40, determine the given probabilities using Table 8.4.2 and the Standard Normal Table in Appendix D.

27. $P(Z > 0)$

28. $P(Z < 0)$

29. $P(Z > 1.30)$

30. $P(Z > 1.60)$

31. $P(Z \leq -1.76)$

32. $P(Z \leq -1.88)$

33. $P(Z \geq -0.73)$

34. $P(Z \geq -0.91)$

35. $P(Z < 1.96)$

36. $P(Z < 2.18)$

37. $P(-0.50 < Z < 0.50)$

38. $P(-0.38 < Z < 0.38)$

39. $P(-0.91 \leq Z \leq -0.12)$

40. $P(-0.87 \leq Z \leq -0.32)$

Applications

In the remainder of the exercises, assume that X represents a normal random variable.

41. The LandTrans Company runs shuttle buses from the airport to the local hotels. Records show that the mean number of runs for each bus each day is $\mu = 15$ with a standard deviation of $\sigma = 0.5$. For an audit, the chief financial officer at LandTrans Company selects a shuttle bus at random.

(a) Determine $P(X > 15)$ and interpret.

(b) Determine $P(X < 15)$ and interpret.

42. The Lock-It-Now Company finds that, on any given day, a mean of $\mu = 24{,}000$ locks is produced with a standard deviation of $\sigma = 2000$. For an audit, the chief financial officer at Lock-It-Now Company selects a manufacturing day at random.

(a) Determine $P(X \geq 24{,}000)$ and interpret.

(b) Determine $P(X \leq 24{,}000)$ and interpret.

43. The We-Make-Tools machine shop manufactures wire and finds that its standard-gauge wire has a mean of $\mu = 0.12$ inch with a standard deviation of $\sigma = 0.02$ inch. For an inspection, a quality control engineer selects a wire at random.

(a) Determine $P(0.12 \leq X \leq 0.16)$ and interpret.

(b) Determine $P(0.08 \leq X \leq 0.12)$ and interpret.

44. The electronic life of the MindReader personal digital assistant is found to have a mean of $\mu = 4.01$ years with a standard deviation of $\sigma = 0.25$ year. For an upcoming article, a computer magazine selects a MindReader owner at random.

(a) Determine $P(4.01 \leq X \leq 4.26)$ and interpret.

(b) Determine $P(3.76 \leq X \leq 4.01)$ and interpret.

45. The weight of players on the 2000 Ohio State University spring training football roster had a mean of

$\mu = 234.7$ pounds with a standard deviation of $\sigma = 48.8$ pounds (*Source:* www.ohiostatebuckeyes.fansonly.com). A nutritionist selects a player from the spring training football roster at random.

(a) Determine $P(X \leq 260)$ and interpret.

(b) Determine $P(200 \leq X \leq 260)$ and interpret.

46. From January 1, 1996, to August 19, 1996, the weekly mean retail price of a gallon of gasoline was \$1.37 with a standard deviation of \$0.07 (*Source:* U.S. Department of Energy, www.eia.doe.gov). For an upcoming story in a motor magazine, a week during this time period is chosen at random.

(a) Determine $P(X \leq 1.40)$ and interpret.

(b) Determine $P(1.30 \leq X \leq 1.40)$ and interpret.

47. The average monthly temperature (in degrees Fahrenheit) in northwestern Pennsylvania during the month of March from 1895 to 1921 had a mean of 33.7 degrees with a standard deviation of 4.8 degrees (*Source:* National Weather Service, www.nws.noaa.gov). In a study on global warming, a climatologist selects a March monthly temperature from this time period at random.

(a) Determine the probability that the average monthly temperature was below 30 degrees.

(b) Determine the probability that the average monthly temperature was between 30 and 40 degrees.

48. In 1998 the mean school district enrollment in the Cuyahoga County, Ohio, school system was 6395.3 students in each school district with a standard deviation of 560.1 students (*Source:* www.nsrs.com). For a new testing procedure, the state superintendent selects a school district at random.

(a) Determine the probability that the school district has fewer than 6000 students.

(b) Determine the probability that the school district has between 6000 and 7500 students.

49. From 1943 to 1974, the U.S. minimum wage had a mean of \$1.00 per hour with a standard deviation of \$0.46 (*Source:* U.S. Department of Labor, www.dol.gov). An economics professor selects a minimum-wage earner from this time period at random.

(a) Determine the probability that the selected minimum-wage earner made between \$1.10 per hour and \$1.50 per hour.

(b) Determine the probability that the selected minimum-wage earner made between \$0.90 per hour and \$1.50 per hour.

50. In 1998 the number of acres per state of nonfederal developed land in the continental United States had a mean of 1.84 million acres with a standard deviation of 1.44 millions acres (*Source:* U.S. Census Bureau, www.census.gov).

Suppose that in 1998 a state in the continental United States is chosen at random by a political science student for a report.

(a) Determine $P(1.5 \leq X \leq 1.7)$ and interpret.

(b) Determine $P(1.3 \leq X \leq 1.7)$ and interpret.

51. In 1995 the mean age of a person receiving outpatient care in the United States was 45.73 years with a standard deviation of 18.68 years (*Source:* National Center for Health Statistics, www.cdc.gov/nchs). An HMO selects an outpatient from 1995 at random.

(a) Determine the probability that the patient is at least 40 years old.

(b) Determine the probability that the patient is at most 50 years old.

52. In 1995 the mean age of a person receiving emergency room care in the United States was 42.92 years with a standard deviation of 19.57 years (*Source:* National Center for Health Statistics, www.cdc.gov/nchs). An HMO selects an emergency room patient from 1995 at random.

(a) Determine the probability that the patient is at least 38 years old.

(b) Determine the probability that the patient is at most 45 years old.

53. In 1996 the mean age of a person getting the flu in the United States was 27.43 years with a standard deviation of 20.98 years (*Source:* National Center for Health Statistics, www.cdc.gov/nchs). An HMO selects a person who had the flu in 1996 at random.

(a) Determine the probability that the patient was between 16 and 19 years old.

(b) Determine the probability that the patient was between 20 and 40 years old.

54. Between 1981 and 1998 the age of a person with AIDS had a mean of 37.26 years with a standard deviation of 10.24 years (*Source:* National Center for Health Statistics, www.cdc.gov/nchs). For a national news report, a person with AIDS during this time period is chosen at random.

(a) Determine the probability that the age of the person was between 30 and 35 years.

(b) Determine the probability that the age of the person was between 38 and 48 years.

55. In 1995 the age of a person in a nursing home in the United States had a mean of 81.77 years with a standard deviation of 7.25 years (*Source:* National Center for Health Statistics, www.cdc.gov/nchs). A psychology professor selects a nursing home resident from 1995 at random.

(a) Determine the probability that the resident was older than 85.

(b) Determine the probability that the resident was older than 72.

 SECTION PROJECT

The distribution of weights of American females aged 40 to 49 from 1988 to 1994 is shown in the table.

Weight (in pounds)	Percent	Weight (in pounds)	Percent
100–109	1.6	230–239	1.7
110–119	3.9	240–249	2.4
120–129	7.6	250–259	0.3
130–139	12.5	260–269	0.5
140–149	12.3	270–279	0.3
150–159	12.0	280–289	0.3
160–169	9.9	290–299	0.3
170–179	10.2	300–309	0.2
180–189	8.6	310–319	0.1
190–199	4.0	320–329	0.1
200–209	4.8	330–339	0.1
210–219	3.1	340–349	0.1
220–229	2.7	350 and over	0.4

SOURCE: U.S. Census Bureau, www.census.gov

(a) According to the table, what percent of American females aged 40 to 49 weighed anywhere between 140 and 170 pounds?

(b) Use the MEANSD program in Appendix C to estimate the mean and the standard deviation of the distribution.

(c) Using μ and σ found in part (b), determine $P(140 \leq X \leq 170)$ and interpret.

(d) How do you account for the difference in the answers in part (a) and in part (c)?

Why We Learned It

In Chapter 8 we saw that random variables are powerful tools that can be used in various disciplines. We saw many situations in which random variables, probability distributions, expected value, and standard deviation were applied. Discrete random variables and some probability distributions associated with them included the uniform distribution and the binomial distribution. We saw that the binomial distribution had many varied applications. For example, a social worker who desires to do a case study on the educational attainment of parents of prisoners can use the binomial distribution to predict the probability that, say, 5 parents in a group of 20 have a high school diploma. The social worker could also compute the expected value (mean) to determine how many are expected to have a high school diploma. Finally, the social worker could compute the standard deviation of the distribution to measure the dispersion. Continuous random variables gave rise to the normal distribution and the standard normal distribution. A quality control engineer in a machine shop could use the standard normal distribution to determine the probability that the diameter of wire produced is between 0.12 and 0.16 inch given that the mean is 0.12 inch and the standard deviation is 0.02 inch.

CHAPTER REVIEW EXERCISES

In Exercises 1–3, determine if the table meets the two criteria of a probability distribution.

1.

$X = x$	$P(X = x)$
$X = 0$	$P(X = 0) = -\frac{1}{2}$
$X = 1$	$P(X = 1) = \frac{3}{4}$
$X = 2$	$P(X = 2) = \frac{3}{4}$

2.

$X = x$	$P(X = x)$
$X = -3$	$P(X = -3) = 0.125$
$X = -2$	$P(X = -2) = 0.138$
$X = -1$	$P(X = -1) = 0.225$
$X = 0$	$P(X = 0) = 0.118$
$X = 1$	$P(X = 1) = 0.206$
$X = 2$	$P(X = 2) = 0.134$
$X = 3$	$P(X = 3) = 0.054$

3.

$X = x$	$P(X = x)$
$X = 5$	$P(X = 5) = \frac{1}{10}$
$X = 10$	$P(X = 10) = \frac{3}{5}$
$X = 15$	$P(X = 15) = \frac{1}{8}$
$X = 20$	$P(X = 20) = \frac{1}{9}$

In Exercises 4–6, a formula for $P(X = x)$ is given. Complete the following.

(a) Write a table of values for $P(X = x)$.

(b) Determine if the table meets the two criteria of a probability distribution. Remember that $P(X = x) = 0$, where the x-values are not defined.

4. $P(X = x) = \frac{1}{7}x, \quad x = 1, 2, 3, 4$

5. $P(X = x) = \frac{1}{30}x^2, \quad x = 1, 2, 3, 4$

6. $P(X = x) = \frac{x + 4}{34}, \quad x = 3, 4, 5, 6$

In Exercises 7 and 8, determine the value of k so that $P(X = x)$ is a probability mass function.

7. $P(X = x) = \frac{4}{k}x, \quad x = 2, 3, 4, 5$

8. $P(X = x) = \frac{2}{k}x^2, \quad x = 1, 2, 3, 4$

In Exercises 9 and 10, a table of frequencies is given. Construct the corresponding probability distribution with $X = x$ in the left column and $P(X = x)$ in the right column.

9.

Value	Frequency
1	10
2	12
3	18
4	14

10.

Value	Frequency
0	220
5	300
10	150
15	170

11. Salem High School has determined that the number of students per class can be represented by a discrete random variable X with the following probability distribution:

$X = x$	$P(X = x)$
$X = 25$	$P(X = 25) = 0.10$
$X = 26$	$P(X = 26) = 0.20$
$X = 27$	$P(X = 27) = 0.35$
$X = 28$	$P(X = 28) = 0.10$
$X = 29$	$P(X = 29) = 0.20$
$X = 30$	$P(X = 30) = 0.05$

(a) Construct a probability histogram of the distribution.

(b) Determine $P(26 \le X \le 28)$ and interpret.

12. Sopris Ski Club has determined that the number of snowboarders that they have on any given night during the ski season can be presented by a discrete random variable X having the following probability distribution:

$X = x$	$P(X = x)$
8	$P(X = 8) = 0.06$
9	$P(X = 9) = 0.28$
10	$P(X = 10) = 0.36$
11	$P(X = 11) = 0.20$
12	$P(X = 12) = 0.10$

(a) Construct a probability histogram of the distribution.

(b) Determine $P(X \ge 10)$ and interpret.

13. Hillsbourough Health Club determines that the number of new memberships during any given month can be represented by a discrete random variable X having the following probability distribution:

$X = x$	$P(X = x)$
$X = 4$	$P(X = 4) = 0.22$
$X = 5$	$P(X = 5) = 0.15$
$X = 6$	$P(X = 6) = 0.26$
$X = 7$	$P(X = 7) = 0.28$
$X = 8$	$P(X = 8) = 0.09$

(a) Determine the probability that there will be at least 6 new memberships during any given month.

(b) According to the distribution, what is the least likely number of new memberships during any given month?

14. The enrollment at Saco River Kayaking School for last June is given in the following frequency table.

Course	Enrollment
Beginning	25
Intermediate	12
Advanced	8
Expert	5

(a) Let X be a discrete random variable, where X represents the level of the course in which a person is enrolled, meaning $x = 1, 2, 3, 4$ for beginning, intermediate, advanced, and expert, respectively. Construct a probability distribution for the data.

(b) Compute $P(X = 4)$ and interpret.

🌐 **15.** The number of unprovoked worldwide shark attacks for the years 1997 to 2000 is shown in the table.

Year	Number of Unprovoked Attacks
1997	56
1998	51
1999	58
2000	79

SOURCE: Florida Museum of Natural History, www.flmnh.ufl.edu

(a) Let X be a discrete random variable, where X represents a year between 1997 and 2000. Construct a probability distribution for the data.

(b) Compute $P(X = 2000)$ and interpret.

(c) Compute $P(X \ge 1999)$ and interpret.

In Exercises 16–18, a table of values for the probability distribution $P(X = x)$ is given. Determine the expected value $E(X)$ of the random variable X having the given discrete probability distribution.

16.

$X = x$	$P(X = x)$
$X = 0$	$P(X = 0) = \frac{1}{8}$
$X = 1$	$P(X = 1) = \frac{1}{4}$
$X = 2$	$P(X = 2) = \frac{1}{4}$
$X = 3$	$P(X = 3) = \frac{3}{8}$

17.

$X = x$	$P(X = x)$
$X = -1$	$P(X = -1) = 0.227$
$X = 0$	$P(X = 0) = 0.314$
$X = 1$	$P(X = 1) = 0.186$
$X = 2$	$P(X = 2) = 0.273$

18.

$X = x$	$P(X = x)$
$X = 5$	$P(X = 5) = 0.2$
$X = 10$	$P(X = 10) = 0.1$
$X = 15$	$P(X = 15) = 0.3$
$X = 20$	$P(X = 20) = 0.1$
$X = 25$	$P(X = 25) = 0.3$

In Exercises 19–21, a probability mass function $P(X = x)$ is given. Determine the expected value.

19. $P(X = x) = \dfrac{1}{6}, \quad x = 1, 2, 3, 4, 5, 6$

20. $P(X = x) = \dfrac{1}{55}x^2, \quad x = 1, 2, 3, 4, 5$

21. $P(X = x) = \dfrac{x - 2}{6}, \quad x = 3, 4, 5$

In Exercises 22–24, determine the standard deviation, σ_X, for the following:

22. For the discrete probability distribution in Exercise 16

23. For the discrete probability distribution in Exercise 17

24. For the probability mass function in Exercise 20

25. Salem High School has determined that the number of students per class can be represented by a discrete random variable X with the following probability distribution:

$X = x$	$P(X = x)$
$X = 25$	$P(X = 25) = 0.10$
$X = 26$	$P(X = 26) = 0.20$
$X = 27$	$P(X = 27) = 0.35$
$X = 28$	$P(X = 28) = 0.10$
$X = 29$	$P(X = 29) = 0.20$
$X = 30$	$P(X = 30) = 0.05$

Determine the expected value $E(X)$ and interpret.

26. Sopris Ski Club has determined that the number of snowboarders that they have on any given night during the ski season can be presented by a discrete random variable X having the following probability distribution:

$X = x$	$P(X = x)$
8	$P(X = 8) = 0.06$
9	$P(X = 9) = 0.28$
10	$P(X = 10) = 0.36$
11	$P(X = 11) = 0.20$
12	$P(X = 12) = 0.10$

(a) Determine the expected value and interpret.

(b) What is the most frequently occurring number of snowboaders on any given night? That is, what x-value yields the greatest $P(X)$-value?

27. Hillsbourough Health Club determines that the number of new memberships during any given month can be represented by a discrete random variable X having the following probability distribution:

$X = x$	$P(X = x)$
$X = 4$	$P(X = 4) = 0.22$
$X = 5$	$P(X = 5) = 0.15$
$X = 6$	$P(X = 6) = 0.26$
$X = 7$	$P(X = 7) = 0.28$
$X = 8$	$P(X = 8) = 0.09$

(a) According to the distribution, how many new memberships can the health club expect during a month?

(b) Determine the standard deviation of the distribution.

28. Chenush Gas is a natural gas drilling company. After testing and analysis by geologists, Chenush Gas is considering drilling on two different sites. The geologists have estimated that Site I will net \$75 million if successful (probability 0.2) and lose \$7 million if not successful (probability 0.8). Site II is estimated to net \$60 million if successful (probability 0.3) and lose \$5 million if not successful (probability 0.7).

(a) Determine the expected return for each site.

(b) Based on part (a), which site would you advise the company to select?

(c) Which site has less risk? (This corresponds to which has a smaller variance.)

29. The number of times that KTTV radio gives the weather each hour is uniformly distributed between 1 and 4, inclusive. Let X be a discrete uniform random variable representing the number of times that the weather is given each hour.

(a) Write a probability mass function for X.

(b) Determine μ_X and interpret.

(c) Compute σ_X for the distribution.

(d) Determine the probability that the weather will be given 3 or more times in a given hour.

In Exercises 30–32, evaluate the binomial probability mass function $P(X = x) = {}_nC_x \cdot p^x(1 - p)^{n-x}$ for the given values of n, p, and x.

30. $n = 4$, $p = \dfrac{1}{4}$, $x = 2$

31. $n = 3$, $p = 0.2$, $x = 0$

32. $n = 10$, $p = 0.40$, $x = 4$

In Exercises 33 and 34, determine if the assumptions of a binomial experiment are reasonably satisfied. If so, identify the values for p and n.

33. In a survey, 100 people are asked which brand of toothpaste they use.

34. Eighty-five percent of the students at Safeway Driving School pass their written driver's examination. A sample of 15 students is asked if they passed the test.

In Exercises 35 and 36, a binomial probability mass function is given. Complete the following:

(a) Make a table of values for $P(X = x)$.

(b) Construct a probability histogram for the distribution in part (a).

35. $P(X = x) = {}_3C_x \left(\dfrac{1}{8}\right)^x \left(1 - \dfrac{1}{8}\right)^{3-x}, \quad x = 0, 1, 2, 3$

36. $P(X = x) = {}_5C_x(0.4)^x(1 - 0.4)^{5-x}, \quad x = 0, 1, 2, 3, 4, 5$

In Exercises 37 and 38, match the probability with the appropriately shaded probability histogram.

(a) $P(X=x)$

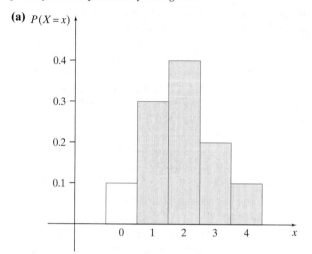

(b) $P(X=x)$

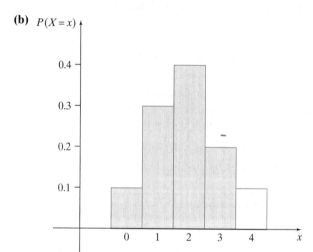

37. $P(X > 0)$

38. $P(X \le 3)$

In Exercises 39 and 40, assume that X represents a binomial random variable. Determine the mean $E(X) = \mu_X$ and the standard deviation σ_X of the distribution.

39. $n = 14$, $p = \dfrac{1}{4}$

40. $P(X = x) = {}_7C_x \cdot \left(\dfrac{1}{5}\right)^x \left(1 - \dfrac{1}{5}\right)^{7-x}$,

$x = 0, 1, 2, 3, 4, 5, 6, 7$

In Exercises 41–45, assume that all applications can be represented by a binomial random variable.

41. A survey at Hawthorne Middle School reveals that 75% of the students pass their physical fitness test. A group of 8 students is selected at random.

(a) Verify that the assumptions of a binomial experiment are reasonably satisfied.

(b) Evaluate $P(X = 5)$ and interpret.

42. The Direct Satisfaction Company knows from experience that 20% of the people who receive a free trial issue of *Backcountry* magazine will sign up for a 1-year subscription. The marketing department selects a group of 14 people at random who received a free trial issue of *Backcounty*.

(a) Determine $P(X = 6)$ and interpret.

(b) Determine $P(6 \le X \le 7)$ and interpret.

43. A 1998 study showed that 49% of Americans dined out at least once within 12 months of the survey (*Source:* U.S. Statistical Abstract, www.census.gov/statab). A sample of 15 Americans is selected at random.

(a) Determine the probability that 8 people dined out at least once within 12 months of the survey.

(b) Determine the probability that at most 7 of them dined out at least once within 12 months of the survey.

44. In a 1998 report on the American teen-age public, 72% of American teen-agers said that they can always trust their parents to be there for them when needed (*Source:* www.publicagenda.org). A poll selects 10 American teen-agers at random.

(a) Construct a probability histogram for the distribution.

(b) Compute μ_X and σ_X for the distribution and interpret each.

45. In 1998, 29% of the residents in Wisconsin lived in counties not meeting U.S. Environmental Protection Agency air quality standards (*Source:* National Center for Chronic Disease Prevention, www.cdc.gov). Eight residents of Wisconsin are selected at random.

(a) How many are expected to live in counties not meeting U.S. Environmental Protection Agency air quality standards?

(b) What is the probability that at most 3 of the residents live in counties not meeting U.S. Environmental Protection Agency air quality standards?

In Exercises 46–60, assume that each exercise can be represented by a normal random variable X.

46. The mean and standard deviation of the final exam scores for Kris's Biology and Physics classes are shown in the table.

	Biology	Physics
μ	78	73
σ	2.1	6.2

(a) Kris scored an 85 on the Biology exam and an 88 on the Physics exam. Compute Kris's z-scores for each exam.

(b) Graph the z-scores from part (a) on the standard normal curve. Relative to the rest of the students taking the Biology and Physics final exams, on which exam did Kris do better?

In Exercises 47–49, find the shaded area using Table 8.4.2 and the Standard Normal Table in Appendix D.

47.

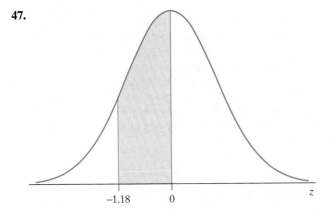

48.

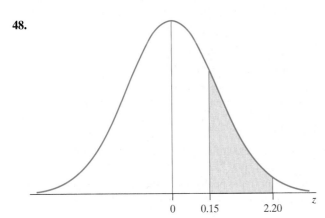

49.

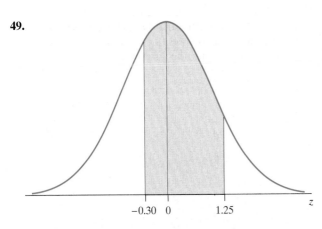

In Exercises 50–52, find the area between the given z-scores using Table 8.4.2 and the Standard Normal Table in Appendix D.

50. Between $z = -1.30$ and $z = 1.46$

51. Between $z = 0.27$ and $z = 2.18$

52. Between $z = -2.24$ and $z = -1.19$

In Exercises 53–55, determine the given probabilities using Table 8.4.2 and the Standard Normal Table in Appendix D.

53. $P(z > 1.28)$

54. $P(-0.45 \le z \le 0.48)$

55. $P(-1.85 < z < -0.35)$

56. The LandTrans Company runs shuttle buses from the ski resort to local hotels. Records show that the mean number of runs for each bus each day is $\mu = 20$ with a standard deviation of $\sigma = 1.4$. For an audit, the chief financial officer at LandTrans Company selects a shuttle bus at random.

(a) Determine $P(X > 17)$ and interpret.

(b) Determine $P(X < 17)$ and interpret.

57. Speedy Printing Company finds that, on any given day, a mean of $\mu = 10{,}000$ copies are made with a standard deviation of $\sigma = 800$ copies. The chief financial officer at Speedy Printing Company selects a business day at random.

(a) Determine $P(8000 \le X \le 12{,}000)$ and interpret.

(b) Determine $P(10{,}000 \le X \le 12{,}000)$ and interpret.

58. The weights of players at a soccer tournament had a mean of $\mu = 165.3$ pounds with a standard deviation of $\sigma = 17.2$ pounds. The athletic trainer selects a player from the roster at random.

(a) Determine the probability that the player weighed at least 150 pounds.

(b) Determine the probability that the player weighed between 150 and 170 pounds.

59. The mean school district enrollment for the 1999–2000 academic year in the Claremont school system was 3485.4 students in each school district with a standard deviation of 285.9 students. The superintendent selects a school district at random.

(a) Determine the probability that the school district has fewer than 3000 students.

(b) Determine the probability that the school district has between 3000 and 4000 students.

60. During the 1999–2000 ski season, the average age of a skier at a local ski resort was 29.7 years with a standard deviation of 14.3 years. The local newspaper selects a skier at random for an interview.

(a) Determine the probability that the skier is at least 18 years old.

(b) Determine the probability that the skier is between 18 and 35 years old.

CHAPTER 8 PROJECT

SOURCE: U.S. Census Bureau, www.census.gov

In this project we will see that we can use the normal distribution to approximate values for the binomial distribution.

According to the U.S. Census Bureau (www.census.gov), about 19% of Americans who are 65 years old and over are taking adult education courses. Suppose that a sample of 10 Americans 65 years old and over is selected at random.

1. Identify n and p.

2. Determine $P(X = 2)$ and interpret.

3. Determine the probability that either 1 or 2 of the 10 Americans in the sample are taking adult education courses.

Now suppose that group of 100 Americans that are 65 years old and over are selected at random. In this case, with $n = 100$, the calculations of the binomial distribution formula can be cumbersome. We can simplify the calculations however if we use the **normal approximation to the binomial distribution**. Statisticians use the following rule of thumb: The normal distribution can be used to approximate the binomial distribution if

$$(i)\ np \geq 5$$

$$(ii)\ n(1 - p) \geq 5$$

The values for the mean and the standard deviation are $\mu = np$ and $\sigma = \sqrt{npq} = \sqrt{np(1 - p)}$, respectively.

4. Verify that the two conditions are satisfied for the values of n and p given in step 1.

5. Use the values of n and p to determine μ and σ for the normal distribution.

Since the binomial distribution is a discrete probability distribution and the normal distribution is a continuous probability distribution, we need to use a **continuity correction**. Let's say that we want to compute $P(X = 5)$ using the normal approximation for the binomial distribution. On a probability histogram, the rectangle over $X = 5$ would begin at 4.5 and end at 5.5. So when we apply the continuity correction to the binomial probability $P(X = 5)$, we get $P(4.5 \leq X \leq 5.5)$. We can then convert these values to z-scores and use the computation techniques from Section 8.4. We can use the following guideline to compute the continuity correction.

Binomial Probability	Normal Probability with Continuity Correction
$P(X = a)$	$P(a - 0.5 \leq X \leq a + 0.5)$
$P(a \leq X \leq b)$	$P(a - 0.5 \leq X \leq b + 0.5)$

6. Compute $P(19.5 \leq X \leq 20.5)$ using the mean and the standard deviation values found in step 5.

7. Now let's determine the probability that anywhere from 10 to 20 of those in the sample of 100 are taking adult education courses. Write a binomial probability to determine this value.

8. Apply the continuity correction to the probability that you wrote in step 7.

9. Compute the probability value that you found in step 8 using the mean and the standard deviation values found in step 5. Interpret.

10. Using statistical software, we can determine that the value of $P(10 \leq X \leq 20)$ using $n = 100$ and $p = 0.19$ is 0.6524. Compare this answer to the one you got in step 9. How do you account for the difference?

Markov Chains

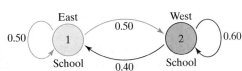

$$S_n = S_0 P^n$$

$$= [9000 \quad 13{,}000] \cdot \begin{bmatrix} \dfrac{6}{11} & \dfrac{5}{11} \\[2mm] \dfrac{6}{11} & \dfrac{5}{11} \end{bmatrix}$$

$$= [12{,}000 \quad 10{,}000]$$

One of the more controversial ideas in the debate over how to improve and fund public education is the use of school vouchers. To analyze the impact of vouchers, economists and public education officials use various models. One such model shows that enrollment changes resulting from school vouchers can be thought of as a system associated with a Markov chain. A state transition diagram is a visual representation of a Markov chain. A matrix product can reveal the long term enrollment pattern in a given school as a result of school vouchers.

What We Know

In Chapters 6, 7, and 8 we studied probability and its applications. This included conditional probability and Bayes' Theorem.

Where Do We Go

In this chapter we will extend the idea of conditional probability and use matrix operations to compute a sequence of experimental trials. This sequence of trials is called a Markov Chain.

Section 9.1   Introduction to Markov Chains

Many real-life applications involve modeling a process that changes from one category to the next, depending on given probabilities. For example, let's say that in a voting precinct in a certain city a voter can cast a vote for a Democrat, Republican, or Independent congressional candidate. These three outcomes associated with the party affiliation can be thought of as **states**, since they are the categories into which a voter can be identified. The voting precinct, relative to all congressional elections now, in the past, and in the future, can be thought of as a **system**. We call an individual voter in the precinct an **element** of the system. If the system changes from one state to another based on chance, then the sequence can be thought of as a **Markov chain**. This is part of a branch of the study of probability called **stochastic theory**, in which we study mathematical models whose experimental outcomes are based on outcomes of previous experiments. Markov chains are named after the Russian mathematician A. A. Markov who began the study of stochastic theory in the early twentieth century. In this section we will learn how to express Markov chains with diagrams and matrices.

From Your Toolbox

In Section 6.4 we saw that the event A does *not* occur is given by the **complement of A**, denoted by A^C. If $P(A) = a$, then $P(A^C) = 1 - a$.

Diagramming Markov Chains

We begin our study of Markov chains by developing ways to express states of systems visually. States of Markov chains are sometimes expressed as given events and complements of given events. To review how complements are expressed in a system, let's flashback to an example from Chapter 6. For a review of the complement of an event, consult the Toolbox to the left.

Flashback

Harris Poll Revisited

In Example 1 of Section 6.7, we saw that a sample of American adults was asked, "By the year 2020, do you think the quality of life in your community will be better or worse?" The percentages of results, separated by gender, are shown in Table 9.1.1.

Table 9.1.1

	Male	Female	Total
Worse	31%	30%	61%
Better	19%	20%	39%
Total	50%	50%	100%

SOURCE: www.harrisinteractive.com

Assume that we just use the results from the female respondents. Answer the following:

(a) Describe the system and the two states associated with the poll.

(b) If A represents the event that a female responded that the quality of life would be better, write the values of $P(A)$ and $P(A^C)$, and draw a tree diagram for the possible outcomes.

Flashback Solution

(a) The system is the opinions of all American females on the quality of life in their community and the two states are

State 1: Responded that the quality of life would be better

State 2: Responded that the quality of life would be worse

(b) To determine the probabilities, we first note that $P(A) = \frac{0.20}{0.50} = 0.40$. This means that $P(A^C) = 1 - 0.40 = 0.60$. The tree diagram for the female's opinion is shown in Figure 9.1.1.

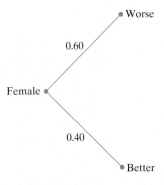

Figure 9.1.1

Note that we had a finite number of outcomes, that is, two, and that the poll was taken so that the respondents in the poll could not change their opinions once they had given their opinions. The Flashback exhibits many of the characteristics of a Markov chain.

■ Markov Chain

A sequence of trials associated with an experiment is a **Markov chain** if the outcome of an experiment meets the following two requirements:

1. The outcome is one of a finite set of defined states.
2. The outcome depends only on the present states and not any previous states.

A distinguishing component of a Markov chain is the **time step**. To see the impact of time steps on a Markov chain, let's examine another scenario. Suppose that a market analyst for Galaxy Communications is interested in whether those customers who sign a 1-year agreement for wireless telephone service will renew their contracts the next year. (Here we consider only those who renew with Galaxy or another company. Those who discontinue service altogether are not included.) A market analysis shows that if a wireless phone customer currently has a 1-year agreement with Galaxy Communications there is a 70% chance that he or she will renew this agreement for the next year. If a wireless phone customer does not have an agreement with Galaxy, then there is a 20% chance that he or she will switch to Galaxy when it comes time to renew the agreement. Here the two states are

State 1: User has a 1-year agreement with Galaxy Communications.

State 2: User has a 1-year agreement with another wireless phone provider.

The percentages given are called **transient probabilities** and correspond to the following conditional probabilities as we transcend from one state to another.

$$P(\text{state } 1 \mid \text{state } 1) = 0.70$$
$$P(\text{state } 1 \mid \text{state } 2) = 0.20$$

To denote this probability, we will use the letter p to represent the probability and use subscripts to identify the states. So we can write the conditional probabilities as

$$p_{1,1} = 0.70$$
$$p_{2,1} = 0.20$$

A graphical representation of these states with their probabilities is called a **state transition diagram**. We use labeled circles for the numbered states and arrows to represent the probabilities. The part of the diagram that represents $p_{1,1}$ is shown in Figure 9.1.2, which shows that there is a 70% chance that an element in state 1 will remain in state 1 when it comes time to renew the agreement next year. Observe that the time step for this system is 1 year. The number inside each circle corresponds to the number of each state. The notation $p_{1,2}$ represents the probability that a wireless user who currently has an agreement with Galaxy will switch to a different provider next year. Since we know that 70% of the current Galaxy customers will renew, this means that 30% of the customers will not. So we can write this complement as

$$p_{1,2} = 1 - p_{1,1} = 1 - 0.70 = 0.30$$

So the probability of going from state 1 to state 2 is 0.30. This is illustrated in Figure 9.1.3.

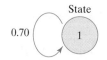

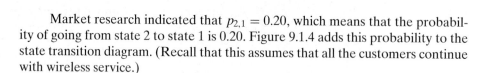

Figure 9.1.2

Figure 9.1.3

Market research indicated that $p_{2,1} = 0.20$, which means that the probability of going from state 2 to state 1 is 0.20. Figure 9.1.4 adds this probability to the state transition diagram. (Recall that this assumes that all the customers continue with wireless service.)

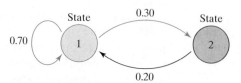

Figure 9.1.4

To complete the diagram, we see that the probability that a wireless customer who has an agreement with a provider other than Galaxy will also have an agreement with a provider other than Galaxy next year is $p_{2,2} = 1 - 0.20 = 0.80$. The complete state transition diagram for the system is shown in Figure 9.1.5.

Before we continue, let's list some properties of state transition diagrams.

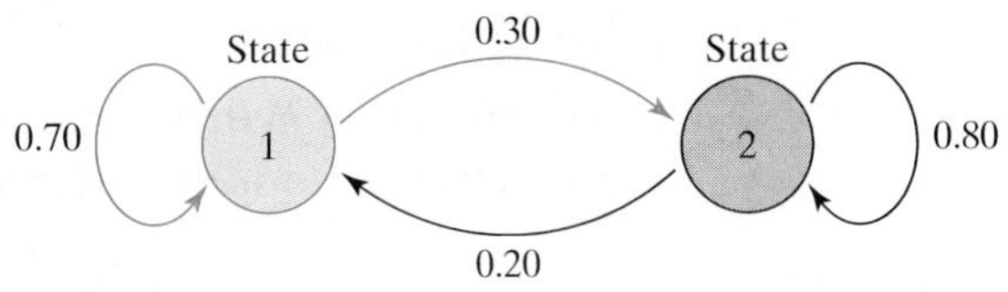

State 1 0.70, State 2 0.80, 0.30 from 1 to 2, 0.20 from 2 to 1

Figure 9.1.5

■ **Properties of State Transition Diagrams**
- There must be one circle to represent each state.
- Every state must have at least one outgoing arrow.
- The sum of the probabilities labeled on the arrows coming from a state must be 1.

Example 1 ## Drawing a State Transition Diagram

At Urbanix College, records show that 15% of all education majors changed their major between their first and second years. Of all the noneducation majors, 10% changed their major to education, while the rest did not.

(a) List the system and the states.

(b) Draw the state transition diagram for the system.

Solution

(a) The system is represented by all students who have attended, are attending, or will attend Urbanix College. The two states are

State 1: The student is an education major.

State 2: The student is not an education major.

(b) The state transition diagram for the system is shown in Figure 9.1.6. Notice that the corresponding probabilities for the system are $p_{1,1} = 0.85$, $p_{1,2} = 0.15$, $p_{2,1} = 0.10$, and $p_{2,2} = 0.90$.

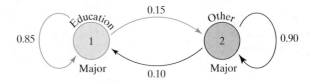

Figure 9.1.6

The subscripts that we are using are not accidental. Notice that they look just like those used when naming elements of a matrix. This means that we can write a matrix corresponding to the Urbanix College system as

$$P = \begin{bmatrix} 0.85 & 0.15 \\ 0.10 & 0.90 \end{bmatrix}$$

We call a matrix such as P a **transition matrix**.

▪ Transition Matrix

The **transition matrix** P for a given system associated with a Markov chain with two states whose corresponding state transition diagram is

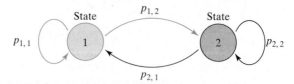

is given by

$$P = \begin{bmatrix} p_{1,1} & p_{1,2} \\ p_{2,1} & p_{2,2} \end{bmatrix} \quad \begin{array}{l} \text{Arrows that come from state 1} \\ \text{Arrows that come from state 2} \end{array}$$

Example 2   Writing Transition Matrices

Write a transition matrix for each of the state transition diagrams in Figures 9.1.7 and 9.1.8.

(a)

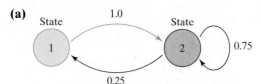

Figure 9.1.7

(b)

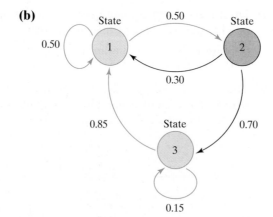

Figure 9.1.8

Solution

(a) Since there is only one arrow originating from state 1, we have $p_{1,2} = 1$. This means that $p_{1,1} = 0$. Knowing that $p_{2,1} = 0.25$ and $p_{2,2} = 0.75$, we get the transition matrix

$$P = \begin{bmatrix} 0 & 1 \\ 0.25 & 0.75 \end{bmatrix}$$

(b) Even though this diagram has three states, the transition matrix is formed using the same method as in part (a). Here the transition matrix will have three rows and three columns. We have two arrows originating from state 1. These arrows indicate that $p_{1,1} = 0.50$, $p_{1,2} = 0.50$, and $p_{1,3} = 0$. From state 2 we get $p_{2,1} = 0.30$, $p_{2,2} = 0$, and $p_{2,3} = 0.70$. For state 3 we get the values $p_{3,1} = 0.85$, $p_{3,2} = 0$, and $p_{3,3} = 0.15$. This gives us the transition matrix

$$P = \begin{bmatrix} 0.50 & 0.50 & 0 \\ 0.30 & 0 & 0.70 \\ 0.85 & 0 & 0.15 \end{bmatrix}$$ ∎

✓ **Checkpoint 1**

Now work Exercise 23.

Example 2 illustrates some properties of transition matrices. We can now generalize the results and list some of the properties of transition matrices for any number of states.

Properties of Transition Matrices

For a system of n states associated with a Markov chain, we have the following:

1. The size of the transition matrix is $n \times n$. That is, the transition matrix has the form

$$P = \begin{bmatrix} p_{1,1} & p_{1,2} & p_{1,3} & \cdots & p_{1,n} \\ p_{2,1} & p_{2,2} & p_{2,3} & \cdots & p_{2,n} \\ \cdot & \cdot & \cdot & \cdot & \cdot \\ \cdot & \cdot & \cdot & \cdot & \cdot \\ \cdot & \cdot & \cdot & \cdot & \cdot \\ p_{n,1} & p_{n,2} & p_{n,3} & \cdots & p_{n,n} \end{bmatrix}$$

2. The elements of a transition matrix are values between 0 and 1, inclusive. That is, $0 \le p_{i,j} \le 1$ for $i = 1, 2, \ldots, n$ and $j = 1, 2, \ldots, n$.
3. The sum of the elements in any row is 1.

Initial State Vectors

Once we get the transition matrix for a system, we can use known values associated with the system to make time step estimations. To see how, let's suppose that the Milson City School System has had a tax levy on the ballot for the last two elections, with the levy failing each time. Exit polls reveal the following facts:

- Sixteen percent of those who opposed the levy in the first election favored the levy in the second election, while the rest voted against the levy in both elections.
- On the other hand, 7% of those who favored the levy in the first election voted against it in the second election, while the rest favored the levy in both elections.

The system associated with this Markov chain has two states:

State 1: Opposed the levy
State 2: Favored the levy

The state transition diagram for the system is shown in Figure 9.1.9. This gives us the transition matrix $P = \begin{bmatrix} 0.84 & 0.16 \\ 0.07 & 0.93 \end{bmatrix}$. Now suppose that in the first election,

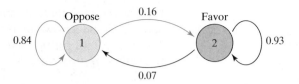

Figure 9.1.9

of the approximately 15,000 votes cast, 9000 were against the levy, while 6000 voted for the levy. To approximate how many voted against the levy in the second election, we can compute

$$\begin{pmatrix} \text{number opposed in} \\ \text{second election} \end{pmatrix} = \begin{pmatrix} \text{number} \\ \text{opposed in} \\ \text{the first} \\ \text{election} \end{pmatrix} \begin{pmatrix} \text{percent who} \\ \text{continued to} \\ \text{oppose in the} \\ \text{second election} \end{pmatrix} + \begin{pmatrix} \text{number who} \\ \text{favored in} \\ \text{first} \\ \text{election} \end{pmatrix} \begin{pmatrix} \text{percent who} \\ \text{switched to being} \\ \text{opposed in} \\ \text{second election} \end{pmatrix}$$

$$= (9000)(0.84) + (6000)(0.07) = 7980$$

The number who favored the levy in the second election is given by

$$\begin{pmatrix} \text{number favored in} \\ \text{second election} \end{pmatrix} = \begin{pmatrix} \text{number} \\ \text{opposed in} \\ \text{the first} \\ \text{election} \end{pmatrix} \begin{pmatrix} \text{percent who} \\ \text{switched to} \\ \text{favoring in} \\ \text{second election} \end{pmatrix} + \begin{pmatrix} \text{number who} \\ \text{favored in} \\ \text{first} \\ \text{election} \end{pmatrix} \begin{pmatrix} \text{percent who} \\ \text{continued to} \\ \text{favor in the} \\ \text{second election} \end{pmatrix}$$

$$= (9000)(0.16) + (6000)(0.93) = 7020$$

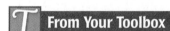

From Your Toolbox

If A is an $m \times n$ matrix and B is an $n \times p$ matrix, then the matrix product is an $m \times p$ matrix. To find the entry of the ith row and jth column of AB, multiply each entry in the ith row of A by the corresponding entry in the jth column of B and add the products.

The pattern used in the calculations is one that we first saw in Chapter 2. (Consult the Toolbox to the left.)

Notice that if we let $[\,9000 \quad 6000\,]$ be a 1×2 matrix and multiply by the 2×2 transition matrix, we get

$$[\,9000 \quad 6000\,]\begin{bmatrix} 0.84 & 0.16 \\ 0.07 & 0.93 \end{bmatrix} = [\,7980 \quad 7020\,]$$

Here we call the matrix $[\,9000 \quad 6000\,]$ an **initial state vector**. A **vector** is a name given to any matrix that is made up of a single row or a single column. If a system associated with a Markov chain has n states, then the size of the initial state vector is $1 \times n$.

■ **Initial State Vector**

For a system associated with a Markov chain with n states, the **initial state vector** S_0 is a $1 \times n$ row vector whose elements are the system's values in its beginning state.

▶ **Notes:** 1. The matrix product $S_0 P = S_1$ gives us the **first state vector**.

　　　　　2. We can also use probability values for initial state vectors.

Using the definition for the initial state vector, the initial vector S_0 for the Milson City School System tax levy is $S_0 = [\,9000 \quad 6000\,]$. The first state vector is given by

$$S_1 = S_0 P$$

$$= [\,9000 \quad 6000\,] \cdot \begin{bmatrix} 0.84 & 0.16 \\ 0.07 & 0.93 \end{bmatrix}$$

$$= [\,7980 \quad 7020\,]$$

Let's assume that the percentages of voters who will either change their minds or vote the same way remain the same in the upcoming election. To approximate how many will oppose or favor the tax levy the next time it is on the ballot, we can repeat the matrix multiplication using the first state vector to get

$$S_2 = S_1 P = [7980 \quad 7020] \begin{bmatrix} 0.84 & 0.16 \\ 0.07 & 0.93 \end{bmatrix} = [7194.6 \quad 7805.4]$$

This means that by the third time the levy is placed on the ballot we can expect the levy to pass with approximately 7195 opposed and approximately 7805 favoring.

Now let's see how probability values can be used in initial state vectors. For example, since 15,000 people voted in the first election and 9000 opposed the levy, we see that $\frac{9000}{15,000} = 60\%$ of the voters opposed the levy and $\frac{6000}{15,000} = 40\%$ favored the levy. This means that we can write $S_0 = [0.60 \quad 0.40]$ as our initial state vector and compute the first state vector as follows:

$$S_1 = S_0 P = [0.60 \quad 0.40] \begin{bmatrix} 0.84 & 0.16 \\ 0.07 & 0.93 \end{bmatrix} = [0.532 \quad 0.468]$$

We can now compute the second state vector, S_2. This is

$$S_2 = S_1 P = [0.532 \quad 0.468] \begin{bmatrix} 0.84 & 0.16 \\ 0.07 & 0.93 \end{bmatrix} \approx [0.480 \quad 0.520]$$

This tells use that by the third time the levy goes on the ballot we can expect about 48% to oppose the levy and about 52% to favor the levy. When we use probabilities as elements in our vectors, we call these **probability distribution vectors**.

Computing State Vectors

In general, we call S_k the **kth state vector**, where $S_k = S_{k-1} P$. But we do not have to compute all the lower state vectors first in order to get S_k. Using the properties of matrix multiplication, we have

$$\text{first state vector} = S_1 = S_0 P$$
$$\text{second state vector} = S_2 = S_1 P = (S_0 P)P = S_0 P^2$$
$$\text{third state vector} = S_3 = S_2 P = (S_0 P^2)P = S_0 P^3$$

This pattern illustrates that we can use powers of the transition matrix to determine the kth state vector.

■ **kth State Vector**

For a system associated with a Markov chain with a transition matrix P and initial state vector S_0, the **kth state vector** is given by

$$S_k = S_0 P^k$$

where $k = 0, 1, 2, \ldots.$

Example 3   Determining the kth State Vector

The zoning commission in Chesterfield identifies properties as being zoned in one of three categories: public, residential, or commercial. During the last 5 years, 85% of the land zoned as public has remained public, while 15% of the land has been rezoned as residential. Seventy percent of the land zoned as residential has remained residential, with 10% rezoned as public and 20% rezoned as commercial. The commission also found that 60% of the land zoned as commercial has remained commercial, while 25% has been rezoned as public and 15% rezoned as residential. Chesterfield initially had 2000 acres zoned public, 24,000 acres zoned residential, and 6000 acres zoned commercial.

(a) Identify the time step, draw a state transition diagram, and write a transition matrix for the zoned properties.

(b) Compute S_4 and interpret.

Solution

(a) The time step for the system is 5 years. The system has three states:

  State 1: Land is zoned public.

  State 2: Land is zoned residential.

  State 3: Land is zoned commercial.

The state transition diagram for the zoned property is shown in Figure 9.1.10.

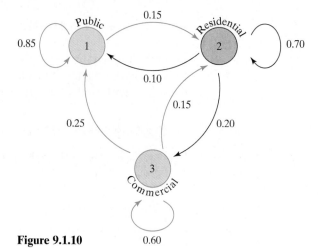

Figure 9.1.10

The diagram indicates that the 3×3 transition matrix for the system is

$$P = \begin{bmatrix} 0.85 & 0.15 & 0 \\ 0.10 & 0.70 & 0.20 \\ 0.25 & 0.15 & 0.60 \end{bmatrix}$$

(b) **Understand the Situation:** The initial state vector is

$$S_0 = [\,2000 \quad 24000 \quad 6000\,]$$

Recall that the time step for the system is 5 years, so S_4 represents the states in 20 years after the land was initially zoned. To determine S_4, we need to compute $S_4 = S_0 P^4$.

So we get

$$S_4 = S_0 P^4 = [\,2000 \quad 24{,}000 \quad 6000\,] \begin{bmatrix} 0.85 & 0.15 & 0 \\ 0.10 & 0.70 & 0.20 \\ 0.25 & 0.15 & 0.60 \end{bmatrix}^4$$

$$= [\,12{,}661.85 \quad 11{,}886.75 \quad 7451.40\,]$$

Interpret the Solution: This means that 20 years after initially zoning the land Chesterfield can expect to have about 12,662 acres zoned public, 11,887 acres zoned residential, and 7451 acres zoned commercial. ■

Technology Option

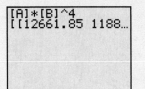

We can compute S_4 on our calculator by assigning S_0 to matrix A and P to matrix B and then computing $A * B\char`^4$. See Figure 9.1.11.

To learn more on using your calculator to perform matrix operations, consult the on-line graphing calculator manual at www.prenhall.com/armstrong

Figure 9.1.11

 **Checkpoint 2**   Now work Exercises 31 and 41.

SUMMARY

- A sequence of trials associated with an experiment is a **Markov chain** if the outcome of an experiment is one of a finite set of defined states and the outcomes depend only on the present states and not any previous states.
- A graphic depiction of a Markov system is a **state transition diagram**.
- A **transition matrix** shows the transitional probabilities from one state to another.

- The **initial state vector** S_0 is a $1 \times n$ row vector whose elements are the system's values in its beginning state.
- For a system associated with a Markov chain with a transition matrix P and initial state vector S_0, the **kth state vector** is given by $S_k = S_0 P^k$.

SECTION 9.1 EXERCISES

In Exercises 1–10, determine if the matrices are transition matrices. Supply an explanation for your answer.

1. $\begin{bmatrix} 0.1 & 0.9 \\ 0.3 & 0.7 \end{bmatrix}$ **2.** $\begin{bmatrix} 0.5 & 0.5 \\ 0.6 & 0.4 \end{bmatrix}$

3. $\begin{bmatrix} 0.4 & 0.6 \\ 0.8 & 0.3 \end{bmatrix}$ **4.** $\begin{bmatrix} 0.1 & 0.8 \\ 0.9 & 0.1 \end{bmatrix}$

5. $\begin{bmatrix} 0 & -0.9 & 0.1 \\ 0 & 1 & 0 \\ 0 & 0 & 1 \end{bmatrix}$ **6.** $\begin{bmatrix} 0.5 & -0.1 & 0.6 \\ 0.4 & 0.3 & 0.3 \\ 0.9 & 0 & 0.1 \end{bmatrix}$

7. $\begin{bmatrix} 0.3 & 0 & 0.7 \\ 0.1 & 0.5 & 0.4 \\ 0.7 & 0.3 & 0 \end{bmatrix}$ **8.** $\begin{bmatrix} 0.6 & 0.4 & 0 \\ 0.9 & 0.1 & 0 \\ 0 & 0 & 1 \end{bmatrix}$

9. $\begin{bmatrix} 0.35 & 0.60 & 0.05 \\ 0.20 & 0.35 & 0.45 \\ 0.50 & 0.15 & 0.35 \end{bmatrix}$ **10.** $\begin{bmatrix} 0.33 & 0.36 & 0.31 \\ 0.64 & 0.20 & 0.16 \\ 0.91 & 0.02 & 0.07 \end{bmatrix}$

In Exercises 11–20, write the transition matrix for the state transition diagrams.

11.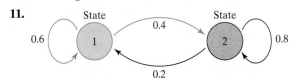

12.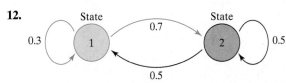

13.

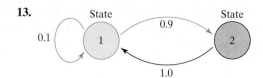

14.

15.

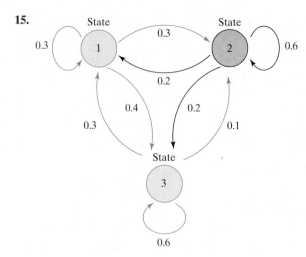

16.

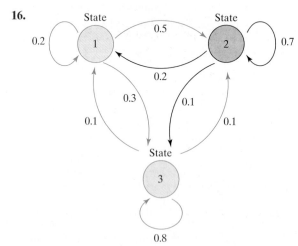

17.

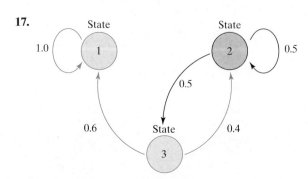

18.

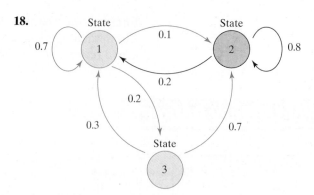

19.

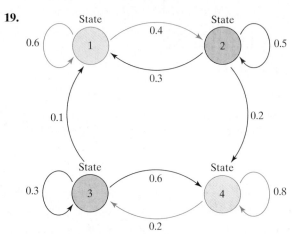

20.

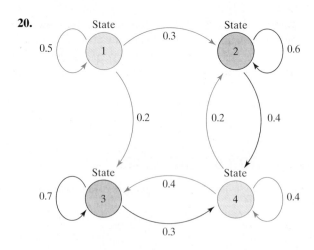

In Exercises 21–30, draw a state transition diagram for the given transition matrix.

21. $P = \begin{bmatrix} 0.3 & 0.7 \\ 0.1 & 0.9 \end{bmatrix}$

22. $P = \begin{bmatrix} 0.5 & 0.5 \\ 0.8 & 0.2 \end{bmatrix}$

✔ **23.** $P = \begin{bmatrix} 0 & 1 \\ 0.4 & 0.6 \end{bmatrix}$

24. $P = \begin{bmatrix} 0.7 & 0.3 \\ 0.1 & 0.9 \end{bmatrix}$

25. $P = \begin{bmatrix} 0.2 & 0.3 & 0.5 \\ 0.8 & 0.1 & 0.1 \\ 0.5 & 0.1 & 0.4 \end{bmatrix}$

26. $P = \begin{bmatrix} 0.7 & 0.2 & 0.1 \\ 0.3 & 0.5 & 0.2 \\ 0.6 & 0.3 & 0.1 \end{bmatrix}$

27. $P = \begin{bmatrix} 0 & 0.5 & 0.5 \\ 0.63 & 0.30 & 0.07 \\ 0 & 0.6 & 0.4 \end{bmatrix}$

28. $P = \begin{bmatrix} 0.69 & 0.23 & 0.08 \\ 0.53 & 0 & 0.47 \\ 0.66 & 0 & 0.34 \end{bmatrix}$

29. $P = \begin{bmatrix} 0.22 & 0.15 & 0.60 & 0.03 \\ 0.48 & 0.11 & 0.03 & 0.38 \\ 0.06 & 0.37 & 0.31 & 0.26 \\ 0.52 & 0 & 0.31 & 0.17 \end{bmatrix}$

30. $P = \begin{bmatrix} 0.36 & 0.05 & 0.08 & 0.51 \\ 0 & 0.62 & 0.33 & 0.05 \\ 0.09 & 0.02 & 0.59 & 0.30 \\ 0.81 & 0 & 0.18 & 0.01 \end{bmatrix}$

For Exercises 31–40, complete the following:

(a) Identify the states.

(b) Draw the state transition diagram.

(c) Write the transition matrix P.

✓ **31.** A market survey on activity-related sunglasses shows that 30% of the owners of Blade Shades did not like the product and switched to Optifilters the next summer, while the rest continued to purchase Blade Shades. Conversely, 40% of the owners of Optifilters did not like their sunglasses and switched to Blade Shades the next summer, while the rest continued to purchase Optifilters.

32. Following a new advertising campaign, a market analysis shows that 20% of the consumers who do not currently drink SportSquirt will purchase it the next time that they buy a sport drink. Ninety percent of the consumers who currently drink SportSquirt will purchase it the next time that they buy a sport drink.

33. The *Suburban Almanac* presumes that the severity of hot temperatures of one summer can be determined by the severity of the previous summer. Records show that if the last year's summer was severely hot, then there is a 45% chance that the next summer will be severely hot. On the other hand, if the last year's summer was moderate, then there is a 40% chance that the next summer will be moderate.

34. The *Suburban Almanac* presumes that the chance of rain the previous day can indicate if it will rain the next day. Records show that if it rained the previous day there is a 40% chance that it will rain again the next day. If it was sunny the previous day, there is an 80% chance that it will be sunny the next day.

35. The town of Minsterton has an election for mayor every 2 years. Voting records show that if a Democrat is in office then there is a 60% chance that a Democrat will be elected in the next election and a 40% chance that a

Republican will be elected. If a Republican is in office, there is a 65% chance that a Republican will be elected in the next election and a 35% chance that a Democrat will be elected.

36. A regional health study shows that, from one year to the next, 75% of the people in the region who smoked will continue to smoke, while 25% of the people who smoked will stop smoking the next year. Eight percent of the people who did not smoke will start smoking the next year, while 92% of those who did not smoke will continue to be nonsmokers.

37. The Move-It-Safe Company rents moving trucks in the cities of Adamstown, Breston, and Carpersville. They have both round-trip and point-to-point rentals. Records show that if a truck is rented in Adamstown, there is a 70% chance that it will be returned to Adamstown, a 15% chance that it will be returned to Breston, and a 15% chance that it will be returned to Carpersville. If a truck is rented in Breston, there is a 65% chance that it will be returned to Breston, a 10% chance that it will be returned to Adamstown, and a 25% chance that it will be returned to Carpersville. If a truck is rented in Carpersville, there is an 85% chance that it will be returned to Carpersville, a 5% chance that it will be returned to Adamstown, and a 10% chance that it will be returned to Breston.

38. In the LaDona metropolitan area, commuters either drive alone, carpool, or take public transportation. A study completed by the transportation department shows that 80% of those who drive alone intend to drive alone a year from now, while 15% intend to switch to carpooling and 5% intend to use public transportation. Ninety percent of those who carpool intend to continue carpooling the next year, while 5% intend to drive alone and 5% will begin to use public transportation. Sixty percent of those who use public transportation will continue to use it next year, while 30% intend to drive alone and 10% will begin to carpool.

39. In a particular county in the northern part of the country, voters can declare themselves as members of the Republican Party, Democrat Party or Green Party. A public opinion poll shows that 85% of current Republicans intend to continue with their party affiliation, while 13% intend to switch to the Democrat Party in the next election and 2% to the Green Party. Seventy percent of current Democrats intend to continue their party affiliation, while 10% intend to switch to the Republican Party in the next election and 20% to the Green Party. Ninety-five percent of current Green Party members plan to continue with their party affiliation, while 1% intend to switch to the Republican party in the next election and 4% to the Democrat Party.

40. The Armont Conservatory has been keeping records on their own Armont carnation and finds that the flowering color of the next generation of plant is related to the current generation. If a carnation is red, there is a 60%

chance that the next generation will be red, a 40% chance that the plant will be pink, and no chance of the plant being white. If a carnation is pink, there is a 60% chance that the next generation will be pink, a 30% chance that it will be red, and a 10% chance that it will be white. If a carnation is white, there is a 50% chance that the next generation will be white, a 50% chance that it will be pink, and no chance that it will be red.

✓ **41.** Refer to Exercise 31.

(a) Currently, Blade Shade holds 60% of the market share and Optifilters holds 40%. Write the probability distribution vector S_0 for the market share.

(b) Determine S_3 and interpret.

42. Refer to Exercise 32.

(a) Currently, about 35% of those who drink sports drinks purchase SportSquirt. Write the probability distribution vector S_0 for the market share.

(b) Determine S_3 and interpret.

43. Refer to Exercise 33.

(a) If the probability distribution vector is given to be $S_0 = [0.4 \quad 0.6]$, interpret the values of these two matrix elements.

(b) Determine S_2 and interpret.

44. Refer to Exercise 34.

(a) If the probability distribution vector is given to be $S_0 = [0.7 \quad 0.3]$, interpret the values of these two matrix elements.

(b) Determine S_2 and interpret.

45. Refer to Exercise 35.

(a) Interpret each coordinate of the probability distribution vector $S_0 = [0.4 \quad 0.6]$.

(b) Determine S_5 and interpret.

46. Refer to Exercise 36.

(a) Currently, there are about 60,000 smokers and 250,000 nonsmokers in the region. Write the corresponding initial state vector S_0.

(b) Determine S_4 and interpret.

47. Refer to Exercise 37.

(a) Currently, 800 Move-It-Safe Company moving trucks have been rented from Adamstown, 1200 from Breston, and 400 from Carpersville. Write the corresponding initial state vector S_0.

(b) Determine S_2 and interpret.

48. Refer to Exercise 38.

(a) Interpret each value of the probability distribution vector $S_0 = [0.8 \quad 0.1 \quad 0.1]$.

(b) Determine S_2 and interpret.

49. Refer to Exercise 39.

(a) Interpret each value of the probability distribution vector $S_0 = [0.62 \quad 0.37 \quad 0.01]$.

(b) Determine S_4 and interpret.

50. Refer to Exercise 40.

(a) Interpret each value of the probability distribution vector $S_0 = [0.3 \quad 0.6 \quad 0.1]$.

(b) Determine S_8 and interpret.

SECTION PROJECT

To explore the computing of powers of matrices, consider the transition matrix

$$P = \begin{bmatrix} \frac{1}{5} & \frac{4}{5} \\ \frac{3}{10} & \frac{7}{10} \end{bmatrix}$$

(a) Compute $10P$ using scalar multiplication. What type of elements does this matrix have?

(b) Compute $(10P)^5$, the fifth power of the scalar product $10P$.

(c) Multiply the result of part (b) by the scalar $\frac{1}{10^5}$, to get the expression $\frac{1}{10^5}(10P)^5$.

(d) Use your calculator to compute P^5. Compare the answer to the answer in part (c). What do you notice?

Section 9.2   Regular Markov Chains

In the previous section we learned that Markov chains can be used to analyze trends of data for a given increment of time called a time step. Now we will turn our attention to the long-range behavior of Markov chains and kth state matrices. We will also see what conditions are necessary to study these long-range behaviors.

Steady-State Transition Matrices

So far we have studied how we can use powers of transition matrices to determine the kth state vector S_k. But suppose that we want to determine the outcome of a Markov chain over a very long period of time. This means we would need to find a kth state vector for a large value of k. For example, consider the transition matrix $P = \begin{bmatrix} 0.2 & 0.8 \\ 0.7 & 0.3 \end{bmatrix}$. If we raise P to higher and higher powers, we get (after rounding to four decimal places)

$$P^2 = \begin{bmatrix} 0.6 & 0.4 \\ 0.35 & 0.65 \end{bmatrix}, \qquad P^4 = \begin{bmatrix} 0.5 & 0.5 \\ 0.4375 & 0.5625 \end{bmatrix}$$

$$P^{16} \approx \begin{bmatrix} 0.4667 & 0.5333 \\ 0.4667 & 0.5333 \end{bmatrix}, \qquad P^{32} \approx \begin{bmatrix} 0.4667 & 0.5333 \\ 0.4667 & 0.5333 \end{bmatrix}$$

Let's make two observations about these matrices:

- The two rows of the higher-power matrices have the same entries.
- As the power of P gets relatively large, we see that no change occurs in the elements of P^n.

For the transition matrix $P = \begin{bmatrix} 0.2 & 0.8 \\ 0.7 & 0.3 \end{bmatrix}$, we can show for large values of n that P^n is approximately $\begin{bmatrix} \frac{7}{15} & \frac{8}{15} \\ \frac{7}{15} & \frac{8}{15} \end{bmatrix}$. Then, for any initial probability vector S_0 and $S = \begin{bmatrix} \frac{7}{15} & \frac{8}{15} \end{bmatrix}$, it can be shown that

$$S_0 P^n \approx S \qquad \text{for large values of } n$$

It is also true that $SP = S$. We call S a **steady-state vector**. The steady-state vector is sometimes called a long-run distribution. The matrix P^n, for large values of n, can also be called a **limiting matrix**. (Observe that the steady-state vector, S, gives the rows of the limiting matrix P^n.) Now let's see how steady-state vectors can be used in applications.

Example 1 **Using Steady-State Vectors in Applications**

Snellburg has two schools on opposite ends of the city, East School and West School. The city school system adopts a voucher system that allows a student to choose the school that he or she attends every 4 years. After the first 4 years, it was found that half of the students who attend East School opted to stay at that school while the other half chose to go to West School. Forty percent of the students who attended West School opted to stay at West, while 60% used their vouchers to enroll at East School. When the voucher system began, 9000 students attended East School and 13,000 students attended West School.

(a) Draw a state transition diagram and write the transition matrix for the East and West School enrollments.

(b) Write the initial state matrix S_0. Given that, for large values of n,

$$P^n \approx \begin{bmatrix} \frac{6}{11} & \frac{5}{11} \\ \frac{6}{11} & \frac{5}{11} \end{bmatrix}, \text{ compute } S_n = S_0 P^n \text{ and interpret.}$$

Solution

(a) The system has two states:

> **State 1**: Student is enrolled at East School.
> **State 2**: Student is enrolled at West School.

The state transition diagram for the system is shown in Figure 9.2.1. This gives us the transition matrix $P = \begin{bmatrix} 0.5 & 0.5 \\ 0.6 & 0.4 \end{bmatrix}$.

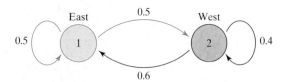

Figure 9.2.1

(b) The initial state matrix is $S_0 = [\,9000 \quad 13{,}000\,]$. Multiplying by

$$P^n \approx \begin{bmatrix} \frac{6}{11} & \frac{5}{11} \\ \frac{6}{11} & \frac{5}{11} \end{bmatrix} \text{ gives us}$$

$$S_n = S_0\, P^n \approx [\,9000 \quad 13{,}000\,] \begin{bmatrix} \frac{6}{11} & \frac{5}{11} \\ \frac{6}{11} & \frac{5}{11} \end{bmatrix} = [\,12{,}000 \quad 10{,}000\,]$$

Interpret the Solution: Assuming that the enrollment trends continue, this means that, enrollment at East School will increase to 12,000, while enrollment at West School will decrease to 10,000, in the long run. ∎

Technology Option

We can use the calculator to help us to determine steady-state vectors. To determine the steady-state vector in Example 1, we store the transition matrix $P = \begin{bmatrix} 0.5 & 0.5 \\ 0.6 & 0.4 \end{bmatrix}$ in our calculator as matrix A. Figure 9.2.2 shows the output of computing P^{16}.

Using the FRAC command gives the result shown in Figure 9.2.3.

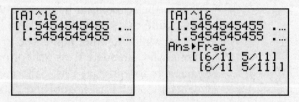

Figure 9.2.2   **Figure 9.2.3**

For more information on performing matrix operations on your calculator, consult the on-line graphing calculator manual at www.prenhall.com/armstrong

Regular Transition Matrices

For a transition matrix to have a steady-state vector, it is sufficient that it be what is called a **regular transition matrix**.

■ Regular Transition Matrix

Let P be a transition matrix corresponding to a Markov chain. P is a **regular transition matrix** if some power of P has only positive elements.

▶ **Note:** Every regular transition matrix has a steady-state vector.

Let's examine some techniques that can be used to determine if a transition matrix is a regular transition matrix.

Example 2   **Determining Regular Matrices**

Determine whether the following transition matrices are regular transition matrices.

(a) $P = \begin{bmatrix} 0.6 & 0.4 \\ 0.1 & 0.9 \end{bmatrix}$ (b) $P = \begin{bmatrix} 0.5 & 0 & 0.5 \\ 0 & 1 & 0 \\ 0 & 0 & 1 \end{bmatrix}$

Solution

(a) Notice that all the elements of P are positive. When we multiply to get higher powers of P, we will always be multiplying and adding positive numbers, which will give us a positive result. This means that P is a regular transition matrix.

(b) Let's compute some powers of P to see if a pattern emerges.

$$P^2 = P \cdot P = \begin{bmatrix} 0.5 & 0 & 0.5 \\ 0 & 1 & 0 \\ 0 & 0 & 1 \end{bmatrix} \begin{bmatrix} 0.5 & 0 & 0.5 \\ 0 & 1 & 0 \\ 0 & 0 & 1 \end{bmatrix} = \begin{bmatrix} 0.25 & 0 & 0.75 \\ 0 & 1 & 0 \\ 0 & 0 & 1 \end{bmatrix}$$

$$P^3 = P^2 \cdot P = \begin{bmatrix} 0.25 & 0 & 0.75 \\ 0 & 1 & 0 \\ 0 & 0 & 1 \end{bmatrix} \begin{bmatrix} 0.5 & 0 & 0.5 \\ 0 & 1 & 0 \\ 0 & 0 & 1 \end{bmatrix} = \begin{bmatrix} 0.125 & 0 & 0.875 \\ 0 & 1 & 0 \\ 0 & 0 & 1 \end{bmatrix}$$

$$P^4 = P^3 \cdot P = \begin{bmatrix} 0.125 & 0 & 0.875 \\ 0 & 1 & 0 \\ 0 & 0 & 1 \end{bmatrix} \begin{bmatrix} 0.5 & 0 & 0.5 \\ 0 & 1 & 0 \\ 0 & 0 & 1 \end{bmatrix} = \begin{bmatrix} 0.0625 & 0 & 0.9375 \\ 0 & 1 & 0 \\ 0 & 0 & 1 \end{bmatrix}$$

Notice that for each power of P the zero elements seem to always be in the same location. It appears that all higher powers of P will also have zero elements in the same location. This means that P is not a regular transition matrix.   ■

✓ Checkpoint 1

Now work Exercise 7.

Example 2 has illustrated some ways to inspect transition matrices in order to determine whether they are regular. Before analyzing a more direct way to determine steady-state matrices, let's first examine some properties of regular transition matrices.

> ■ **Properties of Regular Transition Matrices**
>
> Let P be a regular transition matrix. Then the following properties apply.
>
> 1. There is a unique vector S such that $S_0 P^n \approx S$, for large values of n.
> 2. The rows of the limiting matrix P^n are the same, for large values of n. The elements of S are the row elements of P^n, for large values of n.
> 3. If a transition matrix has all positive elements, then the matrix is a regular transition matrix.
> 4. The steady-state vector, S, corresponding to a regular transition matrix can be determined by solving a system of equations.

To find the steady-state vector for a given regular transition matrix, we can use the equation $SP = S$ and write the corresponding system of equations. We will also use the condition that the sum of the elements in a row of a transition matrix is 1. This allows us to solve the resulting system of equations using the substitution method shown in Chapter 1 or Gauss–Jordan Elimination shown in Chapter 2. The solution to this system gives us the **steady-state vector**. The elements of this vector are the row elements of our limiting matrix P^n, for large values of n.

Example 3 Computing the Steady-State Vector and Limiting Matrix

Solve a system of equations to determine the steady-state vector for the transition matrix

$$P = \begin{bmatrix} 0.1 & 0.9 \\ 0.7 & 0.3 \end{bmatrix}$$

Solution

Note that, since P has only positive elements, P is a regular transition matrix. To determine the steady-state vector, we let $S = \begin{bmatrix} x & y \end{bmatrix}$ and solve the matrix equation

$$\begin{bmatrix} x & y \end{bmatrix} \begin{bmatrix} 0.1 & 0.9 \\ 0.7 & 0.3 \end{bmatrix} = \begin{bmatrix} x & y \end{bmatrix}$$

If we perform the matrix multiplication, we get a system of two equations and two unknowns.

$$0.1x + 0.7y = x$$
$$0.9x + 0.3y = y$$

If we write the equations in standard form, we get

$$-0.9x + 0.7y = 0$$
$$0.9x - 0.7y = 0$$

Since we know that all the row entries must equal 1, we get a third equation:

$$-0.9x + 0.7y = 0$$
$$0.9x - 0.7y = 0$$
$$x + \quad y = 1$$

From the third equation, we know that $x = 1 - y$. Substituting this expression into the second equation gives us

$$0.9(1 - y) - 0.7y = 0$$
$$0.9 - 0.9y - 0.7y = 0$$
$$0.9 - 1.6y = 0$$
$$0.9 = 1.6y$$
$$\frac{9}{16} = y$$

Substituting this value into the third equation gives

$$x + \frac{9}{16} = 1$$
$$x = 1 - \frac{9}{16} = \frac{7}{16}$$

Interactive Activity

Verify the solution to Example 3 by computing P^{16}, P^{32}, and P^{64} on your calculator and using the FRAC key.

So we get the steady-state vector $[\,x \quad y\,] = \left[\,\frac{7}{16} \quad \frac{9}{16}\,\right]$. Since the rows of a limiting matrix are made up of the elements of the steady-state vector $[\,x \quad y\,]$, we have the limiting matrix, P^n, that corresponds to the transition matrix $P = \begin{bmatrix} 0.1 & 0.9 \\ 0.7 & 0.3 \end{bmatrix}$ is

$$P^n \approx \begin{bmatrix} \frac{7}{16} & \frac{9}{16} \\ \frac{7}{16} & \frac{9}{16} \end{bmatrix} \qquad \text{for large } n$$

Applications of Steady-State Vectors

Two key pieces of information result from computing the limiting matrix P^n. First, the steady-state vector gives us the probability that an element will stay in a particular state in the long run. In Example 1 the limiting matrix was

$P^n \approx \begin{bmatrix} \frac{6}{11} & \frac{5}{11} \\ \frac{6}{11} & \frac{5}{11} \end{bmatrix}$. This tells us that the steady-state vector for the system is

$[\,x \quad y\,] = \left[\,\frac{6}{11} \quad \frac{5}{11}\,\right]$. This means that, in the long run, about $\frac{6}{11} \approx 55\%$ of the students will enroll in East School, while about $\frac{5}{11} \approx 45\%$ will attend West School.

The other piece of information is gathered from multiplying the initial state vector S_0 by the limiting matrix P^n to get $S_0 P^n \approx S_n$. This gives us the number of elements that will stay in a particular state in the long run. This is what we computed in part (b) of Example 1. Now let's try some of the techniques that we have learned in an application.

Example 4 Applying the Steady-State Vector

In Suffox County deregulation of gas and electricity utilities has had many customers reevaluating whether they will use electricity, natural gas, or heating oil as their primary heating energy source. Many are switching from one source to another based on their particular needs. A consumer group has surveyed the households of the county and has found that, during the past year:

- 70% of those who use electricity remained using electricity, while 20% have switched to natural gas and 10% to heating oil.
- 80% of those who use natural gas continued using it, while 10% have switched to electricity and 10% to heating oil.
- 60% of those who use heating oil continued to use heating oil, while 30% have switched to natural gas and 10% to electricity.

(a) Draw a state transition diagram and write the transition matrix P for the system.

(b) Let $S = [x \quad y \quad z]$ be a steady-state vector. Solve the matrix equation $SP = S$ to determine the values of the steady-state vector and interpret the results.

(c) Prior to deregulation, about 35,000 households used electricity, 45,000 households used natural gas, and about 15,000 used heating oil. About how many households will be using each of these energy sources in the long run?

Solution

(a) Here we have a system with three states:

State 1: Household uses electricity.

State 2: Household uses natural gas.

State 3: Household uses heating oil.

The state transition diagram for the system is shown in Figure 9.2.4. This gives us the transition matrix $P = \begin{bmatrix} 0.7 & 0.2 & 0.1 \\ 0.1 & 0.8 & 0.1 \\ 0.1 & 0.3 & 0.6 \end{bmatrix}$.

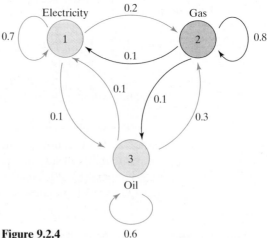

Figure 9.2.4

(b) To determine the steady-state vector, we need to get the values of $[x \quad y \quad z]$ by solving the matrix equation

$$[x \quad y \quad z] \begin{bmatrix} 0.7 & 0.2 & 0.1 \\ 0.1 & 0.8 & 0.1 \\ 0.1 & 0.3 & 0.6 \end{bmatrix} = [x \quad y \quad z]$$

Using matrix multiplication, we get the system of equations

$$0.7x + 0.1y + 0.1z = x$$
$$0.2x + 0.8y + 0.3z = y$$
$$0.1x + .01y + 0.6z = z$$

Knowing that the sum of the rows of the steady-state vector must equal 1 and writing the equations in standard form give us

$$-0.3x + 0.1y + 0.1z = 0$$
$$0.2x - 0.2y + 0.3z = 0$$
$$0.1x + 0.01y - 0.4z = 0$$
$$x + y + z = 1$$

Using the techniques of Chapter 2, we need to row reduce the augmented matrix

$$\begin{bmatrix} -0.3 & 0.1 & 0.1 & | & 0 \\ 0.2 & -0.2 & 0.3 & | & 0 \\ 0.1 & 0.1 & -0.4 & | & 0 \\ 1 & 1 & 1 & | & 1 \end{bmatrix}$$

We leave the details on using Gauss–Jordan to you in the Interactive Activity to the left. We urge you to refer to Chapter 2 to review the Gauss–Jordan method. The row reduced augmented matrix is

$$\begin{bmatrix} 1 & 0 & 0 & | & \frac{1}{4} \\ 0 & 1 & 0 & | & \frac{11}{20} \\ 0 & 0 & 1 & | & \frac{1}{5} \\ 0 & 0 & 0 & | & 0 \end{bmatrix}$$

This gives us the steady-state vector values $\begin{bmatrix} x & y & z \end{bmatrix} = \begin{bmatrix} \frac{1}{4} & \frac{11}{20} & \frac{1}{5} \end{bmatrix}$.

Interpret the Solution: This means that in the long run about $\frac{1}{4} = 25\%$ of the households will use electricity as their heating energy source, about $\frac{11}{20} = 55\%$ will use natural gas, and about $\frac{1}{5} = 20\%$ will use heating oil.

(c) From part (b), we know that the limiting matrix is

$$P^n \approx \begin{bmatrix} 0.25 & 0.55 & 0.20 \\ 0.25 & 0.55 & 0.20 \\ 0.25 & 0.55 & 0.20 \end{bmatrix} \quad \text{for large values of } n$$

From the given information, the initial state vector is $S_0 = \begin{bmatrix} 35{,}000 & 45{,}000 & 15{,}000 \end{bmatrix}$. Multiplying the initial state vector S_0 by the limiting matrix P^n, we get

$$S_n = S_0 P^n = \begin{bmatrix} 35{,}000 & 45{,}000 & 15{,}000 \end{bmatrix} \begin{bmatrix} 0.25 & 0.55 & 0.20 \\ 0.25 & 0.55 & 0.20 \\ 0.25 & 0.55 & 0.20 \end{bmatrix}$$

$$= \begin{bmatrix} 23{,}750 & 52{,}250 & 19{,}000 \end{bmatrix}$$

Interpret the Solution: This means that, in the long run, we would expect about 23,750 to use electricity as their heating energy source, while about 52,250 will use natural gas and about 19,000 will use heating oil. ∎

Technology Option

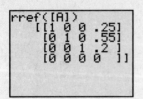

Figure 9.2.5

The rref(command on the calculator takes a matrix that has been entered in the matrix editor and transforms it into reduced row echelon form. For the matrix

$$\left[\begin{array}{ccc|c} -0.3 & 0.1 & 0.1 & 0 \\ 0.2 & -0.2 & 0.3 & 0 \\ 0.1 & 0.1 & -0.4 & 0 \\ 1 & 1 & 1 & 1 \end{array}\right]$$

shown in part (b) of Example 4, we get the row reduced echelon form shown in Figure 9.2.5.

For more information about the rref(command, consult the on-line graphing calculator manual at www.prenhall.com/armstrong

✓ **Checkpoint 2**

Now work Exercise 39.

SUMMARY

In this section, we saw that we could determine high-state Markov chains using the steady-state vector. We saw that we can use these special vectors to determine long-run trends.

- P is a **regular transition matrix** if some power of P has only positive elements.

- Solving $SP = S$ for S gives us the **steady-state vector**.
- If P is a regular transition matrix, then $S_0 P^n \approx S$ for large values of n.

SECTION 9.2 EXERCISES

In Exercises 1–10, compute P^4, P^{16}, P^{32}, and P^{64} on a calculator to determine if the given transition matrix P is a regular transition matrix. If the matrix is regular, approximate the limiting matrix P^n accurate to four decimal places.

1. $P = \begin{bmatrix} 0.2 & 0.8 \\ 0.9 & 0.1 \end{bmatrix}$ **2.** $P = \begin{bmatrix} 0.7 & 0.3 \\ 0.5 & 0.5 \end{bmatrix}$

3. $P = \begin{bmatrix} \frac{1}{3} & \frac{2}{3} \\ \frac{4}{5} & \frac{1}{5} \end{bmatrix}$ **4.** $P = \begin{bmatrix} \frac{1}{2} & \frac{1}{2} \\ \frac{1}{8} & \frac{7}{8} \end{bmatrix}$

5. $P = \begin{bmatrix} 1 & 0 \\ 0.65 & 0.35 \end{bmatrix}$ **6.** $P = \begin{bmatrix} 0.55 & 0.45 \\ 0 & 1 \end{bmatrix}$

✓ **7.** $P = \begin{bmatrix} 0.1 & 0.1 & 0.8 \\ 0.2 & 0.2 & 0.6 \\ 0.1 & 0.2 & 0.7 \end{bmatrix}$

8. $P = \begin{bmatrix} 0.5 & 0.2 & 0.3 \\ 0.1 & 0.4 & 0.5 \\ 0.2 & 0.7 & 0.1 \end{bmatrix}$

9. $P = \begin{bmatrix} 0 & 1 & 0 \\ 0.4 & 0.2 & 0.4 \\ 1 & 0 & 0 \end{bmatrix}$

10. $P = \begin{bmatrix} 0.3 & 0.5 & 0.2 \\ 1 & 0 & 0 \\ 0.5 & 0.4 & 0.1 \end{bmatrix}$

In Exercises 11–18, solve the matrix equation $SP = S$ for S to get the steady-state vector. Solve a system of equations using either the substitution or elimination method to determine the values for the steady-state vector $[x \quad y]$. Also, use the steady-state vector to determine the limiting matrix P^n.

11. $P = \begin{bmatrix} 0.2 & 0.8 \\ 0.1 & 0.9 \end{bmatrix}$ **12.** $P = \begin{bmatrix} 0.5 & 0.5 \\ 0.6 & 0.4 \end{bmatrix}$

13. $P = \begin{bmatrix} 0.1 & 0.9 \\ 0.7 & 0.3 \end{bmatrix}$ **14.** $P = \begin{bmatrix} 0.7 & 0.3 \\ 0.6 & 0.4 \end{bmatrix}$

15. $P = \begin{bmatrix} 0.8 & 0.2 \\ 1 & 0 \end{bmatrix}$ **16.** $P = \begin{bmatrix} 0 & 1 \\ 0.5 & 0.5 \end{bmatrix}$

17. $P = \begin{bmatrix} 0.25 & 0.75 \\ 0.85 & 0.15 \end{bmatrix}$ **18.** $P = \begin{bmatrix} 0.55 & 0.45 \\ 0.65 & 0.35 \end{bmatrix}$

In Exercises 19–26, solve the matrix equation $SP = S$ for S to get the steady-state vector. Use the Gauss–Jordan method to determine the values for the steady-state vector $[\,x \quad y \quad z\,]$. Also, use the steady-state vector to determine the limiting matrix P^n.

19. $P = \begin{bmatrix} 0 & 1 & 0 \\ 0.3 & 0.1 & 0.6 \\ 0.5 & 0.5 & 0 \end{bmatrix}$ **20.** $P = \begin{bmatrix} 0 & 1 & 0 \\ 0 & 0 & 1 \\ 0.5 & 0 & 0.5 \end{bmatrix}$

21. $P = \begin{bmatrix} 0 & \frac{1}{2} & \frac{1}{2} \\ \frac{1}{2} & \frac{1}{2} & 0 \\ 0 & \frac{1}{4} & \frac{3}{4} \end{bmatrix}$ **22.** $P = \begin{bmatrix} \frac{1}{4} & 0 & \frac{3}{4} \\ \frac{1}{2} & \frac{1}{4} & \frac{1}{4} \\ 0 & \frac{1}{4} & \frac{3}{4} \end{bmatrix}$

23. $P = \begin{bmatrix} \frac{1}{2} & \frac{1}{2} & 0 \\ \frac{1}{2} & \frac{1}{4} & \frac{1}{4} \\ \frac{1}{8} & \frac{3}{8} & \frac{1}{2} \end{bmatrix}$ **24.** $P = \begin{bmatrix} \frac{1}{10} & \frac{1}{10} & \frac{4}{5} \\ \frac{1}{2} & 0 & \frac{1}{2} \\ \frac{1}{2} & 0 & \frac{1}{2} \end{bmatrix}$

25. $P = \begin{bmatrix} 0.1 & 0.2 & 0.7 \\ 0.1 & 0.3 & 0.6 \\ 0.1 & 0.4 & 0.5 \end{bmatrix}$ **26.** $P = \begin{bmatrix} 0.09 & 0.05 & 0.86 \\ 0.28 & 0.16 & 0.56 \\ 0.43 & 0.45 & 0.12 \end{bmatrix}$

Applications

In Exercises 27–40, you can use your calculator to approximate the limiting matrix P^n by computing P^{64} or solve the matrix equation $SP = S$ for S using the methods in Exercises 11–26.

27. The Osborn–Wilcox Company manufactures paint pigments and tests the pigments that are made in a quality control lab. Records show that if a sample passed quality test then there is an 80% chance that the next sample will pass and a 20% chance that the next sample will fail. If the sample failed the test the last time it was taken, there is a 30% chance that the next sample will fail and a 70% chance that it will pass.

(a) Label the two states and write the transition matrix.

(b) Determine the limiting matrix P^n, for large values of n.

(c) Use the solution from part (b) to determine the probability that, in the long run, a test will pass.

28. Histronics Incorporated manufactures gauges for uses in aeronautics and calibrates each altimeter before it is shipped to see if the altimeter is within specifications. The calibration department's data show that if an altimeter is not in proper specification then there is a 60% chance that the next altimeter will not be within specification and a 40% chance that the next altimeter will be within specification. If an altimeter is within specification, there is a 75% chance that the next altimeter will be within specification and a 25% chance that the next altimeter will not be within specification.

(a) Label the two states and write the transition matrix.

(b) Determine the limiting matrix P^n, for large values of n.

(c) Use the solution from part (b) to determine the probability that, in the long run, an altimeter will be within specification.

29. The Rock Steady Insurance Company classifies their insured motorists as low risk or high risk based on whether they have had an accident in the last 3 years. Their research shows that 40% of the motorists who are classified as low risk remain classified as low risk 3 years later and 60% are reclassified as high risk. Fifty percent of the motorists who are classified as high risk remain classified as high risk 3 years later and the other 50% are reclassified as low risk.

(a) Draw a state transition diagram and write the transition matrix for the motorist's insurance classifications.

(b) Determine the limiting matrix P^n, for large values of n.

(c) The company currently insures 12,000 low-risk motorists and 8000 high-risk motorists. In the long run, about how many of the motorists will be classified as low risk and how many will be classified as high risk?

30. Metro Area Temps is an employment agency for temporary employees and classifies their temporary employees as either dependable or questionable based on whether they were cited as being late to work or absent without notice from work in their last job. They find that 90% of the employees classified as dependable will be classified the same way in their next job and 10% will be reclassified as questionable. Eighty-five percent of those who are classified as questionable will be classified the same way in their next job, while 15% will be reclassified as dependable.

(a) Draw a state transition diagram and write the transition matrix for the temporary employee classifications.

(b) Determine the limiting matrix P^n, for large values of n.

(c) The company currently has 900 employees classified as dependable and 400 classified as questionable. In the long run, about how many will be classified as dependable and how many will be classified as questionable?

31. (*Refer to Exercise 31 in Section 9.1.*) A market survey on sunglasses shows that 30% of the owners of Blade Shades did not like the product and switched to Optifilters the next summer, while the rest continued to purchase Blade Shades. Conversely, 40% of the owners of Optifilters did not like their sunglasses and switched to Blade Shades the next summer, while the rest continued to purchase Optifilters.

(a) Determine the limiting matrix P^n, for large values of n.

(b) Assuming that Blade Shades and Optifilters are the two primary manufacturers of activity-related sunglasses, how much of the market share will each company have in the long run?

32. (*Refer to Exercise 32 in Section 9.1.*) Following a new advertising campaign, a market analysis shows that 20% of the consumers who did not currently drink SportSquirt will purchase it the next time that they buy a sport drink, and 90% of the consumers who currently drink SportSquirt will purchase it the next time that they buy a sport drink.

(a) Determine the limiting matrix P^n, for large values of n.

(b) How much of the market share would SportSquirt expect to have in the long run?

33. At V. F. Rashford Business College, students are declared as either business majors or general studies majors. The Research and Development Office finds that 63% of the students who are declared business majors switch to general studies majors the next year, while 37% remain business majors. Eighty-two percent of those declared as general studies majors remain general studies majors the next year, while 18% switch to being business majors.

(a) Draw the state transition diagram and write the transition matrix for the student's declared majors.

(b) Determine the limiting matrix P^n, for large values of n, and interpret the result.

34. The city of Galston in Montburg County undergoes a renovation program to deter what they interpret as being urban flight. A year after the program is completed, a consulting firm for the city determines that 89% of the residents in Galston remain in the city, while 11% relocate to the suburbs in Montburg County. They also find that 6% of those who live in a Montburg County suburb relocate in Galston, while 94% remain in the suburbs.

(a) Draw the state transition diagram and write the transition matrix for the public's migration.

(b) Determine the limiting matrix P^n, for large values of n, and interpret the result.

35. (*Refer to Exercise 35 in Section 9.1.*) The town of Minsterton has an election for mayor every 2 years. Voting records show that if a Democrat is in office, there is a 60% chance that a Democrat will be elected in the next election and a 40% chance that a Republican will be elected. If a Republican is in office, there is a 65% chance that a Republican will be elected in the next election and a 35% chance that a Democrat will be elected.

(a) Determine the limiting matrix P^n, for large values of n.

(b) What percentage of Democrats would be expected to win the mayoral election in the long run?

36. (*Refer to Exercise 36 in Section 9.1.*) A regional health study shows that, from one year to the next, 75% of the people who smoked will continue to smoke, while 25% of the people who smoke will stop smoking the next year. Eight percent of the people who did not smoke will start smoking the next year, while 92% of those who did not smoke will continue to be nonsmokers.

(a) Determine the limiting matrix P^n, for large values of n.

(b) What percentage of smokers would the region expect to have in the long run?

37. (*Refer to Exercise 37 in Section 9.1.*) The Move-It-Safe Company rents moving trucks in the cities of Adamstown, Breston, and Carpersville. They have both round-trip and point-to-point rentals. Records show that if a truck is rented in Adamstown, there is a 70% chance that it will be returned to Adamstown, a 15% chance that it will be returned to Breston, and a 15% chance that it will be returned to Carpersville. If a truck is rented in Breston, there is a 65% chance that it will be returned to Breston, a 10% chance that it will be returned to Adamstown, and a 25% chance that it will be returned to Carpersville. If a truck is rented in Carpersville, there is an 85% chance that it will be returned to Carpersville, a 5% chance that it will be returned to Adamstown, and a 10% chance that it will be returned to Breston.

(a) Determine the limiting matrix P^n, for large values of n.

(b) In the long run, what percentage of trucks would be returned to Adamstown, Breston, and Carpersville, respectively?

38. (*Refer to Exercise 38 in Section 9.1.*) In the LaDona metropolitan area, commuters either drive alone, carpool, or take public transportation. A study completed by the transportation department show that 80% of those who drive alone intend to drive alone a year from now, while 15% intend to switch to carpooling and 5% intend to use public transportation. Ninety percent of those who carpool intend to continue carpooling the next year, while 5% intend to drive alone and 5% will begin to use public transportation. Sixty percent of those who use public transportation will continue to use it next year, while 30% intend to drive alone and 10% will begin to carpool.

(a) Determine the limiting matrix P^n, for large values of n.

(b) In the long run, what percentage of LaDona commuters would be expected to either drive alone, carpool, or take public transportation?

✓ **39.** (*Refer to Exercise 39 in Section 9.1.*) In a particular county in the northern part of the country, voters can declare themselves as members of the Republican Party, Democrat Party, or Green Party. A public opinion poll shows that 85% of the Republicans intend to continue with their party affiliation, while 13% intend to switch to the Democrat Party in the next election and 2% to the Green Party. Seventy percent of the Democrats intend to continue their party affiliation, while 10% intend to switch to the Republican Party in the next election and 20% to the Green Party. Ninety-five percent of the Green Party members plan to continue with their party affiliation, while 1% intend to switch to the Republican Party in the next election and 4% to the Democratic Party.

(a) Determine the limiting matrix P^n, for large values of n.

(b) In the long run, what percentage of voters in the county is expected to be members of the Green Party?

40. (*Refer to Exercise 40 in Section 9.1.*) The Armont Conservatory has been keeping records on their own Armont carnations and find that the flowering color of the next generation of plant is related to the current generation. If a carnation is red, there is a 60% chance that the next generation will be red, a 40% chance that the plant will be pink, and no chance of the plant being white. If a carnation is pink, there is a 60% chance that the next generation will be pink, 30% that it will be red, and a 10% chance that it will be white. If a carnation is white, there is a 50% chance that the next generation will be white, a 50% chance that it will be pink, and no chance that it will be red.

(a) Determine the limiting matrix P^n, for large values of n.

(b) In the long run, what percentage of Armont carnations are expected to be pink?

SECTION PROJECT

To generalize some of the properties of regular transition matrices, consider the following:

(a) Let the transition matrix for a Markov chain be

$$P = \begin{bmatrix} p & 1-p \\ q & 1-q \end{bmatrix}$$ for $0 < p < 1$ and $0 < q < 1$. Show that the limiting matrix, P^n, is

$$P^n = \begin{bmatrix} \frac{q}{1+q-p} & \frac{1-p}{1+q-p} \\ \frac{q}{1+q-p} & \frac{1-p}{1+q-p} \end{bmatrix}$$ for large values of n

(b) Determine the limiting matrix P^n for large values of n, if possible, for the transition matrix

$$P = \begin{bmatrix} p & 1-p \\ 1 & 0 \end{bmatrix}$$

for $0 < p < 1$. If the limiting matrix P^n cannot be determined, explain why not.

(c) Determine the limiting matrix P^n for large values of n, if possible, for the transition matrix

$$P = \begin{bmatrix} 0 & 1 & 0 \\ 0 & 0 & 1 \\ p & q & 1-p-q \end{bmatrix}$$

for $0 < p+q < 1$. If the limiting matrix P^n cannot be determined, explain why not.

Section 9.3   Absorbing Markov Chains

In Section 9.2 we saw that if a transition matrix associated with a Markov chain was a regular matrix then the matrix reached a steady state from which we could make long-term estimations. However, some Markov chains involve entering certain states, but never leaving these states. We call these **absorbing states**. In this section we will examine the transition matrix that is associated with these absorbing states and see how we can make long-term estimations of systems that have these states.

Absorbing Systems

To begin to see how these absorbing states can occur, let's examine the state transition diagram in Figure 9.3.1. The transition matrix associated with the system shown is

$$P = \begin{bmatrix} 0.2 & 0.8 & 0 \\ 0.1 & 0.6 & 0.3 \\ 0 & 0 & 1 \end{bmatrix}$$

State 3 is an example of an **absorbing state**, since the arrows in the diagram only enter state 3. There are no arrows leaving this state. In states 1 and 2 we see that there is at least one arrow leaving each of these states. Since states 1 and 2 are not absorbing, we call these states **nonabsorbing states**. We can see in the transition

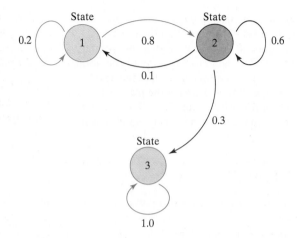

Figure 9.3.1

matrix for the system that the third row contains a 1 on the main diagonal and has 0s elsewhere. This illustrates how absorbing states can be identified in transition matrices.

■ **Absorbing State**

- For a system associated with a Markov chain, an **absorbing state** is a state with the characteristic that, once the state is reached, it is not possible for an element to leave the state. If a state is not an absorbing state, we call it a **nonabsorbing state**.
- A Markov chain is an **absorbing Markov chain** if
  1. The system associated with the Markov chain has at least one absorbing state.
  2. It is possible for an element to go from any nonabsorbing state to an absorbing state in a finite number of time steps.

In an absorbing Markov chain, nonabsorbing states can also be called **transient**.

▶ **Note:** Ways to identify absorbing states in a system include:
- In the state transition diagram, no arrows leave the absorbing state.
- In the transition matrix, a row has a 1 on the main diagonal and 0s elsewhere. In this case the matrix row number and the absorbing state number are the same.

Notice in the second part of the preceding definition that simply having absorbing states in a system does not guarantee that we have an absorbing Markov chain. To see how this can happen, let's examine the state transition diagram in Figure 9.3.2. We see that state 3 is an absorbing state. Yet it is not possible to follow the arrows from state 1 or state 2 to absorbing state 3. To have an absorbing Markov chain, an element in the system must be able to go from each nonabsorbing state to the absorbing state in a finite number of time steps. This means that the system depicted in the diagram in Figure 9.3.2 is not associated with an absorbing Markov chain.

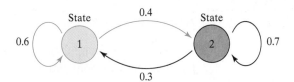

Figure 9.3.2

Example 1 Identifying Absorbing Markov Chains

For each of the following, identify the absorbing states and determine if the associated system represents an absorbing Markov chain.

(a)

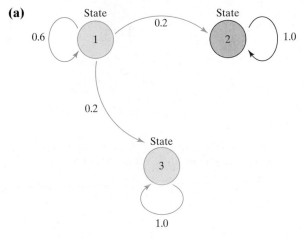

(b) $P = \begin{bmatrix} 0.2 & 0.8 & 0 \\ 0.3 & 0.7 & 0 \\ 0 & 0 & 1 \end{bmatrix}$

Solution

(a) Since states 2 and 3 have arrows entering, but not leaving, the states, we conclude that states 2 and 3 are absorbing states. Since it is possible to go from state 1 (a nonabsorbing state) to state 2 or 3 (the absorbing states), the associated system represents an absorbing Markov chain.

(b) Since the final entry in the third row of the transition matrix is 1 and all the other entries in the row are 0s, we see that state 3 is an absorbing state. Notice that all the other entries in the third column in the transition matrix are zero. That is, $p_{1,3} = 0$ and $p_{2,3} = 0$. This means that it is not possible to go from either state 1 or from state 2 to state 3. Thus, the associated system does not represent an absorbing Markov chain. ∎

Now work Exercise 13.

To be able to make long-term estimates using absorbing Markov chains, we use the following convention:

In an absorbing Markov system, we will always number the absorbing states last.

We follow this convention because the transition matrix for an absorbing system has a particular structure. To see what this structure looks like, consider the state transition diagram in Figure 9.3.3.

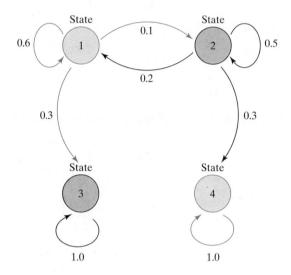

Figure 9.3.3

First, notice that two states, states 3 and 4, are absorbing states. The transition matrix associated with this system is

$$P = \begin{bmatrix} 0.6 & 0.1 & 0.3 & 0 \\ 0.2 & 0.5 & 0 & 0.3 \\ 0 & 0 & 1 & 0 \\ 0 & 0 & 0 & 1 \end{bmatrix}$$

Notice that there is a square 2×2 identity matrix in the lower-right corner resulting from the two absorbing states. Let's now **partition** the transition matrix so that we can see the identity matrix more clearly.

$$P = \left[\begin{array}{cc|cc} 0.6 & 0.1 & 0.3 & 0 \\ 0.2 & 0.5 & 0 & 0.3 \\ \hline 0 & 0 & 1 & 0 \\ 0 & 0 & 0 & 1 \end{array} \right]$$

In addition to the 2×2 identity matrix in the lower-right corner, there is a zero matrix in the lower-left corner. Generally, if a system has n states of which k of them are absorbing states, then the transition matrix has the partitioned form

$$P = \left[\begin{array}{c|c} Q & R \\ \hline 0 & I \end{array} \right]$$

where I is a k by k identity matrix, 0 is a k by $(n-k)$ zero matrix, and Q is an $(n-k)$ by $(n-k)$ matrix. For the transition matrix associated with Figure 9.3.3, the four partitioned matrices are

$$Q = \begin{bmatrix} 0.6 & 0.1 \\ 0.2 & 0.5 \end{bmatrix} \qquad R = \begin{bmatrix} 0.3 & 0 \\ 0 & 0.3 \end{bmatrix}$$

$$0 = \begin{bmatrix} 0 & 0 \\ 0 & 0 \end{bmatrix} \qquad I = \begin{bmatrix} 1 & 0 \\ 0 & 1 \end{bmatrix}$$

Since matrix Q is associated with states 1 and 2, this matrix gives the probabilities of going from one nonabsorbing state to another nonabsorbing state. But we know that the probability exists for us to go from a nonabsorbing to an absorbing state. The number of time steps it is expected to take to go from a nonabsorbing state to an absorbing state is given by the **fundamental matrix**.

▨ Fundamental Matrix

Let P be a transition matrix associated with an absorbing Markov chain, where P has the partitioned form

$$P = \left[\begin{array}{c|c} Q & R \\ \hline 0 & I \end{array} \right]$$

Then the **fundamental matrix** F gives the expected number of time steps that the Markov chain will spend in each nonabsorbing state and is given by

$$F = (I - Q)^{-1}$$

Note that the identity matrix I that is in matrix P and matrix F are not necessarily the same. That is, they may not have the same dimension. In matrix F, I must have the same dimension as Q.

▶ **Note:** We can make the following two interpretations with regard to the fundamental matrix.

1. The i,jth element of F, that is $f_{i,j}$, gives the expected number of time steps that an element that begins in the ith nonabsorbing state will be in the jth nonabsorbing state before it reaches an absorbing state.

2. The sum of the elements in the ith row of the fundamental matrix gives the expected number of time steps that an element that begins in the ith state will take to reach an absorbing state.

Example 2   Computing the Fundamental Matrix

The proline Communications Cable Company has an accounting system in which each customer is given 3 months to pay his or her current bill before cable service is discontinued. The cable company classifies its accounts as current (less than 1 month due), 1 month overdue, 2 months overdue, 3 months overdue, and service discontinued. The company's past records show the following:

- 75% of the accounts are paid within 1 month.
- 60% of the accounts that are 1 month overdue are paid at the end of the month.

- 40% of the accounts that are 2 months overdue are paid at the end of the month.
- 80% of the accounts that are 3 months overdue have service discontinued, while the remaining 20% pay their bill at the end of the month.

(a) Draw a state transition diagram and write a transition matrix P for the system.

(b) Compute the fundamental matrix F and interpret the element in the first row, fourth column (that is, $f_{1,4}$).

Solution

(a) Here we have six states, with two being absorbing states:

   State 1: Account is less than 1 month due.
   State 2: Account is 1 month overdue.
   State 3: Account is 2 months overdue.
   State 4: Account is 3 months overdue.
   State 5: Account is paid.
   State 6: Service is discontinued.

The state transition diagram for the accounts is shown in Figure 9.3.4.

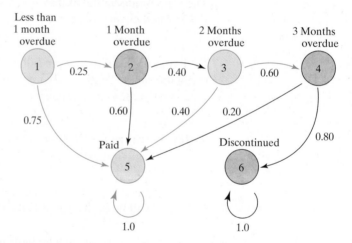

Figure 9.3.4

The corresponding partitioned transition matrix for the system is

$$P = \begin{bmatrix} 0 & 0.25 & 0 & 0 & 0.75 & 0 \\ 0 & 0 & 0.40 & 0 & 0.60 & 0 \\ 0 & 0 & 0 & 0.60 & 0.40 & 0 \\ 0 & 0 & 0 & 0 & 0.20 & 0.80 \\ 0 & 0 & 0 & 0 & 1 & 0 \\ 0 & 0 & 0 & 0 & 0 & 1 \end{bmatrix}$$

So we see that

$$Q = \begin{bmatrix} 0 & 0.25 & 0 & 0 \\ 0 & 0 & 0.40 & 0 \\ 0 & 0 & 0 & 0.60 \\ 0 & 0 & 0 & 0 \end{bmatrix}$$

(b) Using the definition of the fundamental matrix, we get

$$
F = (I - Q)^{-1} = \left(\begin{bmatrix} 1 & 0 & 0 & 0 \\ 0 & 1 & 0 & 0 \\ 0 & 0 & 1 & 0 \\ 0 & 0 & 0 & 1 \end{bmatrix} - \begin{bmatrix} 0 & 0.25 & 0 & 0 \\ 0 & 0 & 0.40 & 0 \\ 0 & 0 & 0 & 0.60 \\ 0 & 0 & 0 & 0 \end{bmatrix} \right)^{-1}
$$

$$
= \begin{bmatrix} 1 & -0.25 & 0 & 0 \\ 0 & 1 & -0.40 & 0 \\ 0 & 0 & 1 & -0.6 \\ 0 & 0 & 0 & 1 \end{bmatrix}^{-1} = \begin{bmatrix} 1 & 0.25 & 0.1 & 0.06 \\ 0 & 1 & 0.40 & 0.24 \\ 0 & 0 & 1 & 0.60 \\ 0 & 0 & 0 & 1 \end{bmatrix}
$$

Interpret the Solution: We see that the element in the first row, fourth column is $f_{1,4} = 0.06$. This means that an account that is less than 1 month due (state 1) will spend about 0.06 time step as an account that is 3 months overdue (state 4) before it reaches an absorbing state (state 5 or state 6). ∎

Technology Option

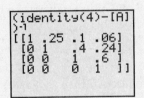

Figure 9.3.5

Graphing calculators with matrix menus can readily compute the inverse of a matrix that has been entered in the matrix editor. If we enter matrix Q,

$$
\begin{bmatrix} 0 & 0.25 & 0 & 0 \\ 0 & 0 & 0.40 & 0 \\ 0 & 0 & 0 & 0.60 \\ 0 & 0 & 0 & 0 \end{bmatrix}
$$

in our calculator as matrix A, we can compute $(I - Q)^{-1}$ by entering $(I - A)^{-1}$ on the calculator. See Figure 9.3.5.

For more information about using your calculator to perform these types of operations, consult the on-line graphing calculator manual at www.prenhall.com/armstrong

Interactive Activity

Verify the matrix inverse to get the fundamental matrix in Example 2 by using the Gauss–Jordan method to write the augmented matrix

$$
\left[\begin{array}{cccc|cccc} 1 & -0.25 & 0 & 0 & 1 & 0 & 0 & 0 \\ 0 & 1 & -0.40 & 0 & 0 & 1 & 0 & 0 \\ 0 & 0 & 1 & -0.6 & 0 & 0 & 1 & 0 \\ 0 & 0 & 0 & 1 & 0 & 0 & 0 & 1 \end{array} \right]
$$

in reduced row echelon form.

In our partitioning of the transition matrix P, it appears that we have ignored the matrix R. However, this matrix can be useful, too, since the matrix product FR gives the probability that, if an element in the system was originally in a nonabsorbing state, it will go to a specific absorbing state. In Example 2 we found matrices F and R to be

$$F = \begin{bmatrix} 1 & 0.25 & 0.1 & 0.06 \\ 0 & 1 & 0.40 & 0.24 \\ 0 & 0 & 1 & 0.60 \\ 0 & 0 & 0 & 1 \end{bmatrix} \quad \text{and} \quad R = \begin{bmatrix} 0.75 & 0 \\ 0.60 & 0 \\ 0.40 & 0 \\ 0.20 & 0.80 \end{bmatrix}$$

The matrix product FR is

$$FR = \begin{bmatrix} 1 & 0.25 & 0.1 & 0.06 \\ 0 & 1 & 0.40 & 0.24 \\ 0 & 0 & 1 & 0.60 \\ 0 & 0 & 0 & 1 \end{bmatrix} \begin{bmatrix} 0.75 & 0 \\ 0.60 & 0 \\ 0.40 & 0 \\ 0.20 & 0.80 \end{bmatrix} = \begin{bmatrix} 0.952 & 0.048 \\ 0.808 & 0.192 \\ 0.520 & 0.480 \\ 0.200 & 0.800 \end{bmatrix}$$

We can see by inspecting row 2 of the matrix product FR, for example, that if an account is 1 month overdue there is an 80.8% chance of the bill being paid and a 19.2% chance that the service will be discontinued. Now let's see what kind of information the sum of the entries of the fundamental matrix can provide.

Example 3 Interpreting the Row Sums of the Fundamental Matrix

The fundamental matrix is often used to analyze enrollment trends in colleges. To illustrate how this is done, consider the following scenario. Bearstate Junior College has found by studying historical data that any one of its students can be classified in one of the following four states:

State 1: Student is a freshman.

State 2: Student is a sophomore.

State 3: Student transfers to a 4-year school or graduates with an associate's degree.

State 4: Student drops out.

The state transition diagram for the college is shown in Figure 9.3.6.

(a) Determine the fundamental matrix F for the system.

(b) Compute the sum of the entries of the rows of the fundamental matrix and interpret.

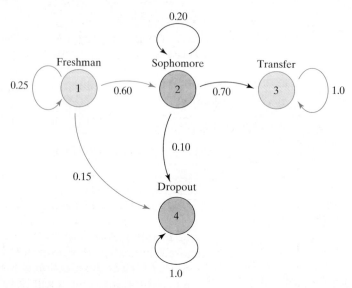

Figure 9.3.6

Solution

(a) From the state transition diagram in Figure 9.3.6, we see that the partitioned transition matrix for the system is

$$P = \begin{bmatrix} 0.25 & 0.60 & 0 & 0.15 \\ 0 & 0.20 & 0.70 & 0.10 \\ 0 & 0 & 1 & 0 \\ 0 & 0 & 0 & 1 \end{bmatrix} = \begin{bmatrix} Q & R \\ 0 & I \end{bmatrix}$$

Knowing that $Q = \begin{bmatrix} 0.25 & 0.60 \\ 0 & 0.20 \end{bmatrix}$, we get the fundamental matrix:

$$F = (I - Q)^{-1} = \left(\begin{bmatrix} 1 & 0 \\ 0 & 1 \end{bmatrix} - \begin{bmatrix} 0.25 & 0.60 \\ 0 & 0.20 \end{bmatrix} \right)^{-1}$$

$$= \begin{bmatrix} 0.75 & -0.60 \\ 0 & 0.80 \end{bmatrix}^{-1} = \begin{bmatrix} \frac{4}{3} & 1 \\ 0 & \frac{5}{4} \end{bmatrix}$$

(b) The sum of the entries in row 1 is $\frac{4}{3} + 1 = 2\frac{1}{3}$, and the sum of the entries in row 2 is $0 + \frac{5}{4} = 1\frac{1}{4}$. This means that we can expect a freshman to be at Bearstate Junior College for about $2\frac{1}{3}$ years before dropping out, transferring, or getting an associate's degree, and we can expect a sophomore to be at the college about $1\frac{1}{4}$ years before dropping out, transferring, or getting an associate's degree.   ∎

SUMMARY

- An **absorbing state** is a state with the characteristic that, once the state is reached, it is not possible for an element to leave the state.
- If a state is not an absorbing state, we call it a **nonabsorbing state**.
- A Markov chain is an **absorbing Markov chain** if the system associated with the Markov chain has at least one absorbing state and it is possible for an element to go from any nonabsorbing state to an absorbing state in a finite number of time steps.

- For the transition matrix P in the partitioned form $P = \begin{bmatrix} Q & R \\ 0 & I \end{bmatrix}$, the **fundamental matrix** $F = (I - Q)^{-1}$ gives the expected number of time steps that the Markov chain will spend in each nonabsorbing state.
- The matrix product FR gives the probability that, if an element in the system was originally in a nonabsorbing state, it will go to a specific absorbing state.

SECTION 9.3 EXERCISES

In Exercises 1–10, a transition matrix is given. Identify the absorbing states, if any, and draw a state transition diagram corresponding to the matrix.

1. $P = \begin{bmatrix} \frac{1}{2} & \frac{1}{2} \\ 1 & 0 \end{bmatrix}$ **2.** $P = \begin{bmatrix} 0 & 1 \\ \frac{1}{2} & \frac{1}{2} \end{bmatrix}$

3. $P = \begin{bmatrix} 0.3 & 0.7 & 0 \\ 0 & 1 & 0 \\ 0 & 0 & 1 \end{bmatrix}$

4. $P = \begin{bmatrix} 0 & 1 & 0 \\ 0.3 & 0.5 & 0.2 \\ 0 & 0 & 1 \end{bmatrix}$

5. $P = \begin{bmatrix} 0.5 & 0.5 & 0 \\ 0 & 0.5 & 0.5 \\ 0 & 0 & 1 \end{bmatrix}$

6. $P = \begin{bmatrix} 0.5 & 0 & 0.5 \\ 0.5 & 0.5 & 0 \\ 0 & 0.6 & 0.4 \end{bmatrix}$

7. $P = \begin{bmatrix} 1 & 0 & 0 \\ 0.4 & 0.4 & 0.2 \\ 0 & 0 & 1 \end{bmatrix}$

8. $P = \begin{bmatrix} 0 & 0 & 1 \\ 1 & 0 & 0 \\ 0 & 1 & 0 \end{bmatrix}$

9. $P = \begin{bmatrix} 0.9 & 0 & 0.1 & 0 \\ 0.3 & 0.3 & 0.4 & 0 \\ 0.5 & 0 & 0 & 0.5 \\ 0 & 0 & 0 & 1 \end{bmatrix}$

10. $P = \begin{bmatrix} 0.25 & 0.25 & 0.3 & 0.2 \\ 0.25 & 0.25 & 0.25 & 0.25 \\ 0 & 0 & 1 & 0 \\ 0 & 0 & 0 & 1 \end{bmatrix}$

In Exercises 11–20, a state transition diagram is given. Identify the absorbing states and determine if the diagram represents an absorbing Markov chain.

11.

12.

✓ **13.**

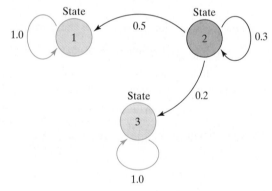

14.

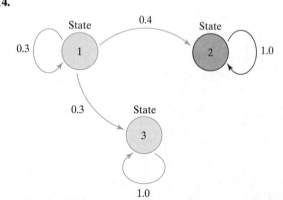

15.

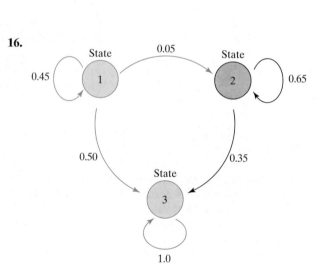

16.

17.

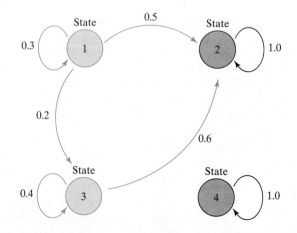

18.

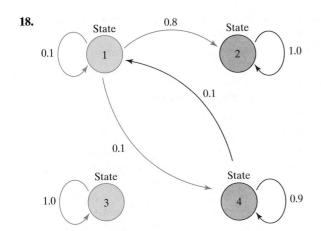

19.

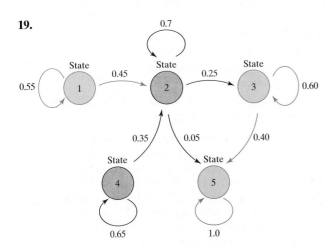

20.

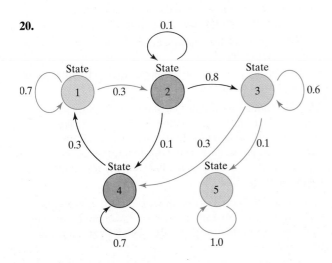

In Exercises 21–36:

(a) Determine the fundamental matrix F, if possible.

(b) Determine FR, if possible.

21. $P = \begin{bmatrix} 0.5 & 0 & 0.5 \\ 0.75 & 0.25 & 0 \\ 0 & 0 & 1 \end{bmatrix}$ **22.** $P = \begin{bmatrix} 0.3 & 0 & 0.7 \\ 0.2 & 0.6 & 0.2 \\ 0 & 0 & 1 \end{bmatrix}$

23. $P = \begin{bmatrix} \frac{1}{2} & 0 & \frac{1}{2} \\ \frac{3}{4} & \frac{1}{4} & 0 \\ 0 & 0 & 1 \end{bmatrix}$ **24.** $P = \begin{bmatrix} \frac{1}{4} & 0 & \frac{3}{4} \\ 0 & \frac{3}{4} & \frac{1}{4} \\ 0 & 0 & 1 \end{bmatrix}$

25. $P = \begin{bmatrix} 0.2 & 0.7 & 0.1 \\ 0.5 & 0 & 0.5 \\ 0 & 0 & 1 \end{bmatrix}$ **26.** $P = \begin{bmatrix} 0.1 & 0.6 & 0.3 \\ 0.9 & 0 & 0.1 \\ 0 & 0 & 1 \end{bmatrix}$

27. $P = \begin{bmatrix} 0.9 & 0.1 & 0 \\ 0.1 & 0.9 & 0 \\ 0 & 0 & 1 \end{bmatrix}$ **28.** $P = \begin{bmatrix} 0.4 & 0.6 & 0 \\ 0.6 & 0.4 & 0 \\ 0 & 0 & 1 \end{bmatrix}$

29. $P = \begin{bmatrix} 0.3 & 0 & 0.7 \\ 1 & 0 & 0 \\ 0 & 0 & 1 \end{bmatrix}$ **30.** $P = \begin{bmatrix} 0.4 & 0 & 0.6 \\ 1 & 0 & 0 \\ 0 & 0 & 1 \end{bmatrix}$

31. $P = \begin{bmatrix} \frac{1}{2} & \frac{1}{2} & 0 & 0 \\ \frac{3}{4} & \frac{1}{4} & 0 & 0 \\ 0 & 0 & 1 & 0 \\ 0 & 0 & 0 & 1 \end{bmatrix}$ **32.** $P = \begin{bmatrix} \frac{2}{3} & \frac{1}{3} & 0 & 0 \\ \frac{1}{5} & \frac{4}{5} & 0 & 0 \\ 0 & 0 & 1 & 0 \\ 0 & 0 & 0 & 1 \end{bmatrix}$

33. $P = \begin{bmatrix} 0.2 & 0.2 & 0.2 & 0.4 \\ 0.5 & 0.2 & 0 & 0.3 \\ 0 & 0 & 1 & 0 \\ 0 & 0 & 0 & 1 \end{bmatrix}$

34. $P = \begin{bmatrix} 0 & 0.5 & 0.1 & 0.4 \\ 0.3 & 0.3 & 0.3 & 0.1 \\ 0 & 0 & 1 & 0 \\ 0 & 0 & 0 & 1 \end{bmatrix}$

35. $P = \begin{bmatrix} 0.15 & 0.25 & 0.10 & 0.3 & 0.2 \\ 0.10 & 0.10 & 0.25 & 0.25 & 0.3 \\ 0.20 & 0.25 & 0.15 & 0.2 & 0.2 \\ 0 & 0 & 0 & 1 & 0 \\ 0 & 0 & 0 & 0 & 1 \end{bmatrix}$

36. $P = \begin{bmatrix} 0.2 & 0.3 & 0.5 & 0 & 0 \\ 0.1 & 0.4 & 0.2 & 0.3 & 0 \\ 0.3 & 0.3 & 0.3 & 0 & 0.1 \\ 0 & 0 & 0 & 1 & 0 \\ 0 & 0 & 0 & 0 & 1 \end{bmatrix}$

Applications

37. The zoning commission in the town of Gresden zones parcels of land as public, commercial, or residential. Over the last 5 years, records show that 75% of the land zoned as residential remained that way, while the remaining 25% was rezoned as commercial. Seventy-five percent of the land zoned as commercial was rezoned that way, with 15% being rezoned as residential and 10% as public. All the public land zoned as public remained zoned as public.

(a) Identify the three states and draw a state transition diagram for the zoning records.

(b) Write the transition matrix, and determine the fundamental matrix F.

(c) Interpret $f_{1,2}$.

38. The CorpNet software company gives each of its employees the title of programmer or project manager. An annual evaluation shows that, during any given year, 70% of the programmers are retitled as programmers, 20% are retitled as project managers, and 10% are asked to leave the company. Ninety percent of the project managers are retitled as project managers, and 10% are asked to leave the company.

(a) Identify the three states and draw a state transition diagram for the zoning records.

(b) Write the transition matrix, and determine the fundamental matrix F.

(c) Interpret $f_{1,2}$.

39. At DeRosa State Community College, records show that 15% of the freshmen at the school will remain freshmen after 1 year, 65% will become sophomores, and 20% will leave the school. During the same year, 25% of the sophomores will remain sophomores and 75% will leave the school (either by dropping out or graduating). Assume that none of the students who leave the school re-enroll.

(a) Identify the three states and draw a state transition diagram for the student enrollment.

(b) Write the transition matrix, and determine the fundamental matrix F.

(c) Interpret $f_{1,2}$.

40. The OilFirst Company employs geologists to conduct their fieldwork. Before being given their field assignment, however, the geologists first must pass a rigorous classroom program. Company records show that, during any given year, 45% of those enrolled in the classroom training will be promoted to field assignments, while the remaining 55% leave the company. Eighty percent of those involved in field assignments continue to do that work, while 20% of the field employees leave the company.

(a) Identify the three states and draw a state transition diagram for the company's training.

(b) Write the transition matrix, and determine the fundamental matrix F.

41. Drive-On Auto Sales gives short-term loans and classifies its clients as being in good standing if they make their payments by the due date or as probationary if they periodically miss payments. The accounting office records show that, during any given month, 75% of the clients in good standing remain in good standing, 20% pay off their loans, and 5% have their cars repossessed. Forty percent of the probationary clients remain probationary, 30% are reclassified as being in good standing, 10% pay off their loans, and 20% have their cars repossessed.

(a) Identify the four states and draw a state transition diagram for the loan classifications.

(b) Write the transition matrix, and determine the fundamental matrix F.

(c) Compute the sum of the entries in each row of the fundamental matrix F and interpret.

(d) Compute FR and interpret the entries in the first row of this matrix product.

42. During any given week, Piloton Metropolitan Library has books that are soon to be due or overdue. Thirty percent of the books that are due remain due, 50% are returned, and 20% are relabeled as overdue. Sixty-five percent of the overdue books remain overdue, while 20% of the books are returned and 15% of the patrons who borrow books have their library cards voided.

(a) Identify the four states and draw a state transition diagram for the library book labelings.

(b) Write the transition matrix, and determine the fundamental matrix F.

(c) Compute the sum of the entries in each row of the fundamental matrix F and interpret.

43. At the Public Water Works Laboratories, employees can have the titles of trainee, technician, or supervisor. During any given year, 20% of the trainees remain trainees, 60% are promoted to technician, and 20% are dismissed or quit. Fifty-five percent of the technicians remain technicians, 15% are promoted to supervisor, and 30% are dismissed or quit. Supervisor is the highest title that an employee can have at the lab.

(a) Identify the four states and draw a state transition diagram for the employee classifications.

(b) Write the transition matrix, and determine the fundamental matrix F.

(c) How many years can an employee expect to hold one of the two lowest positions?

44. Many homeowners who have their mortgages at the Clipston Financial Trust refinance their loans to take advantage of falling interest rates. The banks offers variable-rate loans, 30-year fixed-rate mortgages, and 15-year fixed-rate mortgages. Their records reveal the following facts:

• 55% of the borrowers with variable-rate mortgages remain with variable-rate mortgages, 35% refinance to get a 30-year fixed-rate mortgage, 5% pay off their loan, and 5% foreclose.

• 54% of the borrowers with 30-year fixed-rate loans remain with those loans, 15% refinance to get a variable-rate mortgage, 25% refinance to get a 15-year fixed-rate loan, 5% pay off their loans, and 1% foreclose.

• 75% of those with 15-year fixed-rate mortgages remain with those loans, while 20% refinance to have a variable-rate mortgage, 4% pay off their loans, and 1% foreclose.

(a) Identify the five states and draw a state transition diagram for the bank loans.

(b) Write the transition matrix, and determine the fundamental matrix F.

(c) In the long run, what percentage of borrowers who now have variable-rate loans are expected to have variable-rate loans, 30-year fixed-rate mortgages, or 15-year fixed-rate mortgages?

45. At Edwards and Smith University, past experience shows that students move through the undergraduate ranks as follows:

• 65% of the freshmen will become sophomores after 1 year, 10% will remain freshmen, and 25% will withdraw from the university.

• 65% of the sophomores will become juniors after 1 year, 15% will remain sophomores, and 20% will withdraw from the university.

• 75% of the juniors will become seniors after 1 year, 5% will remain juniors, and 20% will withdraw from the university.

• 80% of the seniors will graduate from the university, 15% will remain as seniors, and 5% will withdraw from the university.

(a) Identify the six states and draw a state transition diagram for the student ranking.

(b) Write the transition matrix, and determine the fundamental matrix F.

(c) Determine the number of years that a freshman, sophomore, junior, and senior can expect to be at the university.

SECTION PROJECT

Let's explore some of the more general properties of absorbing Markov chains.

(a) Consider an absorbing Markov system whose transition matrix is given by

$$P = \begin{bmatrix} p & 1-p \\ 0 & 1 \end{bmatrix}$$

where $0 < p < 1$. Compute the fundamental matrix F and show that, if the Markov chain starts in state 1, then the expected number of times that it is in state 1 before being absorbed is given by $\frac{1}{p}$.

(b) Consider an absorbing Markov system whose transition matrix is given by

$$P = \begin{bmatrix} p & q & 1-p-q \\ 0 & 1 & 0 \\ 0 & 0 & 1 \end{bmatrix}$$

where $0 < p+q < 1$. Compute the fundamental matrix F. If the Markov chain starts in state 1, determine the expected number of time steps needed before an element is absorbed. Determine the probability that it is absorbed in state 3.

Why We Learned It

In this chapter we offered a brief introduction to Markov chains. We analyzed many diverse applications of Markov chains. For example, in the area of education some applications included school officials in a local school district analyzing vouchers, a community college analyzing the length of time that a student spends as a freshman or a sophomore, and a university analyzing the switching of majors by students. Long-term projections can be made by schools officials once a steady-state vector and/or a limiting matrix is determined. We also saw that zoning commissions can analyze how land is being zoned and project the long-term partition of land through a Markov process. Political parties can even use Markov chains to analyze voter registration and project the long-term consequences of voters changing party affiliations. Finally, we saw how botanists could use the Markov process to analyze a long-term projection of floral types. These are just a few of the applications of Markov chains. A more thorough study of Markov chains would yield even more areas of application for this branch of mathematics.

CHAPTER REVIEW EXERCISES

In Exercises 1–3, determine if the matrices are transition matrices.

1. $\begin{bmatrix} 0.4 & 0.6 \\ 0.1 & 0.9 \end{bmatrix}$
2. $\begin{bmatrix} 0.3 & -0.2 & 0.9 \\ 0 & 0.8 & 0.2 \\ 0.1 & 0.3 & 0.6 \end{bmatrix}$

3. $\begin{bmatrix} 0.32 & 0.41 & 0.27 \\ 0.15 & 0.28 & 0.57 \\ 0.05 & 0.78 & 0.17 \end{bmatrix}$

In Exercises 4–6, write the transition matrix for the state transition diagrams.

4.

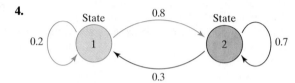

5.

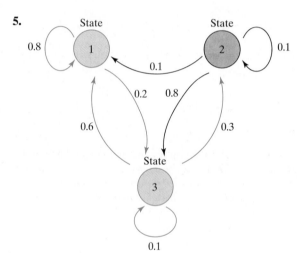

6.

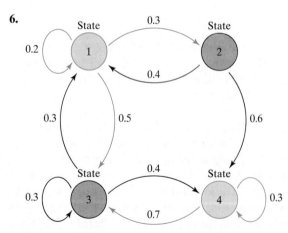

In Exercises 7–9, draw a state transition diagram for the given transition matrix.

7. $\begin{bmatrix} 0.1 & 0.9 \\ 0.2 & 0.8 \end{bmatrix}$
8. $\begin{bmatrix} 0.1 & 0.2 & 0.7 \\ 0.3 & 0.2 & 0.5 \\ 0.2 & 0.2 & 0.6 \end{bmatrix}$

9. $\begin{bmatrix} 0.34 & 0 & 0.28 & 0.38 \\ 0.10 & 0.34 & 0 & 0.56 \\ 0.66 & 0 & 0.19 & 0.15 \\ 0 & 0 & 0.25 & 0.75 \end{bmatrix}$

In Exercises 10–12, complete the following:

(a) Identify the states.

(b) Draw the state transition diagram.

(c) Write the transition matrix P.

10. The town of Plymouth has two elite soccer teams, Metro and Juventus. The director of parks and recreation has determined that there is a 30% chance that a player on Metro will play for Juventus next summer, while the rest continue to play for Metro. Conversely, 15% of the Juventus players will play for Metro next summer, while the rest continue to play for Juventus.

11. Goffstown has a community bike program in which a resident may pick up a bike at one of three locations and ride the bike anywhere provided that the bike is returned to one of the pickup locations the next morning. The pickup locations are the library, the YMCA, and the grocery store. The director of the program has determined that if a bike is picked up at the library there is a 10% chance that it will be returned to the library, a 25% chance that it will be returned to the YMCA, and a 65% chance that it will be returned to the grocery store. If a bike is picked up at the YMCA, there is an 80% chance that it will be returned to the YMCA, a 10% chance that it will be returned to the library, and a 10% chance that it will be returned to the grocery store. If a bike is picked up at the grocery store, there is a 5% chance that it will be returned to the grocery store, a 75% chance that it will be returned to the library, and a 20% chance that it will be returned to the YMCA.

12. In a particular county in New England, voters can declare themselves as Republican, Democrat, or Independent. A public opinion poll shows that 80% of the Republicans intend to continue with their party affiliation, while 12% intend to switch to Democrat in the next election and 8% intend to switch to Independent. Ninety percent of Democrats intend to continue with their party affiliation, while 6% intend to switch to Republican in the next election and 4% intend to switch to Independent. Ninety-six percent of Independents intend to continue with their party affiliation, while 1% intends to switch to Republican in the next election and 3% intend to switch to Democrat.

13. Refer to Exercise 11.

(a) On Monday, 20 bikes were picked up from the library, 24 bikes were picked up from the YMCA, and 18 bikes were picked up from the grocery store. Write the corresponding initial state vector S_0.

(b) Determine S_2 and interpret.

14. Refer to Exercise 12.

(a) Interpret each value of the probability distribution vector $S_0 = [0.58 \quad 0.40 \quad 0.02]$.

(b) Determine S_4 and interpret.

In Exercises 15–17, compute P^4, P^{16}, P^{32}, and P^{64} on a calculator to determine if the given transition matrix P is a regular transition matrix. If the matrix is regular, approximate the limiting matrix P^n accurate to four decimal places.

15. $P = \begin{bmatrix} 0.3 & 0.7 \\ 0.1 & 0.9 \end{bmatrix}$ **16.** $P = \begin{bmatrix} 0.65 & 0.35 \\ 0 & 1 \end{bmatrix}$

17. $P = \begin{bmatrix} 0.2 & 0.2 & 0.6 \\ 0.1 & 0.3 & 0.6 \\ 0.5 & 0.2 & 0.3 \end{bmatrix}$

In Exercises 18–20, solve the matrix equation $SP = S$ for S to determine the steady-state vector. Solve a system of equations using either the substitution or elimination method to determine the values for the steady state vector $[x \quad y]$. Also, use the steady-state vector to determine the limiting matrix P^n.

18. $P = \begin{bmatrix} 0.4 & 0.6 \\ 0.7 & 0.3 \end{bmatrix}$ **19.** $P = \begin{bmatrix} 0 & 1 \\ 0.4 & 0.6 \end{bmatrix}$

20. $P = \begin{bmatrix} 0.22 & 0.78 \\ 0.34 & 0.66 \end{bmatrix}$

In Exercises 21–23, solve the matrix equation $SP = S$ for S to determine the steady-state vector. Use the Gauss–Jordan method to determine the values for the steady-state vector $[x \quad y \quad z]$. Also, use the steady-state vector to determine the limiting matrix P^n.

21. $\begin{bmatrix} 0 & 1 & 0 \\ 0 & 0 & 1 \\ 0.8 & 0 & 0.2 \end{bmatrix}$ **22.** $\begin{bmatrix} \frac{1}{2} & 0 & \frac{1}{2} \\ \frac{1}{2} & \frac{1}{2} & 0 \\ 0 & \frac{1}{5} & \frac{4}{5} \end{bmatrix}$

23. $\begin{bmatrix} 0.10 & 0 & 0.90 \\ 0.55 & 0.15 & 0.30 \\ 0.45 & 0.25 & 0.30 \end{bmatrix}$

In Exercises 24–28, you can use your calculator to approximate the limiting matrix P^n or to solve the matrix equation $SP = S$ for S using the methods of Exercises 18–23.

24. A professional sports league gives its players a monthly drug test. The league has determined that if a player passes a monthly test there is an 80% chance that the player will pass the test next month and a 20% chance that the player will fail the test next month. If a player fails a monthly test, there is a 40% chance that the player will fail the test next month and a 60% chance that the player will pass the test.

(a) Label the two states and write the transition matrix.

(b) Determine the limiting matrix P^n, for large values of n.

(c) Use the solution from part (b) to determine the probability that, in the long run, a player will pass the test.

25. Students at Hillsbourough High School are put on probation if they are truant in any of their classes a total of five or more times a week. The attendance office has found that

90% of the students who are not on probation will remain off probation the next week, while 10% will go on probation. Eighty-five percent of the students who are on probation will remain on probation the next week, while 15% will go off probation.

(a) Draw a state transition diagram and write the transition matrix.

(b) Determine the limiting matrix P^n, for large values of n.

(c) Hillsbourough High School currently has 900 students not on probation and 300 students on probation. In the long run, about how many students will be off probation and how many will be on probation?

26. The town of Plymouth has two elite soccer teams, Metro and Juventus. The director of parks and recreation has determined that there is a 30% chance that a player on Metro will play for Juventus next summer, while the rest continue to play for Metro. Conversely, 15% of Juventus players will play for Metro next summer, while the rest continue to play for Juventus.

(a) Determine the limiting matrix P^n, for large values of n.

(b) If Metro originally has a roster of 24 players and Juventus has a roster of 18 players, in the long run how many players will be on each team?

27. Goffstown has a community bike program in which a resident may pick up a bike at one of three locations and ride the bike anywhere provided that the bike is returned to one of the pickup locations the next morning. The pickup locations are the library, the YMCA, and the grocery store. The director of the program has determined that if a bike is picked up at the library there is a 10% chance that it will be returned to the library, a 25% chance that it will be returned to the YMCA, and a 65% chance that it will be returned to the grocery store. If a bike is picked up at the YMCA, there is an 80% chance that it will be returned to the YMCA, a 10% chance that it will be returned to the library, and a 10% chance that it will be returned to the grocery store. If a bike is picked up at the grocery store, there is a 5% chance that it will be returned to the grocery store, a 75% chance that it will be returned to the library, and a 20% chance that it will be returned to the YMCA.

(a) Determine the limiting matrix P^n, for large values of n.

(b) In the long run, what percentage of the bikes will be at the YMCA?

28. In a particular county in New England, voters can declare themselves as Republican, Democrat, or Independent. A public opinion poll shows that 80% of the Republicans intend to continue with their party affiliation, while 12% intend to switch to Democrat in the next election and 8% intend to switch to Independent. Ninety percent of Democrats intend to continue with their party affiliation, while 6% intend to switch to Republican in the next election and 4% intend to switch to Independent. Ninety-six percent of Independents intend to continue with their party affiliation, while 1% intends to switch to

Republican in the next election and 3% intend to switch to Democrat.

(a) Determine the limiting matrix P^n, for large values of n.

(b) In the long run, what percentage of voters will be declared as Independents?

In Exercises 29–31, a transition matrix is given. Identify the absorbing states, if any, and draw a state transition diagram corresponding to the matrix.

29. $P = \begin{bmatrix} \frac{1}{4} & \frac{3}{4} \\ 0 & 1 \end{bmatrix}$

30. $P = \begin{bmatrix} 0.4 & 0 & 0.6 \\ 0.3 & 0.7 & 0 \\ 0 & 0.2 & 0.8 \end{bmatrix}$

31. $P = \begin{bmatrix} 0.6 & 0.3 & 0.1 & 0 \\ 0.2 & 0.4 & 0.2 & 0.2 \\ 0 & 0 & 1 & 0 \\ 0 & 0 & 0 & 1 \end{bmatrix}$

In Exercises 32–35, a state transition diagram is given. Identify the absorbing states and determine if the diagram represents an absorbing Markov chain.

32.

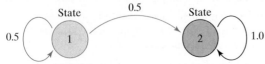

33.

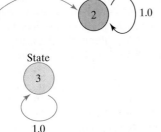

34.

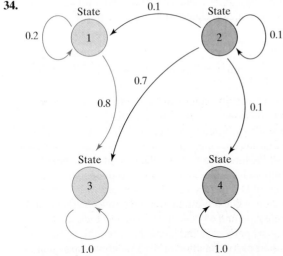

35.

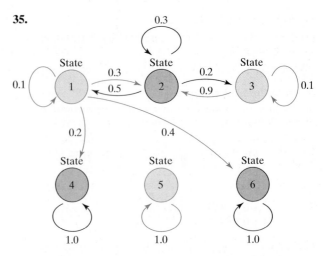

In Exercises 36–40, determine the fundamental matrix F.

36. $P = \begin{bmatrix} \frac{2}{5} & 0 & \frac{3}{5} \\ \frac{3}{10} & \frac{7}{10} & 0 \\ 0 & 0 & 1 \end{bmatrix}$

37. $P = \begin{bmatrix} 0.2 & 0 & 0.8 \\ 0.1 & 0.6 & 0.3 \\ 0 & 0 & 1 \end{bmatrix}$

38. $P = \begin{bmatrix} 0.3 & 0.7 & 0 \\ 0.7 & 0.3 & 0 \\ 0 & 0 & 1 \end{bmatrix}$

39. $P = \begin{bmatrix} \frac{1}{10} & \frac{3}{5} & \frac{1}{5} & \frac{1}{10} \\ \frac{3}{10} & \frac{2}{5} & \frac{3}{10} & 0 \\ 0 & 0 & 1 & 0 \\ 0 & 0 & 0 & 1 \end{bmatrix}$

40. $P = \begin{bmatrix} 0.10 & 0.20 & 0.35 & 0.10 & 0.25 \\ 0.25 & 0.15 & 0.10 & 0.25 & 0.25 \\ 0.30 & 0.10 & 0.20 & 0.15 & 0.25 \\ 0 & 0 & 0 & 1 & 0 \\ 0 & 0 & 0 & 0 & 1 \end{bmatrix}$

41. Players on Metro's soccer team are classified as first string or second string. Records from past seasons show that there is a 75% chance that a first-string player will remain first string the next season, a 20% chance that the player will be on the second string next season, and a 5% chance that the player will leave the team. There is a 40% chance that a second-string player will remain second string the next season, a 20% chance that the player will become first string the next season, and a 40% chance that the player will leave the team. Assume that none of the players who leave the team return.

(a) Identify the three states and draw a state transition diagram for a player's classification.

(b) Write the transition matrix, and determine the fundamental matrix F.

(c) Interpret $f_{1,2}$.

42. At Hillsbourough Community College, records show that 10% of the freshmen at the school will remain freshmen after 1 year, 70% will become sophomores, and 20% will leave the school. During the same year, 20% of sophomores will remain sophomores and 80% will leave the school (either by dropping out or graduating). Assume that none of the students who leave the school re-enroll.

(a) Identify the three states and draw a state transition diagram for the student enrollment.

(b) Write the transition matrix, and determine the fundamental matrix F.

(c) Compute FR and interpret the entries in the first row of this matrix product.

43. At Bedford Hospital's Physical Therapy Center patients are placed in a two-part program. Results of a weekly strength test determine in which stage a patient is placed. Stage I is for patients who test at less than half of their normal strength levels, and Stage II is for patients who test at greater than half of their normal strength levels. When a patient tests at 100%, the patient completes the program. Past records show that there is a 40% chance that a patient in Stage I remains in Stage I after the next testing, a 40% chance that the patient will move to Stage II, and a 20% chance that the patient will transfer to another program. There is a 45% chance that a patient in Stage II will remain in Stage II after the next testing, a 15% chance that the patient returns to Stage I, indicating reinjury, a 5% chance that the patient will transfer to another program, and a 35% chance that the patient will test 100% and complete the program. Assume that a patient that transfers or completes the program does not return.

(a) Identify the four states and draw a state transition diagram for the patient's status.

(b) Write the transition matrix, and determine the fundamental matrix F.

(c) Compute the sum of the entries in each row of the fundamental matrix F and interpret.

44. Vertical Climbing Gym offers beginner, intermediate, and advanced courses. There is a 20% chance that a student in the beginner course will remain in that course next session, a 60% chance that the student moves into the intermediate course, and a 20% chance that the student does not take another course. There is a 30% chance that a student in the intermediate course remains there next session, a 50% chance that the student moves into the advanced course, and a 20% chance that the student does not take another course. The advanced course is the highest course offered, and there is a 25% chance that a student remains in this course next session.

(a) Identify the four states and draw a state transition diagram for the student's course.

(b) Write the transition matrix, and determine the fundamental matrix F.

(c) If a student begins in the intermediate course, about how many sessions will the student take?

45. At Keller and Harris College, past records show that ranks of undergraduates progress as follows:

• 70% of the freshmen will become sophomores after 1 year, 15% will remain freshmen, and 15% will withdraw from the college.

• 70% of the sophomores will become juniors after 1 year, 10% will remain sophomores, and 20% will withdraw from the college.

• 65% of the juniors will become seniors after 1 year, 25% will remain juniors, and 10% will withdraw from the college.

• 85% of the seniors will graduate from the university, 10% will remain seniors, and 5% will withdraw from the university.

(a) Identify the six states and draw a state transition diagram for the ranks of undergraduates.

(b) Write the transition matrix, and determine the fundamental matrix F.

(c) Determine the number of years that a freshman, sophomore, junior, or senior can expect to be at the college.

(d) Determine the probability that a freshman will eventually graduate.

(e) Determine the probability that a sophomore will eventually withdraw.

CHAPTER 9 PROJECT

SOURCE: U.S. Census Bureau, www.census.gov

A Markov chain is a way to describe and predict travel through *states*. In this project, we'll consider the mobility of Americans within the United States. Consider the following information (in thousands):

178 moved from the West to the Midwest.

101 moved from the West to the Northeast.

400 moved from the West to the South.

98 moved from the Midwest to the Northeast.

556 moved from the Midwest to the South.

253 moved from the Midwest to the West.

383 moved from the Northeast to the South.

146 moved from the Northeast to the West.

94 moved from the Northeast to the Midwest.

334 moved from the South to the West.

463 moved from the South to the Midwest.

262 moved from the South to the Northeast.

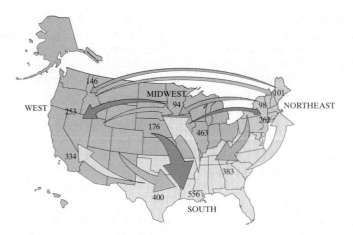

Figure R.1 Census Regions and Divisions of the United States

Let $x_1 =$ population in the West, $x_2 =$ population in the Midwest, $x_3 =$ population in the Northeast, and $x_4 =$ population in the South. Assume that no one enters the United States and no one leaves.

1. In 1998 there were 60,263,000 people in the West, 62,951,000 people in the Midwest, 51,686,000 people in the Northeast, and 95,349,000 people in the South. Find the percentage of each transition. Round to five decimal places.

2. Determine the percentage of people who stayed within their own region. Use the decimals from question 1 to find the total percentage of people that left each region. Subtract this answer from 1 to get the percentage of people who stayed in the region.

3. Draw a transition diagram.

4. Use the diagram in question 3 to form a transition matrix, P. Make sure that your rows add up to 1.

5. Given that the initial state vector is $S_0 = [\,60,263 \quad 62,951 \quad 51,686 \quad 95,349\,]$, find and interpret S_2. Round answers to the nearest whole number.

6. Solve $SP = S$ for S using a system of equations in which P is the matrix found in question 4 and S is the steady-state matrix. Round to five decimal places.

7. Interpret the entries in S found in question 6.

8. Assuming that the U. S. population maintains its 1998 level of 270,249 (in thousands), multiply each entry of S by 270,249 and interpret the results.

9. Use the result from question 6 to determine the limiting matrix, P^n.

10. Find $S_0 P^n$ and compare your answer to that of question 8.

10 Game Theory

	Economy strong	Recession
Remodel	300,000	−100,000
Don't Remodel	−150,000	−125,000

The decision to remodel a store can be thought of as a game between the store and the economy. Although a dated looking store can discourage customers, spending too much money on the remodel can bankrupt the store. Managers can use matrices and game theory as an aid in determining which course of action to follow. The above payoff matrix gives the expected payoff or loss in dollars. The −100,000 is the saddle point which gives the store its best strategy in this strictly determined game.

What We Know

We have learned many things in this text. Linear systems, matrices, linear programming, the simplex method, probability, and expected value are a few topics that we have covered so far.

Where Do We Go

In this chapter, we study game theory. Game theory provides an exceptional review, similar to a capstone, for all the topics listed above. At times, you may need to refer to other chapters in the text to review specific items.

Section 10.1   Strictly Determined Games

When we hear the word **game**, we tend to think of competitors working to determine a strategy that allows one of the competitors to win the game. Many games, such as a chess match, a football game, or a wrestling match, are played by only two competitors. We call these types of games **two-person games**. For our purposes, the number of actual players does not matter since we may be able to think of two-person games as involving two competing forces. However, some games may have several competitors. For example, a jewelry store in a mall might have several jewelry stores in the mall competing for business.

In this chapter we offer a brief exposure to competitive **game theory**, a branch of mathematics invented by John von Neumann. This exposure is restricted to games involving only two competitors, that is, two-person games. Also, we restrict our work in this chapter to games where the amount "gained" or "won" by one competitor is equal to the amount "lost" by the opposing competitor. This means that at the end of one play of the game the sum of the winnings of the two players is zero. Hence, we denote a loss by one competitor with a negative number. The types of games that we have just described are known as **zero-sum games**. In this chapter we analyze **two-person**, **zero-sum games**. Example 1 introduces us to one such game.

Example 1   Introducing a Two-Person, Zero-Sum Game

The city of Bundyville has two digital camera stores, Digicity and ClearVu. Both stores offer discounts of 25% or 35%, depending on how aggressively they want business. When both stores offer the same discount, Digicity gets 65% of the business. When Digicity discounts more than ClearVu, Digicity gets 75% of the business. When ClearVu discounts more than Digicity, Digicity gets 55% of the business. Determine the percentage of business that ClearVu "loses":

(a) When both stores have the same discount.

(b) When Digicity discounts more than ClearVu.

Solution

(a) When both stores offer the same discount, Digicity gets 65% of the business. This means that ClearVu loses 65% of the business.

(b) When Digicity discounts more than ClearVu, it gets 75% of the business. Hence, ClearVu loses 75% of the business. ∎

Example 1 illustrates a **two-person, zero-sum game**, since the sales gain of one store comes at the expense of the other store. (The game in Example 1 can actually be viewed as a *constant-sum* game, since the sum of Digicity's percentage and ClearVu's percentage is 100%.) In the game in Example 1, each store has a choice of how much to discount. Each choice leads to different percentages of the market share for the stores. For example, if Digicity's strategy is to discount 25% and ClearVu's strategy is to discount 35%, then Digicity would have a market share of 55%, while ClearVu would have a market share of 45%. This scenario outlines succinctly what this chapter covers:

1. Each player has a choice on a course of action, and these choices affect the outcome of the game.

2. Any choice made by a player is called a **strategy** for that player.

3. How does each player decide on a best strategy after considering all the choices that the other player has?

Determining the best or **optimal strategy** for each player is called **solving the game** and this is, of course, our ultimate goal. To help us in solving a game, it is advantageous to organize information in what is known as a **payoff matrix**.

Payoff Matrix

Continuing with the scenario in Example 1, the choice made by Digicity or ClearVu is called a **strategy** and each strategy has a consequence, which is called a **payoff**. We can summarize the payoff for each strategy in what is known as a **payoff matrix**. Each entry in the payoff matrix gives the payoff for a particular pair of choices by Digicity and ClearVu. To facilitate our organization, we arbitrarily select one store, Digicity, and use the rows of the matrix to represent its gains, as a percentage of the market share written as a decimal. In this setup, Digicity is called the **row player** and ClearVu is called the **column player**.

$$
\begin{array}{c}
\text{ClearVu}\\
25\%35\%\\
\text{Digicity}\begin{array}{c}25\%\\35\%\end{array}\begin{bmatrix}0.65 & 0.55\\0.75 & 0.65\end{bmatrix}
\end{array}
$$

Payoff matrix where Digicity is the row player.

Example 2 Constructing a Two-Person, Zero-Sum Payoff Matrix

Construct a payoff matrix for Digicity and ClearVu in which ClearVu is the row player.

Solution

Using the information given in Example 1, the payoff matrix is given by

$$
\begin{array}{c}
\text{Digicity}\\
25\%35\%\\
\text{ClearVu}\begin{array}{c}25\%\\35\%\end{array}\begin{bmatrix}0.35 & 0.25\\0.45 & 0.35\end{bmatrix}
\end{array}
$$

∎

✓ **Checkpoint 1** Now work Exercise 9.

Example 3 Understanding Entries in a Payoff Matrix

Zach and Josh are playing a penny-matching game. They each choose a side of a coin without knowing the other's choice. Then, when their friend Diane blows a whistle, both players show their choices simultaneously. The payoff matrix associated with this two-person, zero-sum game is given by

$$
\begin{array}{c}
\text{Josh's choice}\\
\text{H}\text{T}\\
\text{Zach's choice}\begin{array}{c}\text{H}\\\text{T}\end{array}\begin{bmatrix}4 & -5\\3 & 2\end{bmatrix}
\end{array}
$$

Given that the entries are in dollars, supply an interpretation of this payoff matrix.

Solution

Understand the Situation: Since this is a two-person, zero-sum game, a gain for one player equals a loss of the same size for the other player.

If Zach chooses heads (row 1 of the payoff matrix), he wins $4 if Josh chooses heads (column 1), whereas Zach loses $5 if Josh chooses tails (column 2). If Zach chooses tails (row 2), he wins $3 if Josh chooses heads (column 1), and Zach wins $2 if Josh chooses tails (column 2). ∎

Strictly Determined Games and Optimal Strategies

Looking more closely at the game and payoff matrix in Example 3, we may wonder how Zach and Josh should play the game. If Zach is greedy, he may choose row 1 because of the $4 payoff. However, we assume that Josh is not stupid and as a result would try to avoid column 1 to avoid losing $4. If Josh plays column 2, then he might win $5. But Zach, who we also assume is not stupid, might anticipate Josh's rationale and in turn he might play row 2, preventing Josh from winning at all!

To help us to determine a best play for Josh and Zach, as well as illustrate a general approach for a two-person, zero-sum game with an $m \times n$ payoff matrix, we utilize a worst-case scenario mentality. For Zach, the row player, the worst that could happen in row 1 is a $5 loss and in row 2 a $2 gain. We circle these entries as shown below.

$$\begin{array}{c} & \text{Josh's choice} \\ & \begin{array}{cc} \text{H} & \text{T} \end{array} \\ \text{Zach's choice} \begin{array}{c} \text{H} \\ \text{T} \end{array} & \begin{bmatrix} 4 & \textcircled{-5} \\ 3 & \textcircled{2} \end{bmatrix} \end{array}$$

Zach's **best strategy** is to select row 2, since it has the *largest* of these minimum values. (Some call this the **maximin strategy**.) By choosing row 2, Zach is guaranteed a $2 win no matter what Josh does.

For Josh, the column player, the worst that could happen in column 1 is a $4 loss and in column 2 a $2 loss. (Remember, for Josh, the column player, a positive number indicates a loss to Zach, the row player.) We place squares around these entries as shown below.

$$\begin{array}{c} & \text{Josh's choice} \\ & \begin{array}{cc} \text{H} & \text{T} \end{array} \\ \text{Zach's choice} \begin{array}{c} \text{H} \\ \text{T} \end{array} & \begin{bmatrix} \boxed{4} & \textcircled{-5} \\ 3 & \boxed{\textcircled{2}} \end{bmatrix} \end{array}$$

Josh's **best strategy** is to select column 2, since it has the *smallest* of the maximum values. (Some call this the **minimax strategy**.) By choosing column 2, Josh is guaranteed no worse than a $2 loss no matter what Zach does. Unfortunately for Josh, this is the best he can do as long as Zach continues to make his best move.

Notice that the 2 in row 2, column 2 is enclosed by a circle and a square. This means that it is both the smallest entry in its row and the largest entry in its column. This entry is called a **saddle value** or a **saddle point**. Just like a horse saddle, the saddle point is a minimum in one direction and a maximum from the other direction. Since saddle points are crucial in game theory, we now formalize the procedure on how to find them.

■ **Locating Saddle Points**

1. Circle the minimum value in each row. This may occur in more than one place.
2. Place squares around the maximum value in each column. This may occur in more than one place.
3. Any entry enclosed by a circle and a square is a **saddle point**. If more than one occurs, they will be equal.

Example 4 **Finding a Saddle Point**

Locate the saddle point for the following payoff matrix from a two-person, zero-sum game.

$$\begin{bmatrix} 2 & -2 & -6 \\ -1 & 1 & -2 \\ 4 & 10 & -8 \end{bmatrix}$$

Solution

To locate the saddle point, we use the three steps in the Locating Saddle Points box.

Step 1: Here we circle the minimum value in each row to get

$$\begin{bmatrix} 2 & -2 & -6 \\ -1 & 1 & -2 \\ 4 & 10 & -8 \end{bmatrix}$$

Step 2: Using the matrix that resulted from step 1, we now place squares around the maximum value in each column to get

$$\begin{bmatrix} 2 & -2 & -6 \\ -1 & 1 & -2 \\ 4 & 10 & -8 \end{bmatrix}$$

Step 3: The resulting matrix in step 2 shows that there is one entry enclosed by a circle and a square. This means that there is one saddle point and its value is −2.   ■

✓ **Checkpoint 2**

Now work Exercise 13a.

Let's return to Zach and Josh and the payoff matrix for their game.

Josh's choice

H   T

Zach's choice   H $\begin{bmatrix} 4 & -5 \\ 3 & 2 \end{bmatrix}$
                 T

We know that 2 is the saddle point. We also know that this means that it is the smallest entry in row 2 and the largest entry in column 2. This means that Zach should always play row 2 and Josh should always play column 2. This results in a win of $2 for Zach every time. This is the **optimal strategy** for both Zach and Josh. Notice that if Zach keeps playing row 2 and Josh changes to column 1 Zach would win more money (bad for Josh). If Josh keeps playing column 2 and Zach changes to row 1, then Josh wins $5 (bad for Zach). This type of game is said to be **strictly determined**, since Zach must always play row 2 to maximize his winnings and Josh must always play column 2 to minimize his losses. Optimal strategy and strictly determined games can be determined by the payoff matrix and saddle points as follows:

■ **Strictly Determined Games**

A game is called **strictly determined** if its payoff matrix has a saddle point. In a strictly determined game, the **optimal strategy** for row player R and column player C is as follows:

1. R should choose any row that contains a saddle point.
2. C should choose any column that contains a saddle point.

A saddle point is also called the **value of the game**.

▶ **Notes:**
- If the value of the game is positive, then the game favors the row player.
- If the value of the game is negative, then the game favors the column player.
- If the value of the game is zero, then the game is called a **fair game**.

Example 5   Determining Optimal Strategies

Sam's Shoes and Louie's Loafers are two competing shoe stores. Each must decide how to price a popular brand of hiking boots. Each store will charge either $45 or $49 for the boots. If Sam's charges $45 and Louie's charges $49, Sam's will get 70% of the business for these boots. If Sam's charges $49 and Louie's charges $45, Sam's will get 40% of the business. If both stores charge the same amount, each will get 50% of the business.

(a) Construct the payoff matrix for this game writing percentages as decimals.

(b) Verify that the game is strictly determined.

(c) Determine the optimal strategy for each company.

Solution

From Your Toolbox

Recall that an entry in a matrix is given by $a_{i,j}$, where i is the row and j is the column.

(a) Arbitrarily assigning Sam's Shoes as the row player, the payoff matrix is

$$
\begin{array}{cc}
 & \text{Louie's Loafers} \\
 & \begin{array}{cc} \$45 & \$49 \end{array} \\
\text{Sam's Shoes } \begin{array}{c} \$45 \\ \$49 \end{array} & \begin{bmatrix} 0.50 & 0.70 \\ 0.40 & 0.50 \end{bmatrix}
\end{array}
$$

(b) A game is strictly determined if it has a saddle point. Using the three steps in locating a saddle point, we have

$$
\begin{array}{cc}
 & \text{Louie's Loafers} \\
 & \begin{array}{cc} \$45 & \$49 \end{array} \\
\text{Sam's Shoes } \begin{array}{c} \$45 \\ \$49 \end{array} & \begin{bmatrix} 0.50 & 0.70 \\ 0.40 & 0.50 \end{bmatrix}
\end{array}
$$

The payoff matrix has a saddle point (enclosed by both a circle and a square) and hence is a strictly determined game.

(c) The entry $a_{1,1} = 0.50$ is the saddle point, so Sam's Shoes' optimal strategy is to choose row 1 and Louie's Loafers' optimal strategy is to choose column 1. In other words, Sam's Shoes and Louie's Loafers should both charge $45 for the hiking boots. ∎

✓ **Checkpoint 3**   Now work Exercise 29.

SUMMARY

In this section we introduced **two-person, zero-sum games**. Central to our introduction to game theory was the **payoff matrix**. The payoff matrix allows us to determine if the game is **strictly determined**, to locate **saddle points**, and to find an **optimal strategy** for each player. The saddle point also gives the value of the game, which in turn tells us if the game favors one player over the other or if it is a fair game.

- **Locating Saddle Points**
  1. Circle the minimum value in each row. This may occur in more than one place.
  2. Place squares around the maximum value in each column. This may occur in more than one place.

3. Any entry enclosed by a circle and a square is a **saddle point**. If more than one occurs, they will be equal.

- **Strictly Determined Games** A game is called **strictly determined** if its payoff matrix has a saddle point. In a strictly determined game, the **optimal strategy** for row player R and column player C is as follows:
  1. R should choose any row that contains a saddle point.
  2. C should choose any column that contains a saddle point.

A saddle point is also called the **value of the game**.

SECTION 10.1 EXERCISES

Exercises 1–8, use the following payoff matrix for a two-person, zero-sum game. The entries represent dollars.

$$\text{Jenna} \begin{array}{c} & \text{Elle} \\ \begin{bmatrix} 2 & -2 & 3 \\ -1 & 4 & 6 \\ 1 & -3 & 2 \end{bmatrix} \end{array}$$

Determine the payoff when the following strategies are used.

1. Jenna plays row 1; Elle plays column 1.

2. Jenna plays row 1; Elle plays column 3.

3. Jenna plays row 1; Elle plays column 2.

4. Jenna plays row 2; Elle plays column 1.

5. Jenna plays row 2; Elle plays column 3.

6. Jenna plays row 3; Elle plays column 1.

7. Jenna plays row 3; Elle plays column 2.

8. Jenna plays row 3; Elle plays column 3.

In Exercises 9–12, you are given a payoff matrix for a game. Construct a new payoff matrix for the game where the row player becomes the column player and the column player becomes the row player.

✓ **9.** $\text{Bill} \begin{array}{c} \text{Don} \\ \begin{bmatrix} 3 & -3 \\ -2 & -4 \end{bmatrix} \end{array}$

10. $\text{Lisa} \begin{array}{c} \text{Melissa} \\ \begin{bmatrix} -3 & 4 \\ 7 & 1 \end{bmatrix} \end{array}$

11. $\text{Eric} \begin{array}{c} \text{Patrice} \\ \begin{bmatrix} -3 & 1 & 2 \\ 4 & -6 & 3 \\ 4 & 6 & -8 \end{bmatrix} \end{array}$

12. $\text{Ann} \begin{array}{c} \text{Sally} \\ \begin{bmatrix} 3 & -2 & -1 \\ -1 & 2 & -7 \\ -2 & -3 & 5 \end{bmatrix} \end{array}$

In Exercises 13–24, you are given a payoff matrix for a two-person, zero-sum game. If the game is strictly determined (if it is not, say so) determine:

(a) The saddle point(s).

(b) The optimal strategy for each player.

(c) The value of the game, and state whether the game favors one player over the other.

✓ **13.** $\text{Bill} \begin{array}{c} \text{Al} \\ \begin{bmatrix} 2 & 3 \\ -2 & 5 \end{bmatrix} \end{array}$

14. $\text{George} \begin{array}{c} \text{Dick} \\ \begin{bmatrix} 3 & -4 \\ -7 & -5 \end{bmatrix} \end{array}$

Monica

15. Hillary $\begin{bmatrix} 3 & -7 \\ -2 & 1 \end{bmatrix}$

Linda

16. Ken $\begin{bmatrix} 7 & -6 \\ -3 & -1 \end{bmatrix}$

Scoonie

17. Khalid $\begin{bmatrix} 2 & 4 & 2 \\ 0 & 3 & 0 \\ -1 & -2 & 1 \end{bmatrix}$

Maria

18. Nina $\begin{bmatrix} 1 & 2 & 3 \\ -1 & 2 & -1 \\ 1 & 6 & 3 \end{bmatrix}$

Dylan

19. Austin $\begin{bmatrix} 4 & 3 & 0 \\ 5 & 7 & 8 \\ 3 & -1 & 9 \end{bmatrix}$

Rusty

20. Randy $\begin{bmatrix} 1 & 0 & 7 \\ -3 & 2 & 0 \\ 2 & -3 & 8 \end{bmatrix}$

Ash

21. Brock $\begin{bmatrix} -5 & 1 \\ -2 & 0 \\ 4 & 3 \\ -3 & 1 \end{bmatrix}$

Misty

22. Erika $\begin{bmatrix} 3 & 1 & -4 & 5 \\ -2 & 5 & 0 & 3 \end{bmatrix}$

Mike

23. Brian $\begin{bmatrix} 5 & -2 & 3 & -8 \\ 3 & 0 & 2 & 1 \\ 1 & -1 & -7 & 2 \\ 4 & 0 & 1 & 3 \end{bmatrix}$

Luther

24. Denyse $\begin{bmatrix} 1 & 4 & 0 & 1 \\ -8 & 3 & -3 & 2 \\ 4 & 1 & 0 & 4 \\ -2 & 3 & -7 & 0 \end{bmatrix}$

Applications

25. The Chasm, a clothing retail store, is contemplating remodeling to modernize its physical appearance. A market research firm determines that the main determining factor in The Chasm's decision is the economy and whether it will remain strong or slip into recession. The firm determines that if The Chasm remodels and the economy is strong The Chasm will earn $300,000. However, if The Chasm remodels and the economy slips into recession, the business will lose $100,000. If The Chasm does not remodel and the economy remains strong, the business will lose $150,000 due to customers going to other stores. Finally, if The Chasm does not remodel and there is a recession, the business will lose $125,000. The market research firm summarizes this information in the following payoff matrix, where the two-person game is The Chasm against the economy. Find the saddle point and determine the best strategy for The Chasm.

Economy

The Chasm $\quad$ Strong $\quad$ Recession

$\begin{array}{c} \text{Remodel} \\ \text{Don't remodel} \end{array} \begin{bmatrix} 300{,}000 & -100{,}000 \\ -150{,}000 & -125{,}000 \end{bmatrix}$

26. You Do It and I Did It, two home improvement stores, are each exploring opening new stores at one of two possible locations: Eastown Mall and Westown Mall. A market research firm has developed the following payoff matrix, where each entry represents the annual amounts (in millions of dollars) either gained or lost by one store from or to the other as a result of the location selected. Find the saddle point and determine the best strategy for the management of each store.

I Did It

You Do It $\quad$ Eastown $\quad$ Westown

$\begin{array}{c} \text{Eastown} \\ \text{Westown} \end{array} \begin{bmatrix} 1 & -1 \\ 1.5 & -2 \end{bmatrix}$

27. Big Lakes Bank and Bank Two are two banks that serve a certain region. Each bank is exploring opening a new branch that will be located in one of the three following towns: Jasonville (J), Kerrton (K), or Heathton (H). Depending on where each bank locates its new branch, Bank Two will either gain or lose deposits at the expense of Big Lakes Bank as indicated in the following payoff matrix (entries are in thousands of dollars):

Big Lakes Bank

Bank Two $\quad$ $J \quad K \quad H$

$\begin{array}{c} J \\ K \\ H \end{array} \begin{bmatrix} 80 & 0 & -20 \\ 50 & 25 & 40 \\ -30 & -20 & 100 \end{bmatrix}$

(a) Is this a strictly determined game?

(b) Find the saddle point and determine the best strategy for each bank.

(c) Determine the value of the game. Does this game favor one bank over the other? If so, which bank has the advantage?

28. C Me Play and R You Game are two competing sporting goods stores. On a weekly basis, each store selects one (and only one) of the following types of advertisement: television, radio, or mail circulars. A survey by a market research company yields the following payoff matrix. Entries represent the percentages of the market gain or loss for each choice of action by the stores. (Assume that a gain by C Me Play is a loss for R You Game, and vice versa.)

$$
\begin{array}{cc}
 & \text{C Me Play} \\
 & \begin{array}{ccc} \text{TV} & \text{Radio} & \text{Mail} \end{array} \\
\text{R You Game} \begin{array}{c} \text{TV} \\ \text{Radio} \\ \text{Mail} \end{array} & \left[\begin{array}{ccc} 2 & 2 & 3 \\ -3 & -2 & 0 \\ 2 & 2 & -2 \end{array}\right]
\end{array}
$$

(a) Is this a strictly determined game?

(b) Find the saddle point and determine the best strategy for each store.

(c) Determine the value of the game. Does this game favor one store over the other? If so, which store has the advantage?

✓ **29.** Two competing auto parts stores, Redan and Erog, are each planning to add another store in a certain city. It is possible for each store to be located centrally in the city or peripherally in a large suburb of the city. If both stores decide to build in the city, Redan will have a monthly profit of $200 more than the profit of Erog. If both build in the suburb, then Redan's monthly profit will be $400 less than Erog's. If Redan builds in the suburb while Erog builds in the city, Redan's monthly profit will be $800 more than Erog's. If Redan builds in the city and Erog builds in the suburb, then Redan's monthly profit will be $600 less than Erog's. Let Redan be the row player and Erog be the column player.

(a) Construct the payoff matrix for this game, where the entries are in hundreds of dollars.

(b) Verify that the game is strictly determined.

(c) Determine the optimal strategy for each store.

30. (*continuation of Exercise 29*)

(a) Determine the value of the game in Exercise 29.

(b) Does the game in Exercise 29 favor one store over the other? If so, which store has the advantage?

31. Two health clubs, Gutbuster and Noflab, are considering expanding the special fitness programs that they offer. Gutbuster is considering offering a middle-age spread buster program, a senior citizen toning program, and a twenty-something fit-for-life program. Noflab is considering a middle-age spread buster program and a senior citizen toning program. A survey by a market research firm determines the following. If Gutbuster starts

a middle-age spread buster program, it will gain 5% of the market if Noflab starts a middle-age spread buster program, but will lose 5% of the market if Noflab starts a senior citizen toning program. If Gutbuster starts a senior citizen toning program, it will gain 5% of the market if Noflab starts a middle-age spread buster program, but see no gain if Noflab starts a senior citizen toning program. If Gutbuster starts a twenty-something program, it will gain 10% of the market if Noflab starts a middle-age spread buster program and gain 15% of the market if Noflab starts a senior citizen toning program. (Assume that a gain by Gutbuster is a loss for Noflab, and vice versa.) Let Gutbuster be the row player and Noflab be the column player.

(a) Construct the payoff matrix for this game, where the entries are percentages of the market.

(b) Verify that the game is strictly determined.

(c) Determine the optimal strategy for each health club.

32. (*continuation of Exercise 31*)

(a) Determine the value of the game in Exercise 31.

(b) Does the game in Exercise 31 favor one health club over the other? If so, which health club has the advantage?

33. A candidate for governor of a certain state has informed his campaign manager about a prior, potentially damaging indiscretion. The campaign manager has to decide whether to reveal this "skeleton in the closet." The campaign manager has determined that if the candidate reveals the skeleton and the news media are critical, the candidate will lose 35,000 votes. However, if the news media are not critical, the candidate will gain 7000 votes. If the candidate does not reveal the skeleton and the media learns of it first and are critical, the candidate will lose 150,000 votes. Yet if the candidate does not reveal it and the media does not learn about it (it is not revealed during the campaign), the candidate will neither gain nor lose votes. This is a two-person game between the candidate and the media. Let the candidate be the row player and the media be the column player.

(a) Construct the payoff matrix for this game, where the entries are votes gained or lost in thousands.

(b) Verify that the game is strictly determined.

(c) Determine the optimal strategy for the candidate.

SECTION PROJECT

Brian and Jo play a game of matching fingers. When Jamie blows a whistle, both players extend, at the same time, 1, 2 or 3 fingers. If the sum of the extended fingers is 2 or 3, then Brian receives from Jo an amount in dollars equal to that sum. If the sum of the extended fingers is 5 or 6, then Jo receives from Brian an amount in dollars equal to that sum. If the sum of the extended fingers is 4, then neither player wins any money. Let Brian be the row player and Jo the column player.

(a) Construct the payoff matrix for this game.

(b) Is the game strictly determined?

(c) What is the optimal strategy for Brian and Jo?

(d) What is the value of the game?

(e) Does the game favor Brian or Jo or is it a fair game?

Section 10.2   Mixed Strategies

In Section 10.1 we studied strictly determined games and saw that, in a strictly determined game, both the row player and the column player choose a row and a column, respectively, that contains a saddle point. This is the optimal strategy for each player. To review how to find saddle points and determine the corresponding optimal strategy, let's take a brief Flashback.

Flashback

Penny-Matching Game Revisited

In Section 10.1, Example 3, we saw that Zach and Josh are playing a penny-matching game. They each choose a side of a coin without knowing the other's choice. Then, when their friend Diane blows a whistle, both players show their choices simultaneously. The payoff matrix associated with this two-person, zero-sum game is given by

$$
\begin{array}{cc}
 & \text{Josh's choice} \\
 & \begin{array}{cc} \text{H} & \text{T} \end{array} \\
\text{Zach's choice} \begin{array}{c} \text{H} \\ \text{T} \end{array} & \begin{bmatrix} 4 & -5 \\ 3 & 2 \end{bmatrix}
\end{array}
$$

Recall that entries are in dollars. Find the saddle point and optimal strategy for each player.

Flashback Solution

Using the circle and square method given in Section 10.1, we have

$$
\begin{array}{cc}
 & \text{Josh's choice} \\
 & \begin{array}{cc} \text{H} & \text{T} \end{array} \\
\text{Zach's choice} \begin{array}{c} \text{H} \\ \text{T} \end{array} & \begin{bmatrix} \boxed{4} & \boxed{-5} \\ 3 & \boxed{2} \end{bmatrix}
\end{array}
$$

The saddle point is 2; hence the optimal strategy for Zach is row 2 and the optimal strategy for Josh is column 2. Also recall that the saddle point gives the value of the game. Here the value of the game is 2. Since the value of the game is a positive number, the game favors the row player, Zach.

In repeated plays of the game in the Flashback, Zach or Josh having knowledge of the other's strategy is of no advantage provided that the opponent is using an optimal strategy. If Josh figures out that Zach is always playing row 2, it does him no good to change to column 1, since he would then lose more money to Zach. Likewise, if Zach discovers that Josh is always playing column 2, it would do him no good to change to row 1, since he would then lose money to Josh. So the optimal strategy for both in repeated plays of the game is to make the same move over and over again. Such strategies are called **pure strategies**.

Nonstrictly Determined Games

Let's consider the following variation of the payoff matrix for Josh and Zach in the Flashback, while keeping everything else the same.

Josh's choice

$$\begin{array}{cc} & H \quad T \end{array}$$

Zach's choice $\begin{array}{c} H \\ T \end{array} \begin{bmatrix} 4 & -5 \\ -3 & 2 \end{bmatrix}$

To see if Zach or Josh has an advantage, as well as to see if there exists a pure strategy, we begin by checking to see if the game is strictly determined. The circle and square method allows us to determine this.

Josh's choice

$$\begin{array}{cc} & H \quad T \end{array}$$

Zach's choice $\begin{array}{c} H \\ T \end{array} \begin{bmatrix} \boxed{4} & \text{(−5)} \\ \text{(−3)} & \boxed{2} \end{bmatrix}$

Since no entry is enclosed by both a circle and a square, there are no saddle points. This means that the game is *not* strictly determined. So how should Zach and Josh play this game?

If Zach consistently plays row 1 because of its large payoff of $4, it would not take Josh too long to figure out Zach's strategy, and as a result Josh would then play column 2. But then Zach would play row 2. This would cause Josh to then play column 1. This brief analysis indicates that neither player should consistently play the same row or column. Instead, each player should play each row and each column in some mixed pattern unknown to the other player. Strategies such as this are called **mixed strategies**.

Mixed Strategies

How should Zach or Josh choose a mixed pattern? The best way to select a mixed pattern is to use a probability distribution and some sort of chance device to determine the distribution. For example, Zach might choose row 1 with probability $\frac{1}{2}$ and row 2 with probability $\frac{1}{2}$. This could be determined by flipping an unbiased coin before each move, with heads corresponding to playing row 1 and tails corresponding to playing row 2. Similarly, Josh might choose column 1 with probability $\frac{3}{13}$ and column 2 with probability $\frac{10}{13}$. These could be determined by drawing a card at random from a standard 52-card deck, where selecting a Jack, Queen, or King corresponds to playing column 1 and selecting any other card corresponds to playing column 2. (The selected card would need to be replaced after each play.)

Since payoffs for a game are expressed in a payoff matrix, it makes mathematical sense to use matrices for mixed strategies. For Zach, the row player, and the aforementioned strategy of flipping a coin, his strategy is represented by the **row matrix**

$$P = \begin{bmatrix} \frac{1}{2} & \frac{1}{2} \end{bmatrix}$$

For Josh, the column player, and the aforementioned strategy of drawing a card, his strategy is given by the **column matrix**

$$Q = \begin{bmatrix} \frac{3}{13} \\ \frac{10}{13} \end{bmatrix}$$

Before we illustrate exactly why row and column matrices are used for strategies, let's summarize what we have done so far.

■ Strategies for the Row Player and Column Player

Given the payoff matrix

$$A = \begin{bmatrix} a & b \\ c & d \end{bmatrix}$$

the row player strategy is

$$P = \begin{bmatrix} p_1 & p_2 \end{bmatrix}$$

where $p_1 \geq 0$, $p_2 \geq 0$.

$$p_1 + p_2 = 1$$

the column player strategy is

$$Q = \begin{bmatrix} q_1 \\ q_2 \end{bmatrix}$$

where $q_1 \geq 0$, $q_2 \geq 0$.

$$q_1 + q_2 = 1$$

In Zach and Josh's game we saw that pure strategies were not the best choices since it is a **nonstrictly determined game**. This led us to using mixed strategies. However, is there an optimal mixed strategy for each player? The answer is yes and it uses the idea of **expected value** discussed in Section 8.2.

Expected Value of a Game

Let's assume for Zach (row player) and Josh (column player) the payoff matrix A, row matrix P, and column matrix Q introduced earlier.

$$A = \begin{bmatrix} a & b \\ c & d \end{bmatrix} = \text{Zach's choice} \quad \begin{array}{cc} & \begin{array}{cc} \text{H} & \text{T} \end{array} \\ \begin{array}{c} \text{H} \\ \text{T} \end{array} & \begin{bmatrix} 4 & -5 \\ -3 & 2 \end{bmatrix} \end{array}$$

Josh's choice

Suppose that in repeated plays of the game Zach uses the mixed strategy

$$P = \begin{bmatrix} p_1 & p_2 \end{bmatrix} = \begin{bmatrix} \frac{1}{2} & \frac{1}{2} \end{bmatrix}$$

and Josh uses mixed strategy

$$Q = \begin{bmatrix} q_1 \\ q_2 \end{bmatrix} = \begin{bmatrix} \frac{3}{13} \\ \frac{10}{13} \end{bmatrix}$$

This means that Zach plays row 1 with probability $\frac{1}{2}$ and row 2 with probability $\frac{1}{2}$, while Josh plays column 1 with probability $\frac{3}{13}$ and column 2 with probability $\frac{10}{13}$. Each play of the game yields four possible outcomes or payoffs, as given by the payoff matrix A. Each of these payoffs has a probability of occurring that we can determine. For example, the probability of $4 occurring is simply the probability associated with Zach playing row 1 and Josh playing column 1. Since the choice of moves is made by a player without knowing the other's choice, each outcome (or payoff) makes up a pair of independent events. Reviewing the information in the Toolbox to the left, the probability of $4 occurring is simply the

product $\dfrac{1}{2}\left(\dfrac{3}{13}\right) = \dfrac{3}{26}$. That is,

$$P(\text{row 1 and column 1}) = P(\text{row 1}) \cdot P(\text{column 1})$$
$$P(a) = p_1 \cdot q_1$$
$$P(4) = \frac{1}{2} \cdot \frac{3}{13} = \frac{3}{26}$$

In a similar fashion, we can compute the probability of the other three payoffs. This is summarized in Table 10.2.1.

Table 10.2.1

Payoff	Probability
$a = 4$	$p_1 q_1 = \frac{1}{2} \cdot \frac{3}{13} = \frac{3}{26}$
$b = -5$	$p_1 q_2 = \frac{1}{2} \cdot \frac{10}{13} = \frac{10}{26}$
$c = -3$	$p_2 q_1 = \frac{1}{2} \cdot \frac{3}{13} = \frac{3}{26}$
$d = 2$	$p_2 q_2 = \frac{1}{2} \cdot \frac{10}{13} = \frac{10}{26}$

From Your Toolbox

Expected value is computed as follows:
$$E(X) = x_1 \cdot P(X = x_1)$$
$$+ x_2 \cdot P(X = x_2)$$
$$+ \cdots + x_n \cdot P(X = x_n)$$

Now using the concept of expected value from Section 8.2 (consult the Toolbox to the left), the **expected value of the game E** is

$$E = ap_1q_1 + bp_1q_2 + cp_2q_1 + dp_2q_2$$

$$= 4\left(\frac{1}{2}\right)\left(\frac{3}{13}\right) + (-5)\left(\frac{1}{2}\right)\left(\frac{10}{13}\right) + (-3)\left(\frac{1}{2}\right)\left(\frac{3}{13}\right) + 2\left(\frac{1}{2}\right)\left(\frac{10}{13}\right)$$

$$= -\frac{27}{26} \approx -1.04$$

This means that in the long run, when Zach and Josh use the indicated mixed strategies, Zach will lose about \$1.04 per game on average. Hence, Josh will win about \$1.04 per game on average.

Before we address whether this is the best that Zach and Josh can do, let's summarize what we have done.

■ **Expected Value of a Game**

Given the payoff matrix

$$A = \begin{bmatrix} a & b \\ c & d \end{bmatrix}$$

and strategies

$$P = \begin{bmatrix} p_1 & p_2 \end{bmatrix} \quad \text{and} \quad Q = \begin{bmatrix} q_1 \\ q_2 \end{bmatrix}$$

for the row player and the column player, respectively, the **expected value of the game**, E, is given by

$$E = PAQ$$

▶ **Note:** To show that $E = PAQ$, use matrix multiplication to multiply PAQ and show that $PAQ = ap_1q_1 + bp_1q_2 + cp_2q_1 + dp_2q_2$. This is left for you to do in Exercise 29.

Example 1 **Computing an Expected Value of a Game**

Given the payoff matrix A for Zach and Josh

Josh's choice

$$A = \text{Zach's choice} \quad \begin{matrix} & \text{H} \quad \text{T} \\ \text{H} \\ \text{T} \end{matrix} \begin{bmatrix} 4 & -5 \\ -3 & 2 \end{bmatrix}$$

and mixed strategies

$$P = \begin{bmatrix} \frac{1}{2} & \frac{1}{2} \end{bmatrix} \quad \text{and} \quad Q = \begin{bmatrix} \frac{1}{2} \\ \frac{1}{2} \end{bmatrix}$$

determine the expected value, E, of the game and interpret.

Solution

Since $E = PAQ$, we have

$$E = \begin{bmatrix} \frac{1}{2} & \frac{1}{2} \end{bmatrix} \cdot \begin{bmatrix} 4 & -5 \\ -3 & 2 \end{bmatrix} \cdot \begin{bmatrix} \frac{1}{2} \\ \frac{1}{2} \end{bmatrix}$$

$$= \begin{bmatrix} \frac{1}{2} & -\frac{3}{2} \end{bmatrix} \cdot \begin{bmatrix} \frac{1}{2} \\ \frac{1}{2} \end{bmatrix}$$

$$= \begin{bmatrix} -\frac{1}{2} \end{bmatrix} \quad \text{or simply} \quad -\frac{1}{2}$$

Interpret the Solution: This means that in the long run, when Zach and Josh use the indicated mixed strategies, Zach will lose about $0.50 per game on average. ■

 Checkpoint 1

Now work Exercise 3.

Technology Option

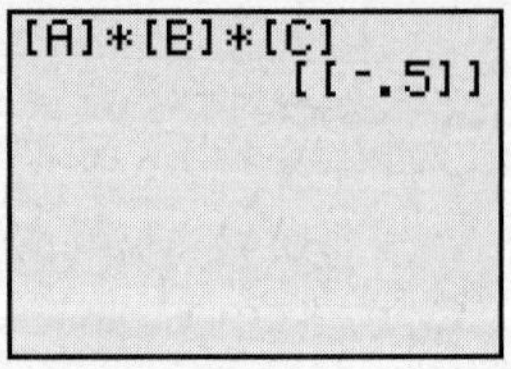

$$\begin{array}{l} [A]*[B]*[C] \\ \qquad\qquad [[-.5]] \end{array}$$

Figure 10.2.1

Recall from Chapter 2 that matrix multiplication can be performed on a graphing calculator. In Figure 10.2.1, we have entered matrices P, A, and Q into a calculator and, for the calculator's sake, we have renamed them matrices A, B, and C. Figure 10.2.1 shows the result of the matrix multiplication. Note that it is the same result that we obtained in Example 1.

For more information on how to use your calculator with matrices, consult the on-line graphing calculator manual at www.prenhall.com/armstrong

Solutions to a Game

In Section 10.1 we saw that the optimal strategy (pure strategy) for a strictly determined game was the row and column associated with a saddle point. This optimal strategy gave us the **solution** to a strictly determined game. For **nonstrictly determined games** a player needs a mixed strategy, since a pure strategy would be detected by the opponent. Unfortunately, there are an infinite number of mixed strategies for each player in a nonstrictly determined game. So how can an optimal mixed strategy be determined for each player? (An optimal mixed strategy would maximize the expected payoff for the row player, while the column player simultaneously wants to minimize it).

Before we supply a formula to determine the optimal mixed strategy, let's consider the thought processes of each player. The row player assumes that any mixed strategy he uses will be countered by the column player using a strategy that minimizes the row player's payoff. Hence, the row player will use the mixed strategy for which his expected payoff is maximized when the column player offers his best counterstrategy.

Likewise, the column player assumes that the row player will use a counterstrategy that maximizes the row player's payoff. Hence, the column player will use the mixed strategy that minimizes the payoff to the row player, when the row player uses his best counterstrategy.

The following formulas, which give the optimal strategies and therefore the solution to a nonstrictly determined game, are supplied without proof. In Section 10.3 we show how to solve a game by converting it to a linear programming problem and using the simplex method, along with duality.

■ **Optimal Mixed Strategy for Nonstrictly Determined Games**

For payoff matrix

$$A = \begin{bmatrix} a & b \\ c & d \end{bmatrix}$$

let $S = (a + d) - (b + c)$.

The **optimal mixed strategy for the row player** is

$$P_{\text{opt}} = [\, p_1 \quad p_2 \,]$$

where $p_1 = \dfrac{d - c}{S}$ and $p_2 = 1 - p_1$.

The **optimal mixed strategy for the column player** is

$$Q_{\text{opt}} = \begin{bmatrix} q_1 \\ q_2 \end{bmatrix}$$

where $q_1 = \dfrac{d - b}{S}$ and $q_2 = 1 - q_1$.

The **value of the game**, v, is equivalent to the expected value $E = P_{\text{opt}} A Q_{\text{opt}}$. This can be simplified to $v = \dfrac{ad - bc}{S}$.

▶ **Note:** In Section 2.4 we defined the determinant of a 2×2 matrix, det (A), as det $(A) = ad - bc$. Notice that the value of the game, v, can be restated as $v = \dfrac{\det(A)}{S}$.

Example 2 **Finding Optimal Mixed Strategies**

Solve the Zach and Josh penny-matching game where the payoff matrix is

$$A = \text{Zach's choice} \quad \begin{array}{c} \\ H \\ T \end{array} \begin{bmatrix} 4 & -5 \\ -3 & 2 \end{bmatrix}$$

Josh's choice
H   T

Solution

Understand the Situation: To solve this game means to determine the optimal mixed strategy for both players.

From the payoff matrix, we have $a = 4, b = -5, c = -3$, and $d = 2$. So S is

$$S = (a + d) - (b + c) = (4 + 2) - (-5 + -3) = 14$$

So p_1 is

$$p_1 = \frac{d-c}{S} = \frac{2-(-3)}{14} = \frac{5}{14}$$

and p_2 is

$$p_2 = 1 - p_1 = 1 - \frac{5}{14} = \frac{9}{14}$$

This gives us Zach's optimal mixed strategy:

$$P_{\text{opt}} = [\, p_1 \quad p_2 \,] = \left[\, \tfrac{5}{14} \quad \tfrac{9}{14} \,\right]$$

For Josh we have

$$q_1 = \frac{d-b}{S} = \frac{2-(-5)}{14} = \frac{7}{14} = \frac{1}{2}$$

and q_2 is

$$q_2 = 1 - q_1 = 1 - \frac{1}{2} = \frac{1}{2}$$

Josh's optimal mixed strategy is

$$Q_{\text{opt}} = \begin{bmatrix} q_1 \\ q_2 \end{bmatrix} = \begin{bmatrix} \tfrac{1}{2} \\ \tfrac{1}{2} \end{bmatrix}$$ ■

✓ **Checkpoint 2**

Now work Exercise 13.

Example 3   **Finding the Value of a Game**

Find the value of the penny-matching game in Example 2 and determine if it favors one player over the other.

Solution

Recall from Example 2 that $a = 4$, $b = -5$, $c = -3$, and $d = 2$. We also determined that $S = 14$. Since the value of a game, v, is given by $v = \dfrac{ad - bc}{S}$, we determine that $ad - bc = 4(2) - (-5)(-3) = -7$. So the value of the game is

$$v = \frac{ad - bc}{S} = \frac{-7}{14} = -\frac{1}{2}$$

Interpret the Solution: Since the value of the game is negative, this game favors Josh, the column player, over Zach, the row player. Over the long run, in repeated plays of this game, when each player uses his optimal mixed strategy, Josh is expected to win $0.50 per game on average.   ■

✓ **Checkpoint 3**

Now work Exercise 23.

When doing the applications in the exercises, be sure to write down the payoff matrix (if it is not supplied) and employ the techniques of this section if the payoff matrix represents a nonstrictly determined game. If the payoff matrix represents a strictly determined game, use the techniques from Section 10.1.

SUMMARY

In this section we saw that **pure strategies** are used in strictly determined games and **mixed strategies** are used in non-strictly determined games. Mixed strategies saw us revisit some old topics: probability, expected value, and independent events. We also revisited row matrices and column matrices, as well as matrix multiplication. We provided formulas for determining the optimal strategy (solution) for nonstrictly determined games.

- **Optimal Mixed Strategy for Nonstrictly Determined Games** For payoff matrix

$$A = \begin{bmatrix} a & b \\ c & d \end{bmatrix}$$

let $S = (a + d) - (b + c)$.

The **optimal mixed strategy for the row player** is

$$P_{opt} = [\, p_1 \quad p_2 \,]$$

where $p_1 = \dfrac{d - c}{S}$ and $p_2 = 1 - p_1$.

The **optimal mixed strategy for the column player** is

$$Q_{opt} = \begin{bmatrix} q_1 \\ q_2 \end{bmatrix}$$

where $q_1 = \dfrac{d - b}{S}$ and $q_2 = 1 - q_1$.

The **value of the game**, v, is equivalent to the expected value $E = P_{opt} A Q_{opt}$. This can be simplified to $v = \dfrac{ad - bc}{S}$.

SECTION 10.2 EXERCISES

In Exercises 1–10, determine the expected value E for each game whose payoff matrix and strategies P and Q (row and column player, respectively) are given.

1. $\begin{bmatrix} 1 & -2 \\ -2 & 3 \end{bmatrix}$, $P = [\, \frac{1}{2} \quad \frac{1}{2} \,]$, $Q = \begin{bmatrix} \frac{1}{3} \\ \frac{2}{3} \end{bmatrix}$

2. $\begin{bmatrix} -4 & 2 \\ 3 & 1 \end{bmatrix}$, $P = [\, \frac{1}{3} \quad \frac{2}{3} \,]$, $Q = \begin{bmatrix} \frac{3}{4} \\ \frac{1}{4} \end{bmatrix}$

✓ **3.** $\begin{bmatrix} 5 & -4 \\ -3 & 6 \end{bmatrix}$, $P = [\, \frac{3}{8} \quad \frac{5}{8} \,]$, $Q = \begin{bmatrix} \frac{1}{2} \\ \frac{1}{2} \end{bmatrix}$

4. $\begin{bmatrix} -3 & 2 \\ 5 & -4 \end{bmatrix}$, $P = [\, \frac{2}{3} \quad \frac{1}{3} \,]$, $Q = \begin{bmatrix} \frac{1}{4} \\ \frac{3}{4} \end{bmatrix}$

5. $\begin{bmatrix} 4 & 1 & -3 \\ 3 & -2 & 0 \\ 4 & 2 & -1 \end{bmatrix}$, $P = [\, 0.3 \quad 0.4 \quad 0.3 \,]$, $Q = \begin{bmatrix} 0.2 \\ 0.5 \\ 0.3 \end{bmatrix}$

6. $\begin{bmatrix} 5 & -3 & 1 \\ 1 & 6 & -2 \\ 2 & -4 & 3 \end{bmatrix}$, $P = [\, 0.1 \quad 0.4 \quad 0.5 \,]$, $Q = \begin{bmatrix} 0.6 \\ 0.1 \\ 0.3 \end{bmatrix}$

7. $\begin{bmatrix} 3 & -1 & 4 \\ -2 & 2 & 5 \end{bmatrix}$, $P = [\, 0.75 \quad 0.25 \,]$, $Q = \begin{bmatrix} 0.2 \\ 0.2 \\ 0.6 \end{bmatrix}$

8. $\begin{bmatrix} -1 & 3 & -3 \\ 1 & -2 & 4 \end{bmatrix}$, $P = [\, 0.15 \quad 0.85 \,]$, $Q = \begin{bmatrix} 0.25 \\ 0.35 \\ 0.4 \end{bmatrix}$

9. $\begin{bmatrix} -1 & 3 \\ 2 & -2 \\ 3 & -3 \end{bmatrix}$, $P = [\, 0.2 \quad 0.4 \quad 0.4 \,]$, $Q = \begin{bmatrix} 0.7 \\ 0.3 \end{bmatrix}$

10. $\begin{bmatrix} 3 & -2 \\ -1 & 1 \\ -3 & 2 \end{bmatrix}$, $P = [\, 0.3 \quad 0.4 \quad 0.3 \,]$, $Q = \begin{bmatrix} 0.2 \\ 0.8 \end{bmatrix}$

11. A certain game has a payoff matrix

$$\begin{bmatrix} 2 & -3 \\ -2 & 4 \end{bmatrix}$$

Compute the expected value E for the pairs of strategies given in (a) to (d). Which of these strategies is most favorable to the row player?

(a) $P = [\, \frac{1}{2} \quad \frac{1}{2} \,]$, $Q = \begin{bmatrix} \frac{1}{2} \\ \frac{1}{2} \end{bmatrix}$

(b) $P = [\, \frac{1}{3} \quad \frac{2}{3} \,]$, $Q = \begin{bmatrix} \frac{1}{4} \\ \frac{3}{4} \end{bmatrix}$

(c) $P = [\, \frac{2}{5} \quad \frac{3}{5} \,]$, $Q = \begin{bmatrix} \frac{2}{7} \\ \frac{5}{7} \end{bmatrix}$

(d) $P = [\, \frac{3}{8} \quad \frac{5}{8} \,]$, $Q = \begin{bmatrix} \frac{2}{9} \\ \frac{7}{9} \end{bmatrix}$

12. A certain game has a payoff matrix

$$\begin{bmatrix} 4 & 1 & -3 \\ 3 & -2 & 0 \\ 4 & 2 & -1 \end{bmatrix}$$

Compute the expected value E for the pairs of strategies given in (a) to (d). Which of these strategies is most favorable to the row player?

(a) $P = [\, 0.3 \quad 0.4 \quad 0.3 \,]$, $Q = \begin{bmatrix} 0.2 \\ 0.5 \\ 0.3 \end{bmatrix}$

(b) $P = [0.4 \quad 0.2 \quad 0.4], \quad Q = \begin{bmatrix} 0.6 \\ 0.3 \\ 0.1 \end{bmatrix}$

(c) $P = [1 \quad 0 \quad 0], \quad Q = \begin{bmatrix} 1 \\ 0 \\ 0 \end{bmatrix}$

(d) $P = [0.25 \quad 0.45 \quad 0.3], \quad Q = \begin{bmatrix} 0.55 \\ 0.25 \\ 0.2 \end{bmatrix}$

In Exercises 13–22, determine the optimal strategies P_{opt} and Q_{opt} for the row and column players, respectively. Be sure to check for saddle points since some may be strictly determined games.

✓ **13.** $\begin{bmatrix} -4 & 5 \\ 3 & -4 \end{bmatrix}$ **14.** $\begin{bmatrix} -1 & 3 \\ 2 & 1 \end{bmatrix}$

15. $\begin{bmatrix} 3 & -4 \\ -7 & -5 \end{bmatrix}$ **16.** $\begin{bmatrix} 3 & -7 \\ -2 & 1 \end{bmatrix}$

17. $\begin{bmatrix} 0 & 3 \\ 4 & 0 \end{bmatrix}$ **18.** $\begin{bmatrix} 4 & 0 \\ 0 & 5 \end{bmatrix}$

19. $\begin{bmatrix} 5 & -4 \\ -3 & 6 \end{bmatrix}$ **20.** $\begin{bmatrix} -3 & 2 \\ 5 & -4 \end{bmatrix}$

21. $\begin{bmatrix} -\frac{1}{2} & \frac{1}{2} \\ \frac{2}{3} & -\frac{1}{3} \end{bmatrix}$ **22.** $\begin{bmatrix} \frac{5}{4} & -\frac{3}{4} \\ -\frac{7}{8} & \frac{7}{4} \end{bmatrix}$

In Exercises 23–28:

(a) Determine the value of the game.

(b) Determine which player (if any) the game favors, assuming that the row player and the column player use their optimal strategies.

✓ **23.** The game in Exercise 13

24. The game in Exercise 14

25. The game in Exercise 17

26. The game in Exercise 18

27. The game in Exercise 21

28. The game in Exercise 22

29. Let $A = \begin{bmatrix} a & b \\ c & d \end{bmatrix}$ be a payoff matrix and $P = [\, p_1 \quad p_2 \,]$

and $Q = \begin{bmatrix} q_1 \\ q_2 \end{bmatrix}$ be the strategies for the row player and the column player, respectively.

(a) Use matrix multiplication to multiply PA.

(b) Use matrix multiplication to multiply the result from part (a) by Q.

(c) Part (b) is really the product PAQ. Verify that this is equal to E, the expected value, given in the text.

30. Let $A = \begin{bmatrix} a & b \\ c & d \end{bmatrix}$ be a payoff matrix.

(a) Compute $P_{opt} = [\, p_1 \quad p_2 \,]$, the optimal mixed strategy for the row player.

(b) Compute $Q_{opt} = \begin{bmatrix} q_1 \\ q_2 \end{bmatrix}$, the optimal mixed strategy for the column player.

(c) Compute $E = P_{opt} A Q_{opt}$, the expected value of the game. Simplify the result to verify that $E = P_{opt} A Q_{opt} = \dfrac{ad - bc}{(a + d) - (b + c)}$.

Applications

31. The Stumpf Company introduces its new product on the market with a major television and newspaper advertising campaign. Dave, the CEO of the Stumpf Company, finds out that its major competitor, the Stitz Company, is planning a major advertising campaign for a comparable product. A market research firm has determined the following payoff matrix, where each entry shows the sales (in millions of dollars) either gained or lost by one company to the other.

$$\begin{array}{cc} & \text{Stumpf} \\ & \text{TV} \quad\; \text{Paper} \\ \text{Stitz} \begin{array}{c} \text{TV} \\ \text{Paper} \end{array} & \begin{bmatrix} 1.5 & -0.8 \\ -0.7 & 1.2 \end{bmatrix} \end{array}$$

(a) Find the optimum strategies for the Stumpf Company and the Stitz Company.

(b) Find the value of the game assuming that both use the optimal mixed strategy. Does the game favor anyone? If so, who?

32. The Stumpf Company introduces its new product on the market with a radio and mail circular advertising campaign. Dave, the CEO of the Stumpf Company, finds out that its major competitor, the Stitz Company, is planning a major advertising campaign for a comparable product. A market research firm has determined the following payoff matrix, where each entry shows the sales (in millions of dollars) either gained or lost by one company to the other.

$$\begin{array}{cc} & \text{Stumpf} \\ & \text{Radio} \quad \text{Mail} \\ \text{Stitz} \begin{array}{c} \text{Radio} \\ \text{Mail} \end{array} & \begin{bmatrix} 1.5 & -0.8 \\ -0.7 & 1.2 \end{bmatrix} \end{array}$$

(a) Find the optimum strategies for the Stumpf Company and the Stitz Company.

(b) Find the value of the game, assuming that both use the optimal mixed strategy. Does the game favor anyone? If so, who?

33. The Bridge, a clothing retail store, is contemplating remodeling to modernize its physical appearance. A market research firm determines that the main determining factor in The Bridge's decision is the economy and whether it will remain strong or slip into recession. The firm determines that if The Bridge remodels and the economy is strong The Bridge will earn $400,000. However, if The Bridge remodels and the economy slips into recession, the business will lose $150,000. If The Bridge does not remodel and the economy remains strong, the business will lose $125,000 due to customers going to other stores. Finally, if The Bridge does not remodel and there is a recession, the business will lose $100,000. Let The Bridge be the row player and the economy the column player.

(a) Construct the payoff matrix for this game, where the entries are in thousands of dollars.

(b) Determine the optimal strategy for The Bridge.

(c) If you were the CEO of The Bridge, what would you do?

34. (*continuation of Exercise 33*)

(a) Assuming the optimal strategies in Exercise 33, determine the value of the game.

(b) Does the game in Exercise 33 favor anyone? If so, who?

35. The Riblets have decided to invest $50,000 in the stocks of two companies listed on the New York Stock Exchange (NYSE). One of the companies gets its revenue mainly from the technology sector, whereas the other is a pharmaceutical company. Market analysts believe that if the economy is growing the technology stock will have an annual gain of 35%, whereas the pharmaceutical stock will have an annual gain of 15%. However, if the economy experiences a recession, the technology stock will have an annual loss of 15% and the pharmaceutical stock will have an annual gain of 16%. Let each stock choice be the row player and the economy be the column player.

(a) Construct the payoff matrix for this game, where entries are percentages of increase or decrease in the value of each investment.

(b) Determine the optimal strategy for the Riblet's investment of $50,000.

(c) If the Riblet's use their optimal investment strategy, what profit can they expect to make on their investments after 1 year?

36. The Hannuses have decided to invest $80,000 into two types of mutual funds: a stock fund and a bond fund. Their financial planner believes that if the economy is growing the stock fund will have an annual gain of 30%, whereas the bond fund will have an annual gain of 10%. However, if the economy experiences a recession, the stock fund will have an annual loss of 15% and the bond fund will have

an annual gain of 12%. Let the fund choice be the row player and the economy be the column player.

(a) Construct the payoff matrix for this game, where entries are percentages of increase or decrease in the value of each investment.

(b) Determine the optimal strategy for the Hannuses investment of $80,000.

(c) If the Hannuses use their optimal investment strategy, what profit can they expect to make on their investments after 1 year?

37. Beth Toth and Mike Williams are running for governor of a certain state. The state has several urban areas and rural areas. If both candidates campaign only in urban areas, Toth is expected to get 65% of the vote, and if both campaign only in rural areas, Toth is expected to get 60% of the vote. If Toth campaigns solely in the urban areas and Williams campaigns solely in the rural areas, Toth is expected to get 35% of the vote. If Toth campaigns solely in the rural areas and Williams campaigns solely in the urban areas, Toth is expected to get 40% of the vote. Let Toth be the row player and Williams the column player.

(a) Construct the payoff matrix for this game, where the entries are percentages written as decimals.

(b) Determine the optimal strategy for Toth and Williams.

38. (*continuation of Exercise 37*) A scandal in the Toth versus Williams governor race has been reported and it is affecting the Toth campaign. If both candidates campaign only in urban areas, Toth is expected to get 55% of the vote, and if both campaign only in rural areas, Toth is expected to get 65% of the vote. If Toth campaigns solely in urban areas and Williams campaigns solely in rural areas, Toth is expected to get 45% of the vote. If Toth campaigns solely in rural areas and Williams campaigns solely in urban areas, Toth is expected to get 35% of the vote. Let Toth be the row player and Williams the column player.

(a) Construct the payoff matrix for this game, where the entries are percentages written as decimals.

(b) Determine the optimal strategy for Toth and Williams.

SECTION PROJECT

Maria and Gloria are playing a game called **matching fingers** in which they show either 1 or 2 fingers at the same time. If there is a match (both show 1 or 2 fingers), Maria wins an amount of dollars equal to the total number of fingers shown. If there is no match, Gloria wins an amount of dollars equal to the number of fingers shown. Let Maria be the row player and Gloria the column player.

(a) Write the payoff matrix for this game.

(b) Find the optimal strategies for Maria and Gloria.

(c) Determine the value of this game, assuming that both players use their optimal strategies.

(d) Does the game favor one player over the other? If so, who?

(e) Repeat parts (a) to (d), except now the game is one in which Maria and Gloria show either 1 or 3 fingers at the same time. Assume the same type of money payoff as described earlier.

Section 10.3   Game Theory and Linear Programming

From Your Toolbox

A matrix is square if the number of rows equals the number of columns.

All games that we discussed in Section 10.2 used mixed strategies and had a square matrix for the payoff matrix. (Consult the Toolbox to the left.) Specifically, all games had 2×2 payoff matrices. In this section we will see how the techniques of linear programming and the simplex method can be utilized to solve games in which the row player and column player have an arbitrary (but finite) number of choices.

The Fundamental Theorem of Game Theory

Before we get to the heart of this section, we state without proof the **fundamental theorem of game theory**.

> **Fundamental Theorem of Game Theory**
>
> For a two-person, zero-sum game with a finite payoff matrix A, there exist strategies P_{opt} and Q_{opt} (not necessarily unique) for the row player and column player, respectively, and a unique number v such that
>
> $$P_{\text{opt}} A Q \geq v$$
>
> for every strategy Q of the column player and
>
> $$P A Q_{\text{opt}} \leq v$$
>
> for every strategy P of the row player.

Some observations on the fundamental theorem of game theory follow:

- v is the **value of the game**.
- P_{opt} and Q_{opt} are the optimal strategies for the row player and the column player, respectively.
- $P_{\text{opt}} A Q$ is the expected value of the game when the row player uses the optimal strategy and the column player uses any strategy. $P A Q_{\text{opt}}$ is the expected value of the game when the column player uses the optimal strategy and the row player uses any strategy.
- The theorem says that every game has optimal strategies for the row player and the column player. Finding these strategies, as we did in Section 10.2, is called **solving the game**.
- As we saw in Section 10.2, when both players use optimal strategies, the expected value of the game is v.

We delayed introducing the fundamental theorem of game theory until now since it has some structures in it (the $\geq$ and $\leq$) that lead us to linear programming.

Linear Programming and Game Theory

Now that we have introduced the fundamental theorem of game theory, let's look at an example in which the optimal strategy for the row player can be determined by rewriting it as a linear programming problem.

Example 1 Using Linear Programming to Find an Optimal Strategy

Determine a linear programming problem that will yield the row player's optimal strategy for the payoff matrix

$$A = \begin{bmatrix} 1 & 4 \\ 3 & 2 \end{bmatrix}$$

Solution

In general, any strategy for the row and column player, respectively, is given by

$$P = [\, p_1 \quad p_2 \,] \quad \text{and} \quad Q = \begin{bmatrix} q_1 \\ q_2 \end{bmatrix}$$

This means that the expected value of the game, in general, is

$$E = PAQ = [\, p_1 \quad p_2 \,] \cdot \begin{bmatrix} 1 & 4 \\ 3 & 2 \end{bmatrix} \cdot \begin{bmatrix} q_1 \\ q_2 \end{bmatrix}$$

$$= [\, p_1 + 3p_2 \quad 4p_1 + 2p_2 \,] \cdot \begin{bmatrix} q_1 \\ q_2 \end{bmatrix}$$

$$= (p_1 + 3p_2)q_1 + (4p_1 + 2p_2)q_2$$

If the column player always chooses column 1, then $q_1 = 1$ and $q_2 = 0$, and the expected value is

$$E = p_1 + 3p_2 \tag{I}$$

However, if the column player always chooses column 2, then $q_1 = 0$ and $q_2 = 1$ and the expected value is

$$E = 4p_1 + 2p_2 \tag{II}$$

The goal of the row player is to find values for p_1 and p_2 (where $p_1 + p_2 = 1$) that maximize the expected value of the game. Let v be the smaller of the two expected values in Equations (I) and (II). In either case, we have $E \geq v$. This means that if we can maximize v then we maximize E. Also, since all entries in the payoff matrix A are positive, we can claim that v is positive, that is, $v > 0$. We can now write Equations (I) and (II) as

$$p_1 + 3p_2 \geq v$$
$$4p_1 + 2p_2 \geq v$$

Since $v > 0$, we can divide both sides of each inequality by v and not have to worry about the inequality symbol reversing direction. Doing this gives

$$\frac{p_1}{v} + 3\left(\frac{p_2}{v}\right) \geq 1$$

$$4\left(\frac{p_1}{v}\right) + 2\left(\frac{p_2}{v}\right) \geq 1$$

To simplify notation, we substitute $x_1 = \dfrac{p_1}{v}$ and $x_2 = \dfrac{p_2}{v}$ ($x_1 \geq 0$ and $x_2 \geq 0$). The inequalities can be written more simply as

$$x_1 + 3x_2 \geq 1 \tag{III}$$
$$4x_1 + 2x_2 \geq 1 \tag{IV}$$
$$x_1 \geq 0, \ x_2 \geq 0 \tag{V}$$

Inequalities (III), (IV), and (V) look like the constraints for a linear programming problem. All we need is to determine the objective function. The objective function comes from the fact that $p_1 + p_2 = 1$. Dividing both sides of $p_1 + p_2 = 1$ by v yields

$$\frac{p_1}{v} + \frac{p_2}{v} = \frac{1}{v} \quad \text{or} \quad x_1 + x_2 = \frac{1}{v}$$

Remember that we are trying to **maximize** v. This happens if $\frac{1}{v}$ is **minimized**. (As v gets larger, $\frac{1}{v}$ gets smaller.) Hence, we now have the linear programming problem as

$$\text{Minimize:} \qquad y = \frac{1}{v} = x_1 + x_2$$

$$\text{Subject to:} \qquad x_1 + 3x_2 \geq 1$$
$$4x_1 + 2x_2 \geq 1$$
$$x_1 \geq 0, x_2 \geq 0$$

Notice that once the values for x_1 and x_2 have been found we can use $\frac{1}{v} = x_1 + x_2$ to find v. Quite simply, v is

$$v = \frac{1}{x_1 + x_2}$$

Also, the substitutions $x_1 = \frac{p_1}{v}$ and $x_2 = \frac{p_2}{v}$ allow us to write the optimal row strategy as

$$p_1 = vx_1 \quad \text{and} \quad p_2 = vx_2 \qquad \blacksquare$$

Using Example 1 as a guide, we now summarize the procedure for finding the optimal row strategy when given a 2×2 payoff matrix. This can be further generalized to larger payoff matrices, as we show later in Example 3.

■ **Using Linear Programming to Determine Optimal Row Strategy**

Given a game with payoff matrix

$$A = \begin{bmatrix} a & b \\ c & d \end{bmatrix}$$

where $a, b, c,$ and d are positive numbers, we can find $P_{\text{opt}} = [\, p_1 \quad p_2 \,]$, the optimal row strategy, by solving the linear programming problem

$$\text{Minimize:} \qquad y = \frac{1}{v} = x_1 + x_2$$

$$\text{Subject to:} \qquad ax_1 + cx_2 \geq 1$$
$$bx_1 + dx_2 \geq 1$$
$$x_1 \geq 0, x_2 \geq 0$$

The expected value of the game is $v = \dfrac{1}{x_1 + x_2}$. The optimal row strategy is $P_{\text{opt}} = [\, p_1 \quad p_2 \,]$, where $p_1 = vx_1$ and $p_2 = vx_2$.

T **From Your Toolbox**

For a given matrix A, if we interchange the rows and columns, we produce the transpose of A, denoted A^T.

Before continuing, notice how the coefficients of x_1 and x_2 in the main constraints are the rows of the **transpose** of the original payoff matrix A. (Consult the Toolbox to the left.)

Going through a similar analysis as we did in Example 1, except for the column player, we would reach the following for finding the optimal column strategy.

■ Using Linear Programming to Determine Optimal Column Strategy

Given a game with payoff matrix

$$A = \begin{bmatrix} a & b \\ c & d \end{bmatrix}$$

where a, b, c, and d are positive numbers, we can find $Q_{opt} = \begin{bmatrix} q_1 \\ q_2 \end{bmatrix}$, the optimal column strategy, by solving the linear programming problem

Maximize: $\quad z = \dfrac{1}{v} = y_1 + y_2$

Subject to: $\quad ay_1 + by_2 \leq 1$

$\qquad\qquad\quad cy_1 + dy_2 \leq 1$

$\qquad\qquad\quad y_1 \geq 0,\ y_2 \geq 0$

The expected value of the game is $v = \dfrac{1}{y_1 + y_2}$. The optimal column strategy

is $Q_{opt} = \begin{bmatrix} q_1 \\ q_2 \end{bmatrix}$, where $q_1 = vy_1$ and $q_2 = vy_2$.

Here observe that the coefficients of y_1 and y_2 in the main constraints are simply the rows of the payoff matrix A. Since the coefficients of x_1 and x_2 in the main constraints in the Optimal Row Strategy box are the transpose of payoff matrix A and the coefficients of y_1 and y_2 in the main constraints in the Optimal Column Strategy box are simply the payoff matrix A, it appears that the linear programming problems for row and column strategies are actually **dual problems** (see Section 4.3). This means that we can determine both optimal strategies by using the simplex method to solve the linear programming problem for the column player's optimal strategy. The concept of duality then allows us to read the row player's optimal strategy from the bottom row of the final tableau. Example 2 illustrates.

Example 2 Solving a Game Using the Simplex Method

Solve the game that has payoff matrix

$$A = \begin{bmatrix} 1 & 4 \\ 3 & 2 \end{bmatrix}$$

Solution

Understand the Situation: This is the same payoff matrix given in Example 1. The row player's optimal strategy is found by solving

Minimize: $\quad y = \dfrac{1}{v} = x_1 + x_2$

Subject to: $\quad x_1 + 3x_2 \geq 1$

$\qquad\qquad\quad 4x_1 + 2x_2 \geq 1$

$\qquad\qquad\quad x_1 \geq 0,\ x_2 \geq 0$

The column player's optimal strategy is found by solving

$$\text{Maximize:} \qquad z = \frac{1}{v} = y_1 + y_2$$

$$\text{Subject to:} \qquad y_1 + 4y_2 \le 1$$
$$3y_1 + 2y_2 \le 1$$
$$y_1 \ge 0,\ y_2 \ge 0$$

From Your Toolbox

The column with the most negative indicator is the pivot column. If there are two, use the leftmost indicator. To find the pivot row, divide each number from the rightmost column by the corresponding number from the pivot column. The smallest nonnegative quotient gives the pivot row. For more details, see Section 4.2.

We use the simplex method on the column player's linear programming problem. The initial tableau, with first pivot element (consult the Toolbox to the left) circled, is

$$\begin{array}{ccccc} y_1 & y_2 & s_1 & s_2 & z \\ \end{array}$$
$$\left[\begin{array}{ccccc|c} 1 & 4 & 1 & 0 & 0 & 1 \\ ③ & 2 & 0 & 1 & 0 & 1 \\ \hline -1 & -1 & 0 & 0 & 1 & 0 \end{array}\right] \begin{array}{c} \text{Quotients:} \\ 1/1 \\ 1/3 \end{array}$$

$$\underbrace{\phantom{\begin{array}{ccccc} 1 & 4 & 1 & 0 & 0 \end{array}}}_{\text{Indicators}}$$

Pivoting gives us

$$\begin{array}{ccccc} y_1 & y_2 & s_1 & s_2 & z \\ \end{array}$$
$$\left[\begin{array}{ccccc|c} 1 & 4 & 1 & 0 & 0 & 1 \\ ③ & 2 & 0 & 1 & 0 & 1 \\ \hline -1 & -1 & 0 & 0 & 1 & 0 \end{array}\right] \frac{1}{3}R_2 \to R_2$$

$$\begin{array}{ccccc} y_1 & y_2 & s_1 & s_2 & z \\ \end{array}$$
$$\left[\begin{array}{ccccc|c} 1 & 4 & 1 & 0 & 0 & 1 \\ 1 & \frac{2}{3} & 0 & \frac{1}{3} & 0 & \frac{1}{3} \\ \hline -1 & -1 & 0 & 0 & 1 & 0 \end{array}\right]$$

$$\begin{array}{cc} -1R_2 + R_1 \to R_1 \\ R_2 + R_3 \to R_3 \end{array}$$
$$\begin{array}{ccccc} y_1 & y_2 & s_1 & s_2 & z \\ \end{array}$$
$$\left[\begin{array}{ccccc|c} 0 & \frac{10}{3} & 1 & -\frac{1}{3} & 0 & \frac{2}{3} \\ 1 & \frac{2}{3} & 0 & \frac{1}{3} & 0 & \frac{1}{3} \\ \hline 0 & -\frac{1}{3} & 0 & \frac{1}{3} & 1 & \frac{1}{3} \end{array}\right]$$

The next pivot element is the $\dfrac{10}{3}$ in row 1, column 2.

$$\begin{array}{ccccc} y_1 & y_2 & s_1 & s_2 & z \\ \end{array}$$
$$\left[\begin{array}{ccccc|c} 0 & ⑩⁄③ & 1 & -\frac{1}{3} & 0 & \frac{2}{3} \\ 1 & \frac{2}{3} & 0 & \frac{1}{3} & 0 & \frac{1}{3} \\ \hline 0 & -\frac{1}{3} & 0 & \frac{1}{3} & 1 & \frac{1}{3} \end{array}\right] \frac{3}{10}R_1 \to R_1$$

$$\begin{array}{ccccc} y_1 & y_2 & s_1 & s_2 & z \\ \end{array}$$
$$\left[\begin{array}{ccccc|c} 0 & 1 & \frac{3}{10} & -\frac{1}{10} & 0 & \frac{1}{5} \\ 1 & \frac{2}{3} & 0 & \frac{1}{3} & 0 & \frac{1}{3} \\ \hline 0 & -\frac{1}{3} & 0 & \frac{1}{3} & 1 & \frac{1}{3} \end{array}\right]$$

$$\begin{array}{cc} -\frac{2}{3}R_1 + R_2 \to R_2 \\ \frac{1}{3}R_1 + R_3 \to R_3 \end{array}$$
$$\begin{array}{ccccc} y_1 & y_2 & s_1 & s_2 & z \\ \end{array}$$
$$\left[\begin{array}{ccccc|c} 0 & 1 & \frac{3}{10} & -\frac{1}{10} & 0 & \frac{1}{5} \\ 1 & 0 & -\frac{1}{5} & \frac{2}{5} & 0 & \frac{1}{5} \\ \hline 0 & 0 & \frac{1}{10} & \frac{3}{10} & 1 & \frac{2}{5} \end{array}\right]$$

We now have our final tableau. From this we find that $y_1 = \dfrac{1}{5}$ and $y_2 = \dfrac{1}{5}$. This means that $y_1 + y_2 = \dfrac{2}{5}$, giving the expected value of the game to be $v = \dfrac{5}{2}$. Also, we have

$$q_1 = vy_1 \quad \text{and} \quad q_2 = vy_2$$

$$= \frac{5}{2}\left(\frac{1}{5}\right) \qquad\qquad = \frac{5}{2}\left(\frac{1}{5}\right)$$

$$= \frac{1}{2} \qquad\qquad\qquad = \frac{1}{2}$$

Hence the column player's optimal strategy is $Q_{\text{opt}} = \begin{bmatrix} q_1 \\ q_2 \end{bmatrix} = \begin{bmatrix} \frac{1}{2} \\ \frac{1}{2} \end{bmatrix}$.

Now the solution to the minimum linear programming problem (the row player's strategy) is obtained from the bottom row of the final tableau, precisely in the slack variables columns. In other words, the final tableau shows us that $x_1 = \frac{1}{10}$ and $x_2 = \frac{3}{10}$. This means that $x_1 + x_2 = \frac{4}{10} = \frac{2}{5}$, from which we again find that the expected value of the game is $v = \frac{5}{2}$. Also, we have

$$p_1 = vx_1 \quad \text{and} \quad p_2 = vx_2$$

$$= \frac{5}{2}\left(\frac{1}{10}\right) \qquad\qquad = \frac{5}{2}\left(\frac{3}{10}\right)$$

$$= \frac{1}{4} \qquad\qquad\qquad = \frac{3}{4}$$

So the row player's optimal strategy is $P_{\text{opt}} = [\, p_1 \quad p_2 \,] = \begin{bmatrix} \frac{1}{4} & \frac{3}{4} \end{bmatrix}$. As a final check, the value of the game $v = \frac{5}{2}$, should be the result of $P_{\text{opt}} \cdot A \cdot Q_{\text{opt}}$. Evaluating, we get

$$P_{\text{opt}} \cdot A \cdot Q_{\text{opt}} = \begin{bmatrix} \frac{1}{4} & \frac{3}{4} \end{bmatrix} \cdot \begin{bmatrix} 1 & 4 \\ 3 & 2 \end{bmatrix} \cdot \begin{bmatrix} \frac{1}{2} \\ \frac{1}{2} \end{bmatrix}$$

$$= \begin{bmatrix} \frac{5}{2} & \frac{5}{2} \end{bmatrix} \cdot \begin{bmatrix} \frac{1}{2} \\ \frac{1}{2} \end{bmatrix} = 2.5 = \frac{5}{2} \qquad\blacksquare$$

✓ **Checkpoint 1**   Now work Exercise 1.

Before considering another example, we need to comment on the requirement for all entries in payoff matrix A to be positive. Recall that in Example 1 this was necessary to preserve the direction of the inequalities. However, we can relax this restriction from this point forward. If some of the entries in a payoff matrix A are negative, we can add a positive number n to each entry so that we create a new payoff matrix B in which all entries are positive. Matrix B and matrix A will have exactly the same optimal strategies. To find the value of the game with matrix A, we find the value of the game with matrix B and then subtract n. All this can be summarized as follows:

Optimal strategies of a game do not change if a positive constant n is added to each entry in the payoff matrix. If w is the value of the new game, then the value of the original game is $v = w - n$.

In the Flashback that follows, we illustrate how to use this technique.

Flashback

Zach and Josh Revisited

Recall the penny-matching game that Zach and Josh are playing with the payoff matrix

$$
\begin{array}{cc}
 & \text{Josh's choice} \\
 & \begin{array}{cc} \text{H} & \text{T} \end{array} \\
\text{Zach's choice} \quad \begin{array}{c} \text{H} \\ \text{T} \end{array} & \begin{bmatrix} 4 & -5 \\ -3 & 2 \end{bmatrix}
\end{array}
$$

Use the simplex method and duals to find the optimal strategy for each player and the value of the game.

Flashback Solution

We begin by adding 6 to each entry in the payoff matrix to yield an alternative payoff matrix B with all positive entries.

$$
B = \begin{bmatrix} 10 & 1 \\ 3 & 8 \end{bmatrix}
$$

From Josh's viewpoint, the column player, the linear programming problem has the following standard maximum form:

$$
\text{Maximize:} \quad z = \frac{1}{v} = y_1 + y_2
$$

$$
\text{Subject to:} \quad 10y_1 + y_2 \le 1
$$
$$
3y_1 + 8y_2 \le 1
$$
$$
y_1 \ge 0, \, y_2 \ge 0
$$

The initial simplex tableau with pivot element circled is

$$
\begin{array}{ccccc}
y_1 & y_2 & s_1 & s_2 & z
\end{array}
$$
$$
\left[\begin{array}{ccccc|c}
⑩ & 1 & 1 & 0 & 0 & 1 \\
3 & 8 & 0 & 1 & 0 & 1 \\
\hline
-1 & -1 & 0 & 0 & 1 & 0
\end{array}\right]
$$

Pivoting gives us

$$
\begin{array}{ccccc}
y_1 & y_2 & s_1 & s_2 & z
\end{array}
$$
$$
\left[\begin{array}{ccccc|c}
⑩ & 1 & 1 & 0 & 0 & 1 \\
3 & 8 & 0 & 1 & 0 & 1 \\
\hline
-1 & -1 & 0 & 0 & 1 & 0
\end{array}\right] \quad \tfrac{1}{10}R_1 \to R_1 \quad
\begin{array}{ccccc}
y_1 & y_2 & s_1 & s_2 & z
\end{array}
$$
$$
\left[\begin{array}{ccccc|c}
1 & \frac{1}{10} & \frac{1}{10} & 0 & 0 & \frac{1}{10} \\
3 & 8 & 0 & 1 & 0 & 1 \\
\hline
-1 & -1 & 0 & 0 & 1 & 0
\end{array}\right]
$$

$$
\begin{array}{l}
-3R_1 + R_2 \to R_2 \\
R_1 + R_3 \to R_3
\end{array}
\quad
\begin{array}{ccccc}
y_1 & y_2 & s_1 & s_2 & z
\end{array}
$$
$$
\left[\begin{array}{ccccc|c}
1 & \frac{1}{10} & \frac{1}{10} & 0 & 0 & \frac{1}{10} \\
0 & \frac{77}{10} & -\frac{3}{10} & 1 & 0 & \frac{7}{10} \\
\hline
0 & -\frac{9}{10} & \frac{1}{10} & 0 & 1 & \frac{1}{10}
\end{array}\right]
$$

Next we pivot on the $\dfrac{77}{10}$ in row 2, column 2.

$$
\begin{array}{ccccc}
y_1 & y_2 & s_1 & s_2 & z \\
\end{array}
$$

$$
\left[\begin{array}{ccccc|c}
1 & \frac{1}{10} & \frac{1}{10} & 0 & 0 & \frac{1}{10} \\
0 & \boxed{\frac{77}{10}} & -\frac{3}{10} & 1 & 0 & \frac{7}{10} \\ \hdashline
0 & -\frac{9}{10} & \frac{1}{10} & 0 & 1 & \frac{1}{10}
\end{array}\right]
\quad \frac{10}{77}R_2 \to R_2 \quad
\left[\begin{array}{ccccc|c}
1 & \frac{1}{10} & \frac{1}{10} & 0 & 0 & \frac{1}{10} \\
0 & 1 & -\frac{3}{77} & \frac{10}{77} & 0 & \frac{1}{11} \\ \hdashline
0 & -\frac{9}{10} & \frac{1}{10} & 0 & 1 & \frac{1}{10}
\end{array}\right]
$$

$$
\begin{array}{ccccc}
y_1 & y_2 & s_1 & s_2 & z \\
\end{array}
$$

$$
\begin{array}{l}
-\frac{1}{10}R_2 + R_1 \to R_1 \\[4pt]
\frac{9}{10}R_2 + R_3 \to R_3
\end{array}
\left[\begin{array}{ccccc|c}
1 & 0 & \frac{8}{77} & -\frac{1}{77} & 0 & \frac{1}{11} \\
0 & 1 & -\frac{3}{77} & \frac{10}{77} & 0 & \frac{1}{11} \\ \hdashline
0 & 0 & \frac{5}{77} & \frac{9}{77} & 1 & \frac{2}{11}
\end{array}\right]
$$

From the final tableau, we find that $y_1 = \dfrac{1}{11}$ and $y_2 = \dfrac{1}{11}$. This means that $y_1 + y_2 = \dfrac{2}{11}$ and the value of the game is $v = \dfrac{11}{2}$. Also

$$
\begin{array}{lcl}
q_1 = vy_1 & \quad\text{and}\quad & q_2 = vy_2 \\[6pt]
 = \dfrac{11}{2}\left(\dfrac{1}{11}\right) & & = \dfrac{11}{2}\left(\dfrac{1}{11}\right) \\[10pt]
 = \dfrac{1}{2} & & = \dfrac{1}{2}
\end{array}
$$

So Josh's optimal strategy is $Q_{\text{opt}} = \begin{bmatrix} \frac{1}{2} \\ \frac{1}{2} \end{bmatrix}$. The row player Zach's optimal strategy is found by noting that $x_1 = \dfrac{5}{77}$ and $x_2 = \dfrac{9}{77}$. So $x_1 + x_2 = \dfrac{14}{77} = \dfrac{2}{11}$, which means again that the value of the game is $v = \dfrac{11}{2}$. Also,

$$
\begin{array}{lcl}
p_1 = vx_1 & \quad\text{and}\quad & p_2 = vx_2 \\[6pt]
 = \dfrac{11}{2}\left(\dfrac{5}{77}\right) & & = \dfrac{11}{2}\left(\dfrac{9}{77}\right) \\[10pt]
 = \dfrac{5}{14} & & = \dfrac{9}{14}
\end{array}
$$

Interpret the Solution: So Zach's optimal strategy is $P_{\text{opt}} = \begin{bmatrix} \frac{5}{14} & \frac{9}{14} \end{bmatrix}$. The value of the original game is $v = \dfrac{11}{2}$ minus the 6, which was added to the original payoff matrix. This gives $v = -\dfrac{1}{2}$.

Interactive Activity

Compare the result of the Flashback with Example 2 and Example 3 in Section 10.2.

✔ **Checkpoint 2**

Now work Exercise 11.

Our final example illustrates how the linear programming techniques and the simplex method can be applied to payoff matrices of any size.

Example 3   **Analyzing Larger Payoff Matrices**

Use linear programming and the simplex method to find the optimal strategy for the column player for a game that has payoff matrix

$$A = \begin{bmatrix} 1 & 4 & 7 \\ 5 & 3 & 8 \end{bmatrix}$$

Solution

The linear programming problem for the column player is

$$\text{Maximize:} \qquad z = \frac{1}{v} = y_1 + y_2 + y_3$$

$$\text{Subject to:} \qquad y_1 + 4y_2 + 7y_3 \leq 1$$
$$5y_1 + 3y_2 + 8y_3 \leq 1$$
$$y_1 \geq 0,\ y_2 \geq 0,\ y_3 \geq 0$$

The initial tableau with pivot element circled is

$$\begin{array}{cccccc} y_1 & y_2 & y_3 & s_1 & s_2 & z \end{array}$$
$$\begin{bmatrix} 1 & 4 & 7 & 1 & 0 & 0 & | & 1 \\ ⑤ & 3 & 8 & 0 & 1 & 0 & | & 1 \\ -1 & -1 & -1 & 0 & 0 & 1 & | & 0 \end{bmatrix}$$

Pivoting gives us

$$\begin{array}{ccccccc} & y_1 & y_2 & y_3 & s_1 & s_2 & z \end{array}$$
$$\begin{array}{c} \frac{1}{5}R_2 \rightarrow R_2 \\ \text{Then } -1R_2 + R_1 \rightarrow R_1 \\ R_2 + R_3 \rightarrow R_3 \end{array} \begin{bmatrix} 0 & \frac{17}{5} & \frac{27}{5} & 1 & -\frac{1}{5} & 0 & | & \frac{4}{5} \\ 1 & \frac{3}{5} & \frac{8}{5} & 0 & \frac{1}{5} & 0 & | & \frac{1}{5} \\ 0 & -\frac{2}{5} & \frac{3}{5} & 0 & \frac{1}{5} & 1 & | & \frac{1}{5} \end{bmatrix}$$

Pivoting on the $\dfrac{17}{5}$ in row 1, column 2 yields

$$\begin{array}{ccccccc} & y_1 & y_2 & y_3 & s_1 & s_2 & z \end{array}$$
$$\begin{array}{c} \frac{5}{17}R_1 \rightarrow R_1 \\ \text{Then } -\frac{3}{5}R_1 + R_2 \rightarrow R_2 \\ \frac{2}{5}R_1 + R_3 \rightarrow R_3 \end{array} \begin{bmatrix} 0 & 1 & \frac{27}{17} & \frac{5}{17} & -\frac{1}{17} & 0 & | & \frac{4}{17} \\ 1 & 0 & \frac{11}{17} & -\frac{3}{17} & \frac{4}{17} & 0 & | & \frac{1}{17} \\ 0 & 0 & \frac{21}{17} & \frac{2}{17} & \frac{3}{17} & 1 & | & \frac{5}{17} \end{bmatrix}$$

From our final tableau, we find that $y_1 = \dfrac{1}{17}$, $y_2 = \dfrac{4}{17}$, and $y_3 = 0$. So $y_1 + y_2 + y_3 = \dfrac{5}{17}$, which means that $v = \dfrac{17}{5}$. This gives us

$$\begin{array}{ccc} q_1 = vy_1 & q_2 = vy_2 & q_3 = vy_3 \\[4pt] = \frac{17}{5}\left(\frac{1}{17}\right) & = \frac{17}{5}\left(\frac{4}{17}\right) & = \frac{17}{5}(0) \\[4pt] = \frac{1}{5} & = \frac{4}{5} & = 0 \end{array}$$

So the column player's optimal strategy is

$$Q_{\text{opt}} = \begin{bmatrix} q_1 \\ q_2 \\ q_3 \end{bmatrix} = \begin{bmatrix} \frac{1}{5} \\ \frac{4}{5} \\ 0 \end{bmatrix}$$

Interactive Activity

Find the row player's optimal strategy in Example 3.

SUMMARY

We introduced this section with the fundamental theorem of game theory and used it to help to set up a linear programming problem to determine optimal strategies. We saw how the linear programming problems to determine P_{opt} and Q_{opt} are actually **dual problems**. The techniques outlined in Section 4.3 allowed us to solve and use duals to determine each player's optimal strategy. We also saw how to modify a payoff matrix so that it has no negative entries and how this has no effect on the optimal strategies. The section

concluded with an example of how to use linear programming to solve a game with any size of payoff matrix.

- **Fundamental Theorem of Game Theory** For a game with payoff matrix A, there exist strategies P_{opt} and Q_{opt} (not necessarily unique) for the row player and column player, respectively, and a unique number v such that $P_{opt} AQ \geq v$ for every strategy Q of the column player and $PAQ_{opt} \leq v$ for every strategy P of the row player.

SECTION 10.3 EXERCISES

In Exercises 1–10, the payoff matrix for a two-person, zero-sum game is given (some are strictly determined).

(a) Solve each game. For those that are nonstrictly determined, solve using the techniques of this section.

(b) Determine the value of each game.

✓ **1.** $\begin{bmatrix} 1 & 3 \\ 7 & 2 \end{bmatrix}$ **2.** $\begin{bmatrix} 2 & 6 \\ 4 & 1 \end{bmatrix}$

3. $\begin{bmatrix} 8 & 11 \\ 9 & 7 \end{bmatrix}$ **4.** $\begin{bmatrix} 3 & 5 \\ 9 & 4 \end{bmatrix}$

5. $\begin{bmatrix} 2 & 6 \\ 1 & 3 \end{bmatrix}$ **6.** $\begin{bmatrix} 3 & 2 \\ 1 & 1 \end{bmatrix}$

7. $\begin{bmatrix} 4 & 2 \\ 5 & 1 \\ 3 & 4 \end{bmatrix}$ **8.** $\begin{bmatrix} 5 & 1 \\ 7 & 3 \\ 2 & 6 \end{bmatrix}$

9. $\begin{bmatrix} 3 & 6 & 2 \\ 2 & 1 & 4 \end{bmatrix}$ **10.** $\begin{bmatrix} 2 & 4 & 6 \\ 4 & 5 & 3 \end{bmatrix}$

In Exercises 11–20, the payoff matrix for a two-person, zero-sum game is given. Use the duality of a linear programming problem to determine the row and column player's optimal strategies.

✓ **11.** $\begin{bmatrix} -3 & 1 \\ 2 & -2 \end{bmatrix}$ **12.** $\begin{bmatrix} 3 & -2 \\ -3 & 4 \end{bmatrix}$

13. $\begin{bmatrix} 3 & -5 \\ -9 & 4 \end{bmatrix}$ **14.** $\begin{bmatrix} -4 & 3 \\ 5 & -2 \end{bmatrix}$

15. $\begin{bmatrix} 5 & 3 & -2 \\ 2 & -2 & 3 \end{bmatrix}$ **16.** $\begin{bmatrix} -3 & 1 & 2 \\ -1 & -2 & 4 \end{bmatrix}$

17. $\begin{bmatrix} 2 & -2 & 4 \\ 3 & 4 & -3 \\ -3 & 5 & 3 \end{bmatrix}$ **18.** $\begin{bmatrix} 3 & -1 & 8 \\ 1 & 2 & -2 \\ -3 & 4 & 5 \end{bmatrix}$

19. $\begin{bmatrix} 5 & 2 \\ -3 & 4 \\ 2 & 6 \end{bmatrix}$ **20.** $\begin{bmatrix} -4 & 3 \\ 5 & -1 \\ -3 & -2 \end{bmatrix}$

Applications

21. The Hoffman Company introduces its new product on the market with a major television and radio advertising campaign. Ty, the CEO of the Hoffman Company, finds out

that its major competitor, the Baker Company, is planning a major advertising campaign for a comparable product. A market research firm has determined the following payoff matrix, where each entry shows the sales (in millions of dollars) either gained or lost by one company to the other.

$$\text{Baker} \quad \begin{array}{c} \\ \text{TV} \\ \text{Radio} \end{array} \begin{array}{c} \text{Hoffman} \\ \begin{array}{cc} \text{TV} & \text{Radio} \end{array} \\ \begin{bmatrix} 2 & -1 \\ -1 & 1 \end{bmatrix} \end{array}$$

(a) Use the techniques of this section to determine the optimal strategy for each company.

(b) Does the game favor anyone? If so, who?

22. The Hoffman Company introduces its new product on the market with a newspaper and mail circular advertising campaign. Ty, the CEO of the Hoffman Company, finds out that its major competitor, the Baker Company, is planning a major advertising campaign for a comparable product. A market research firm has determined the following payoff matrix, where each entry shows the sales (in millions of dollars) either gained or lost by one company to the other.

$$\text{Baker} \quad \begin{array}{c} \\ \text{Paper} \\ \text{Mail} \end{array} \begin{array}{c} \text{Hoffman} \\ \begin{array}{cc} \text{Paper} & \text{Mail} \end{array} \\ \begin{bmatrix} 1 & -1 \\ -2 & 3 \end{bmatrix} \end{array}$$

(a) Use the techniques of this section to determine the optimal strategy for each company.

(b) Does the game favor anyone? If so, who?

23. Jessie and Jamie play the following game: Jessie hides a coin in one of her hands. If Jamie guesses which hand the coin is in, Jessie gives Jamie $4. If Jamie does not guess correctly, Jamie gives Jessie $3 if the coin is in the right hand and $6 if the coin is in the left hand. Let Jessie be the row player and Jamie the column player.

(a) Write the payoff matrix associated with this game.

(b) Use the techniques of this section to determine the optimal strategy for each player.

(c) Does this game favor anyone? If so, who?

24. Two-finger morra is a game in which two players, at a predetermined signal, simultaneously show 1 or 2 fingers. Jose and Maria are playing a version of two-finger morra in which if the total number of fingers shown is even Maria pays Jose $2. However, if the total number of fingers shown is odd, Jose pays Maria $3. Let Maria be the row player and Jose the column player.

(a) Write the payoff matrix associated with this game.

(b) Use the techniques of this section to determine the optimal strategy for each player.

(c) Does this game favor anyone? If so, who?

25. Based on preliminary tests, a physician determines that Jacques has one of two possible diseases, labeled 1 and 2. The doctor can prescribe one of three treatments, labeled A, B, and C. Each treatment is effective on certain diseases but less so for others. The payoff matrix for this scenario is

$$
\begin{array}{c}
\quad\quad\quad\quad \text{Diseases} \\
\quad\quad\quad\quad 1 \quad\; 2 \\
\text{Treatment}\;\; \begin{array}{c} A \\ B \\ C \end{array} \left[\begin{array}{rr} 4 & 2 \\ -1 & 1 \\ -2 & 3 \end{array}\right]
\end{array}
$$

Determine the optimal strategy for the physician.

SECTION PROJECT

The game rock, paper, scissors is well known in many parts of the world. In this game, two players simultaneously show a hand in the form of a fist (rock), open hand (paper), or two fingers extended (scissors). The payoff for this game is one unit according to the following rules: paper beats rock, scissors beats paper, and rock beats scissors. If the players present the same choice, the game is a tie and the payoff is 0.

(a) Set up the payoff matrix for this game.

(b) Using the methods of this section, determine the optimum strategies for the row player and the column player.

(c) Determine the value of this game. Are you surprised?

■ Why We Learned It

In Chapter 10 we saw that game theory was an interesting topic that reviewed many topics in this text. We saw that probability, expected value, matrices, linear programming, and the simplex method are used in the study of game theory. Situations in which game theory arises are many and varied. A political campaign manager may employ game theory and consider the game to be the candidate for office and the media. This is especially true if there is some potentially damaging information in the candidate's past. The decision on whether to remodel a store can become a game between the store manager and the economy. If market research can determine expected profits, or losses, for the store based on the decision of remodeling or not and whether the economy remains strong or slips into recession, techniques of game theory can help the store owner to make a more informed decision. Game theory can also be applied to games such as matching fingers or two-finger morra.

■ CHAPTER REVIEW EXERCISES

In Exercises 1–4, use the following payoff matrix for a two-person, zero-sum game. The entries represent dollars.

$$
\begin{array}{c}
\quad\quad\quad \text{Grace} \\
\text{Amy}\; \left[\begin{array}{rrr} -3 & -2 & 4 \\ 1 & 6 & 2 \\ -5 & 1 & 4 \end{array}\right]
\end{array}
$$

Determine the payoff when the following strategies are used.

1. Amy plays row 1, Grace plays column 1.

2. Amy plays row 1, Grace plays column 3.

3. Amy plays row 2, Grace plays column 2.

4. Amy plays row 3, Grace plays column 1.

In Exercises 5 and 6, you are given a payoff matrix for a game. Construct a new payoff matrix for the game where the row player becomes the column player and the column player becomes the row player.

$$
\begin{array}{c}
\quad\quad\quad \text{Sharon} \\
\textbf{5.}\;\; \text{Ron}\; \left[\begin{array}{rr} 2 & -5 \\ -1 & -3 \end{array}\right]
\end{array}
$$

Nelle

6. Quinn $\begin{bmatrix} -2 & 1 & -3 \\ 3 & -4 & 2 \\ 2 & 3 & -5 \end{bmatrix}$

In Exercises 7–10, you are given a payoff matrix for a two-person, zero-sum game that is strictly determined.

(a) Determine the saddle point(s).

(b) Determine the optimal strategy for each player.

(c) Determine the value of the game and state which person the game favors.

Jen

7. Mike $\begin{bmatrix} 3 & -5 \\ 4 & 2 \end{bmatrix}$

Larry

8. Fran $\begin{bmatrix} 5 & 4 & 3 \\ 1 & 4 & 1 \\ -1 & -3 & 2 \end{bmatrix}$

Mark

9. Lindsay $\begin{bmatrix} -2 & -1 \\ -1 & 0 \\ 0 & -3 \\ 3 & 5 \end{bmatrix}$

Nicole

10. Troy $\begin{bmatrix} -1 & -3 & 6 & 8 \\ 0 & -2 & 1 & 4 \\ 4 & -5 & -5 & 7 \\ 3 & -2 & -1 & 0 \end{bmatrix}$

11. A professional indoor soccer team is contemplating moving to another city. The team owner determines that the main determining factor in the team's decision is whether they will have a winning or losing season. If the team moves and has a winning season, they will generate $120,000 more than last season due to increased ticket sales. If the team moves and has a losing season, they will generate $40,000 less than last season due to decreased ticket sales. If the team does not move and they have a winning season, they will generate $50,000 more than last season due to increased ticket sales. If the team does not move and they have a losing season, they will generate $8000 less than last season due to decreased ticket sales. The team's owner summarizes this information in the following payoff matrix, where the two-person game is the team against their season. Find the saddle point and determine the best strategy for the team.

Season

Winning Losing

Team $\begin{matrix} \text{Move} \\ \text{Don't move} \end{matrix}$ $\begin{bmatrix} 120{,}000 & -40{,}000 \\ 50{,}000 & -8000 \end{bmatrix}$

12. Mark's Auto and Jim's Auto are two competing used car dealerships. On a weekly basis each dealership selects one (and only one) of the following types of advertisement: television, radio, or newspaper. A survey by a market research company yields the following payoff matrix. Entries represent the percentage of the market gain or loss for each choice of action by the dealerships. (Assume that a gain by Mark's Auto is a loss for Jim's Auto, and vice versa.)

Jim's Auto

TV   Radio Newspaper

Mark's Auto $\begin{matrix} \text{TV} \\ \text{Radio} \\ \text{Newspaper} \end{matrix}$ $\begin{bmatrix} 3 & 2 & 1 \\ -2 & -1 & 0 \\ -1 & 2 & -3 \end{bmatrix}$

(a) Is this a strictly determined game?

(b) Find the saddle point and determine the best strategy for each dealership.

(c) Determine the value of the game. Does this game favor one dealership over the other? If so, which dealership has the advantage?

13. Two competing ski resorts, Willard Mountain and Gorham Mountain, are considering expanding the type of daily ski passes that they offer. Both mountains currently only offer full-day passes. Willard Mountain is considering offering a 2-hour, 3-hour, or 4-hour pass and Gorham Mountain is considering offering a 2-hour or 3-hour pass. Each day both mountains decide on one of the passes to offer. If Willard Mountain offers a 2-hour pass, it will lose 2% of the market if Gorham Mountain offers a 2-hour pass and gain 4% of the market if Gorham Mountain offers a 3-hour pass. If Willard Mountain offers a 3-hour pass, it will see no gain if Gorham Mountain offers a 2-hour pass and a 1% gain of the market if Gorham Mountain offers a 3-hour pass. If Willard Mountain offers a 4-hour pass, it will lose 1% of the market if Gorham Mountain offers a 2-hour pass and lose 2% of the market if Gorham Mountain offers a 3-hour pass. (Assume that a gain by Willard Mountain is a loss for Gorham Mountain, and vice versa.) Let Willard Mountain be the row player and Gorham Mountain be the column player.

(a) Construct the payoff matrix for this game, where the entries are percentages of the market.

(b) Verify that the game is strictly determined.

(c) Determine the optimal strategy for each ski resort.

14. (*continuation of Exercise 13*)

(a) Determine the value of the game in Exercise 13.

(b) Does the game in Exercise 13 favor one ski resort over the other? If so, which ski resort has the advantage?

15. A bookstore is considering expanding to include a music section and a coffee bar. A market research firm determines that the main factor in the bookstore's decision is the economy and whether it remains strong or slips into recession. The firm determines that if the bookstore expands and the economy is strong, the store will earn

$50,000. However, if the store expands and the economy slips into recession, the store will lose $100,000. If the store does not expand and the economy is strong, the store will lose $75,000 due to customers going to other stores. If the store does not expand and there is a recession, the store will lose $80,000. This is a two-person game between the store and the economy. Let the store be the row player and the economy be the column player.

(a) Construct the payoff matrix for the game.

(b) Verify that the game is strictly determined.

(c) Determine the optimal strategy for the bookstore.

In Exercises 16–18, determine the expected value E for each game whose payoff matrix and strategies P and Q (row and column player, respectively) are given.

16. $\begin{bmatrix} -2 & 3 \\ 5 & 1 \end{bmatrix}$, $P = \begin{bmatrix} \frac{1}{4} & \frac{3}{4} \end{bmatrix}$, $Q = \begin{bmatrix} \frac{5}{6} \\ \frac{1}{6} \end{bmatrix}$

17. $\begin{bmatrix} -2 & 0 & 1 \\ 3 & -1 & 2 \\ 4 & -5 & 6 \end{bmatrix}$, $P = \begin{bmatrix} 0.1 & 0.3 & 0.6 \end{bmatrix}$,

$Q = \begin{bmatrix} 0.2 \\ 0.3 \\ 0.5 \end{bmatrix}$

18. $\begin{bmatrix} 1 & -1 & -2 \\ -2 & 3 & 2 \end{bmatrix}$, $P = \begin{bmatrix} 0.25 & 0.75 \end{bmatrix}$,

$Q = \begin{bmatrix} 0.15 \\ 0.35 \\ 0.50 \end{bmatrix}$

19. A certain game has a payoff matrix $\begin{bmatrix} 2 & -1 \\ -4 & 3 \end{bmatrix}$.

Compute the expected value E for the pairs of strategies given in parts (a) to (d). Which of these strategies is most favorable to the column player?

(a) $P = \begin{bmatrix} \frac{1}{2} & \frac{1}{2} \end{bmatrix}$, $Q = \begin{bmatrix} \frac{1}{2} \\ \frac{1}{2} \end{bmatrix}$

(b) $P = \begin{bmatrix} \frac{1}{4} & \frac{3}{4} \end{bmatrix}$, $Q = \begin{bmatrix} \frac{1}{6} \\ \frac{5}{6} \end{bmatrix}$

(c) $P = \begin{bmatrix} \frac{1}{3} & \frac{2}{3} \end{bmatrix}$, $Q = \begin{bmatrix} \frac{2}{3} \\ \frac{1}{3} \end{bmatrix}$

(d) $P = \begin{bmatrix} \frac{1}{8} & \frac{7}{8} \end{bmatrix}$, $Q = \begin{bmatrix} \frac{1}{2} \\ \frac{1}{2} \end{bmatrix}$

In Exercises 20–22, determine the optimal strategies P_{opt} and Q_{opt} for the row and column players, respectively. Be sure to check for saddle points since some may be strictly determined games.

20. $\begin{bmatrix} -2 & 0 \\ 3 & -1 \end{bmatrix}$

21. $\begin{bmatrix} 2 & -3 \\ -8 & -6 \end{bmatrix}$

22. $\begin{bmatrix} -\frac{1}{3} & \frac{2}{3} \\ \frac{1}{2} & -\frac{1}{4} \end{bmatrix}$

In Exercises 23–25, use the specified payoff matrix for a two-person, zero-sum game given in Exercises 20–22.

(a) Determine the value of the game.

(b) Determine which player (if any) the game favors assuming that the row player and the column player use their optimal strategies.

23. Use the game in Exercise 20.

24. Use the game in Exercise 21.

25. Use the game in Exercise 22.

26. Sopris Mountaineering is deciding to introduce new guided trips to two new destinations, either Mount Cook or Denali. To advertise, Sopris Mountaineering will launch a major magazine and newspaper campaign. Greg, the CEO of Sopris Mountaineering, finds out that its major competitor, Mountain Adventures, is also planning an advertising campaign for comparable trips. A market research firm has determined the following payoff matrix, where each entry shows the sales (in dollars) either gained or lost by one company to the other.

		Mountain Adventures	
		Magazine	Newspaper
Sopris	Magazine	$20,000$	$-12,000$
Mountaineering	Newspaper	-8000	$15,000$

(a) Find the optimal strategies for Sopris Mountaineering and Mountain Adventures.

(b) Find the value of the game assuming that both use the optimal mixed strategy. Does the game favor anyone? If so, who?

27. A bookstore is considering expanding to include a music section and a coffee bar. A market research firm determines that the main factor in the bookstore's decision is the economy and whether it remains strong or slips into recession. The firm determines that if the bookstore expands and the economy is strong the store will earn $50,000. However, if the store expands and the economy slips into recession, the store will lose $100,000. If the store does not expand and the economy is strong, the store will lose $125,000 due to customers going to other stores. If the store does not expand and there is a recession, the store will lose $80,000. This is a two-person game between the store and the economy. Let the store be the row player and economy be the column player.

(a) Construct the payoff matrix for this game where the entries are in thousands of dollars.

(b) Determine the optimal strategy for the bookstore.

(c) If you were the owner of the bookstore, what would you do?

28. (*continuation of Exercise 27*)

(a) Assuming the optimal strategies in Exercise 27, determine the value of the game.

(b) Does the game in Exercise 27 favor anyone? If so, who?

29. Kris and Rachel are running for senior class president. They may deliver their campaign speech at all school meetings and football games. If both candidates campaign only at all school meetings, Kris is expected to get 55% of the vote, and if both campaign only at football games, Kris is expected to get 60% of the vote. If Kris campaigns solely at all school meetings and Rachel campaigns solely at football games, Kris is expected to get 40% of the vote. If Kris solely campaigns at football games and Rachel solely campaigns at all school meetings, Kris is expected to get 45% of the vote. Let Kris be the row player and Rachel be the column player.

(a) Construct the payoff matrix for this game where the entries are percentages written as decimals.

(b) Determine the optimal strategy for Kris and Rachel.

30. (*continuation of Exercise 29*) The school newspaper has printed some information that has affected Kris's campaign. If both candidates campaign only at all school meetings, Kris is expected to get 40% of the vote, and if both campaign only at football games, Kris is expected to get 55% of the vote. If Kris campaigns solely at all school meetings and Rachel campaigns solely at football games, Kris is expected to get 35% of the vote. If Kris campaigns solely at football games and Rachel solely campaigns at all school meetings, Kris is expected to get 45% of the vote. Let Kris be the row player and Rachel be the column player.

(a) Construct the payoff matrix for this game where the entries are percentages written as decimals.

(b) Determine the optimal strategy for Kris and Rachel.

In Exercises 31– 35, the payoff matrix for a two-person, zero-sum game is given (some are strictly determined).

(a) Solve the game. For those that are not strictly determined, solve using the techniques of Section 10.3.

(b) Determine the value of each game.

31. $\begin{bmatrix} 3 & 7 \\ 8 & 4 \end{bmatrix}$ **32.** $\begin{bmatrix} 3 & 5 \\ 4 & 6 \end{bmatrix}$

33. $\begin{bmatrix} 3 & 1 \\ 4 & 3 \\ 2 & 4 \end{bmatrix}$ **34.** $\begin{bmatrix} 3 & 1 & 4 \\ 2 & 3 & 5 \end{bmatrix}$

35. $\begin{bmatrix} 2 & 3 & 4 \\ 1 & 4 & 6 \end{bmatrix}$

In Exercises 36–40, the payoff matrix for a two-person, zero-sum game is given. Use the duality of a linear programming problem to determine the row and column players' optimal strategies.

36. $\begin{bmatrix} -2 & 1 \\ 5 & -4 \end{bmatrix}$ **37.** $\begin{bmatrix} 1 & -6 \\ -4 & 8 \end{bmatrix}$

38. $\begin{bmatrix} 2 & -3 & 5 \\ -3 & 1 & 4 \end{bmatrix}$

39. $\begin{bmatrix} -2 & 2 & 4 \\ 0 & 1 & -3 \\ 1 & -1 & 3 \end{bmatrix}$

40. $\begin{bmatrix} -3 & -4 \\ -2 & 0 \\ 1 & -3 \end{bmatrix}$

41. The Babcock Company introduces its new product on the market with a major newspaper and magazine campaign. Grace, the CEO of the Babcock Company, finds that its major competitor, the Mayhew Company, is planning a major newspaper and magazine campaign for a comparable product. A market research firm has determined the following payoff matrix, where each entry shows the sales (in millions of dollars) either gained or lost by one company to the other.

$$\begin{array}{cc} & \text{Mayhew} \\ & \begin{array}{cc} \text{Newspaper} & \text{Magazine} \end{array} \\ \text{Babcock} \begin{array}{c} \text{Newspaper} \\ \text{Magazine} \end{array} & \begin{bmatrix} 3 & -2 \\ -1 & 1 \end{bmatrix} \end{array}$$

(a) Use the techniques of Section 10.3 to determine the optimal strategy for each company.

(b) Does the game favor anyone? If so, who?

42. (*continuation of Exercise 41*) The Mayhew Company has decided to launch a television campaign in addition to the newspaper and magazine campaign. Based on this new campaign, a market research firm has determined the following payoff matrix, where each entry shows the sales (in millions of dollars) either gained or lost by one company to the other.

$$\begin{array}{cc} & \text{Mayhew} \\ & \begin{array}{ccc} \text{Newspaper} & \text{Magazine} & \text{TV} \end{array} \\ \text{Babcock} \begin{array}{c} \text{Newspaper} \\ \text{Magazine} \end{array} & \begin{bmatrix} 3 & -2 & -3 \\ -1 & 1 & -5 \end{bmatrix} \end{array}$$

(a) Determine the optimal strategy for each company.

(b) Does the game favor anyone? If so, who?

43. (a) Use the techniques of Section 10.2 to verify the solutions to Exercise 41(a).

(b) What is the value of the game?

44. Two-finger morra is a game in which two players, at a predetermined signal, simultaneously show 1 or 2 fingers. Lissa and Lera are playing a version of the two-finger morra in which, if the total number of fingers shown is odd, Lissa pays Lera $2. However, if the total number of fingers shown is even, Lera pays Lissa $1. Let Lissa be the row player and Lera be the column player.

(a) Write the payoff matrix associated with this game.

(b) Use the techniques of Section 10.3 to determine the optimal strategy for each player.

(c) Verify the solution to part (b) by using the techniques of Section 10.2.

(d) What is the value of the game?

(e) Does the game favor anyone? If so, who?

45. Based on preliminary tests, a physician determines that Mika has one of two possible diseases, labeled 1 and 2. The doctor can prescribe one of three treatments, labeled A, B, and C. Each treatment is effective on certain diseases, but less so for others. The payoff matrix for this scenario is

$$\begin{array}{c} & \text{Disease} \\ & \begin{array}{cc} 1 & 2 \end{array} \\ \text{Treatment} \begin{array}{c} A \\ B \\ C \end{array} & \left[\begin{array}{cc} -1 & 2 \\ -2 & 4 \\ 3 & -2 \end{array}\right] \end{array}$$

Determine the optimal strategy for the physician.

CHAPTER 10 PROJECT

Kids love to receive money, even if they have to do chores to earn it. Consider a two-person game between brothers in which the goal is to make the most money from the completed chores. Suppose that Donald can either do the laundry or wash cars, while Matthew can vacuum, do the dishes, or mow the lawn. Consider the following values for the chores:

If Donald does the laundry and if Matthew vacuums, Donald receives $3.

If Donald does the laundry and if Matthew does the dishes, Matthew receives $5.

If Donald does the laundry and if Matthew mows the lawn, Donald receives $4.

If Donald washes the cars and if Matthew vacuums, Donald receives $6.

If Donald washes the cars and if Matthew does the dishes, Donald receives $7.

If Donald washes the cars and if Matthew mows the lawn, Matthew receives $6.

Suppose that Nicole, the boys' older sister, realizes that this is actually a two-person game between Donald and Matthew. Nicole is in charge of distributing the money between the boys. She knows that if Donald makes any money it is considered a loss to Matthew. Also, if Matthew makes any money, it is considered a loss to Donald. Only one brother can be paid based on whether his chore was more valuable.

1. Determine the payoff matrix, A, for this game for which each entry is the dollar value of the chore. Have Donald be the row player and Matthew the column player. Let L stand for does laundry, W for washes cars, V for vacuums, D for does dishes, and M for mows lawn.

2. Determine the location of the saddle point for the payoff matrix (if any). Is the game strictly determined? Why or why not?

3. Does the payoff matrix have all nonnegative entries? If not, add a positive number, n, to each entry.

4. Find a linear programming problem to determine the optimal (row) strategy for Donald.

5. Find a linear programming problem to determine the optimal (column) strategy for Matthew.

6. If both linear programming problems in questions 4 and 5 were solved, what would be true about their optimal values? Why?

7. Use the Simplex Method to solve the linear programming problem in question 5.

8. Determine the optimal (row) strategy for Donald, the optimal (column) strategy for Matthew, and the expected value of the game. Be sure to subtract n from the expected value of the game (where n is the number added in 3).

9. In the long run, to whom would Nicole pay the most?

Appendices

Appendix A
Prerequisites of Finite Mathematics

Section A.1   Percents, Decimals, and Fractions

In many applications, we are given information in the form of a **percentage**. When we hear facts such as "15.5% of the U.S. military in 1998 was made up of females" or "In 1997, 25.6% of all the lumber exported from the United States went to Japan" we are using percentages. Whenever we perform mathematical calculations involving percentages, we convert them to **decimal** form. Since the word **percent** means "per 100," we convert a percent into a decimal form by dividing by 100. For example, 25.6% in decimal form is determined by

$$25.6\% = \frac{25.6}{100} = 0.256$$

This illustrates the following shortcut used in converting percents to decimals.

■ **Converting Percents to Decimals**

To convert a percent into decimal form, we move the decimal point two places to the left and remove the percent sign.

Example 1   **Converting Percents to Decimals**

Rewrite each of the following in decimal form.

  (a) 73.1%   **(b)** 3.1%

Solution

  (a) When we move the decimal two places to the left and remove the percent sign, we see that

$$73.1\% = 0.731$$

  (b) Since there is only one digit to the left of the decimal place, we insert a zero to get

$$3.1\% = 0.031$$

✓ **Checkpoint 1**

Now work Exercise 3 in the A.1 Exercises.

Let's see how percent to decimal conversion is used in an application.

Example 2   **Applying Percent to Decimal Conversion**

In 1997, about 24% of the doctorates conferred in earth science were awarded to females (*Source:* National Science Foundation, www.nsf.gov). If 862 earth science doctorates were awarded in 1997, how many went to females?

Solution

Understand the Situation: To determine the number of females receiving earth science doctorates, we must take 24% of those receiving doctorates.
   This is determined by performing

$$24\% \text{ of } \left(\begin{array}{c}\text{total earth} \\ \text{science doctorates}\end{array}\right) = 0.24 \cdot 862 = 206.88 \approx 207$$

Interpret the Solution: This means that of the 862 students receiving earth science doctorates in 1997 about 207 of them were female.   ∎

Converting Fractions to Percents

Another often used tool in many of our applications is the conversion of fractions into percents. To do this, we first convert the fraction to a decimal and then convert the decimal into a percent.

▌ **Converting Fractions to Percents**

To convert a fraction into a percent, first convert the fraction into a decimal by dividing the numerator of the fraction by its denominator. Then multiply the decimal number by 100 and add on a percent sign.

▷ **Note:** Converting decimals to percents is equivalent to moving the decimal point to the right two places and then adding on a percent sign.

Example 3   **Converting Fractions to Percents**

Convert each of the following fractions into a percent.

(a) $\dfrac{4}{5}$ (b) $\dfrac{47}{200}$

Solution

(a) We begin by writing the fraction in decimal form as follows.

$$\frac{4}{5} = 4 \div 5 = 0.80$$

Now we multiply the quotient by 100 and add on a percent sign.

$$0.80 \cdot 100 = 80\%$$

(b) Dividing the numerator by the denominator gives us

$$\frac{47}{200} = 47 \div 200 = 0.235$$

Multiplying this decimal number by 100 gives us

$$0.235 \cdot 100 = 23.5\%$$

∎

 Checkpoint 2

Now work Exercise 15 in the A.1 Exercises.

Many of the applications that we study involve finding what percentage of a whole is made up by a small group. In these cases, converting the quantity to a fraction can make the solution easier to understand. Now let's look at an application of fraction to percentage conversion.

Example 4   Converting Fractions to Percents in Applications

Interactive Activity

The information in Example 4 shows that the number of farms in 1992 was about 1,925,000, with 283,000 being 50- to 99-acre farms. Use these numbers to convert to a percent. Is the answer the same as in Example 4? Explain.

In 1992 there were about 1925 thousand farms in the United States and 283 thousand of these had sizes between 50 and 99 acres (*Source:* U.S. Department of Agriculture, www.usda.gov). Convert $\dfrac{283}{1925}$ to a percentage and interpret.

Solution
Dividing this fraction gives us

$$\frac{283}{1925} \approx 0.147$$

Now converting to a percent gives us

$$0.147 \cdot 100 = 14.7\%$$

Interpret the Solution: This means that in 1992 about 14.7% of all U.S. farms had a size between 50 and 99 acres.

∎

SECTION A.1 EXERCISES

In Exercises 1–8, write each of the following in decimal form.

1. 73%   **2.** 56%

✓ **3.** 4%   **4.** 8%

5. 59.1%   **6.** 46.4%

7. 0.7%   **8.** 0.9%

🌐 **9.** In 1990 the average daily cost of hospital care was $1104, of which 47.5% of the charges were reimbursed by private health insurance. How much per day was covered by private health insurance? (*Source:* U.S. Census Bureau, www.census.gov).

🌐 **10.** In 1997 the average daily cost of hospital care was $2131, of which 49.9% of the charges were reimbursed by private health insurance. How much per day was covered by private health insurance? (*Source:* U.S. Census Bureau, www.census.gov).

🌐 **11.** In 1980 about 9866 million board feet of lumber was imported to the United States from Canada. If this represents 97.5% of all the lumber imported that year, how many board feet does this represent? (*Source:* U.S. Census Bureau, www.census.gov).

🌐 **12.** In 1998 there were about 272 trillion cubic feet of natural gas produced worldwide. If the United States produced 23.1% of all the natural gas in 1998, how many trillion cubic feet of natural gas does this represent? (*Source:* U.S. Census Bureau, www.census.gov).

In Exercises 13–20, convert each of the following fractions to a percent.

13. $\dfrac{1}{5}$ **14.** $\dfrac{3}{4}$

✓ **15.** $\dfrac{1}{8}$ **16.** $\dfrac{7}{8}$

17. $\dfrac{267}{500}$ **18.** $\dfrac{117}{250}$

19. $\dfrac{3}{7}$ **20.** $\dfrac{4}{11}$

21. The 2008 projected school enrollment of students in kindergarten through eighth grade is about 38 million, of which 4.6 million are projected to be in private schools. Convert $\dfrac{4.6}{38}$ to a percent and interpret. (*Source:* U.S. National Center for Education Statistics, www.nces.ed.gov).

22. In 1988 the school enrollment of students in kindergarten through eighth grade was about 38.4 million, of which about 4 million were enrolled in private schools. Convert $\dfrac{4}{38.4}$ to a percent and interpret. (*Source:* U.S. National Center for Education Statistics, www.nces.ed.gov).

23. In 1985 the United States spent about $49,887 million on research and development, of which $2725 million was spent on technology research and development. Convert $\dfrac{2725}{49,887}$ to a percent and interpret. (*Source:* U.S. Census Bureau, www.census.gov).

24. In 1999 the United States spent about $77,225 million on research and development, of which $8037 million was spent on technology research and development. Convert $\dfrac{8037}{77,225}$ to a percent and interpret. (*Source:* U.S. Census Bureau, www.census.gov).

Section A.2   Evaluating Expressions and Order of Operations

In mathematics an **expression** can be thought of as a combination of arithmetic operations (addition, subtraction, multiplication, and division), along with exponents and parentheses. An expression can contain numbers and variables. By **evaluating** an expression, we mean to perform as many of these operations as possible. To evaluate expressions in a consistent manner, we use the **order of operations** to perform the simplifications.

■ Order of Operations

Given a mathematical expression, we perform all operation in the following order.

1. **Parentheses**: Simplify all expressions in parentheses or other grouping symbols such as fraction bars. Calculate from the innermost parentheses out. In the case of a fraction bar, calculate the numerator and denominator separately and then perform the division.
2. **Exponents**: Simplify all expressions containing exponents by raising all numbers to their indicated powers.
3. **Multiplication and Division**: Perform all multiplications and divisions, reading from left to right, in the order that they appear.
4. **Addition and Subtraction**: Perform all additions and subtractions, reading from left to right, in the order that they appear.

▶ **Note:** Some recall the order of operations by remembering the pneumonic **P**lease **E**xcuse **M**y **D**ear **A**unt **S**ally.

Example 1   Performing the Order of Operations

Evaluate the following expressions:

(a) $4 - 36 \div 4 + 3$

(b) $(2 + 13) \cdot 2 - 3^3 + 6$

(c) $\dfrac{6 - 3}{4}$

(d) $-2^2 + 3(5)$

Solution

(a) Since no parentheses or exponents are involved, we start by performing the division.

$$4 - \mathbf{36 \div 4} + 3 = 4 - 9 + 3$$

Now we can perform the subtraction and the addition, going from left to right.

$$\mathbf{4 - 9} + 3 = -5 + 3 = -2$$

(b) Completing the operation within the parentheses gives

$$(\mathbf{2 + 13}) \cdot 2 - 3^3 + 6 = 15 \cdot 2 - 3^3 + 6$$

Then we simplify the exponential part of the expression to get

$$15 \cdot 2 - \mathbf{3^3} + 6 = 15 \cdot 2 - 27 + 6$$

Multiplying yields

$$\mathbf{15 \cdot 2} - 27 + 6 = 30 - 27 + 6$$

Now we go from left to right, performing the addition and subtraction.

$$\mathbf{30 - 27} + 6 = 3 + 6 = 9$$

(c) For this fraction, we first simplify the numerator and then divide.

$$\frac{6 - 3}{4} = \frac{3}{4} = 0.75$$

(d) For this expression, we begin by squaring the 2. Since we can write -2 as $-1 \cdot 2$, -2^2 really means $-1 \cdot 2^2$. Notice that the result is still negative.

$$-\mathbf{2^2} + 3(5) = -4 + 3(5)$$

Now we perform the multiplication implied by the parentheses and finally the addition to get

$$-4 + \mathbf{3(5)} = -4 + 15 = 11 \qquad \blacksquare$$

Technology Option

We can evaluate expressions on a graphing calculator fairly easily. Figures A.2.1 through A.2.4 show the results of Example 1, parts (a)–(d), respectively.

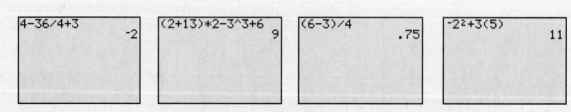

| Figure A.2.1 | Figure A.2.2 | Figure A.2.3 | Figure A.2.4 |

To see how expressions are evaluated on your calculator, consult the on-line calculator manual at www.prenhall.com/armstrong.

✓ **Checkpoint 1**   Now work Exercise 5 in the A.2 Exercises.

Many applications involve evaluating an expression that comes from a given formula. Example 2 illustrates how the order of operations can be used in an application.

Example 2   Using the Order of Operations in an Application

In 1998 about 77% of all Asian Americans living in the United States were anywhere from $31.9 - 1.2(20.1)$ to $31.9 + 1.2(20.1)$ years old. (*Source:* U.S. Census Bureau, www.census.gov). Simplify these expressions and interpret.

Solution

When simplifying these expressions, we must first perform the multiplication implied by the parentheses and then complete the addition and subtraction.

$$31.9 - 1.2(20.1) = 31.9 - 24.12 \qquad 31.9 + 1.2(20.1) = 31.9 + 24.12$$
$$= 7.78 \qquad\qquad\qquad\qquad\qquad = 56.02$$

Interpret the Solution: This means that in 1997 about 77% of all Asian Americans living in the United States were anywhere from 7.78 to 56.02 years old. ∎

SECTION A.2 EXERCISES

In Exercises 1–8, evaluate the following expressions without the use of a calculator.

1. $3 + 4 \cdot 7$

2. $2 + 5 \cdot 8$

3. $9 - 8 \div 4 + 1$

4. $5 + 8 \div 2 + 7$

✓ **5.** $(5 - 2)^2 - 5$

6. $(1 + 3)^2 - 1$

7. $2(3) + 4(1)$

8. $6(8) + 2(4)$

In Exercises 9–18, evaluate the following expressions using a calculator.

9. $\dfrac{7 - 3}{0.5}$

10. $\dfrac{10 - 6}{0.2}$

11. $3(-2)^2 - 4(-3)^2$

12. $2(-1)^2 - 6(-3)^2$

13. $12(0.6)^2(0.4)^3$

14. $8(0.7)^4(0.3)^2$

15. $5^2 + 2 \cdot 9 - (4 - 6 \div 3)$

16. $2^3 + 3 \cdot 4 - (6 - 15 \div 5)$

17. $(0.7)(0.3) + (0.4)(0.6)$

18. $(0.1)(0.9) + (0.5)(0.5)$

19. In 1996 the middle 50% of the unmarried mothers in the United States had ages from $23.3 - 0.67(5.9)$ to $23.3 + 0.67(5.9)$ years old (*Source:* U.S. National Center for Health Statistics, www.cdc.gov/nchs). Simplify these expressions and interpret.

20. The middle 60% of the families who had a child born during 1998 had family incomes anywhere from $29.44 - 0.84(20.33)$ to $29.44 + 0.84(20.33)$ (*Source:* U.S. Census Bureau, www.census.gov). Simplify these expressions and interpret.

Section A.3   Properties of Integer Exponents

Integer exponents are used as an efficient way to write repeated multiplication, just as multiplication is used as an efficient way to write repeated addition. Many of the properties of integer exponents can be derived directly from the way that we define the integer exponent.

From Your Toolbox

- By an **integer**, we mean the set of numbers comprised of the whole numbers and their opposites. The set of integers looks like $\{\ldots, -3, -2, -1, 0, 1, 2, 3, \ldots\}$.
- A **real number** is any number on the real number line.

■ Exponent

If b is a real number and n is a positive integer, then b^n is the product of n factors of b. That is

$$b^n = \underbrace{b \cdot b \cdot \,\cdots\, \cdot b}_{n \text{ factors of } b}$$

where n is the **exponent** and b is the **base**.

To see how we can use this definition to develop some shortcuts for simplifying exponents, let's consider a product such as $3^2 \cdot 3^4$. Using the definition of exponent, we see that

$$3^2 \cdot 3^4 = (3 \cdot 3) \cdot (3 \cdot 3 \cdot 3 \cdot 3) = 3 \cdot 3 \cdot 3 \cdot 3 \cdot 3 \cdot 3 = 3^6$$

Notice that if we simply add the exponents we get

$$3^2 \cdot 3^4 = 3^{2+4} = 3^6$$

This illustrates the **multiplication rule** for exponents. The other basic properties of integer exponents can be verified in a similar manner.

■ Basic Properties of Exponents

Let m and n be integers and let b represent any real number.

Rule	Meaning	Example
1. Multiplication Rule: $b^m \cdot b^n = b^{m+n}$	When multiplying numbers with the same base, add the exponents.	$5^4 \cdot 5^3 = 5^{4+3} = 5^7$
2. Division Rule: $\dfrac{b^m}{b^n} = b^{m-n}, \quad b \neq 0$	When dividing numbers with the same base, subtract the exponents.	$\dfrac{9^{11}}{9^5} = 9^{11-5} = 9^6$
3. Power-to-Power Rule: $(b^m)^n = b^{m \cdot n}$	When an exponential expression is raised to a power, multiply the exponents.	$(6^2)^4 = 6^{2 \cdot 4} = 6^8$

▶ **Note:** We can show that these properties, along with the other properties in this section, also apply to real number exponents. However, we shall restrict our discussion to only integer exponents.

Example 1 Simplifying Expressions with Integer Exponents

Use the basic properties of exponents to simplify the following and write the solution in the form b^n. Then evaluate the simplified expression if possible.

(a) $\dfrac{4^{10}}{4^8}$ 　　　　　 (b) $(1+a)^6(1+a)^3$ 　　　　　 (c) $(0.5^2)^3$

Solution

(a) Using the division rule, we get

$$\frac{4^{10}}{4^8} = 4^{10-8} = 4^2 = 16$$

(b) In this case the base is the expression $1 + a$. The multiplication rule gives us

$$(1+a)^6(1+a)^3 = (1+a)^{6+3} = (1+a)^9$$

(c) To simplify this expression, we use the power rule to get

$$(0.5^2)^3 = 0.5^{2\cdot3} = 0.5^6 = 0.015625$$

∎

✓ Checkpoint 1

Now work Exercises 9 and 13 in the A.3 Exercises.

Now we examine a few special cases of exponents. Let's say that we have the fraction $\frac{3^4}{3^4}$. If we use the definition of exponent to rewrite the numerator and denominator and then cancel the common factors, we get

$$\frac{3^4}{3^4} = \frac{3\cdot3\cdot3\cdot3}{3\cdot3\cdot3\cdot3} = 1$$

However, using the Division Rule, this fraction can be simplified as $\frac{3^4}{3^4} = 3^{4-4} = 3^0$. So it appears that $3^0 = 1$. Since we can use any nonzero base and get a similar result, this illustrates that any nonzero real number raised to the zero power is equal to 1.

To see what other properties can be shown, let's examine the fraction $\frac{6^2}{6^5}$. If we factor and cancel the common factors, the reduced form of the fraction is

$$\frac{6^2}{6^5} = \frac{6\cdot6}{6\cdot6\cdot6\cdot6\cdot6} = \frac{1}{6\cdot6\cdot6} = \frac{1}{6^3}$$

From the division rule, we know that $\frac{6^2}{6^5} = 6^{2-5} = 6^{-3}$. So it appears that $6^{-3} = \frac{1}{6^3}$.

To conclude these properties, let's now consider the expression $(4x)^3$. In this case, the exponent is 3 and the base is $4x$. If we use the definition of exponent to write the factored form, we get

$$(4x)^3 = 4x\cdot4x\cdot4x = (4\cdot4\cdot4)\cdot(x\cdot x\cdot x) = 4^3\cdot x^3 = 64x^3$$

This illustrates the property that if a product, like $4x$, is raised to a power we raise each factor to that power. A similar argument can be made to show that $\left(\frac{5}{x}\right)^3 = \frac{5^3}{x^3} = \frac{125}{x^3}$. Now let's gather this list of integer exponent properties.

■ Properties of Integer Exponents

Let m and n be integers and let a and b represent any real numbers.

1. $b^0 = 1, \quad b \neq 0$

2. $b^{-n} = \frac{1}{b^n}, \quad b \neq 0$

3. $(ab)^n = a^n b^n$

4. $\left(\frac{a}{b}\right)^n = \frac{a^n}{b^n}, \quad b \neq 0$

Example 2   Using the Properties of Integer Exponents

Simplify the following:

(a) 4^{-2} (b) $(5x^3)^2$ (c) $\left(\dfrac{2}{5}\right)^3$

Solution

(a) Here we have a base of 4 and the exponent is negative. Property 2 gives us

$$4^{-2} = \frac{1}{4^2} = \frac{1}{16}$$

(b) Here the base is $5x^3$, so we have to square 5 and square x^3 using Property 3.

$$(5x^3)^2 = 5^2(x^3)^2 = 25x^6$$

Observe how we used the Power-to-Power Rule to simplify $(x^3)^2$.

(c) To simplify this expression, we must take the numerator and the denominator to the third power by using Property 4 to get

$$\left(\frac{2}{5}\right)^3 = \frac{2^3}{5^3} = \frac{8}{125}$$

Many of the applications that we encounter, particularly when studying the mathematics of finance in Chapter 5, make use of integer exponents. We also see these exponents used in the study of probability.

Example 3   Applying Integer Exponents

Shawna buys a big-screen digital television and wants to pay back the $4600 price in 18 months. The electronics store is running a financing promotion and determines that the cost of each monthly payment is given by the expression $4600\dfrac{0.015}{1-(1.015)^{-18}}$. Evaluate this expression and interpret.

Solution

Using the properties of integer exponents, we can rewrite the denominator and simplify.

$$4600\frac{0.015}{1-(1.015)^{-18}} = 4600\frac{0.015}{1-\dfrac{1}{1.015^{18}}} \approx 4600 \cdot \frac{0.015}{0.235} \approx 293.62$$

This means that Shawna will pay $293.62 each month for 18 months to pay off the television.

SECTION A.3 EXERCISES

In Exercises 1–14, simplify using the basic properties of exponents and write the solution in the form b^n. Then evaluate the simplified expression, if possible.

1. $3 \cdot 3^5$

2. $5 \cdot 5^4$

3. $2^{10} \cdot 2^5$

4. $9^8 \cdot 9^2$

5. $\dfrac{10^8}{10^2}$

6. $\dfrac{7^7}{7^5}$

7. $\dfrac{3^8}{3^7}$

8. $\dfrac{8^{12}}{8^{11}}$

✓ **9.** $(6^2)^3$

10. $(8^3)^4$

11. $(-3^3)^5$

12. $(-2)^7$

✓ **13.** $(x+3)^4(x+3)^2$

14. $(5-x)^5(5-x)$

In Exercises 15–30, simplify using the properties of integer exponents.

15. 5^0

16. 3^0

17. $(-5)^0$

18. $(-3)^0$

19. 6^{-2}

20. 3^{-3}

21. $\left(\dfrac{1}{6}\right)^{-2}$

22. $\left(\dfrac{1}{3}\right)^{-3}$

23. $(2x^3)^4$

24. $(4x^5)^3$

25. $(3x^5)^{-1}$

26. $(8x^2)^{-1}$

27. $\left(\dfrac{3}{4}\right)^2$

28. $\left(\dfrac{2}{3}\right)^3$

29. $\left(\dfrac{1}{8}\right)^{-2}$

30. $\left(\dfrac{1}{10}\right)^{-3}$

🌐 **31.** During 1995, statistics show that if 20 women who just had babies were selected at random the chance that 2 of them were African American is given by the expression $190(0.16)^2(0.84)^{20-2}$ (*Source:* U.S. Census Bureau, www.census.gov). Evaluate this expression, convert to a percent, and interpret.

🌐 **32.** U.S. Coast Guard statistics show that if the records of 30 oil spills from 1994 are chosen at random the chance that 27 of them were spills of less than 100 gallons is given by the expression $4060(0.94)^{27}(0.06)^{30-27}$ (*Source:* U.S. Coast Guard, www.uscg.mil). Evaluate this expression, convert to a percent, and interpret.

33. Candy is considering the purchase of a new car from West Side Auto Sales. She chooses a car costing $22,000 and is told by the salesperson that, if monthly payments are made for 5 years, the cost of each payment is given by the expression $22{,}000\dfrac{0.065}{1-(1.065)^{-60}}$. Evaluate this expression and interpret.

34. The Punderson family is shopping for a new dining room table and hutch. They find that the table and hutch that they want costs $9600. They decide to finance the purchase and learn from their banker that, in order to pay off the purchase in 2 years, each monthly payment will be $9600\dfrac{0.08}{1-(1.08)^{-24}}$. Evaluate this expression and interpret.

Section A.4   Rational Exponents and Radicals

Fractions that have integers in the numerator and denominator are called **rational numbers**. Let's see what happens if an expression contains an exponent that is a rational number. Using the multiplication rule for exponents, we know that, for example,

$$3^{\frac{1}{2}} \cdot 3^{\frac{1}{2}} = 3^{\frac{1}{2}+\frac{1}{2}} = 3^1 = 3$$

But we also know that $\sqrt{3} \cdot \sqrt{3} = \sqrt{9} = 3$. So it appears that we can say that $3^{\frac{1}{2}} = \sqrt{3}$. This illustrates that we can take an expression written in radical form and write it as a rational exponent, and vice versa.

■ **Rational Exponent**

Let m and n be integers, $n \neq 0$, and let b represent any real number. Then $\dfrac{m}{n}$ is a **rational exponent** and we can write

$$b^{\frac{m}{n}} = \sqrt[n]{b^m} = \left(\sqrt[n]{b}\right)^m$$

Example 1   Writing Radical Expression in Exponential Form

Write each of the following in the form $b^{\frac{m}{n}}$.

(a) $\sqrt[3]{x^2}$

(b) $\dfrac{1}{\sqrt{x}}$

Solution

(a) Using the definition of rational exponent with $n = 3$ and $m = 2$, we get

$$\sqrt[3]{x^2} = x^{\frac{2}{3}}$$

(b) Here we see that, since the radical expression is in the denominator, the exponent will be negative.

$$\frac{1}{\sqrt{x}} = \frac{1}{x^{\frac{1}{2}}} = x^{-\frac{1}{2}}$$

✓ **Checkpoint 1**

Now work Exercise 9 in the A.4 Exercises.

Example 2 **Writing Expressions in Radical Form**

Write each of the following expressions in radical form.

(a) $3x^{\frac{3}{2}}$ (b) $x^{-\frac{1}{4}}$

Solution

Interactive Activity

Write the expression $(3x)^{\frac{3}{2}}$ in radical form. How does this answer differ from that in part (a) of Example 2?

(a) Here the exponent is $\frac{3}{2}$ and the base is x, so we get

$$3x^{\frac{3}{2}} = 3\sqrt{x^3}$$

(b) We first make the exponent positive to get

$$x^{-\frac{1}{4}} = \frac{1}{x^{\frac{1}{4}}} = \frac{1}{\sqrt[4]{x}}$$

SECTION A.4 EXERCISES

In Exercises 1–10, write each of the following in the form $b^{\frac{m}{n}}$.

1. $\sqrt{x}$ 2. $\sqrt[3]{x}$

3. $\sqrt[7]{x^4}$ 4. $\sqrt[5]{x^3}$

5. $\sqrt[4]{5x^2}$ 6. $\sqrt[6]{10x^5}$

7. $\dfrac{1}{\sqrt[5]{x}}$ 8. $\dfrac{1}{\sqrt[8]{x}}$

✓ 9. $\dfrac{8}{\sqrt[7]{x^2}}$ 10. $\dfrac{11}{\sqrt[5]{x^4}}$

In Exercises 11–20, write each of the following expressions in radical form.

11. $x^{\frac{1}{4}}$ 12. $x^{\frac{1}{6}}$

13. $x^{\frac{5}{8}}$ 14. $x^{\frac{3}{4}}$

15. $(6x^2)^{\frac{1}{4}}$ 16. $(4x^4)^{\frac{1}{2}}$

17. $x^{-\frac{1}{3}}$ 18. $x^{-\frac{1}{10}}$

19. $5x^{-\frac{2}{3}}$ 20. $9x^{-\frac{5}{7}}$

Appendix B
Algebra Review

Multiplying Polynomial Expressions

A **polynomial** in one variable is an algebraic expression that contains terms of the form $a_n x^n$ that are combined using addition and subtraction. The first part of the term, a_n, is called the **coefficient** and can be any real number. The **variable part** of the term is x^n, and the number n is the **degree** of the term and can be any whole number. We name some polynomials by the number of terms that they have. See Table B.1.1.

Table B.1.1

Number of terms	Name	Example
1	Monomial	$4x^2$
2	Binomial	$x^2 - 7x$
3	Trinomial	$3x^3 + 2x - 7$
4	Four-term polynomial	$7x^3 - 0.2x^2 + 6x - 4$
⋮	⋮	⋮

From Your Toolbox

The **Multiplication Rule** for integer exponents tells us that when we are multiplying and the bases are the same we add the exponents. That is, $b^m \cdot b^n = b^{m+n}$.

To see how polynomials are multiplied together, let's consider several cases.

Case 1: Multiplying a Monomial by a Monomial

To Multiply: Use the Multiplication Rule for integer exponents to multiply the two variable parts (consult the Toolbox to the left) and then multiply the coefficients.

Example 1   **Multiplying a Monomial by a Monomial**

Multiply the following monomials.

(a) $4x^3 \cdot 12x$ **(b)** $(-3x^2)(x^4)$ **(c)** $-6(5x^4)$

Solution

(a) Here the respective coefficients are 4 and 12, and the variable parts are x^3 and x. So we get

$$4x^3 \cdot 12x = (4 \cdot 12)(x^3 \cdot x) = 48x^{3+1} = 48x^4$$

(b) We can write the term x^4 as $1x^4$ to get the product

$$(-3x^2)(x^4) = (-3 \cdot 1)(x^2 \cdot x^4) = -3x^6$$

(c) Since we know that $x^0 = 1$, we can write the first terms as $-6x^0$ to give us

$$-6(5x^4) = (-6 \cdot 5)(x^0 \cdot x^4) = -30x^4 \qquad \blacksquare$$

✓ **Checkpoint 1**

Now work Exercise 3 in the B.1 Exercises.

Case 2: Multiplying a Monomial by a Polynomial

To Multiply: Use the Distributive Property.

> ■ **Distributive Property**
> 1. $a(b+c) = ab + ac$
> 2. $a(b-c) = ab - ac$
> 3. $a(b+c+d+\cdots+n) = ab + ac + ad + \cdots + an$

The distributive property means that we can multiply the monomial by each term of the polynomial in the parentheses. After multiplying, we can sometimes combine the **like terms**. Two terms are like terms if their variable parts are the same and they have the same degree. To combine like terms, we combine their coefficients and write the common variable part.

Example 2 Multiplying a Monomial by a Polynomial

Multiply the following and simplify if possible.

(a) $6x(3x^2 - 7x + 4)$ **(b)** $2x(x+3) + (x^2 + 7x)$

Solution

(a) Using the distributive property, we get

$$6x(3x^2 - 7x + 4) = 6x(3x^2) - 6x(7x) + 6x(4)$$
$$= 18x^3 - 42x^2 + 24x$$

(b) We begin by distributing $2x$ over the binomial $x + 3$.

$$2x(x+3) + (x^2+7) = 2x(x) + 2x(3) + (x^2 + 7x)$$
$$= 2x^2 + 6x + (x^2 + 7x)$$

Now let's remove the parentheses and combine the like terms.

$$= 2x^2 + 6x + x^2 + 7x$$
$$= 3x^2 + 13x \qquad \blacksquare$$

Figure B.1.1

Case 3: Multiplying a Binomial by a Binomial

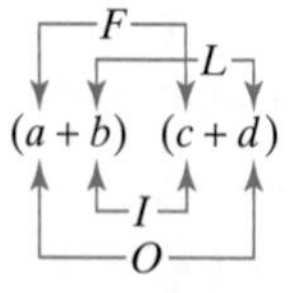

To Multiply: Use the FOIL Method.

The FOIL method reminds us that when we multiply two binomials together we multiply the **F**irst two terms, the **O**utside terms, the **I**nside terms, and the two **L**ast terms. See Figure B.1.1.

Example 3 ## Multiplying Two Binomials

Multiply and combine like terms.

(a) $(x - 3)(2x + 1)$ (b) $(0.3x^2 + 5)(2x^2 + 6)$

Solution

(a) Let's begin by identifying the terms used for the FOIL method.

$$\mathbf{F} \quad\quad \mathbf{O} \quad\quad \mathbf{I} \quad\quad \mathbf{L}$$
$$(x - 3)(2x + 1) = x(2x) + x(1) - 3(2x) - 3(1)$$
$$= 2x^2 + x - 6x - 3$$
$$= 2x^2 - 5x - 3$$

(b) Using the FOIL method, we get

$$(0.3x^2 + 5)(2x^2 + 6) = (0.3x^2)(2x^2) + (0.3x^2)(6) + (5)(2x^2) + (5)(6)$$
$$= 0.6x^4 + 1.8x^2 + 10x^2 + 30$$
$$= 0.6x^4 + 11.8x^2 + 30$$

✓ **Checkpoint 2**

Now work Exercise 15 in the B.1 Exercises.

Case 4: Multiplying a Polynomial by Another Polynomial

To Multiply: Use the general distributive property.

■ **General Distributive Property**

If p is a polynomial expression, then

1. $(b + c)p = bp + cp$
2. $(b + c + d + \cdots + n)p = bp + cp + dp + \cdots + np$

The general distributive property means that we can distribute a polynomial over each term of another polynomial. Let's see how this property is used to multiply two polynomials together.

Example 4 ## Multiplying Two Polynomials

Multiply and combine like terms.

(a) $(3x - 2)(x^2 - 6x + 4)$ (b) $(2x^2 - x + 6)(x^2 + 3x - 1)$

Solution

(a) Using the general distributive property, we start by distributing the terms $3x$ and -2 over the trinomial $x^2 - 6x + 4$. Then we combine the like terms.

$$(3x - 2)(x^2 - 6x + 4) = 3x(x^2 - 6x + 4) - 2(x^2 - 6x + 4)$$
$$= 3x^3 - 18x^2 + 12x - 2x^2 + 12x - 8$$
$$= 3x^3 - 20x^2 + 24x - 8$$

(b) When we multiply these two trinomials, we distribute the terms $2x$, $-x$, and 6 over the trinomial $x^2 + 3x - 1$ to get

$$(2x^2 - x + 6)(x^2 + 3x - 1) = 2x^2(x^2 + 3x - 1) - x(x^2 + 3x - 1) + 6(x^2 + 3x - 1)$$
$$= 2x^4 + 6x^3 - 2x^2 - x^3 - 3x^2 + x + 6x^2 + 18x - 6$$
$$= 2x^4 + 5x^3 + x^2 + 19x - 6 \qquad ▪$$

Interactive Activity

Polynomials can also be multiplied using the vertical method. The solution to part (a) of Example 4 using the vertical method is

$$
\begin{array}{r}
x^2 - 6x + 4 \\
3x - 2 \\
\hline
-2x^2 + 12x - 8 \\
3x^3 - 18x^2 + 12x \\
\hline
3x^3 - 20x^2 + 24x - 8
\end{array}
$$

Verify the solution to part (b) of Example 4 using this method.

SECTION B.1 EXERCISES

In Exercises 1–6, multiply the following monomials.

1. $5x^3 \cdot 2x^2$

2. $7x^3 \cdot 9x^7$

✓ **3.** $(-8x^9)(x^3)$

4. $(-2x^2)(x^2)$

5. $-10(15x^4)$

6. $-9(6x^3)$

In Exercises 7–14, multiply the following:

7. $2x(x - 5)$

8. $3x(x + 1)$

9. $-4x^2(x^2 - 3x)$

10. $-9x^3(x^2 - 2x)$

11. $(4x^5 - 6x)(-3x)$

12. $(2x^3 - 7x)(-6x)$

13. $-2x^2(x^2 - 7x + 5)$

14. $-5x^4(2x^2 + 9x - 1)$

In Exercises 15–22, multiply and combine like terms.

✓ **15.** $(x + 3)(x + 5)$

16. $(x + 4)(x + 7)$

17. $(1.2x - 5)(x - 4)$

18. $(1.4x - 1)(x - 3)$

19. $(3x + 4)(3x - 8)$

20. $(8x - 1)(2x + 6)$

21. $(4x^2 + 5)(2x^2 + 3)$

22. $(6x^2 + 3)(3x^2 + 2)$

In Exercises 23–32, multiply and combine like terms.

23. $(x + 1)(x^2 + 2x + 3)$

24. $(x + 2)(x^2 + 2x + 5)$

25. $(0.2x - 3)(3x^2 - 7x + 1)$

26. $(0.3x - 2)(2x^2 + x - 3)$

27. $(x^2 + x - 2)(4x^2 + 3)$

28. $(x^2 + 2x - 1)(5x^2 + 2)$

29. $(x^2 + 3x - 3)(x^2 + 9x + 2)$

30. $(x^2 - 5x + 2)(x^2 + 2x + 8)$

31. $(2x^3 - 2x^2 + 3)(x^2 + 5x + 2)$

32. $(x^3 + 5x^2 - 6)(x^2 + 9x + 8)$

Section B.2   Factoring Trinomials

Factoring polynomials can be thought of as the opposite of multiplying them. Factoring can be a useful tool for solving equations. When we say factoring, we are referring to factoring polynomials so that their factors contain integer coefficients and constants. Several techniques can be used in factoring, depending on the characteristics of a given polynomial. Let's consider each case by case.

Case 1: Factoring Out the Greatest Common Monomial

To Factor: Use the reverse of the distributive property. That is, we use

Factor out a
$$ab + ac = a(b + c)$$

or, in general,

$$ab + ac + ad + \cdots + an = a(b + c + d + \cdots + n)$$

The **greatest common monomial** is made up of a coefficient part and a variable part. The coefficient part is the greatest common factor of all the coefficients. The variable part contains the smallest power of the variable that appears in every term of the polynomial.

Example 1   Factoring Out the Greatest Common Monomial

Factor the following:

(a) $36x^3 + 24x^2$ **(b)** $26x^5 + 13x^3 - 39x^2$

Solution

(a) The coefficients are 36 and 24, so the greatest common factor of these numbers is 12. The smallest power of x in each term is 2, so the greatest common monomial is $12x^2$.

$$36x^3 + 24x^2 = 12x^2(3x + 2)$$

(b) For this trinomial, the greatest common monomial is $13x^2$. Factoring this monomial from each term yields

$$26x^5 + 13x^3 - 39x^2 = 13x^2(2x^3 + x - 3)$$ ■

Case 2: Difference of Squares

To Factor: Use the formula $a^2 - b^2 = (a + b)(a - b)$

We call the binomial $a^2 - b^2$ a difference of squares because each term is a **perfect square** (such as 4, 9, 16, 25 or $x^2, x^4, x^6, \ldots$), and, by subtracting, we are taking their **difference**. To verify the difference of squares formula, let's use the FOIL method to multiply $(a + b)(a - b)$.

$$(a + b)(a - b) = a \cdot a - ab + ab - b \cdot b = a^2 - b^2$$

Notice that the two middle terms cancel one another.

Example 2   Factoring a Difference of Squares

Factor the following completely.

(a) $x^2 - 16$ **(b)** $x^4 - 81$

Solution

(a) We see that both terms are perfect squares since they can be written in the form $(x)^2$ and $(4)^2$, respectively. So we get

$$x^2 - 16 = (x)^2 - (4)^2 = (x + 4)(x - 4)$$

(b) Knowing that $x^4 = (x^2)^2$ and $81 = (9)^2$, we get

$$x^4 - 81 = (x^2)^2 - (9)^2 = (x^2 + 9)(x^2 - 9)$$

But now notice that the binomial $x^2 - 9$ is also a difference of squares, so we can factor this binomial to give us the completely factored form

$$x^4 - 81 = (x^2 + 9)(x^2 - 9) = (x^2 + 9)(x + 3)(x - 3) \qquad ■$$

Case 3: Factoring by Grouping

To Factor: Using the general distributive property, factor $ap + bp = (a + b)p$, where p is a polynomial.

Factoring by grouping is used when a polynomial expression has four terms. We first factor the greatest common monomial from the two first terms and then repeat the process for the next two terms. Finally, we factor using the general distributive property.

Example 3 **Factoring by Grouping**

Factor the following completely.

(a) $8x^3 - 12x^2 + 6x - 9$ **(b)** $x^3 + 2x^2 - x - 2$

Solution

Interactive Activity

Verify the answer to part (a) of Example 3 by multiplying $(4x^2 + 3)(2x - 3)$ using the FOIL method.

(a) We can factor $4x^2$ from the two first terms and factor 3 from the last two terms. This gives us

$$8x^3 - 12x^2 + 6x - 9 = 4x^2(2x - 3) + 3(2x - 3)$$

Using the general distributive property, we get

$$\underbrace{a}\quad\underbrace{p}\ +\underbrace{b}\quad\underbrace{p}\quad=\ \underbrace{(a + b)}\quad\underbrace{p}$$
$$4x^2(2x - 3) + 3(2x - 3) = (4x^2 + 3)(2x - 3)$$

(b) For this polynomial, we begin by factoring x^2 from the two first terms.

$$x^3 + 2x^2 - x - 2 = x^2(x + 2) - x - 2$$

To complete the factoring by grouping process, we factor -1 from the last two terms.

$$x^2(x + 2) - x - 2 = x^2(x + 2) - 1(x + 2) = (x^2 - 1)(x + 2)$$

Now we see that the first binomial $x^2 - 1$ is a difference of squares. So the complete factored form of the polynomial is

$$x^3 + 2x^2 - x - 2 = (x^2 - 1)(x + 2) = (x + 1)(x - 1)(x + 2) \qquad ■$$

The final two cases involve the techniques that are used to factor **quadratic trinomials**. There are two cases that depend on whether the leading coefficient is 1 or the leading coefficient is some integer other than 1.

Case 4: Factoring Trinomials of the Form $x^2 + bx + c$

To Factor: Factor the expression $x^2 + bx + c$ into $(x + n_1)(x + n_2)$ under the conditions that $n_1 \cdot n_2 = c$ and $n_1 + n_2 = b$.

To see how this factoring method works, let's multiply $(x + n_1)(x + n_2)$ using the FOIL method.

$$(x + n_1)(x + n_2) = x^2 + n_2 x + n_1 x + n_1 \cdot n_2$$

If we factor x from the two middle terms, we get

$$x^2 + \underbrace{(n_1 + n_2)}_{b} x + \underbrace{n_1 \cdot n_2}_{c} = x^2 + bx + c$$

Example 4 Factoring a Trinomial of the Form $x^2 + bx + c$

Factor the following completely.

(a) $x^2 + 3x - 10$ (b) $x^4 - 13x^2 - 48$

Solution

(a) For this trinomial, we need two numbers, n_1 and n_2, so that $n_1 \cdot n_2 = -10$, while $n_1 + n_2 = 3$. That is, we need two numbers such that when they are multiplied together we get -10, and when we add the two numbers, we get 3. Notice that if we let $n_1 = 5$ and $n_2 = -2$ we get $n_1 \cdot n_2 = (5)(-2) = -10$ and $n_1 + n_2 = 5 + (-2) = 3$. This means that we can write the factored form of the polynomial as

$$x^2 + bx + c = (x + n_1)(x + n_2)$$
$$x^2 + 3x - 10 = (x + 5)\ (x - 2)$$

This solution can be checked using the FOIL method.

(b) Even though this is a fourth-degree polynomial, we can rewrite it in quadratic form by letting $x^2 = y$. This gives us

$$x^4 - 13x^2 - 48 = y^2 - 13y - 48$$

Now we need to two numbers n_1 and n_2 such that their product is $n_1 \cdot n_2 = -48$ and their sum is $n_1 + n_2 = -13$. We can find these values by listing the factors of -48. See Table B.2.1.

Table B.2.1

Factors of -48	Sum of Factors	
$-48(1)$	$-48 + 1 = -47$	
$-24(2)$	$-24 + 2 = -22$	
$-16(3)$	$-16 + 3 = -13$	Desired values
$-12(4)$	$-12 + 4 = -8$	
$-8(6)$	$-8 + 6 = -2$	
$-6(8)$	$-6 + 8 = 2$	
$-4(12)$	$-4 + 12 = 8$	
$-3(16)$	$-3 + 16 = 13$	
$-2(24)$	$-2 + 24 = 22$	
$-1(48)$	$-1 + 48 = 47$	

Now we see that $n_1 = -16$ and $n_2 = 3$. This gives us the factored form

$$y^2 - 13y - 48 = (y - 16)(y + 3)$$

Rewriting the factored form in terms of x (recall that we let $x^2 = y$) gives us

$$x^4 - 13x^2 - 48 = (x^2 - 16)(x^2 + 3)$$

But notice that the binomial $x^2 - 16$ is a difference of squares. Factoring this binomial, we get

$$x^4 - 13x^2 - 48 = (x^2 - 16)(x^2 + 3) = (x + 4)(x - 4)(x^2 + 3)$$

The expression is now factored completely.

∎

✓ Checkpoint 1

Now work Exercise 31 in the B.2 Exercises.

Case 5: Factoring a Trinomial of the Form $ax^2 + bx + c, a \neq 1$

To Factor: One can use the following process:

1. Determine numbers n_1 and n_2 so that their product is $n_1 \cdot n_2 = a \cdot c$ and their sum is $n_1 + n_2 = b$.
2. Rewrite the trinomial as a four-term polynomial of the form

$$ax^2 + n_1 x + n_2 x + c$$

where $n_1 x + n_2 x = bx$.
3. Factor by grouping.

Let's see how this process works in the next example.

Example 5 **Factoring a Trinomial of the Form $ax^2 + bx + c, a \neq 1$**

Factor the following completely.

(a) $2x^2 - 5x - 3$ (b) $6x^2 + 11x - 10$

Solution

Interactive Activity

In part (b) of Example 5, suppose that we let the two numbers be $n_1 = -4$ and $n_2 = 15$. Will we get the same answer? Try it and see.

(a) For this part of the example, we will outline the three steps.

1. We need two numbers such that $n_1 \cdot n_2 = a \cdot c = (2)(-3) = -6$ and $n_1 + n_2 = b = -5$. We can see that both of these conditions are satisfied when $n_1 = -6$ and $n_2 = 1$.
2. Rewriting the trinomial as a four-term polynomial using the number that we found in step 1 gives us

$$2x^2 - 5x - 3 = 2x^2 - 6x + x - 3$$

3. Using the factoring by grouping technique, we begin by factoring $2x$ from the two first terms and 1 from the last two terms.

$$2x^2 - 6x + x - 3 = 2x(x - 3) + 1(x - 3)$$
$$= (2x + 1)(x - 3)$$

(b) For this trinomial we need $n_1 \cdot n_2 = (6)(-10) = -60$ and $n_1 + n_2 = 11$. We can check to see that the two numbers are $n_1 = 15$ and $n_2 = -4$. Using these values, we rewrite the trinomial and factor by grouping to give us

$$6x^2 + 11x - 10 = 6x^2 + 15x - 4x - 10$$
$$= 3x(2x + 5) - 2(2x + 5)$$
$$= (3x - 2)(2x + 5)$$

∎

We have this final comment about factoring. Many students opt to use a trial and error method to factor trinomials, once any common factors have been factored. This method is perfectly fine and will yield the same completely factored form.

▮ SECTION B.2 EXERCISES

In Exercises 1–8, factor out the greatest common monomial.

1. $2x - 2$

2. $5x + 5$

3. $35x - 10$

4. $28x - 7$

5. $12x^2 - 26x^3$

6. $2x^3 + 3x^2$

7. $21x^3 + 4x^2 - 14$

8. $15x^3 + 3x - 5$

In Exercises 9–18, factor the difference of squares completely.

9. $x^2 - 81$

10. $x^2 - 49$

11. $9x^2 - 1$

12. $4x^2 - 25$

13. $(2x - 1)^2 - 4$

14. $(x + 1)^2 - 100$

15. $x^4 - 1$

16. $x^4 - 16$

17. $16x^4 - 81$

18. $256x^4 - 1$

In Exercises 19–26, factor by grouping.

19. $x^2 + 4x + x + 4$

20. $x^2 + 5 + x + 5$

21. $2x^3 + x^2 + 6x + 1$

22. $3x^3 + 5x^2 + 3x + 5$

23. $2x^3 - 2x^2 + 4x - 4$

24. $9x^3 + 18x^2 - 5x - 10$

25. $6x^3 - 24x - 3x^2 + 12$

26. $5x^3 - 45x + x^2 - 9$

In Exercises 27–36, factor completely. In Exercises 35 and 36, first factor out the greatest common monomial.

27. $x^2 + 8x + 15$

28. $x^2 + x + 9$

29. $x^2 + 2x + 1$

30. $x^2 + 12x + 36$

✓ **31.** $x^2 - 4x - 5$

32. $x^2 - 5x - 6$

33. $x^2 - 9x + 81$

34. $x^2 - 10x - 200$

35. $2x^2 + 10x - 48$

36. $4x^2 - 4x - 8$

In Exercises 37–46, factor completely using the three-step process or trial and error.

37. $2x^2 + 3x + 1$

38. $3x^2 + 7x + 2$

39. $3x^2 - 8x + 4$

40. $2x^2 - 9x + 4$

41. $7x^2 - 12x + 5$

42. $5x^2 + 36x + 7$

43. $7x^2 + 4x - 11$

44. $5x^2 - 54x - 11$

45. $15x^2 + 10x - 40$

46. $3x^2 - 21x + 36$

Section B.3   Solving Linear Equations

A **linear equation** is an equation that can be written in the form $ax + b = c$, where a is the real number coefficient of the variable x ($a \neq 0$), and b and c represent any real numbers. Once a linear equation is in this form, we can solve for x using the following steps:

$$ax + b = c \qquad \text{Given equation}$$

$$ax = b - c \qquad \text{Subtract } c \text{ from both sides}$$

$$x = \frac{b - c}{a} \qquad \text{Divide both sides by } a$$

Example 1   Solving a Linear Equation

Solve the following equations for x.

(a) $3x - 4 = 6$

(b) $0.25x + 1 = 3$

Solution

(a) Using the steps for solving a linear equation, we get

$$
\begin{array}{ll}
3x - 4 = 6 & \text{Given equation} \\
3x = 6 + 4 & \text{Add 4 to both sides.} \\
3x = 10 & \text{Simplify.} \\
x = \dfrac{10}{3} & \text{Divide both sides by 3.}
\end{array}
$$

(b) Even though the coefficient of x is a decimal, we solve the equation in the same manner.

$$
\begin{array}{l}
0.25x + 1 = 3 \\
0.25x = 3 - 1 \\
0.25x = 2 \\
x = \dfrac{2}{0.25} \\
x = 8
\end{array}
$$

■

There are times when a linear equation must be rewritten or simplified so that it is in the form $ax + b = c$. This could involve simplifying each side of the equation, collecting all the variable terms on one side of the equation, or the use of the distributive property. Let's see how we can solve these kinds of linear equations.

Example 2 Simplifying and Solving a Linear Equation

Simplify and solve the following equations for x.

(a) $5x - 2(3x + 12) = 10$ **(b)** $6x - (5x + 1) = 2(x + 3)$

Solution

(a) We begin by distributing -2 over the binomial $3x + 12$.

$$
\begin{array}{ll}
5x - 2(3x + 12) = 10 & \text{Given equation} \\
5x - 6x - 24 = 10 & \text{Distribute } -2. \\
-x - 24 = 10 & \text{Combine like terms.} \\
-x = 34 & \text{Add 24 to both sides.} \\
x = -34 & \text{Divide both sides by } -1.
\end{array}
$$

(b) In this case the variable x appears on both sides of the equation.

$$
\begin{array}{ll}
6x - (5x + 1) = 2(x + 3) & \text{Given equation} \\
6x - 5x - 1 = 2x + 6 & \text{Distribute } -1 \text{ (left side) and 2 (right side).} \\
x - 1 = 2x + 6 & \text{Combine like terms.} \\
-x - 1 = 6 & \text{Subtract } 2x \text{ from both sides.} \\
-x = 7 & \text{Add 1 to both sides.} \\
x = -7 & \text{Divide both sides by } -1.
\end{array}
$$

■

✓ **Checkpoint 1**

Now work Exercise 11 in the B.3 Exercises.

When studying applications of linear equations, we often solve equations that are based on **linear models**. A linear model is a mathematical expression derived from real-world data. When solving equations that involve linear models, we write a sentence that **interprets** the mathematical solution. Let's examine one of these applications.

Example 3 **Solving an Equation Based on a Linear Model**

The year that the number of violent crimes in the United States was about 7.7 million can be determined by solving the equation

$$-0.78x + 13.94 = 7.7$$

where x represents the number of years since 1990 and the number of violent crimes is expressed in millions (*Source:* U.S. Bureau of Justice Statistics, www.ojp.usdoj.gov/bjs). Solve for x and interpret.

Solution

Solving this equation for x gives us

$$-0.78x + 13.94 = 7.7$$
$$-0.78x = 7.7 - 13.94$$
$$-0.78x = -6.24$$
$$x = \frac{-6.24}{-0.78}$$
$$x = 8$$

Since x represents the number of years since 1990, we see that $x = 8$ corresponds to the year 1998.

Interpret the Solution: This means that in 1998 about 7.7 million violent crimes were reported in the United States. ∎

SECTION B.3 EXERCISES

In Exercises 1–10, solve the equation for x.

1. $x + 6 = 8$

2. $x - 7 = 3$

3. $2x = 6$

4. $3x = 15$

5. $\frac{1}{4}x = 7$

6. $\frac{1}{6}x = 8$

7. $3x + 1 = 3$

8. $8x - 2 = 13$

9. $0.2x + 1.3 = 0.8$

10. $0.7x - 1.2 = 0.9$

In Exercises 11–20, simplify and solve the following:

✓ **11.** $3x - 11 = 7 - 2x$

12. $x + 18 = 6x - 3$

13. $2(2x + 1) = 5 + 3x$

14. $3(x - 4) = y + 6$

15. $4(x + 1) = -2(4 - x)$

16. $5(x + 4) = 2(3 - x)$

17. $3(x - 5) + 10 = 2(x + 4)$

18. $2(5x + 2) + 1 = 3(3x - 2)$

19. $0.22(4x - 3) = 1.37$

20. $0.85(5x + 10) = 2.55$

21. The year that the annual expenditures on pollution abatement in the United States totaled 60.17 billion dollars can be determined by solving the equation

$$2x + 48.17 = 60.17$$

where x represents the number of years since 1972 (*Source:* U.S. Census Bureau, www.census.gov). Solve the equation and interpret.

22. The year that United States industries spent 3 billion dollars on research and development can be determined by solving the equation

$$0.19x + 1.67 = 3$$

where x represents the number of years since 1972 (*Source:* U.S. Census Bureau, www.census.gov). Solve the equation and interpret.

23. The year that worldwide military expenditures were 887.92 billion dollars can be determined by solving

the equation

$$-50.77x + 1091 = 887.92$$

where x represents the number of year since 1990 (*Source:* U.S. Census Bureau, www.census.gov). Solve the equation and interpret.

24. The year that the total expenditures for colleges and universities in the United States was 221.7 billion dollars can be determined by solving the equation

$$5.05x + 191.4 = 221.7$$

where x represents the number of years since 1990 (*Source:* U.S. National Center for Education Statistics, www.nces.ed.gov). Solve the equation and interpret.

Section B.4   Solving Quadratic Equations

So far we have examined solutions to linear equations. Now we turn our attention to **quadratic equations**. These are equations that include a second-degree polynomial and have the form

$$ax^2 + bx + c = 0, \qquad \text{where } a \neq 0$$

We call this form the **standard form** of a quadratic equation. Quadratic equations are useful in such applications as physical motion problems, geometry problems, and mathematical modeling. We will study two methods for solving quadratic equations. These are solving by factoring and solving by the quadratic formula.

Solving Quadratic Equations by Factoring

If a trinomial in the form $ax^2 + bx + c$ is factorable, then its corresponding standard form equation $ax^2 + bx + c = 0$ can be solved by using the **zero product property**.

■ **Zero Product Property**

Let a and b represent any two real numbers. If the product

$$a \cdot b = 0$$

then either $a = 0$, $b = 0$, or both equal zero.

▶ **Note:** The zero product property tells us that if a product is equal to zero then at least one of its factors is equal to zero.

Let's say that we have an equation such as $(3x - 1)(x - 7) = 0$. We can use the zero product property to set each linear factor equal to zero and solve.

$$(3x - 1)(x - 7) = 0 \qquad \text{Given equation}$$

$$3x - 1 = 0 \quad \text{or} \quad x - 7 = 0 \qquad \text{By the zero product property}$$

$$3x = 1 \quad \text{or} \qquad x = 7 \qquad \text{Solving each equation for } x$$

$$x = \frac{1}{3} \quad \text{or} \qquad x = 7 \qquad \text{The solution to the equation}$$

So the solutions to the equation $(3x - 1)(x - 7) = 0$ are $x = \frac{1}{3}$ and $x = 7$. If a quadratic equation is factorable over the real numbers, then the solution consists of either two real number solutions or one repeated number solution.

■ **Solving a Quadratic Equation by Factoring**

1. If needed, write the quadratic equation in the standard form $ax^2 + bx + c = 0$ by setting one side equal to zero.
2. Factor the polynomial completely.
3. Use the zero product property to set each linear factor equal to zero.
4. Solve the linear equations found in step 3.

Example 1   **Solving a Quadratic Equation by Factoring**

Solve the following equations by factoring.

(a) $x^2 + x = 6$ **(b)** $2x^2 - 7x - 4 = 0$

Solution

(a) We begin by writing the equation in standard form.

$$x^2 + x = 6$$

$$x^2 + x - 6 = 0$$

Next we factor the trinomial on the left side of the equation.

$$x^2 + x - 6 = 0 \qquad \text{Equation in standard form}$$

$$(x + 3)(x - 2) = 0 \qquad \text{Factor the polynomial completely.}$$

$$x + 3 = 0 \quad \text{or} \quad x - 2 = 0 \qquad \text{Use the zero product property.}$$

$$x = -3 \quad \text{or} \qquad x = 2 \qquad \text{Solve the linear factors for } x.$$

So the solutions are $x = -3$ and $x = 2$.

(b) We can see that there is no greatest common monomial to factor out. So we factor and solve as in part (a).

$$2x^2 - 7x - 4 = 0 \qquad \text{Given equation}$$

$$(2x + 1)(x - 4) = 0 \qquad \text{Factor by trial and error.}$$

$$2x + 1 = 0 \quad \text{or} \quad x - 4 = 0 \qquad \text{Use the zero product property.}$$

$$x = -\frac{1}{2} \quad \text{or} \qquad x = 4 \qquad \text{Solve the linear factors for } x.$$

So the solutions are $x = -\frac{1}{2}$ and $x = 4$.

Technology Option

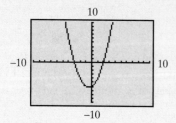

Figure B.4.1

We can verify the solution to part (a) of Example 1 by graphing $y = x^2 + x - 6$ on the calculator. See Figure B.4.1. Notice that the graph crosses the x-axis at the points $(-3, 0)$ and $(2, 0)$. Explain why this shows that $x = -3$ and $x = 2$ are the solutions to $x^2 + x - 6 = 0$.

Solving Quadratic Equations Using the Quadratic Formula

If a quadratic equation is not factorable, then we must use another method to determine the solution. The general solution to the quadratic equation $ax^2 + bx + c = 0$ for $a \neq 0$ is given by the **quadratic formula**

$$x = \frac{-b \pm \sqrt{b^2 - 4ac}}{2a}$$

This formula has two important features. First is the symbol $\pm$ (read "plus or minus"), which means that the quadratic formula is really two formulas that represent the two solutions of a quadratic equation.

$$x = \frac{-b - \sqrt{b^2 - 4ac}}{2a} \quad \text{or} \quad x = \frac{-b + \sqrt{b^2 - 4ac}}{2a}$$

The other important feature is the expression under the radical sign. This expression, $b^2 - 4ac$, is called the **discriminant** and can be used to determine what type of solution a quadratic equation will have. See Table B.4.1.

Table B.4.1

Value of Discriminant $b^2 - 4ac$	Type of Solution
Positive, meaning $b^2 - 4ac > 0$	Two distinct real number solutions
Zero, meaning $b^2 - 4ac = 0$	One repeated real number solution
Negative, meaning $b^2 - 4ac < 0$	No real number solutions

■ **Solving a Quadratic Equation Using the Quadratic Formula**

1. If needed, write the quadratic equation in the standard form $ax^2 + bx + c = 0$ by setting one side equal to zero.
2. Identify the values of a, b, and c.
3. Substitute the values from step 2 into the quadratic formula

$$x = \frac{-b \pm \sqrt{b^2 - 4ac}}{2a}$$

4. Evaluate and simplify to obtain the solutions.

▷ **Note:** If after step 4 we see that the discriminant is negative, then we write *no real solutions*. The solutions are not real-number solutions because we cannot take the square root of a negative number. Such solutions are complex-number solutions and involve the imaginary unit i.

Example 2 ### Using the Quadratic Formula

Find the real-number solutions of the following equations, if possible, using the quadratic formula.

(a) $x^2 + 4x + 5 = 0$ **(b)** $4x^2 = 11 - 4x$

Solution

(a) The equation is given in standard form, so we can see that $a = 1$, $b = 4$, and $c = 5$. Substituting these values into the quadratic equation yields

$$x = \frac{-b \pm \sqrt{b^2 - 4ac}}{2a} = \frac{-4 \pm \sqrt{4^2 - 4(1)(5)}}{2(1)}$$

$$= \frac{-4 \pm \sqrt{16 - 20}}{2}$$

$$= \frac{-4 \pm \sqrt{-4}}{2}$$

Interactive Activity
For the two solutions in Example 2(b), convert to decimal approximations. Check the results in the original equation.

Since the discriminant $b^2 - 4ac = -4$ is negative, this equation has *no real solutions*.

(b) We begin by writing the equation in standard form.

$$4x^2 = 11 - 4x$$
$$4x^2 + 4x - 11 = 0$$

This gives $a = 4$, $b = 4$, and $c = -11$. Substituting these values in the quadratic formula gives us

$$x = \frac{-4 \pm \sqrt{4^2 - 4(4)(-11)}}{2(4)} = \frac{-4 \pm \sqrt{16 + 176}}{8}$$

$$= \frac{-4 \pm \sqrt{192}}{8}$$

Since $192 = 64 \cdot 3 = 8^2 \cdot 3$, we can simplify the solutions by writing

$$x = \frac{-4 \pm \sqrt{192}}{8} = \frac{-4 \pm \sqrt{8^2 \cdot 3}}{8} = \frac{-4 \pm 8\sqrt{3}}{8}$$

$$= \frac{4(-1 \pm 2\sqrt{3})}{8} = \frac{-1 \pm 2\sqrt{3}}{2}$$

So the solutions to the equation $4x^2 = 11 - 4x$ are $x = \dfrac{-1 - 2\sqrt{3}}{2}$ and $x = \dfrac{-1 + 2\sqrt{3}}{2}$. ∎

Quadratic equations are often used in applications that involve **mathematical models**. When solving a quadratic equation based on a mathematical model, we must be sure that the solution is part of the **reasonable values of x**.

These values, denoted in the form $l \le x \le u$, are the values on which the model is based.

Example 3 Solving Quadratic Models

The year that the number of calories consumed each day per person in the United States was 3361 can be represented by the equation

$$0.89x^2 - 1.93x + 3306.27 = 3361, \qquad 0 \le x \le 15$$

where x represents the number of years since 1974 (*Source:* U.S. Census Bureau, www.census.gov).

(a) Interpret the meaning of $0 \le x \le 15$.

(b) Solve the equation for x and interpret.

Solution

(a) The statement $0 \le x \le 15$ gives the reasonable values of x on which the model is based. Since x represents the number of years since 1974, this model is valid for the years 1974 to 1989.

(b) Writing the equation in standard form, we get

$$0.89x^2 - 1.93x - 54.73 = 0$$

Knowing that $a = 0.89$, $b = -1.93$, and $c = -54.73$, we have

$$x = \frac{1.93 \pm \sqrt{(-1.93)^2 - 4(0.89)(-54.73)}}{2(0.89)}$$

$$= \frac{1.93 \pm \sqrt{198.5637}}{1.78}$$

Using a calculator to approximate these solutions, we get

$$x = \frac{1.93 - \sqrt{198.5637}}{1.78} \approx -6.83 \quad \text{or} \quad x = \frac{1.93 + \sqrt{198.5637}}{1.78} \approx 9.00$$

Since -6.83 is not between $0 \le x \le 15$, the only approximate solution is $x \approx 9$.

Interpret the Solution: This means that in 1983 the number of calories consumed each day per person in the United States was 3361. ∎

✓ Checkpoint 1

Now work Exercise 23 in the B.4 Exercises.

SECTION B.4 EXERCISES

In Exercises 1–12, solve the equation by factoring.

1. $4x^2 + 8x = 0$

2. $5x^2 - 10x = 0$

3. $x^2 - 9 = 0$

4. $x^2 - 36 = 0$

5. $x^2 + 8x + 15 = 0$

6. $x^2 + 7x + 12 = 0$

7. $x^2 = 5x - 6$

8. $x^2 = 3x - 2$

9. $2x^2 + 3x + 1 = 0$

10. $3x^2 + 10x + 3 = 0$

11. $3x^2 = x + 4$

12. $2x^2 = 7x + 4$

*In Exercises 13–20, solve the equation by using the quadratic formula. If the discriminant is negative, write **no real solutions**.*

13. $4x^2 + x - 2 = 0$

14. $3x^2 + 7x + 3 = 0$

15. $x^2 - 3x + 3 = 0$

16. $2x^2 + 3x + 3 = 0$

17. $4x^2 + 7 = 12x$ **18.** $3x^2 = 2 - 6x$

19. $2x^2 = 6x - 1$ **20.** $5x^2 = 1 - 2x$

🌐 **21.** The year that the membership in the Boy Scouts of America was 5433 thousand can be found by solving the equation

$$21.82x^2 - 143.75x + 5555.57 = 5433, \qquad 0 \le x \le 7$$

where x represents the number of years since 1989. (*Source:* U.S. Census Bureau, www.census.gov). Solve the equation for x and interpret.

🌐 **22.** The year that the annual total U.S. exports was $442.5 billion can be found by solving the equation

$$5.18x^2 + 5.86x + 410.6 = 442.5, \qquad 1 \le x \le 6$$

where x represents the number of years since 1990 (*Source:* U.S. Census Bureau, www.census.gov). Solve the equation for x and interpret.

🌐 ✓ **23.** The year that the number of lawyers in private practice was 195 thousand can be found by solving the equation

$$-0.37x^2 - 1.43x + 190.34 = 195, \qquad 0 \le x \le 30$$

where x represents the number of years since 1960. (*Source:* U.S. Census Bureau, www.census.gov). Solve the equation for x and interpret. (*Hint:* Begin by multiplying each side of the equation by -1.)

🌐 **24.** The year that the number of aggravated assaults in the United States was 297 per 100,00 people can be found by solving the equation

$$-3.08x^2 + 40.35x + 305.89 = 297, \qquad 0 \le x \le 10$$

where x represents the number of years since 1986 (*Source:* U.S. Census Bureau, www.census.gov). Solve the equation for x and interpret. (*Hint:* Begin by multiplying each side of the equation by -1.)

Section B.5   Properties of Logarithms

We can think of subtraction as a way to undo addition and division as a way to undo multiplication. A logarithm can be thought of as a way to undo exponentiation. To see how, let's take a look at the definition of logarithm.

> ■ **Logarithm**
>
> We can define a **logarithm base b of x** by saying that
>
> $$y = \log_b x \quad \text{if and only if} \quad b^y = x, \qquad b \ne 1, \; x > 0$$

▶ **Note:** From the definition we see that we cannot take the logarithm of a negative number.

This definition shows that an expression written in logarithmic form can be written in exponential form, and vice versa. Table B.5.1 lists some examples.

Table B.5.1

Logarithmic Form	Exponential Form
$\log_2 8 = 3$	$2^3 = 8$
$\log_5 625 = 4$	$5^4 = 625$
$\log_{10} 10,000 = 4$	$10^4 = 10,000$
$\log_{\frac{1}{2}} \frac{1}{32} = 5$	$\left(\frac{1}{2}\right)^5 = \frac{1}{32}$
$\log_{\frac{1}{4}} 64 = -3$	$\left(\frac{1}{4}\right)^{-3} = 64$

Some logarithms have special bases. If $b = 10$, then $\log_{10} x = \log x$ is a **common logarithm**. Many applications in the area of finance use the number e as the base. This is the **natural logarithm** and is denoted by $\log_e x = \ln x$.

The definition of logarithm can be used to prove many of the properties of logarithms. Let's take a look at some of these properties.

> ■ **Properties of Logarithms**
>
> Let $b > 0$ with $b \neq 1$ and let m and n represent positive real numbers with r being any real number.
>
> 1. $\log_b 1 = 0$
> 2. $\log_b b = 1$
> 3. $\log_b(m \cdot n) = \log_b m + \log_b n$
> 4. $\log_b\left(\frac{m}{n}\right) = \log_b m - \log_b n$
> 5. $\log_b m^r = r \log_b m$
> 6. $\log_b b^r = r$

Example 1   Using the Properties of Logarithms

Use the properties of logarithms to write the following as a sum, difference, or product of logarithms.

(a) $\log \dfrac{(x-3)y}{x}$ **(b)** $\ln((x+7)y)^4$

Solution

(a) $\log \dfrac{(x-3)y}{x}$ Given expression

$= \log(x-3)y - \log x$ Logarithm property 4

$= \log(x-3) + \log y - \log x$ Logarithm property 3

(b) $\ln((x+7)y)^4$ Given expression

$= 4\ln((x+7)y)$ Logarithm property 5

$= 4(\ln(x+7) + \ln y)$ Logarithm property 3   ■

✓ **Checkpoint 1**   Now work Exercise 5 in the B.5 Exercises.

At times we need to solve **logarithmic equations**. Sometimes we can use the definition of a logarithm to solve equations. Other times we need to use the method based on the following property.

$$\text{If } \log_b m = \log_b n, \quad \text{then } m = n.$$

Here m and n are called the **arguments** of the logarithmic expressions. Another useful property is

$$\text{If } m = n, \quad \text{then } \log_b m = \log_b n.$$

In mathematics, when we have this "two-way street," we can express both properties by writing

$$\log_b m = \log_b n \quad \text{if and only if} \quad m = n.$$

▶ **Note:** This assumes, from the definition, that $m > 0$ and $n > 0$.

Part (a) of the next example illustrates using the definition of a logarithm for solving, while part (b) shows how the aforementioned property is used.

Example 2   **Solving Logarithmic Equations**

Solve each of the following for x.

(a) $\log(2x + 1) = 4$

(b) $\ln(x + 6) - \ln(x + 2) = \ln x$

Solution

(a) Here we use the definition of logarithm to solve. This gives us

$\log(2x + 1) = 4$	Given equation
$10^4 = 2x + 1$	Definition of logarithm
$10{,}000 = 2x + 1$	Simplification
$9999 = 2x$	Subtracting 1 from both sides
$\dfrac{9999}{2} = x$	Dividing both sides by 2

(b) For this equation we need to rewrite the left side as a single logarithmic term using property 4 in order to apply the property $\log_b m = \log_b n$ if and only if $m = n$.

$\ln(x + 6) - \ln(x + 2) = \ln x$	Given equation
$\ln\left(\dfrac{x + 6}{x + 2}\right) = \ln x$	Logarithm property 4
$\dfrac{x + 6}{x + 2} = x$	$\log_b m = \log_b n$ if and only if $m = n$
$x + 6 = x(x + 2)$	Multiply both sides by $x + 2$
$x + 6 = x^2 + 2x$	Distribute x.
$0 = x^2 + x - 6$	The quadratic equation in standard form
$0 = (x - 2)(x + 3)$	Factor the right side of the equation.
$x = 2 \quad \text{or} \quad x = -3$	Solve for x.

Since we cannot take the logarithm of a negative number, we see that $x = -3$ is not a solution. Thus, $x = 2$ is the only solution to the equation. ∎

Logarithms are a useful tool for solving **exponential equations**. These are equations that have a form such as $b^y = x$, where the variable that we are solving for is the exponent. Since the property $\log_b m = \log_b n$ if and only if $m = n$ is a two-way street, we can solve exponential equations by taking the logarithm of each side of the equation and using logarithm property 5. Since it does not matter which base of logarithm we use when solving these equations, we usually use the natural logarithm.

Example 3 **Solving Exponential Equations**

Solve the equation $3(5^x) = 21$ for x.

Solution

$$3(5^x) = 21 \qquad \text{Given equation}$$
$$5^x = 7 \qquad \text{Divide both sides by 3.}$$
$$\ln(5^x) = \ln 7 \qquad \log_b m = \log_b n \text{ if and only if } m = n$$
$$x \ln 5 = \ln 7 \qquad \text{Logarithm property 5}$$
$$x = \frac{\ln 7}{\ln 5} \approx 1.21 \qquad \text{Divide both sides by } \ln 5.$$ ∎

SECTION B.5 EXERCISES

In Exercises 1–8, use the properties of logarithms to write the following as a sum, difference, or product of logs.

1. $\log(2x \cdot y)$

2. $\log\left(\dfrac{3x}{5y}\right)$

3. $\log\left(\dfrac{2y}{5x}\right)^3$

4. $\log\left(\dfrac{10x}{y}\right)^7$

✓ **5.** $\ln\left(\dfrac{xy^2}{x+2}\right)$

6. $\ln\left(\dfrac{x^3 y}{x-10}\right)$

7. $\ln \sqrt[3]{\dfrac{5x^3}{9y}}$

8. $\ln \sqrt{\dfrac{y^4}{x^3 y}}$

In Exercises 9–16, solve the logarithmic equation for x.

9. $\log(x-2) = 2$

10. $\log(x+5) = 3$

11. $\ln(4x-2) = 3$

12. $\ln(5x+1) = 5$

13. $\log(x-2) = \log(x-7) + \log 4$

14. $\log_2 x - \log_2(x+1) = \log_2 5$

15. $\ln(4x+5) - \ln(x+3) = \ln 3$

16. $\ln(4x-2) + \ln 4 = \ln(x-2)$

In Exercises 17–22, solve the exponential equation for x.

17. $3^x = 6$

18. $4^x = 12$

19. $2^{5x+2} = 8$

20. $10^{3x-7} = 5$

21. $5(2^{3x}) = 8$

22. $0.4(3^x) = 0.8$

🌐 **23.** The year that 18.89% of all waste generated in the United States was yard waste can be determined by solving the exponential equation

$$19.92(0.994^x) = 18.89, \qquad 1 \le x \le 6$$

where x represents the number of years since 1990 (*Source:* U.S. Census Bureau, www.census.gov). Solve the equation for x and interpret.

🌐 **24.** The year that the pupil–teacher ratio in the United States was 19 students for each teacher can be determined by solving the exponential equation

$$25.34(0.987^x) = 19, \qquad 1 \le x \le 36$$

where x represents the number of years since 1959 (*Source:* U.S. National Center for Education Statistics, www.nces.ed.gov). Solve the equation for x and interpret.

Section B.6 **Geometric Sequence and Series**

In our everyday use of the word *sequence*, we usually mean that there is a collection of some objects with an order to the collection. For example, if we were to list in order the U.S. presidents since the end of World War II, our list would have an identified first member (Harry S. Truman), an identified second member

(Dwight D. Eisenhower), an identified third member (John F. Kennedy), and so on.

In mathematics, we use the word *sequence* in a manner similar to our everyday usage of the word. The main difference is that in mathematics, our collection of objects is usually numbers. At this time, we limit our study to **geometric sequences**.

Geometric Sequence

A finite **geometric sequence** has a finite number of terms where each successive term is found by multiplying the preceding term by a number, denoted r. (Think of r as being short for *ratio*.) In general, a finite geometric sequence looks like the following:

■ **Finite Geometric Sequence**

A **finite geometric sequence** has the form

$$a, ar, ar^2, ar^3, \ldots, ar^{n-1}$$

where a is the first term and r is the **common ratio** ($r \neq 1$), and there are n terms.

Example 1 **Identifying Elements of a Finite Geometric Sequence**

For the following geometric sequences, identify a, r, and n.

(a) $3, 9, 27, 81$

(b) $3, \dfrac{3}{10}, \dfrac{3}{10^2}, \dfrac{3}{10^3}, \dfrac{3}{10^4}$

Solution

(a) The first term is $a = 3$. The common ratio can be found by dividing any two consecutive terms. This gives us $r = \dfrac{9}{3} = 3$. The number of terms is $n = 4$.

(b) Here it may be beneficial to rewrite the sequence as follows:

$$3, \frac{3}{10}, \frac{3}{10^2}, \frac{3}{10^3}, \frac{3}{10^4}$$

$$3, 3\left(\frac{1}{10}\right), 3\left(\frac{1}{10^2}\right), 3\left(\frac{1}{10^3}\right), 3\left(\frac{1}{10^4}\right)$$

$$3, 3\left(\frac{1}{10}\right), 3\left(\frac{1}{10}\right)^2, 3\left(\frac{1}{10}\right)^3, 3\left(\frac{1}{10}\right)^4$$

The first term is $a = 3$. The common ratio is $r = \dfrac{1}{10}$. The number of terms is $n = 5$. Notice how rewriting the sequence made determining r relatively easy. Also notice that the number of terms, $n = 5$, is one more than the last exponent. ■

Geometric Series

When we sum the terms of a sequence, we form what is called a **series**. A finite **geometric series** has the following definition:

■ **Finite Geometric Series**

A **finite geometric series** has the form

$$a + ar + ar^2 + ar^3 + \cdots + ar^{n-1}$$

where a is the first term and r is the common ratio ($r \neq 1$), and there are n terms.

▶ **Note:** 1. Notice that we are simply adding the terms of a finite geometric sequence.

2. The $\cdots$ means that there may be other terms that follow the indicated pattern.

Example 2 **Identifying Elements of a Finite Geometric Series**

For the following geometric series, identify a, r, and n.

(a) $3 + 9 + 27 + 81$

(b) $3 + \dfrac{3}{10} + \dfrac{3}{10^2} + \cdots + \dfrac{3}{10^{50}}$

Solution

(a) As in Example 1a, we have $a = 3, r = 3$ and $n = 4$.

(b) As in Example 1b, we can rewrite the series as follows.

$$3 + \frac{3}{10} + \frac{3}{10^2} + \cdots + \frac{3}{10^{50}}$$

$$3 + 3\left(\frac{1}{10}\right) + 3\left(\frac{1}{10}\right)^2 + \cdots + 3\left(\frac{1}{10}\right)^{50}$$

We have $a = 3$ and $r = \dfrac{1}{10}$. The number of terms is 51, one more than the last exponent. ■

Each finite geometric series in Example 2 has a sum. We can quickly compute that the sum in Example 2(a) is 120, whereas the sum in Example 2(b) would require more effort. At this time we would like to develop a formula to find the sum of **any** finite geometric series. (This formula is used in Chapter 5.) To do this, we will call the sum S_n. This gives us

$$S_n = a + ar + ar^2 + ar^3 + \cdots + ar^{n-1} \tag{I}$$

Next we multiply both sides of the equation by r to give us

$$r\,S_n = ar + ar^2 + ar^3 + ar^4 + \cdots + ar^n \tag{II}$$

Now we subtract Equation (II) from Equation (I).

$$S_n - rS_n = (a + ar + ar^2 + ar^3 + \cdots + ar^{n-1}) - (ar + ar^2 + ar^3 + ar^4 + \cdots + ar^n)$$
$$= a - ar^n$$

Finally, we factor and solve for S_n to get

$$S_n(1 - r) = a(1 - r^n)$$
$$S_n = \frac{a(1 - r^n)}{1 - r}, \qquad r \neq 1$$

■ **Sum of Finite Geometric Series**

A finite geometric series has a sum of

$$a + ar + ar^2 + ar^3 + \cdots + ar^{n-1} = \frac{a(1 - r^n)}{1 - r}, \qquad r \neq 1$$

Example 3   **Using the Sum of a Finite Geometric Series Formula**

Which would produce more money? $10,000 each day in the month of April or 1¢ the first day, 2¢ the second day, 4¢ the third day, and so on, with the money received doubling each day.

Solution

Since April has 30 days, the first choice would generate $10,000(30) = 300,000$ or $300,000. The second choice would generate, in cents,

$$\text{Total} = 1 + 2 \cdot 1 + 2^2 \cdot 1 + 2^3 \cdot 1 + 2^4 \cdot 1 + \cdots + 2^{29} \cdot 1$$

This is a finite geometric series with $a = 1, r = 2$, and $n = 30$. Using our sum formula yields

$$\text{total} = \frac{a(1 - r^n)}{1 - r} = \frac{1(1 - 2^{30})}{1 - 2} = \frac{-1,073,741,823}{-1} = 1,073,741,823\text{¢}$$

In dollars, this is equal to $10,737,418.23. Clearly, the second choice is much better! ■

SECTION B.6 EXERCISES

In Exercises 1–8, find the sum of the given finite geometric series.

1. $3 + 3\left(\frac{1}{10}\right) + 3\left(\frac{1}{10}\right)^2 + \cdots + 3\left(\frac{1}{10}\right)^{50}$

2. $4 + 4\left(\frac{1}{10}\right) + 4\left(\frac{1}{10}\right)^2 + \cdots + 4\left(\frac{1}{10}\right)^{50}$

3. $1 + 2 + 2^2 + 2^3 + \cdots + 2^{20}$

4. $1 + 3 + 3^2 + 3^3 + \cdots + 3^{20}$

5. $2 + 2(5) + 2(5)^2 + 2(5)^3 + \cdots + 2(5)^9$

6. $5 + 5(2) + 5(2)^2 + 5(2)^3 + \cdots + 5(2)^9$

7. $2 - 2(3) + 2(3)^2 - 2(3)^3 + \cdots + 2(3)^8$

8. $3 - 3(2) + 3(2)^2 - 3(2)^3 + \cdots + 3(2)^8$

Applications

9. An NBA player has recently been fined for fighting with a fan during a game. The league office has decided to fine him according to the following: $1000 for the first day after its decision; $2000 for the second day after its decision; $4000 for the third day, and so on, with the amount doubling each day until the player pays the fines. Find the total fine for 12 days, when the player decides to pay the fine.

10. The NBA player in Exercise 9 decides to appeal his fine. The appeal board gave the player the option of paying the total fine as determined in Exercise 9 or to pay $100 to every fan in attendance at the team's next home game (maximum capacity of the basketball arena is 19,000). As his agent, what would you advise him to do?

11. Your very wealthy and very eccentric Aunt Liz has offered you the following choices as a graduation present: She picks the month and gives you $100,000 each day of the month *or* you pick the month and she gives you 1¢ the first day, 2¢ the second day, 4¢ the third day, and so on, doubling the amount received each day. Which would you choose and why?

12. Suppose that Aunt Liz in Exercise 11 slightly changes your choices as follows: You pick the month and she gives you $100,000 each day of the month *or* she picks the month and gives you 1¢ the first day, 2¢ the second day, 4¢ the third day, and so on, doubling the amount received each day. Which would you choose and why?

Appendix C
Calculator Programs

Simplex Method Program

The following program, SIMPLEX, utilizes the simplex method to solve linear programming problems as examined in Chapter 4. As with all programs in Appendix C, this program is written for the Texas Instruments TI-83 family of calculators. If you have a calculator different from this, consult the text's companion Web site, www.prenhall.com/armstrong, for the program for your model of calculator.

```
PROGRAM:SIMPLEX
ClrHome
dim([A])→L₁
L₁(1)-1→dim(L₂)
rowSwap([A],1,L₁(1))→[A]
For(J,1,L₁(2)-1)
0→C:0→M
For(I,2,L₁(1))
C+abs([A](I,J))→C
If [A] (I,J)=1
Then
I→R:1→M
End:End
If C=1 and [A](1,J)>0 and M=1
Then
*row+(-[A] (1,J), [A], R, 1)→[A]
End:End
Lbl LP
-1→C
Matr▶list([A]ᵀ,1,L₃)
L₁(2)-1→dim(L₃)
For(I,1,L₁(2)-1)
If L₃(I)=min(L₃) and L₃(I)<0
Then
I→C
End:End
If C=-1
Then
Goto QT
End
For(I,2,L₁(1))
If [A] (I,C)>0
```

```
Then
[A] (I,L₁(2))/[A] (I,C)→L₂(I-1)
Else:1E6→L₂(I-1)
End:End
For(I,1,dim(L₂))
If L₂(I)=min(L₂)
Then
I+1→R
End:End
*row(1/[A] (R,C),[A],R)→[A]
For(I,1,L₁(1))
If I≠R
Then
*row+(-[A] (I,C),[A],R,I)→[A]
End:End
Goto LP
Lbl QT
round([A],4)→[A]
[A] (1,L₁(2))→Z
Disp "Z="
Output(1,5,Z)
Disp ""
rowSwap([A],1,L₁(1))→[A]
L₁(2)-1→dim(L₄)
For(J,1,L₁(2)-1)
Matr▶list([A],J,L₂)
If sum(abs(L₂))>1
Then
0→L₄(J)
Else
For(I,1,L₁(1))
If [A] (I,J)=1
Then
[A] (I,L₁(2))→L₄(J)
End:End:End:End
L₄
```

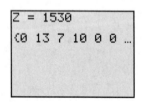

Figure C.1

Example Executing the Simplex Program

For the situation in Section 4.2, Example 3, we enter the initial simplex tableau into matrix A and execute the SIMPLEX program. The output of the program is shown in Figure C.1. The optimal solution of 1530 is shown, as well as the values of all variables. You can scroll to the right to see the value of all decision and slack variables. ∎

Future Value of an Annuity Program

The following program calculates the future value of an annuity.

```
PROGRAM:ANNFV
Prompt P,I,N
P*((1+I)^N-1)/I→F
Disp "FUTURE VALUE IS"
Disp F
```

Example Executing the ANNFV Program

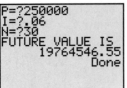

For the situation in Section 5.3, Example 3, the input and output for executing the future value of an annuity program are shown in Figure C.2. Notice that the I must be entered as the periodic rate in decimal form. ■

Figure C.2

Payment to Fund an Annuity Program

The following program determines the periodic payment amount needed to fund an annuity. Classic examples are sinking funds.

```
PROGRAM:ANNPMT
Prompt F,I,N
(F*I)/((1+I)^N-1)→P
Disp "PAYMENT IS"
Disp P
```

Example Executing the ANNPMT Program

For the situation in Section 5.3, Example 4, the input and output for executing the payment to fund an annuity program are shown in Figure C.3. Pay particular attention to how I and N were entered. ■

Figure C.3

Present Value of an Annuity Program

The following program calculates the present value of an annuity.

```
PROGRAM:ANNPV
Prompt P,I,N
P*((1-(1+I)^-N)/I)→A
Disp "PRESENT VALUE IS"
Disp A
```

Example Executing the ANNPV Program

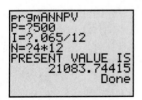

For the situation in Section 5.4, Example 1, the input and output for executing the present value of an annuity program are shown in Figure C.4. Pay particular attention to how I and N are entered. ■

Figure C.4

Program for Payment to Amortize a Loan

The following program calculates the periodic payment necessary to amortize a loan.

```
PROGRAM:AMORTPAY
Prompt P,I,N
P*(I/(1-(1+I)^-N))→T
Disp "PAYMENT IS"
Disp T
```

Example **Executing the AMORTPAY Program**

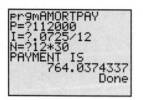

Figure C.5

For the situation in Section 5.4, Example 4, the input and output for executing the payment to amortize a loan program are shown in Figure C.5. Pay particular attention to how I and N are entered. ∎

Bayes Theorem Program

This program is used to compute

$$P(A) = P(B_1) \cdot P(A \mid B_1) + P(B_2) \cdot P(A \mid B_2) + \cdots + P(B_k) \cdot P(A \mid B_k)$$

and

$$P(B_i \mid A) = \frac{P(B_i) \cdot P(A \mid B_i)}{P(B_1) \cdot P(A \mid B_1) + P(B_2) \cdot P(A \mid B_2) + \cdots + P(B_k) \cdot P(A \mid B_k)}$$

for any value i, where $1 \le i \le k$.

The values of $P(B_1), P(B_2), \ldots, P(B_k)$ are entered in L_1 and the values of $P(A \mid B_1), P(A \mid B_2), \ldots, P(A \mid B_k)$ are entered in L_2 before executing the program.

```
PROGRAM: BAYES
Input "NO. BRANCHES ",K
0→T
For (J,1,K)
L₁(J)*L₂(J)→S
T+S→T
End
Disp "P(A)= ",T
Input "ITH BRANCH ",I
(L₁(I)*L₂(I))→N
N/T→P
Disp "PROBABILITY",P
```

Example **Executing the BAYES Program**

For the situation in Section 6.7, Example 4, the input and output for executing the Bayes program are shown in Figures C.6 to C.8.

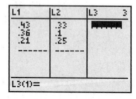

Figure C.6

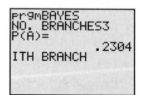

Figure C.7

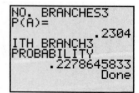

Figure C.8 ∎

Mean and Standard Deviation Program

For a given frequency distribution, this program computes the mean and standard deviation where the midpoints of the class are in L_1 and the frequencies are in L_2.

```
PROGRAM:MEANSD
sum(L₁*L₂)/sum(L₂)→M
ClrList L₃
(L₁-M)²→L₃
√((sum(L₃*L₂))/(sum(L₂)-1))→S
Disp "MEAN", M
Disp "STD DEV", S
```

To see the input and output for this program, consult the Technology Option box after Example 3 in Section 7.4.

Binomial Probability Distribution Program

This program displays a table of $X = x$ and $P(X = x)$ values of the binomial distribution for given values of n and p.

```
PROGRAM: BINOMPD
Input "N ",N
Input "P ",P
ClrList L₁,L₂
For (I, 0, N)
I→L₁(I+1)
binompdf(N,P,I)→L₂(I+1)
End
```

Example   ### Executing the BINOMPD Program

For the situation in Section 8.3, Example 3, the input for the program is in Figure C.9 and the probability distribution is in Figure C.10. Recall that for this example $n = 7$ and $p = 0.71$.

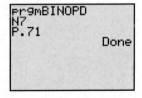

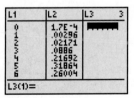

Figure C.9   **Figure C.10**

Appendix D
Standard Normal Table

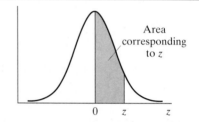

Area corresponding to z

Area Under the Standard Normal Curve

z	.00	.01	.02	.03	.04	.05	.06	.07	.08	.09
0.0	0.0000	0.0040	0.0080	0.0120	0.0160	0.0199	0.0239	0.0279	0.0319	0.0359
0.1	0.0398	0.0438	0.0478	0.0517	0.0557	0.0596	0.0636	0.0675	0.0714	0.0753
0.2	0.0793	0.0832	0.0871	0.0910	0.0948	0.0987	0.1026	0.1064	0.1103	0.1141
0.3	0.1179	0.1217	0.1255	0.1293	0.1331	0.1368	0.1406	0.1443	0.1480	0.1517
0.4	0.1554	0.1591	0.1628	0.1664	0.1700	0.1736	0.1772	0.1808	0.1844	0.1879
0.5	0.1915	0.1950	0.1985	0.2019	0.2054	0.2088	0.2123	0.2157	0.2190	0.2224
0.6	0.2257	0.2291	0.2324	0.2357	0.2389	0.2422	0.2454	0.2486	0.2517	0.2549
0.7	0.2580	0.2611	0.2642	0.2673	0.2704	0.2734	0.2764	0.2794	0.2823	0.2852
0.8	0.2881	0.2910	0.2939	0.2967	0.2995	0.3023	0.3051	0.3078	0.3106	0.3133
0.9	0.3159	0.3186	0.3212	0.3238	0.3264	0.3289	0.3315	0.3340	0.3365	0.3389
1.0	0.3413	0.3438	0.3461	0.3485	0.3508	0.3531	0.3554	0.3577	0.3599	0.3621
1.1	0.3643	0.3665	0.3686	0.3708	0.3729	0.3749	0.3770	0.3790	0.3810	0.3830
1.2	0.3849	0.3869	0.3888	0.3907	0.3925	0.3944	0.3962	0.3980	0.3997	0.4015
1.3	0.4032	0.4049	0.4066	0.4082	0.4099	0.4115	0.4131	0.4147	0.4162	0.4177
1.4	0.4192	0.4207	0.4222	0.4236	0.4251	0.4265	0.4279	0.4292	0.4306	0.4319
1.5	0.4332	0.4345	0.4357	0.4370	0.4382	0.4394	0.4406	0.4418	0.4429	0.4441
1.6	0.4452	0.4463	0.4474	0.4484	0.4495	0.4505	0.4515	0.4525	0.4535	0.4545
1.7	0.4554	0.4564	0.4573	0.4582	0.4591	0.4599	0.4608	0.4616	0.4625	0.4633
1.8	0.4641	0.4649	0.4656	0.4664	0.4671	0.4678	0.4686	0.4693	0.4699	0.4706
1.9	0.4713	0.4719	0.4726	0.4732	0.4738	0.4744	0.4750	0.4756	0.4761	0.4767
2.0	0.4772	0.4778	0.4783	0.4788	0.4793	0.4798	0.4803	0.4808	0.4812	0.4817
2.1	0.4821	0.4826	0.4830	0.4834	0.4838	0.4842	0.4846	0.4850	0.4854	0.4857
2.2	0.4861	0.4864	0.4868	0.4871	0.4875	0.4878	0.4881	0.4884	0.4887	0.4890
2.3	0.4893	0.4896	0.4898	0.4901	0.4904	0.4906	0.4909	0.4911	0.4913	0.4916
2.4	0.4918	0.4920	0.4922	0.4925	0.4927	0.4929	0.4931	0.4932	0.4934	0.4936
2.5	0.4938	0.4940	0.4941	0.4943	0.4945	0.4946	0.4948	0.4949	0.4951	0.4952
2.6	0.4953	0.4955	0.4956	0.4957	0.4959	0.4960	0.4961	0.4962	0.4963	0.4964
2.7	0.4965	0.4966	0.4967	0.4968	0.4969	0.4970	0.4971	0.4972	0.4973	0.4974
2.8	0.4974	0.4975	0.4976	0.4977	0.4977	0.4978	0.4979	0.4979	0.4980	0.4981
2.9	0.4981	0.4982	0.4982	0.4983	0.4984	0.4984	0.4985	0.4985	0.4986	0.4986
3.0	0.4987	0.4987	0.4987	0.4988	0.4988	0.4989	0.4989	0.4989	0.4990	0.4990
3.1	0.4990	0.4991	0.4991	0.4991	0.4992	0.4992	0.4992	0.4992	0.4993	0.4993
3.2	0.4993	0.4993	0.4994	0.4994	0.4994	0.4994	0.4994	0.4995	0.4995	0.4995
3.3	0.4995	0.4995	0.4995	0.4996	0.4996	0.4996	0.4996	0.4996	0.4996	0.4997
3.4	0.4997	0.4997	0.4997	0.4997	0.4997	0.4997	0.4997	0.4997	0.4997	0.4998
3.5	0.4998	0.4998	0.4998	0.4998	0.4998	0.4998	0.4998	0.4998	0.4998	0.4998
3.6	0.4998	0.4998	0.4999	0.4999	0.4999	0.4999	0.4999	0.4999	0.4999	0.4999
3.7	0.4999	0.4999	0.4999	0.4999	0.4999	0.4999	0.4999	0.4999	0.4999	0.4999
3.8	0.4999	0.4999	0.4999	0.4999	0.4999	0.4999	0.4999	0.4999	0.4999	0.4999
3.9	0.5000	0.5000	0.5000	0.5000	0.5000	0.5000	0.5000	0.5000	0.5000	0.5000

Table entries represent the area under the standard normal curve from 0 to z, $z \geq 0$.

Taken from Barnett/Ziegler/Byleen   Finite Math 8/e (Prentice Hall)

Answers

Chapter 1

Section 1.1

1. $x + 10 = 14$ **3.** $x - 7 = 28$ **5.** $5x = 45$

7. $6x - 1 = 29$ **9.** $3x + 11 = 44$

11. (a) It will cost \$267 to produce ten picture frames.
   (b) $C = 1.50x + 252$

13. (a) It costs \$11,250 to produce 110 portable cribs.
   (b) $C = 100x + 250$

15. (a) $220x + 80y = 1400$ **(b)** $6x + 4y = 40$

17. (a) $x + y = 640$ **(b)** $0.75x + 2y = 1080$

19. (a) Let x = number of student tickets sold and let y = number of adult tickets sold.

$$x + y = 1600$$
$$3x + 5y = 5760$$

   (b) $x + y = 1600$ **(c)** $x + y = 1600$
         $3x + 5y = 5980$ $3x + 5y = 5500$

21. (a) Let C = monthly checking costs of City account and let x = fee per check.

$$C = 5 + 0.15x$$

   (b) Let C = monthly checking costs of Urban account and let x = fee per check.

$$C = 4 + 0.2x$$

23. (a) Let C = attorney costs at Frank & Earnst Law Firm and let x = amount of settlement.

$$C = 1500 + 0.25x$$

   (b) Let C = attorney costs at Mackenzie Law Firm and let x = amount of settlement.

$$C = 2000 + 0.20x$$

25. (a) $0.5x + 0.8y + 1.1z = 450$
   (b) $0.6x + 0.7y + 0.9z = 390$
   (c) $0.2x + 0.3y + 0.6z = 190$

27. (a) $x + y + z = 50{,}000$ **(b)** $z = \frac{1}{2}x + 1000$
   (c) $0.06x + 0.07y + 0.08z = 3410$

29. (a) $240x + 215y + 336z = 65{,}790$
   (b) $148x + 98y + 171z = 31{,}860$
   (c) $42x + 69y + 147z = 21{,}630$

31. For parts (a), (b), and (c), let x = amount of Basic!, let y = amount of Vaca, and let z = amount of Secra.
   (a) $20x + 10y + 30z = 340$ **(b)** $10x + 5y + 20z = 205$
   (c) $20x + 15y + 10z = 225$

33. For parts (a), (b), and (c), let x = quantity produced at Plant I, let y = quantity produced at Plant II, and let z = quantity produced at Plant III.
   (a) $18x + 20y + 16z = 5900$
   (b) $12x + 9y + 11z = 3580$
   (c) $22x + 18y + 17z = 6350$

Section 1.2

1. Ordered pairs are $(0, 1)$, $(3, 5)$, and $(6, 7)$.

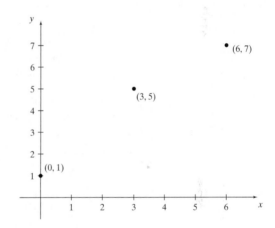

3. Ordered pairs are $(-2, 3)$, $(-1, 5)$, and $(4, -4)$.

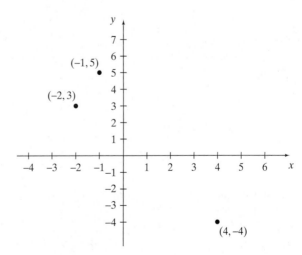

5. Ordered pairs are $(-4, -5), (-3.5, -6), (-2, -8)$, and $(2, -11)$.

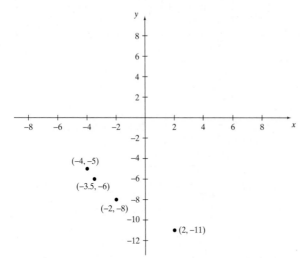

7. Ordered pairs are $(0, 1.1), (1, 1.5), (2, 1.8)$, and $(3, 2)$.

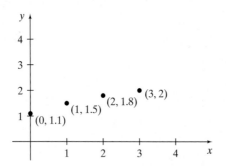

9. Domain: $\{0, 3, 6\}$
Range: $\{1, 5, 7\}$

11. Domain: $\{-2, -1, 4\}$
Range: $\{-4, 3, 5\}$

13. Domain: $\{-4, -3.5, -2, 2\}$
Range: $\{-11, -8, -6, -5\}$

15. Domain: $\{0, 1, 2, 3\}$
Range: $\{1.1, 1.5, 1.8, 2\}$

17. (a) Independent variable: x
Dependent variable: $f(x)$

(b)

x	$f(x)$
2	-3
-3	-8

19. (a) Independent variable: x
Dependent variable: $g(x)$

(b)

x	$g(x)$
-1	-2.5
0	0.5
5	15.5

21. (a) Independent variable: x
Dependent variable: y

(b)

x	y
0	6
5	7.5
10	9

23. (a) Independent variable: t
Dependent variable: $f(t)$

(b)

t	$f(t)$
0	3.5
1	3.35
3	3.05
6	2.6

25. (a) Independent variable: x
Dependent variable: $f(x)$

(b)

x	$f(x)$
1	1
4	2.2
6	3

27. The table represents a function; there is a unique y-value for every x-value.

29. The table does not represent a function; there are two y-values for the x-value of -2.

31. The graph represents a function.

33. The graph does not represent a function.

35. (a)

x	Expression	y
0	$3(0) + 1$	1
2	$3(2) + 1$	7
1	$3(1) + 1$	4

(b) $f(x) = 3x + 1$

37. (a)

x	Expression	y
0	$6 - 3(0)$	6
1	$6 - 3(1)$	3
2	$6 - 3(2)$	0
3	$6 - 3(3)$	-3

(b) $f(x) = 6 - 3x$

39. $m = \frac{3}{10}$; increasing

41. $m = -3$; decreasing

43. $m = 0$; horizontal

45. m is undefined; vertical

47. $m = -\frac{3}{2}$; decreasing

49. y-intercept: $(0, -10)$
x-intercept: $(5, 0)$

51. y-intercept: $(0, -2)$
x-intercept: $(3, 0)$

53. y-intercept: $(0, -4)$
x-intercept: $(8, 0)$

55. y-intercept: $\left(0, -\frac{5}{2}\right)$
x-intercept: $\left(\frac{1}{2}, 0\right)$

57. y-intercept: $(0, 10.3)$
x-intercept: $(-51.5, 0)$

59. $C(x) = 70x + 10{,}000$

61. $C(x) = 134x + 14{,}500$

63. (a) Fixed costs are \$1250 and the cost per unit is \$3.
(b) $C(400) = 2450$. The cost to produce 400 quick-replace spokes is \$2450.
(c) 90 quick-replace spokes can be produced for \$830.

65. (a) The fixed costs are \$10 and the cost per unit is \$32.
(b) $C(15) = 490$
To arrange and deliver 15 get-well bouquets, the cost is \$490.
(c) 250 get-well bouquets can be arranged and delivered for \$8010.

67. $m = 2.42$
From 1989 to 1999, public expenditures at private universities and colleges in the United States increased at an average rate of 2.42 billion dollars per year.

Section 1.3

1. $\frac{2}{3}x - y = \frac{19}{3}$ **3.** $3x + y = 12$ **5.** $4x - y = 7$

7. $1.375x - y = -3.725$ **9.** $x = 0$ **11.** $x = 9$

13. $\frac{2}{3}x + y = 2$ **15.** $f(x) = 3x - 20$ **17.** $f(x) = -\frac{5}{8}x + \frac{29}{8}$

19. $f(x) = 5x - 13$ **21.** $f(x) = \frac{2}{11}x + \frac{31}{11}$

23. $f(x) = 0.1825x + 0.0875$ **25.** $f(x) = -\frac{1}{4}x + \frac{7}{4}$

27. $f(x) = \frac{1}{8}x + \frac{3}{2}$ **29.** $f(x) = -7.25x + 32.55$

31. $y = \frac{1}{6}x + \frac{8}{3}$ **33.** $y = -\frac{2}{7}x + \frac{22}{7}$

35. (a) Let x = age of photocopier, in years. Let y = value of photocopier, in dollars.

$y = 25{,}000 - 1250x$

(b) The photocopier will be worth $10,000 in 12 years.

37. (a) Let x = age of machine press, in years. Let y = value of machine press, in dollars.

$y = 90{,}000 - 6000x$

(b) The machine press will be worth half of its original value in 7.5 years.

39. (a) The amount of money spent on pollution abatement is generally increasing at an average rate of $2 billion per year.

(b) $f(12) = 72.17$

In 1984, $72.17 billion were spent on pollution abatement.

(c) In 1992.

41. (a) $f(50) = 163.15$

In 1950, the population of the United States was about 163.15 million people.

(b) $x \approx 43.15$

In 1943, the population of the United States was about 150 million people.

43. (a) Spending in college bookstores increases $190,000,000 each year, on average.

(b) In 1985, $2.24 billion was spent in college bookstores in the United States.

(c) In 1990, $3.19 billion was spent in college bookstores in the United States.

(d) $0.95 billion more was spent in 1990 than in 1985.

45. (a) $g(x) = y = 0.12x + 4.24$

(b) In 1994, there were 4.72 million children in the United States with physical disabilities.

47. (a) $y = 1776.71(x - 5) + 25{,}483.95$

(b) During the year 1997, the average price for an import car was $30,000.

49. (a) $m = 4.36$

The average rate of change in the median price from 1989 to 1999 was $4360 per year.

(b) $y = 4.36x + 96.4$

(c) During 1994 the median price of a single family house in the southern United States was $118.2 thousand.

51. (a) $m = 1.52$

The average rate of change in the adult civilian labor force in the United States from 1990 to 2000 was increasing at a rate of 1.52 million people per year.

(b) $f(x) = 1.52x + 125.50$

(c) $f(4) = 131.58$

In 1994, the size of the adult civilian labor force in the United States was 131.58 million people.

53. (a) $m = 2.94$

The number of motor vehicle registrations in the United States from 1990 to 2000 was increasing at an average rate of 2.94 million vehicles per year.

(b) $f(x) = 2.94x + 188.34$

(c) The y-intercept is (0,188.34). In 1990, the number of motor vehicle registrations was 188.34 million.

55. (a) $11x - 100y = -1375$

(b) $x = 5$

There were 14.3 million students enrolled in higher education institutions in the United States in 1995.

Section 1.4

1. Yes **3.** No **5.** No **7.** No **9.** No

11. The solution to the system is the ordered pair $(3, -1)$.

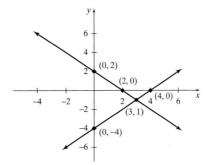

13. The solution to the system is the ordered pair $(0, 4)$.

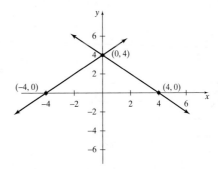

15. The solution to the system is the ordered pair $(0, -3)$.

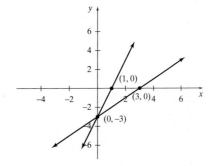

17. Parallel lines; no solution.

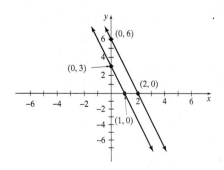

19. Graphs coincide; infinitely many solutions.

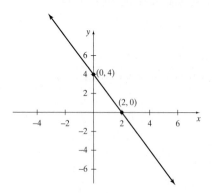

21. The intersection is at $(2, -1)$.

23. The intersection is at $(1, 0)$.

25. $(6, -2)$

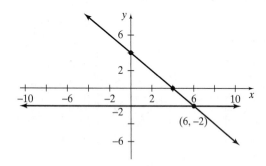

27. $(4, -4)$

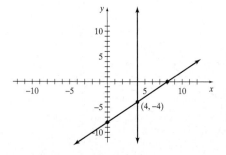

29. $(3, -2)$

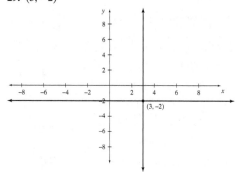

31. $(2, -1)$

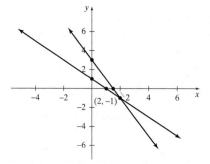

33. The solution to the system is the ordered pair $(-3, -2)$.

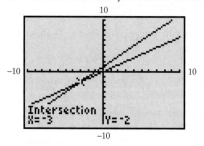

35. The system has no solution.

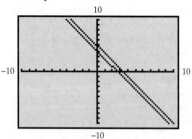

37. The solution to the system of equations is the ordered pair $(0, 3)$.

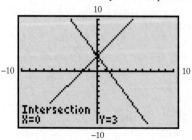

39. $k = 3$. **41.** $k = 6$.

43. Answers vary. They are in the form $k = z - \frac{1}{2}y$, where z is any number, $z \neq 2$.

45. (a) x-intercept: $(133.33, 0)$ **(b)** $s(0) = 0$
y-intercept: $(0, 80)$ $s(130) = 52$

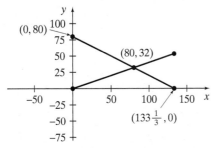

(c) The intersection is at the point $(80, 32)$. The supply will meet the demand when 80 teleconference mini-suites are supplied at a price of $32 per unit.

47. (a) x-intercept: $(750, 0)$ **(b)** $s(0) = 0$
y-intercept: $(0, 150)$ $s(400) = 160$

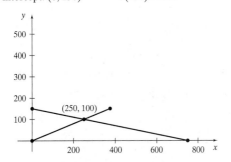

(c) The intersection is at the point $(250, 100)$. The supply will meet the demand when 250 TK-2 model color televisions are supplied at a cost of $100 per unit.

49. (a) $d(x) = y = 75 - \frac{1}{4}x$ $s(x) = y = \frac{1}{2}x$

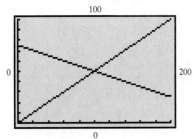

(b) The intersection point is $(100, 50)$.

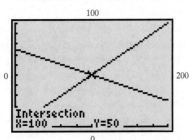

49. (b) *(continued)* The daily supply will meet demand when 100 pup tents are supplied at a cost of $50 per tent.

51. (a) $C(x) = y = 252 + 1.5x$
$R(x) = y = 5.5x$

(b) The lines intersect at about $(60, 350)$. This is the break-even point. The x-coordinate is the quantity of frames produced and the y-coordinate is the amount, in dollars, of the cost and revenue.

53. Let x = the quantity of cribs produced, $C(x)$ = the cost, in dollars, to produce the cribs, and $R(x)$ = the revenue, in dollars, from sales of the cribs.
$$C(x) = 250 + 100x$$
$$R(x) = 125x$$

55. Let x = quantity of bumper stickers sold, $C(x)$ = the cost, in dollars, to produce the bumper stickers, and $R(x)$ = the revenue, in dollars, from sales.
$$C(x) = 200 + 3x$$
$$R(x) = 5x$$

57. Let x = the quantity of radios, $C(x)$ = the cost to produce the radios, and $R(x)$ = the revenue, in dollars, from sales.
$$C(x) = 20{,}000 + 30x$$
$$R(x) = 50x$$

59. Let x = the quantity of pressure washers, $C(x)$ = the cost, in dollars, to produce the pressure washers, and $R(x)$ = the revenue, in dollars, from sales.
$$C(x) = 100{,}000 + 150x$$
$$R(x) = 250x$$

61. (a) $f(0) = 2.51$
$f(8) = 3.47$
$g(0) = 2.84$
$g(8) = 2.68$

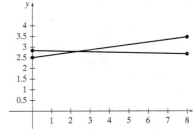

(b) The point of intersection is approximately $(2.5, 3)$. In 1992, there were approximately 3 million children under the age of 5 who were of Hispanic origin and 3 million children under the age of 5 who were of African-American origin in the United States.

63. (a) $f(0) = 2.335$
$f(8) = 2.527$
$g(0) = 2.629$
$g(8) = 2.445$

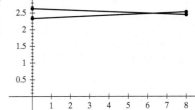

(b) The intersection point is approximately $(6.5, 2.5)$. In 1996, there were approximately 2.5 million people 25 to 29 years of age who were of Hispanic origin and 2.5 million people 25 to 29 years of age who were of African-American origin in the United States.

Section 1.5

1. $(x, y) = (2, 2)$ **3.** $(x, y) = (1, 3)$ **5.** $(x, y) = (5, 1)$

7. $(x, y) = (4, 1)$ **9.** $(x, y) = (-1, -2)$

11. There is no solution to the system.   **13.** $(x, y) = (3, 7)$

15. $(x, y) = (x, 3x - 5)$ **17.** $(x, y) = \left(\frac{8}{3}, \frac{7}{3}\right)$

19. $(x, y) = (0.2, -0.3)$ **21.** $(x, y) = (2, 4)$

23. $(x, y) = (5, 2)$ **25.** $(x, y) = (-4, -2)$

27. $(x, y) = (5, -6)$ **29.** $(x, y) = (1, 3)$

31. There is no solution to the system of equations.

33. $(x, y) = \left(-1, \frac{5}{2}\right)$ **35.** $(x, y) = \left(\frac{12}{5}, \frac{3}{5}\right)$

37. $(x, y) = (2, 3)$ **39.** $(x, y) = \left(\frac{7}{25}, -\frac{1}{25}\right)$

41. $(x, y, z) = \left(-\frac{11}{27}, \frac{79}{27}, \frac{22}{27}\right)$ **43.** $(x, y, z) = (1, -3, 3)$

45. $(x, y, z) = (-3, 0, 4)$

47. There is no solution to the system of equations.

49. $(x, y, z) = (4, -5, 8)$

51. (a) Let x = price of a coat and y = price of the slacks.

$2x + y = 120$
$x + 2y = 120$

(b) $(x, y) = (40, 40)$

The price for the coats was \$40 and the price of the slacks were also \$40.

53. $(x, y) \approx (4.4, 250.29)$

In 1984, both the District of Columbia and South Dakota had about 250 thousand households.

55. $(x, y) \approx (4.78, 555.29)$

In 1994, population of the District of Columbia and the population of Alaska were equal at about 555.29 thousand people.

57. (a) $x + y + z = 100$

(b) $4x + 6y + 7z = 520$
         $6x + 9y + 12z = 810$

(c) $(x, y, z) = (50, 30, 20)$

The company should manufacture 50 boxes of standard clips, 30 boxes of large clips, and 20 boxes of jumbo clips.

59. (a) $2x + 3y + z = 14$
           $x + 2y + z = 9$
           $2x + y + 2z = 9$

(b) $(x, y, z) = (2, 3, 1)$

Two servings of Food A, three servings of Food B, and one serving of Food C are needed to provide the necessary number of grams of fat, carbohydrates and protein.

Chapter 1 Review Exercises

1. $x + 7 = 12$ **3.** $3x = 36$ **5.** $9x + 5 = 41$

7. (a) $7x + 5y = 70$ **(b)** $4x + 4y = 48$

9. (a) $x + y + z = 20{,}000$ **(b)** $y = \frac{1}{4}x + 8000$

(c) $0.06x + 0.07y + 0.05z = 1280$

11.

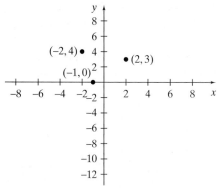

13. Domain: $\{-2, -1, 2\}$
Range: $\{0, 3, 4\}$

15. (a) x is the independent variable.

$g(x)$ is the dependent variable.

(b)

x	$g(x)$
-2	$-\frac{13}{4}$
4	$\frac{35}{4}$
10	$\frac{83}{4}$

17. This table does not represent a function.

19. This graph represents a function.

21. (a)

x	Expression	y
0	$2(0) - 1$	-1
4	$2(4) - 1$	7
$\frac{1}{2}$	$2(\frac{1}{2}) - 1$	0
2	$2(2) - 1$	3

(b) $y = 2x - 1$

23. $m = \frac{1}{4}$ **25.** The x-intercept is the point $(7, 0)$.

Increasing   The y-intercept is the point $(0, -14)$.

27. $C(x) = 17.33x + 210$

29. (a) The fixed cost is \$700, and the per unit cost is \$1.50.

(b) $C(50) = 775$

It costs \$775 to produce 50 baseballs.

(c) 75 baseballs can be produced for \$812.50.

31. $3x - 4y = -19$ **33.** $x = -5$ **35.** $y = 2x - 6$

37. $g(x) = -3x - 5$ **39.** $y = -3x + 2$

41. (a) Let y represent the value of the car in dollars after x years.

$y = -800x + 15{,}525$

(b) In four years the car will have a value of \$12,325.

43. (a) The total pounds of garbage produced per year is increasing at a rate of 50 pounds per year.

(b) 5,200 pounds

(c) $f(5) = 5450$

In the year 2000 the housing community was producing 5450 pounds of garbage per year.

45. (a) $y = 2.07(x - 2) + 20.15$

(b) The average price of a fishing license exceeded \$24.50 during the year 1994.

47. The given ordered pair is a solution to the system.

49. There are infinite solutions because the graphs are the same line.

51. The solution is the point of intersection $(2, -3)$.

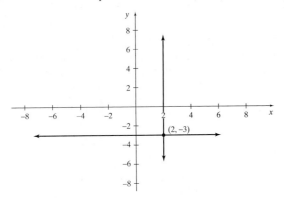

53. The lines are parallel; there is no solution to the system.

55. $k = 12$

57. **(a)** $d(x) = 100 - \frac{1}{5}x$
$s(x) = \frac{3}{5}x$

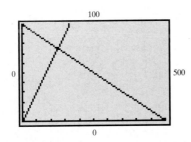

(b) The point of intersection is $(125, 75)$.
The supply will meet the demand when 125 shoes are supplied at a price of $75 per shoe.

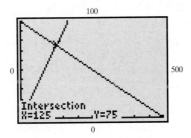

59. **(a)** Let x = the quantity of sweaters produced,
$C(x)$ = the cost to produce x sweaters, and
$R(x)$ = the revenue from producing x sweaters.

$C(x) = 14x + 65$
$R(x) = 30x$

(b) The intersection point is approximately $(4, 122)$. Molly needs to produce and sell approximately 4 sweaters for $122 in order to break even.

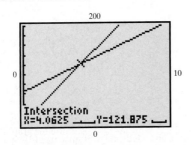

61. The solution to the system is $(-2, 1)$.

63. The solution to the system is $(0, -3)$.

65. The solution to the system is $(10, -2)$.

67. There are infinitely many solutions. The solution in parameter form is $(3y + y, y)$.

69. The solution to the system of equations is $(1, 1, -2)$.

71. **(a)** Let x = the cost of the T-shirts and y = the cost of the jeans.

$3x + 2y = 90$
$x + 4y = 130$

(b) $x, y = (10, 30)$
The T-shirts were $10 each and the jeans were $30 each.

73. The point of intersection is approximately $(7.22, 3.91)$. Approximately 7.22 years after July 1990 (about September 1997) the populations of Colorado and Kentucky were about equal at 3.91 million people in each state.

75. **(a)** $4x + 2y + z = 14$
$3x + y + 2z = 15$
$2x + 2y + 3z = 18$

(b) $x, y, z = (2, 1, 4)$
Two servings of food A, one serving of food B, and four servings of food C are required to satisfy requirements.

Chapter 2

Section 2.1

1. 2×3 **3.** 3×2 **5.** 3×4 **7.** 4×1 **9.** $b_{3,1} = 5$

11. $d_{2,3} = -2$ **13.** $f_{2,3} = 7$ **15.** $h_{1,3} = 2$

17. $\begin{bmatrix} 3 & -7 & | & 4 \\ 2 & 5 & | & 8 \end{bmatrix}$ **19.** $\begin{bmatrix} 1 & 3 & | & -11 \\ 2 & 5 & | & -17 \end{bmatrix}$

21. $\begin{bmatrix} 1 & 1 & 1 & | & 1 \\ 3 & 1 & -1 & | & -2 \\ 2 & -1 & 3 & | & 4 \end{bmatrix}$ **23.** $\begin{bmatrix} 3 & 1 & 10 & | & 38 \\ 5 & 10 & 1 & | & 22 \\ 7 & -1 & -12 & | & -15 \end{bmatrix}$

25. $2x - 3y = 7$ **27.** $x = 8$ **29.** $3x - y + 4z = 7$
$-2x + 4y = 3$ $\quad\quad y = -3$ $\quad\quad 2x + 8y - 3z = 2$
$\quad\quad\quad\quad\quad\quad\quad\quad\quad\quad\quad\quad\quad\quad -x + 4y + 5z = -3$

31. $x = 8$ **33.** $\begin{bmatrix} 1 & -1 & | & -7 \\ 3 & -1 & | & -4 \end{bmatrix}$ **35.** $\begin{bmatrix} 1 & 3 & | & -2 \\ 0 & 10 & | & -1 \end{bmatrix}$
$y = -2$
$z = 7$

37. $\begin{bmatrix} 1 & -1 & -\frac{3}{4} & | & -\frac{1}{2} \\ -5 & 18 & 4 & | & 22 \\ 7 & 31 & -2 & | & 37 \end{bmatrix}$ **39.** $\begin{bmatrix} 1 & -1 & -1 & | & -2 \\ 0 & 3 & 2 & | & 5 \\ 0 & -6 & 3 & | & 9 \end{bmatrix}$

41. The solution is $x = 13$, $y = 11$.

43. The solution is $x = 11$, $y = 5$.

45. The solution is $x = 4$, $y = 3$.

47. The solution is $x = -\frac{41}{27}$, $y = \frac{22}{27}$.

49. The solution is $x = 1$, $y = 0$, $z = -1$.

51. The solution is $x = 6$, $y = -2$, $z = 3$.

53. The solution is $x = \frac{21}{4}$, $y = 0$, $z = -\frac{1}{4}$.

55. The solution is $x = \frac{39}{35}$, $y = -\frac{44}{35}$, $z = -\frac{8}{5}$.

57. The solution is $x = -\frac{1}{2}$, $y = \frac{11}{10}$, $z = -\frac{13}{10}$, $w = -\frac{9}{5}$

59. $x = 1.5738$
$\quad\;\; y = -1.5229$

61. $x = 1.9668$
$\quad\;\; y = -0.0922$

63. $x = 1.2752$
$\quad\;\; y = -1.0035$
$\quad\;\; z = 0.1508$

65. $x = -0.1178$
$\quad\;\; y = 0.2013$
$\quad\;\; z = -1.0281$
$\quad\;\; w = -0.1270$

67. (a) $5x + 4y = 40$
$\qquad\; 200x + 80y = 1200$

(b) Max can make 4 end tables and 5 night stands.

69. The company should make 544 two-person models and 96 four-person models to operate at full capacity.

71. Six boxes should be ordered from Tidy Trader wholesaler and 2 boxes from the Trade-it-Now wholesaler.

73. (a) $0.5x + 0.8y + 1.1z = 450$
$\qquad\; 0.6x + 0.7y + 0.9z = 390$
$\qquad\; 0.2x + 0.3y + 0.6z = 190$

(b) Twenty-two-person tents, 396 four-person tents, and 112 six-person tents can be made in one week.

75. (a) $x + y + z = 50{,}000$
$\qquad\; 0.06x + 0.07y + 0.08z = 3410$
$\qquad\; -0.5x + z = 1000$

(b) The corporation borrowed $20,000 at 6%, $19,000 at 7%, and $11,000 at 8%.

77. $\frac{8}{7}$ quarts of the 21% acid solution and $\frac{6}{7}$ quarts of the 14% acid solution should be used to make 2 quarts of an 18% acid solution.

79. The patient should eat 4 ounces of Basic!, 5 ounces of Vaca, and 7 ounces of Secra.

81. (a) It is not possible for him to use all his resources completely.

(b) With the additional amount of nitrogen, Juan can use all resources completely. He should plant 30 acres of asparagus, 210 acres of cauliflower, and 40 acres of celery.

83. $\left[\begin{array}{ccc|c} 0.8 & 1.1 & 1.7 & 32.1 \\ 0.6 & 1.3 & 1.5 & 29.7 \\ 1.2 & 0.8 & 1.3 & 33.0 \end{array}\right]$

In reduced row echelon form:

$\left[\begin{array}{ccc|c} 1 & 0 & 0 & 15 \\ 0 & 1 & 0 & 9 \\ 0 & 0 & 1 & 6 \end{array}\right]$

Tammy K's should admit 15 students to the medical training, 9 students to the small business training, and 6 students to the executive secretary training.

Section 2.2

1. $\begin{bmatrix} 2 & -1 & 0 \\ 1 & 4 & 5 \end{bmatrix}$ **3.** $A + B$ cannot be added.   **5.** $\begin{bmatrix} 5 & -4 \\ 1 & -7 \end{bmatrix}$

7. $\begin{bmatrix} 4 & 0 & -6 \\ -2 & 6 & -8 \end{bmatrix}$ **9.** $[\,4 \;\; -4 \;\; 16 \;\; -12\,]$

11. $\begin{bmatrix} 2 & -2 & 3 \\ 3 & 5 & 14 \end{bmatrix}$ **13.** $\begin{bmatrix} -13 & -4 \\ -21 & 25 \\ 23 & -14 \end{bmatrix}$ **15.** $\begin{bmatrix} 14 & 0 \\ 18 & -18 \\ -18 & 16 \end{bmatrix}$

17. B and C.   **19.** E.

21. D and G cannot be added because they do not have the same dimension. D is a 2×3 matrix and G is a 3×2 matrix.

23. $D^T = \begin{bmatrix} 4 & 0 \\ -1 & 7 \\ -3 & 1 \end{bmatrix}$; 3×2 **25.** $C^T = \begin{bmatrix} -4 & 3 \\ 1 & 5 \end{bmatrix}$; 2×2

27. $\begin{bmatrix} 0 & 0 & 0 \\ 0 & 0 & 0 \end{bmatrix}$ **29.** $\begin{bmatrix} 0 & 0 \\ 0 & 0 \\ 0 & 0 \end{bmatrix}$ **31.** $x = -1$; $y = 4$; $z = 3$

33. $x = 5$, $y = 3$, $z = 1$ **35.** $x = 5$, $y = 13$, $z = 3$

37. $x = 27$, $y = 4$, $z = 2$, $w = 3$, $p = \frac{1}{2}$

39. $\begin{bmatrix} 9.43 & 13.9 & 2.6 \\ 7.29 & 5.92 & 10.2 \end{bmatrix}$ **41.** $\begin{bmatrix} 26.97 & 37.03 & 6.61 \\ 18.73 & 14.98 & 24.99 \end{bmatrix}$

43. $\begin{bmatrix} 1.716 & 6.071 & 1.547 \\ 4.082 & 3.614 & 7.293 \end{bmatrix}$ **45.** $\begin{bmatrix} 1.7497 & 9.4988 & 2.5806 \\ 6.7363 & 6.0548 & 12.5154 \end{bmatrix}$

47. No answer; verification.

49. (a) $J = \begin{bmatrix} 13 & 7 \\ 5 & 12 \end{bmatrix}$, $Y = \begin{bmatrix} 12 & 9 \\ 7 & 14 \end{bmatrix}$, $A = \begin{bmatrix} 14 & 11 \\ 3 & 19 \end{bmatrix}$

(b) $J + Y + A = \begin{bmatrix} 39 & 27 \\ 15 & 45 \end{bmatrix}$

51. $\begin{bmatrix} 13 & 9 \\ 5 & 15 \end{bmatrix}$

This represents the average that was sold at Krizan's Northside dealership for the months of June, July, and August.

53. (a) $\begin{bmatrix} 26.5 & 37 \\ 12.5 & 26 \end{bmatrix}$

This represents the average of material and labor costs for the East and West Coast plants.

(b) $1.1E = \begin{bmatrix} 27.5 & 38.5 \\ 11 & 27.5 \end{bmatrix}$, $0.95W = \begin{bmatrix} 26.6 & 37.05 \\ 14.25 & 25.65 \end{bmatrix}$

55. (a) $I = 0.08A = \begin{bmatrix} 1600 & 2000 \\ 1200 & 2800 \end{bmatrix}$

(b) $A + I = \begin{bmatrix} 21{,}600 & 27{,}000 \\ 16{,}200 & 37{,}800 \end{bmatrix}$

57. (a) $MCR = \begin{bmatrix} 32.3 \\ 33.2 \\ 33.1 \\ 33.9 \\ 34.7 \\ 35.2 \\ 35.6 \end{bmatrix}$ **(b)** $MCD = \begin{bmatrix} 24.3 \\ 29.4 \\ 31.7 \\ 31.6 \\ 31.9 \\ 31.5 \\ 29.0 \end{bmatrix}$

(c) $TG = MCR + MCD = \begin{bmatrix} 56.6 \\ 62.6 \\ 64.8 \\ 65.5 \\ 66.6 \\ 66.7 \\ 64.6 \end{bmatrix}$

This represents the number of people, in millions, covered by Medicare and Medicaid insurance from 1990 through 1997.

59. (a) $m_{3,2} = 31.4$; 31.4% of males in the United States between the ages of 25 and 34 are smokers. $f_{3,2} = 28.8$; 28.8% of females in the United States between the ages of 25 and 34 are smokers.

(b) $M - F = \begin{bmatrix} -2.4 & 6.2 & 6.1 & 3.5 & 6.1 \\ 4.1 & 3.4 & 9.7 & 4.5 & 3.1 \\ 4.6 & 2.6 & 6.4 & 5.5 & 2.1 \end{bmatrix}$

61. (a) $M = \begin{bmatrix} 57.2 & 51.3 \\ 58.0 & 47.3 \\ 54.0 & 41.6 \\ 53.4 & 42.6 \end{bmatrix}$ **(b)** $F = \begin{bmatrix} 48.5 & 36.8 \\ 45.0 & 34.7 \\ 40.7 & 26.4 \\ 35.7 & 21.9 \end{bmatrix}$

(c) $M - F = \begin{bmatrix} 8.7 & 14.5 \\ 13.0 & 12.6 \\ 13.3 & 15.2 \\ 17.7 & 20.7 \end{bmatrix}$

Section 2.3

1. (a) 3×3 **(b)** 2×2 **3. (a)** 2×2 **(b)** 4×4

5. (a) Does not exist. **(b)** 5×4

7. (a) Does not exist. **(b)** Does not exist.

9. $\begin{bmatrix} -3 \\ -2 \end{bmatrix}$ **11.** $\begin{bmatrix} 2 \\ 10 \end{bmatrix}$ **13.** $\begin{bmatrix} 10 & -10 \\ 7 & -9 \end{bmatrix}$ **15.** $\begin{bmatrix} 28 & 1 \\ -8 & 17 \end{bmatrix}$

17. Matrix product does not exist.

19. $\begin{bmatrix} 5 & -6 \\ -2 & 1 \end{bmatrix}$ **21.** $\begin{bmatrix} 6 & -1 & 4 \\ 0 & 3 & 2 \\ 1 & 4 & 5 \end{bmatrix}$ **23.** $\begin{bmatrix} 440 \\ 212 \\ 620 \\ 694 \end{bmatrix}$

25. $\begin{bmatrix} 19.48 & 12.83 \\ 43.6 & 28.28 \end{bmatrix}$ **27.** $\begin{bmatrix} 1 & 0 \\ 0 & 1 \end{bmatrix}$ **29.** $\begin{bmatrix} 93,300 \\ 43,600 \\ 78,700 \end{bmatrix}$

31. $\begin{bmatrix} 1 & 0 & 0 \\ 0 & 1 & 0 \\ 0 & 0 & 1 \end{bmatrix}$

33. (a) $A^2 = \begin{bmatrix} 0.671 & 0.329 \\ 0.6674 & 0.3326 \end{bmatrix}$

$A^5 = \begin{bmatrix} 0.669811064 & 0.330188936 \\ 0.6698118416 & 0.3301881584 \end{bmatrix}$

$A^{10} = \begin{bmatrix} 0.6698113208 & 0.3301886792 \\ 0.6698113208 & 0.3301886792 \end{bmatrix}$

$A^{20} = \begin{bmatrix} 0.6698113208 & 0.3301886792 \\ 0.6698113208 & 0.3301886792 \end{bmatrix}$

(b) The values of $a_{1,1}$ and $a_{2,1}$ are becoming equal as the matrix is raised to increasingly higher powers. The same thing is happening to the entries $a_{2,1}$ and $a_{2,2}$.

35. $\begin{bmatrix} 3 & -2 \\ 2 & -5 \end{bmatrix} \begin{bmatrix} x \\ y \end{bmatrix} = \begin{bmatrix} 14 \\ 8 \end{bmatrix}$ **37.** $\begin{bmatrix} 1 & 1 \\ -3 & -2 \end{bmatrix} \begin{bmatrix} x \\ y \end{bmatrix} = \begin{bmatrix} 4 \\ -5 \end{bmatrix}$

39. $\begin{bmatrix} 1 & 2 & 2 \\ 1 & -1 & -1 \\ 2 & 5 & 9 \end{bmatrix} \begin{bmatrix} x \\ y \\ z \end{bmatrix} = \begin{bmatrix} 11 \\ -4 \\ 39 \end{bmatrix}$

41. $\begin{bmatrix} -2 & 3 & 4 \\ 3 & -5 & 2 \\ 4 & 2 & -3 \end{bmatrix} \begin{bmatrix} x \\ y \\ z \end{bmatrix} = \begin{bmatrix} 10 \\ 7 \\ -2 \end{bmatrix}$

43. (a) $MP = \begin{bmatrix} 965,300 & 1,436,200 \\ 1,000,100 & 1,487,800 \end{bmatrix}$

(b) $mp_{1,1}$ is the wholesale value of the three most popular selling SUVs at the East dealership. $mp_{1,2}$ is the retail value of the three most popular selling SUVs at the East dealership. $mp_{2,1}$ is the wholesale value of the three most popular selling SUVs at the West dealership. $mp_{2,2}$ is the retail value of the three most popular selling SUVs at the West dealership.

45. (a) $LW = \begin{bmatrix} 17.5 & 18.2 \\ 23.9 & 25.0 \\ 33.8 & 35.5 \end{bmatrix}$ **(b)** $18.20 per hour

(c) $33.80 per hour

(d) It costs $17.50 per hour to manufacture a two-person tent on the East coast, and $18.20 per hour to manufacture a two-person tent on the West coast. For the four-person tent, it costs $23.90 per hour to manufacture it on the East coast and $25.00 per hour to manufacture it on the West coast. For the six-person tent, the manufacturing costs are $33.80 per hour on the East coast and $35.50 per hour on the West coast.

47. (a) $A = \begin{bmatrix} 8 \\ 5 \\ 9 \end{bmatrix}$ **(b)** $B = \begin{bmatrix} 4 \\ 5 \\ 7 \end{bmatrix}$

(c) $FA = \begin{bmatrix} 480 \\ 285 \\ 325 \end{bmatrix}$

At lunchtime, the patient consumed 480 units of calcium, 285 units of vitamin A, and 325 units of vitamin C

(d) $FB = \begin{bmatrix} 340 \\ 205 \\ 225 \end{bmatrix}$

At dinner, the patient consumed 340 units of calcium, 205 units of vitamin A, and 225 units of vitamin C.

49. (a) $FP = \begin{bmatrix} 46,400 \\ 36,770 \\ 15,520 \end{bmatrix}$

(b) He will need 46,400 pounds of nitrogen, 36,770 pounds of phosphates, and 15,520 pounds of potash for the entire growing season.

51. $BA = [\, 21,429.92 \quad 21,545.72 \quad 18,728.24 \,]$

In 1985, there were 21,429.92 thousands of tons of these pollutants released into the atmosphere. In 1990, there were 21,545.72 thousands of tons of these pollutants released into the atmosphere, and in 1995, there were 18,728.24 thousands of tons of these pollutants released into the atmosphere. In all 3 years, pollutants caused by electric companies.

53. (a) $AP = \begin{bmatrix} 41,100 \\ 49,850 \end{bmatrix}$

(b) On Dec. 31, 2000, Denyse's stock holdings were valued at $41,100; Jamie's were valued at $49,850.

55. (a) $\frac{1}{3}(SC) = \begin{bmatrix} 82.67 \\ 87.0 \\ 91.3 \\ 86.67 \\ 81.67 \end{bmatrix}$

This matrix represents the average grade that each student received for the three tests given. Josh had an average score of 82.67, Juan's average score was 87.0, Maria's average score was 91.3, Diane's average was 86.67, and Zach's average score for the three tests was 81.67.

(b) $[87.4 \quad 82 \quad 88]$

This matrix represents the class average for each test given. For the first test, the class was 87.4. On the second test, the class earned an average of 82, and for the third test the average grade for the class was 88.

57. (a) $NF = \begin{bmatrix} 220 & 150 & 140 \\ 570 & 405 & 390 \\ 750 & 525 & 500 \end{bmatrix}$

(b) There are 150 grams per ounce of protein in Mix II.

(c) There are 570 grams per ounce of carbohydrates in Mix I.

(d) In Mix I there are 220 grams per ounce of protein, 570 grams per ounce of carbohydrates, and 750 grams per ounce of fat. In Mix II, there are 150 grams per ounce of protein, 405 grams per ounce of carbohydrates, and 525 grams per ounce of fat. In Mix III, there are 140 grams per ounce of protein, 390 grams per ounce of carbohydrates, and 500 grams per ounce of fat.

59. $BA = [89.778 \quad 18.531 \quad 79.3835]$

Of the eligible voters, 89,778,000 will vote Democratic, 18,531,000 will vote Independent, and 79,383,500 will vote Republican.

Section 2.4

1. They are inverses of each other.

3. They are not inverses of each other.

5. They are inverses of each other.

7. They are not inverses of each other.

9. $\begin{bmatrix} -\frac{1}{5} & \frac{3}{10} \\ \frac{2}{5} & -\frac{1}{10} \end{bmatrix}$ **11.** $\begin{bmatrix} -\frac{1}{30} & -\frac{7}{30} \\ \frac{2}{15} & -\frac{1}{15} \end{bmatrix}$

13. $\begin{bmatrix} -\frac{11}{10} & -\frac{1}{5} & \frac{7}{10} \\ -\frac{9}{10} & \frac{1}{5} & \frac{3}{10} \\ \frac{3}{5} & \frac{1}{5} & -\frac{1}{5} \end{bmatrix}$ **15.** $\begin{bmatrix} -\frac{1}{9} & -\frac{1}{3} & \frac{2}{9} \\ \frac{11}{9} & \frac{2}{3} & -\frac{4}{9} \\ -\frac{2}{3} & 0 & \frac{1}{3} \end{bmatrix}$

17. $\begin{bmatrix} -1 & 1 & 1 & 0 \\ -11 & \frac{54}{5} & \frac{38}{5} & \frac{7}{5} \\ -6 & \frac{28}{5} & \frac{21}{5} & \frac{4}{5} \\ 4 & -\frac{19}{5} & -\frac{13}{5} & -\frac{2}{5} \end{bmatrix}$ **19.** $\begin{bmatrix} \frac{2}{7} & \frac{3}{28} \\ -\frac{1}{7} & \frac{1}{14} \end{bmatrix}$

21. The matrix does not have an inverse; it is singular.

23. $\begin{bmatrix} \frac{2}{11} & \frac{3}{11} \\ -\frac{1}{11} & -\frac{7}{11} \end{bmatrix}$ **25.** $\begin{bmatrix} \frac{7}{10} & \frac{1}{10} \\ \frac{2}{5} & \frac{1}{5} \end{bmatrix}$

27. No answer, verification.

29. $\begin{bmatrix} -\frac{21}{52} & 1 & -\frac{21}{52} \\ \frac{41}{52} & 0 & -\frac{11}{52} \\ -\frac{5}{26} & -1 & \frac{21}{26} \end{bmatrix}$ **31.** $\begin{bmatrix} -\frac{42}{113} & \frac{19}{113} & -\frac{31}{226} & \frac{79}{226} \\ \frac{22}{113} & \frac{7}{226} & \frac{27}{226} & -\frac{18}{113} \\ -\frac{13}{113} & \frac{1}{226} & \frac{10}{113} & \frac{11}{226} \\ \frac{38}{113} & -\frac{29}{226} & -\frac{15}{226} & \frac{10}{113} \end{bmatrix}$

33. $\begin{bmatrix} \frac{5}{22} & -\frac{5}{44} & \frac{3}{11} & -\frac{1}{44} \\ \frac{1}{11} & -\frac{13}{44} & \frac{31}{22} & \frac{37}{44} \\ \frac{3}{22} & -\frac{7}{22} & \frac{19}{22} & \frac{7}{11} \\ -\frac{5}{22} & \frac{4}{11} & -\frac{17}{22} & -\frac{5}{22} \end{bmatrix}$

35. $\begin{bmatrix} \frac{159}{1213} & -\frac{757}{1213} & -\frac{325}{1213} & -\frac{385}{1213} & \frac{428}{1213} \\ \frac{293}{1213} & -\frac{716}{1213} & -\frac{309}{1213} & -\frac{534}{1213} & \frac{392}{1213} \\ \frac{86}{1213} & -\frac{28}{1213} & \frac{137}{1213} & \frac{13}{1213} & -\frac{5}{1213} \\ -\frac{229}{1213} & \frac{2021}{1213} & \frac{552}{1213} & \frac{1531}{1213} & -\frac{1242}{1213} \\ -\frac{34}{1213} & -\frac{807}{1213} & \frac{167}{1213} & \frac{795}{1213} & \frac{679}{1213} \end{bmatrix}$

37. The solution is $x = \frac{13}{10}$, $y = \frac{19}{10}$.

39. The solution is $x = \frac{16}{15}$, $y = \frac{11}{15}$.

41. The solution is $x = -\frac{17}{80}$, $y = -\frac{133}{80}$, $z = \frac{31}{40}$.

43. The solution is $x = \frac{11}{9}$, $y = \frac{5}{9}$, $z = -\frac{2}{3}$.

45. The solution is $x = -3$, $y = -\frac{153}{5}$, $z = -\frac{81}{5}$, $w = \frac{58}{5}$.

47. The solution is $x = -\frac{45}{13}$, $y = \frac{112}{13}$, $z = -\frac{14}{13}$.

49. The solution is $x = \frac{53}{22}$, $y = \frac{70}{11}$, $z = \frac{111}{22}$, $w = -\frac{97}{22}$.

51. The solution is $x = \frac{3890}{1213}$, $y = \frac{4277}{1213}$, $z = \frac{357}{1213}$, $v = -\frac{12,728}{1213}$, $w = \frac{6347}{1213}$.

53. (a) $\begin{bmatrix} 1 & 2 & 2 \\ 3 & 2 & 3 \\ 2 & 2 & 1 \end{bmatrix} \cdot \begin{bmatrix} x \\ y \\ z \end{bmatrix} = \begin{bmatrix} 24 \\ 45 \\ 27 \end{bmatrix}$

(b) $\begin{bmatrix} x \\ y \\ z \end{bmatrix} = \begin{bmatrix} 8 \\ 3 \\ 5 \end{bmatrix}$

Daily orders are 8 loaves of Lemon Poppyseed glaze, 3 loaves of Pumpkin glaze, and 5 loaves of Banana Walnut glaze.

55. (a) Thursday: $\begin{aligned} x + y &= 1600 \\ 3x + 5y &= 5760 \end{aligned}$

Friday: $\begin{aligned} x + y &= 1600 \\ 3x + 5y &= 5980 \end{aligned}$

Saturday: $\begin{aligned} x + y &= 1600 \\ 3x + 5y &= 5500 \end{aligned}$

(b) Thursday: $\begin{bmatrix} 1 & 1 \\ 3 & 5 \end{bmatrix}\begin{bmatrix} x \\ y \end{bmatrix} = \begin{bmatrix} 1,600 \\ 5,760 \end{bmatrix}$

Friday: $\begin{bmatrix} 1 & 1 \\ 3 & 5 \end{bmatrix}\begin{bmatrix} x \\ y \end{bmatrix} = \begin{bmatrix} 1,600 \\ 5,980 \end{bmatrix}$

Saturday: $\begin{bmatrix} 1 & 1 \\ 3 & 5 \end{bmatrix}\begin{bmatrix} x \\ y \end{bmatrix} = \begin{bmatrix} 1600 \\ 5500 \end{bmatrix}$

(c) On Thursday night, 1120 three dollar tickets were sold and 480 five dollar tickets were sold. On Friday night, 1010 three dollar tickets were sold and 590 five dollar tickets were sold. On Saturday night, 1250 three dollar tickets were sold and 350 five dollar tickets were sold.

57. (a) $\begin{aligned} 18x + 20y + 16z &= 5900 \\ 12x + 9y + 11z &= 3580 \\ 22x + 18y + 17z &= 6350 \end{aligned}$

(b) $\begin{bmatrix} 18 & 20 & 16 \\ 12 & 9 & 11 \\ 22 & 18 & 17 \end{bmatrix} \cdot \begin{bmatrix} x \\ y \\ z \end{bmatrix} = \begin{bmatrix} 5900 \\ 3580 \\ 6350 \end{bmatrix}$

(c) $\begin{bmatrix} x \\ y \\ z \end{bmatrix} = \begin{bmatrix} 130 \\ 90 \\ 110 \end{bmatrix}$

To fill the order, Plant I should schedule 130 hours, Plant II should schedule 90 hours, and Plant III should schedule 110 hours.

59. $\begin{bmatrix} x \\ y \\ z \end{bmatrix} = \begin{bmatrix} 20 \\ 396 \\ 112 \end{bmatrix}$

The company can produce 20 two-person tents, 396 four-person tents, and 112 six-person tents operating at full capacity.

61. (a) $\begin{bmatrix} 1 & 1 & 1 \\ 0.06 & 0.075 & 0.085 \\ -\frac{1}{3} & 0 & 1 \end{bmatrix} \cdot \begin{bmatrix} x \\ y \\ z \end{bmatrix} = \begin{bmatrix} 120,000 \\ 8,350 \\ 5,000 \end{bmatrix}$

(b) $\begin{bmatrix} x \\ y \\ z \end{bmatrix} = \begin{bmatrix} 60,000 \\ 35,000 \\ 25,000 \end{bmatrix}$

The corporation borrowed \$60,000 at 6.0%, \$35,000 at 7.5%, and 25,000 at 8.5%.

63. $\begin{bmatrix} x \\ y \\ z \end{bmatrix} = \begin{bmatrix} 5 \\ 4 \\ 7 \end{bmatrix}$

To exactly meet the dietary requirements, Mona should use 5 ounces of Lacto, 4 ounces of Fracto, and 7 ounces of Hima.

65. To use all of his resources, Zach should plant 120 acres of head lettuce, 100 acres of cantaloupe, and 80 acres of bell peppers.

Section 2.5

1. 1 unit of agriculture requires 0.27 units of agriculture, 0.05 units of manufacturing, 0.06 units of trade, and 0.12 units of service.

3. 1 unit of trade requires 0 units of agriculture, 0 units of manufacturing, 0.36 units of trade, and 0.2 units of service.

5. $TX = \begin{bmatrix} 18,250 \\ 10,500 \end{bmatrix}$

The internal demands for utilities and transportation are \$18,250 and \$10,500, respectively.

7. $TX = \begin{bmatrix} 4,800 \\ 15,850 \\ 20,650 \end{bmatrix}$

The internal demands for agriculture, construction, and utilities are \$4,800, \$15,850 and \$20,650, respectively.

9. To produce \$1.00 worth of construction, 0.2 units of construction and 0.4 units of services are required.

11. $I - T = \begin{bmatrix} 0.8 & -0.1 \\ -0.4 & 0.4 \end{bmatrix}$ **13.** $X = \begin{bmatrix} \frac{145}{14} \\ \frac{160}{7} \end{bmatrix}$

$(I-T)^{-1} = \begin{bmatrix} \frac{10}{7} & \frac{5}{14} \\ \frac{10}{7} & \frac{20}{7} \end{bmatrix}$

15. To produce \$1.00 worth of M, 0.2 units of A, 0.5 units of M, and 0 units of T are required.

17. $(I-T) = \begin{bmatrix} 0.8 & -0.2 & -0.1 \\ -0.4 & 0.5 & -0.2 \\ -0.1 & 0 & 0.8 \end{bmatrix}$ **19.** $\begin{bmatrix} \frac{6150}{247} \\ \frac{12,020}{247} \\ \frac{5400}{247} \end{bmatrix}$

21. The total output for agriculture and energy is \$1.1 billion and \$4.75 billion respectively.

23. The total output is \$1955.88 million in agriculture and \$293,227.26 million in manufacturing.

25. (a) $\begin{array}{c} \\ A \\ T = M \\ T \end{array} \begin{array}{ccc} A & M & T \\ \begin{bmatrix} 0.27 & 0.39 & 0 \\ 0.05 & 0.15 & 0 \\ 0.06 & 0.07 & 0.36 \end{bmatrix} \end{array}$ **(b)** $D = \begin{bmatrix} 8 \\ 23 \\ 10 \end{bmatrix}$

(c) \$26.23 billion, \$28.47 billion, and \$21.09 billion for agriculture, manufacturing, and trade, respectively.

27. \$73,894.20 million, \$755,004.27 million, and \$117,761.22 million for agriculture, manufacturing, and trade, respectively.

Chapter 2 Review Exercises

1. 3×2 **3.** -1 **5.** $\begin{bmatrix} -2 & 3 & -1 & | & 5 \\ 1 & 1 & 1 & | & 13 \\ 4 & -5 & 2 & | & -1 \end{bmatrix}$

7. $x = -1, \ y = 2$ **9.** $(-5z+4, \ 2z-1, \ z)$

11. $x = -3.8248, \ y = -0.4023$

13. (a) $\begin{array}{l} 6x + 8y = 40 \\ 60x + 80y = 400 \end{array}$ **(b)** Four 4-wt rods, two 5-wt rods

15. (a) $\begin{array}{l} 0.3x + 0.4y + 0.6z = 310 \\ 0.4x + 0.5y + 0.8z = 405 \\ 0.1x + 0.2y + 0.3z = 150 \end{array}$

(b) 100 small, 250 medium, 300 large

17. 10 oz Alive, 8 oz Gday, 5 oz Cardio

19. $\begin{bmatrix} 3 & 4 & 0 \\ -1 & -8 & 0 \\ 5 & 0 & -1 \end{bmatrix}$ **21.** $\begin{bmatrix} 3 & 9 & -15 \\ 0 & -24 & -12 \\ 6 & -3 & 9 \end{bmatrix}$

23. $\begin{bmatrix} -1 & 7 & -25 \\ 2 & -24 & -20 \\ 0 & -5 & -13 \end{bmatrix}$ **25.** $\begin{bmatrix} 1 & -1 & 1 \\ -2 & 8 & 2 \\ 10 & 8 & 5 \end{bmatrix}$

27. $x = 5, \ y = 3, \ z = -7, \ w = 7, \ p = 4$

29. $\begin{bmatrix} -0.76 & -3.19 \\ -36.95 & -3.67 \\ 3.23 & -12.14 \end{bmatrix}$

31. (a) $I = \begin{bmatrix} 2250 & 1800 \\ 1350 & 2700 \end{bmatrix}$ **(b)** $\begin{bmatrix} 27,250 & 21,800 \\ 16,350 & 32,700 \end{bmatrix}$

(c) CD: 6%; TN: 8%

33. $A_{2,2} = 30,000$ and represents the amount of money in treasury notes at Hilltop Bank; $I_{1,2} = 1800$ and represents the amount of interest earned after 1 year on \$20,000 worth of treasury notes at First Century.

35. (a) 5×5 **(b)** 2×2 **37.** $\begin{bmatrix} 9 \\ 2 \end{bmatrix}$

39. $\begin{bmatrix} -13 & -8 & 7 \\ 32 & 40 & 4 \\ 19 & 10 & -12 \end{bmatrix}$ **41.** $\begin{bmatrix} 49 \\ -21 \\ 3 \\ -27 \end{bmatrix}$

43. (a) $\begin{bmatrix} 0.3895 & 0.6105 \\ 0.3774 & 0.626 \end{bmatrix}, \begin{bmatrix} 0.3820324245 & 0.6179675755 \\ 0.3820163194 & 0.61798336806 \end{bmatrix},$

$\begin{bmatrix} 0.3820224721 & 0.6179775279 \\ 0.3820224718 & 0.6179775282 \end{bmatrix},$

$\begin{bmatrix} 0.3820224719 & 0.6179775281 \\ 0.3820224719 & 0.6179775281 \end{bmatrix}$

(b) As n increases, A^n approaches $\begin{bmatrix} 0.3820224719 & 0.6179975281 \\ 0.3820224719 & 0.6179975281 \end{bmatrix}$.

45. $\begin{bmatrix} 1 & 3 & -2 \\ -2 & 4 & -5 \\ -1 & -1 & 1 \end{bmatrix}\begin{bmatrix} x \\ y \\ z \end{bmatrix} = \begin{bmatrix} 4 \\ -1 \\ 7 \end{bmatrix}$

47. **(a)** $A = \begin{bmatrix} 5 \\ 8 \\ 12 \end{bmatrix}$ **(b)** $B = \begin{bmatrix} 6 \\ 10 \\ 16 \end{bmatrix}$

(c) $\begin{bmatrix} 251 \\ 285 \\ 129 \end{bmatrix}$; $fa_{1,1}$ represents the total units of iron consumed at breakfast from all three food types. $fa_{2,1}$ represents the total units of vitamin C consumed at breakfast from all three food types. $fa_{3,1}$ represents the total units of vitamin B consumed at breakfast from all three food types.

(d) $\begin{bmatrix} 318 \\ 360 \\ 164 \end{bmatrix}$; $fb_{1,1}$ represents the total units of iron consumed at lunch from all three food types. $fb_{2,1}$ represents the total units of vitamin C consumed at lunch from all three food types. $fb_{3,1}$ represents the total units of vitamin B consumed at lunch from all three food types.

49. **(a)** $\begin{bmatrix} 500 \\ 400 \\ 600 \end{bmatrix}$ **(b)** $\begin{bmatrix} 860 \\ 640 \end{bmatrix}$ **(c)** 860 people **(d)** 640 people

51. Not inverses

53. $\begin{bmatrix} -\frac{2}{5} & \frac{3}{10} & \frac{3}{20} \\ \frac{2}{5} & \frac{1}{5} & -\frac{2}{5} \\ \frac{1}{5} & \frac{1}{10} & \frac{1}{20} \end{bmatrix}$ **55.** $\begin{bmatrix} \frac{3}{13} & -\frac{2}{13} \\ \frac{1}{13} & \frac{3}{26} \end{bmatrix}$

57. $\begin{bmatrix} \frac{93}{239} & -\frac{40}{239} & -\frac{13}{239} \\ \frac{40}{239} & \frac{119}{239} & -\frac{59}{239} \\ -\frac{123}{239} & -\frac{109}{239} & \frac{102}{239} \end{bmatrix}$ **59.** $x = -1,\ y = 4$

61. $x = \frac{80}{239},\ y = -\frac{479}{239},\ z = -\frac{260}{239}$

63. **(a)** $X = \begin{bmatrix} x \\ y \\ z \end{bmatrix}$; $\begin{bmatrix} 1 & 3 & 3 \\ 2 & 3 & 4 \\ 1 & 2 & 3 \end{bmatrix}\begin{bmatrix} x \\ y \\ z \end{bmatrix} = \begin{bmatrix} 135 \\ 180 \\ 125 \end{bmatrix}$

(b) $\begin{bmatrix} x \\ y \\ z \end{bmatrix} = \begin{bmatrix} 15 \\ 10 \\ 30 \end{bmatrix}$ **(c)** 30 large pizzas

65. **(a)** $10x + 12y + 8z = 2800$
$15x + 7y + 12z = 2880$
$6x + 12y + 15z = 2970$

(b) $\begin{bmatrix} 10 & 12 & 8 \\ 15 & 7 & 12 \\ 6 & 12 & 15 \end{bmatrix}\begin{bmatrix} x \\ y \\ z \end{bmatrix} = \begin{bmatrix} 2800 \\ 2880 \\ 2970 \end{bmatrix}$

(c) $\begin{bmatrix} x \\ y \\ z \end{bmatrix} = \begin{bmatrix} 80 \\ 120 \\ 70 \end{bmatrix}$; plant I should operate 80 hr, plant II 120 hr, and plant III 70 hr.

67. **(a)** $\begin{bmatrix} 0.511 & 0.48 & 0.474 \\ 0.382 & 0.423 & 0.41 \\ 0.07 & 0.091 & 0.107 \end{bmatrix}\begin{bmatrix} x \\ y \\ z \end{bmatrix} = \begin{bmatrix} 9815 \\ 7930 \\ 1669 \end{bmatrix}$

(b) $\begin{bmatrix} x \\ y \\ z \end{bmatrix} \approx \begin{bmatrix} 10{,}020 \\ 5312 \\ 4525 \end{bmatrix}$, approximately 10,020,000, 5,312,000, and 4,525,000 people voted in California, Florida, and Ohio, respectively.

69. 0.18 unit of agriculture, 0.15 unit of energy, 0 units of chemicals

71. Util.: \$13,500; Trans.: \$45,000

73. 0.1 unit of C, 0.2 unit of M, 0.1 unit of T

75. $\begin{bmatrix} 0.9 & -0.3 & -0.1 \\ -0.2 & 0.9 & -0.3 \\ -0.4 & -0.1 & 0.5 \end{bmatrix}$

77. C: 19.6 million; M: 22.2 million; T: 40.1 million

79. A: 12.3 million; E: 21.7 million

81. **(a)** $T = \begin{matrix} & A & C & E & \\ & \begin{bmatrix} 0.15 & 0.02 & 0.04 \\ 0.04 & 0.12 & 0.18 \\ 0.05 & 0.25 & 0.10 \end{bmatrix} & & & \begin{matrix} A \\ C \\ E \end{matrix} \end{matrix}$ **(b)** $D = \begin{bmatrix} 12 \\ 8 \\ 15 \end{bmatrix}$

(c) A: 15.5 billion; C: 14.2 billion; E: 21.5 billion

Chapter 3

Section 3.1

1.

x	$x > 6$	Statement
−12	$-12 > 6$	False
2	$2 > 6$	False
6	$6 > 6$	False
10	$10 > 6$	True

3.

x	$x \leq -4$	Statement
−8	$-8 \leq -4$	True
−6	$-6 \leq -4$	True
−4	$-4 \leq -4$	True
0	$0 \leq -4$	False

5.

x	$x \geq -1$	Statement
−2	$-2 \geq -1$	False
−1	$-1 \geq -1$	True
0	$0 \geq -1$	True
1	$1 \geq -1$	True

7. $x + 5 > 4$ **9.** $x - 8 < -18$ **11.** $9x \geq 54$

13. $3x - 15 \leq -20$ **15.** $8x + 1 \neq 14$

17. $x > 7$

19. $x < 5$

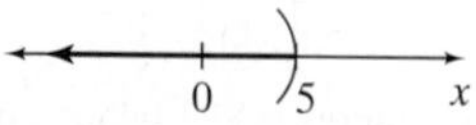

21. $x \le 8$

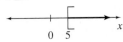

23. $x \ge 5$

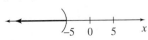

25. $x < -5$

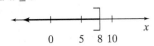

27. $x \ge 3$

29. $2x + 3y \le 32$
$2.5x + 2.5y \le 30$
$x \ge 0, y \ge 0$

31. $4x + 3y \le 240$
$x + 2y \le 140$
$x \ge 0, \ y \ge 0$

33. $x + y \le 200$
$110x \ge 11,000$
$35y \ge 2,205$

35. **(a)** Let $x =$ quantity of two-person tents produced and $y =$ quantity of four-person tents produced.
$x \ge 0, \ y \ge 0$

(b) $6x + 8y \le 580$
$2x + 4y \le 260$

37. **(a)** Let $x =$ quantity of WOW 1000s produced and $y =$ quantity of WOW 1500s produced.
$x \ge 0, \ y \ge 0$

(b) $3x + 6y \le 150$
$3x + 305y \le 100$

39. **(a)** Let $x =$ acres of cauliflower to plant and $y =$ acres of head lettuce to plant.
$x \ge 0, \ y \ge 0$

(b) $x + y \le 210$
$215x + 173y \le 38,380$
$98x + 122y \le 22,240$

41. **(a)** Let $x =$ number of students in the medical field training program, $y =$ number of students in the small business program, and $z =$ number of students in the executive secretary program.
$x \ge 0, \ y \ge 0, \ z \ge 0$

(b) $0.8x + 1.1y + 1.7z \le 32.1$
$0.6x + 1.3y + 1.5z \le 29.7$
$1.2x + 0.8y + 1.3z \le 33$

Section 3.2

1. The correct graph is (b).

3. The correct graph is (d).

5.

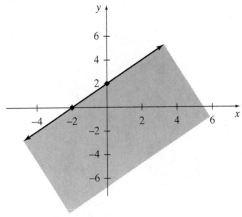

7.

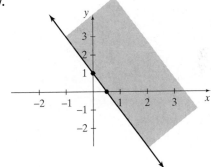

9.

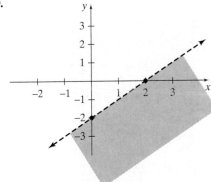

11.

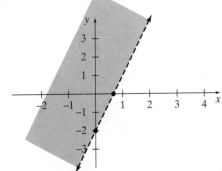

13.

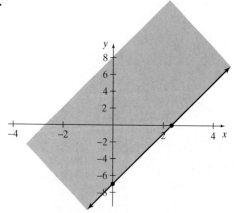

15.

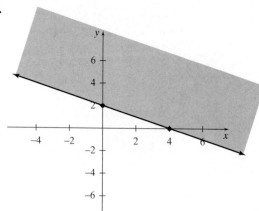

17.

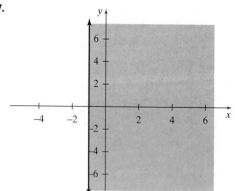

19.

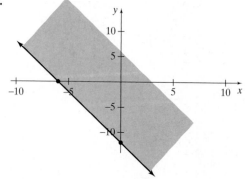

21.

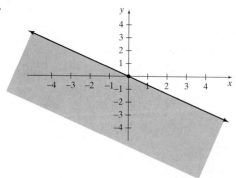

23.

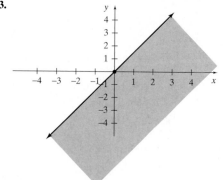

25. The system of inequalities corresponds to region II.

27. The system of inequalities corresponds to region III.

29. The region is unbounded.

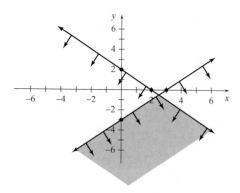

31. The region is unbounded.

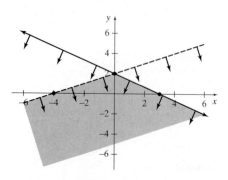

33. The lines are parallel, and there is no feasible solution to the system of inequalities.

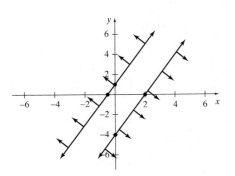

35. The region is bounded.

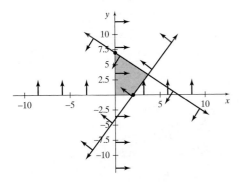

37. The region is bounded.

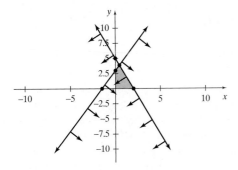

39.

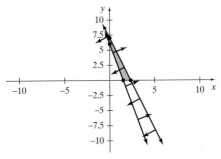

The region is bounded.

41. The region is bounded.

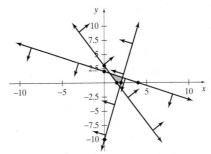

43.

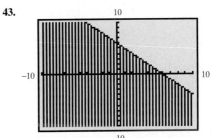

45.

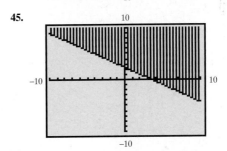

47.

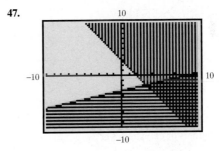

49. (a) $2x + 3y \leq 32$
$2.5x + 2.5y \leq 30$
$x \geq 0, \ y \geq 0$

(b) Every point in the feasible region satisfies the system of inequalities.

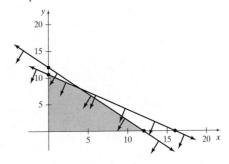

51. (a) Let x = the quantity of switch A produced and y = the quantity of switch B produced.

$$4x + 3y \le 240$$
$$x + 2y \le 140$$
$$x \ge 0, y \ge 0$$

(b) Every point in the feasible region satisfies the system of inequalities.

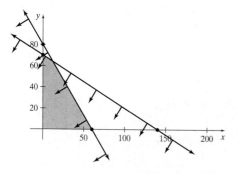

53. (a) Let x = acres of corn to plant and y = acres of wheat to plant.

$$x + y \le 200$$
$$110x \ge 11{,}000$$
$$35y \ge 2205$$

(b) Every point in the feasible region satisfies the system of inequalities.

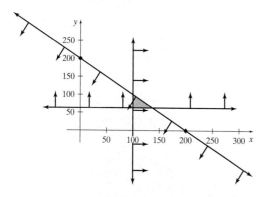

55. (a) Let x = small orders of French fries and y = garden salads.

$$3x + 7y \ge 12$$
$$x \ge 0, y \ge 0$$

(b) Every point in the feasible region satisfies the system of inequalities.

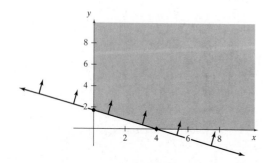

57. (a) $3x + 7y \ge 12$
$10x + 6y \le 25$
$x \ge 0, y \ge 0$

(b) Every point in the feasible region satisfies the system of inequalities.

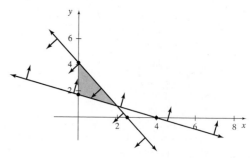

59. (a) No answer given; verification.

(b)

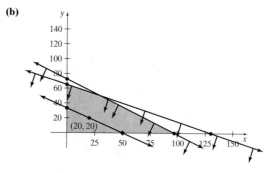

(c) $40x + 60y = 2400$

(d) The slope of the line $40x + 60y = 2000$ is $-\frac{2}{3}$, which is equal to the slope of the line $40x + 60y = 2400$. The lines are parallel.

Section 3.3

1. The minimum value of $z = 2x + 3y$ is 8 at the corner point $(1, 2)$; the maximum value is 39 at the corner point $(6, 9)$.

3. The minimum value of $z = 2x + 5y$ is 0 at the corner point $(0, 0)$; the maximum value is 30 at the corner point $(0, 6)$.

5. The minimum value of $z = x + 4y$ is 5 at the corner point $(5, 0)$; the region is unbounded and no maximum exists.

7. The minimum value of $z = 0.4x + 0.3y$ is 0 at the corner point $(0, 0)$; the maximum value is 4.5 at the corner point $(3, 11)$.

9. The maximum value of $z = 4x + 3y$ is 11 at the corner point $(2, 1)$.

11. The minimum value of $z = x + 2y$ is 0 at the corner point $(0, 0)$.

13. The maximum value of $z = 2x + 2y$ is 14 at the corner points $(7, 0)$ and $(0, 7)$. This is a multiple optimal solution, and all points on the line segment connecting $(7, 0)$ and $(0, 7)$ will produce a maximum value of 14.

15. The minimum value of $z = 3x + 2y$ is 0 at the corner point $(0, 0)$, and the maximum value is 11 at the corner point $(1, 4)$.

17. The minimum value of $z = x + 4y$ is 0 at the corner point $(0, 0)$, and the maximum value is 48 at the corner point $(0, 12)$.

19. The minimum value of $z = 2x + 5y$ is 0 at the corner point $(0, 0)$, and the maximum value is 20 at the corner point $(0, 4)$.

21. The minimum value of $z = 3x + 5y$ is 7 at the corner point $\left(\frac{3}{2}, \frac{1}{2}\right)$, and the maximum value is 47 at the corner point $(4, 7)$.

Section 3.4

1. (a) Let P = profit, x = quantity of two-person tents produced and y = quantity of four-person tents produced.

Maximize: $P = 40x + 60y$
Subject to: $6x + 8y \leq 580$
$2x + 4y \leq 260$
$x \geq 0, y \geq 0$

(b) To maximize weekly profit and stay within department hour restriction, the Yuen Company should produce and sell 30 two-person tents and 50 six-person tents. This generates a maximum weekly profit of $4200.

3. (a) Let P = profit, x = quantity Zap 1000s produced, and y = quantity of Zap 1200s produced.

Maximize: $P = 300x + 450y$
Subject to: $3x + 6y \leq 150$
$3x + 3.5y \leq 100$
$x \geq 0, y \geq 0$

(b) To maximize weekly profit and stay within department hour restrictions, Denyse and Mike should produce and sell 10 Zap 1000s and 20 Zap 1200s. This will generate a maximum weekly profit of $12,000.

5. (a) Let C = cost, x = quantity of slick tires produced, and y = quantity of rain tires produced.

Minimize: $C = 50x + 60y$
Subject to: $4x + 12y \geq 18,000$
$x + y \geq 2000$
$y \geq 800$

(b) To minimize daily costs and stay within labor union restrictions, HotRubber, Inc., should produce and sell 750 slick tires and 1250 rain tires per day. The minimum daily costs will be $112,500.

7. To maximize return on investment, Kevan should invest approximately $16,666.67 in the growth mutual fund and $33,333.33 in the income fund. Kevan will realize a $5000 return on his investment.

9. (a) To fulfill the order while minimizing daily operating costs, the Findlay refinery should operate for 60 days and the Ft. Wayne refinery should operate for 150 days.

(b) The minimum cost will be $6,630,000.

11. (a) To maximize profit while following EPA guidelines, the Roskos Chemical Co. should produce 500 gallons using the old process and 5500 gallons using the new process.

(b) The maximum daily profit is $1080.

13. (a) To maximize revenue, 39 hardback books and 95 softback books should be shipped.

(b) The maximum revenue is $1521.66.

15. (a) To maximize revenue, one 1200 series laptop and seven 1700 series laptops should be shipped.

(b) The maximum potential revenue is $11,592.

17. (a) To maximize revenue and have enough ingredients, Dominic should produce and sell 8 pancakes and 14 crepes.

(b) The maximum revenue is $101.

19. (a) To minimize daily costs and meet minimum nutrition requirements, Anita should use 50 ounces of HiGlo and 300 ounces of Sheen.

(b) The minimum daily cost is $11.50.

21. (a) To minimize daily food costs and meet minimum nutritional requirements, Mike should eat 3 cheeseburgers and 10 small orders of onion rings daily.

(b) Mike's minimum daily food costs are $13.47.

23. (a) To minimize daily fat consumption while meeting minimum daily nutritional requirements, Austin should eat 6 Filet-o-Fish™ sandwiches and 4 large orders of French fries daily.

(b) Minimum daily total fat intake is 260 grams of fat.

25. (a) To maximize profit, Frances should plant 40 acres of artichokes and 100 acres of asparagus.

(b) Maximum profit is $21,600.

Chapter 3 Review Exercises

1.

x	$x > 5$	Statement
-14	$-14 > 5$	False
5	$5 > 5$	False
7	$7 > 5$	True
10	$10 > 5$	True

3. $x + 8 = 12$ **5.** $5x - 7 \leq 9$

7. $x > 8$

9. $x \geq 4$

11. $x \geq 0, y \geq 0$
$2x + 3y \leq 37$
$3x + 4y \leq 51$

13. (a) Let x = number of day packs that can be made each week and y = number of multiday packs that can be made each week; $x \geq 0, y \geq 0$.

(b) $4x + 8y \leq 340$
$2x + 6y \leq 230$

15. (a) Let x = number of students admitted each month for the drum workshop, y = number of students admitted each month for the guitar workshop, and z = number of students admitted each month for the piano workshop; $x \geq 0, y \geq 0, z \geq 0$.

(b) $2x + 3y + 5z \leq 57$
$4x + 5y + 2z \leq 54$
$x + 3y + z \leq 24$

17. (a)

19.

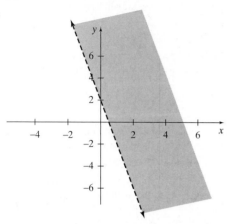

21.

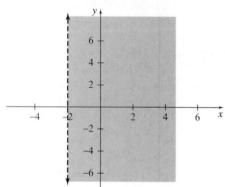

23. Region I

25. Unbounded

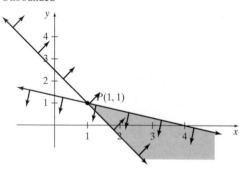

27. Unbounded

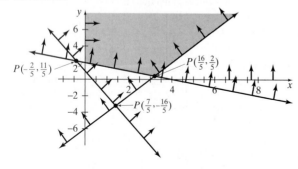

29.

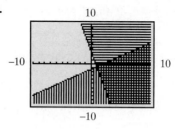

31. (a) $x \geq 0, \ y \geq 0$
$2x + 3y \leq 37$
$3x + 4y \leq 51$
$x + y \leq 14$

(b) Every point in the feasible region is a feasible solution to the problem.

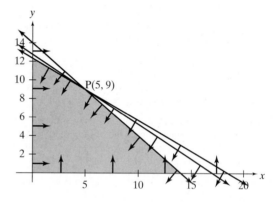

33. (a) Let x = number of 6-inch turkey subs, and y = number of 6-inch veggie delite subs.

$x \geq 0, \ y \geq 0$
$16x + 7y \geq 25$

(b) Every point in the feasible region is a feasible solution to the problem.

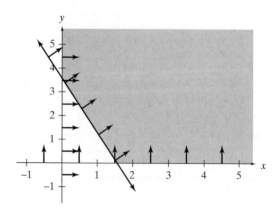

35. (a) $x \geq 0, y \geq 0$
$16x + 7y \geq 25$
$3.5x + 2.5y \leq 8$

(b) Every point in the feasible region is a feasible solution to the problem.

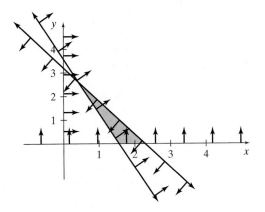

37. Maximum: 23; minimum: 0

39. Maximum: none; minimum: 10

41. Maximum: 6 at $(2, 0)$ **43.** Minimum: $\frac{86}{5}$ at $\left(\frac{8}{5}, \frac{14}{5}\right)$

45. Minimum: 6 at $\left(\frac{3}{2}, 0\right)$; maximum: 32 at $\left(\frac{14}{3}, \frac{8}{3}\right)$

47. (a) 17 drum students and 0 guitar students   **(b)** \$425

49. (a) Aztec: 10 days; Buena Vista: 282 days   **(b)** \$8,780,000

Chapter 4

Section 4.1

1. $x_1 + x_2 + s_1 = 8$
$2x_1 - x_2 + s_2 = 5$

3. $2x_1 + x_2 - x_3 + s_1 = 5$
$-3x_1 + 2x_2 + x_3 + s_2 = 7$
$4x_1 - x_2 + 2x_3 + s_3 = 9$

5. $2x_1 - x_2 + s_1 = 4$
$-3x_1 + 2x_2 + s_2 = 7$
$x_1 - 4x_2 + s_3 = 8$

7. $2x_1 + 3x_2 - x_3 + s_1 = 8$
$-4x_1 + x_2 + x_3 + s_2 = 5$

9.

x_1	x_2	s_1	s_2	z	
1	1	1	0	0	6
1	-1	0	1	0	2
-2	-1	0	0	1	0

11.

x_1	x_2	s_1	s_2	z	
1	1	1	0	0	7
2	3	0	1	0	16
-1	-2	0	0	1	0

13.

x_1	x_2	s_1	s_2	z	
1	1	1	0	0	3
1	-2	0	1	0	0
-2	-2	0	0	1	0

15.

x_1	x_2	s_1	s_2	s_3	z	
3	-2	1	0	0	0	0
1	1	0	1	0	0	5
2	1	0	0	1	0	6
-2	-2	0	0	0	1	0

17.

x_1	x_2	s_1	s_2	s_3	s_4	z	
-1	2	1	0	0	0	0	10
2	1	0	1	0	0	0	15
5	1	0	0	1	0	0	8
-1	1	0	0	0	1	0	1
-5	-1	0	0	0	0	1	0

19.

x_1	x_2	x_3	s_1	s_2	z	
1	-1	2	1	0	0	4
2	1	-3	0	1	0	7
-2	-1	-3	0	0	1	0

21.

x_1	x_2	x_3	s_1	s_2	s_3	z	
2	1	4	1	0	0	0	8
4	-1	3	0	1	0	0	15
12	2	7	0	0	1	0	12
-3	-1	-5	0	0	0	1	0

23.

x_1	x_2	x_3	s_1	s_2	z	
2	0	3	1	0	0	8
3	2	0	0	1	0	5
-4	-2	-3	0	0	1	0

25.

x_1	x_2	x_3	s_1	s_2	z	
15	3	0	1	0	0	9
0	6	1	0	1	0	7
-5	-1	-3	0	0	1	0

27. (a) Basic variables: x_1, s_1, z
Nonbasic variables: x_2, s_2

(b) $x_1 = 5$, $x_2 = 0$, $s_1 = 2$,
$s_2 = 0$, $z = 10$

29. (a) Basic variables: s_1, s_2, z
Nonbasic variables: x_1, x_2

(b) $x_1 = 0$, $x_2 = 0$, $s_1 = 3$, $s_2 = 7$, $z = 9$

31. (a) Basic variables: s_1, s_2, s_3, z
Nonbasic variables: x_1, x_2

(b) $x_1 = 0$, $x_2 = 0$, $s_1 = 3$,
$s_2 = 8$, $s_3 = 15$, $z = 0$

33. (a) Basic variables: x_1, x_2, z
Nonbasic variables: x_3, s_1, s_2

(b) $x_1 = 7$, $x_2 = 12$, $x_3 = 0$,
$s_1 = 0$, $s_2 = 0$, $z = 27$

35. (a) Basic variables: x_2, x_3, s_2, z
Nonbasic variables: x_1, x_4, s_1, s_3

(b) $x_1 = 0$, $x_2 = 15$, $x_3 = 20$, $x_4 = 0$,
$s_1 = 0$, $s_2 = 10$, $s_3 = 0$, $z = 100$

37. (a)

x_1	x_2	s_1	s_2	z	
1	0	$-\frac{1}{2}$	$\frac{5}{2}$	0	4
0	1	$\frac{1}{2}$	$\frac{1}{2}$	0	1
0	0	$\frac{5}{2}$	$\frac{13}{2}$	1	15

(b) $x_1 = 4$, $x_2 = 1$, $s_1 = 0$, $s_2 = 0$, $z = 15$

39. (a)

x_1	x_2	s_1	s_2	z	
1	$-\frac{3}{2}$	$\frac{1}{2}$	0	0	$\frac{3}{2}$
0	12	-2	1	0	1
0	$\frac{7}{2}$	$\frac{1}{2}$	0	1	$\frac{21}{2}$

(b) $x_1 = \frac{3}{2}$, $x_2 = 0$, $s_1 = 0$, $s_2 = 1$, $z = \frac{21}{2}$

41. (a)

$$
\begin{array}{c c c c c c}
x_1 & x_2 & x_3 & s_1 & s_2 & z \\
\end{array}
$$

$$
\left[
\begin{array}{cccccc|c}
-\frac{7}{4} & 0 & -\frac{1}{4} & 1 & 0 & 0 & \frac{3}{2} \\
\frac{3}{8} & 1 & \frac{1}{8} & 0 & 0 & 0 & \frac{3}{4} \\
\frac{41}{8} & 0 & -\frac{5}{8} & 0 & 1 & 0 & \frac{17}{4} \\
\hline
\frac{5}{8} & 0 & \frac{7}{8} & 0 & 0 & 1 & \frac{21}{4}
\end{array}
\right]
$$

(b) $x_1 = 0$, $x_2 = \frac{3}{4}$, $x_3 = 0$, $s_1 = \frac{3}{2}$, $s_2 = \frac{17}{4}$, $z = \frac{21}{4}$

43. (a)

$$
\begin{array}{c c c c c c}
x_1 & x_2 & x_3 & s_1 & s_2 & z \\
\end{array}
$$

$$
\left[
\begin{array}{cccccc|c}
7 & \frac{5}{2} & 0 & 1 & -\frac{1}{2} & 0 & 70 \\
\frac{3}{2} & \frac{3}{4} & 1 & 0 & \frac{1}{4} & 0 & 10 \\
\hline
10 & 3 & 0 & 0 & 2 & 1 & 80
\end{array}
\right]
$$

(b) $x_1 = 0$, $x_2 = 0$, $x_3 = 10$, $s_1 = 70$, $s_2 = 0$, $z = 80$

45. (a)

$$
\begin{array}{c c c c c c c c}
x_1 & x_2 & x_3 & x_4 & s_1 & s_2 & s_3 & z \\
\end{array}
$$

$$
\left[
\begin{array}{cccccccc|c}
1 & 0 & -\frac{5}{2} & 0 & -\frac{11}{2} & -\frac{25}{2} & -\frac{39}{2} & 0 & -\frac{117}{2} \\
0 & 0 & -\frac{3}{2} & 1 & -\frac{11}{2} & -\frac{13}{2} & -\frac{25}{2} & 0 & -\frac{55}{2} \\
0 & 1 & \frac{1}{2} & 0 & \frac{3}{2} & \frac{7}{2} & \frac{9}{2} & 0 & \frac{27}{2} \\
\hline
0 & 0 & 1 & 0 & 4 & 7 & 13 & 1 & 107
\end{array}
\right]
$$

(b) $x_1 = -\frac{117}{2}$, $x_2 = \frac{27}{2}$, $x_3 = 0$, $x_4 = -\frac{55}{2}$,

$s_1 = 0$, $s_2 = 0$, $s_3 = 0$, $z = 107$

47. (b) $x_1 = 0$, $x_2 = 2.2$, $s_1 = 0.6$, $s_2 = 0$, $z = 21.6$

49. (b) $x_1 = 0$, $x_2 = 0$, $x_3 = 13$, $s_1 = 0$, $s_2 = -11$, $z = 130$

51. (a) Let x_1 = gallons of chemical produced using older process and x_2 = gallons of chemical produced using newer process.

Maximize: $z = 0.40x_1 + 0.16x_2$

Subject to: $4x_1 + 2x_2 \le 10{,}000$

$3x_1 + x_2 \le 7000$

$x_1 \ge 0, x_2 \ge 0$

(b) $4x_1 + 2x_2 + s_1 = 10{,}000$

$3x_1 + x_2 + s_2 = 7000$

$-0.4x_1 - 0.16x_2 + z = 0$

(c)

$$
\begin{array}{c c c c c}
x_1 & x_2 & s_1 & s_2 & z \\
\end{array}
$$

$$
\left[
\begin{array}{ccccc|c}
4 & 2 & 1 & 0 & 0 & 10{,}000 \\
3 & 1 & 0 & 1 & 0 & 7000 \\
\hline
-0.4 & -0.16 & 0 & 0 & 1 & 0
\end{array}
\right]
$$

53. (a) Let x_1 = quantity of WOW 1000s produced and x_2 = quantity of WOW 1500s produced.

Maximize: $P = 100x_1 + 150x_2$

Subject to: $3x_1 + 6x_2 \le 150$

$3x_1 + 3.5x_2 \le 100$

$x_1 \ge 0, x_2 \ge 0$

(b) $3x_1 + 6x_2 + s_1 = 150$

$3x_1 + 3.5x_2 + s_2 = 100$

$-100x_1 - 150x_2 + z = 0$

(c)

$$
\begin{array}{c c c c c}
x_1 & x_2 & s_1 & s_2 & z \\
\end{array}
$$

$$
\left[
\begin{array}{ccccc|c}
3 & 6 & 1 & 0 & 0 & 150 \\
3 & 3.5 & 0 & 1 & 0 & 100 \\
\hline
-100 & -150 & 0 & 0 & 1 & 0
\end{array}
\right]
$$

55. (a) Let x_1 = acres of artichokes and x_2 = acres of asparagus.

Maximize: $z = 140x_1 + 160x_2$

Subject to: $197x_1 + 240x_2 \le 31{,}880$

$123x_1 + 42x_2 \le 9120$

$x_1 \ge 0, x_2 \ge 0$

(b) $197x_1 + 240x_2 + s_1 = 31{,}880$

$123x_1 + 42x_2 + s_2 = 9120$

$-140x_1 - 160x_2 + z = 0$

(c)

$$
\begin{array}{c c c c c}
x_1 & x_2 & s_1 & s_2 & z \\
\end{array}
$$

$$
\left[
\begin{array}{ccccc|c}
197 & 240 & 1 & 0 & 0 & 31{,}880 \\
123 & 42 & 0 & 1 & 0 & 9120 \\
\hline
-140 & -160 & 0 & 0 & 1 & 0
\end{array}
\right]
$$

57. (a) Let x_1 = quantity of two-person tents, x_2 = quantity of four-person tents, and x_3 = quantity of six-person tents.

Maximize: $z = 30x_1 + 50x_2 + 80x_3$

Subject to: $0.5x_1 + 0.8x_2 + 1.1x_3 \le 420$

$0.6x_1 + 0.7x_2 + 0.9x_3 \le 310$

$0.2x_1 + 0.3x_2 + 0.6x_3 \le 180$

$x_1 \ge 0, x_2 \ge 0, x_3 \ge 0$

(b) $0.5x_1 + 0.8x_2 + 1.1x_3 + s_1 = 420$

$0.6x_1 + 0.7x_2 + 0.9x_3 + s_2 = 310$

$0.2x_1 + 0.3x_2 + 0.6x_3 + s_3 = 180$

$-30x_1 - 50x_2 - 80x_3 + z = 0$

(c)

$$
\begin{array}{c c c c c c c}
x_1 & x_2 & x_3 & s_1 & s_2 & s_3 & z \\
\end{array}
$$

$$
\left[
\begin{array}{ccccccc|c}
0.5 & 0.8 & 1.1 & 1 & 0 & 0 & 0 & 420 \\
0.6 & 0.7 & 0.9 & 0 & 1 & 0 & 0 & 310 \\
0.2 & 0.3 & 0.6 & 0 & 0 & 1 & 0 & 180 \\
\hline
-30 & -50 & -80 & 0 & 0 & 0 & 1 & 0
\end{array}
\right]
$$

59. (a) Let x_1 = acres of asparagus, x_2 = acres of cauliflower, and x_3 = acres of celery.

Maximize: $z = 160x_1 + 90x_2 + 80x_3$

Subject to: $240x_1 + 215x_2 + 336x_3 \le 65{,}970$

$148x_1 + 98x_2 + 171x_3 \le 31{,}860$

$42x_1 + 69x_2 + 147x_3 \le 21{,}630$

$x_1 \ge 0, x_2 \ge 0, x_3 \ge 0$

(b) $240x_1 + 215x_2 + 336x_3 + s_1 = 65{,}970$

$148x_1 + 98x_2 + 171x_3 + s_2 = 31{,}860$

$42x_1 + 69x_2 + 147x_3 + s_3 = 21{,}630$

$-160x_1 - 90x_2 - 80x_3 + z = 0$

(c)

$$
\begin{array}{c c c c c c c}
x_1 & x_2 & x_3 & s_1 & s_2 & s_3 & z \\
\end{array}
$$

$$
\left[
\begin{array}{ccccccc|c}
240 & 215 & 336 & 1 & 0 & 0 & 0 & 65{,}970 \\
148 & 98 & 171 & 0 & 1 & 0 & 0 & 31{,}860 \\
42 & 69 & 147 & 0 & 0 & 1 & 0 & 21{,}630 \\
\hline
-160 & -90 & -80 & 0 & 0 & 0 & 1 & 0
\end{array}
\right]
$$

Section 4.2

1. $x_1 = 2$, $x_2 = 0$, $s_1 = 3$, $s_2 = 0$, $z = 18$

3. $x_1 = 5$, $x_2 = 0$, $s_1 = 1$, $s_2 = 0$, $z = 29$

5. $x_1 = 0, x_2 = \frac{3}{4}$,

$s_1 = \frac{3}{2}, s_2 = 0, s_3 = \frac{17}{4}$,

$z = 5.25$

7. $x_1 = 0$, $x_2 = 0$, $x_3 = 10$, $s_1 = 70$, $s_2 = 0$, $z = 80$

9. $x_1 = 0$, $x_2 = \frac{9}{5}$, $x_3 = \frac{117}{5}$, $x_4 = \frac{38}{5}$,

$s_1 = 0$, $s_2 = \frac{38}{5}$, $s_3 = 0$, $z = 107$

11. (a)

x_1	x_2	s_1	s_2	z	
1	1	1	0	0	6
1	−1	0	1	0	2
−2	−1	0	0	1	0

(b) $x_1 = 4$, $x_2 = 2$, $s_1 = 0$, $s_2 = 0$, $z = 10$

13. (a)

x_1	x_2	s_1	s_2	z	
3	2	1	0	0	8
1	−1	0	1	0	7
−2	−3	0	0	1	0

(b) $x_1 = 0$, $x_2 = 4$, $s_1 = 0$, $s_2 = 11$, $z = 12$

15. (a)

x_1	x_2	s_1	s_2	s_3	z	
2	1	1	0	0	0	32
1	1	0	1	0	0	18
1	3	0	0	1	0	36
−1	−4	0	0	0	1	0

(b) $x_1 = 0$, $x_2 = 12$, $s_1 = 20$, $s_2 = 6$, $s_3 = 0$, $z = 48$

17. (a)

x_1	x_2	s_1	s_2	s_3	z	
3	−2	1	0	0	0	0
1	1	0	1	0	0	5
2	1	0	0	1	0	6
−2	−2	0	0	0	1	0

(b) $x_1 = 1$, $x_2 = 4$, $s_1 = 5$, $s_2 = 0$, $s_3 = 0$, $z = 10$
(solution not unique; multiple optimal solution)

19. (a)

x_1	x_2	s_1	s_2	s_3	s_4	z	
−1	2	1	0	0	0	0	10
2	1	0	1	0	0	0	15
5	1	0	0	1	0	0	8
−1	1	0	0	0	1	0	1
−5	−1	0	0	0	0	1	0

(b) $x_1 = \frac{8}{5}$, $x_2 = 0$, $s_1 = \frac{58}{5}$,

$s_2 = \frac{59}{5}$, $s_3 = 0$, $s_4 = \frac{13}{5}$,

$z = 8$

21. (a)

x_1	x_2	x_3	s_1	s_2	z	
2	1	1	1	0	0	10
5	−1	2	0	1	0	8
−3	−2	−4	0	0	1	0

(b) $x_1 = 0$, $x_2 = 4$, $x_3 = 6$,
$s_1 = 0$, $s_2 = 0$, $z = 32$

23. (a)

x_1	x_2	x_3	s_1	s_2	s_3	z	
5	2	1	1	0	0	0	20
1	3	5	0	1	0	0	13
4	1	3	0	0	1	0	15
−2	−5	−1	0	0	0	1	0

(b) $x_1 = \frac{34}{13}$, $x_2 = \frac{45}{13}$, $x_3 = 0$,

$s_1 = 0$, $s_2 = 0$, $s_3 = \frac{14}{13}$, $z = \frac{293}{13}$

25. (a)

x_1	x_2	x_3	s_1	s_2	z	
2	0	3	1	0	0	8
3	2	0	0	1	0	5
−4	−2	−3	0	0	1	0

(b) $x_1 = 0$, $x_2 = \frac{5}{2}$, $x_3 = \frac{8}{3}$,

$s_1 = 0$, $s_2 = 0$, $z = 13$

27. To maximize profit, the Young Furniture Company should produce 30 tables and 40 chairs. This yields a maximum profit of $4100.

29. To maximize weekly profit and stay within EPA regulations, the Henderson Chemical Company should produce 2000 gallons using the older process and 1000 gallons using the newer process. This will yield a maximum weekly profit of approximately $960.

31. To maximize weekly profit, 10 units of the WOW 1000 and 20 units of the WOW 1500 should be produced. This will yield a maximum profit of $4000.

33. (a) To maximize profit, Franco should plant 40 acres of artichokes and 100 acres of asparagus.

(b) Maximum profit is $21,600.

35. To maximize profit, the Smolko Tent Company should produce and sell 0 two-person tents, 160 four-person tents, and 220 six-person tents.

37. To maximize profit, Eldrick should plant about 215 acres of asparagus, 0 acres of cauliflower, and 0 acres of celery. This will yield a maximum profit of approximately $34,443. The nonzero values for s_1 and s_2 indicate that there will be approximately 14,305 pounds of nitrogen and 12,589 pounds of potash, respectively, that will not be used.

39. To maximize total grams of protein, Meagan should eat 7.2 servings of Cookie's 'n' Creme candies, and no servings of Reese's Cups or York Peppermint. She will consume 14.4 grams of protein. The nonzero value for s_1 indicates that she will consume 52 fewer calories than the maximum amount (700). The nonzero value for s_2 indicates that she will consume 8 fewer grams of carbohydrates than the maximum amount (80 grams).

41. To maximize profit, Julia should make 8 loaves of Lemon Poppy Seed Glaze bread, 3 loaves of Pumpkin Glaze bread and 5 loaves of Banana Walnut Glaze bread. This will yield a maximum profit of $20.

Section 4.3

1. $A^T = \begin{bmatrix} 2 & 4 \\ 1 & 1 \\ 3 & 2 \end{bmatrix}$
3. $A^T = \begin{bmatrix} 5 & 3 & 0 \\ 7 & 2 & 1 \\ 6 & 1 & 1 \\ 1 & 0 & 2 \end{bmatrix}$
5. $A^T = \begin{bmatrix} 4 & 1 & 8 \\ 1 & 2 & 1 \\ 3 & 7 & 9 \end{bmatrix}$

7. Maximize: $z = 4x_1 + 7x_2$
   Subject to: $5x_1 + 3x_2 \leq 2$
   $2x_1 + x_2 \leq 3$
   $x_1 \geq 0, x_2 \geq 0$

9. Maximize: $z = 7x_2 + 11x_3$
   Subject to: $2x_1 + 8x_2 + 5x_3 \leq 1$
   $x_1 + x_2 + 2x_3 \leq 3$
   $4x_1 + x_2 + 7x_3 \leq 4$
   $x_1 \geq 0, x_2 \geq 0, x_3 \geq 0$

11. Maximize: $z = 8x_1 + 7x_2 + 11x_3$
    Subject to: $5x_1 + 3x_2 \leq 2$
    $x_1 + x_3 \leq 1$
    $8x_2 + x_3 \leq 3$
    $x_1 \geq 0, x_2 \geq 0, x_3 \geq 0$

13. (a) Minimum is $w = 6$ at $y_1 = 6$, $y_2 = 0$.
    (b) Maximum is $z = 6$ at $x_1 = 1$, $x_2 = 0$.

15. (a) Minimum is $w = 24$ at $y_1 = 12$, $y_2 = 0$.
    (b) Maximum is $z = 24$ at $x_1 = 2$, $x_2 = 0$, $x_3 = 0$.

17. (a) Maximize: $z = 15x_1 + 4x_2$
    Subject to: $8x_1 + x_2 \leq 7$
    $10x_1 + x_2 \leq 3$
    $x_1 \geq 0, x_2 \geq 0$
    (b) Minimum is $w = 12$ at $y_1 = 0$, $y_2 = 4$.

19. (a) Maximize: $z = 5x_1 + 6x_2$
    Subject to: $x_1 + x_2 \leq 1$
    $2x_1 + 3x_2 \leq 4$
    $x_1 \geq 0, x_2 \geq 0$
    (b) Minimum is $w = 6$ at $y_1 = 6$, $y_2 = 0$.

21. (a) Maximize: $z = 8x_1 + 4x_2$
    Subject to: $x_1 + x_2 \leq 2$
    $x_1 + 2x_2 \leq 1$
    $x_1 \geq 0, x_2 \geq 0$
    (b) Minimum is $w = 8$ at $y_1 = 0$, $y_2 = 8$.

23. (a) Maximize: $z = 29x_1 + 55x_2 + 18x_3$
    Subject to: $2x_1 + 5x_2 + 2x_3 \leq 4$
    $3x_1 + 4x_2 + x_3 \leq 5$
    $x_1 \geq 0, x_2 \geq 0, x_3 \geq 0$
    (b) Minimum is $w = 53$ at $y_1 = 7$, $y_2 = 5$.

25. (a) Maximize: $z = 29x_1 + 20x_2 + 78x_3$
    Subject to: $15x_1 + x_2 + 8x_3 \leq 2$
    $x_1 + 6x_2 + 7x_3 \leq 3$
    $x_1 \geq 0, x_2 \geq 0, x_3 \geq 0$
    (b) Minimum is $w = 22$ at $y_1 = 8$, $y_2 = 2$.

27. (a) Maximize: $z = 6x_1 + 9x_2$
    Subject to: $x_1 + 2x_2 \leq 14$
    $x_1 + x_2 \leq 8$
    $3x_1 + x_2 \leq 20$
    $x_1 \geq 0, x_2 \geq 0$
    (b) Minimum is $w = 66$ at $y_1 = 3$, $y_2 = 3$, $y_3 = 0$.

29. (a) Maximize: $z = 6x_1 + 9x_2 + 3x_3$
    Subject to: $x_1 + x_2 + 3x_3 \leq 4$
    $3x_1 + x_2 + x_3 \leq 10$
    $x_1 + 3x_2 + x_3 \leq 9$
    $x_1 \geq 0, x_2 \geq 0, x_3 \geq 0$
    (b) Minimum is $w = \frac{63}{2}$ at $y_1 = \frac{9}{2}$, $y_2 = 0$, $y_3 = \frac{3}{2}$.

31. (a) Minimize: $C = 0.08y_1 + 0.09y_2$
    Subject to: $5y_1 + 10y_2 \geq 50$
    $y_1 + 2y_2 \geq 5$
    $y_1 \geq 0, y_2 \geq 0$
    (b) Maximize: $z = 50x_1 + 5x_2$
    Subject to: $5x_1 + x_2 \leq 0.08$
    $10x_1 + 2x_2 \leq 0.09$
    $x_1 \geq 0, x_2 \geq 0$
    (c) The solution to the primal problem is
    $C = 0.45$ at $y_1 = 0$, $y_2 = 5$.

    To minimize cost, Richie should take 0 pills of the AllDay
    supplement and 5 pills of the Live! supplement. Minimum
    daily cost is $0.45.

33. (a) Minimize: $C = 0.30y_1 + 0.40y_2$
    Subject to: $210y_1 + 240y_2 \geq 2115$
    $150y_1 + 380y_2 \geq 3460$
    $y_1 \geq 0, y_2 \geq 0$
    (b) Maximize: $z = 2115x_1 + 3460x_2$
    Subject to: $210x_1 + 150x_2 \leq 0.30$
    $240x_1 + 380x_2 \leq 0.40$
    $x_1 \geq 0, x_2 \geq 0$
    (c) The solution to the primal problem is
    $C = \frac{346}{95} \approx 3.64$ at $y_1 = 0$, $y_2 = \frac{173}{19} \approx 9.11$.

    To minimize cost, Elle should feed her beagles 0 pounds of
    Shine dog food and approximately 9.11 pounds of Lustre
    dog food. Minimum daily cost will be about $3.64.

35. (a) Let $y_1 = $ number of days Mine I should operate,
    $y_2 = $ number of days Mine II should operate.

    Minimize: $C = 12y_1 + 10y_2$
    Subject to: $3y_1 + 2y_2 \geq 13$
    $3y_1 + y_2 \geq 9$
    $4y_1 + 6y_2 \geq 24$
    $y_1 \geq 0, y_2 \geq 0$
    (b) Maximize: $z = 13x_1 + 9x_2 + 24x_3$
    Subject to: $3x_1 + 3x_2 + 4x_3 \leq 12$
    $2x_1 + x_2 + 6x_3 \leq 10$
    $x_1 \geq 0, x_2 \geq 0, x_3 \geq 0$
    (c) The solution to the primal problem is $C = 56$ at
    $y_1 = 3$, $y_2 = 2$.

    To minimize daily cost, AJA Mining Co. should operate
    Mine I for 3 days and Mine II for 2 days. Minimum daily
    cost will be $56,000.

37. (a) To minimize cost, and fulfill the minimum order, the
    refinery in Findlay should operate for 60 days and the
    refinery in Ft. Wayne should operate for 150 days.
    (b) The minimum cost is $6,630,000.

39. (a) To minimize cost, and meet dietary requirements, Anita should use 50 ounces of Flakes and 300 ounces of Woday.

(b) The minimum daily cost is $11.50.

41. (a) To minimize fat intake, and meet dietary requirements, Warren should eat 2 servings of Burrito Supreme^R-Beef and 3 servings of Double Burrito Supreme^R-Steak.

(b) Minimum daily fat intake is 90 grams of fat.

43. (a) The cost of a milligram of vitamin C is $0.009 and the cost of a milligram of niacin is 0.

(b) New total cost is about $0.46.

45. (a) A unit of protein is worth zero and a unit of fat is worth $\frac{1}{950}$ of a dollar.

(b) New total cost is about $3.64.

Section 4.4

1. $3x_1 + 2x_2 + s_1 = 12$
$x_1 + 4x_2 - s_2 = 7$

3. $x_1 + 4x_2 + 3x_3 + s_1 = 9$
$2x_1 + x_2 + 5x_3 - s_2 = 10$
$3x_1 + 5x_2 + 2x_3 + s_3 = 8$

5. $x_1 + 2x_2 + x_3 + s_1 = 60$
$2x_1 + 3x_3 - s_2 = 18$
$3x_2 + 2x_3 - s_3 = 30$

7. (a) Maximize: $\quad z = -w = -8x_1 - 3x_2$
Subject to: $\quad 4x_1 + 2x_2 \le 8$
$\qquad\qquad\quad x_1 + 4x_2 \ge 7$
$\qquad\qquad\quad x_1 \ge 0, x_2 \ge 0$

(b) $4x_1 + 2x_2 + s_1 = 8$
$x_1 + 4x_2 - s_2 = 7$
$8x_1 + 3x_2 + z = 0$

(c)
$$\begin{array}{c} \begin{array}{ccccc} x_1 & x_2 & s_1 & s_2 & z \end{array} \\ \left[\begin{array}{ccccc|c} 4 & 2 & 1 & 0 & 0 & 8 \\ 1 & 4 & 0 & -1 & 0 & 7 \\ \hline 8 & 3 & 0 & 0 & 1 & 0 \end{array}\right] \end{array}$$

9. (a) Maximize: $\quad z = -w = -2x_1 - x_2 - 3x_3$
Subject to: $\quad 5x_1 + 6x_2 + x_3 \le 20$
$\qquad\qquad\quad 2x_1 + x_2 + 5x_3 \ge 10$
$\qquad\qquad\quad 7x_2 + 4x_3 \ge 20$
$\qquad\qquad\quad x_1 \ge 0, x_2 \ge 0, x_3 \ge 0$

(b) $5x_1 + 6x_2 + x_3 + s_1 = 20$
$2x_1 + x_2 + 5x_3 - s_2 = 10$
$0x_1 + 7x_2 + 4x_3 - s_3 = 20$
$2x_1 + x_2 + 3x_3 + z = 0$

(c)
$$\begin{array}{c} \begin{array}{ccccccc} x_1 & x_2 & x_3 & s_1 & s_2 & s_3 & z \end{array} \\ \left[\begin{array}{ccccccc|c} 5 & 6 & 1 & 1 & 0 & 0 & 0 & 20 \\ 2 & 1 & 5 & 0 & -1 & 0 & 0 & 10 \\ 0 & 7 & 4 & 0 & 0 & -1 & 0 & 20 \\ \hline 2 & 1 & 3 & 0 & 0 & 0 & 1 & 0 \end{array}\right] \end{array}$$

11. The solution is $x_1 = 18$, $x_2 = 0$, $s_1 = 7$, $s_2 = 12$, $s_3 = 0$, $z = 90$.

13. The solution is $x_1 = 0, x_2 = 10, s_1 = 5, s_2 = 0, s_3 = 18, z = 30$.

15. The solution is $x_1 = 0$, $x_2 = 15$, $x_3 = 45$, $s_1 = 45$, $s_2 = 0$, $s_3 = 0$, $z = 150$.

17. The solution is $x_1 = 2$, $x_2 = 0$, $s_1 = 8$, $s_2 = 0$, $w = 2$.

19. The solution is $x_1 = 0$, $x_2 = 1$, $s_1 = 16$, $s_2 = 9$, $s_3 = 0$, $w = 4$.

21. The solution is $x_1 = 0, x_2 = \frac{8}{5}, x_3 = \frac{13}{5}, s_1 = 7, s_2 = 0$, $s_3 = 0, w = \frac{34}{5}$.

23. The solution is $x_1 = 0, x_2 = 12, x_3 = 0, s_1 = 10, s_2 = 30, s_3 = 0$, $w = -12$.

25. (a) To maximize profit, the Tentsmith Co. should make and sell 30 two-person tents and 50 six-person tents. This will yield a profit of $4200.

(b) $s_1 = 0, s_2 = 0$ indicate that all hours in the fabric cutting and assembly and packaging departments have been used. $s_3 = 60$ indicates that the company has exceeded the minimum number of tents required by 60.

27. (a) To maximize profit, Mike and Denyse should produce and sell 10 Deluxe 1000 models and 20 Deluxe 1200 models. Maximum profit will be $12,000.

(b) $s_1 = s_2 = 0$ indicate that all hours in the assembly and testing department are being used. $s_3 = 22$ indicates that they will exceed the minimum 8 microcomputers required by 22.

29. (a) To minimize daily food cost and meet minimum nutritional requirements, Mike should eat 3 cheeseburgers and 10 small orders of onion rings.

(b) Minimum daily cost will be $13.47.

31. (a) To minimize operating cost and fulfill the minimum order, the Findlay refinery should operate for 60 days and the Ft. Wayne refinery should operate for 150 days. The minimum cost is $6,630,000.

(b) $s_1 = 138,000$ indicates that they will exceed the minimum order of 60,000 barrels of low-octane gasoline by 138,000 barrels. $s_4 = 90$ indicates that the refineries will operate 90 days less than the 300-day maximum restriction, and the zero values for s_2 and s_3 indicate that they will exactly meet the minimum requirements of high-octane and diesel, respectively.

33. (a) To minimize cost and meet nutritional requirements, Anita should use 50 ounces of Goli and 300 ounces of Zarobila. Minimum daily costs are $11.50.

(b) $s_2 = 150$ indicates that the minimum requirement of 500 units of minerals will be exceeded by 150 units. $s_4 = 100$ indicates that Anita will use 100 ounces less than the maximum 450 ounces allowed. The zero values for the remaining slack and surplus variables indicate that these restrictions are being met exactly.

35. To minimize transportation costs, warehouse A should ship 110 DVD players to retailer I and zero DVD players to retailer II. Warehouse B should ship 150 DVD players to retailer II and zero DVD players to retailer I. The minimum transportation costs will be $3860.

Chapter 4 Review Exercises

1.
$$x_1 + x_2 + s_1 = 14$$
$$-2x_1 + 5x_2 + s_2 = 6$$

3.

x_1	x_2	s_1	s_2	z	
1	2	1	0	0	8
1	−1	0	1	0	4
−1	−3	0	0	1	0

5.

x_1	x_2	x_3	s_1	s_2	z	
1	2	3	1	0	0	14
−1	4	−2	0	1	0	3
−2	−3	−1	0	0	1	0

7. (a) Basic: x_1, s_1, z; nonbasic: x_2, s_2

(b) $x_1 = 14, x_2 = 0, s_1 = 6, s_2 = 0, z = 5$

9. (a)

x_1	x_2	s_1	s_2	s_3	z	
$\frac{11}{4}$	0	1	$-\frac{3}{4}$	0	0	0
$-\frac{1}{4}$	1	0	$\frac{1}{4}$	0	0	1
$\frac{1}{2}$	0	0	$\frac{1}{2}$	1	0	14
$\frac{17}{4}$	0	0	$-\frac{5}{4}$	0	1	−3

(b) $x_1 = 0, x_2 = 1, s_1 = 0, s_2 = 0, s_3 = 14, z = -3$

11. (a) Let x_1 = number of students admitted each month for the drum workshop, and let x_2 = number of students admitted each month for the guitar workshop.

Maximize: $P = 25x_1 + 30x_2$
Subject to: $2x_1 + 4x_2 \le 84$
$4x_1 + 2x_2 \le 78$
$x_1 \ge 0, x_2 \ge 0$

(b)
$$2x_1 + 4x_2 + s_1 = 84$$
$$4x_1 + 2x_2 + s_2 = 78$$
$$-25x_1 - 30x_2 + P = 0$$

(c)

x_1	x_2	s_1	s_2	P	
2	4	1	0	0	84
4	2	0	1	0	78
−25	−30	0	0	1	0

13. (a) Let x_1 = number of +15 bags made each week, and let x_2 = number of −15 bags made each week.

Maximize: $P = 60x_1 + 70x_2$
Subject to: $3x_1 + 6x_2 \le 240$
$3x_1 + 3x_2 \le 180$
$x_1 \ge 0, x_2 \ge 0$

(b)
$$3x_1 + 6x_2 + s_1 = 240$$
$$3x_1 + 3x_2 + s_2 = 180$$
$$-60x_1 - 70x_2 + P = 0$$

(c)

x_1	x_2	s_1	s_2	P	
3	6	1	0	0	240
3	3	0	1	0	180
−60	−70	0	0	1	0

15. (a) Let x_1 = number of acres allotted for head lettuce, x_2 = number of acres allotted for cantaloupe, and x_3 = number of acres allotted for bell peppers.

Maximize: $P = 120x_1 + 100x_2 + 140x_3$
Subject to: $x_1 + x_2 + x_3 \le 220$
$173x_1 + 175x_2 + 212x_3 \le 34{,}700$
$122x_1 + 161x_2 + 108x_3 \le 27{,}845$
$76x_1 + 39x_2 + 72x_3 \le 13{,}350$
$x_1 \ge 0, x_2 \ge 0, x_3 \ge 0$

(b)
$$x_1 + x_2 + x_3 + s_1 = 220$$
$$173x_1 + 175x_2 + 212x_3 + s_2 = 34{,}700$$
$$122x_1 + 161x_2 + 108x_3 + s_3 = 27{,}845$$
$$76x_1 + 39x_2 + 72x_3 + s_4 = 13{,}350$$
$$-120x_1 - 100x_2 - 140x_3 + P = 0$$

(c)

x_1	x_2	x_3	s_1	s_2	s_3	s_4	P	
1	1	1	1	0	0	0	0	220
173	175	212	0	1	0	0	0	34,700
122	161	108	0	0	1	0	0	27,845
76	39	72	0	0	0	1	0	13,350
−120	−100	−140	0	0	0	0	1	0

17. $x_1 = 5, x_2 = 0, s_1 = 23, s_2 = 0, s_3 = 17, z = 20$

19. $x_1 = 0, x_2 = 1, x_3 = \frac{12}{5}, x_4 = \frac{41}{5}, s_1 = 0, s_2 = 0, s_3 = 0, z = \frac{149}{5}$

21. (a)

x_1	x_2	s_1	s_2	s_3	z	
4	−2	1	0	0	0	0
1	1	0	1	0	0	7
2	1	0	0	1	0	8
−2	−2	0	0	0	1	0

(b) Maximum is 14 at $x_1 = 1, x_2 = 6, s_1 = 8$, and $s_2 = s_3 = 0$ (may be multiple solutions).

23. (a)

x_1	x_2	x_3	s_1	s_2	z	
1	−1	1	1	0	0	5
2	1	4	0	1	0	4
−3	−1	−1	0	0	1	0

(b) Maximum is $z = 6$ at $x_1 = 2, x_2 = x_3 = s_2 = 0$, and $s_1 = 3$.

25. (a)

x_1	x_2	s_1	s_2	z	
1	1	1	0	0	9
2	2	0	1	0	4
−8	−2	0	0	1	0

(b) Maximum is $z = 16$ at $x_1 = 2, x_2 = s_2 = 0$, and $s_1 = 7$.

27. (a) A maximum monthly profit of $660 is obtained by admitting 8 drum students and 12 guitar students each month. The value of 2 for s_3 indicates that 2 hours of available music recording instruction are unused.

(b) Answers will vary.

29. A maximum profit of $23,000 is obtained by allotting 150 acres for head lettuce and 50 acres for cantaloupe. The value of 20 for s_1 indicates that there are 20 available acres unused. The value of 1495 for s_3 indicates that there are 1495 available pounds of phosphate unused.

31.
$$\begin{bmatrix} 3 & 2 & 4 \\ 1 & 0 & 1 \\ 0 & 3 & 8 \\ 2 & 7 & 9 \end{bmatrix}$$

33. Maximize: $z = 10x_1 + 5x_2$
Subject to: $4x_1 + x_2 \leq 6$
$2x_1 + 3x_2 \leq 4$
$x_1 \geq 0, x_2 \geq 0$

35. (a) Minimum is 4 at $y_1 = 4$, $y_2 = 0$.

(b) Maximum is 4 at $x_1 = 1$, $x_2 = 0$.

37. (a) Maximize: $z = 6x_1 + 9x_2$
Subject to: $4x_1 + 3x_2 \leq 3$
$2x_1 + 4x_2 \leq 4$
$x_1 \geq 0, x_2 \geq 0$

(b) Minimum is 9 at $y_1 = 3$, $y_2 = 0$.

39. (a) Maximize: $z = 5x_1 + 12x_2$
Subject to: $x_1 + 2x_2 \leq 16$
$2x_1 + x_2 \leq 7$
$3x_1 + x_2 \leq 22$
$x_1 \geq 0, x_2 \geq 0$

(b) Minimum is 84 at $y_1 = 0$, $y_2 = 12$, $y_3 = 0$.

41. (a) Minimize: $C = 0.08y_1 + 0.07y_2$
Subject to: $6y_1 + 6y_2 \geq 60$
$2y_1 + y_2 \geq 12$
$y_1 \geq 0, y_2 \geq 0$

(b) Maximize: $z = 60x_1 + 12x_2$
Subject to: $6x_1 + 2x_2 \leq 0.08$
$6x_1 + x_2 \leq 0.07$
$x_1 \geq 0, x_2 \geq 0$

(c) A minimum cost of $0.72 is obtained by taking 2 Everyday pills and 8 Allday pills daily.

43. (a) Aztec: 50 days; Buena Vista: 125 days

(b) $5,800,000

45. (a) $x_1 =$ the value of each mg of vitamin C, $x_2 =$ the value of each mg of Niacin; $x_1 = 0.01$, $x_2 = 0.01$.

(b) $0.74

47. $4x_1 + x_2 + 3x_3 + s_1 = 12$
$x_1 + 2x_2 + 5x_3 - s_2 = 9$
$3x_1 + x_2 + 7x_3 + s_3 = 4$

49. (a) Maximize: $z = -w = -x_1 - 3x_2 - 8x_3$
Subject to: $x_1 + 2x_2 + 5x_3 \leq 7$
$x_1 + x_2 + 8x_3 \geq 10$
$4x_2 + 2x_3 \geq 24$
$x_1 \geq 0, x_2 \geq 0, x_3 \geq 0$

(b) $x_1 + 2x_2 + 5x_3 + s_1 = 7$
$x_1 + x_2 + 8x_3 - s_2 = 10$
$4x_2 + 2x_3 - s_3 = 24$
$x_1 + 3x_2 + 8x_3 + z = 0$

(c)
$$\begin{array}{ccccccc|c} x_1 & x_2 & x_3 & s_1 & s_2 & s_3 & z & \\ 1 & 2 & 5 & 1 & 0 & 0 & 0 & 7 \\ 1 & 1 & 8 & 0 & -1 & 0 & 0 & 10 \\ 0 & 4 & 2 & 0 & 0 & -1 & 0 & 24 \\ \hline 1 & 3 & 8 & 0 & 0 & 0 & 1 & 0 \end{array}$$

51. Maximum is $z = 64$ at $x_1 = 16$, $x_2 = s_3 = 0$, $s_1 = 6$, $s_2 = 22$.

53. Minimum is $w = 4$ at $x_1 = s_2 = 0$, $x_2 = 2$, $s_1 = 20$.

55. Minimum is $w = 10$ at $x_1 = 5$, $x_2 = x_3 = s_2 = 0$, $s_1 = 25$, $s_3 = 15$.

57. (a) Day packs: 90; Multi-day packs: 0; $3150

(b) $s_1 = 0$ indicates that no available hours in fabric cutting are unused, $s_2 = 80$ indicates that 80 available hours in assembly and packaging are unused, $s_3 = 75$ indicates that 75 more backpacks are made weekly than the minimum requirement of 15.

59. (a) 6-inch roasted chicken sandwiches: 3; 6-inch Subway Club® sandwiches: 2

(b) $19.40

(c) Answers will vary.

Chapter 5

Section 5.1

1. $187.50 **3.** $2750 **5.** $318.75 **7.** $11,000

9. $2280 **11.** $13,250 **13.** $7350 **15.** $32,675

17. 7.5% **19.** 9% **21.** 9.51% **23.** 10.52%

25. $4310.34 **27.** $7920.79 **29.** $3814.06 **31.** $13,300.08

33. (a) $A = 8000 + 520t$

(b)

$m = 520$
Each year, Sue earns $520 in interest on her $8000 deposit.

35. (a) $A = 4000 + 210t$

(b)

$m = 210$
Each year, Rich earns $210 in interest on his $4000 deposit.

37. It will take 10 years for a deposit to double if invested at 10% simple interest.

39. It will take approximately 11.76 years for a deposit to double if invested at 8.5% simple interest.

41. It will take approximately 17.39 years for a deposit to double if invested at 5.75% simple interest.

43. The discount is $46.67 and the proceeds are $953.33.

45. The discount is $650 and the proceeds are $4350.

47. The discount is $2887.50 and the proceeds are $7112.50.

49. $1048.95   **51.** $5747.13   **53.** $14,059.75

55. The amount of interest due on the loan is $18,000.

57. The amount of interest due on the loan is $160.

59. The amount of interest due on the loan is $18 million.

61. Ardra repaid $25,630 to her aunt.

63. Denyse repaid $23,625 to her grandfather.

65. The discount is $1190 and the proceeds are $5810.

67. Brian should borrow $8433.73.

69. Melissa should invest approximately $2643.17.

71. Bill should invest approximately $16,293.28.

73. Francine will pay 8.25% interest.

75. Lenny will pay 7.3% interest.

77. Dylan should invest $3948.06.

79. Franco should invest $1864.80.

81. The annual simple interest rate is 3.7%.

83. Alicia should pay $4773.27.

85. (a) The loan to value ratio is $32,800.
  (b) Bill will be able to take a $2788 tax deduction in a given year due to interest on the home equity loan.

Section 5.2

1. Future value of approximately $3828.84 after 5 years and a future value of approximately $4432.37 after 8 years.

3. Future value of approximately $2623.30 after 5 years and a future value of approximately $3087.02 after 8 years.

5. Future value of approximately $11,458.08 after 5 years and a future value of approximately $14,214.29 after 8 years.

7. Future value of approximately $21,001.72 after 5 years and a future value of approximately $25,701.19 after 8 years.

9. Future value of approximately $14,560.77 after 5 years and a future value of approximately $18,579.40 after 8 years.

11. Future value of $9424.55 after 3 years and $11,102.76 after 6 years.

13. Future value of $11,877.82 after 3 years and $14,108.27 after 6 years.

15. The present value is approximately $4113.51.

17. The present value is approximately $2287.19.

19. The present value is approximately $5523.22.

21. The present value is approximately $8115.48.

23. The present value is approximately $3639.06.

25. The present value is approximately $6995.44.

27. The present value is approximately $5817.44.

29. Graph (c)   **31.** Graph (a)

33. It will take approximately 4 years.

35. It will take approximately 18 compounding periods (about 9 years).

37. It will take approximately 37 compounding periods (about $9\frac{1}{4}$ years).

39. It will take approximately 92 compounding periods (about $7\frac{1}{2}$ years).

41. The effective rate is 8.775%.

43. The effective rate is 6.168%.

45. The effective rate is 5.319%.

47. The effective rate is 6.963%.

49. The effective rate is 7.714%.

51. The effective rate is 7.229%.

53. The effective rate is 6.348%.

55. The effective rate is 5.904%.

57. (a) The nominal rate of 6.25% compounded quarterly is 6.109%.
  (b) The nominal rate of 6.25% compounded monthly is 6.078%.

59. (a) The nominal rate of 5.5% compounded quarterly is 5.39%.
  (b) The nominal rate of 5.5% compounded monthly is 5.366%.

61. (a) The nominal rate of 5.75% compounded quarterly is 5.63%.
  (b) The nominal rate of 5.75% compounded monthly is 5.604%.

63. (a) The nominal rate of 7.5% compounded quarterly is 7.298%.
  (b) The nominal rate of 7.5% compounded monthly is 7.254%.

65. 8% compounded monthly is the better investment.

67. 8.25% compounded quarterly is the better investment.

69. If the rate holds true, in 5 years a gallon of milk will cost about $2.62.

71. The account will be worth about $2962.37 in 4 years.

73. If the rate holds true, in 7 years the house will be worth about $166,033.

75. Alec and Lexi should invest about $16,074.45 now at 5.5% interest compounded quarterly to have $20,000 in 4 years.

77. Sam and Diane should invest $16,368.75 now at 8.5% interest compounded monthly to have $25,000 in 5 years.

79. It will take 37 compounding periods, or 37 quarters, for the $10,000 to grow to at least $16,000.

81. (a) The effective rate of 4.5% compounded monthly is 4.594%.

(b) The effective rate of 4.4% compounded daily is 4.498%.

(c) The account at First Internet bank paying 4.50% interest compounded monthly is a better investment; the effective rate is higher.

83. Paige's grandparents should pay $11,546.49 for the bond.

85. The annual compound rate of return is 9.746%.

87. The interest rate, compounded annually, is 25.137%.

Section 5.3

1. The future value of the annuity is $24,581.90.

3. The future value of the annuity is $206,798.81.

5. The future value of the annuity is $191,086.80.

7. The future value of the annuity is $79,240.57.

9. The future value of the annuity is $105,997.82.

11. The future value of the annuity is $110,943.44.

13. The amount of interest is $6581.90.

15. The amount of interest is $146,798.81.

17. The amount of interest is $131,086.80.

19. The periodic payment amount is $1517.36 annually.

21. The periodic payment amount is $263.55 monthly.

23. The periodic payment amount is $700 quarterly.

25. The annual payment needed is $1193.84.

27. The monthly payment needed is $258.14.

29. The quarterly payment needed is $1129.35.

31. (a) $15,173.60   **(b)** $4,826.40

33. (a) $47,439   **(b)** $32,561

35. (a) $22,400   **(b)** $7600

37. The annual interest rate is 6.66%.

39. The annual interest rate is 6.92%.

41. The annual interest rate is 8.12%.

43. The account will be worth $15,322.29 after 10 years.

45. Teri deposited $12,000 in the account and earned $3322.29 in interest.

47. The account will be worth $71,924.44 after 30 years.

49. The account will be worth $41,693.54 after 10 years.

51. The account will be worth $18,838.56 after 4 years.

53. They should deposit $356.76 monthly.

55. The quarterly deposit needed is $1014.92.

57. Irene should invest $208.81 each month.

59. (a) $5355

(b) 4 years of college tuition at $5355 per year will be $21,420.

(c) $80.14 should be invested monthly at 5% interest compounded monthly to have $21,420 after 15 years.

61. (a) $188,921.57   **(b)** $272,615.08   **(c)** $398,041.76

63. (a) On her 65th birthday, Debbie's account will be worth $367,264.71.

(b) Bubba's account will be worth $291,901.24 on his 65th birthday.

(c) Debbie deposited $20,000 into her account, and earned $347,264.71 in interest. Bubba deposited $66,000 into his account and earned $225,901.24 in interest. The amount that Debbie deposited was considerably less than the amount that Bubba deposited, but she had more in her account on her 65th birthday. The moral is to begin making regular deposits into a savings account as early as possible!

65. (a) $13,286.89   **(b)** Monthly investment is $3286.89.

(c) The value of the account after 30 years is $3,301,640.04.

Section 5.4

1. $60,876.17   **3.** $111,718.92   **5.** $21,994.72

7. $66,753.57   **9.** $103,671.89   **11.** $15,586.87

13. $296.75   **15.** $716.41   **17.** $761.87   **19.** $410.33

21. $820.95   **23.** $707.29

25. (a) The total amount paid over the life of the loan is $35,610.

(b) The total amount of interest paid over the life of the loan is $10,610.

27. (a) The total amount paid over the life of the loan is $257,907.60.

(b) The total amount of interest paid over the life of the loan is $157,907.60.

29. (a) The total amount paid over the life of the loan is $15,237.40.

(b) The total amount of interest paid over the life of the loan is $2237.40.

31. Teri needs to deposit a lump sum of $9537.64 now at 4.75% interest compounded monthly for 10 years to yield the same amount as the annuity.

33. (a) The monthly payment will be $359.45.

(b) The total interest paid over the life of the loan is $15,934.

35. Assuming that the account continues to earn 6.5% interest compounded monthly, Irene can withdraw $1493.80 each month for 25 years.

37. (a) Irene deposited $72,000 into her account.

(b) Nick deposited $72,000 into his account.

(c) Irene receives $17,925.60 annually while Nick receives $16,994.73 annually. Irene's account earned more interest than Nick's due to the more frequent deposits and the monthly compounding of interest.

39. Malika needs to deposit $310.26 into the annuity each month for 35 years to be able to withdraw $3000 each month for 25 years.

41. (a) The lump sum payment will be $1,835,310.77.

(b) The lump sum payment will be $1,654,538.78.

43. The cost of the house is $138,566.64.

45. (a) The down payment is $36,000. The financed amount is $144,000.

    (b) The monthly payment is $995.06.

47. (a) They will pay $214,221.60 in interest if they take out the 30-year mortgage at 7.38%.

    (b) They will pay $88,975.80 in interest if they take out the 15-year mortgage at 7.0%.

49. (a) After 10 years of payments, the unpaid balance is $124,651.82.

    (b) After 15 years of payments, the unpaid balance is $108,134.44.

    (c) After 15 years of payments, the unpaid balance is $84,272.55.

51. (a) $y = 995.06 \left[\dfrac{1-(1.00615)^{-(360-x)}}{0.00615} \right]$

    (b)

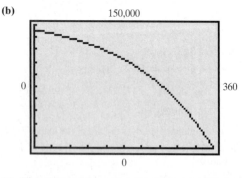

    (c) The calculator approximation of the unpaid balance is $124,651.82, which is the same result as in Exercise 49a.

    (d) The calculator approximation of the unpaid balance is $108,134.44, which is the same result as in Exercise 49b.

    (e) The calculator approximation of the unpaid balance is $84,272.55, which is the same result as in Exercise 49c.

53.

Payment Number	Amount of Payment	Interest for Period	Reduction of Unpaid Balance	Unpaid Balance
0				10,000
1	874.51	75.00	799.51	9,200.49
2	874.51	69.00	805.51	8,394.98
3	874.51	62.96	811.55	7,583.43
4	874.51	56.88	817.63	6,765.80
5	874.51	50.74	823.77	5,942.03
6	874.51	44.57	829.94	5,112.09
7	874.51	38.34	836.17	4,275.92
8	874.51	32.07	842.44	3,433.48
9	874.51	25.75	848.76	2,548.72
10	874.51	19.39	855.12	1,693.60
11	874.51	12.70	861.81	831.79
12	838.03	6.24	831.79	0

Notice that the final month's payment will usually need to be adjusted to bring the unpaid balance to zero.

Chapter 5 Review Exercises

1. (a) $2970   **(b)** $14,970   **3.** 7.5%   **5.** $6439.39

7. (a) $A = 6000 + 375t$

    (b)

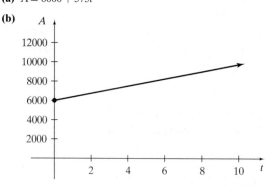

    (c) $m = 375$; Amara earns $375 in interest each year on her $6000 deposit.

9. 16 years   **11.** $D = \$4200$, $p = \$7800$ **13.** $29,937.63

15. $20,860   **17.** 10%   **19.** $9128.98, $12,722.65

21. $10,064.76, $11,902.67   **23.** $5008.01   **25.** $8929.13

27. 10 years   **29.** 7.978%   **31. (a)** 5.630%   **(b)** 5.604%

33. $1.87   **35.** $6997.61   **37.** $14,280.11   **39.** $270,949.97

41. $180,949.97   **43.** $861.97

45. (a) $11,606.40   **(b)** $8393.60   **47.** 4.47%

49. $6918.28   **51.** $312.38   **53.** $94,891.55

55. $46,269.22   **57.** $265.47   **59.** $511.63

61. (a) $47,784.60   **(b)** $17,784.60

63. (a) $291.04   **(b)** $22,387.20

65. (a) $1849.27

    (b) If Lindsay deposits $240 a month for 25 years, she deposits a total of $72,000 and receives equal monthly payments of $1421.11 for the last 20 years. Whereas, if Lindsay deposits $400 each month for 15 years, she deposits a total of $73,000 and receives monthly payments of $1849.27 for the last 20 years.

67.

Payment Number	Amount of Payment	Interest for Period	Reduction of Unpaid Balance	Unpaid Balance (principal)
0				$6000
1	$521.93	$40.00	$481.93	$5518.07
2	521.93	36.79	485.14	5032.93
3	521.93	33.55	488.38	4544.55
4	521.93	30.30	491.63	4052.92
5	521.93	27.02	494.91	3558.01
6	521.93	23.72	498.21	3059.80
7	521.93	20.40	501.53	2558.27
8	521.93	17.06	504.87	2053.40
9	521.93	13.69	508.24	1545.16
10	521.93	10.30	511.63	1033.53
11	521.93	6.89	515.04	518.49
12	521.95	3.46	518.49	0

69. (a) Simple interest   **(b)** 4.44%

71. (a) Compound interest   **(b)** $9971.65

Chapter 6

Section 6.1

1. This is an example of probability.

3. This is an example of statistical inference.

5. This is an example of probability.

7. This is an example of probability.

9. This process represents a probability experiment.

11. This process does not represent a probability experiment.

13. All the questions that have an oval to fill in as a response represent a probability experiment.

15. Order is important.   **17.** Order is important.

19. Order is not important.   **21.** Order is not important.

23. Order is not important.

25. The events can happen at the same time.

27. The events cannot happen at the same time.

29. The events can happen at the same time.

31. The events cannot happen at the same time.

33. The events cannot happen at the same time.

In Exercises 35–45, the answers will vary.

35. The outcome is tails.   **37.** It will be light outside.

39. The person has perfect hearing.

41. The person has blonde hair.

43. The student receives a passing grade in geography.

45. The card is a queen.

47. The outcome of the flood influences the postponement of the picnic.

49. The outcome of the first event does not influence the second event.

51. The outcome of the first event influences the second event.

53. The outcome of the first event does not influence the second event.

55. The outcome of the first event influences the second event.

In Exercises 57–67, the answers will vary.

57. There will be a full moon on Wednesday night.

59. The child watches Sesame Street in the afternoon.

61. The outcome of the next roll is an odd number.

63. The event that the person will have a sunburn.

65. The event that Tammy gives a speech at the ceremony.

67. The event that Ron fails the political science exam.

Section 6.2

1. $A = \{1, 2, 3, 4, 5, 6, 7\}$

3. $C = \{2, 4, 6, 8, 10, 12, 14, 16, 18, 20, 22, 24\}$

5. $E = \{$September, October, November, December$\}$

7. $A = \{x | x$ is a whole number and $2 \leq x \leq 6\}$

9. $C = \{x | x$ is a multiple of 5 and $1 \leq x \leq 30\}$

11. $E = \{x | x$ is a day of the week that starts with the letter T$\}$

13. $\emptyset, \{2\}$ **15.** $\emptyset, \{$Honda$\}, \{$Ford$\}, \{$Honda, Ford$\}$

17. $\emptyset, \{8\}, \{9\}, \{10\}, \{8, 9\}, \{8, 10\}, \{9, 10\}, \{8, 9, 10\}$

19. $\supset$ **21.** $\subseteq$ **23.** $\subseteq$ **25.** $A \cap B = \{2, 4\}$

27. $B \cap C = \emptyset$ **29.** $A \cup B = \{1, 2, 3, 4, 5, 6, 8, 10\}$

31. $B \cup C = \{2, 4, 6, 8, 10, 11, 12, 13, 14, 15\}$

33. $A^C = \{6, 7, 8, 9, 10, 11, 12, 13, 14, 15\}$

35. $B \cap C^C = \{2, 4, 6, 8, 10\}$ **37.** $U^C = \emptyset$

39. $(A \cap C)^C = \{1, 2, 3, 4, 5, 6, 7, 8, 9, 10, 11, 12, 13, 14, 15\}$

41. $A^C \cup C = \{6, 7, 8, 9, 10, 11, 12, 13, 14, 15\}$

43. $A \cup \emptyset = \{1, 2, 3, 4, 5\}$ **45.** $\emptyset \cap B^C = \emptyset$

47. $A = \{a, b, c, d, e\}$

49. $U = \{a, b, c, d, e, f, g, h, i, j, k, l, m, n\}$

51. $A \cap B = \{b, e\}$ **53.** $A^C = \{f, g, h, i, j, k, l, m, n\}$

55. $(A \cup B)^C = \{h, j, k, l, m, n\}$

57. $A \cup (B \cap C) = \{2, 4, 5, 6, 8\}$

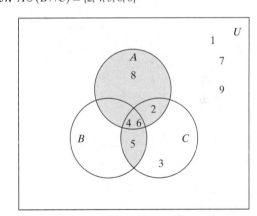

59. $A \cap (B \cup C)^C = \{8\}$

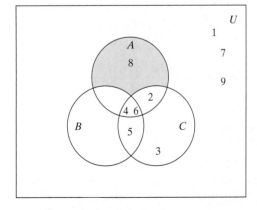

61. $(A \cup B) \cap (A \cup C) = \{2, 4, 5, 6, 8\}$

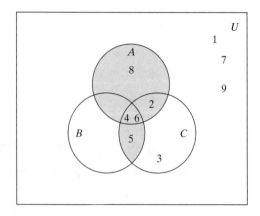

63. $A^C \cup B^C = \{1, 2, 3, 5, 7, 8, 9\}$

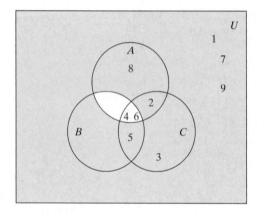

65. $(A \cap B)^C = \{1, 2, 3, 5, 7, 8, 9\}$

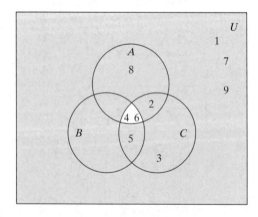

67. $n(A) = 250$

During 1996, the value of arm transfers from the United States to India and Pakistan was $250 million.

69. $n(B \cap D) = 70$

During 1996, the value of arms transfers from the United Kingdom to Pakistan was $70 million.

71. $n(B \cup D) = n(B) + n(D) - n(B \cap D)$
$= 70 + 210 - 70 = 210$

During 1996, the values of arms transfers either sent by the United Kingdom or received by Pakistan was $210 million.

73. $n(A^C) = n(U) - n(A)$
$= 84.9 - 53.8 = 31.1$

There are about 31.1 million households in the Midwest and southern United States whose household income is not under $35,000.

75. $n(A \cup D) = 71.8$

For the southern and midwestern United States, there are about 71.8 million households who are in the southern U.S. or whose household income is under $35,000.

77. $n(A^C \cup D) = 49.6$

For the southern and midwestern United States, there are 49.6 million households located in the southern U.S or whose household income is not under $35,000.

Section 6.3

1. (a)

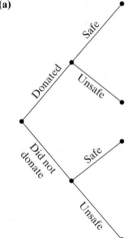

(b) There are four possible ways a respondent could answer the questions.

3. (a)

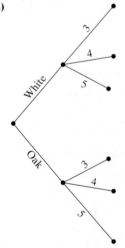

(b) There are six possible ways to build the bookshelf.

5. (a)

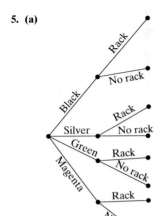

(b) There are eight possible ways to create a motorcycle package.

7. There are 192 ways that a person can construct a package deal.

9. There are 1512 ways that the awards can be given.

11. There are 720 possible ways to order the meal.

13. The textbooks can be chosen in 4550 ways.

15. The car can be ordered in 60 ways.

17. There are 30,240 possible combinations for the lock.

19. There are 1,125, 899, 906, 842, 624 different ways to answer the 25 questions.

21. 5040   **23.** 60   **25.** 5040   **27.** 1

29. $_{50}P_{20} \approx 1.15 \times 10^{32}$ **31.** $_{40}P_{15} \approx 5.26 \times 10^{22}$

33. $_{90}P_{30} \approx 1.79 \times 10^{56}$

35. There are 1320 different ways that the company officers can be selected.

37. There are 5814 ways that the winner, first runner-up, and second runner-up can be selected from the 19 entrants.

39. There are 303,600 different ways that the restaurants could be ranked.

41. 95,040 different messages are possible.

43. There are 362,880 ways that the batting order could be arranged.

45. 10   **47.** 210   **49.** 792   **51.** 7   **53.** 1   **55.** 1

57. $_{50}C_{20} \approx 4.71 \times 10^{13}$ **59.** $_{40}C_{15} \approx 4.02 \times 10^{10}$

61. $_{90}C_{30} \approx 6.73 \times 10^{23}$

63. There are 20 ways that the judges can be selected.

65. There are 15,504 different ways that the research team can be formed.

67. There are 118,755 different ways that five of the companies could be chosen.

69. In the Maine state lottery *Megabucks*, numbers can be chosen in 5,245,786 different ways.

71. 76,904,685 different sampler cups are possible.

73. There are 350 different ways that the subcommittee could be formed with 2 Democrats and 4 Republicans.

75. There are 8,953,560 ways that a sample of 10 could be chosen so that 2 are defective and 8 are not.

77. There are 8400 different ways that the committee of 7 could be formed with 2 Republicans, 2 Democrats, and 3 Green Party members.

Section 6.4

1. $S = \{1, 2, 3, 4, 5, 6, 7, 8, 9\}$ **3.** $S = \{2, 3, 4, 5, 6, 7, 8, 9, 10\}$
$n(S) = 9$ $n(S) = 9$

5. $S = \{\text{conservative, moderate, liberal}\}$
$n(S) = 3$

7. $S = \{(\text{good}, 1), (\text{good}, 2), (\text{good}, 3)$
$(\text{good}, 4), (\text{excellent}, 1),$
$(\text{excellent}, 2), (\text{excellent}, 3),$
$(\text{excellent}, 4), (\text{superior}, 1),$
$(\text{superior}, 2), (\text{superior}, 3),$
$(\text{superior}, 4)\}$
$n(S) = 12$

9. $S = \{(\text{better, AM}), (\text{better, FM}),$
$(\text{worse, AM}), (\text{worse, FM})\}$
$n(S) = 4$

11. $P(A) = \frac{5}{9}, \ P(B) = \frac{4}{9}$ **13.** $P(A) = \frac{5}{9}, \ P(B) = \frac{4}{9}$

15. $P(A) = 0, \ P(B) = 0.2$

17. $P(A) = \frac{4}{12} \approx 0.33, \ P(B) = \frac{3}{12} = 0.25$

19. $P(A) = \frac{4}{16} = 0.25, \ P(B) = \frac{7}{16} = 0.4375$

21. (a) $_{30}C_2 = 435$ **(b)** 105

(c) $P(A) = \dfrac{105}{435} \approx 0.24$

There is about a 24% chance that both cookies will be oatmeal raisin.

23. (a) 4060   **(b)** 455

(c) $P(A) = \dfrac{455}{4,060} \approx 0.112$

There is about an 11.2% chance that the 3 marbles will be blue.

25. (a) 924   **(b)** 350

(c) $P(A) = \dfrac{350}{924} \approx 0.3788$

There is about a 37.9% probability that the subcommittee will be formed with 2 Democrats and 4 Republicans.

27. (a) 30,045,015   **(b)** 8,953,560

(c) $P(A) = \dfrac{8,953,560}{30,045,015} \approx 0.298$

There is about a 29.8% probability that in a sample of 10 car stereos 2 will be defective.

29. $P(A) = \dfrac{8400}{77,520} \approx 0.108$

There is about a 10.8% chance that a committee of 7 will be formed with 2 Republicans, 2 Democrats, and 3 Green Party members.

31. (a) Let A represent the event that the periodical was a quarterly.

$$P(A) = \frac{1444}{8904} \approx 0.162$$

The probability that the periodical was a quarterly is approximately 16.2%

(b) Let B represent the event that the periodical was a weekly.

$$P(B) = \frac{1716}{8904} \approx 0.193$$

The probability that the periodical was a weekly is approximately 19.3%

33. (a) Let A represent the event that the crime involved a gas station.

$$P(A) = \frac{13}{459} \approx 0.0283$$

The probability that the crime involved a gas station is about 2.83%.

(b) Let B represent the event that the crime involved a bank.

$$P(B) = \frac{11}{459} \approx 0.024$$

The probability that the crime involved a bank is about 2.4%.

35. (a) Let A represent the event that the person was in the Army.

$$P(A) = \frac{162.3}{380.1} \approx 0.427$$

The probability that a randomly selected military personnel was in the Army is about 42.7%.

(b) Let B represent the event that the person was in the Navy.

$$P(B) = \frac{93.3}{380.1} \approx 0.2455$$

The probability that a randomly selected military personnel was in the Navy is about 24.55%.

37. (a) Let A represent the event that the scientist is a life scientist.

$$P(A) = \frac{180.0}{665.8} \approx 0.270$$

The probability that a randomly selected scientist, employed in the United States in 1996, is a life scientist is about 27.0%.

(b) The probability that a randomly selected scientist, employed in the United States in 1996, is not a life scientist is about 73.0%.

39. (a) Let A represent the event that the house sold for anywhere between $120,000 and $149,000.

$$P(A) = \frac{183,000}{891,000} = 0.205$$

The probability that a house sold in 1998, selected at random, sold for anywhere between $120,000 and $149,000 is about 20.5%.

(b) Let B represent the event that the house sold for $120,000 or more.

$$P(B) = \frac{639,000}{891,000} \approx 0.7172$$

The probability that a house sold in 1998, selected at random, sold for $120,000 or more is about 20.5%.

41. Subjective   **43.** Empirical   **45.** Classical

Section 6.5

1. The events are not mutually exclusive.

3. The events are mutually exclusive.

5. The events are not mutually exclusive.

7. The events are mutually exclusive.

9. The events are mutually exclusive.

11. $\frac{1}{5}$ **13.** $\frac{1}{10}$ **15.** $\frac{11}{15}$ **17.** $\frac{1}{3}$

19. $P(A) = \frac{175}{500} = 0.35$

The probability that a Canal Junction water treatment plant worker, chosen at random, took a 3-week vacation last year is 35.0%.

21. $P(A \cap B) = \frac{50}{500} = 0.10$

The probability that a Canal Junction water treatment plant worker, chosen at random, took a 3-week vacation and worked more than 5 weeks of overtime last year is 10.0%.

23. $P(A \cup B) = 0.50$

The probability that a worker, selected at random, took a 3-week vacation or worked more than 5 weeks of overtime is 50%.

25. $P(B) = 0.35$

The probability that a senior, selected at random, had been placed on the Dean's List for at least one semester is 35.0%.

27. $P(A \cap B)^C = 0.95$

The probability that a senior, selected at random, was not placed on the dean's list and was not placed on academic probation for at least one semester last year is 95.0%.

29. $P(A \cup B) = 0.55$

The probability that a senior, selected at random, was either placed on academic probation or was placed on the dean's list is 55.0%.

31. $P(A) = \frac{3113}{6639} \approx 0.469$

The probability that the doctorate is in social sciences is about 46.9%.

33. $P(A)^C = 0.531$

The probability that the doctorate is not in social sciences is about 53.1%.

35. $P(A \cap C) = \frac{2353}{6639} \approx 0.354$

The probability that the doctorate is in social sciences and is conferred on a U.S. citizen is about 35.4%.

37. $P(A \cap B) = 0$

The events are mutually exclusive.

39. $P(A \cup D) \approx 0.636$

The probability that the doctorate is in social sciences or is conferred on foreign-born citizen is about 63.6%.

41. $P(B) = 52.4\%$

In 1996, the probability that a U.S. adult, selected at random, exercised regularly is 52.4%.

43. $P(B^C) = 47.6\%$

In 1996, the probability that a U.S. adult, selected at random, did not exercise regularly is 47.6%.

45. $P(B \cap D) = 26.8$

The probability that, in 1996, a U.S. adult, selected at random, exercised regularly and is a female is 26.8%.

47. $P(C \cap D) = 0$

The events are mutually exclusive.

49. $P(B \cup C) = 74.4\%$.

The probability that in 1996 a U.S. adult, selected at random, either exercised regularly or is male is 74.4%.

51. (a) The odds in favor of event A occurring are 4 to 1.

(b) The odds against event A occurring are 1 to 4.

53. (a) The odds in favor of event A occurring are 3 to 7.

(b) The odds against event A occurring are 7 to 3.

55. $P(A) = \frac{5}{8}$

57. The probability that event A will not occur is $\frac{2}{7}$.

59. The odds that a tail will come up in a single toss of a coin is 1 to 1.

61. The odds that at least one tail will come up in two coin tosses is 3 to 1.

63. The odds against a head coming up in a single toss of a coin are 1 to 1.

65. The odds against 2 tails coming up in two tosses of a coin are 1 to 3.

67. (a) The odds that it will rain tomorrow are 3 to 7.

(b) The odds that it will not rain tomorrow are 7 to 3.

69. The probability that the Dallas Cowboys will win the 2002 Super Bowl is $\frac{1}{76}$ or about 1.32%.

71. The probability that the Los Angeles Dodgers will win the 2001 World Series is $\frac{1}{26}$ or about 3.84%.

73. The probability that the Los Angeles Lakers will win the 2002 NBA Championship is $\frac{5}{12}$ or about 41.7%.

75. No answer given; verification

Section 6.6

1. The events are dependent.

3. The events are independent.

5. The events are dependent.

7. The events are independent.

9. The events are dependent.

11. $\frac{1}{9}$ **13.** $\frac{25}{81}$

15. (a) The events are independent. The opinion of one randomly selected person does not influence the opinion of another randomly selected person.

(b) $P(A \cap B) = 0.0196$

There is a 1.96% probability that, in the year 2000, two Americans selected at random believe that the most important problem facing the country is the decline of ethics, morals, and family values.

(c) $P(A^C) = P(B^C) = 0.86$

The probability that, in 2000, the first person does not believe that the most important problem facing the county is the decline of ethic, morals, and family values is 86%. (Same probability for the second person also.)

(d) $P(A^C \cap B^C) = 0.7396$

The probability that, in 2000, two randomly selected Americans did not believe that the biggest problem facing the country is the decline of ethic, morals, and family values is 73.96%.

17. $(0.25)^4 \approx 0.004$

The probability that, in March 2000, four randomly selected Americans all reported their doctor had made an error is about 0.4%.

19. (a)

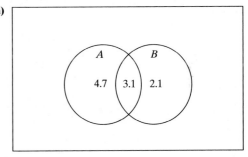

(b) $P(A \cap B) = \frac{3.1}{9.9} \approx 0.31$

The probability that a randomly selected American, who uses her computer for more than 16 hours per week, uses the computer for personal and business use is about 31%.

(c) $P(B|A) = \frac{\frac{3.1}{9.9}}{\frac{7.8}{9.9}} = \frac{3.1}{7.8} \approx 0.397$

The probability that a randomly selected computer owner, who uses her computer for more than 16 hours a week, uses the computer for business use, given that the computer is for personal use, is about 39.7%.

21. $P(A \cap B) = \frac{4}{27}$

$P(B \mid A) = \frac{4}{9} \approx 0.44$

23. $P(A \cap B) = \frac{12}{40} = 0.3$

$P(B \mid A) = \frac{12}{20} = 0.6$

25. $P(A \cap B) = \frac{1}{18}$

$P(B \mid A) = \frac{1}{3}$

27. $P(A \cap B) = 0.38$

$P(B \mid A) = \frac{0.38}{0.61} \approx 0.623$

29. $P(A \cap C) = \frac{40}{200} = 0.2$

$P(C \mid A) = \frac{40}{95} \approx 0.421$

31. $P(A \cap C) = \frac{87}{250} = 0.348$

$P(C \mid A) = \frac{87}{148} \approx 0.588$

33. $P(B \cap D) = \frac{2}{5}$

$P(D \mid B) = \frac{4}{7}$

35. $P(B \cap D) = 0.39$

$P(D \mid B) = \frac{0.39}{0.63} \approx 0.619$

37. (a) $\frac{4}{15}$ **(b)** $\frac{2}{15}$ **39. (a)** $\frac{1}{120}$ **(b)** $\frac{7}{40}$

41. $P(C \cap D) = 0$

The events are mutually exclusive.

43. $P(B \cap D) = \dfrac{1111}{6639} \approx 0.167$

45. $P(D \mid B) = \dfrac{\frac{1111}{6639}}{\frac{3526}{6639}} = \dfrac{1111}{3526} \approx 0.315$

There is 31.5% probability that the doctorate was conferred on a foreign-born citizen, given that the doctorate is in physical sciences.

47. $P(A \mid D) = \dfrac{\frac{760}{6639}}{\frac{1871}{6639}} = \dfrac{760}{1871} \approx 0.406$

49. $P(A \cap B) = 0$

The events are mutually exclusive.

51. $P(A \cap C) = 22.0$

The probability that the person is male and did not exercise regularly is 22.0%.

53. $P(C \mid A) = \dfrac{22.0}{47.6} \approx 0.462$

The probability that the person is male, given that the person did not exercise regularly, is about 46.2%.

55. $P(B \mid C) = \dfrac{25.6}{47.6} \approx 0.538$

The probability that the person exercised regularly, given that the person is male, is about 53.8%.

Section 6.7

1. $P(D \mid B) = \dfrac{1}{5}$ **3.** $P(D \mid B) = \dfrac{16}{27}$

$P(B \mid D) = \dfrac{1}{4}$ $P(B \mid D) = \dfrac{8}{11}$

5. $P(C \mid A) = \dfrac{2}{3}$ **7.** $P(C \mid A) = \dfrac{0.22}{0.64} = 0.34375$

$P(A \mid C) = \dfrac{1}{2}$ $P(A \mid C) = \dfrac{0.22}{0.37} \approx 0.595$

9. (a) $P(D \mid B) = \dfrac{1111}{3526} \approx 0.315$

(b) $P(B \mid D) = \dfrac{1111}{1871} \approx 0.594$

11. $P(A) = 0.17$

$P(B \mid A) = \dfrac{(0.3)(0.1)}{0.17} \approx 0.176$

$P(B^C \mid A) = \dfrac{(0.7)(0.2)}{0.17} \approx 0.824$

13. $P(A) = \dfrac{1}{4}$

$P(B \mid A) = \dfrac{(\frac{1}{4})(\frac{2}{5})}{\frac{1}{4}} = \dfrac{2}{5}$

$P(B^C \mid A) = \dfrac{(\frac{3}{4})(\frac{1}{5})}{\frac{1}{4}} = \dfrac{3}{5}$

15. $P(A) = 0.1454$

$P(B \mid A) = \dfrac{(0.23)(0.13)}{0.1454} \approx 0.206$

$P(B^C \mid A) = \dfrac{(0.77)(0.15)}{0.1454} \approx 0.794$

17. (a) $P(A) = \left(\dfrac{300}{500}\right)\left(\dfrac{150}{300}\right) + \left(\dfrac{200}{500}\right)\left(\dfrac{150}{200}\right) = \dfrac{3}{5}$

The probability that a randomly selected factory worker favored St Patrick's day for a paid holiday is 60%.

(b) $P(B \mid A) = \dfrac{1}{2}$

The probability that a randomly selected factory worker works day shift, given that the worker favors St. Patrick's day as a paid holiday, is 50%.

19. (a) $P(A) = 0.2586$

There is a 25.86% probability that a randomly selected Ketterson College student favors switching from quarters to semesters.

(b) $P(B \mid A) \approx 0.842$

The probability that a Ketterson College student is an underclassperson, given that he favors switching from quarters to semesters is about 84.2%.

21. (a) $P(A) = 0.6842$

The probability that a randomly selected respondent thought that Tiger Woods was the greatest active golfer is about 68.4%.

(b) $P(B \mid A) \approx 0.308$

There is about a 30.8% probability that the respondent is a golf fan, given that he/she felt that Tiger Woods is the greatest active golfer.

(c) $P(B^C \mid A) \approx 0.692$

There is about a 69.2% probability that the respondent is not a golf fan, given that he/she felt that Tiger Woods is the greatest active golfer.

23. (a) $P(A) = 0.3795$

There is a 37.95% probability that a randomly selected respondent felt that Shoeless Joe Jackson should be eligible for the Hall of Fame.

(b) $P(B \mid A) \approx 0.522$

The probability that the respondent is a baseball fan, given that he/she felt that Shoeless Joe Jackson should be eligible for the Hall of Fame, is about 52.2%.

(c) $P(B^C \mid A) \approx 0.478$

There is about a 47.8% probability that the respondent is not a baseball fan, given that he/she feels that Shoeless Joe Jackson should be eligible for the Hall of Fame.

25. (a) $P(A) = 0.4826$

The probability that the respondent is willing to pay higher prices for organic foods is 48.26%.

(b) $P(B \mid A) \approx 0.62$

The probability that the respondent eats organic foods regularly, given that he/she is willing to pay higher prices for organic foods, is about 62%.

(c) $P(B^C \mid A) \approx 0.38$

The probability that the respondent does not eat organic foods regularly, given that he/she is willing to pay higher prices for organic foods, is about 38%.

27. (a) $P(A) = 0.612$

There is a 61.2% probability that the respondent felt that kids whose parents are not involved in school sometimes get shortchanged.

(b) $P(B \mid A) \approx 0.624$

There is about a 62.4% probability that the respondent is a parent, given that he/she felt that kids whose parents are not involved in school sometimes get shortchanged.

(c) $P(B^C \mid A) \approx 0.353$

There is about a 35.3% probability that the respondent is a teacher, given that the respondent felt that kids whose parents are not involved is school sometimes get shortchanged.

29. $P(B_1) + P(B_2) = 1$

The events are collectively exhaustive.

31. $P(B_1) + P(B_2) \neq 1$

The events are not collectively exhaustive.

33. $P(B_1) + P(B_2) + P(B_3) = 1$

The events are collectively exhaustive.

35. $P(A) = 0.24$
$P(B_1|A) \approx 0.333$
$P(B_3|A) \approx 0.167$

37. $P(A) = \dfrac{37}{240}$
$P(B_1|A) = \dfrac{20}{37}$
$P(B_3|A) = \dfrac{12}{37}$

39. $P(A) = 0.22$
$P(B_1|A) = \dfrac{9}{22}$
$P(B_3|A) = \dfrac{1}{11}$

41. (b) $P(A) = 0.023$

The probability that the light fixture is defective is about 2.3%

(c) $P(B_2|A) \approx 0.217$

The probability that the light fixture came from factory II, given that the fixture is defective, is about 21.7%.

(d) $P(B_3|A) \approx 0.609$

The probability that the light fixture came from factory III, given that the light fixture is defective, is about 60.9%.

43. (b) $P(A) = 0.277$

There is a 27.7% probability that the randomly selected woman has never married or cohabitated.

(c) $P(B_2|A) \approx 0.196$

There is about a 19.6% probability that the woman is between the ages of 25 and 34, given that she has never married or cohabitated.

(d) $P(B_3|A) \approx 0.078$

There is about a 7.8% probability that the woman is between the ages of 35 and 44, given that she has never married or cohabited.

Chapter 6 Review Exercises

1. Statistical inference   **3.** Yes   **5.** Yes   **7.** Yes

9. No   **11.** No   **13.** No

15. (a) Answers may vary.   **(b)** Answers may vary.

17. $B = \{$New Hampshire, New Jersey, New Mexico, New York$\}$

19. $B = \{x \mid x$ is a state that starts with the letter A$\}$

21. Ø, {McKinley}, {Pikes Peak}, {Washington},
{McKinley, Pikes Peak}, {McKinley, Washington},
{Pikes Peak, Washington},
{McKinley, Pikes Peak, Washington}

23. $A \cap B = \{17, 19\}$

25. $(A \cup C)^C = \{21, 22, 23, 24, 25\}$ **27.** $A^C \cap B = \{6, 7, 8\}$

29. $A^C \cap (B \cup C) = \{2, 4\}$

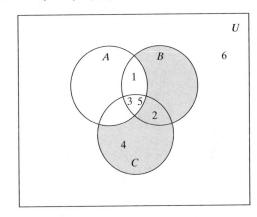

31. $n(A \cup C) = 25{,}508{,}000$; in 1997, 25,508,000 students were either in grades 9 to 12 at public and private schools or were at a public school in grades 9 to 12 and college.

33. 528 ways   **35.** 1024 ways   **37.** 30   **39.** 2.74×10^{17}

41. 70   **43.** 4.19×10^{15} **45.** 28 ways   **47.** 5,565,120 ways

49. $S = \{$A, B, C, D, E, F, G, H, I, J, K, L$\}$; $n(S) = 12$

51. $S = \{$(Yes, public), (Yes, private), (No, public), (No, private)$\}$;
$n(S) = 4$

53. $P(A) = \dfrac{1}{3}$; $P(B) = \dfrac{1}{6}$ **55.** $P(A) = \dfrac{1}{6}$; $P(B) = \dfrac{5}{36}$

57. (a) 2002 ways   **(b)** 560

(c) $\dfrac{560}{2002} \approx 0.2797$; there is about a 28% probability that a subcommittee is formed with 3 Democrats and 2 Republicans.

59. (a) About 0.763   **(b)** About 0.237   **61.** Classical

63. Not mutually exclusive   **65.** $\dfrac{25}{47} \approx 0.532$

67. $\dfrac{210}{400} = 0.525$; if we select a student from the survey group at random, the probability that the student is an upperclassman is 0.525 or 52.5%.

69. $\dfrac{320}{400} = 0.80$; if we select a student from the survey group at random, the probability that the student is not an upperclassman with season football tickets is 80%.

71. $\dfrac{150}{400} = 0.375$ **73.** 0.557   **75. (a)** $\dfrac{2}{3}$ **(b)** $\dfrac{3}{2}$

77. $\dfrac{2}{3}$ **79.** $\dfrac{1}{6}$

81. Dependent; if Rebecca's car breaks down, the chances that she will need to ride her bike to work are increased, so the events are dependent.

83. (a)

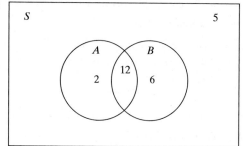

(b) $\frac{12}{25} = 0.48$; if one of the sampled students is selected at random, the probability that the student studied for at least 3 hours and received a grade of 80% or higher is 0.48 or 48%.

(c) $\frac{12}{14} \approx 0.857$; if one of the sampled students is selected at random, the probability that the student received a grade of 80% or higher, given that the student studied at least 3 hours, is about 85.7%.

85. $P(A \cap C) = \frac{15}{105} \approx 0.143$; $P(C|A) = \frac{15}{40} = 0.375$

87. (a) $\frac{24}{91}$ **(b)** $\frac{15}{91}$

89. 0; if a person living in California or Texas is selected at random, the probability that the person lives in California and Texas is zero.

91. 0.5   **93.** $P(D|B) = \frac{10}{40} = 0.25$; $P(B|D) = \frac{10}{25} = 0.40$

95. (a) About 0.519; if a person who participated in the survey is selected at random, the probability that the person is a man, given that the person is afraid of heights, is about 0.519 or 51.9%.

(b) 0.624; if a person who participated in the survey is selected at random, the probability that the person is afraid of heights, given that the person is a man, is 0.624 or 62.4%.

97. $P(A) = \frac{2}{5}$; $P(B|A) = \frac{1}{4}$; $P(B^C|A) = \frac{3}{4}$

99. (a) About 0.570; if a respondent is selected at random, the probability that the respondent supports the attacks is about 0.570 or 57%.

(b) About 0.739

(c) About 0.261; if a respondent is selected at random, the probability that the respondent was not aware of the attacks prior to the survey, given that the respondent supports the attacks after being made aware of them, is about 0.261 or 26.1%.

101. Collectively exhaustive

103. $P(A) = 0.26$; $P(B_1|A) \approx 0.077$; $P(B_3|A) \approx 0.615$

105. (a)

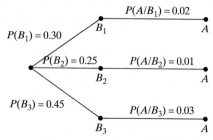

(b) 0.022; if a ski binding is selected at random, the probability that the binding has a defect is 0.022 or 2.2%.

(c) About 0.114

(d) About 0.614; if a ski binding is selected at random, the probability that the binding came from factory III, given that the binding is defective, is about 0.614 or 61.4%.

Chapter 7

Section 7.1

1. The data set is a population.   **3.** The data set is a sample.

5. The data set is a population.   **7.** The data set is a sample.

9. The population is all adult Americans; the sample is the 568 adult Americans who were surveyed.

11. The population is all baseball fans; the sample is the 467 baseball fans that were surveyed.

13. The population is all Internet users; the sample is the 19,026 respondents.

15. Qualitative   **17.** Quantitative   **19.** Qualitative

21.

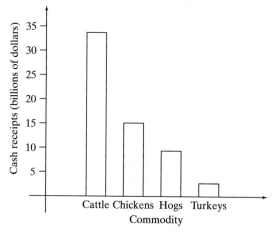

23.

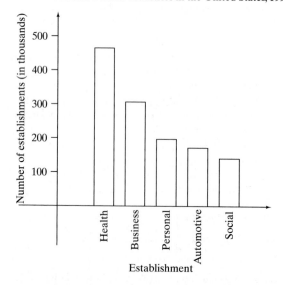

25.

Recorded Audio Media Shipped in the United States, 1990.

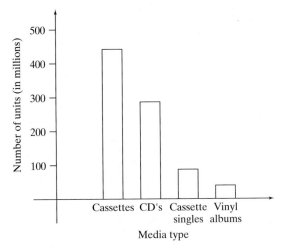

27. Books in various Subjects Imported to the United States, 1990.

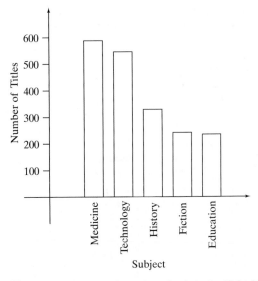

29. Foreign Students who came to Study in the United States, 1995.

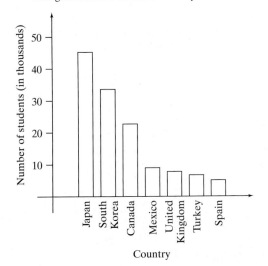

31. (a) United States, United Kingdom
   (b) Singapore, Germany

33. (a) Sulfur dioxide, carbon monoxide, nitrogen oxide and lead
   (b) There are about 740 more monitoring stations for particulates than there are for lead.

35. (a) About 65,000 households took trips for rest and relaxation.
   (b) About 40,000 more households took trips for rest and relaxation than households that took trips for sightseeing.

37. (a) In 1995, 58.6% of family debt was from home secured debt and installment.
   (b) In 1995, 63.6% of family debt was from credit card balances and installment.

39. (a) In 1996, buses were the least used mode of motorized transportation in France.
   (b) About 8.9% were motorcycles.

41. (a) Life scientist
   (b) About 39.6% of the scientists were social scientists.

43. (a) Premium cable
   (b) About 23.8% of the time was spent watching basic cable.

45. (a) About 45,000 gas station and convenience store crimes were reported in 1996.
   (b) About 12.4% of crimes reported in 1996 were residence robbery crimes.

Section 7.2

1. (a) The class width is 500.
   (b) In 1997, there were 10,680,000 single family homes with square footage from 2000 to 2499 square feet.

3. (a) The class width is 10.
   (b) The 75 to 84 age group had the smallest amount of financial debt in 1995 and the 45 to 54 age group had the largest amount of financial debt in 1995.

5. (a) In 1992, 908,000 farms were owned by people who were 55 or older.
   (b) In 1992, 990,000 farms were owned by people who were between 25 and 54 years of age.

7. (a) 204,700,000 people, aged 15 to 74 years were covered by private or government health insurance in 1997.
   (b) The age group from 35 to 44 years had the highest frequency of Americans covered by private or government health insurance in 1997.

9. (a) 4,755,000 northeastern homeowners had monthly housing costs of less than $500.
   (b) 4,648,000 northeastern homeowners had monthly housing costs of $300 or more but less than $700.

11. (a)

Class Boundaries
500.5 to 999.5
999.5 to 1499.5
1499.5 to 1999.5
1999.5 to 2499.5
2499.5 to 2999.5
2999.5 to 3499.5
3499.5 to 3999.5

(b)

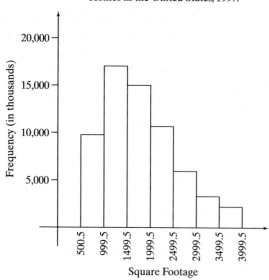

Square Footage for Single-Family Homes in the United States, 1997.

(b)

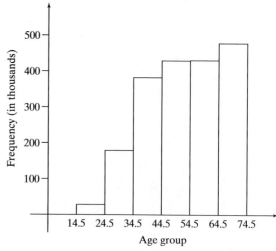

Number of Farms Owned by Americans, 1992.

13. (a)

Class Boundaries
24.5 to 34.5
34.5 to 44.5
44.5 to 54.5
54.5 to 64.5
64.5 to 74.5
74.5 to 84.5

17. (a)

Class Boundaries
14.5 to 24.5
24.5 to 34.5
34.5 to 44.5
44.5 to 54.5
54.5 to 64.5
64.5 to 74.5

(b)

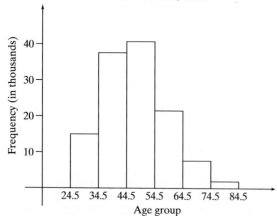

Financial Debt Held by Families in the United States, 1995.

(b)

Class	Frequency	Relative Frequency
15 to 24	32.3	0.16
25 to 34	39.3	0.19
35 to 44	44.5	0.22
45 to 54	34.1	0.17
55 to 64	22.3	0.11
65 to 74	32.1	0.16

(c)

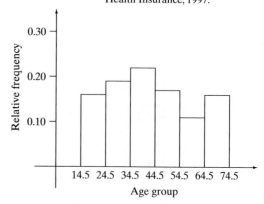

Relative Frequency of Americans, Aged 15 to 74, Covered by either Private or Government Health Insurance, 1997.

15. (a)

Class Boundaries
14.5 to 24.5
24.5 to 34.5
34.5 to 44.5
44.5 to 54.5
54.5 to 64.5
64.5 to 74.5

19. (a)

Class Boundaries
199.5 to 299.5
299.5 to 399.5
399.5 to 499.5
499.5 to 599.5
599.5 to 699.5
699.5 to 799.5
799.5 to 899.5
899.5 to 999.5

(b)

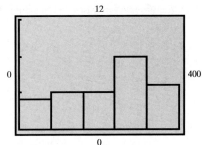

(b)

Class	Frequency	Relative Frequency
200 to 299	1923	0.23
300 to 399	1493	0.17
400 to 499	1339	0.16
500 to 599	967	0.11
600 to 699	849	0.10
700 to 799	765	0.09
800 to 899	482	0.06
900 to 999	722	0.08

(c)

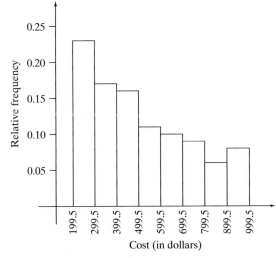

Relative Frequency of Monthly Housing Costs under 1000 for Homeowners in the Northeastern United States

23. (a)

Class Limits	Frequency
75 to 197	18
198 to 320	6
321 to 443	0
444 to 566	1
567 to 689	2
690 to 813	2

(b)

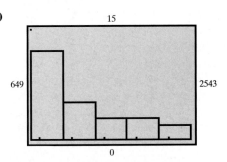

25. (a)

Class Limits	Frequency
649 to 1027	12
1028 to 1406	5
1407 to 1785	3
1786 to 2164	3
2165 to 2543	2

(b)

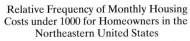

21. (a)

Class Limits	Frequency
0 to 79	4
80 to 159	5
160 to 239	5
240 to 319	10
320 to 400	6

27. (a)

Class	Midpoint	Frequency
20 to 29	24.5	15
30 to 39	34.5	50
40 to 49	44.5	56
50 to 59	54.5	30
60 to 69	64.5	18

(b)

Employee Ages at New Finish Corporation

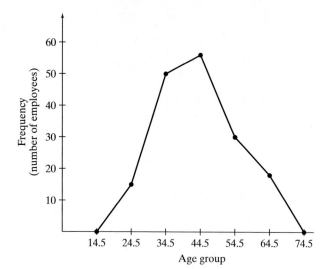

(b)

Square Footage of Single-family Homes in the United States, 1997.

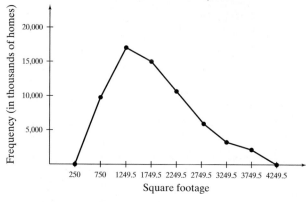

29. (a)

Class	Midpoint	Frequency
59 to 61	60	4
62 to 64	63	8
65 to 67	66	12
68 to 70	69	13
71 to 73	72	21
74 to 76	75	15
77 to 79	78	12
80 to 82	81	9
83 to 85	84	4
86 to 88	87	2

33. (a)

Class	Midpoint
25 to 34	29.5
35 to 44	39.5
45 to 54	49.5
55 to 64	59.5
65 to 74	69.5
75 to 84	79.5

(b)

Weights of 100 Sixth-graders at P. H. Johnson Elementary School

(b)

Average Amount of United States Family Debt, 1995.

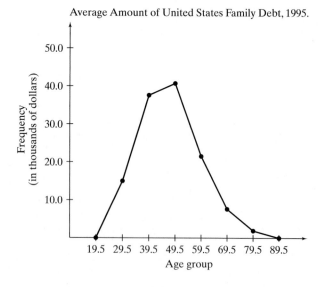

31. (a)

Class	Midpoint
501 to 999	750
1000 to 1499	1249.5
1500 to 1999	1749.5
2000 to 2499	2249.5
2500 to 2999	2749.5
3000 to 3499	3249.5
3500 to 3999	3749.5

35. (a)

Class	Cumulative Frequency
20 to 29	15
30 to 39	65
40 to 49	121
50 to 59	151
60 to 69	169

(b)

Employees Ages at New Finish Corporation

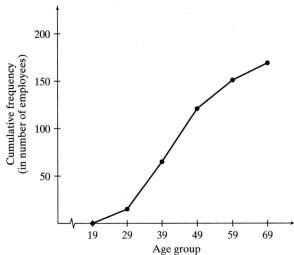

37. (a)

Class	Cumulative Frequency
59 to 61	4
62 to 64	12
65 to 67	24
68 to 70	37
71 to 73	58
74 to 76	73
77 to 79	85
80 to 82	94
83 to 85	98
86 to 88	100

(b)

Weights of 100 Sixth-graders at P. H. Johnson Elementary School

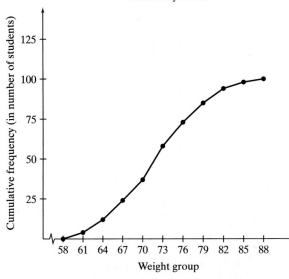

39. (a)

Class (age group)	Cumulative Frequency
15 to 24	21
25 to 34	149
35 to 44	520
45 to 54	987
55 to 64	1414
65 to 74	1911

(b)

Farm Ownership by Americans, 1997.

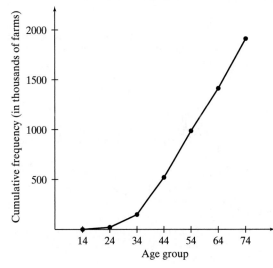

41. The 24.5 to 29.5 age group had the most newborns.

43. The class width is 5.

45. The 24.5 to 29.5 age group had the most newborns.

47. The class width is 5.   **49.** The class width is 3.

51. The second class (parents of children ages 4 to 6) spent over $1300 annually on child care and education.

53. The class width is 10.

55. In the seventh and eighth classes, the number of deaths caused by cancer is close to 160,000; this corresponds to the age groups of 65 to 74 and 75 to 84, respectively.

57. The lower limit of the first class is 15 and the upper limit is 24.

59. About 25,300 drivers aged 34.5 years and below were involved in alcohol-related fatal crashes in 1998.

61. The lower limit of the first class is 5 and the upper limit is 14.

63. About 54,300 individuals aged 54.5 years and below died from accidental and adverse effects in 1997.

Section 7.3

1. $\bar{x} \approx 6.17$ **3.** $\bar{x} = 3.6$ **5.** $\bar{x} \approx 11.3$
$M_0 = 6$ $M_0 = 1$ and 4   There is no mode.

7. $\bar{x} = 15.16$ **9.** $\bar{x} = 0.5$ **11.** $\mu = 5.7$
$M_0 = 11.2$ and 15.8   There is no mode.

13. $\mu \approx 154.7$ **15.** $\mu = 54.0375$ **17.** $M = 5$
$MR = 7$

19. $M = 5$ **21.** $M = 19.5$ **23.** $M = 0.44$
$MR = 6$ $MR = 20$ $MR = 0.51$

25. $M = 205$ **27.** The mode is medium.
$MR = 353$

29. Blue and brown are the modes.   **31.** $\bar{x} \approx 6.64$

33. $\bar{x} = 30$ **35.** $\bar{x} = \frac{94.5}{25} = 3.78$ **37.** $\bar{x} \approx 93.3$

39. Symmetric   **41.** Skewed left

43. Skewed left   **45.** Symmetric

47. (a) $\bar{x} = 40.3$
In May 2000, the average number of visits to the top six (most visited) Web sites was 40.3 million.
(b) $M = 40.7$
In May 2000, the median number of visits to the top six (most visited) web sites was 40.7 million.

49. (a) $\bar{x} \approx 93.2$
There was an average of 93.2 billion barrels of oil reserves in the top nine countries (with the largest oil reserves) in 1999.
(b) $M_0 = 93$
The most frequently occurring amount of oil reserves is 93 million barrels.

51. (a) $M = 8.3$
In 1998, in the top 15 countries in terms of personal computer usage, the median number of PCs in use was 8.3 million.
(b) $MR = \frac{129 + 4.6}{2} = 66.8$
In 1998, in the top 15 countries in terms of personal computer usage, the midrange number of computers was 66.8 million.

53. (a) $n = 95.5$
This means that in 1997 the number of emergency room visits in the United States by people aged 5 to 84 totaled about 95.5 million.
(b) The mean is about 38.73. This means that the average age of people going to emergency rooms in the United States in 1997 was about 38.73 years.

55. (a) $\bar{x} \approx 37.76$
In 1997, the average age of a U.S. prisoner sentenced to death was about 37.76 years.
(b) In general, the data are skewed right.

Section 7.4

1. $R = 6$ **3.** $R = 8$ **5.** $R = 17$ **7.** $R = 10.5$

9. $R = 0.82$ **11.** $\sigma \approx 3.07$ **13.** $\sigma \approx 22.63$

15. $\sigma \approx 14.42$ **17.** $s \approx 3.11$ **19.** $s \approx 2.34$

21. $s \approx 6.48$ **23.** $s \approx 0.247$ **25.** $s \approx 92.3$

27. $s \approx 3.65$ **29.** $s \approx 10.3$ **31.** $s \approx 0.55$

33. $s \approx 36.3$

35. (a) $R = 691$
The range of selected bodies of water in the Pacific Coast region is 691 square miles.
(b) $s \approx 326.71$
This means that the areas of these selected bodies of water varied on average by about 326.71 square miles from the mean area of 328.5 square miles.

37. $\sigma \approx 9$
The ages of the U.S. prisoners sentenced to death in 1980 varied on average by about 9 years from the mean age of about 30.5 year.

39. $\sigma \approx 21.94$
This means that the ages of people who visited emergency rooms in the United States in 1997 varied on average by about 21.94 years from the mean age of 38.73 years.

41. $SK \approx -6.43$
The distribution is skewed left.

43. $SK = 0$
The distribution is symmetric.

45. $SK = 0.25$
The distribution is skewed right.

Chapter 7 Review Exercises

1. The population consists of the responses, or possible responses, of all American adults. The sample is made up of the 1060 American adults who participated in the survey.

3. Quantitative   **5.** Qualitative

7.

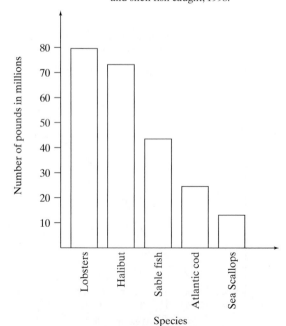

Number of pounds of domestic fish and shell fish caught, 1998.

9. (a) Bachelor's, master's, doctorate
   (b) About $40,000

11. (a) About 200,000
   (b) About 650,000

13. (a) Dangerous nonnarcotic drugs
   (b) 77.0%

15. (a) 10   **(b)** 21.9 million adults

17. (a) 24.5 to 34.5, 34.5 to 44.5, 44.5 to 54.5, 54.5 to 64.5, 64.5 to 74.5
   (b)

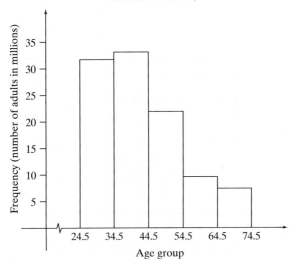

Number of adults who attended
at least one movie, 1996.

19. (a)

Class (number of minutes exercising)	Frequency, f (number of adults)
0 to 17	10
18 to 35	7
36 to 53	6
54 to 71	3
72 to 89	1
90 to 107	1
108 to 125	2

   (b)

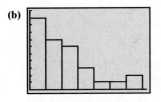

21. (a)

Class (age group)	Class midpoint, m
25 to 34	29.5
35 to 44	39.5
45 to 54	49.5
55 to 64	59.5
65 to 74	69.5

   (b)

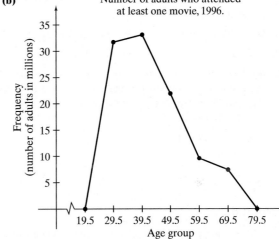

Number of adults who attended
at least one movie, 1996.

23. (a)

Class (age group)	Cumulative Frequency
25 to 34	31.7
35 to 44	64.8
45 to 54	86.7
55 to 64	96.3
65 to 74	103.7

   (b)

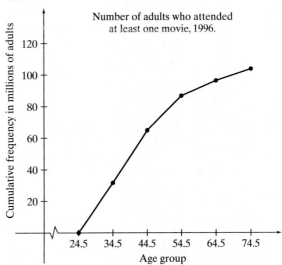

Number of adults who attended
at least one movie, 1996.

25. 10   **27.** 55 to 64   **29.** 10   **31.** 5.1; 2, 8   **33.** 133

35. 18.5; 20.5   **37.** First   **39.** Approximately 8.39

41. Skewed right

43. (a) Approximately 381.9; the average domestic gross, in millions, per film of the top six summer films of all time was about $381.9 million.

(b) 378.45; the median domestic gross, in millions, of the top six summer films of all time was $378.45 million. Half of the top six summer films grossed more than $378.45 million and half grossed less than $378.45 million.

45. (a) Approximately 15.67; an estimate of the average age of a 10 to 18 year-old arrested in 1999 was 15.67 years old.

(b) Skewed left

47. 38   **49.** Approximately 4.20   **51.** Approximately 2.87

53. Approximately 4.87   **55.** Approximately 4.48

57. (a) 334; of the seven counties listed in the table, the difference between the most and least populated counties is 334,000.

(b) Approximately 116.26; the population of the seven counties varied on average by about 116,260 from the mean.

59. Approximately 19.46; skewed right

Chapter 8

Section 8.1

1. $P(X = x)$ meets both criteria; therefore, it is a probability mass function.

3. Since $P(X = x)$ meets both criteria, it is a probability mass function.

5. Both criteria are met; therefore, it is a probability mass function.

7. Since $P(X = 10) = -0.1$, $P(X = x)$ is not a probability mass function.

9. Since $P(X = x)$ meets both criteria, it is a probability mass function.

11. (a)

$X = x$	$P(X = x) = \frac{1}{6}x$
$X = 0$	$P(X = 0) = 0$
$X = 1$	$P(X = 1) = \frac{1}{6}$
$X = 2$	$P(X = 2) = \frac{1}{3}$
$X = 3$	$P(X = 3) = \frac{1}{2}$

(b) Both of the criteria have been met, so $P(X = x)$ is a probability mass function.

13. (a)

$X = x$	$P(X = x) = \frac{1}{4}$
$X = 1$	$P(X = 1) = \frac{1}{4}$
$X = 2$	$P(X = 2) = \frac{1}{4}$
$X = 3$	$P(X = 3) = \frac{1}{4}$
$X = 4$	$P(X = 4) = \frac{1}{4}$

(b) Both criteria have been met, so $P(X = x)$ is a probability mass function.

15. (a)

$X = x$	$P(X = x) = \frac{1}{15}x^2$
$X = 0$	$P(X = 0) = 0$
$X = 1$	$P(X = 1) = \frac{1}{15}$
$X = 2$	$P(X = 2) = \frac{4}{15}$
$X = 3$	$P(X = 3) = \frac{9}{15}$
$X = 4$	$P(X = 4) = \frac{16}{15}$

(b) Since $P(X = 4) > 1$, $P(X = x)$ is not a probability mass function.

17. (a)

$X = x$	$P(X = x) = 0.05x$
$X = 2$	$P(X = 2) = 0.10$
$X = 3$	$P(X = 3) = 0.15$
$X = 4$	$P(X = 4) = 0.20$
$X = 6$	$P(X = 6) = 0.30$

(b) Both criteria have not been met, so $P(X = x)$ is not a probability mass function.

19. (a)

$X = x$	$P(X = x) = \frac{x+2}{10}$
$X = 0$	$P(X = 0) = \frac{1}{5}$
$X = 1$	$P(X = 1) = \frac{3}{10}$
$X = 2$	$P(X = 2) = \frac{2}{5}$
$X = 3$	$P(X = 3) = \frac{1}{2}$

(b) Both criteria have not been met, so $P(X = x)$ is not a probability mass function.

21. $k = 6$ **23.** $k = \frac{1}{50}$ **25.** $k = 21$

27.

$X = x$	$P(X = x)$
$X = 1$	$P(X = 1) = \frac{5}{20} = 0.25$
$X = 2$	$P(X = 2) = \frac{5}{20} = 0.25$
$X = 3$	$P(X = 3) = \frac{10}{20} = 0.50$

29.

$X = x$	$P(X = x)$
$X = 1$	$P(X = 1) = \frac{10}{75} \approx 0.13$
$X = 2$	$P(X = 2) = \frac{20}{75} \approx 0.27$
$X = 3$	$P(X = 3) = \frac{15}{75} = 0.20$
$X = 4$	$P(X = 4) = \frac{5}{75} \approx 0.07$
$X = 5$	$P(X = 5) = \frac{25}{75} \approx 0.33$

31.

$X = x$	$P(X = x)$
$X = 0$	$P(X = 0) = \frac{150}{1350} \approx 0.11$
$X = 1$	$P(X = 1) = \frac{600}{1350} \approx 0.44$
$X = 2$	$P(X = 2) = \frac{400}{1350} \approx 0.30$
$X = 3$	$P(X = 3) = \frac{200}{1350} \approx 0.15$

33. (a)

$P(X=x)$

(b) $P(1 \leq X \leq 2) = 0.80$

The probability that a customer will rent one or two videos is 80%.

35. (a)

$P(X=x)$

(b) $P(2 \leq X \leq 4) = 0.7$

The probability that 2, 3, or 4 women are hired is 70%.

37. (a)

$P(X=x)$

(b) $P(X \leq 3) = 0.80$

There is an 80% probability that the number of commercials aired per show during after school television will be 3 or fewer.

39. (a) $P(X \leq 14) = 0.65$

The probability that 12, 13, or 14 students will drop out of introductory psychology during any given quarter is 65%.

(b) 12 is the least likely number of students to drop out of introductory psychology during any given quarter.

41. (a) $P(X \geq 4) = 0.62$

The probability that the size of the family will be at least 4 is 62%.

(b) 7 is the least likely size that the chosen family will be.

43. (a)

$X = x$	$P(X = x)$
0	$P(X=0) = \frac{7800}{19,000} \approx 0.41$
1	$P(X=1) = \frac{5500}{19,000} \approx 0.29$
2	$P(X=2) = \frac{3200}{19,000} \approx 0.17$
3	$P(X=3) = \frac{2500}{19,000} \approx 0.13$

(b) $P(X=3) \approx 0.13$

There is about a 13% probability that a randomly selected student has completed 3 years of college.

45. (a)

$X = x$	$P(X = x)$
$X=0$	$P(X=0) = \frac{53}{665} \approx 0.08$
$X=1$	$P(X=1) = \frac{127}{665} \approx 0.19$
$X=2$	$P(X=2) = \frac{280}{665} \approx 0.42$
$X=3$	$P(X=3) = \frac{128}{665} \approx 0.19$
$X=4$	$P(X=4) = \frac{77}{665} \approx 0.12$

(b) $P(X=3) \approx 0.19$

There is about a 19% probability that 3 passengers will miss the overnight flight from Tokyo to Honolulu.

47. (a)

$X = x$	$P(X = x)$
$X=16$	$P(X=16) = \frac{1.6}{24.2} \approx 0.07$
$X=17$	$P(X=17) = \frac{2.2}{24.2} \approx 0.09$
$X=18$	$P(X=18) = \frac{2.4}{24.2} \approx 0.10$
$X=19$	$P(X=19) = \frac{2.7}{24.2} \approx 0.11$
$X=20$	$P(X=20) = \frac{2.8}{24.2} \approx 0.12$
$X=21$	$P(X=21) = \frac{2.9}{24.2} \approx 0.12$
$X=22$	$P(X=22) = \frac{2.9}{24.2} \approx 0.12$
$X=23$	$P(X=23) = \frac{3.1}{24.2} \approx 0.13$
$X=24$	$P(X=24) = \frac{3.6}{24.2} \approx 0.15$

(b) $P(X=20) \approx 0.12$

There is about a 12% probability that the driver involved in the accident was 20 years old.

(c) $P(X \leq 20) = 0.49$

There is a 49% probability that the driver involved in the accident was 20 years old or younger.

49. (a) $P(X = 2) = 0.197$

The probability that at least one parent participated in 2 school activities is 19.7%.

(b) $P(X \geq 2) = 0.891$

There is an 89.1% probability that at least one parent participated in 2 or more school activities.

Section 8.2

1. $E(X) = 0.75$ **3.** $E(X) = 2.65$ **5.** $E(X) = 0.34$

7. $E(X) = 29$ **9.** $E(X) = 10.25$ **11.** $E(X) = 1$

13. $E(X) = 4.2$ **15.** $E(X) = \frac{7}{3}$ **17.** $E(X) = 3.75$

19. $E(X) = 2.5$ **21.** $E(X) = \frac{18}{7}$ **23.** $E(X) = 4.5$

25. $E(X) = 1$ **27.** $\sigma_X = \sqrt{0.6875} \approx 0.829$

29. $\sigma_X = \sqrt{1.4275} \approx 1.195$ **31.** $\sigma_X = \sqrt{69} \approx 8.31$

33. $\sigma_X = \sqrt{0.556} \approx 0.745$ **35.** $\sigma_X = \sqrt{1.25} \approx 1.12$

37. $\sigma_X = \sqrt{0.388} \approx 0.623$ **39.** $\sigma_X = \sqrt{\frac{73}{70}} \approx 1.02$

41. $E(X) = 1.55$

The average number of video rentals per customer is 1.55 videos.

43. (a) $E(X) = 2.25$

The average number of commercials per show aired during after school television is 2.25.

(b) The most frequently occurring number of commercials on any given show is 2.

45. The number of students expected to drop out of introductory psychology is 14.14 students.

47. $E(X) = -8$

The expected value of one ticket is $-\$8$.

49. (a) $E(X) \approx 1.53$

The number of inspections required to discover a defective spring is expected to be 1.53.

(b) $\sigma_X \approx 0.72$

The number of inspections required to discover a defective spring varied an average by about 0.72 from the mean of 1.53.

51. (a) $E(X) = 178.81$

The expected average weight of an American male aged 70 to 79, from 1988 to 1994, is 178.81 pounds.

(b) $\sigma_X \approx 31$

The expected average weight of an American male aged 70 to 79 varied on average by approximately 31 pounds from the mean of 178.81 pounds.

53. The insurance company's expected gain is $-\$1624$.

55. (a) The expected value of Site I is $7 million. The expected value of Site II is $4 million.

(b) Based on part (a), advise the company to select Site I.

(c) Site I has less risk.

57. (a) $P(X = x) = \frac{1}{10}, x = 1, 2, 3, 4, 5, 6, 7, 8, 9, 10$

(b) $\mu_X = 5.5$

The expected number of times that the printing machine is expected to break down during a shift is 5.5 times.

(c) $\sigma_X \approx 2.87$

The expected number of times that the machine breaks down varied on average by about 2.87 from the mean of 5.5.

(d) $P(X \leq 3) = 0.3$

There is a 30% probability that during any given shift the machine will break down 3 or fewer times.

59. (a) $P(X = x) = \frac{1}{25}, \quad 1 \leq x \leq 25$

(b) $\mu_X = 13$

The expected hardness of the plastic pieces manufactured is 13 on the Rockford hardness scale.

(c) $\sigma_X \approx 7.21$

The expected hardness of the plastic pieces varied on average by about 7.21 from the mean of 13.

(d) $P(X \geq 20) = 0.24$

The probability that a plastic piece, selected at random, will have a rating on the Rockford hardness scale of more than 20 is 24%.

Section 8.3

1. $P(X = 2) = 0.375$ **3.** $P(X = 0) = 0.32768$

5. $P(X = 6) \approx 0.2731$ **7.** $P(X = 5) \approx 0.0264$

9. $P(X = 10) \approx 0.1762$

11. This is a binomial experiment.

$p = 0.50 \qquad n = 30$

13. This is not a binomial experiment. There are not exactly two outcomes.

15. This is a binomial experiment.

$p = 0.90 \qquad n = 12$

17. This is not a binomial experiment. There is not a definite number of trials.

19. This is a binomial experiment.

$p = 0.20 \qquad n = 150$

21. (a)

$X = x$	$P(X = x)$
$X = 0$	$_3C_0\left(\frac{1}{2}\right)^0\left(1 - \frac{1}{2}\right)^{3-0} = 0.125$
$X = 1$	$_3C_1\left(\frac{1}{2}\right)^1\left(1 - \frac{1}{2}\right)^{3-1} = 0.375$
$X = 2$	$_3C_2\left(\frac{1}{2}\right)^2\left(1 - \frac{1}{2}\right)^{3-2} = 0.375$
$X = 3$	$_3C_3\left(\frac{1}{2}\right)^3\left(1 - \frac{1}{2}\right)^{3-3} = 0.125$

(b) $P(X = x)$

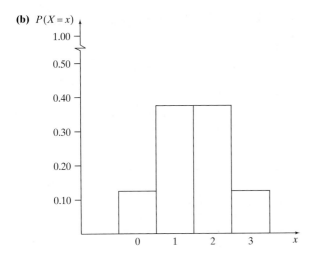

23. (a)

$X = x$	$P(X = x)$
$X = 0$	$_5C_0(0.3)^0(1 - 0.3)^{5-0} = 0.16807$
$X = 1$	$_5C_1(0.3)^1(1 - 0.3)^{5-1} = 0.36015$
$X = 2$	$_5C_2(0.3)^2(1 - 0.3)^{5-2} = 0.3087$
$X = 3$	$_5C_3(0.3)^3(1 - 0.3)^{5-3} = 0.1323$
$X = 4$	$_5C_4(0.3)^4(1 - 0.3)^{5-4} = 0.02835$
$X = 5$	$_5C_5(0.3)^5(1 - 0.3)^{5-5} = 0.00243$

(b) $P(X = x)$

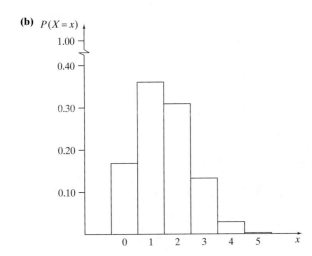

25. (a)

$X = x$	$P(X = x)$
$X = 0$	$_5C_0(0.7)^0(1 - 0.7)^{5-0} = 0.00243$
$X = 1$	$_5C_1(0.7)^1(1 - 0.7)^{5-1} = 0.02835$
$X = 2$	$_5C_2(0.7)^2(1 - 0.7)^{5-2} = 0.1323$
$X = 3$	$_5C_3(0.7)^3(1 - 0.7)^{5-3} = 0.3087$
$X = 4$	$_5C_4(0.7)^4(1 - 0.7)^{5-4} = 0.36015$
$X = 5$	$_5C_5(0.7)^5(1 - 0.7)^{5-5} = 0.16807$

(b) $P(X = x)$

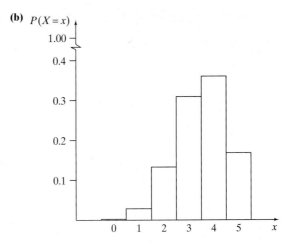

27. b   **29.** c   **31.** f

33. $E(X) = \mu_X = 6$
$\sigma_x = \sqrt{3} \approx 1.732$

35. $E(X) = \mu_X = 5$
$\sigma_x = \sqrt{3.75} \approx 1.94$

37. $E(X) = \mu_X = 1.5$
$\sigma_x = \sqrt{1.125} \approx 1.06$

39. $E(X) = \mu_X = 2.25$
$\sigma_x = \sqrt{\frac{63}{32}} \approx 1.40$

41. (a) There are two outcomes: a prospective employee either passes or does not pass the exam. The probability is the same (0.40). There are a definite number of trials (7), which are independent of each other.

(b) $n = 7$, $p = 0.40$
$P(X = 5) \approx 0.077$
There is a 7.7% probability that 5 prospective employees will pass the exam.

43. (a) $P(X = 4) \approx 0.213$
The probability that 4 of the 12 sweepstakes entrants also ordered a magazine is about 21.3%.

(b) $P(4 \leq X \leq 5) \approx 0.44$
The probability that 4 or 5 of the 12 sweepstakes entrants also ordered a magazine is about 44%.

45. (a) $P(X = 3) \approx 0.24$
There is about a 24% probability that 3 of the 15 Republicans responded Abraham Lincoln.

(b) $P(X \leq 3) \approx 0.535$
The probability that 3 or fewer of the 15 Republicans responded Abraham Lincoln is about 53.5%.

47. (a) $P(X = 7) \approx 0.267$
The probability that 7 of the 8 households had cable is abut 26.7%.

(b) $P(X \leq 5) \approx 0.320$
The probability that at most 5 of the 8 households had cable is about 32%.

49. (a) There is about a 34.8% probability that one of the 50 people was unemployed.

(b) There is about a 59% probability that at most one of the 50 people was unemployed.

51. (a) The probability that 3 of the 13 were cigar smokers is about 20.6%.

(b) The probability that at most 3 of the 13 were cigar smokers is about 38%.

53. (a) $P(X = x)$

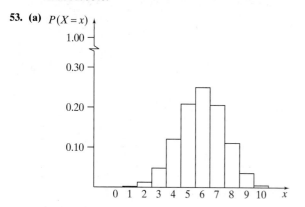

(b) $\mu_X = 5.9$

Out of 10 African Americans in New York City it is expected that 5.9 will not have an automobile.

$\sigma_X = \sqrt{2.419} \approx 1.56$

As the experiment is repeated, in samples of 10, the number of African Americans in New York without a car varies on average by about 1.56 from the expected value of 5.9.

55. (a) $\mu_X = 9.8$

It is expected, out of 20 prisoners, that 9.8 of them will not have high school diplomas.

(b)

$X = x$	$P(X = x)$
$X = 0$	$P(X = 0) \approx 0.0000014$
$X = 1$	$P(X = 1) \approx 0.000027$
$X = 2$	$P(X = 2) \approx 0.00025$
$X = 3$	$P(X = 3) \approx 0.00143$
$X = 4$	$P(X = 4) \approx 0.00585$
$X = 5$	$P(X = 5) \approx 0.01799$
$X = 6$	$P(X = 6) \approx 0.04321$
$X = 7$	$P(X = 7) \approx 0.08302$
$X = 8$	$P(X = 8) \approx 0.12962$
$X = 9$	$P(X = 9) \approx 0.16605$
$X = 10$	$P(X = 10) \approx 0.17549$
$X = 11$	$P(X = 11) \approx 0.15328$
$X = 12$	$P(X = 12) \approx 0.11045$
$X = 13$	$P(X = 13) \approx 0.06531$
$X = 14$	$P(X = 14) \approx 0.03137$
$X = 15$	$P(X = 15) \approx 0.01206$
$X = 16$	$P(X = 16) \approx 0.00363$
$X = 17$	$P(X = 17) \approx 0.00082$
$X = 18$	$P(X = 18) \approx 0.00013$
$X = 19$	$P(X = 19) \approx 0.000013$
$X = 20$	$P(X = 20) \approx 0.00000064$

The most likely number of prisoners to not have a high school diploma is 10.

57. (a) $E(X) = \mu_X = 9.78$

Out of 30 women, selected at random, who gave birth in 1994, 9.78 of them are expected to be unmarried.

(b) $P(X \le 10) \approx 0.618$

There is about a 61.8% probability that at most 10 of the 30 women were unmarried.

Section 8.4

1. (a) For the Jupiter, $z = 1$. For the Atlantis, $z \approx 0.608$

(b) The Jupiter got better mileage relative to the rest of the cars of the same make.

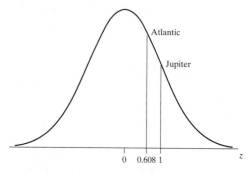

3. (a) 1999–2000: $z \approx -1.51$

2000–2001: $z \approx -1.26$

(b) The student who took the test in 2000–2001 scored better on the test relative to the rest of the students who took the test that year.

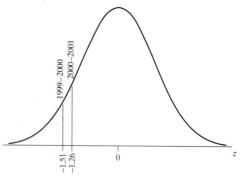

5. (a) MegaStar: $z = 2.5$. HemelGraph: $z = 2.3$

(b) HemelGraph's sale division traveled less, relative to the rest of the months.

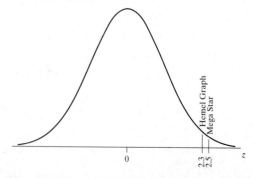

7. 0.4842 **9.** 0.2371 **11.** 0.0721 **13.** 0.5759

15. 0.3894 **17.** 0.4332 **19.** 0.2422 **21.** 0.6198

23. 0.2807 **25.** 0.3191 **27.** $P(Z > 0) = 0.5000$

29. $P(Z > 1.30) = 0.0968$ **31.** $P(Z \leq -1.76) = 0.0392$

33. $P(Z \geq -0.73) = 0.7673$ **35.** $P(Z < 1.96) = 0.9750$

37. $P(-0.50 < Z < 0.50) = 0.3830$

39. $P(-0.91 \leq Z \leq -0.12) = 0.2708$

41. (a) $z = \frac{15-15}{0.15} = 0$

$P(X > 15) = P(Z > 0) = 0.50$

There is a 50% probability that the shuttle bus will have made more than 15 runs from the airport to the local hotels.

(b) $P(X < 15) = P(Z < 0) = 0.50$

There is a 50% probability that the shuttle bus will have made less than 15 runs from the airport to the local hotels.

43. (a) $P(0.12 \leq X \leq 0.16) = P(0 \leq Z \leq 2) = 0.4772$

The probability that the wire is between 0.12 and 0.16 inch is about 47.72%.

(b) $P(0.08 \leq X \leq 0.12) = P(-2 \leq Z \leq 0) = 0.4772$

The probability that the wire is between 0.08 and 0.12 inch is about 47.72%.

45. (a) $P(X \leq 260) \approx P(Z \leq 0.52) = 0.50 + 0.1985 = 0.6985$

The probability that the player will weigh less than 260 pounds is about 69.85%.

(b) $P(200 \leq X \leq 260) \approx P(-0.71 \leq Z \leq 0.52)$
$= 0.2611 + 0.1985 = 0.4596$

The probability that the player will weigh between 200 and 260 pounds is about 45.96%.

47. (a) $P(X \leq 30) \approx P(Z \leq -0.77) = 0.2206$

The probability that the average monthly temperature was below 30°F is about 22%.

(b) $P(30 \leq X \leq 40) \approx P(-0.77 \leq Z \leq 1.31) = 0.6843$

The probability that the average monthly temperature was between 30° and 40°F is about 68.43%.

49. (a) $P(1.10 \leq X \leq 1.50) \approx P(0.22 \leq Z \leq 1.09) = 0.275$

There is about a 27.5% probability that the selected minimum wage earner made between $1.10 and $1.50 per hour.

(b) $P(0.90 \leq X \leq 1.50) \approx P(-0.22 \leq Z \leq 1.09) = 0.4492$

There is about a 45% probability that the selected minimum wage earner made between $0.90 and $1.50 per hour.

51. (a) $P(X \geq 40) \approx P(Z \geq -0.31) = 0.6217$

The probability that the patient is at least 40 years old is about 62.17%.

(b) $X = 50:$ $z = \frac{50-45.73}{18.68} \approx 0.23$

$P(0 \leq X \leq 50) \approx P(-2.45 \leq Z \leq 0.23) = 0.5839$

The probability that the patient is at most 50 years old is about 58.39%.

53. (a) $P(16 \leq X \leq 19) \approx P(-0.54 \leq Z \leq -0.40) = 0.05$

The probability that the patient was between 16 and 19 years old is about 5%.

(b) $P(20 \leq X \leq 40) \approx P(-0.35 \leq Z \leq 0.60) = 0.3625$

The probability that the patient was between 20 and 40 years old is about 36.25%.

55. (a) $P(X \geq 85) \approx P(Z \geq 0.45) = 0.3264$

The probability that the resident was older than 85 is about 32.64%.

(b) $X = 72:$ $z = \frac{72-81.77}{7.25} \approx -1.35$

$P(X \geq 72) \approx P(Z \geq -1.35) = 0.9155$

The probability that the resident was older than 72 is about 91.55%.

Chapter 8 Review Exercises

1. Does not meet criteria

3. Does not meet criteria

5. (a)

$X = x$	$P(X = x)$
$X = 1$	$P(X = 1) = \frac{1}{30}$
$X = 2$	$P(X = 2) = \frac{2}{15}$
$X = 3$	$P(X = 3) = \frac{3}{10}$
$X = 4$	$P(X = 4) = \frac{8}{15}$

(b) Does meet criteria

7. $k = 56$

9.

$X = x$	$P(X = x)$
$X = 1$	$P(X = 1) = \frac{10}{54} \approx 0.185$
$X = 2$	$P(X = 2) = \frac{12}{54} \approx 0.222$
$X = 3$	$P(X = 3) = \frac{18}{54} \approx 0.333$
$X = 4$	$P(X = 4) = \frac{14}{54} \approx 0.259$

11. (a)

(b) 0.65; the probability that a randomly selected class at Salem High School will have between 26 and 28 students inclusive is 0.65 or 65%.

13. (a) 0.63   **(b)** 5 memberships

15. (a)

$X = x$	$P(X = x)$
$X = 1997$	$P(X = 1997) = \frac{56}{244} \approx 0.230$
$X = 1998$	$P(X = 1998) = \frac{51}{244} \approx 0.209$
$X = 1999$	$P(X = 1999) = \frac{58}{244} \approx 0.238$
$X = 2000$	$P(X = 2000) = \frac{79}{244} \approx 0.324$

(b) 0.324; the probability that a person selected at random from the group of people who were involved in unprovoked shark attacks between 1997 and 2000 was involved in an attack in 2000 is 0.324 or 32.4%.

(c) 0.561; the probability that a person selected at random from the group of people who were involved in unprovoked shark attacks between 1997 and 2000 was involved in an attack in 1999 or 2000 is 0.561 or 56.1%.

17. $E(X) = 0.505$ **19.** $E(X) = 3.5$

21. $E(X) \approx 4.33$ **23.** $\sigma_X \approx 1.12$

25. $E(X) = 27.25$; if classes at Salem High School are selected at random, in the long run, the average number of students per class is 27.25.

27. (a) 5.87   **(b)** Approximately 1.29

29. (a) $P(X = x) = \frac{1}{4}, x = 1, 2, 3, 4$

(b) $\mu_X = 2.5$; the average number of times KTTV radio gives the weather each hour is 2.5 times.

(c) Approximately 1.19   **(d)** 0.5

31. $\frac{64}{125}$ **33.** Not satisfied

35. (a)

$X = x$	$P(X = x)$
$X = 0$	$P(X = 0) \approx 0.670$
$X = 1$	$P(X = 1) \approx 0.287$
$X = 2$	$P(X = 2) \approx 0.041$
$X = 3$	$P(X = 3) \approx 0.002$

(b)

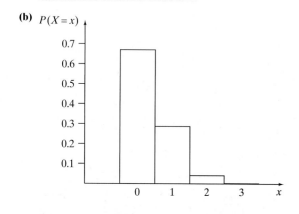

37. b   **39.** $E(X) = 3.5; \sigma_X \approx 1.62$

41. (a) 1. Success: member of sample passes exam; failure: member of sample fails exam

2. $P = 0.75$ **3.** $n = 8$ **4.** Trials are independent.

(b) $P(X = 5) \approx 0.21$; if 8 students at Hawthorne Middle School are selected at random, the probability that 5 of them pass their physical fitness test is about 0.21 or 21%.

43. (a) Approximately 0.19   **(b)** Approximately 0.53

45. (a) 2.32   **(b)** Approximately 0.82   **47.** 0.3810

49. 0.5123   **51.** 0.3790   **53.** 0.1003   **55.** 0.3310

57. (a) 0.9876; if a business day is selected at random, there is about a 98.8% chance that between 8000 and 12,000 copies are made.

(b) 0.4938; if a business day is selected at random, there is about a 49.4% chance that between 10,000 and 12,000 copies are made.

59. (a) 0.0446   **(b)** 0.9195

Chapter 9

Section 9.1

1. It is a transition matrix. It is a 2×2 matrix, with the properties that the sum of each row is equal to 1 and all the elements of the matrix are between 0 and 1.

3. It is not a transition matrix. The sum of the second row is greater than 1.

5. It is not a transition matrix. All the elements are not between 0 and 1.

7. It is a transition matrix. It is a 3×3 matrix, with the properties that the sum of each row is equal to 1 and all the elements of the matrix are between 0 and 1.

9. It is a transition matrix. It is a 3×3 matrix, with the properties that the sum of each row is equal to 1 and all the elements of the matrix are between 0 and 1.

11. $P = \begin{bmatrix} 0.6 & 0.4 \\ 0.2 & 0.8 \end{bmatrix}$ **13.** $P = \begin{bmatrix} 0.1 & 0.9 \\ 1.0 & 0.0 \end{bmatrix}$

15. $P = \begin{bmatrix} 0.3 & 0.3 & 0.4 \\ 0.2 & 0.6 & 0.2 \\ 0.3 & 0.1 & 0.6 \end{bmatrix}$ **17.** $P = \begin{bmatrix} 1.0 & 0 & 0 \\ 0 & 0.5 & 0.5 \\ 0.6 & 0.4 & 0 \end{bmatrix}$

19. $P = \begin{bmatrix} 0.6 & 0.4 & 0 & 0 \\ 0.3 & 0.5 & 0 & 0.2 \\ 0.1 & 0 & 0.3 & 0.6 \\ 0 & 0 & 0.2 & 0.8 \end{bmatrix}$

21.

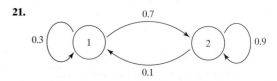

23.

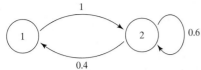

25.

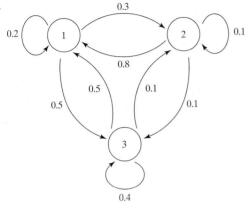

27.

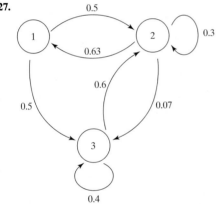

29.

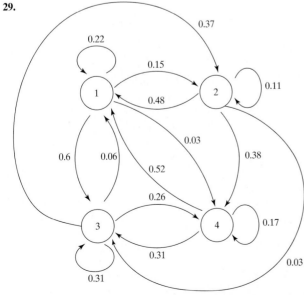

31. (a) State 1: owns Blade Shades
State 2: owns Optifilters

(b)

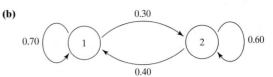

(c) $P = \begin{bmatrix} 0.70 & 0.30 \\ 0.40 & 0.60 \end{bmatrix}$

33. (a) State 1: hot summer
State 2: moderate summer

(b)

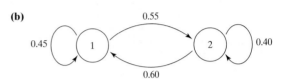

(c) $P = \begin{bmatrix} 0.45 & 0.55 \\ 0.60 & 0.40 \end{bmatrix}$

35. (a) State 1: Democrat in office
State 2: Republican in office

(b)

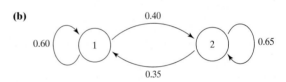

(c) $P = \begin{bmatrix} 0.60 & 0.40 \\ 0.35 & 0.65 \end{bmatrix}$

37. (a) State 1: rented in Adamstown
State 2: rented in Breston
State 3: rented in Carpersville

(b)

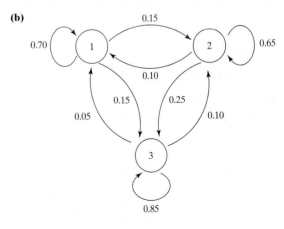

(c) $P = \begin{bmatrix} 0.70 & 0.15 & 0.15 \\ 0.10 & 0.65 & 0.25 \\ 0.05 & 0.10 & 0.85 \end{bmatrix}$

39. (a) State 1: Republican Party
State 2: Democratic Party
State 3: Green Party

(b)

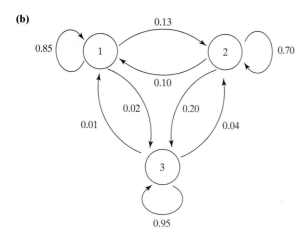

(c) $P = \begin{bmatrix} 0.85 & 0.13 & 0.02 \\ 0.10 & 0.70 & 0.20 \\ 0.01 & 0.04 & 0.95 \end{bmatrix}$

41. (a) $S_0 = [\,0.60 \quad 0.40\,]$

(b) $S_3 = [\,0.5722 \quad 0.4278\,]$

After 3 years, Blade Shades will have about 57% of the market share and Optifilter will have about 43% of the market share.

43. (a) Currently, a 40% chance of a severely hot summer and a 60% chance of a moderate summer.

(b) $S_2 = [\,0.519 \quad 0.481\,]$

After two years there will be about 52% chance of a severely hot summer and about a 48% chance of a moderate summer.

45. (a) A 40% chance that the next mayor is a Democrat and a 60% chance that the next mayor is a Republican.

(b) $S_5 \approx [\,0.467 \quad 0.533\,]$

After 10 years about a 47% chance that the mayor is a Democrat and about a 53% chance that the mayor is a Republican.

47. (a) $S_0 = [\,800 \quad 1200 \quad 400\,]$

(b) $S_2 = [\,622 \quad 792 \quad 986\,]$

After 2 years the Adamstown store will have 622 trucks to rent, the Breston store will have 792 trucks to rent, and the Carpersville store will have 986 trucks to rent.

49. (a) Initially, 62% of the voters were members of the Republican Party, 37% were members of the Democratic Party, and 1% were members of the Green Party.

(b) $S_4 \approx [\,0.4317 \quad 0.2766 \quad 0.2917\,]$

After 4 elections about 43% of the voters will declare themselves as members of the Republican Party, about 28% as members of the Democratic Party, and about 29% as members of the Green Party.

Section 9.2

1. $P^4 = \begin{bmatrix} 0.6424 & 0.3576 \\ 0.4023 & 0.5977 \end{bmatrix}$

$P^{16} \approx \begin{bmatrix} 0.5310 & 0.4690 \\ 0.5277 & 0.4723 \end{bmatrix}$

$P^{32} \approx \begin{bmatrix} 0.5294 & 0.4706 \\ 0.5294 & 0.4706 \end{bmatrix}$

$P^{64} \approx \begin{bmatrix} 0.5294 & 0.4706 \\ 0.5294 & 0.4706 \end{bmatrix}$

This is a regular transition matrix

$P^n = \begin{bmatrix} 0.5294 & 0.4706 \\ 0.5294 & 0.4706 \end{bmatrix}$ for large values of n.

3. $P^4 \approx \begin{bmatrix} 0.5670 & 0.4330 \\ 0.5196 & 0.4804 \end{bmatrix}$

$P^{16} \approx \begin{bmatrix} 0.5455 & 0.4545 \\ 0.5455 & 0.4545 \end{bmatrix}$

$P^{32} \approx \begin{bmatrix} 0.5455 & 0.4545 \\ 0.5455 & 0.4545 \end{bmatrix}$

$P^{64} \approx \begin{bmatrix} 0.5455 & 0.4545 \\ 0.5455 & 0.4545 \end{bmatrix}$

This is a regular transition matrix.

$P^n = \begin{bmatrix} 0.5455 & 0.4545 \\ 0.5455 & 0.4545 \end{bmatrix}$ for large values of n.

5. $P^4 \approx \begin{bmatrix} 1 & 0 \\ 0.985 & 0.015 \end{bmatrix}$

$P^{16} \approx \begin{bmatrix} 1 & 0 \\ 1 & 0 \end{bmatrix}$

$P^{32} \approx \begin{bmatrix} 1 & 0 \\ 1 & 0 \end{bmatrix}$

$P^{64} \approx \begin{bmatrix} 1 & 0 \\ 1 & 0 \end{bmatrix}$

This is a regular transition matrix.

$P^n = \begin{bmatrix} 1 & 0 \\ 1 & 0 \end{bmatrix}$ for large values of n.

7. The matrix is a regular transition matrix.

$P^n = \begin{bmatrix} 0.1188 & 0.1881 & 0.6931 \\ 0.1188 & 0.1881 & 0.6931 \\ 0.1188 & 0.1881 & 0.6931 \end{bmatrix}$ for large values of n.

9. The matrix is a regular transition matrix.

$P^n = \begin{bmatrix} 0.3636 & 0.4545 & 0.1818 \\ 0.3636 & 0.4545 & 0.1818 \\ 0.3636 & 0.4545 & 0.1818 \end{bmatrix}$ for large values of n.

11. The steady-state vector is

$[\,x \quad y\,] = \left[\,\frac{1}{9} \quad \frac{8}{9}\,\right]$

$P^n = \begin{bmatrix} \frac{1}{9} & \frac{8}{9} \\ \frac{1}{9} & \frac{8}{9} \end{bmatrix}$ for large values of n.

13. The steady-state vector is

$[\,x \quad y\,] = \left[\,\frac{7}{16} \quad \frac{9}{16}\,\right]$

$P^n = \begin{bmatrix} \frac{7}{16} & \frac{9}{16} \\ \frac{7}{16} & \frac{9}{16} \end{bmatrix}$ for large values of n.

15. The steady-state vector is

$$S = \begin{bmatrix} \frac{5}{6} & \frac{1}{6} \end{bmatrix}$$

$$P^n = \begin{bmatrix} \frac{5}{6} & \frac{1}{6} \\ \frac{5}{6} & \frac{1}{6} \end{bmatrix} \text{ for large values of } n.$$

17. The steady-state vector is

$$S = \begin{bmatrix} \frac{17}{32} & \frac{15}{32} \end{bmatrix}$$

$$P^n = \begin{bmatrix} \frac{17}{32} & \frac{15}{32} \\ \frac{17}{32} & \frac{15}{32} \end{bmatrix} \text{ for large values of } n.$$

19. The steady-state vector is

$$\begin{bmatrix} x & y & z \end{bmatrix} = \begin{bmatrix} \frac{3}{11} & \frac{5}{11} & \frac{3}{11} \end{bmatrix}$$

$$P^n = \begin{bmatrix} \frac{3}{11} & \frac{5}{11} & \frac{3}{11} \\ \frac{3}{11} & \frac{5}{11} & \frac{3}{11} \\ \frac{3}{11} & \frac{5}{11} & \frac{3}{11} \end{bmatrix} \text{ for large values of } n.$$

21. The steady-state vector is

$$\begin{bmatrix} x & y & z \end{bmatrix} = \begin{bmatrix} \frac{1}{5} & \frac{2}{5} & \frac{2}{5} \end{bmatrix}$$

$$P^n = \begin{bmatrix} \frac{1}{5} & \frac{2}{5} & \frac{2}{5} \\ \frac{1}{5} & \frac{2}{5} & \frac{2}{5} \\ \frac{1}{5} & \frac{2}{5} & \frac{2}{5} \end{bmatrix} \text{ for large values of } n.$$

23. The steady-state vector is

$$\begin{bmatrix} x & y & z \end{bmatrix} = \begin{bmatrix} \frac{3}{7} & \frac{8}{21} & \frac{4}{21} \end{bmatrix}$$

$$P^n = \begin{bmatrix} \frac{3}{7} & \frac{8}{21} & \frac{4}{21} \\ \frac{3}{7} & \frac{8}{21} & \frac{4}{21} \\ \frac{3}{7} & \frac{8}{21} & \frac{4}{21} \end{bmatrix} \text{ for large values of } n.$$

25. The steady-state vector is

$$\begin{bmatrix} x & y & z \end{bmatrix} = \begin{bmatrix} \frac{1}{10} & \frac{19}{55} & \frac{61}{110} \end{bmatrix}$$

$$P^n = \begin{bmatrix} \frac{1}{10} & \frac{19}{55} & \frac{61}{110} \\ \frac{1}{10} & \frac{19}{55} & \frac{61}{110} \\ \frac{1}{10} & \frac{19}{55} & \frac{61}{110} \end{bmatrix} \text{ for large values of } n.$$

27. (a) State 1: the sample passes the quality test
State 2: the sample fails the quality test

$$P = \begin{bmatrix} 0.80 & 0.20 \\ 0.70 & 0.30 \end{bmatrix}$$

(b) $P^n = \begin{bmatrix} \frac{7}{9} & \frac{2}{9} \\ \frac{7}{9} & \frac{2}{9} \end{bmatrix}$ for large values of n.

(c) In the long run, the probability that the sample will pass the quality test is 7/9 (approx, 78%).

29. (a)

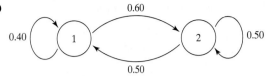

$$P = \begin{bmatrix} 0.40 & 0.60 \\ 0.50 & 0.50 \end{bmatrix}$$

(b) $P^n = \begin{bmatrix} \frac{5}{11} & \frac{6}{11} \\ \frac{5}{11} & \frac{6}{11} \end{bmatrix}$

(c) In the long run about 9091 motorists will be classified as low risk and about 10,909 will be classified as high risk.

31. (a) $P^n = \begin{bmatrix} \frac{4}{7} & \frac{3}{7} \\ \frac{4}{7} & \frac{3}{7} \end{bmatrix} \approx \begin{bmatrix} 0.5714 & 0.4286 \\ 0.5714 & 0.4286 \end{bmatrix}$ for large values of n.

(b) In the long run Blade Shades will have about 57% of the market share and Optifilters will have about 43% of the market share.

33. (a)

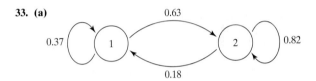

(b) $P^n = \begin{bmatrix} \frac{2}{9} & \frac{7}{9} \\ \frac{2}{9} & \frac{7}{9} \end{bmatrix} \approx \begin{bmatrix} 0.2222 & 0.7778 \\ 0.2222 & 0.7778 \end{bmatrix}$ for large values of n.

In the long run the college can expect to have about 22% of the students declared business majors and about 78% declared general studies majors.

35. (a) $P^n = \begin{bmatrix} \frac{7}{15} & \frac{8}{15} \\ \frac{7}{15} & \frac{8}{15} \end{bmatrix} \approx \begin{bmatrix} 0.4667 & 0.5333 \\ 0.4667 & 0.5333 \end{bmatrix}$ for large values of n.

(b) About 47% of mayoral election Democrats are expected to win in the long run.

37. (a) $P^n \approx \begin{bmatrix} 0.1774 & 0.2419 & 0.5806 \\ 0.1774 & 0.2419 & 0.5806 \\ 0.1774 & 0.2419 & 0.5806 \end{bmatrix}$ for large values of n.

(b) In the long run about 18% of the trucks will be returned to Adamstown, about 24% of the trucks to Breston, and about 58% to Carpersville.

39. (a) $P^n \approx \begin{bmatrix} 0.1511 & 0.1577 & 0.6911 \\ 0.1511 & 0.1577 & 0.6911 \\ 0.1511 & 0.1577 & 0.6911 \end{bmatrix}$ for large values of n.

(b) In the long run about 69% of the voters are expected to be members of the Green Party.

Section 9.3

1. There is not an absorbing state in this matrix.

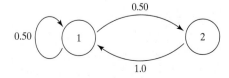

3. States 2 and 3 are absorbing states.

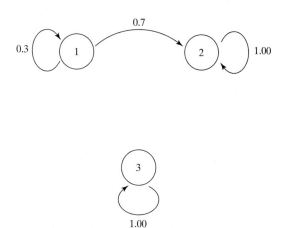

5. State 3 is an absorbing state.

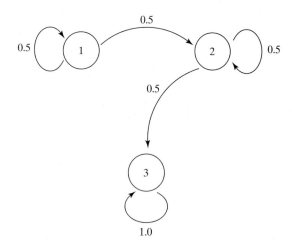

7. States 1 and 3 are absorbing states.

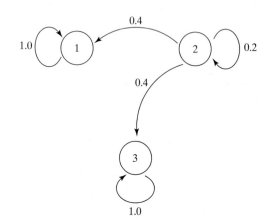

9. State 4 is an absorbing state.

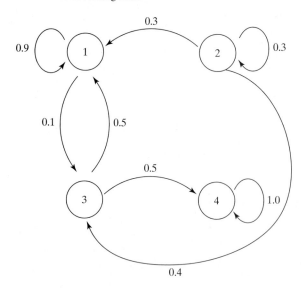

11. State 1 is an absorbing state. The diagram represents an absorbing Markov chain.

13. States 1 and 3 are absorbing states. The diagram represents an absorbing Markov chain.

15. State 2 is an absorbing state. The diagram represents an absorbing Markov chain.

17. States 2 and 4 are absorbing states. The diagram represents an absorbing Markov chain.

19. State 5 is an absorbing state. The diagram represents an absorbing Markov chain.

21. **(a)** $F = \begin{bmatrix} 2 & 0 \\ 2 & \frac{4}{3} \end{bmatrix}$ **(b)** $FR = \begin{bmatrix} 1 \\ 1 \end{bmatrix}$

23. **(a)** $F = \begin{bmatrix} 2 & 0 \\ 2 & \frac{4}{3} \end{bmatrix}$ **(b)** $FR = \begin{bmatrix} 1 \\ 1 \end{bmatrix}$

25. **(a)** $F = \begin{bmatrix} \frac{20}{9} & \frac{14}{9} \\ \frac{10}{9} & \frac{16}{9} \end{bmatrix}$ **(b)** $FR = \begin{bmatrix} 1 \\ 1 \end{bmatrix}$

27. Singular matrix, therefore no inverse exists.

29. **(a)** $F = \begin{bmatrix} \frac{10}{7} & 0 \\ \frac{10}{7} & 1 \end{bmatrix}$ **(b)** $FR = \begin{bmatrix} 1 \\ 1 \end{bmatrix}$

31. Singular matrix, therefore no inverse exists.

33. **(a)** $F = \begin{bmatrix} \frac{40}{27} & \frac{10}{27} \\ \frac{25}{27} & \frac{40}{27} \end{bmatrix}$ **(b)** $FR = \begin{bmatrix} \frac{8}{27} & \frac{19}{27} \\ \frac{5}{27} & \frac{22}{27} \end{bmatrix}$

35. **(a)** $F = \begin{bmatrix} \frac{5620}{4343} & \frac{1900}{4343} & \frac{1220}{4343} \\ \frac{1080}{4343} & \frac{5620}{4343} & \frac{1780}{4343} \\ \frac{1640}{4343} & \frac{2100}{4343} & \frac{5920}{4343} \end{bmatrix}$ **(b)** $FR = \begin{bmatrix} \frac{2405}{4343} & \frac{1938}{4343} \\ \frac{2085}{4343} & \frac{2258}{4343} \\ \frac{2201}{4343} & \frac{2142}{4343} \end{bmatrix}$

37. **(a)** State 1: zoned residential
State 2: zoned commercial
State 3: zoned public

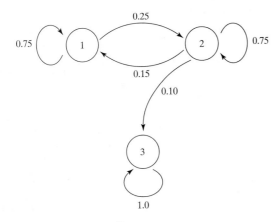

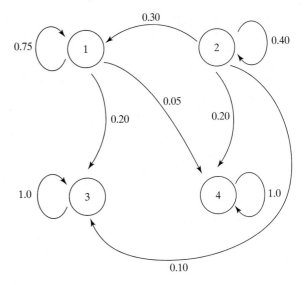

(b) $P = \begin{bmatrix} 0.75 & 0.25 & 0 \\ 0.15 & 0.75 & 0.10 \\ 0 & 0 & 1 \end{bmatrix}$

$F = \begin{bmatrix} 10 & 10 \\ 6 & 10 \end{bmatrix}$

(c) $f_{1,2} = 10$; land that was initially zoned as residential will spend about 10 years zoned as commercial until it is eventually zoned for public use.

39. (a) State 1: freshman
State 2: sophomore
State 3: leave the school

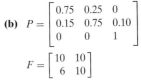

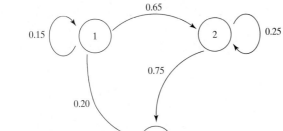

(b) $P = \begin{bmatrix} 0.15 & 0.65 & 0.20 \\ 0 & 0.25 & 0.75 \\ 0 & 0 & 1 \end{bmatrix}$

$F = \begin{bmatrix} \frac{20}{17} & \frac{52}{51} \\ 0 & \frac{4}{3} \end{bmatrix}$

(c) $f_{1,2} = \frac{52}{51}$; this means that a student who is initially a freshman will spend a little more than 1 year as a sophomore before eventually leaving school.

41. (a) State 1: account in good standing
State 2: probationary
State 3: pay off loan
State 4: repossession

(b) $P = \begin{bmatrix} 0.75 & 0 & 0.20 & 0.05 \\ 0.30 & 0.40 & 0.10 & 0.20 \\ 0 & 0 & 1 & 0 \\ 0 & 0 & 0 & 1 \end{bmatrix}$ $\quad F = \begin{bmatrix} 4 & 0 \\ 2 & \frac{5}{3} \end{bmatrix}$

(c) The sum of the entries in row 1 is 4. This indicates that we can expect an account in good standing to be either paid off or enter into repossession in four months.

The sum of the entries in row 2 is about 3.7. This indicates that we can expect an account that is in a probationary state to be either paid off or enter into repossession in about 3.7 months.

(d) $FR = \begin{bmatrix} \frac{4}{5} & \frac{1}{5} \\ \frac{17}{30} & \frac{13}{30} \end{bmatrix}$

An account in good standing has an 80% chance of remaining in good standing and a 20% chance of becoming probationary.

43. (a) State 1: trainee   State 3: supervisor
State 2: technician   State 4: leave the company

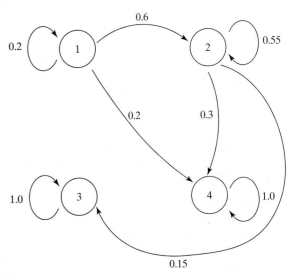

(b) $P = \begin{bmatrix} 0.20 & 0.60 & 0 & 0.20 \\ 0 & 0.55 & 0.15 & 0.30 \\ 0 & 0 & 1 & 0 \\ 0 & 0 & 0 & 1 \end{bmatrix}$ $\quad F = \begin{bmatrix} \frac{5}{4} & \frac{5}{3} \\ 0 & \frac{20}{9} \end{bmatrix}$

(c) The sum of the entries in row 1 is approximately 2.92. The sum of the entries in row 2 is approximately 2.22. This indicates that a trainee can expect to be with the company about 3 years and a technician can expect to be with the company about 2.2 years before becoming a supervisor.

45. (a) State 1: freshman
State 2: sophomore
State 3: junior
State 4: senior
State 5: withdraw
State 6: graduate

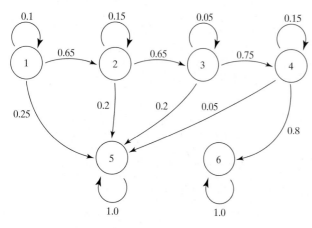

(b) $P = \begin{bmatrix} 0.10 & 0.65 & 0 & 0 & 0.25 & 0 \\ 0 & 0.15 & 0.65 & 0 & 0.20 & 0 \\ 0 & 0 & 0.05 & 0.75 & 0.20 & 0 \\ 0 & 0 & 0 & 0.15 & 0.05 & 0.80 \\ 0 & 0 & 0 & 0 & 1 & 0 \\ 0 & 0 & 0 & 0 & 0 & 1 \end{bmatrix}$

$F \approx \begin{bmatrix} \frac{10}{9} & \frac{130}{153} & \frac{1690}{2907} & 0.513 \\ 0 & \frac{20}{17} & \frac{260}{323} & \frac{3900}{5491} \\ 0 & 0 & \frac{20}{19} & \frac{300}{323} \\ 0 & 0 & 0 & \frac{20}{17} \end{bmatrix}$

The sum of row 1 is approximately 3.055.
The sum of row 2 is approximately 2.692.
The sum of row 3 is approximately 1.981.
The sum of row 4 is approximately 1.176.

This means that a freshman, sophomore, junior, and senior can expect to be at the university 3.055, 2.692, 1.981, and 1.176 years, respectively.

Chapter 9 Review Exercises

1. Transition matrix **3.** Transition matrix

5. $\begin{bmatrix} 0.8 & 0 & 0.2 \\ 0.1 & 0.1 & 0.8 \\ 0.6 & 0.3 & 0.1 \end{bmatrix}$

7.

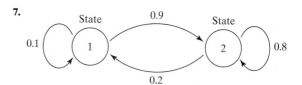

9.

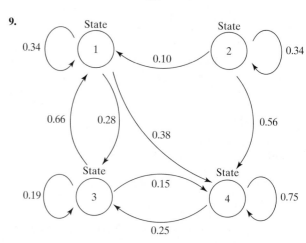

11. (a) State 1: bike is returned to the library; state 2: bike is returned to the YMCA, state 3: bike is returned to the grocery store.

(b)

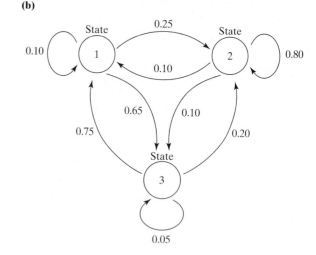

(c) $\begin{bmatrix} 0.10 & 0.25 & 0.65 \\ 0.10 & 0.80 & 0.10 \\ 0.75 & 0.20 & 0.05 \end{bmatrix}$

13. (a) $S_0 = \begin{bmatrix} 20 & 24 & 18 \end{bmatrix}$

(b) $\begin{bmatrix} 16.795 & 29.975 & 15.230 \end{bmatrix}$; on Wednesday morning there will be about 17 bikes at the library, 30 bikes at the YMCA, and 15 bikes at the grocery store.

15. Regular; $P^n = \begin{bmatrix} 0.1250 & 0.8750 \\ 0.1250 & 0.8750 \end{bmatrix}$ for large values of n.

17. Regular; $P^n = \begin{bmatrix} 0.3162 & 0.2222 & 0.4615 \\ 0.3162 & 0.2222 & 0.4615 \\ 0.3162 & 0.2222 & 0.4615 \end{bmatrix}$ for large values of n.

19. $\left[\begin{array}{cc} \frac{2}{7} & \frac{5}{7} \end{array}\right]$; $\left[\begin{array}{cc} \frac{2}{7} & \frac{5}{7} \\ \frac{2}{7} & \frac{5}{7} \end{array}\right]$ **21.** $\left[\begin{array}{ccc} \frac{4}{13} & \frac{4}{13} & \frac{5}{13} \end{array}\right]$; $\left[\begin{array}{ccc} \frac{4}{13} & \frac{4}{13} & \frac{5}{13} \\ \frac{4}{13} & \frac{4}{13} & \frac{5}{13} \\ \frac{4}{13} & \frac{4}{13} & \frac{5}{13} \end{array}\right]$

23. $\left[\begin{array}{ccc} \frac{52}{151} & \frac{45}{302} & \frac{153}{302} \end{array}\right]$; $\left[\begin{array}{ccc} \frac{52}{151} & \frac{45}{302} & \frac{153}{302} \\ \frac{52}{151} & \frac{45}{302} & \frac{153}{302} \\ \frac{52}{151} & \frac{45}{302} & \frac{153}{302} \end{array}\right]$

25. (a) State 1: off probation; state 2: on probation;

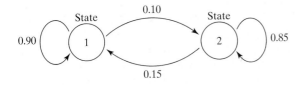

$$\left[\begin{array}{cc} 0.90 & 0.10 \\ 0.15 & 0.85 \end{array}\right]$$

(b) $\left[\begin{array}{cc} \frac{3}{5} & \frac{2}{5} \\ \frac{3}{5} & \frac{2}{5} \end{array}\right]$

(c) Off probation: 720 students; on probation: 480 students

27. (a) $\left[\begin{array}{ccc} \frac{68}{277} & \frac{147}{277} & \frac{62}{277} \\ \frac{68}{277} & \frac{147}{277} & \frac{62}{277} \\ \frac{68}{277} & \frac{147}{277} & \frac{62}{277} \end{array}\right]$ **(b)** About 0.53 or 53%

29. Absorbing state: state 2

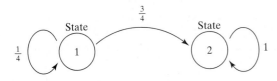

31. Absorbing states: states 3 and 4

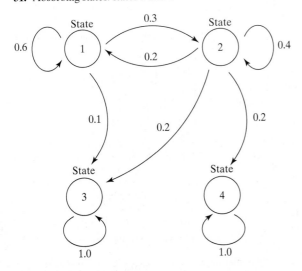

33. Absorbing states: states 2 and 3; does not represent an absorbing Markov chain

35. Absorbing states: states 4, 5, and 6; does not represent an absorbing Markov chain

37. $\left[\begin{array}{cc} 1.25 & 0 \\ 0.3125 & 2.5 \end{array}\right]$

39. $\left[\begin{array}{cc} \frac{5}{3} & \frac{5}{3} \\ \frac{5}{6} & \frac{5}{2} \end{array}\right]$

41. (a) State 1: first string, State 2: second string, State 3: leave the team;

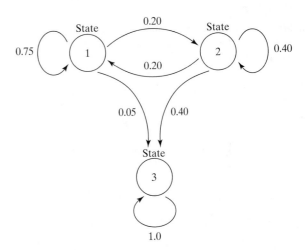

(b) $\left[\begin{array}{ccc} 0.75 & 0.20 & 0.05 \\ 0.20 & 0.40 & 0.40 \\ 0 & 0 & 1 \end{array}\right]$; entries rounded to the nearest hundredth, $\left[\begin{array}{cc} 5.45 & 1.82 \\ 1.82 & 2.27 \end{array}\right]$

(c) A first-string player (state 1) will spend about 1.82 seasons as a second-string player (state 2) before leaving the team (state 3).

43. (a) State 1: stage I; state 2: stage II; state 3: patient transfers; state 4: patient completes program;

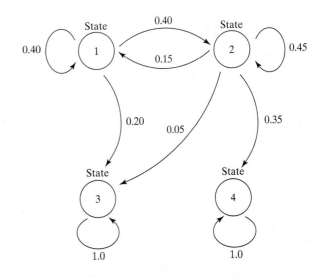

(b) $\begin{bmatrix} 0.40 & 0.40 & 0.20 & 0 \\ 0.15 & 0.45 & 0.05 & 0.35 \\ 0 & 0 & 1 & 0 \\ 0 & 0 & 0 & 1 \end{bmatrix}$; entries rounded to the nearest

hundredth: $\begin{bmatrix} 2.04 & 1.48 \\ 0.56 & 2.22 \end{bmatrix}$

(c) Row 1: about 3.52, a patient in stage I can be expected to be in the program for about 3.52 weeks before transferring or completing the program; row 2: about 2.78, a patient in stage II can be expected to be in the program for about 2.78 weeks before transferring or completing the program.

45. (a) State 1: freshman; state 2: sophomore; state 3: junior; state 4: senior; state 5: withdraw; state 6: graduate;

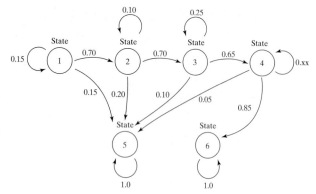

(b) $\begin{bmatrix} 0.15 & 0.70 & 0 & 0 & 0.15 & 0 \\ 0 & 0.10 & 0.70 & 0 & 0.20 & 0 \\ 0 & 0 & 0.25 & 0.65 & 0.10 & 0 \\ 0 & 0 & 0 & 0.10 & 0.05 & 0.85 \\ 0 & 0 & 0 & 0 & 1 & 0 \\ 0 & 0 & 0 & 0 & 0 & 1 \end{bmatrix}$; entries rounded

to the nearest hundredth:

$\begin{bmatrix} 1.18 & 0.92 & 0.85 & 0.62 \\ 0 & 1.11 & 1.04 & 0.75 \\ 0 & 0 & 1.33 & 0.96 \\ 0 & 0 & 0 & 1.11 \end{bmatrix}$

(c) Freshman: about 3.57 years; sophomore: about 2.9 years; junior: about 2.29 years; senior: about 1.11 years

(d) About 0.53 or 53%

(e) About 0.36 or 36%

Chapter 10

Section 10.1

1. Jenna wins $2.   **3.** Elle wins $2.

5. Jenna wins $6.   **7.** Elle wins $3.

9. Don $\begin{array}{c} \\ \\ \end{array}\overset{\text{Bill}}{\begin{bmatrix} -3 & 2 \\ 3 & 4 \end{bmatrix}}$

11. Patrice $\overset{\text{Quincy}}{\begin{bmatrix} 3 & -4 & -4 \\ -1 & 6 & -6 \\ -2 & -3 & 8 \end{bmatrix}}$

13. (a) The saddle point is the entry $a_{1,1} = 2$.

(b) The optimal strategy for Bill is to play row 1 and the optimal strategy for Al is to play column 1.

(c) The value of the game is 2 and the game favors Bill.

15. There is no saddle point in this game. This is not a strictly determined game.

17. (a) The saddle points are the entries $a_{1,1}$ and $a_{1,3}$, which each have a value of 2.

(b) The optimal strategy for Khalid is to play row 1. The optimal strategy for Scoonie is to choose column 1 or 3.

(c) The value of the game is 2 and the game favors Khalid.

19. (a) The saddle point is the entry $a_{2,1} = 5$.

(b) The optimal strategies are for Austin to play row 2 and for Dylan to play column 1.

(c) The value of the game is 5 and the game favors Austin.

21. (a) The saddle point is the entry $a_{3,2} = 3$.

(b) The optimal strategies are for Brock to play row 3 and for Ash to play column 2.

(c) The value of the game is 3 and the game favors Brock.

23. (a) The saddle points are the entries $a_{2,2}$ and $a_{4,2}$, which both are equal to zero.

(b) The optimal strategies are for Brian to play row 2 or 4 and for Mike to play column 2.

(c) The value of the game is 0 and it is a fair game.

25. (a) The saddle point is the entry $a_{1,2} = -100,000$. The best strategy for The Chasm is to remodel (row 1).

27. (a) This is a strictly determined game.

(b) The saddle point is the entry $a_{2,2} = 25$. The optimal strategy for each bank is to open a branch in Kerrton.

(c) The value of the game is 25. The game favors Bank Two.

29. (a) Redan $\begin{array}{c} \\ \text{City} \\ \text{Suburb} \end{array}\overset{\begin{array}{cc} & \text{Erog} \\ \text{City} & \text{Suburb} \end{array}}{\begin{bmatrix} 2 & -6 \\ 8 & -4 \end{bmatrix}}$

(c) Optimal strategy for each store is to build in the suburbs.

31. (a) Gutbuster $\begin{array}{c} \\ \text{MA} \\ \text{SR} \\ \text{TW} \end{array}\overset{\begin{array}{cc} & \text{Noflab} \\ \text{MA} & \text{SR} \end{array}}{\begin{bmatrix} 5 & -5 \\ 5 & 0 \\ 10 & 15 \end{bmatrix}}$

(c) The optimal strategy for Gutbuster is to start a twenty-something fit-for-life program. The optimal strategy for Noflab is to start a middle-age spread program.

33. (a) Candidate $\begin{array}{c} \\ \text{Reveals} \\ \text{Doesn't reveal} \end{array}\overset{\begin{array}{cc} & \text{Media} \\ \text{Critical} & \text{Noncritical} \end{array}}{\begin{bmatrix} -35 & 7 \\ -150 & 0 \end{bmatrix}}$

(c) The optimal strategy is for the candidate to reveal the skeleton.

Section 10.2

1. The expected value is $\frac{1}{6}$.

3. The expected value is $\frac{9}{8}$.

5. The expected value is 0.41.

7. The expected value is 2.85.

7. The expected value is 2.85.

9. The expected value is 0.84.

11. (a) $E = [0.25]$ **(b)** $E = \left[\frac{13}{12}\right]$ **(c)** $E = \left[\frac{26}{35}\right]$

(d) $E = \left[\frac{23}{24}\right]$

Strategy (b) is the most favorable to the row player.

13. $p_{opt} = \left[\frac{7}{16} \quad \frac{9}{16}\right]$ $q_{opt} = \begin{bmatrix} \frac{9}{16} \\ \frac{7}{16} \end{bmatrix}$

15. This is a strictly determined game. Optimal strategy for the row player is row 1; for the column player it is column 2.

17. $p_{opt} = \left[\frac{4}{7} \quad \frac{3}{7}\right]$ **19.** $p_{opt} = \left[\frac{1}{2} \quad \frac{1}{2}\right]$ **21.** $p_{opt} = \left[\frac{1}{2} \quad \frac{1}{2}\right]$

$q_{opt} = \begin{bmatrix} \frac{3}{7} \\ \frac{4}{7} \end{bmatrix}$ $q_{opt} = \begin{bmatrix} \frac{5}{9} \\ \frac{4}{9} \end{bmatrix}$ $q_{opt} = \begin{bmatrix} \frac{5}{12} \\ \frac{7}{12} \end{bmatrix}$

23. (a) $v = -\frac{1}{16}$ **(b)** The game favors the column player.

25. (a) $v = \frac{12}{7}$ **(b)** The game favors the row player.

27. (a) $v = \frac{1}{12}$ **(b)** The game favors the row player.

31. (a) $p_{opt} = \left[\frac{19}{42} \quad \frac{23}{42}\right]$

This is the optimal strategy for the Stitz Co.

$q_{opt} = \begin{bmatrix} \frac{10}{21} \\ \frac{11}{21} \end{bmatrix}$

This is the optimal strategy for the Stumpf Co.

(b) $v = \frac{31}{105}$

The game favors the Stitz Co.

33. (a) Bridge

	Economy	
	Strong	Recession
Remodel	400	−150
Don't remodel	−125	−100

(b) $p_{opt} = \left[\frac{1}{23} \quad \frac{22}{23}\right]$

35. (a) Stock choice

	Economy	
	Growing	Recession
Tech.	0.35	−0.15
Pharm.	0.15	0.16

(b) $p_{opt} = \left[\frac{1}{51} \quad \frac{50}{51}\right]$

The optimal strategy for the Riblets is to invest about $980 in technology stock and about $49,020 in pharmaceutical stock.

(c) The Riblets can expect about a 15.39% annual gain. So after one year the Riblets can expect a profit of about $7695.

37. (a) Toth

	Williams	
	Rural	Urban
Rural	0.60	0.40
Urban	0.35	0.65

(b) $p_{opt} = [0.6 \quad 0.4]$

The optimal strategy for Toth is to campaign 60% of the time in rural areas and 40% of the time in urban areas.

$q_{opt} = [0.5 \quad 0.5]$

The optimal strategy for Williams is to campaign 50% of the time in rural areas and 50% of the time in urban areas.

Section 10.3

1. (a) Optimal strategy for the column player is $Q_{opt} = \begin{bmatrix} \frac{1}{7} \\ \frac{6}{7} \end{bmatrix}$.

Optimal strategy for the row player is $P_{opt} = \left[\frac{5}{7} \quad \frac{2}{7}\right]$.

(b) The value of the game is $\frac{19}{7}$.

3. (a) Column player's optimal strategy is $Q_{opt} = \begin{bmatrix} \frac{1}{5} \\ \frac{4}{5} \end{bmatrix}$.

Row player's optimal strategy is $p_{opt} = \left[\frac{2}{5} \quad \frac{3}{5}\right]$.

(b) The value of the game is $\frac{43}{5}$.

5. (a) Row player's optimal strategy is row 1; column player's is column 1. (This is a strictly determined game).

(b) Value of game is 2.

7. (a) Column player's optimal strategy is $Q_{opt} = \begin{bmatrix} \frac{3}{5} \\ \frac{2}{5} \end{bmatrix}$.

The row player's optimal strategy is $P_{opt} = \left[0 \quad \frac{1}{5} \quad \frac{4}{5}\right]$.

(b) The value of the game is $\frac{17}{5}$.

9. (a) The column player's optimal strategy is $Q_{opt} = \begin{bmatrix} \frac{2}{3} \\ 0 \\ \frac{1}{3} \end{bmatrix}$.

The row player's optimal strategy is $P_{opt} = \left[\frac{2}{3} \quad \frac{1}{3}\right]$.

(b) The value of the game is $\frac{8}{3}$.

11. (a) The column player's optimal strategy is $Q_{opt} = \begin{bmatrix} \frac{3}{8} \\ \frac{5}{8} \end{bmatrix}$.

The row player's optimal strategy is $P_{opt} = \left[\frac{1}{2} \quad \frac{1}{2}\right]$.

13. The column player's optimal strategy is $Q_{opt} = \begin{bmatrix} \frac{3}{7} \\ \frac{4}{7} \end{bmatrix}$.

The row player's optimal strategy is $P_{opt} = \left[\frac{13}{21} \quad \frac{8}{21}\right]$.

15. The column player's optimal strategy is $Q_{opt} = \begin{bmatrix} 0 \\ \frac{1}{2} \\ \frac{1}{2} \end{bmatrix}$.

The row player's optimal strategy is $P_{opt} = \left[\frac{1}{2} \quad \frac{1}{2}\right]$.

17. The column player's optimal strategy is $Q_{opt} = \begin{bmatrix} \frac{43}{116} \\ \frac{9}{29} \\ \frac{37}{116} \end{bmatrix}$.

The row player's optimal strategy is $P_{opt} = \begin{bmatrix} \frac{27}{58} & \frac{10}{29} & \frac{11}{58} \end{bmatrix}$.

19. The column player's optimal strategy is $Q_{opt} = \begin{bmatrix} \frac{4}{7} \\ \frac{3}{7} \end{bmatrix}$.

The row player's optimal strategy is $P_{opt} = \begin{bmatrix} \frac{4}{7} & 0 & \frac{3}{7} \end{bmatrix}$.

21. (a) The column player's optimal strategy is $Q_{opt} = \begin{bmatrix} \frac{2}{5} \\ \frac{3}{5} \end{bmatrix}$.

The optimal strategy for the Hoffman company is to have $\frac{2}{5}$ of their advertising on television and $\frac{3}{5}$ on radio.

The row player's optimal strategy is $P_{opt} = \begin{bmatrix} \frac{2}{5} & \frac{3}{5} \end{bmatrix}$.

The optimal strategy for the Baker Company is to have $\frac{2}{5}$ of their advertising on television and $\frac{3}{5}$ on radio.

(b) The value of the game is $\frac{1}{5}$. The game favors the Baker Company.

23. (a)

Jamie

		L	R
Jessie	L	-4	3
	R	6	-4

(b) Jamie's optimal strategy will be to guess the left hand 7 out of 17 times and the right hand 10 out of 17 times. Jessie's optimal strategy is to hide the coin in the left hand 10 out of 17 times and in the right hand 7 out of 17 times.

(c) The value of the game is $\frac{2}{17}$. The game favors Jessie.

25. The doctor's optimum strategy would be to prescribe treatment A.

Chapter 10 Review Exercises

1. -3, Amy loses \$3, Grace wins \$3.

3. 6, Amy wins \$6, Grace loses \$6.   **5.** Sharon $\begin{bmatrix} -2 & 1 \\ 5 & 3 \end{bmatrix}$ Ron

7. (a) $a_{2,2} = 2$ **(b)** Mike: row 2; Jen: column 2

(c) 2; favors Mike

9. (a) $a_{4,1} = 3$ **(b)** Lindsay: row 4; Mark: column 1

(c) 3; favors Lindsay

11. -8000; don't move

13. (a)

Gorham Mountain

		2-hour pass	3-hour pass
Willard Mountain	2-hour pass	-2	4
	3-hour pass	0	1
	4-hour pass	-1	-2

(b) $a_{2,1} = 0$ is a saddle point.

(c) Willard Mountain: offer a 3-hour pass; Gorham Mountain: offer a 2-hour pass

15. (a) Bookstore

		Strong	Recession
	Expand	50,000	$-100,000$
	Don't expand	$-75,000$	$-80,000$

Economy

(b) $a_{2,2} = -80,000$ is a saddle point.   **(c)** Don't expand

17. 1.78

19. (a) 0   **(b)** $\frac{5}{4} = 1.25$ **(c)** $-\frac{7}{9} \approx -0.78$ **(d)** $-\frac{3}{8} = -0.375$

Strategy (c) is the most favorable to the column player.

21. $\begin{bmatrix} 1 & 0 \end{bmatrix}$; $\begin{bmatrix} 0 \\ 1 \end{bmatrix}$

23. (a) $-\frac{1}{3} \approx -0.33$ **(b)** Column player

25. (a) $\frac{1}{7} \approx 0.14$ **(b)** Row player

27. (a) Bookstore

		Strong	Recession
	Expand	50	-100
	Don't expand	-125	-80

Economy

(b) $\begin{bmatrix} \frac{3}{13} & \frac{10}{13} \end{bmatrix}$ **(c)** Don't expand

29. (a) Kris

	All school meetings	Football games
All school meetings	0.55	0.40
Football games	0.45	0.60

Rachel

(b) $\begin{bmatrix} \frac{1}{2} & \frac{1}{2} \end{bmatrix}$; $\begin{bmatrix} \frac{2}{3} \\ \frac{1}{3} \end{bmatrix}$

31. (a) $\begin{bmatrix} \frac{1}{2} & \frac{1}{2} \end{bmatrix}$; $\begin{bmatrix} \frac{3}{8} \\ \frac{5}{8} \end{bmatrix}$ **(b)** $\frac{11}{2}$

33. (a) $\begin{bmatrix} 0 & \frac{2}{3} & \frac{1}{3} \end{bmatrix}$; $\begin{bmatrix} \frac{1}{3} \\ \frac{2}{3} \end{bmatrix}$ **(b)** $\frac{10}{3}$

35. (a) $\begin{bmatrix} 1 & 0 \end{bmatrix}$; $\begin{bmatrix} 1 \\ 0 \\ 0 \end{bmatrix}$ **(b)** 2   **37.** $\begin{bmatrix} \frac{12}{19} & \frac{7}{19} \end{bmatrix}$; $\begin{bmatrix} \frac{14}{19} \\ \frac{5}{19} \end{bmatrix}$

39. $\begin{bmatrix} \frac{2}{21} & \frac{10}{21} & \frac{3}{7} \end{bmatrix}$; $\begin{bmatrix} \frac{10}{21} \\ \frac{19}{42} \\ \frac{1}{14} \end{bmatrix}$

41. (a) $\begin{bmatrix} \frac{2}{7} & \frac{5}{7} \end{bmatrix}$; $\begin{bmatrix} \frac{3}{7} \\ \frac{4}{7} \end{bmatrix}$ **(b)** Yes; the Babcock Company

43. (a) $S = (a + d) - (b + c) = (3 + 1) - [-2 + (-1)] = 4 + 3 = 7$;

$p_1 = \frac{d - c}{S} = \frac{1 - (-1)}{7} = \frac{2}{7}$; $p_2 = 1 - p_1 = 1 - \frac{2}{7} = \frac{5}{7}$;

$q_1 = \frac{d - b}{S} = \frac{1 - (-2)}{7} = \frac{3}{7}$; $q_2 = 1 - q_1 = 1 - \frac{3}{7} = \frac{4}{7}$

(b) $\frac{1}{7}$

45. $\begin{bmatrix} 0 & \frac{5}{11} & \frac{6}{11} \end{bmatrix}$; treatment C

Appendices

Appendix A.1

1. 0.73 **3.** 0.04 **5.** 0.591 **7.** 0.007

9. \$524.40 was covered by private health insurance in 1990.

11. About 10,119 million board feet. **13.** 20% **15.** 12.5%

17. 53.4% **19.** About 42.9%

21. It is projected that in 2008 approximately 12.1% of the students will be enrolled in private schools.

23. In 1985 approximately 5.46% of the budget was spent on technology research and development.

Appendix A.2

1. 31 **3.** 8 **5.** 4 **7.** 10 **9.** 8 **11.** -24

13. 0.27648 **15.** 41 **17.** 0.45

19. $23.3 - 0.67(5.9) = 19.347$
$23.3 + 0.67(5.9) = 27.253$

The ages of the middle 50% of the unmarried mothers in the United States in 1996 ranged from about 19 to 27.

Appendix A.3

1. $3^6 = 729$ **3.** $2^{15} = 32{,}768$ **5.** $10^6 = 1{,}000{,}000$

7. $3^1 = 3$ **9.** $6^6 = 46{,}656$ **11.** $-3^{15} = -14{,}348{,}907$

13. $(x+3)^6$ **15.** 1 **17.** 1 **19.** $\frac{1}{36}$ **21.** 36 **23.** $16x^{12}$

25. $\frac{1}{3x^5}$ **27.** $\frac{9}{16}$ **29.** 64

31. ≈ 0.2109
There is about a 21.1% chance that 2 out of 20 were African American.

33. The payment will be \$1,463.45 each month for 5 years.

Appendix A.4

1. $x^{\frac{1}{2}}$ **3.** $x^{\frac{4}{7}}$ **5.** $5^{\frac{1}{4}}x^{\frac{2}{4}} = 5^{\frac{1}{4}}x^{\frac{1}{2}}$ **7.** $x^{-\frac{1}{3}}$ **9.** $8x^{-\frac{2}{7}}$

11. $\sqrt[4]{x}$ **13.** $\sqrt[8]{x^5}$ **15.** $\sqrt[4]{6x^2}$ **17.** $\frac{1}{\sqrt[3]{x}}$ **19.** $\frac{5}{\sqrt[3]{x^2}}$

Appendix B.1

1. $10x^5$ **3.** $-8x^{12}$ **5.** $-150x^4$ **7.** $2x^2 - 10x$

9. $-4x^4 + 12x^3$ **11.** $-12x^6 + 18x^2$ **13.** $-2x^4 + 14x^3 - 10x^2$

15. $x^2 + 8x + 15$ **17.** $1.2x^2 - 9.8x + 20$ **19.** $9x^2 - 12x - 32$

21. $8x^4 + 22x^2 + 15$ **23.** $x^3 + 3x^2 + 5x + 3$

25. $0.6x^3 - 10.4x^2 + 21.2x - 3$ **27.** $4x^4 + 4x^3 - 5x^2 + 3x - 6$

29. $x^4 + 12x^3 + 26x^2 - 21x - 6$

31. $2x^5 + 8x^4 - 6x^3 - x^2 + 15x + 6$

Appendix B.2

1. $2(x-1)$ **3.** $5(7x-2)$ **5.** $2x^2(6-13x)$

7. This expression cannot be factored. **9.** $(x+9)(x-9)$

11. $(3x+1)(3x-1)$ **13.** $(2x+1)(2x-3)$

15. $(x^2+1)(x+1)(x-1)$ **17.** $(4x^2+9)(2x+3)(2x-3)$

19. $(x+1)(x+4)$ **21.** Expression cannot be factored.

23. $(2x^2+4)(x-1)$ **25.** $(6x-3)(x+2)(x-2)$

27. $(x+3)(x+5)$ **29.** $(x+1)^2$ **31.** $(x-5)(x+1)$

33. Expression is not factorable. **35.** $2(x+8)(x-3)$

37. $(2x+1)(x+1)$ **39.** $(3x-2)(x-2)$ **41.** $(7x-5)(x-1)$

43. $(7x+11)(x-1)$ **45.** $(5)(3x-4)(x+2)$

Appendix B.3

1. $x = 2$ **3.** $x = 3$ **5.** $x = 28$ **7.** $x = \frac{2}{3}$

9. $x = -\frac{0.5}{0.2} = -\frac{5}{2}$ **11.** $x = \frac{18}{5}$ **13.** $x = 3$ **15.** $x = -6$

17. $x = 13$ **19.** $x = \frac{203}{88}$

21. $x = 6$. In 1978 the annual expenditure on pollution abatement in the United States totaled \$60.17 billion.

23. $x = 4$. In 1994, worldwide military expenditures were \$887.92 billion.

Appendix B.4

1. $x = 0$ or $x = -2$ **3.** $x = -3$ or $x = 3$

5. $x = -5$ or $x = -3$ **7.** $x = 6$ or $x = -1$

9. $x = -\frac{1}{2}$ or $x = -1$ **11.** $x = \frac{4}{3}$ or $x = -1$

13. $x = \dfrac{-1 \pm \sqrt{33}}{8}$ **15.** No real solution

17. $x = \dfrac{3 \pm \sqrt{2}}{2}$ **19.** $x = \dfrac{3 \pm \sqrt{7}}{2}$

21. In 1990 and 1994, membership in the Boy Scouts of America was 5,433,000.

23. No real solution.

Appendix B.5

1. $\log 2 + \log x + \log y$ **3.** $3(\log 2 + \log y - \log 5 - \log x)$

5. $\ln x + 2\ln y - \ln(x+2)$ **7.** $\frac{1}{3}(\ln 5 + 3\ln x - \ln 9 - \ln y)$

9. $x = 102$ **11.** $x = \frac{e^3+2}{4} \approx 5.52$ **13.** $x = \frac{26}{3}$

15. $x = 4$ **17.** $x \approx 1.63$ **19.** $x = \frac{1}{5}$ **21.** $x \approx 0.23$

23. $x \approx 8.82$; in 1998, 18.89% of all waste generated in the United States was yard waste.

Appendix B.6

1. $\frac{10}{3}$ **3.** 2,097,151 **5.** 4,882,812 **7.** 9842

9. \$4,095,000

11. You pick the month choice. Pick a 31-day month and the doubling effect would amount to \$21,474,836.47. A 31-day month during which you receive \$100,000 each day results in a total of \$3,100,000.

Index

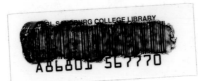
Carl Sandburg College
Learning Resources Center
2400 Tom L. Wilson
Galesburg, Illinois 61401